W9-CHC-924

NOISE REDUCTION

Noise Reduction

PREPARED FOR
A SPECIAL SUMMER PROGRAM AT
THE MASSACHUSETTS INSTITUTE OF TECHNOLOGY

Edited by
LEO L. BERANEK
Lecturer, Massachusetts Institute of Technology
President, Bolt Beranek and Newman Inc.

McGraw-Hill Book Company
NEW YORK TORONTO LONDON
1960

NOISE REDUCTION

 Library of Congress Catalog Card Number 59-15454

8 9 10 11 12-MAMM-1

ISBN 07-004832-0

FOREWORD

With the accelerated pace of technological change, industrial engineers in almost every specialized domain need periodic opportunity to become up to date on the fundamental knowledge that underlies their field. Likewise, the engineering academicians in the fields related to these domains need periodic enlightenment concerning the multitude of industrial uses that stem from their teaching in order that their activities may have appropriate relevance. Each needs to exchange information with the other in order to sharpen man's basic understanding of the technology, to expand the base of its useful exploitation, and to keep abreast of the consequences of both new understanding and new application.

The Special Summer Programs conducted by the Massachusetts Institute of Technology on the subject of noise reduction in industry were intended to serve precisely these functions. The subject of noise in industry, by its very nature, offers a unique opportunity for the profitable blending of academic talent and industrial talent. The noise in this instance is largely an undesirable consequence of our industrial advance. The great increase in size of energy conversion units, such as the jet engine, and the widespread use of machines, with their attendant noise in the home, in the shop, in the office, and even in entertainment, have created many problems.

This book is a contribution to our better understanding of these problems and to means for their solution. It has been compiled from the lectures and discussions presented at the Conferences by both academic and industrial participants.

As the director of the Conferences and as editor of this book, Dr. Beranek serves the dual role of academician and practicing engineer. He has here effectively coupled the academic environment of MIT with the industrial complex that it aims to serve.

Gordon S. Brown
Dean of Engineering, MIT

PREFACE

The field of acoustics has many branches, but none is developing more rapidly than noise control. Noise has assumed an importance in national thinking that could hardly have been believed two decades ago. The control of noise must be considered at all stages of the design and engineering of airports, aircraft, buildings, home appliances, industrial machinery, and automotive vehicles. Even cities are adopting specifications on allowable noise levels in residential and industrial areas.

In the United States and Germany, the demand for information has led to the inauguration of semitechnical magazines that are devoted exclusively to noise and its control. In nearly all of the major nations, symposia and technical meetings on noise are now commonplace.

Special Summer Programs on Noise Reduction were offered at the Massachusetts Institute of Technology in 1953, 1955, 1957, and 1960. Much of the material in this book served as the text for those programs.

This book, which is intended to be readable by graduate engineers in nearly any technical field, is divided into four parts as follows:

One. Sound Waves and Their Measurement

Two. Fundamentals Underlying Noise and Vibration Control

Three. Criteria for Noise and Vibration Control

Four. Practical Noise Control

The material is graded in technical level, with the simpler concepts, apparatus and techniques appearing first, and the more specialized and complex techniques occurring later.

No effort has been made by the authors to produce a handbook or an all-inclusive compendium. Rather, this text seeks to lead the reader by gradual steps from the beginning of the subject on into the more advanced aspects. Each man with a noise problem should find assistance. The text contains many numerical examples and frequent comparison of measured with calculated data and gives practical details of construction.

The editor wishes to express his warm thanks to Mr. R. J. Wells of the General Electric Company and Dr. Peter A. Franken of Bolt Beranek and Newman Inc. who have assisted in a major way in the details of the editing; to Professor R. F. Lambert of the University of Minnesota; to Drs. R. Plunkett and D. Muster of the General Engineering Laboratory of the General Electric Company; and to his co-authors at MIT and Bolt Beranek and Newman Inc. for their patience and for their considerable respect for deadlines. He also wishes to thank Professors J. R. Cox, C. E. Crede, R. B. Newman, W. A. Rosenblith, H. W. Rudmose, K. N. Stevens; Drs. J. J. Baruch, H. Davis, I. Dyer, R. O. Fehr, H. O. Parrack, J. Sataloff, C. R. Williams, J. J. Zwislocki; and Messrs. D. Apps, G. H. Ashley, A. L. Cudworth, C. W. Dietrich, V. H. Hill, and S. E. Pihl for participating in the lectures of previous MIT special summer programs on noise reduction.

I also wish to thank Helene M. Morgan, Elizabeth M. Donnelly, and Polly A. Horan for assistance in preparing the manuscript and the figures.

Leo L. Beranek

CONTENTS

Chapter 1
SOME HISTORY AND EARLIER REFERENCES

Leo L. Beranek

Noise reduction in machines and buildings is approaching the state of an engineering science. Only a few years ago, acoustical engineers generally controlled noise by trial and error and in some cases entirely by exercising ordinary common sense. Although the use of common sense is always advisable, the number of cases in acoustics where simple reasoning alone can lead one astray is so great that noise control has often been said to be an art. The transition from art to near-science has taken about 25 years—starting from before World War II and continuing until now.

Science usually begins with the collection of known information in a given field. Typical of such collections are the bulletins on sound-absorption coefficients which were first issued in 1934 and are now published annually[1] by the Acoustical Materials Association (AMA); and the reports on sound insulation which were first published in 1939[2] by the National Bureau of Standards (NBS). As we shall see later in this text, the methods of test used in acoustical investigations can produce such variations in results that rank orderings of materials and structures made in laboratories may not correspond to rank orderings determined from measurements made on actual installations in the field. Thus, collections of data alone are not sufficient to a science, although they generally come first.

For an area of interest to be a science, we must be able to analyze phenomena in that area and to predict future situations quantitatively. Thus a science needs well-developed instrumentation for measurement, terminology for expressing one's findings to others, generalized formulations for explaining and predicting the behavior of systems when their parameters and stimulus are known, and handbooks of tabulated measured data to use in design.

Much of the support necessary to the development of noise reduction as a science has come from the transportation industry and the military. Between 1930 and 1940, noise control was already being applied actively to buildings, automobiles, some aircraft, ships, and public transportation. A typical selection of papers published in that period is given in the references.[3–18]

Also between 1930 and 1940, university and government laboratories were becoming interested in the internal processes involved in the absorption of sound by porous acoustical materials. These results were presented in the *Journal of the Acoustical Society of America* and in foreign scientific journals.[19–29]

Then came World War II. Military personnel had to live for long periods of time in airplanes, tanks, and ships. What was more important, they had to communicate by voice. So critical was the noise problem that in the United States the National Defense Research Committee (later the Office of Scientific Research and Development) set up two laboratories at Harvard University especially to carry on research on sound control in combat vehicles and research on communication equipment for operation in noise.

One of those Harvard laboratories, the Psycho-Acoustic Laboratory, under the direction of S. S. Stevens, was charged with investigating the psychological aspects of noise control and speech communication. The other laboratory, the Electro-Acoustic Laboratory, under the direction of L. L. Beranek, was charged with investigating the physical aspects of those problems. In addition to reporting on many facets of speech communication in noise, those two laboratories issued reports on the effects of noise on man; on noise control in airplanes, airships, tracked landing vehicles, tanks, ships, and outboard motors; on the properties of acoustical materials; and on apparatus for measuring noise and vibration.

One of the projects of the Electro-Acoustic Laboratory was concerned with developing structures and materials that were light in weight for a given noise reduction. It was soon observed that knowledge of a material's flow resistance as well as its thickness and density was of first-order importance, particularly in determining the transmission loss (TL). It was learned that interleaving porous blankets with impervious septa increased the TL of a structure without negating the surface absorption. It was also observed that it was advantageous to add vibration-damping materials to thin panels to increase their TL. The group at the Electro-Acoustic Laboratory discovered the value of very fine-fibered-glass acoustic blankets in aircraft noise control. This material was initially developed for possible use as filters in gas masks and was being manufactured in small lots when its application to noise reduction was discovered. Pertinent publications on noise—its effect on man and its control—resulting from the war-time work of the Psycho-Acoustic and the Electro-Acoustic Laboratories are given in Refs. 30 through 41.

During World War II unbeknown to American acousticians, an important series of papers on sound absorption in acoustical materials was being published in Europe by Zwikker, Kosten, and v. den Eijk.[42] Those papers and later papers by Scott,[43] Beranek,[36] and Kosten and Janssen[44] showed that it is necessary to know the structure factor, the dynamic stiffness of the gas, the dynamic stiffness of the skeleton of the material, the density, the flow resistance, the thickness, and the porosity of the material in order to determine the acoustical behavior of flexible and rigid porous acoustic structures.

After World War II, noise problems in architecture and in industry received more emphasis than at any other earlier period. Organizations such as the Massachusetts Institute of Technology and the Armour Research Foundation sponsored research and engineering projects pertinent to these demands. Each group, of course, developed its own specialities and its own philosophy of operation. For example, in the period from 1947 to 1952, the Armour group published papers on industrial transportation and residential noise in Chicago, on jet engine test cell noise control, on control of noise in mechanical equipment, and on noise reduction in piping.[45–50]

The MIT group, influenced in part by contacts with students, attempted to bring a new concept to the solution of noise-control problems—namely, the *systems concept*. Each noise-control problem was visualized as being a system with three components; the *source*, the *path*, and the *receiver*. No problem could be said to be satisfactorily handled unless the characteristics of the source were known and a criterion, expressing the desired noise levels at the receiver, was established. Only then should one decide how much control need be put in the intervening path. These concepts were described in a paper by R. H. Bolt[51] and are detailed in three volumes of the "Handbook of Acoustic Noise Control"[52] prepared under a contract with the consulting firm of Bolt Beranek and Newman Inc. (BBN). These volumes are based on a review of the literature, particularly the excellent work of the National Advisory Committee for Aeronautics (NACA), and on published and unpublished works of the MIT and BBN groups.[53–89]

In the summer of 1953, with these volumes as the basis, the Massachusetts Institute of Technology, Summer Session, offered for the first time a special summer course in noise reduction. This course took two weeks, and was attended by 120 men from industry, government, and other academic institutions. Attendance was stimulated in part by the growing awareness of industry that quiet products had sales appeal. More important, perhaps, was the "damage to hearing scare" brought on by several successful lawsuits by workmen against industry in the states of New York and Wisconsin.[90,91]

A greater quantity of papers, too numerous to list here, started to appear in the literature after 1953. Management, industrial hygiene, and engineering organizations of many types held special symposiums on damage to hearing, industrial audiometric programs, and noise control in factories.

The new material of the 1953 special summer program at MIT was incorporated into a text along with other basic material on acoustics.[92] The General Radio Company's "Handbook of Noise Measurement" (1st ed.)[93] was published just ahead of the program. It used the systems concept and the newer terminology of the MIT and BBN groups. Because it was used as one of the texts, in effect, it is a partial record of the 1953 program.

In the years from 1953 to 1955 many publications on noise control appeared in the *Journal of the Acoustical Society of America*.[94] Then, in January, 1955, the magazine *Noise Control*[95] was started by the Acoustical Society of America. This magazine serves a need in the United States for a medium for disseminating information on noise control that others besides scientists and acoustical engineers can understand. The Proceedings of the Annual National Noise Abatement Symposium (NNAS) and of the West Coast Noise Symposium are carried in its pages, thus reducing the number of places that one needs go to find the American literature on this subject.

Foreign publications are also very active in disseminating information on noise and its control. Principal among these periodicals are *Acustica*,[96] *Akustische Beihefte*,[97] *Larmbekampfung*,[98] the *Journal of the Acoustical Society of Japan*,[99] and the *Journal of Acoustics of the Academy of Sciences, U.S.S.R.*[100]

In 1955 and again in 1957 two-week special summer programs on noise reduction were held at MIT. The enrollments were 165 and 185 respectively, and much new material was introduced into the curriculum. Liberal use was made of references to *Noise Control* and to existing texts on acoustics.[92–93,101–104] Lecture notes containing new material were issued. It became obvious that a text on noise reduction could now be written with the material from these lectures as a start.

The book you are reading is partly from those programs, upgraded to serve the 1960 program. Because it covers only the material of a two-week course, we recommend strongly three other publications as supplementary reading.[105–107]

Where does noise reduction go from here? It strikes us that six areas of the field will receive heavy attention in the next few years. These areas are (1) noise of aerodynamic origin, (2) structure-borne sound and vibration, (3) vibration and shock isolation, (4) failure of structures due to fatigue induced by sound and vibration, (5) processes for quietly pumping liquids and transforming and transporting mechanical power, and (6) instrumentation for more rapid and complete measurement and analysis of sound and vibration and for measurement of acoustic and mechanical impedances. References to current literature on these and other topics appear at the end of the chapters of this book.

Without reservation, the authors are fully aware of their great indebtedness to the work of other organizations and individuals in the United States and particularly to the influence of research done on noise control in the fine laboratories of Germany, England, France, Switzerland, Holland, Denmark, and other countries. We hope that the inevitable emphasis placed on our own work in this text will not prove offensive. Our explanation is that the essence of university education is personal contact by students with particular teachers and their particular ideas. We hope, however, that we have included solid, usable material pertinent to many aspects of noise reduction.

REFERENCES

1. "Sound Absorption Coefficients of Architectural and Acoustical Materials" (issued annually), Acoustical Materials Association, 335 E. 45th St., N.Y. 17, N.Y.

2. Sound Insulation of Wall and Floor Constructions, *U.S. Natl. Bur. Standards Rept. BMS* 144, Feb. 25, 1955, and supplement February, 1956.

3. Bassett, P. R., and S. J. Zand: Noise Reduction in Cabin Airplanes, *Trans. ASME,* vol. 56, p. 49, 1934.

4. Chrisler, V. L., and W. F. Snyder: Soundproofing of Airplane Cabins, *J. Research Natl. Bur. Standards,* vol. 2, p. 897, 1929.

5. Davis, A. H.: Silencing Aircraft, *Engr.,* vol. 156, p. 316, 1933.

6. Zand, S. J., and G. Perot: Étude du comfort à bord des avions de transport et application pratique à un appareil, *L'Aéronaut.,* vol. 18, p. 16, 1936.

7. Bruderlin, H. H.: Developments in Aircraft Sound Control, *J. Acoust. Soc. Am.,* vol. 8, pp. 181–184, 1937.

8. Rettinger, M.: Sound Transmission in Materials, *J. Acoust. Soc. Am.,* vol. 8, pp. 172–175, 1937.

9. Norris, R. F.: Automotive Quieting, *J. Acoust. Soc. Am.,* vol. 8, pp. 100–103, 1936.

10. Sivian, L. J.: Sound Propagation in Ducts, *J. Acoust. Soc. Am.,* vol. 9, pp. 135–140, 1937.

11. Attenuation of Noises of Diesel Engines in Ships, *Z. Ver. dent. Ing.,* vol. 81, pp. 770–776, 1937.

12. Morreau, C. J.: Insulation of Airborne Sound, *J. Acoust. Soc. Am.*, vol. 10, pp. 45–49, 1938.

13. Cary, B. C.: Some Acoustical Problems of the Automotive Fan, *J. Acoust. Soc. Am.*, vol. 10, pp. 63–67, 1938.

14. Trevor, J. S.: Silencing London's Tubes, *Transit J.*, vol. 82, pp. 482–483 and 508–509, 1938.

15. Robinson, R. S.: Noise Reduction in Ships, *Engr.*, vol. 166, pp. 568–569, 1938.

16. Henshaw, H., and W. A. Johnson: Reduction of Noise in Coal Mines, *Colliery Guardian*, vol. 157, pp. 148–151, 1938.

17. Leedy, H. A.: Noise and Vibration Isolation, *J. Acoust. Soc. Am.*, vol. 11, pp. 341–345, 1940.

18. Smith, P. H.: Noise Control, *Sci. American*, vol. 161, pp. 204–207, 1939.

19. Chrisler, V. L.: Sound Absorption Coefficients, *J. Acoust. Soc. Am.*, vol. 6, p. 115, 1934.

20. Germant, A. G.: Resistance to Airflow in Sound Absorbers, *Sitzber. preuss. Akad. Wiss. Physik.-math. Kl.*, vol. 17, p. 579, 1933.

21. Paris, E. T.: Reflection of Sound from a Porous Surface, *Proc. Phys. Soc. London*, vol. 115, pp. 407–419, 1927.

22. Paris, E. T.: Oblique Reflection of Sound, *Nature*, vol. 9, p. 126, 1930.

23. Penman, H. L., and E. G. Richardson: The Absorption by Porous Materials at Normal Incidence, *J. Acoust. Soc. Am.*, vol. 4, p. 322–330, 1933.

24. Rettinger, M.: The Theory of Sound Absorption of Porous Materials, *J. Acoust. Soc. Am.*, vol. 8, p. 53–59, 1936.

25. Morse, P. M., R. H. Bolt, and R. L. Brown: Acoustic Impedance and Sound Absorption, *J. Acoust. Soc. Am.*, vol. 12, pp. 217–227, 1940.

26. Wintergerst, E.: Theorie der Schalldurchlässigkeit von einfachen und zusammengesetzen Wänden, *Schalltechnick*, vol. 4, p. 85, 1931, and vol. 5, p. 1, 1932.

27. Kuhl, V., and E. Meyer: Untersuchungen über die Winkel- und Frequenzabhängigkeit der Schallschluckung von porösen Stoffen, *Sitzber. preuss. Akad. Wiss. Physik.-math. Kl.*, vol. 16, p. 416. 1932.

28. Beranek, L. L.: Precision Measurement of Acoustic Impedance, *J. Acoust. Soc. Am.*, vol. 12, pp. 3–13, 1940.

29. Beranek, L. L.: Acoustic Impedance of Commercial Materials, *J. Acoust. Soc. Am.*, vol. 12, pp. 14–23, 1940.

30. Stevens, S. S., J. P. Egan, T. H. Waterman, J. Miller, R. H. Knapp, and S. C. Rowe: Part I, The Effects of Noise on Psychomotor Efficiencies, Part II, Noise Reduction in Aircraft as Related to Communication Annoyance and Aural Injury, *Natl. Research Council Comm. on Sound Control, O.S.R.D. Rept.* 247, 1941. Copies may be purchased from the Office of Technical Service, Dept. of Commerce; refer to *Rept.* PB 8334.

31. Beranek, L. L.: Acoustic Impedance of Porous Materials, *J. Acoust. Soc. Am.*, vol. 13, pp. 248–260, 1942.

32. Wiener, F. M., and R. J. Marquis: Noise Levels Due to an Airplane Passing Overhead, *J. Acoust. Soc. Am.*, vol. 18, pp. 450–452, 1946.

33. Beranek, L. L.: Some Notes on the Measurement of Acoustic Impedance, *J. Acoust. Soc. Am.*, vol. 19, pp. 420–427, 1947.

34. Nichols, R. H., Jr., H. P. Sleeper, Jr., R. L. Wallace, Jr., and H. L. Ericson: Acoustical Materials and Acoustical Treatments for Aircraft, *J. Acoust. Soc. Am.*, vol. 19, pp. 428–443, 1947.

35. Beranek, L. L., and H. W. Rudmose: Sound Control in Airplanes, *J. Acoust. Soc. Am.,* vol. 19, pp. 357–364, 1947.

36. Beranek, L. L.: Acoustical Properties of Homogeneous, Isotropic Rigid Tiles and Flexible Blankets, *J. Acoust. Soc. Am.,* vol. 19, pp. 556–568, 1947.

37. Nichols, R. H., Jr.: Flow Resistance Characteristics of Fibrous Acoustical Materials, *J. Acoust. Soc. Am.,* vol. 19, pp. 866–871, 1947.

38. Beranek, L. L., and H. W. Rudmose: Airplane Quieting I—Measurement of Sound Levels in Flight, *Trans. ASME,* vol. 69, pp. 89–96, 1947.

39. Beranek, L. L.: Airplane Quieting II—Specification of Acceptable Noise Levels, *Trans. ASME,* vol. 69, pp. 97–100, 1947.

40. Rudmose, H. W., and L. L. Beranek: Noise Reduction in Aircraft, *J. Aeronaut. Sci.,* vol. 14, pp. 79–96, 1947.

41. Beranek, L. L., and G. A. Work: Sound Transmission Through Multiple Structures Containing Flexible Blankets, *J. Acoust. Soc. Am.,* vol. 21, pp. 419–428, 1949.

42. Zwikker, C., C. W. Kosten, and J. v. den Eijk: Absorption of Sound by Porous Material, *Physica,* vol. 8, pp. 149, 469, 1094, and 1102, 1941; C. W. Kosten and C. Zwikker: Extended Theory of the Absorption of Sound by Compressible Wall Coverings, *Physica,* vol. 8, p. 968, 1941.

43. Scott, R. A.: The Absorption of Sound In a Homogeneous Porous Medium, *Proc. Phys. Soc. London,* vol. 58, p. 165, 1946.

44. Kosten, C. W., and J. H. Janssen: Acoustic Properties of Flexible and Porous Materials, *Acustica,* vol. 7, pp. 372–378, 1957.

45. Bonvallet, G. L.: Levels and Spectra of Transportation Vehicle Noise, *J. Acoust. Soc. Am.,* vol. 22, pp. 201–205, March, 1950.

46. Hardy, H. C.: Control of Noise in Mechanical Equipment, *Proc. 1st Annual NNAS,* pp. 38–52, 1950.

47. Bonvallet, G. L.: Levels and Spectra of Traffic, Industrial and Residential Area Noise, *J. Acoust. Soc. Am.,* vol. 23, pp. 435–439, July, 1951.

48. Hardy, H. C.: Design Characteristics for Noise Control of Jet Engine Test Cells, *J. Acoust. Soc. Am.,* vol. 24, pp. 185–190, 1952.

49. Potter, S. M., and H. C. Hardy: Quieting of the Jet Engine Testing Facilities in the Willgoos Turbine Laboratory, *J. Acoust. Soc. Am.* (*Abstr.*), vol. 24, p. 116, 1952.

50. Callaway, D. B., F. G. Tyzzer, and H. C. Hardy: Techniques for Evaluation of Noise-reducing Piping Components, *J. Acoust. Soc. Am.,* vol. 24, pp. 725–730, 1952.

51. Bolt, R. H.: The Aircraft Noise Problem, *J. Acoust. Soc. Am.,* vol. 25, pp. 363–366, 1953.

52. Bolt Beranek and Newman Inc., "Handbook of Acoustic Noise Control," vol. I, Physical Characteristics, and vol. II, Noise and Man, *WADC Tech. Rept.* 52–204, Wright-Patterson Air Force Base, Ohio, 1952–53. Order from Office of Technical Services, Dept. of Commerce, vol. I, PB 111,200; vol. II, PB 111,274; and vol. I, suppl. 1, PB 111,200S.

53. Wathen-Dunn, W.: Audible P–80 Jet Aircraft Noise, *Naval Research Lab. Rept. S*–3266, March, 1948.

54. Gutin, L.: Über das Schallfeld einer Rotierenden Luftschraube, *Phys. Z. Sowjetunion,* vol. 9, pp. 57–71, 1936; English translation, *NACA Tech. Mem.* 1195, 1948.

55. Cullum, D. J. W.: "The Practical Application of Acoustic Principles," E. and F. N. Spon, Ltd., London, 1949.

56. Knudsen, V. O., and C. M. Harris: "Acoustical Designing in Architecture," John Wiley & Sons, New York, 1950.

57. Cremer, L.: "Die Wissenschaftlichen Grundlagen der Raumakustik," Band III, Wellentheoretische Raumakustik, S. Hirzel Verlag, Leipzig, 1950.

58. Beranek, L. L., F. Elwell, J. P. Roberts, and C. F. Taylor: Experiments in External Noise Reduction of Light Airplanes, *NACA Tech. Note* 2079, 1950.

59. Hubbard, H. H., and A. A. Regier: Free-space Oscillating Pressures Near the Tips of Rotating Propellers, *NACA Tech. Rept.* 996, 1950.

60. Roberts, J. P., and L. L. Beranek: Experiments in External Noise Reduction of a Small Pusher-type Amphibian Airplane, *NACA Tech. Note* 2727, 1952.

61. Hubbard, H. H., and L. W. Lassiter: Sound from a Two-blade Propeller at Supersonic Tip Speeds, *NACA Research Tech. Rept.* 1079, 1952.

62. von Gierke, H. E., H. O. Parrack, W. J. Gannon, and R. G. Hansen: The Noise Field of a Turbo-jet Engine, *J. Acoust. Soc. Am.,* vol. 24, pp. 169–174, 1952.

63. Hubbard, H. H.: Propeller Noise Charts for Transport Airplanes, *NACA Tech. Rept.* 2968, 1953.

64. Regier, A. A., and H. H. Hubbard: Status of Research on Propeller Noise and Its Reduction, *J. Acoust. Soc. Am.,* vol. 25, pp. 395–404, 1953.

65. Veneklasen, P. S.: Noise Control for Operation of the F–89 Airplane, *J. Acoust. Soc. Am.,* vol. 25, pp. 417–422, 1953.

66. Beranek, L. L.: "Acoustic Measurements," John Wiley & Sons, Inc., New York, 1949.

67. Goldstein, S.: "Modern Developments in Fluid Mechanics," vols. I and II, Oxford University Press, New York, 1950.

68. Numerous articles in the field of psychoacoustics.

69. Stevens, S. S., and H. Davis: "Hearing," John Wiley & Sons, Inc., New York, 1938.

70. Davis, H.: "Hearing and Deafness," Rinehart & Company, Inc., New York, 1947.

71. Watson, L. S., and T. Tolan: "Hearing Tests and Hearing Instruments," The Williams & Wilkins Company, Baltimore, 1949.

72. Kryter, K. D.: The Effects of Noise on Man, *J. Speech and Hearing Disorder,* monograph suppl. 1, 1950.

73. Miller, G. A.: "Language and Communication," McGraw-Hill Book Company, Inc., New York, 1951.

74. Stevens, S. S. (ed.): "Handbook of Experimental Psychology," John Wiley & Sons, Inc., New York, 1951.

75. Hirsch, I. H.: "The Measurement of Hearing," McGraw-Hill Book Company, Inc., New York, 1952.

76. Fletcher, H.: "Speech and Hearing in Communication," D. Van Nostrand Company, Inc., Princeton, N.J., 1953.

77. Beranek, L. L., J. L. Reynolds, and K. E. Wilson: Apparatus and Procedure for Predicting Ventilation System Noise, *J. Acoust. Soc. Am.,* vol. 25, pp. 313–321, 1953.

78. Peistrup, C. F., and J. E. Wesler: Noise of Ventilating Fans, *J. Acoust. Soc. Am.,* vol. 25, pp. 322–326, 1953.

79. Cremer, L.: Theorie de Schalldämmung dünner Wände bei schrägem Einfall, *Akust. Z.,* vol. 7, pp. 81–104, 1942.

80. Bolt, R. H.: On the Design of Perforated Facings of Acoustic Material, *J. Acoust. Soc. Am.,* vol. 19, pp. 917–921, 1947.

81. Cremer, L., and A. Eisenberg: Verbesserung der Schalldämmung dünner Wände durch Verringerung ihrer Biegesteifigkeit, *Bauplanung u. Bautechnik,* vol. 2, p. 235, 1948.

82. London, A.: Transmission of Reverberant Sound through Double Walls, *J. Acoust. Soc. Am.,* vol. 22, pp. 270–279, 1950.

83. Ingard, U., and R. H. Bolt.: Absorption Characteristics of Acoustic Material with Perforated Facings, *J. Acoust. Soc. Am.,* vol. 23, pp. 533–540, 1951.

84. Fehr, R. O.: The Reduction of Industrial Machine Noise, *Proc. 2d Annual NNAS,* pp. 93–103, 1951.

85. Cremer, L.: Theory of Impact Noise Transmission in Floors with Floating Coverings, *Acustica,* vol. 2, pp. 167–178, 1952.

86. Ingard, U.: On the Theory and Design of Acoustic Resonators, *J. Acoust. Soc. Am.,* vol. 25, pp. 1037–1061, 1953.

87. Bolt, R. H., S. Labate, and U. Ingard: The Acoustic Reactance of Small Circular Orifices, *J. Acoust. Soc. Am.,* vol. 21, pp. 94–97, 1949.

88. Lawhead, R. B., and I. Rudnick: Acoustic Wave Propagation along a Constant Normal Impedance Boundary, *J. Acoust. Soc. Am.,* vol. 23, pp. 546–549, 1951.

89. Ingard, U., and D. Pridmore-Brown: Propagation of Sound in a Duct with Constrictions, *J. Acoust. Soc. Am.,* vol. 23, pp. 689–694, 1951.

90. Symons, N. S.: Legal Aspects of Noise: Legal and Legislative Developments in New York State, *Noise Control,* vol. 1, pp. 72–74, January, 1955.

91. Fox, M. S.: The Wisconsin Story of the Industrial Noise Problem, *Noise Control,* vol. 1, pp. 74–76 and 89, January, 1955.

92. Beranek, L. L.: "Acoustics," McGraw-Hill Book Company, Inc., New York, 1954.

93. Peterson, A. P. G., and L. L. Beranek: "Handbook of Noise Measurement," General Radio Company, Cambridge, Mass., 1953, 1954, 1956.

94. *J. Acoust. Soc. Am.,* published by the Acoustical Society of America, 335 E. 45th St., N.Y. 17, N.Y.

95. *Noise Control,* a magazine devoted to the control of noise and vibration, also published by the Acoustical Society of America.

96. *Acustica,* an international journal in acoustics jointly sponsored by (1) the Acoustics Group of the Physical Society of Great Britain, (2) the Groupement des Acousticiens de Langue Française, and (3) the Verbandes Deutscher Physikalischer Gesellschaften. Published by S. Hirzel Verlag, Stuttgart, Germany. (Papers are accepted in three languages.)

97. *Akust. Beih.,* a supplementary publication, in German, indexed and furnished with *Acustica.*

98. *Larmbekampfung* (the German equivalent of the American *Noise Control*), Verlag für Angewandte Wissenschaften Gmbh, Baden-Baden, Germany.

99. *J. Acoust. Soc. Japan* (*Nippon Onkyô Gakkaishi*), published by Acoustical Society of Japan, Institute of Science and Technology, Komaba, MeguroKu, Tokyo.

100. *J. Acoustics,* quarterly, Academy of Science, U.S.S.R. Available in English by subscription from American Institute of Physics, 335 E. 45th St., N.Y. 17, N.Y.

101. Morse, P. M.: "Vibration and Sound," 2d ed., McGraw-Hill Book Company, Inc., New York, 1948.

102. Kinsler, L. E., and A. R. Frey: "Fundamentals of Acoustics," John Wiley & Sons, Inc., New York, 1950.

103. Swenson, G. W., Jr.: "Principles of Modern Acoustics," D. Van Nostrand Company, Inc., Princeton, N.J., 1953.

104. Hunter, J. L.: "Acoustics," Prentice-Hall, Inc., Englewood Cliffs, N.J., 1957.

105. Harris, C. M. (ed.): "Handbook of Noise Control," McGraw-Hill Book Company, Inc., New York, 1957.

106. Noise, chap. 40 of "Heating, Ventilating, and Air Conditioning Guide," American Society of Heating and Air-Conditioning Engineers, Inc., 62 Worth St., N.Y. 13, N.Y., 1957. (Much of the material in this chapter came from the researches of Prof. L. L. Beranek and his students at MIT.)

107. Doelling, N., and R. H. Bolt: Noise Control for Aircraft Engine Test Cells and Ground Run-up Suppressors, *WADC Tech. Rept.* 58-202, Wright-Patterson Air Force Base, Ohio, 1958.

Part One

SOUND WAVES AND THEIR MEASUREMENT

Chapter 2

BEHAVIOR OF SOUND WAVES

William J. Galloway and Leo L. Beranek

2.1 Basic Properties of Waves

We all have some intuitive ideas about what makes up a wave. Almost everyone can visualize some occurrence in nature in which wave motion has been apparent. Probably the most frequently recalled wave motion is that of water waves: ocean waves striking the beach, or ripples in a pond when a stone strikes the surface of a lake. In other ways we are constantly being reminded of the notion of a wave—radio waves, heat waves, shock waves, even waves of enthusiasm or dejection in the stock market.

From physics we learn that all forms of wave motion satisfy the statement that: A disturbance initiated at some point is transmitted (propagated) to other points in a predictable manner determined by the physical properties of the medium existing between the points of observation.

Sound waves are a particular form of a general class of waves known as elastic waves. Elastic waves can occur in media having the properties of mass (inertia) and elasticity. The elasticity tends to pull a displaced particle back to its original position as a spring would. Because of its inertia the displaced particle can transfer momentum to an adjoining particle.

A simple example of the transmission of an elastic wave is shown in Fig. 2.1. An edge-wound steel coil spring is stretched to a length of approximately 2 m. Inertia is associated with the mass of the coils. Elasticity exists in the spring and tends to keep the adjacent coils evenly spaced. When a sudden movement is imparted to one end of the stretched spring in the direction of its axis, the coils are compressed and energy is stored in their movement. The resulting momentum causes the displaced coils to impart force to coils farther down the line and the elasticity tends to return the displaced coils to their initial position. A wave is propagated along the spring and would continue indefinitely to the right if the spring were infinitely long. This type of wave is longitudinal because all motions take place in the direction that the wave is traveling.

Most often when we think of sound waves we are considering air or some other gas as our elastic medium. Although sound propaga-

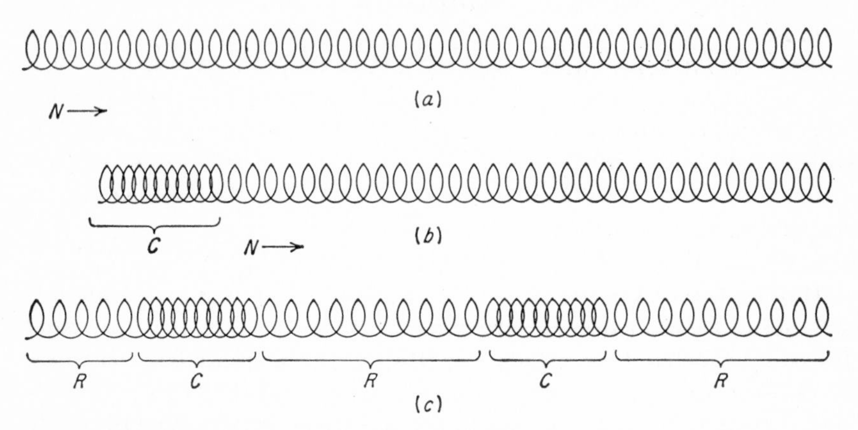

FIG. 2.1 Transmission of an elastic wave along an edge-wound steel coil spring. (*a*) Quiescent state, (*b*) displacement of the end of the spring by sudden compression, (*c*) wave propagation following a series of sudden compressions imparted at the left end. *N*, normal state; *C*, compressed state; *R*, rarefied or extended state.

tion in other media is considered in later chapters, we shall restrict our attention to sound waves in air for the remaining portion of this chapter.

2.2 Air and Its Properties

In our definition of an elastic medium we stated that it must have inertia and elasticity. How do we determine that air satisfies these requirements?

In considering inertia we are implying that the air has mass or that it "weighs" so much. For our purposes here let us assume that air consists of a large number of molecules, which we will think of as very small spheres distributed uniformly within the space in question, each molecule moving in random motion. If we were to enclose a portion of space in a box, first with no air and then with the box filled with air at normal room *temperatures* and *pressures,* we would find that the weight of the air would be about 1 lb for each 13 ft^3 or about 1 kg for each m^3 we had contained. We would also find that the air exerts a force on the surfaces of the box, amounting to about 14.7 lb/$in.^2$, or in mks units, about 100,000 newtons/m^2. This force per unit area is what we normally term *atmospheric pressure* and is directly related to the *density,* or number of molecules, of the air in the box. We find that we have now described the gross properties of our air by specifying the three parameters: (1) pressure, (2) density, and (3) temperature.

We can show that air has elasticity by a number of means. For example, if we take a basketball in its uninflated state and drop it to the floor, it stays there without rebounding. Inflating the ball with air imparts a resiliency to it which permits it to rebound as expected when dropped to the floor. Even more simply, suppose while preparing to inflate the ball we had held a finger over the air outlet of our hand pump so that no air could come out. Pushing on the handle to compress the trapped air in the pump produces the same sensation as pushing on a spring. In fact, by knowing the physical dimensions of the pump and the amount of force we had applied to its handle, we could determine the "spring constant" or compressibility of the gas, thus giving a quantitative description of the elasticity of the air.

In summary, because of its weight, air has inertia. Because of its "springiness," it has elasticity. Because air possesses both inertia and elasticity, a wave can be propagated in air.

2.3 Simple Harmonic Motion—The Sine Wave

Now that we have talked about density, elasticity, pressure, and temperature of air, we are in a position to describe how we can propagate a sound wave through it. While we could discuss very complex sounds, it seems more useful to consider first the simplest form of sound, the pure tone. In an analytical sense, the pure tone originates from a type of motion known as *simple harmonic.* The sound wave produced by this motion is known as a *sine wave.* Let us consider how this wave is produced.

Simple harmonic motion is often described as the projection on a plane of the path of a point moving around a circle at uniform speed. For example, in Fig. 2.2*a* assume that the moving point P_1 travels around the circle at constant speed v. The projection of this point onto a plane would trace back and forth along a line as the point went around the circle (see Fig. 2.2*b*). Now suppose that we moved the plane on which we were plotting our projection to the left at a uniform speed c (as indicated in Fig. 2.2*d*) instead of allowing the trace to pile up on the line as Fig. 2.2*b* indicates. The resulting trace has the form of the graph of the trigonometric function known as the *sine.* That this is so is evident when one recalls that the definition of the sine of an angle is the ratio of the side opposite the angle in a right triangle to the hypotenuse of the triangle. Simple harmonic motion is motion which changes in magnitude and time in the manner prescribed by Fig. 2.2*c* and *d* and which repeats itself exactly. A body is said to be moving in simple harmonic motion if the trace of the displacement of the body provides a plot of displacement as a function of time similar to the sine wave indicated in Fig. 2.2*d*.

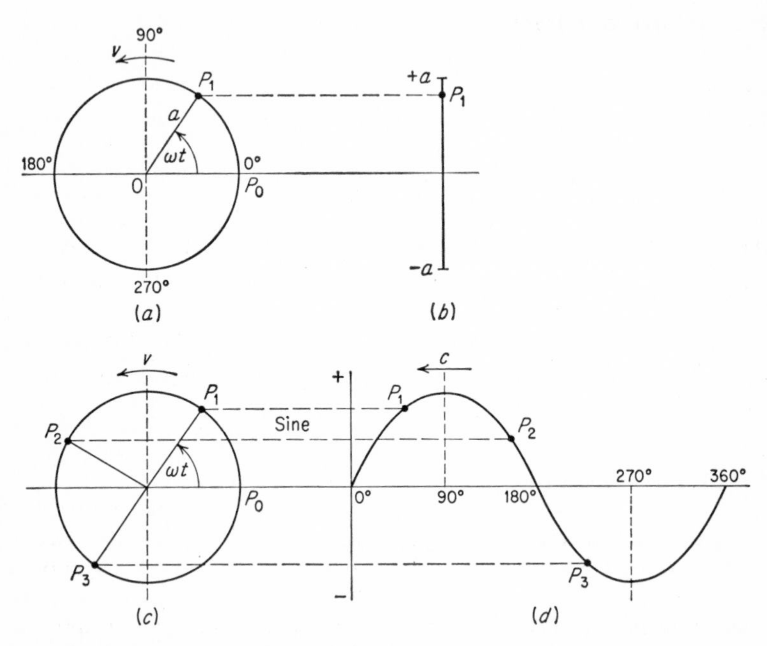

FIG. 2.2 Simple harmonic motion and the sine wave. In (*a*) the point P_1 moves around the circle at a constant tangential speed v and in (*b*) its projection moves up and down between the limits $+a$ and $-a$. In (*c*) the same is true except that the plane on which the projection is plotted (*d*) is traveling to the left at a speed c, thereby generating a sine wave. Note that the angular speed is ω(radians per second) and that the total angle traced out from the 0° position equals ωt in radians.

Consider a long tube containing air, with a movable piston at one end (see Fig. 2.3). Assume that the piston is made to move back and forth in simple harmonic motion. This to-and-fro motion of the piston will cause the air molecules adjacent to the piston to alternately crowd together, or compress, and then move apart or become rarefied. This action of alternate compression and rarefaction moves down the tube due to the elasticity of the medium. A sine-shaped wave is thus generated whose frequency is the rate (number of times per second) at which the piston moves back and forth. The strength of the wave is determined by the magnitude of the displacement of the piston.

2.4 Sound Waves in Air

In the previous section we have indicated how a simple sound wave of a single frequency might be generated. We should now consider how this wave moves through the air and what parameters might be used to characterize its progress.

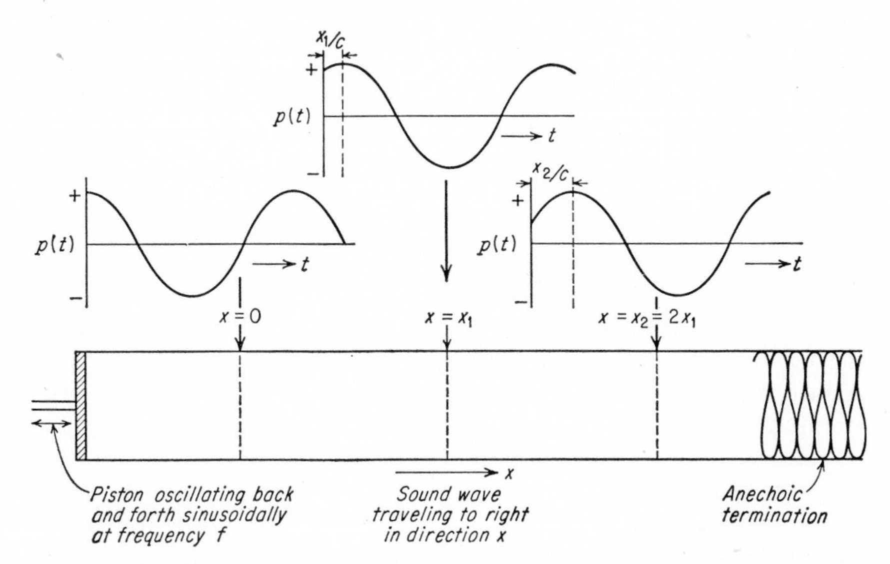

FIG. 2.3 Plane-wave propagation. A plane wave generated by the piston at the left of the tube travels to the right and is absorbed by the anechoic termination. The waves at the top give the variation in sound pressure with time at the three points indicated, $x = 0$, $x = x_1$, and $x = x_2 = 2x_1$.

We have indicated that the strength, or amplitude, of a sound wave is directly determined by the net displacement that the molecules undergo in moving in one direction and in return. Other parameters one can employ in describing the sound wave are particle velocity and sound pressure.

Particle velocity is the rate at which displacement takes place, or displacement per unit time. If the displacement per unit time increases, the amplitude of the sound wave increases, and the particle velocity increases accordingly. One of the major drawbacks to the use of particle velocity in describing the magnitude of a sound wave lies in the relative scarcity of suitable instrumentation of small size and sufficient durability for field use to measure it.

The fact that the molecules are in a type of motion that changes direction implies that particle acceleration or momentum change is present. Since force is the rate of change of momentum, measurement of the forces causing the air molecules to displace from their equilibrium positions provides a direct indication of the amplitude of the sound wave. Measurement of the force exerted on a few molecules is difficult. However, by measuring the force exerted on a relatively large area one can obtain the force per unit area, or pressure. The instantaneous pressure recorded in a sound wave is usually a small variation from the normal atmospheric pressure. This small change from atmospheric pressure is called the *excess sound pressure*, or simply *sound pressure*. The unit most often used in reporting sound pressure is the *microbar*, about 1/1,000,000 atm.*

In the previous section we said that frequency is the rate at which the piston causing the sound wave moved back and forth. By this we mean the number of times per second that the piston displaced from its equilibrium position, rebounded through the equilibrium position to a maximum displacement opposite in direction to the initial displacement, and then returned to its equilibrium position. In other words, the *frequency* of our wave is the number of times per second it traces out one complete cycle of a sine wave. The time required for each cycle is known as the *period* of the wave, and is simply the reciprocal of the frequency. If the piston were vibrating at a frequency of 100 cycles/sec (cps), the period of the vibration would be 0.01 sec.

We have indicated that the motion of the piston in the tube caused the air molecules to compress and expand in such a way that a sound wave is started down the tube. If we were to position ourselves down the tube some distance from the piston and then have the piston started in motion, by suitable instruments we would be able to measure that it would take a definite period of time before the sound wave starting at the piston would reach our observation point. The length of time necessary to transmit the wave down the tube is a function of the elasticity of the air and its density.

By experiment we can determine that the elasticity of the air is equal to a constant times the static pressure of the air, i.e., atmospheric pressure. The constant turns out to be the ratio of the specific heat of the air at constant pressure to the specific heat at constant volume. At most temperatures with which we deal this ratio is 1.4. Thus the speed of sound in air is given by the equation

* The microbar is exactly 1.0 $dyne/cm^2$ or 0.1 $newton/m^2$.

$$c = \sqrt{\frac{1.4P_0}{\rho_0}} \tag{2.1}$$

where P_0 is the atmospheric pressure and ρ_0 is the density of the air.

If we make the assumption that air behaves as an ideal gas, a perfectly reasonable assumption at most temperatures and densities with which we deal, we can show that the speed of sound is really dependent only on the absolute temperature of the air.

In English units, the equation for the speed of sound in feet per second becomes

$$c = 49.03\sqrt{R} \qquad \text{ft/sec} \tag{2.2}$$

where R is the temperature in degrees Rankine, i.e., 459.7° plus the temperature in degrees Fahrenheit.

Example 2.1. Determine the speed of sound at 70°F.

Solution. The Rankine temperature is $459.7 + 70 = 529.7$°R. The speed of sound is then about

$$c = 49\sqrt{R} = 49\sqrt{530}$$

$$= 1{,}128 \text{ ft/sec}$$

In mks units, the equation for the speed of sound in meters per second becomes

$$c = 20.05\sqrt{T} \qquad \text{m/sec} \tag{2.3}$$

where T is the absolute temperature in degrees Kelvin, i.e., 273.2° plus the temperature in degrees centigrade.

Example 2.2. Determine the speed of sound at the same temperature as in Example 2.1, i.e., 21.1°C.

Solution. The absolute temperature is about $273.2 + 21.1 = 294.3$°K. The speed of sound is then

$$c = 20.05\sqrt{294} = 344 \text{ m/sec}$$

Frequency and the speed of sound are related through the important quantity wavelength. The wavelength of a sound wave is the distance the wave travels during one period or cycle. A 100-cps wave would move a distance of 11.3 ft in 1 cycle when the speed of sound is 1,130 ft/sec. Thus, the speed of sound is equal to the product of the frequency and the wavelength:

$$c = f\lambda \tag{2.4}$$

where c = speed of sound, ft/sec or m/sec
f = frequency, cps
λ = wavelength, ft or m

It is important to note that the speed of sound is dependent only upon the properties of the air and is constant at a given temperature. On the other hand, the wavelength decreases as the frequency increases. If the frequency were 1,000 cps in Example 2.1 in the previous paragraphs, the wavelength would be 1.13 ft.

2.5 Free Progressive Waves

Qualitative Aspects. In our discussion so far we talked about a particular sound wave moving down a tube with its motion imparted by a piston vibrating at one end of the tube (Fig. 2.3). We assumed that the tube was infinitely long or else had a nonreflecting (anechoic) termination so that the effects of what was at the end of the tube would not enter into the discussion. The sound wave started by the piston merely progressed on down the tube. We could define all the properties of the wave in terms of the distance down the tube away from the piston and the action of the piston itself. This form of wave is known as a *one-dimensional, plane, free progressive wave.* The one-dimensional aspect relates to the specification of the parameters in terms of a single distance; the planar aspect to the fact that the wavefronts are parallel to each other; and the free progressive aspect to the advancement of the wave without interference from other objects or changes in the medium. In actuality, the wave is not truly free in that it is bounded by the walls of the tube. For practical purposes, however, the motion is essentially the same as if the tube and the piston were infinite in cross section.

We also noted that at any point along the tube we can measure a varying pressure above and below the atmospheric pressure, that is to say we can measure the *sound pressure.* Also we can measure a varying air *particle velocity* associated with the to-and-fro motions of the air molecules. The units of sound pressure are generally microbars (or dynes per square centimeter or newtons per square meter), although they could be pounds per square foot. The units of particle velocity are generally centimeters or meters per second (or feet per second).

In a similar manner we can describe another one-dimensional, free progressive wave—the spherical wave. Assume that a sphere, say a balloon, is sinusoidally pulsating about some equilibrium size, that is to say, the surface of the balloon is uniformly contracting and expanding radially in simple harmonic motion. In a similar manner as that for the piston, the molecules next to the balloon are compressed and expanded in such a way as to form a sound wave that propagates away from the balloon, in this case in the form of a spherical wave. The speed at which the wave propagates is the same as for the plane wave, and the same relations between frequency and wavelength hold. Although the wave is now moving outward with a spherical wavefront, it is still one-dimensional since all the parameters of the wave can be related to one distance, the radial distance of the wavefront from the center of the sphere. This form of wave is known as a *one-dimensional, spherical, free progressive wave.*

A free progressive sound wave transmits energy. That a sound wave transmits energy from the source to more remote points is apparent from the fact that the molecules in the air along the path of the sound wave are forced to move in a prescribed way. Energy is the ability to do work, and work is the product of force times distance. The fact that the sound wave forces the molecules to move implies that the wave has an energy associated with it.

The usual way in which the energy propagation is described is in terms of *intensity,* defined as the energy that flows through a unit area in unit time. The unit for intensity is watts per square centimeter or per square meter (or per square foot). In terms of the parameters describing a free progressive sound wave, the average intensity in the direction of the wave propagation is the time average of the product of the sound pressure and the particle velocity measured in the direction of the wave propagation.

Quantitative Aspects.[1–5] We will soon find that a qualitative description of the behavior of a sound wave is not sufficient for our needs. We must utilize a mathematical model as a means for guiding our reasoning and preventing us from taking wrong steps when carrying out the practical aspects of acoustics.

Plane Wave. We have already said that sound waves travel through an elastic medium whose particles have inertia and elasticity. Obviously, the force required to move a portion of the gas is equal to the mass times the acceleration of the particles (Newton's second law). Furthermore, the pressure required to compress or rarefy a portion of the gas must follow the Boyle-Charles law for a perfect gas ($PV = mRT$). Next we add to this the fact that sound waves oscillate rapidly enough so that the ideal gas compressions are essentially adiabatic ($PV^{1.4} = \text{const}$). Finally, if we note that the law of conservation of mass must hold, we have the material to set up three simultaneous equations (Newton's, adiabatic–Boyle-Charles, and conservation).[1-5] When combined these give us the wave* equation for a plane, one-dimensional sound-pressure wave of angular frequency ω.

$$\frac{\partial^2 p}{\partial x^2} = -\frac{\omega^2}{c^2} p \tag{2.5}$$

where p = sound pressure
c = speed of sound in air
$\omega = 2\pi f$, radians/sec
f = frequency of wave, cps
x = coordinate system along which wave is traveling

Forward-traveling Wave. Now imagine that the piston in our tube is generating a sound wave that travels to the right and is never reflected back (see Fig. 2.3). In other words, the right end of our tube is anechoic (echo-free). Therefore, all of the energy produced by the piston is absorbed by the termination.

Arbitrarily, let us designate a plane along the tube as the $x = 0$ plane. At this plane the sound pressure is varying sinusoidally at the same frequency as the piston and will have a maximum value equal to P_R, so that

$$[p(t)]_{x=0} = P_R \cos \omega t \tag{2.6}$$

Refer now to Fig. 2.4. We observe four diagrams that plot sound pressure vs. the distance variable x. We note that each diagram has vertical lines drawn at 20 places along the x direction. These lines help us to focus our attention on particular places along the path of an outwardly traveling free progressive wave. Practically, such a traveling wave might exist for frequencies below 500 cps in a long tube with cross-sectional dimensions of 5 by 5 cm (2 by 2 in.). A loudspeaker would be located at the left end of the tube and an anechoic (nonreflecting) termination at the right end.

Let us say that the frequency of the source is 100 cps. Then the wavelength is

$$\lambda = \frac{c}{100} = \frac{344}{100} = 3.44 \text{ m} \qquad (11.28 \text{ ft}) \tag{2.7}$$

where c is the speed of sound in air equal to 344 m/sec (1,128 ft/sec). The time it takes for the wave to travel 1 wavelength (3.44 m), i.e., *the period,* is

* Although it is not standard American practice in the field of acoustics, the following types of combinations of words are hyphenated in this book for ease in reading: *sound-pressure level, sound-power level, sound-absorption coefficient, sound-pressure-level measurement,* and *one-third-octave-band analyzer.*

$$T = \frac{\lambda}{c} = \frac{3.44}{c} = \frac{3.44}{344} = 0.01 \text{ sec} = \frac{1}{f} \tag{2.8}$$

First, let us look at the upper graph of Fig. 2.4. At each of the 20 places along the wave, the position of a dot shows the sound pressure.

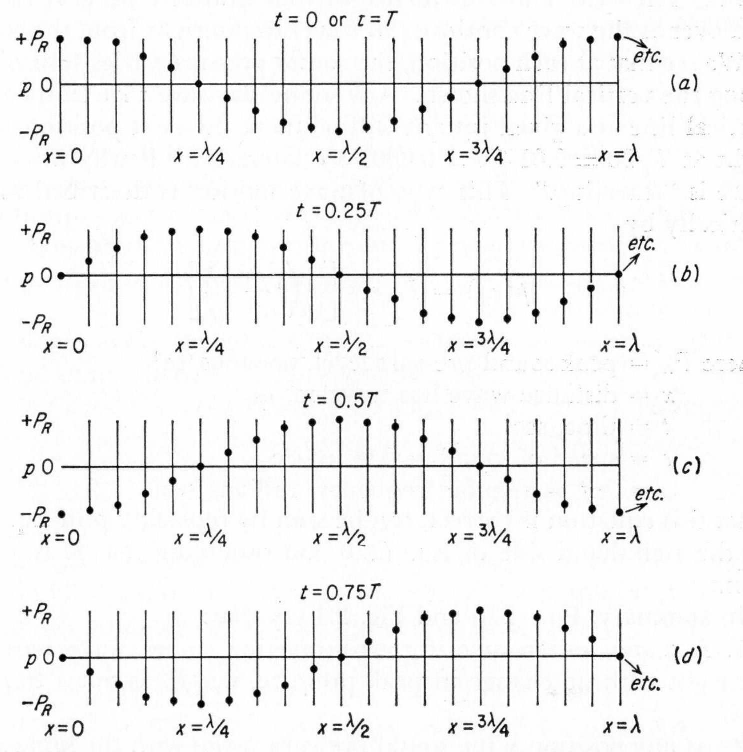

FIG. 2.4 Graphs showing the sound pressure in a plane free progressive forward-traveling wave at 20 places in space at five instants of time t, namely, $t = 0$, $T/4$, $T/2$, $3T/4$, and T sec. The wave is produced by a source at the left and travels to the right with a speed $c = 344$ m/sec. The length of time it takes a wave to travel a distance equal to a wavelength is called the *period* T. Forward-traveling wave: $p(x,t) = P_R \cos[\omega(t - x/c)]$; $T = 1/f = 2\pi/\omega$.

This sound pressure is above the zero line near the left and right ends and is below the zero line in between. The zero line is actually the barometric-pressure line. Graph (*a*) of Fig. 2.4 is a "photo flash" at time $t = 0$ of the sound-pressure distribution in space over a distance that is arbitrarily chosen to be a wavelength.

Now at a short time later equal to 0.0025 sec, i.e., $T/4$ sec, this wave will have traveled to the right a distance equal to ¼ wavelength [see graph (*b*) in Fig. 2.4]. In other words, we have the same sound-pressure distribution except that it is displaced in space 0.86 m to the right.

Each 0.0025 sec later [see graphs (*c*) and (*d*) in Fig. 2.4] the wave will have traveled another 0.86 m, and, finally after 0.01 sec $= T$, the wave will have traveled a full wavelength, and graph (*a*) applies again. Then the entire sequence of four graphs repeats itself over and over as the wave continues to travel to the right from the source.

We see that at each position, the sound pressure varies sinusoidally along the vertical line shown. Any sound pressure that exists at one vertical line at a given instant will occur at the next position to the right at $T/20 = 0.01/20 = 0.0005$ sec later. This is why we say the wave is "traveling." This type of wave motion is described mathematically by

$$p(x,t) = P_R \cos\left[\omega\left(t - \frac{x}{c}\right)\right] \tag{2.9}$$

where P_R = peak sound-pressure level, newtons/m²
x = distance wave has traveled, m
t = time, sec
c = speed of sound = 344 m/sec
$\omega = 2\pi f$ = angular frequency, radians/sec

That this equation is correct, can be seen by replacing p in Eq. (2.5) by the right-hand side of Eq. (2.9) and observing that it is a solution.

In summary, Eq. (2.9) and Fig. 2.4 say that:

1. A plane, progressive, forward-traveling sound wave moves to the right with no change in peak pressure, i.e., P_R is not a function of x.

2. At any position x the sound pressure varies with the same angular frequency ω as at any other position.

3. The only effect of distance is to delay the time at which the pressure at any point is at the same value it is at $x = 0$.

Let us look further at Fig. 2.4. Above the point $x = 0$ we see that the pressure varies as given by Eq. (2.6). At $x = \lambda/4$ the same function is shown except that the pressure reaches its maximum at a time $t = 0.25T = (\lambda/4)/c$ sec later (Eq. 2.9). We also may say that a *phase shift* of $\omega x/c = \omega(\lambda/4)/c = \pi/2$ radians has occurred. At $x = \lambda/2$ the sound pressure reaches its maximum at $t = (\lambda/2)/c = 0.5T$. The phase shift is now π radians.

In the particular example of Fig. 2.4, where $f = 100$ cps,

$$p(x,t) = P_R \cos\left[628\left(t - \frac{x}{344}\right)\right] \tag{2.10}$$

Backward-traveling Wave. A backward-traveling wave is one that travels in the negative direction of x. It can be produced either by placing a sound source in the right-hand end of our 5- by 5-cm tube and having it radiate to the left, or it can be produced by putting a reflecting wall in the right-hand end of the tube in place of the anechoic termination and thereby turning the forward-traveling wave around.

A backward-traveling wave is given by the equation

$$p(x,t) = P_L \cos \left[\omega \left(t + \frac{x}{c} \right) \right] \tag{2.11}$$

The only change from Eq. (2.9) is in the sign following the t in the formula.

The backward-traveling case is diagramed in Fig. 2.5. We see that

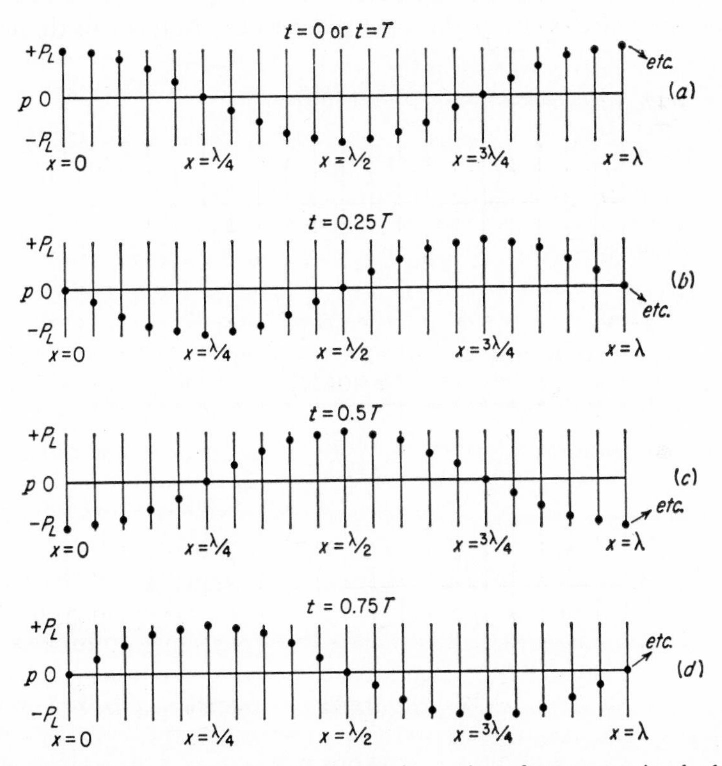

FIG. 2.5 Graphs showing the sound pressure in a plane free progressive backward-traveling wave at 20 places in space at five instants of time t, namely, $t = 0$, $T/4$, $T/2$, $3T/4$, and T sec. The wave is produced by a source at the right (or by being reflected from a boundary) and travels to the left with a speed $c = 344$ m/sec. The period T is defined the same as for Fig. 2.4. Backward-traveling wave: $p(x,t) = P_L \cos [\omega(t + x/c)]$; $T = 1/f = 2\pi/\omega$.

any sound pressure that exists at one of the lines will exist at the next line to the *left* $T/20$ sec later. We say that a phase shift of 18° ($2\pi/20$ radians) exists between adjacent lines. This was also true in Fig. 2.4.

Combined Forward- and Backward-traveling Waves (Standing Wave). As the next step, let us simply add Figs. 2.4 and 2.5 together, i.e., let us perform the operation of Eq. (2.12) for $f = 100$ cps. We have

$$p(x,t) = P_R \cos\left[628\left(t - \frac{x}{344}\right)\right] + P_L \cos\left[628\left(t + \frac{x}{344}\right)\right] \quad (2.12)$$

The graphical combination of these two waves is given in Fig. 2.6. In other words, Fig. 2.6 is the simple algebraic sum of Figs. 2.4 and 2.5.

A surprising change has occurred! No longer does a sound pressure at one place occur at the next place to the right or to the left of

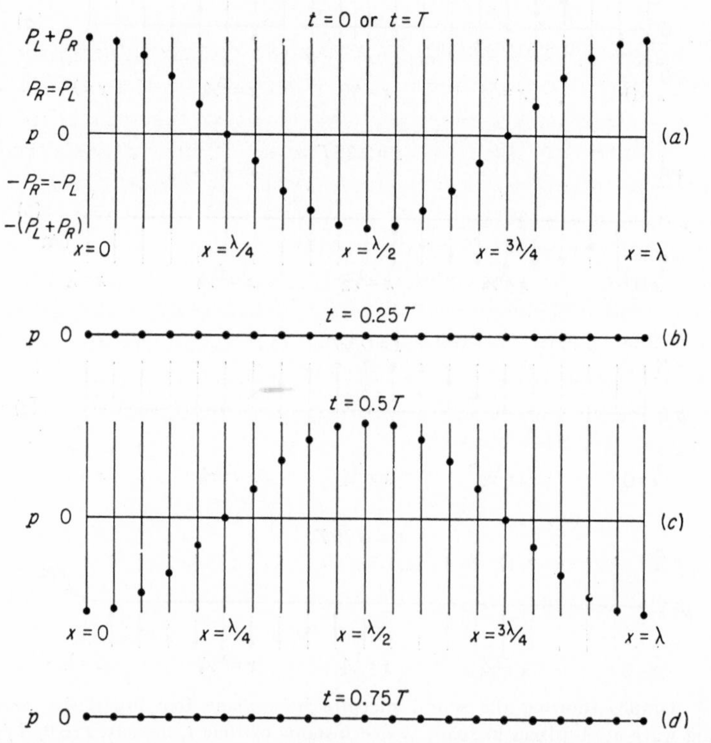

FIG. 2.6 Graphs giving the sound pressure in a plane standing wave at 20 places in space at five instants of time t, $t = 0$, $T/4$, $T/2$, $3T/4$, and T sec. The wave is produced by two sources equal in strength at the right and left of the page. Standing wave: $p(x,t) = P_R \cos[\omega(t - x/c)] + P_L \cos[\omega(t + x/c)]$, where $P_R = P_L$; $T = 1/f$.

it at the next instant. The wave simply does not travel any longer! We see that at each point in space the sound pressure increases and decreases sinusoidally with time, except at the points $x = \lambda/4$ and $x = 3\lambda/4$. There the pressure is always zero. The maximum value of the pressure variation at different points is different, being greatest at $x = 0$, $x = \lambda/2$, and $x = \lambda$. The sound pressures at the nine points between the points $x = \lambda/4$ and $x = 3\lambda/4$ always vary together, i.e., increase or decrease in phase, although the maximum amount to which they increase or decrease is different, as we just said. The sound pressures for the five places to the left of $x = \lambda/4$ and to the right of $x = 3\lambda/4$ also increase or decrease together (*in phase*), but the changes are ½ cps or *180°* out of phase with the sound pressures at the nine points between $x = \lambda/4$ and $x = 3\lambda/4$.

Assuming that $P_R = P_L = P$, we can show mathematically that the sum of the two terms in Eq. (2.12) is

$$p(x,t) = 2P\left[\cos 2\pi\left(1 - \frac{x}{\lambda}\right)\right]\cos \omega t \tag{2.13}$$

Here, we see very clearly the difference between a standing and a traveling wave. In the traveling wave [see Eqs. (2.9) and (2.11)] time t and x/c occur as a sum or difference inside the argument of the cosine. Hence, for the traveling wave, by adjusting both time and distance (according to the speed of sound) in the argument of the cosine, we can always keep the argument and hence the magnitude of the cosine the same. In Eq. (2.13) time and x/c no longer appear together inside the argument of a single cosine. So the same sound pressure cannot occur at an adjacent point in space at a later time.

Particle Velocity. Now what is the particle velocity? From the Newtonian equation we can solve for the particle velocity u in terms of the sound pressure p. We find[1] for the free-progressive-plane-wave case of Figs. 2.3 and 2.4 that

$$u = \frac{p}{\rho_0 c} \tag{2.14}$$

where u is the particle velocity and ρ_0 is the density of air.

As a result, from Eq. (2.9) we see that

$$u(x,t) = \frac{P_R}{\rho_0 c}\cos\left[\omega\left(t - \frac{x}{c}\right)\right] \tag{2.15}$$

The average power per unit area, i.e., the intensity, transmitted to the right in Fig. 2.3 is equal to

$$\text{Average intensity } I = \overline{pu} \tag{2.16}$$

where the overline indicates the time average of the product. Substitution of Eqs. (2.9) and (2.15) in Eq. (2.16) yields the instantaneous intensity (remembering that P_R is the peak or maximum value of the sound pressure)

$$I(t) = \frac{P^2_{\max}}{\rho_0 c}\left[\frac{1}{2} + \frac{1}{2}\cos 2\omega\left(t - \frac{x}{c}\right)\right] \tag{2.17}$$

Because the time average of the cosine is zero, the average intensity is

$$I = \left(\frac{P_{\max}}{\sqrt{2}}\right)^2 \frac{1}{\rho_0 c} \tag{2.18}$$

Taking cognizance of the fact that the root-mean-square (rms) value of a sine wave equals $1/\sqrt{2}$ times its peak value we obtain

$$I = \frac{p^2_{\mathrm{rms}}}{\rho_0 c} \tag{2.19}$$

All sound-level meters are calibrated to read the rms value (also called the "effective" value) of a sound wave so that the intensity I may easily be computed from p_{rms} for this type of wave.

Spherical Wave. Without going into the details, we find that for a spherical, one-dimensional, free progressive wave

$$p(r,t) = \frac{A_{\max}}{r}\cos\omega\left(t - \frac{r}{c}\right) \tag{2.20}$$

This equation is to be compared with Eq. (2.9). We find it too is very interesting. It says that

1. As a spherical, progressive wave travels outward from a source, its pressure amplitude *decreases* in inverse proportion to r.
2. At any position r the sound wave has the same frequency as the spherical source producing it.
3. Besides its effect on amplitude, the distance r also produces a time delay (*phase shift*) in the wave.

In the case of a spherical wave, it can also be shown that the average intensity at a distance r from the source is given by Eq. (2.19), where p_{rms} is the rms pressure at the distance r. Note, however, that here p_{rms} decreases in inverse proportion to r, whereas, by contrast, p_{rms} was constant for a plane wave.

Because we usually desire the intensity I to be in *watts* per unit area, it is customary to work in mks units. Hence, p is in newtons per square meter, ρ_0 is in kilograms per cubic meter, c is in meters per second, and I is in watts per square meter. In the cgs system, I comes out in ergs per second per square centimeter, and multiplication of the result by 10^{-7} is necessary to obtain watts per square centimeter. In English units (foot-pound-second or fps system), p is in pounds per square foot, ρ_0 is in pounds per cubic foot, c is in feet per second, and I is in foot-pounds per second per square foot. It is necessary to multiply the result by 0.1383 to get watts per square foot.

Example 2.3. Consider the case of a sinusoidal, free progressive plane wave having a peak sound pressure of 1 newton/m².

Solution. The effective sound pressure would be $1/\sqrt{2} = 0.707$ newton/m². If $\rho_0 c$ is taken as 406 mks (meter-kilogram-second) rayls, which holds approximately at room temperature near sea level, the intensity is given by

$$I = \frac{(0.707)^2}{406} = \frac{1}{812} \text{ watt/m}^2$$

The analogy between the expression for acoustic power flow and electrical power flow is apparent if one substitutes electromotive force (volts) for sound pressure. The expression $I = p^2/\rho_0 c$, where I is acoustic intensity in watts per square meter, is analogous to $P = E^2/R$, where P is electrical power in watts, E is in volts, and R is resistance in ohms. If we identify $\rho_0 c$ as a kind of acoustic resistance, we have analogous expressions for power in each case.

An important difference between plane waves and spherical waves becomes apparent when we consider the intensity of the sound wave at any given point. If we assume that our source is imparting a constant energy to the sound wave, ideally this energy will be transmitted by the wave without diminution. In the plane wave, intensity remains constant regardless of how far we get from the source, since the area of the wavefront is constant (the tube has constant cross section). In the spherical wave, however, the area of the wavefront increases as the wave progresses farther and farther from the source. In fact, the area increases with the square of the distance from the source since the area of a sphere is equal to 4π times the square of its radius. In the spherical wave, therefore, with a source of constant energy per second (i.e., constant power), the intensity must be decreasing as the square of the distance from the source, since the product of the intensity and the area through which it passes is equal to the power transmitted by the wave. This is the origin of the famous "inverse square law." Simply stated,

$$I \text{ (at radius } r) = \frac{W}{4\pi r^2} = \frac{p_{\text{rms}}^2 \text{ (at } r)}{\rho_0 c} \tag{2.21}$$

where W is the total acoustic power radiated by the source of sound

Example 2.4. Consider a sinusoidal, free progressive spherical wave having a peak sound pressure of 1 newton/m² at the distance r equal to 1 m from a source of constant power. Find the intensity at 1 m, the power radiated by the source, the intensity, and the rms and max sound pressures at 10 m.

Solution. The effective (rms) sound pressure is $1/\sqrt{2} = 0.707$ newton/m². Taking $\rho_0 c$ as 406 mks units, the intensity is given by

$$I = \frac{(0.707)^2}{406} = \frac{1}{812} \text{ watt/m}^2$$

The power radiated by the source is given by

$$W = 4\pi r^2 I = \frac{4\pi}{812} = \frac{1}{65} \text{ watt}$$

At $r = 10$ m, since W remains constant,

$$I \text{ (at } r = 10 \text{ m)} = \frac{1}{4\pi(10)^2}\,\frac{1}{65} = \frac{1}{81{,}200} \text{ watt/m}^2$$

$$p_{\text{rms}} \text{ (at } r = 10 \text{ m)} = \sqrt{\frac{1}{81{,}200} \times 406} = 0.0707 \text{ newton/m}^2$$

$$P_{\text{max}} \text{ (at } r = 10 \text{ m)} = \sqrt{2} \times 0.0707 = 0.1 \text{ newton/m}^2$$

We see that for 10 times the distance the sound pressure has decreased by a factor of 10 and the sound intensity has decreased by a factor of 100. Compare this example with Example 2.3 for the plane free progressive wave.

In a similar manner to the formation of plane and spherical waves, we can generate waves of an even more complex nature. An example of another one-dimensional, free progressive wave can be obtained by considering a very long cylinder, say a straight hose filled with air. Suppose that we cause the surface of the hose to pulsate uniformly in a radial direction. A sound wave will now propagate outward from the cylinder. This wave is in a sense a cross between the plane wave and the spherical wave. At any point the wave will have the same properties as at any other point at the same distance from the cylinder, but at two points having different distances from the cylinder the wave will be different. We can show that the intensity at any distance from the source is inversely proportional to the first power of the radius, rather than the second power as in the case of the spherical wave. This type of wave is called a one-dimensional cylindrical wave.

As an example of a two-dimensional wave suppose that, in addition to the uniform radial motion of the cylinder, we imposed another motion along the axis of the cylinder itself. The resulting sound wave radiated from the cylinder would now depend not only on the radial distance from the source, but also on the distance along the axis of the source.

An approximate picture of the two waves described above is indicated by the sketches in Fig. 2.7. The action of the waves is indicated by a succession of surfaces called "wavefronts." The wavefronts are said to be surfaces having equal phase. This implies that the sound pressure at each surface bears the same basic relation to the sound pressures at each of the other surfaces at any given instant. That is, if the sound pressure is at a maximum at one surface, it will also be at a maximum at each of the other surfaces having the same phase.

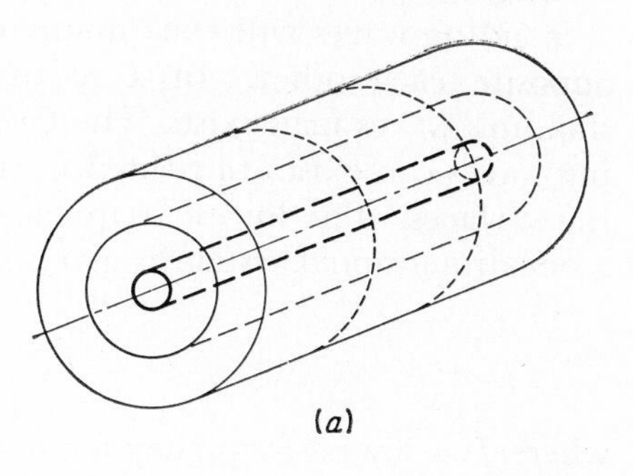

(a)

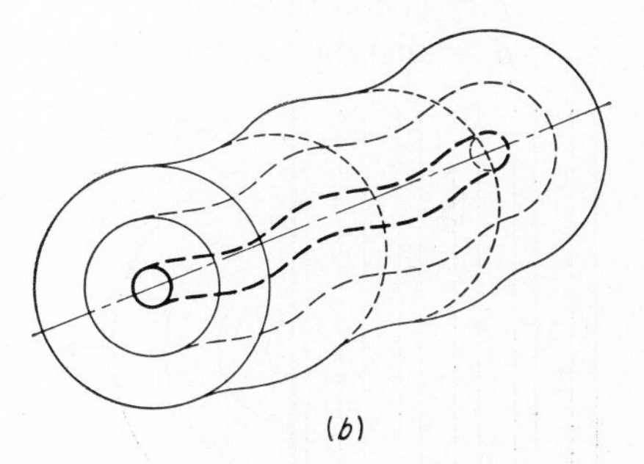

(b)

FIG. 2.7 (*a*) One- and (*b*) two-dimensional waves.

2.6 Diffraction—Standing Waves

Free progressive waves were defined as existing where there were no obstacles or changes in the propagating medium. These waves can exist only in free space. As soon as we enclose the space, or place an obstacle in the path of the wave, the shape of the wave is altered. The mathematical analysis of the interaction of the wave and the obstacle is exceedingly complex in all but the simplest of cases. An indication of how a wavefront can be altered in a simple case may be seen by examining the case of a plane wave striking a very large plane surface containing a small hole. Consider the plane wave advancing on the barrier indicated in Fig. 2.8. As the wave strikes the barrier, a portion of it is reflected to form a plane wave traveling in a direction opposite to the incident wave. The sound passing through the small hole progresses in the form of a spherical wave. That the sound behaves in the manner indicated can be shown both analytically and experimentally. The analysis which studies this phenomenon is known as diffraction theory, and the new sound waves formed are known as *diffraction waves.*

Standing waves will exist in any enclosure having parallel surfaces opposite each other. In a rectangular room six distinct sets of standing waves may exist. The frequencies at which resonant standing waves can exist are related to the separation between the reflecting surfaces. The lowest frequency for a resonant standing wave in a one-dimensional system is given by

$$f = \frac{c}{2d} \tag{2.22}$$

where f = lowest frequency for resonant standing wave
c = speed of sound
d = distance separating two reflecting surfaces

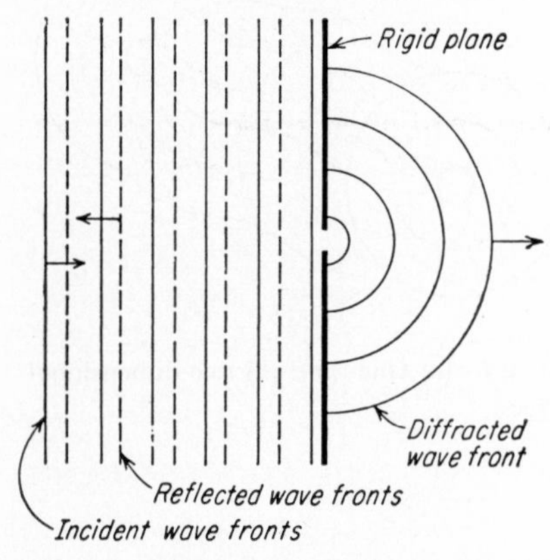

FIG. 2.8 Diffraction through a small hole.

Resonant standing waves can also exist at every integral multiple of this frequency. That is to say,

$$f = \frac{nc}{2d} \tag{2.23}$$

where n is an integer: 1, 2, 3, ...

Example 2.5. If two reflecting surfaces are 10 ft apart, what is the lowest-frequency resonant standing wave that can exist between the surfaces when the speed of sound is 1,120 ft/sec?

Solution

$$f = \frac{1{,}120}{2 \times 10} = 56 \text{ cps}$$

Higher-frequency resonant standing waves can exist at frequencies of $2 \times 56 = 112$, $3 \times 56 = 168$, $4 \times 56 = 224$ cps, etc.

Similar relations can be developed for any enclosure which permits the formation of standing waves. These will be presented in more detail in Chap. 10.

2.7 Complex Waves

Harmonically and Inharmonically Related Wave Combinations. In our discussion so far, we have considered solely waves having a single frequency—so-called "pure tones." In nature, pure tones usually do not exist by themselves. Instead, the sound wave is composed of a number of component waves combining to yield a resultant complex wave.

By the mathematical technique known as "Fourier analysis," it can be shown that any sound wave that repeats itself periodically can be considered as a suitable combination of sine waves whose frequencies are integral multiples of some frequency called the "fundamental frequency." That component with *twice the frequency* of the fundamental is called the *second harmonic,* and so forth. Given the component waves, the waveform of the resultant periodic complex wave can be obtained by the process of "superposition." This merely means that, at any given time, the amplitudes of the individual components are added linearly to give the amplitude of the periodic complex wave. The combined wave will repeat itself exactly after each time period T equal to $1/f_1$, where f_1 is the fundamental frequency.

Combination of several sine waves introduces the concept of the relative *phase* among the individual components. Specification of the phase relationships between a sine wave and its harmonics can be accomplished in a number of ways analytically. One of the simplest is to reference the relative phases of the harmonics, at some fixed point in time (referred to as "$t = 0$"), to the phase of the fundamental. If the time point is selected so that the fundamental has either zero value (the fundamental then can be represented as $C \sin \omega t$), or a positive maximum value (the fundamental then can be represented by $C \cos \omega t$), the harmonics can then be written as $C_n \sin n(\omega t + \theta_n)$, or $C_n \cos n(\omega t + \theta_n)$.* Here n is the order of the harmonic, i.e., 2, 3, 4, etc., C_n is the amplitude of the "nth" harmonic, and θ_n is the phase of the nth harmonic relative to the fundamental.

An example of the combination of one sine wave with another to produce a complex wave is indicated in Fig. 2.9 where a sine wave and its second harmonic are combined with a zero phase difference at $t = 0$ to produce a third wave. The combination of these same two sines with a phase difference of $\pi/4$ radians is shown in Fig. 2.10. A third example plotted in Fig. 2.11 shows the result when a phase difference of $\pi/2$ radians is introduced between the two waves.

The addition of component waves to produce a complex wave raises the question of specifying other properties of the wave. The *peak value,* for example, of complex waves composed of harmonically related components is dependent not only on the amplitude of the

* Note that a sine wave is equally valid as a cosine wave, as a solution to the wave equation because the two are simply related to each other by a phase shift of $\pi/2$ radians.

individual components, but also upon the relative phases between the components. This can be observed in Figs. 2.9, 2.10, and 2.11. In general, the peak-to-peak value will be a maximum when all the higher harmonics are in phase with the fundamental, i.e., $\theta_n = 0$. This assumes that all are expressed as cosine functions.

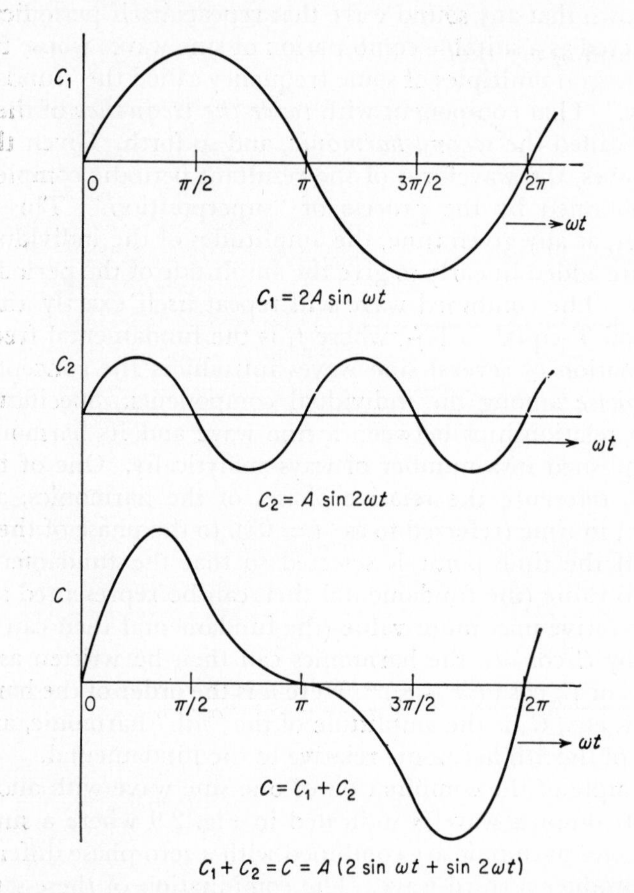

FIG. 2.9 Combination of two harmonically related sine waves to produce a complex wave. Note that a different complex wave would be produced if the phase (lateral displacement) of one of the component waves were changed. The effect of phase is illustrated in Figs. 2.10 and 2.11.

Previously, we indicated that the *effective* (rms) *value* of the pressure of a sound wave was a most useful quantity, since the power transmitted by the wave is generally related to the mean-square pressure in the wave [see Eq. (2.19)]. Let us dwell on this point for a moment.

Assume that we are measuring, with a sound-level meter capable of determining the effective (rms) sound pressure, two sound waves emanating from two sources. Let the instantaneous pressures of the waves be

$$p_1(t) = P_1 \cos (\omega_1 t + \theta_1) \tag{2.24a}$$

$$p_2(t) = P_2 \cos (\omega_2 t + \theta_2) \tag{2.24b}$$

where ω_1 and ω_2 are their angular frequencies and θ_1 and θ_2 are their

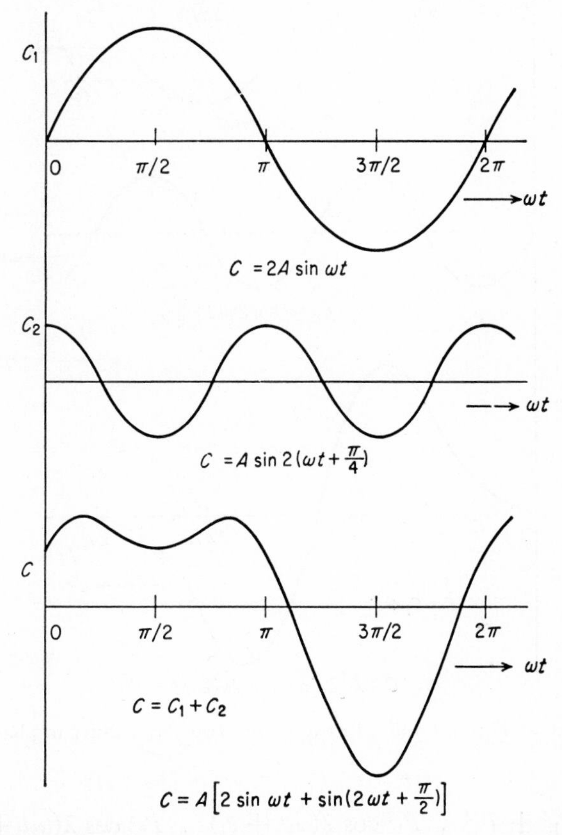

FIG. 2.10 Combination of two harmonically related sine waves, differing in phase by $\omega t = \pi/4$ radians. Note the different appearance of the combined wave from those shown in Figs. 2.9 and 2.11.

respective phases. At any instant of time, the pressure measured by our microphone is

$$p(t) = p_1(t) + p_2(t) = P_1 \cos (\omega_1 t + \theta_1) + P_2 \cos (\omega_2 t + \theta_2) \tag{2.25}$$

In determining the effective value of this summation, the circuiting in an *ideal* sound-level meter will square $p(t)$, take the time average, then display the square root of the result (converted to decibels) on the indicating meter. When $p(t)$ is squared, we obtain

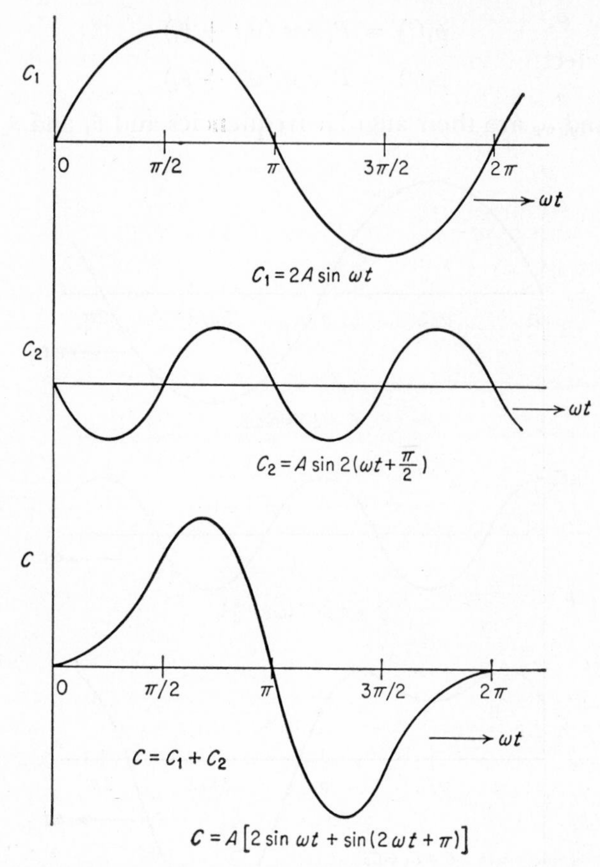

FIG. 2.11 Same as Figs. 2.9 and 2.10 except the two waves differ in phase by $\omega t = \pi/2$ radians.

$$[p(t)]^2 = \frac{P_1^2 + P_2^2}{2} + \frac{P_1^2 \cos 2(\omega_1 t + \theta_1)}{2} + \frac{P_2^2 \cos 2(\omega_2 t + \theta_2)}{2} + 2P_1P_2 \cos (\omega_1 t + \theta_1) \cos (\omega_2 t + \theta_2) \tag{2.26}$$

We know that the time average of each of the single cosines is zero. It can be shown mathematically that the time average of the product of two cosines is also zero in all cases except where they have the same frequency. As a result we can write the following relations for the mean-square pressure (with $\omega_1 \neq \omega_2$),

$$p^2 = \left(\frac{P_1}{\sqrt{2}}\right)^2 + \left(\frac{P_2}{\sqrt{2}}\right)^2 \tag{2.27}$$

We see that $P_1/\sqrt{2}$ = effective sound pressure of the first wave and $P_2/\sqrt{2}$ = effective sound pressure of the second wave because the rms value of a sine wave is known to be $1/\sqrt{2}$ times its peak value. So, for two waves having *different* frequencies

$$p = \sqrt{p_1^2 + p_2^2} \tag{2.28}$$

where p, p_1, and p_2 are effective (rms) magnitudes.

Under the special condition that $\omega_1 = \omega_2 =$ (say) ω, the two quantities p_1 and p_2 combine to produce a third sine wave of the same frequency but of different amplitude and phase, vis.,

$$\begin{aligned} p(t) &= p_1(t) + p_2(t) \\ &= \sqrt{P_1^2 + P_2^2 + 2P_1P_2 \cos(\theta_1 - \theta_2)} \cos(\omega t + \theta_3) \\ &= P_3 \cos(\omega t + \theta_3) \end{aligned} \tag{2.29}$$

where θ_3 is the new phase angle. The rms pressure is

$$p = \frac{P_3}{\sqrt{2}} = \sqrt{p_1^2 + p_2^2 + 2p_1p_2 \cos(\theta_1 - \theta_2)} \tag{2.30}$$

where p, p_1, and p_2 are rms values. Comparison of Eqs. (2.30) and (2.28) reveals the importance of phase when combining two sine waves of the *same* frequency. If the phase difference $(\theta_1 - \theta_2)$ is zero, the two waves are said to be in phase and the total pressure is at its maximum. If $\theta_1 - \theta_2 = \pm 180°$, the total pressure is at its minimum.

In practice, if one wishes to find the effective sound pressure of a number of waves all of which have different frequencies except, say, two, these two are added together according to Eq. (2.30) to obtain a new effective pressure. Then this pressure and the pressure of the remainder of the components are summed according to Eq. (2.28).

One further quantity of interest is the *rectified average* value of a wave. By rectified average we mean that the negative loops of the wave are made positive and the average value under all of the loops is determined. For a sine wave this value is 0.636 of its peak value. Because the effective value of a sine wave is 0.707 of the peak, the rectified average value is about 0.9 of the effective value.

For a complex wave of several harmonically related components, the rectified average depends upon the relative amplitudes and phases of the individual components.

We have shown that phase is not important in determining the effective (rms) value of a combination of sine waves of different frequency. As a result, we usually obtain all the amplitude information that we need from a graph in one of the forms of Fig. 2.12. Four categories of spectra are shown.

Line Spectrum Sound. The uppermost graph (*a*) shows the effective sound-pressure spectrum for a group of four components

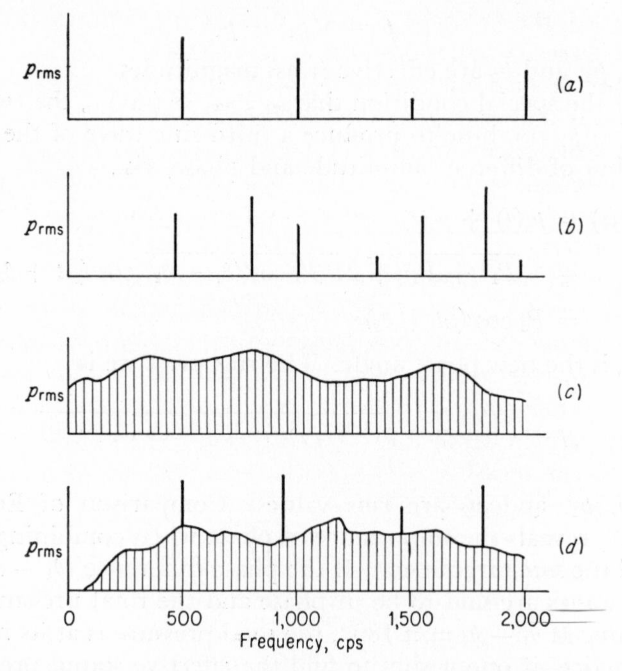

FIG. 2.12 (*a*) Line spectrum harmonically related; (*b*) line spectrum inharmonically related; (*c*) continuous spectrum; (*d*) combination line and continuous spectrum (complex spectrum).

harmonically related with a fundamental frequency of 500 cps. The effective sound pressure of each component is indicated by the vertical height of the line above its frequency. Such a graph is called a *line spectrum.*

The combination (summation) of a number of inharmonically related waves produces a resultant wave that is not periodic. In other words, one can predict the amplitudes of the wave only on a statistical basis (see Ref. 6, pp. 457–464). Of course, the peak amplitude can never exceed the sums of the amplitudes of all the components. But, as the number of components is increased, the probability of achieving the maximum peak amplitude becomes very small. For example, with 10 inharmonically related components of equal amplitude, the maximum possible peak amplitude is 10 times that of one component. However, only 0.02 per cent of the time will the amplitude of the combination exceed 8 times that of one component.

As we showed in the previous section, the *effective* (rms) *value* of a combination of waves of different frequencies is computed very simply by taking the square root of the sum of the squares of the rms pressures of the individual components.

In general, the *rectified average value* of the combination is about 0.9 of the effective value (see Ref. 6, pp. 465–473).

An example of a line spectrum made up of inharmonically related pure tones is shown in graph (*b*) of Fig. 2.12.

Continuous Spectrum Sound (Noise). A very common type of sound is one that is built up from a continuum of components. Such a noise is produced by the exhaust of a jet engine or by the water at the base of Niagara Falls or by the hiss of air and steam jets or by the dumping of a truck load of rocks over a cliff onto a pile of rocks, and so on. To produce such a noise from pure tones would require an infinite number of waves, each with an infinitesimal amplitude.

An example of a continuous spectrum is shown in graph (*c*) of Fig. 2.12. When combined with a line spectrum, a complex spectrum of the type shown in (*d*) is produced.

How do we measure and specify a continuous spectrum sound? Most simply we could measure the over-all sound-pressure level. The answer that we would get would depend, however, on the frequency range of the meter, and no two types of meter would yield the same result. Thus, this method is not so good. It has been agreed, therefore, that we should describe a continuous spectrum noise by plotting (as a function of frequency) the effective sound pressure (or the intensity) existing in increments of frequency along the frequency scale, *each increment* 1 cps *in width.* Therefore, graphs (*c*) and (*d*) of Fig. 2.12 indicate, as a function of frequency, the effective sound pressure in 1-cps-wide frequency bands.

There is one difficulty with this otherwise acceptable method of describing a continuous spectrum noise. There are no instruments available for measuring effective sound pressure that have bandwidths as narrow as 1 cps. Hence, we must measure with a wider bandwidth and convert to the standard method of presenting the data.

To develop the philosophy of converting from a wider bandwidth to a narrower one, let us imagine that we have a machine that produces an intensity of 10^{-6} watt/m^2 in a 1-cps-wide band between 999 and 1,000 cps (see Fig. 2.13*a*). Now imagine that we start a second machine that produces the same intensity between 1,000 and 1,001 cps (see Fig. 2.13*b*). When the noise from the two machines is combined, we get the spectrum of Fig. 2.13*c*. Because the two machines produce twice as much power as one, the intensity must double.

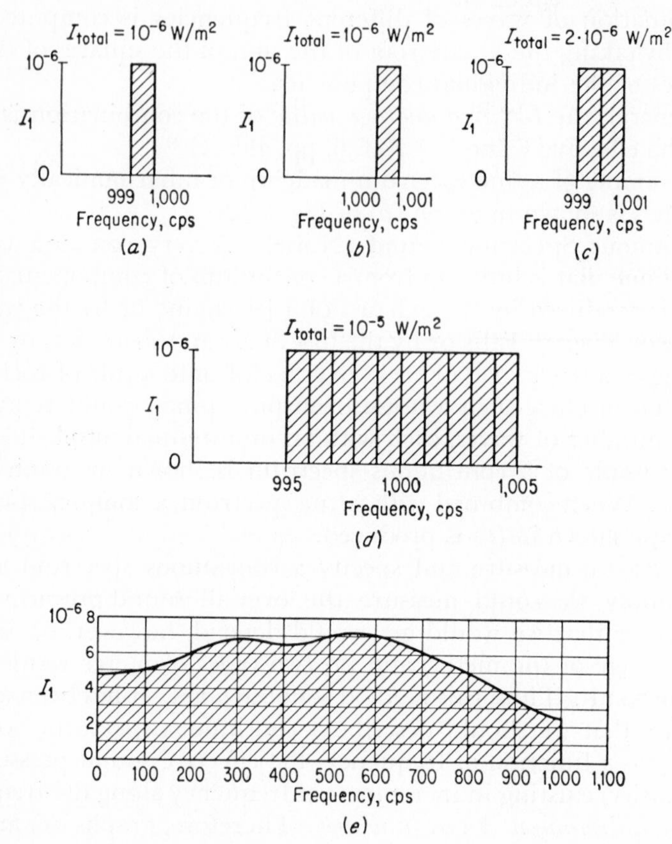

FIG. 2.13 Graphs showing the addition of intensity when the bandwidth of noise is increased.

Similarly in Fig. 2.13*d* 10 such machines would produce 10 times the intensity.

In other words, if the continuous spectrum is "flat" (has the same intensity in every 1-cps band) the total intensity is given by

$$I_{\text{total}} = I_1 \, \Delta f \tag{2.31}$$

or, in words, the total intensity equals the product of the intensity in a 1-cps band multiplied by the width of the total band in cycles per second.

This discussion is all very well if the spectrum is flat. But what is done if the spectrum is different at different frequencies, say like that in Fig. 2.13*e*? We simply substitute for the actual I_1 of that figure, the average value of I_1 in the band in which we are interested.

Example 2.6. What is the total intensity in the 700- to 800-cps band of Fig. 2.13*e*?

Solution. The average intensity for a 1-cps band within the 700- to 800-cps band is seen to be $5.5 \times 10^{-6}\ W/m^2$. Hence, the total intensity in the 700- to 800-cps band is, from Eq. (2.31),

$$I_{\text{total}} = (5.5 \times 10^{-6}) \times 10^2 = 5.5 \times 10^{-4} \frac{W}{m^2}$$

Conversely, if we have measured an intensity I_T with an analyzer having a bandwidth Δf, the average spectrum intensity I_1 (1-cps band) in that band is obtained from the formula

$$I_1 = \frac{I_T}{\Delta f} \tag{2.32}$$

One remaining question needs answering. Because we can measure effective sound pressure easily, but not sound intensity, how do Eqs. (2.31) and (2.32) read when expressed in effective sound pressure? Substitution of Eq. (2.19) in Eq. (2.31) yields

$$\left(\frac{p_{\text{rms}}^2}{\rho_0 c}\right)_{\text{total}} = \left(\frac{p_{\text{rms}}^2}{\rho_0 c}\right)_1 \Delta f \tag{2.33}$$

Cancellation of $\rho_0 c$ and extracting the square root lead to

$$(p_{\text{rms}})_{\text{total}} = (p_{\text{rms}})_1 \sqrt{\Delta f} \tag{2.34}$$

where $(p_{\text{rms}})_1$ is the effective sound pressure in a 1-cps band.

In words, doubling the bandwidth of a flat continuous spectrum noise increases the intensity by a factor of 2, and the effective sound pressure by a factor of 1.41.

2.8 Summary

Sound waves are one means by which energy can be propagated through an elastic medium. We have indicated how the properties of the medium determine the speed at which the wave propagates, using air as our example. The concepts of frequency and wavelength have been considered in their relation to the speed of the wave, and the notions of sound pressure, particle velocity, and intensity have been discussed.

We have introduced the concept of simple harmonic motion, and we have indicated how the sine wave is related to this motion. Employing the sine wave as an example, we have considered the simpler analytic aspects of free progressive plane and spherical waves and how they are used to propagate energy. We discussed standing waves as the sum of forward- and backward-traveling waves.

The construction of a complex wave as a series of sine waves was discussed. We have considered the phase relations of the component waves and how this affects the waveshape, and the average and the rms values of the complex wave when the components are harmonically or inharmonically related. Finally, we have indicated how these concepts can be extended to the case of noise, in which all frequency components are present in the bandwidth under consideration, and how the intensity of the wave depends upon this bandwidth.

REFERENCES

1. Beranek, L. L.: "Acoustics," McGraw-Hill Book Company, Inc., New York, 1954.

2. Hunter, J. L.: "Acoustics," Prentice-Hall, Inc., Englewood Cliffs, N.J., 1957.

3. Kinsler, L. E., and A. R. Frey: "Fundamentals of Acoustics," John Wiley & Sons, Inc., New York, 1950.

4. Swenson, G. W., Jr.: "Principles of Modern Acoustics," D. Van Nostrand Company, Inc., Princeton, N.J., 1953.

5. Morse, P. M.: "Vibration and Sound," 2d ed., McGraw-Hill Book Company, Inc., New York, 1948.

6. Beranek, L. L.: "Acoustic Measurements," John Wiley & Sons, Inc., New York, 1949.

Chapter 3

DECIBELS AND LEVELS

Edward M. Kerwin, Jr.

The field of acoustics makes use of a number of specialized terms for convenience and clarity. Perhaps the most common of these terms is the decibel (db). In this chapter we discuss the nature of the decibel, reasons for using decibels, and some applications of the decibel notation in common acoustical practice.

3.1 Some Physical Relations

As we read in Chap. 2, one of the principal characteristics of a source of sound is its ability to radiate power in the form of acoustic waves. In Fig. 3.1 a simple source of sound is shown, a source that radiates uniformly in all directions. We shall designate the acoustic *power* that it is radiating by W (watts). If we assume that there are no losses in the air surrounding the source, we realize that *all of the power radiated* must pass through any surface that completely encloses the source. As we move away from the source it is easy to see that the power per unit area passing our position decreases because the total power must diverge with distance. We call the power passing through a unit area the *intensity* (see Fig. 3.1).

It is clear that for a spherical surface S (square meters) surrounding our simple source, the intensity is the same at every point on the surface. Thus, the total power W (watts) is equal to the intensity (watts per square meter) multiplied by the total area of the surface S. That is,

$$W = IS \qquad \text{watts} \tag{3.1}$$

It is difficult to measure intensity directly; we do not have any handy acoustical instruments for this purpose. However, most of our acoustical instruments are able to measure the *effective sound pressure* p associated with a propagating acoustic wave. This sound pressure is related directly to sound intensity and sound-energy density.

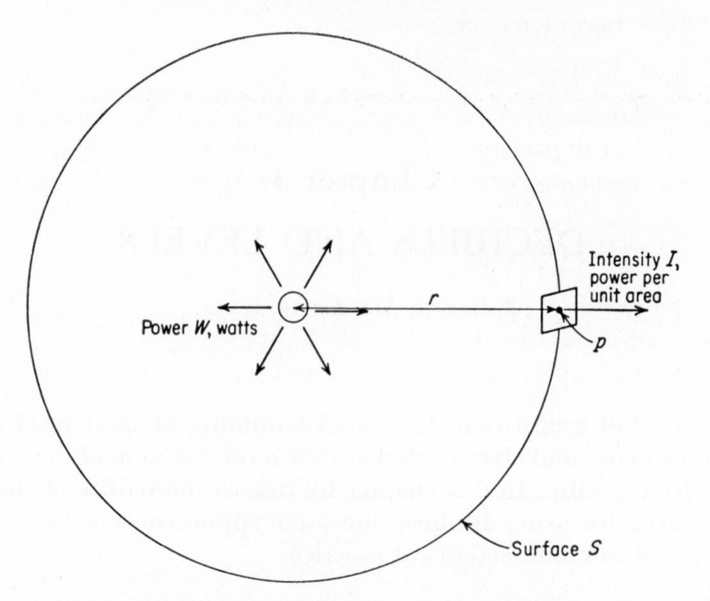

FIG. 3.1 A simple nondirectional sound source radiating W watts producing a sound intensity I in watts per unit area. $I = W/S$, where $S = 4\pi r^2$ is the area of the spherical surface S of radius r centered on the source.

For free progressive plane and spherical waves there is a unique and simple relation between the mean-square sound pressure and the intensity. This relation is

$$I = \frac{p^2}{\rho c} \quad \text{watts/m}^2 \tag{3.2}$$

where p^2 = mean-square sound pressure. (We learned in Chap. 2 that the mean-square sound pressure is the time average of the square of the instantaneous sound pressure in the acoustic wave.)

p = square root of mean-square sound pressure (commonly called *effective* or *rms* sound pressure), newtons/m^2

ρ = density of air, kg/m^3

c = speed of sound in air, m/sec

In standing-wave systems, such as resonators and rooms, the average of p^2 in space can be determined from data obtained as a microphone is moved around the room. This space-average mean-square pressure in air is related to the average sound-energy density by

$$\text{Average sound-energy density} = \frac{p^2}{\rho c^2} = \frac{p^2}{1.4P_0} \quad \text{watt-sec/m}^3 \tag{3.3}$$

where P_0 = barometric pressure

$\doteq 10^5$ newtons/m^2

The sound-energy density is the energy stored in a small volume of gas due to the presence of the sound field. It is the quantity used in classical room-acoustics calculations (see Chap. 11) and elsewhere.

3.2 Some Decibel Scales in Acoustics

Most of us, perhaps, do not realize the tremendous range of sound power covered by common sources of sound. A very soft whisper, for example, involves a total radiated power of about 0.000,000,001 watt,

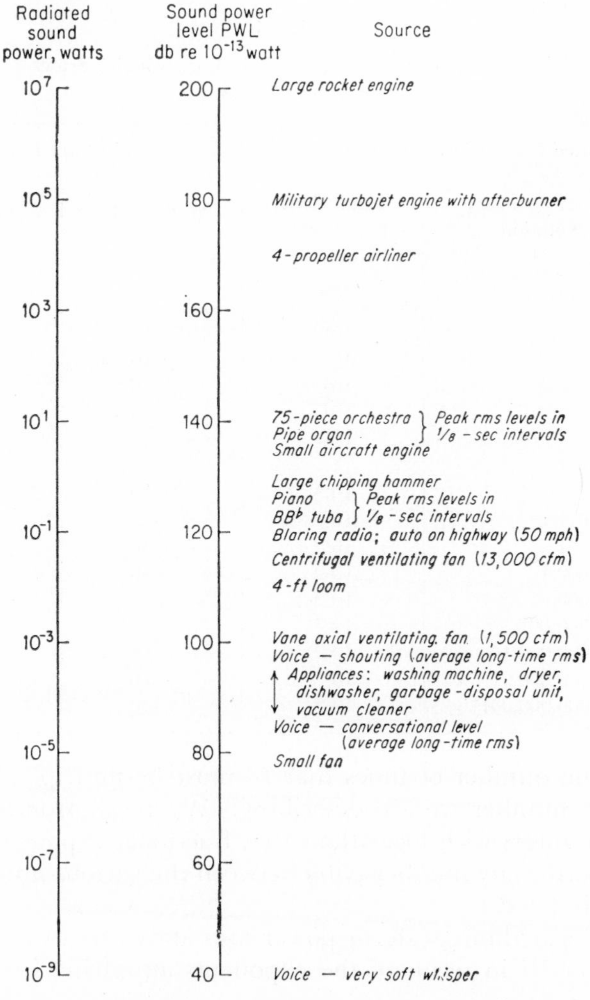

FIG. 3.2 Sound-power levels of some common sound sources.

while for an average shout the power radiated is about 0.001 watt. A large orchestra may radiate a maximum of 10 watts, and a large jet aircraft at takeoff can radiate almost 100,000 watts. A chart showing a number of typical powers radiated by various sources is given in Fig. 3.2.[1]

Decibel Scales for Sound Power. The range of acoustic powers that we have just named is also tabulated in column 1 of Table 3.1. We can see immediately that the numbers as we usually write them are quite clumsy to handle in calculations or in describing an acoustic source.

A much more compact notation is the equivalent exponential notation presented in column 2 of Table 3.1. This notation simply

Table 3.1

Linear, Exponential, and Decibel (Logarithmic) Scales for Sound Power

Radiated Sound Power W, watts		Sound-power Level, db	
Usual Notation (1)	Equivalent Exponential Notation (2)	Relative to 1 watt (3)	Relative to 10^{-13} watt (4)
100,000	10^5	50	180
10,000	10^4	40	170
1,000	10^3	30	160
100	10^2	20	150
10	10^1	10	140
1	1	0	130
0.1	10^{-1}	−10	120
0.01	10^{-2}	−20	110
0.001	10^{-3}	−30	100
0.000,1	10^{-4}	−40	90
0.000,01	10^{-5}	−50	80
0.000,001	10^{-6}	−60	70
0.000,000,1	10^{-7}	−70	60
0.000,000,01	10^{-8}	−80	50
0.000,000,001	10^{-9}	−90	40

expresses the number of times that 10 must be multiplied by itself to give the number we are describing. We recall from our high-school encounters with logarithms that fractional exponents may be used to describe any number lying between the various numbers that we have tabulated.

Now, a logarithmic scale in power also allows us to describe the sound power W in terms of the exponents appearing in column 2 of Table 3.1. Such a scale then describes the ratio of a particular power W to a power of 1 watt, which is called a "reference power." (Recall that the rules governing the use of logarithms require that we always take the logarithm of a *dimensionless quantity*—in the case just mentioned, the ratio of a particular power to a reference power of one watt.) In general for a reference power W_0

$$\text{Sound-power level} = \text{PWL} = 10 \log_{10} \frac{W}{W_0} \qquad \text{db } re\ W_0 \quad (3.4)$$

We use the term *level* to indicate that the scale is logarithmic, not linear. We call the "units" used in our logarithmic scale decibels (db) and Eq. (3.4) may be regarded as a definition of the decibel.

The decibel is basically a unit implying a given *ratio between two powers*. A ratio of 10 in power corresponds to a level difference of 10 db. A ratio of 100 in power corresponds to a level difference of 20 db. Power ratios less than 1 are perfectly allowable; they simply lead to negative numbers of decibels. For example, a power ratio of 0.1 corresponds to a level difference of −10 db, and so forth.

Table 3.2 indicates some typical power ratios and the corresponding level differences in decibels. Obviously, level differences between those shown as well as those beyond the range indicated are possible. The level differences for fractional power ratios as shown in Table 3.2 can be obtained, alternatively, from those for power ratios greater than 1. For example, for a power ratio of 0.5 we need merely note that 0.5 equals the inverse of 2, and we may thus determine the level difference as the negative of the level difference for a power ratio of 2.0, i.e., −3 db.

The Need for a Reference Quantity. We have pointed out that, because a decibel scale must be based on a power ratio, a reference quantity must necessarily be involved whenever we describe sound power by a given number on a decibel scale. For example, let us choose, for the moment, a reference power of 1 watt. In this case, we may express the sound-power level of a power W simply as

$$\text{PWL} = 10 \log_{10} W \qquad \text{db } re \text{ 1 watt} \qquad (3.5)$$

where W is the sound power in watts.

Column 3 of Table 3.1 shows such a power-level scale. On this scale the acoustic power level corresponding to a power of 1 watt is 0 db (that is, zero level difference from the reference quantity). For a power of 10 watts we have +10 db, and so forth.

Example 3.1. Determine the sound-power level of a source radiating 16.3 watts of sound power, using a reference power of 1 watt.

Solution

$$\text{PWL} = 10 \log_{10} 16.3 = 12.1 \text{ db } re \text{ 1 watt}$$

Table 3.2

Some Sound-power Ratios and the Corresponding Power-level Differences

Sound-power Ratio (Dimensionless), W/W_0	*Sound-power-level Difference, db,** $10 \log W/W_0$
1,000	30
100	20
10	10
9	9.5
8	9.0
7	8.5
6	7.8
5	7.0
4	6.0
3	4.8
2	3.0
1	0.0
0.9	−0.5
0.8	−1.0
0.7	−1.5
0.6	−2.2
0.5	−3.0
0.4	−4.0
0.3	−5.2
0.2	−7.0
0.1	−10
0.01	−20
0.001	−30

* To the nearest 0.1 db.

The actual number used as the reference quantity is usually chosen for convenience in calculation. For reasons that we shall discuss later in this chapter, *a reference power of 10^{-13} watt is used in this book.* The power-level scale for this reference power is shown in column 4 of Table 3.1.

It is of greatest importance that *the magnitude of a sound-power level never be stated without a clear indication of the associated reference power.* (We note incidentally that increments in power level do not depend on the reference power. For example, a power ratio of 10 corresponds to a 10-db change in power level regardless of the reference quantity.)

For a reference power of 10^{-13} watt we write

$$\text{PWL} = 10 \log_{10} W - 10 \log 10^{-13}$$

or

$$\text{PWL} = 10 \log_{10} W + 130 \qquad \text{db } re \ 10^{-13} \text{ watt} \tag{3.6}$$

where W is the sound power in watts.

Example 3.2. Determine the sound-power level of a source radiating 0.067 watt.

Solution

$$\begin{aligned} \text{PWL} &= 10 \log_{10} 0.067 + 130 \\ &= -10 \log_{10} \frac{1}{0.067} + 130 \\ &= -10 \log 14.9 + 130 \\ &= 118.3 \text{ db } re \ 10^{-13} \text{ watt} \end{aligned}$$

For orientation we show in Fig. 3.2[1] the sound-power levels associated with familiar acoustic sources. Sound-power levels should not be confused with sound-pressure levels, which are also expressed in decibels (see Sec. 3.2). We can see this distinction readily by remembering that the sound-power level describes the total acoustic power radiated by a source. On the other hand, the sound-pressure level at a given point depends upon the distance from the source, losses in the intervening air, room effects (if indoors), etc. (see Sec. 3.2).

A Decibel Scale for Sound Intensity. In a manner entirely analogous to that used in describing the decibel scale for sound power, we may define a decibel scale for sound intensities as follows:

$$\text{Intensity level} = \text{IL} = 10 \log_{10} \frac{I}{I_{\text{ref}}} \qquad \text{db } re \ I_{\text{ref}} \tag{3.7}$$

where I = sound intensity, watts/m^2 (power passing through a unit area) whose level is being specified

I_{ref} = reference intensity, usually 10^{-12} watt/m^2 (equivalent to 10^{-16} watt/cm^2)

The reason for the choice of 10^{-12} watt/m^2 as reference intensity is discussed subsequently. In this case

$$\text{IL} = 10 \log_{10} I + 120 \qquad \text{db } re \ 10^{-12} \text{ watt/m}^2 \tag{3.8}$$

Example 3.3. Determine the sound-intensity level at a distance of 10 m from a uniformly radiating source in free space, assuming that the source radiates a power of 0.1 watt.

Solution. First calculate the intensity at 10 m:

$$I = \frac{W}{S} = \frac{0.1}{4\pi r^2} = 0.1/4\pi(10)^2 = 7.95 \times 10^{-5} \text{ watt/m}^2$$

Then calculate the intensity level

$$\text{IL} = 10 \log_{10} 7.95 \times 10^{-5} + 120 = 79 \text{ db } re\ 10^{-12}\ W/\text{m}^2$$

A Decibel Scale for Sound Pressures. We have already said that under some circumstances sound intensity or sound power or sound-energy density is proportional to p^2, the mean-square sound pressure. Furthermore, most of our acoustic measurements are made in terms of effective sound pressure. Therefore, it is convenient for us to have a decibel scale for sound pressures.

The range of sound pressures with which we may be concerned runs from about 0.0002 μbar (microbar*) at the threshold of hearing under ideal conditions, to perhaps more than 10,000 to 100,000 μbars for noises associated with jet or rocket propulsion systems or with blast phenomena.

In column 1 of Table 3.3 we list the range of effective sound pressures discussed above. Column 2 expresses these same pressures in the exponential notation as used in Table 3.1.

In defining a decibel scale for effective sound pressure, we follow our earlier procedure of using a ratio of quantities proportional to acoustic power; that is, we use a ratio of mean-square sound pressures as follows:

$$\text{Sound-pressure level (SPL)} = 10 \log_{10} \frac{p^2}{p_{\text{ref}}^2}$$

$$= 20 \log \frac{p}{p_{\text{ref}}} \qquad \text{db } re\ p_{\text{ref}} \qquad (3.9)$$

where p_{ref} is the reference sound pressure.

Columns 3 and 4 of Table 3.3 show the sound-pressure-level scale for reference pressures of 1 μbar and 0.0002 μbar. A reference of 0.0002 μbar (0.00002 newton/m^2) is most commonly used in air. The standard sound-level meter is calibrated in decibels relative to 0.0002 μbar.

Therefore, we may write

$$\text{SPL} = 20 \log_{10} p + 94 \qquad \text{db } re\ 0.0002\ \mu\text{bar} \qquad (3.10)$$

* A microbar is about 1/1,000,000 atm. Specifically, it is equal to 0.1 newton/m^2 or 1.0 dyne/cm^2.

Table 3.3

Linear, Exponential, and Decibel (Logarithmic) Scales for Effective Sound Pressure

Effective Sound Pressure		Sound-pressure Level, db	
Descriptive Term, μbar (1)	Equivalent Notation, μbar (2)	Relative to 1 μbar (3)	Relative to 0.0002 μbar (4)
1,000,000	10^6	120	194
100,000	10^5	100	174
10,000	10^4	80	154
1,000	10^3	60	134
100	10^2	40	114
10	10^1	20	94
1	1	0	74
0.1	10^{-1}	−20	54
0.01	10^{-2}	−40	34
0.001	10^{-3}	−60	14
0.0002	2×10^{-4}	−74	0
0.0001	10^{-4}	−80	−6
0.00001	10^{-5}	−100	−26

where p is the effective sound pressure in *newtons per square meter*, or

$$\text{SPL} = 20 \log_{10} p + 74 \qquad \text{db } re \text{ } 0.0002 \text{ } \mu\text{bar} \tag{3.11}$$

where p is the effective sound pressure in *dynes per square centimeter*.

Example 3.4. Calculate the sound-pressure level *re* 0.0002 μbar for a sound with an effective sound pressure of 2.8 newtons/m².

Solution. Substitution in Eq. (3.10) gives

$$\text{SPL} = 20 \log 2.8 + 94 = 103 \text{ db } re \text{ } 0.0002 \text{ } \mu\text{bar}$$

For convenience we present in Fig. 3.3 four nomograms relating SPL in db *re* 0.0002 μbar to the corresponding effective pressure in (1) microbars (dynes per square centimeter), (2) newtons per square meter, (3) pounds per square inch, and (4) pounds per square foot.

An important difference between the decibel scales for sound pressure and for sound power is that the pressure ratio appears squared rather than to the first power in the logarithm. Equivalently, one uses the expression $20 \log_{10} (p/p_{\text{ref}})$ as opposed to $10 \log_{10} (W/W_{\text{ref}})$.

Thus, as is indicated in Table 3.3, a ratio of 10 in sound pressure corresponds to a sound-pressure-level difference of 20 db, not 10 db.

This difference between sound-pressure level and sound-power level scales need not trouble us if we recall the basic requirement that a decibel scale involves *a ratio of powers or quantities proportional to power*. The following rules should be helpful in avoiding confusion:

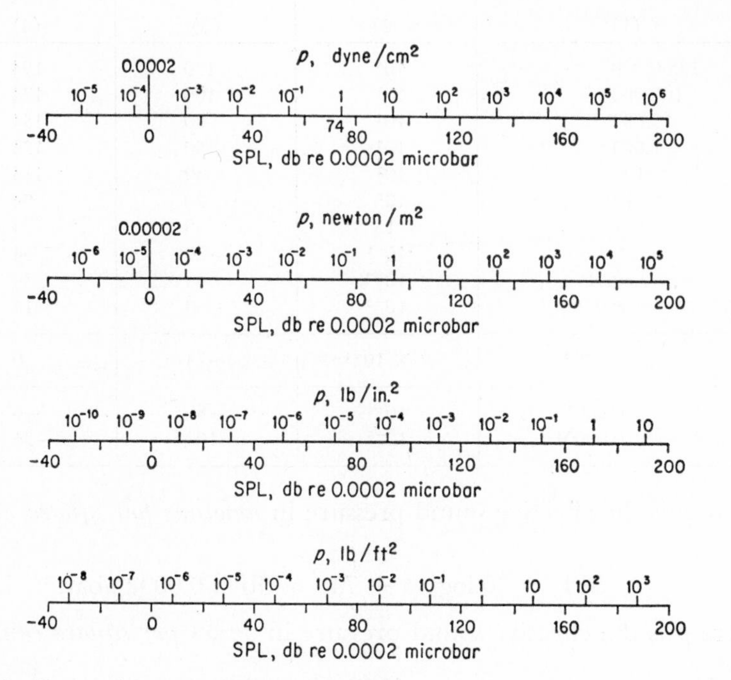

FIG. 3.3 Charts relating SPL (db *re* 0.0002 μbar) to p in dynes/cm², newtons/m², lb/in.², and lb/ft². For example, 1.0 newton/m² equals 1.435×10^{-4} lb/in.² equals 94 db.

Some Rules regarding Decibel Scales

(1) Decibel scales are basically for use with ratios of two powers, W/W_{ref} or two intensities I/I_{ref}, yielding

$$\text{PWL} = 10 \log_{10} \frac{W}{W_{\text{ref}}} \qquad \text{db } re\ W_{\text{ref}}$$

$$\text{IL} = 10 \log_{10} \frac{I}{I_{\text{ref}}} \qquad \text{db } re\ I_{\text{ref}}$$

(2) In certain common cases, sound power is proportional to the mean-square sound pressure, i.e., to p^2.

(3) Thus, by extension of the original definition of the decibel

scale, we *arbitrarily* define a decibel scale for ratios of *pressure squared*, p^2/p^2_{ref}, yielding

$$\text{SPL} = 10 \log_{10} \frac{p^2}{p^2_{\text{ref}}} = 20 \log_{10} \frac{p}{p_{\text{ref}}} \qquad \text{db } re\ P_{\text{ref}}$$

Choice of Reference Quantities for Sound Power, Intensity, and Pressure. As a practical matter, reference quantities for acoustic power, intensity, and pressure have been chosen so that the decibel scales for sound-power level, intensity level, and sound-pressure level are interrelated in a convenient way under certain circumstances.

The threshold of hearing at 1,000 cps for young men with good hearing corresponds to a sound pressure of about 0.0002 μbar. It was thus deemed convenient to choose this approximate threshold pressure, $p_{\text{ref}} = 0.0002$ μbar, as a reference pressure for the decibel scale for sound pressures.

The reference sound intensity $I_{\text{ref}} = 10^{-12}$ watt/m² (10^{-16} watt/cm²) was chosen so that the intensity level and sound-pressure level would be nearly numerically equal for plane or spherical waves in air at room temperature and sea-level pressure. Under these conditions we have the following relation:

$$\text{SPL} = \text{IL} + 0.2 \qquad \text{db } re\ 0.0002\ \mu\text{bar} \tag{3.12}$$

A correction must be made at other than room temperature and sea-level pressure. Figure 3.4[2] shows the correction to be added to SPL as given in Eqs (3.12), (3.13), and (3.14) for various air temperatures and barometric pressures.

We saw in Eq. (3.1) that for a simple source the sound intensity is directly related to the acoustic source power and the area of a surface through which this power is being radiated. The reference sound power used here, $W_{\text{ref}} = 10^{-13}$ watt, chosen so that sound-power level and sound-pressure level (as related to power level through intensity level) would be approximately but simply related to each other when the area of the surface being considered is expressed in *square feet*. Thus, at room temperature and sea-level pressure we have the following relation:

$$\text{SPL} = \text{PWL} - 10 \log_{10} \frac{S}{S_0} + 0.5 \qquad \text{db } re\ 0.0002\ \mu\text{bar} \tag{3.13}$$

or *approximately*

$$\text{SPL} \doteq \text{PWL} - 10 \log_{10} S \qquad \text{db } re\ 0.0002\ \mu\text{bar} \tag{3.14}$$

where S = surface area through which sound power is radiated, ft²
S_0 = reference area (1 ft²)

When W ref is chosen to be 10^{-13} watt, the sound-pressure level and the sound-power level are alike at a distance of 0.283 ft from a "point" source. (At this distance the surface area of a sphere enclosing the source is 1 ft².) Because sound pressure decreases linearly with distance from a point source, then at 1,000 ft (for instance) the sound-pressure levels would be down 71 db from the power levels (20 log 0.283/1,000 = − 20 log 3,540 = −71 db). Losses due to atmospheric effects would reduce further the sound-pressure levels at such a distance.

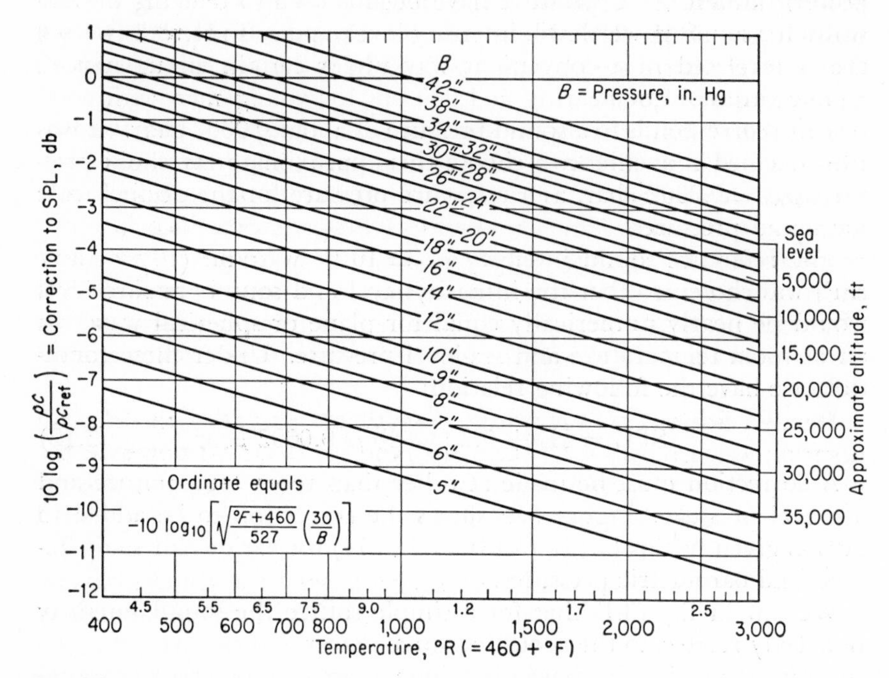

FIG. 3.4 Correction for ambient temperature and pressure, to be added to SPL for conditions of constant sound power or intensity. See Eqs. (3.12), (3.13), and (3.14).

Although the reference power 10^{-13} watt is the most widely used in the United States, some writers prefer a reference power of 10^{-12} watt because, when S and S_0 have the units of square meters,[3] this choice of reference leads to the same sort of result as Eq. (3.13). Certainly in those parts of the world where the metric system is used, a reference of 10^{-12} watt makes better sense than 10^{-13} watt. One should check carefully the reference power used by the person reporting power-level data before using the data in calculations.

Decibel Notation for Transmission Loss, Attenuation, and Noise Reduction. An important part of noise control deals with the amount of sound power or intensity transmitted through a structure. Acoustic barriers (walls, etc.) and mufflers can both be described in terms of a transmission coefficient. This coefficient is related to the *ratio* of the transmitted intensity (or power) to the incident intensity (or power).

Now, because a ratio of powers or intensities is involved, we may use a decibel scale to describe the transmission properties of an acoustic structure. Thus we might describe a wall as having a transmission loss of 36 db at 1,000 cps. Referring to Table 3.2, we see that a level difference (in this case a TL) of 36 db corresponds to a power ratio of 4,000.

The definitions of transmission loss, noise reduction, and attenuation (e.g., of a muffler) are presented in other chapters where they are used. It is sufficient here to point out that these quantities deal with power-like ratios and are therefore properly described in terms of decibels. No specific reference quantity is required for these scales; a reference to the incident or source quantity is implicit. We also note that the word "level" does not happen to appear in the names of the scales. However, each of the quantities implies a level difference of a particular type.

Decibel Scales for Electrical Quantities. Actually the decibel notation was originated for use with ratios of electrical power in the field of electrical communications. For example, a signal power of 10 watts is a factor of 10^4 or 40 db above a reference power of 1 milliwatt.

In just the way that we defined a decibel scale for sound pressure, scales may be used for voltage (e) or current (i). Decibel differences in level are computed as

$$10 \log \frac{e^2}{e_{\text{ref}}^2} = 20 \log \frac{e}{e_{\text{ref}}}$$

and

$$10 \log \frac{i^2}{i_{\text{ref}}^2} = 20 \log \frac{i}{i_{\text{ref}}},$$

respectively.

Such scales for electrical quantities are useful in acoustics because nearly all of our measurement apparatus is electronic in nature.

3.3 Multiple Sources

Very often in noise-control problems we are concerned with contributions from more than one source of noise. Naturally we wish to be able to use our decibel scales to take account of multiple contributions. In this section we shall develop a few rules which will help us avoid the confusion that can result from the incorrect summation of decibel levels. This discussion depends in part on material in Sec. 2.7 of the preceding chapter. It is expected that the reader will be familiar with that section.

Pure Tones. For the moment let us confine our attention to sounds that are "pure tone" (i.e., single frequency) in nature. Such

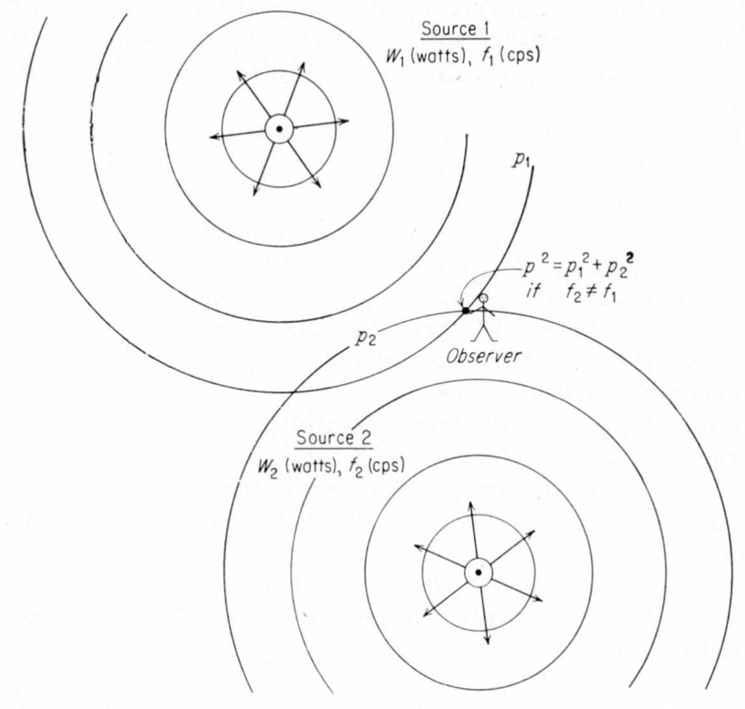

FIG. 3.5 Two sound sources contribute to the mean-square sound pressure p^2 at an observer's position.

a consideration is important, because it can be shown that any sound, no matter how complex, can be built up by summing a number of pure-tone components. Figure 3.5 shows two sound sources contributing to the total effective sound pressure p at the observer's position.

We can best understand the process of combining decibels by working an example.

Example 3.5. Two sound sources having different frequencies produce sound-pressure levels of (*a*) 88 and (*b*) 85 db *re* 0.0002 μbar at an observer's position. Find the total sound-pressure level *re* 0.0002 μbar.

Solution. Perform the operations:

1. Mean-square sound-pressure ratio of (*a*) is

$$\text{antilog}_{10} \frac{88}{10} = 6.3 \times 10^8$$

2. Mean-square sound-pressure ratio of (*b*) is

$$\text{antilog}_{10} \frac{85}{10} = 3.16 \times 10^8$$

3. Total mean-square sound-pressure ratio is

$$9.46 \times 10^8$$

4. Total sound-pressure level is

$$10 \log_{10} 9.46 \times 10^8 = 89.8 \text{ db}$$

Note that we did not actually need to find the pressures, but only their ratios to the reference pressure.

We may also make use of the information presented in Table 3.2 to combine contributions from two sources. In doing this, we may prepare Fig. 3.6 by interpreting column 1 in Table 3.2 as mean-

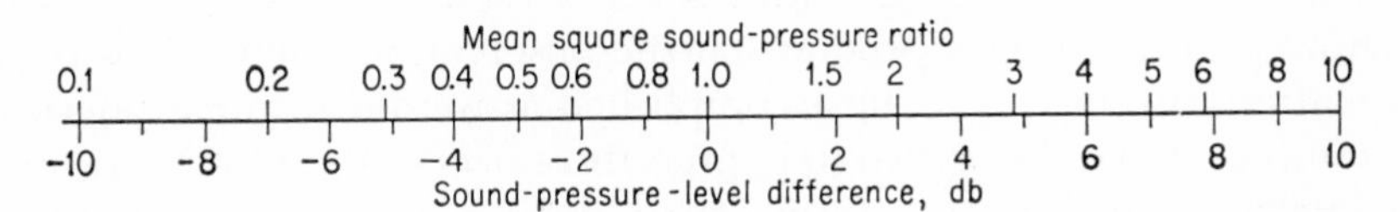

FIG. 3.6 Relation between mean-square sound-pressure ratio (p_T^2/p_1^2) and sound-pressure-level difference in decibels ($\text{SPL}_T - \text{SPL}_1$). This figure is based on Table 3.2.

square-pressure ratio and column 2 as sound-pressure-level difference, respectively. We assume again that the sound waves from two sources have different frequencies. The mean-square sound-pressure ratio when source 2 is combined with source 1 is

$$\left(\frac{p_T}{p_{\text{ref}}}\right)^2 = \left(\frac{p_1}{p_{\text{ref}}}\right)^2 + \left(\frac{p_2}{p_{\text{ref}}}\right)^2 \tag{3.15}$$

For example, if the second source makes an intensity contribution equal to that from the first source, then by Eq. (3.15) the combined mean-square pressure is twice that due to the first source alone. Referring to Fig. 3.6 (or Table 3.2) we find that for a power ratio*

* In this case we are interpreting "ratio" as the ratio of the combined mean-square pressure from the two sources to the mean-square pressure from one source alone. This interpretation is a valid one for Table 3.2 and Fig. 3.6.

of 2 we expect a level difference of 3.0 db. That is to say, the sound-pressure level of the two sources combined will be 3 db above that for either of the sources separately.

Likewise, if the second source is contributing twice as much (in intensity or mean-square pressure) as the first source, the combination will have 3 times the mean-square pressure of the first source. For this case, Fig. 3.6 indicates an SPL increase of 4.8 db over the level of the first source. (Note that this is exactly the result that was obtained in Example 3.5.)

It follows from our earlier discussion that Fig. 3.6 may also be used where more than two sources are combined, provided again no two have the same frequency. For example, 10 sources of equal intensity will give a resultant sound-pressure level (and intensity level) 10 db higher than that for any one of the sources individually.

If any two of the sources have the same frequency, their combined pressure must be determined by Eq. (2.30)* before combining with the others. For example, two sources of the same frequency, the same phase, and of equal strength give us a mean-square pressure ratio of $2^2 = 4$. The resultant sound-pressure-level increase is 6.0 db instead of 3.0 db.

Very often in noise-control work we wish to combine the contributions from two or more sound sources each characterized by the sound-pressure level it produces at the observation point. This combination can, of course, be accomplished as we did in the numerical example above, by a two-step procedure wherein we convert first to mean-square pressure ratios, then add together the mean-square pressure ratios, and then convert back by taking $10 \log_{10}$ of the sum.

However, the conversion process can be done once and for all, and a nomogram prepared. Figure 3.7 presents such a nomogram, which can be used to combine any number of levels taken two at a time. As an example, let us determine the total sound-pressure level for two sources that produce 70 and 75 db *re* 0.0002 μbar SPL respectively, at an observer's position. We choose the larger number as L_1, obtaining a level difference of $L_1 - L_2 = 5$ db. Figure 3.7 tells us that the level of the combination is 1.2 db above L_1 or 76.2 db (*re* 0.0002 μbar). If more than two sound-pressure levels are to be combined, we combine the first two, then combine the third with our first result, and so forth.

Continuous-spectrum Sounds. Many sounds that we encounter in noise control are continuous-spectrum sounds. That is, they contain acoustic power distributed continuously over a range of fre-

* $p = \sqrt{p_1^2 + p_2^2 + 2p_1p_2 \cos(\theta_1 - \theta_2)}$.

quencies (see Figs. 2.12 and 2.13). The roar of a waterfall and the hiss of an air jet are examples of continuous-spectrum noises.

As we showed in Chap. 2, it is customary, when dealing with continuous-spectrum sounds, to consider the mean-square sound pressure in a given frequency band. Such a frequency band may be divided up into a number of increments, and a mean-square pressure may be assigned to each increment. Because no two of these increments have the same frequency, the incremental contributions are combined like pure tones of different frequencies. Therefore, we determine the total mean-square pressure simply by adding the ap-

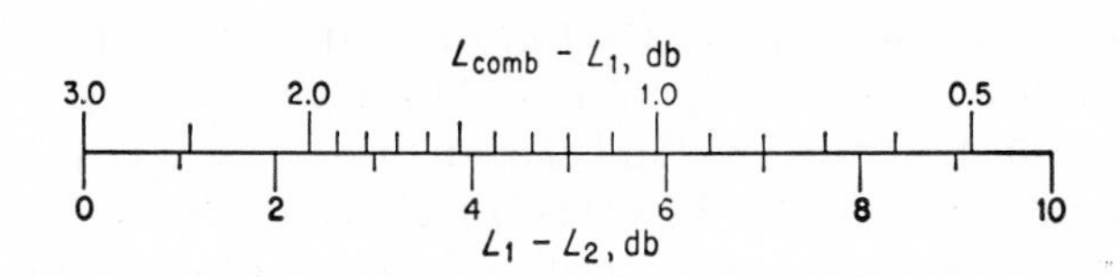

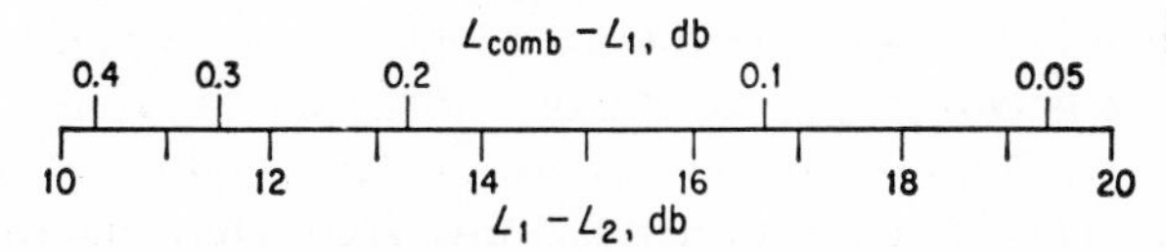

FIG. 3.7 Chart for combining two acoustic levels, L_1 and L_2. Levels may be power levels, sound-pressure levels, or intensity levels.

$$L_1 > L_2, \quad L_{\text{comb}} = (L_{\text{comb}} - L_1) + L_1$$

Example: $L_1 = 88$ db, $L_2 = 85$ db, $L_1 - L_2 = 3$ db

Solution: $L_{\text{comb}} = 88 + 1.8 = 89.8$

propriate values of mean-square pressures for the individual frequency increments.

Thus, as we showed in Sec. 2.7, the mean-square pressure for a frequency band composed of a number of subbands is,

$$p^2_{\text{band}} = p_A{}^2 + p_B{}^2 + p_C{}^2 + p_D{}^2 + \cdots \tag{3.16}$$

where $p_A{}^2$, $p_B{}^2$, etc., are the mean-square pressures for the subbands.

A special case of interest occurs when each subband has a width of 1 cps. If the mean-square pressure for the subbands is $p_1{}^2$ on the average, then the mean-square pressure for the parent band is [see Eq. (2.31)]

$$p^2_{\text{band}} = p_1{}^2 \, \Delta f \tag{3.17}$$

where Δf is the width in cps of the parent band.

Expressed in decibels *re* 0.0002 μbar, this relation becomes

$$10 \log \frac{p_{\text{band}}^2}{p_{\text{ref}}^2} = 10 \log \frac{p_1^2}{p_{\text{ref}}^2} + 10 \log \Delta f \qquad (3.18)$$

or

$$\text{SPL}_{\text{band}} = S(f) + 10 \log \Delta f \qquad \text{db } re \text{ } 0.0002 \text{ } \mu\text{bar} \qquad (3.19)$$

where $S(f)$ is defined as the *spectrum level* (db *re* 0.0002 μbar). In other words, *the spectrum level is that average sound-pressure level in decibels (referred to a band* 1 *cps wide) which when added to* 10 *log* Δf *yields the sound-pressure level of the parent band.* The spectrum level $S(f)$ is seldom measured directly with an analyzer of unity bandwidth, but rather is determined from the SPL_{band} by the mathematical relation of Eq. (3.19). More will be said about the practical use of spectrum level in Chap. 4.

Determination of Over-all Levels from Band Levels. Frequently we measure the sound-pressure levels in a series of contiguous bands that include the entire frequency range of the source. We then desire the level as though it had been measured in a single band covering all of this same frequency range. The level in the all-inclusive band is referred to as an *over-all level* in decibels for the noise.

One very common set of contiguous frequency bands used in noise control is known as *octave bands.* The term "octave" is used to indicate that the upper limit of frequency of the band is just twice the lower limit of frequency. Such bands are examples of so-called constant-percentage frequency bands, in which the bandwidth (i.e., the number of cycles per second included within the band) is proportional to the geometric mean frequency of the band. Other examples of constant-percentage bands are: one-half-octave bands, one-third-octave bands, etc.

As an example of the combination of levels in different frequency bands, to produce an over-all level, let us start with the band levels shown across the top of Fig. 3.8. The frequency limits of the particular bands are not important to the method of calculation. It is sufficient to know that the bands are contiguous and cover the frequency range of interest.

Figure 3.8 indicates schematically the step-by-step computation of the over-all sound-pressure level accomplished with the help of Fig. 3.7. The order in which we combine the various band levels is unimportant. In Fig. 3.8 the levels were taken in increasing order of magnitude. We see that the seventh and eighth bands combine to give a level of 73. That combination and the sixth band give a level of 77.1, etc.

We may make an interesting observation simply by changing the order in which we combine the levels presented in Fig. 3.8. In Fig. 3.9 we show the same sound-pressure-level spectrum with the levels in bands 1 through 4 combined separately from the levels in bands 5 through 8. The result of this revised procedure is interesting in that the over-all level is seen to be determined entirely by the levels in the first four bands.*

The example in Figs. 3.8 and 3.9 points up the fact that characterizing a noise spectrum simply by the over-all level may be completely

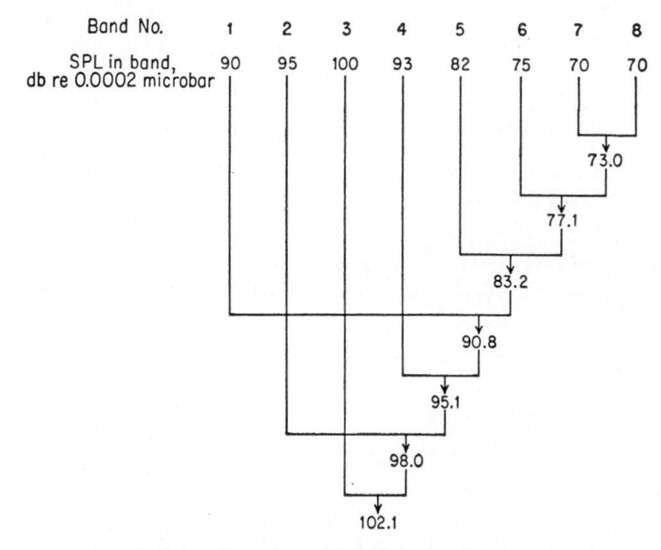

FIG. 3.8 Determination of an over-all sound-pressure level from levels in frequency bands.

inadequate for some noise-control purposes, because it ignores a large portion of the frequency spectrum. For example, in a given noise-control problem, the sound-pressure levels in bands 5 through 8 of Figs. 3.8 and 3.9 might be quite important. They might be closely associated with annoyance or with ability to communicate by voice. However, as we have seen, their influence on the over-all level is negligible. Thus, it is generally useful to describe a sound spectrum in terms of the levels in a number of frequency

* In almost all noise-control problems it makes no sense to deal with small fractions of decibels. Rarely does one require a precision as great as 0.1 db in his calculations, and in many cases the results of an analysis—for example, an estimated sound-pressure-level spectrum—is best stated only to the nearest decibel.

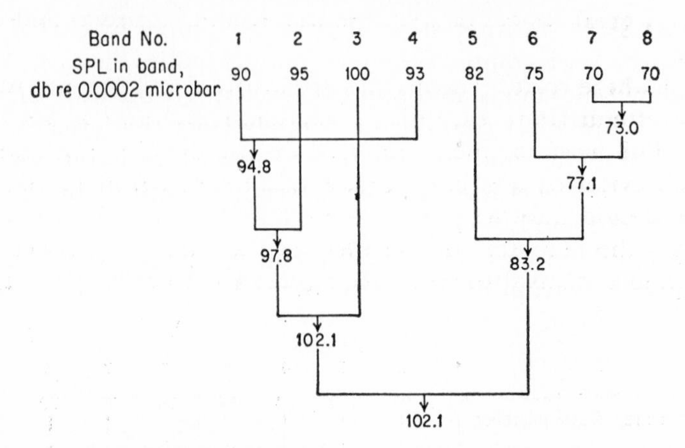

FIG. 3.9 Alternate determination of an over-all sound-pressure level from levels in frequency bands (see Fig. 3.8).

bands. This subject will be treated more fully in many chapters of this book, particularly those dealing with criteria for noise control.

3.4 Summary

We have seen that decibel scales are useful in a number of ways in noise control. Such scales provide a convenient notation to de-

Table 3.4

Decibel Scales for Sound Power, Intensity, and Pressure, with Definitions and Reference Quantities

Decibel Scale	Abbreviation	Definition	Reference Quantity
Sound-power level	PWL	$10 \log_{10} \frac{W}{W_{ref}}$	$W_{ref} = 10^{-13}$ watt
Sound-intensity level	IL	$10 \log_{10} \frac{I}{I_{ref}}$	$I_{ref} = 10^{-12}$ watt/m² $= 10^{-16}$ watt/cm²
Sound-pressure level	SPL	$10 \log \frac{p^2}{p_{ref}^2} = 20 \log \frac{p}{p_{ref}}$	$p_{ref} = 0.00002$ newton/m² $= 0.0002$ μbar $= 0.0002$ dyne/cm²

Note: For two sounds having different frequencies, decibel levels are usually combined on an "energy" or "power" basis, as indicated in Fig. 3.7. Sounds having the same frequency are combined according to Eq. (2.30).

scribe very great ranges of sound power, sound intensity, and sound pressure.

Except where relative quantities are involved (as they are in level changes, attenuation, TL, etc.) *a reference quantity is always required.* The most common decibel scales are defined, together with their abbreviations and appropriate reference quantities, in Table 3.4. We should indicate the reference quantity (as "db *re* reference quantity") the first time the decibel scale is used in a report or set of notes and as many times thereafter as required for clarity.

REFERENCES

1. Peterson, A. P. G., and L. L. Beranek: "Handbook of Noise Measurement," 3d ed., General Radio Company, Cambridge, Mass., 1956.
2. Beranek, L. L.: "Acoustics," p. 315, McGraw-Hill Book Company, Inc., New York, 1954.
3. Harris, C. M. (ed.): "Handbook of Noise Control," McGraw-Hill Book Company, Inc., New York, 1957.

Chapter 4
THE SELECTION OF INSTRUMENTATION

A. C. Pietrasanta

4.1 Introduction

A wide variety of instrumentation is available today for the measurement of noise and vibration. The material presented in Chap. 5 and 6 makes this fact abundantly clear. A number of different microphones, vibration pickups, frequency analyzers, and read-out instruments are described. Confronted with these various items of equipment, the reader may easily become confounded. Which type of microphone is right for the job at hand? Is a detailed frequency analysis required; if so, what type of analyzer should be employed? Would a tape recorder be useful? Answers to these and similar questions can be found in this chapter, which aims to serve, in a general way, as a guide for the selection of instrumentation.

First, in order to relegate each piece of equipment to its proper place in a generalized instrumentation system, the fundamental elements of a generalized system will be described. Second, the type of problem at hand must be considered. Since noise and vibration problems vary widely in their nature, the pertinent factors of the problem at hand influence the choice of equipment. Finally, the various types of instrumentation that come under each of the elements in the instrumentation system are discussed in some detail.

The Instrumentation System. The various elements in an instrumentation system are (1) transducer (microphone or vibration pickup*), (2) electronic amplifier and calibrated attenuator (gain

* Basically, instrumentation systems for the measurement of sound or vibration are essentially alike except for the transducer. For air-borne sound the transducer is a microphone; for structure-borne vibration, the transducer is a vibration pickup. Although the inputs to both kinds of transducers are different, the output of each is the same—an electrical signal that is then fed to the rest of the instrumentation system.

Therefore, since the instrumentation systems for the measurement of noise or vibra-

control), (3) data storage, (4) frequency analyzer, and (5) read-out. Not all of the elements shown are utilized in every instrumentation system. Some possible combinations of the basic elements are shown in Fig. 4.1 and are discussed in the next few paragraphs.

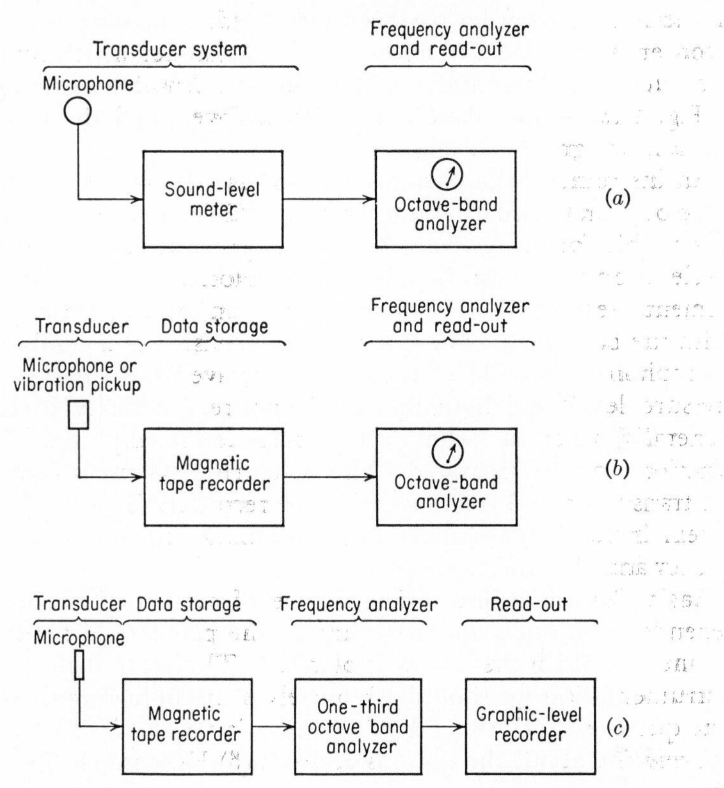

Fig. 4.1 Examples of instrumentation systems: (*a*) basic measuring system, (*b*) basic system with data storage added, (*c*) system with data storage and with graphic recorder used as an automatic read-out device.

The basic measuring system is composed of a microphone, a sound-level meter, and an octave-band analyzer (see Fig. 4.1*a*). The transducer system comprises the microphone and the sound-level meter, which in this case is simply acting as an amplifier. The octave-band analyzer is acting both as a frequency analyzer and as a read-out device (the decibel meter). This system could have been simpli-

tion are essentially the same except for the transducer, the text, for the sake of simplicity, will refer primarily to the measurement of noise. In most instances the discussion will also apply to the measurement of vibration. As a matter of fact, the reader could easily substitute the word *vibration* for *noise* throughout most of the text.

fied even more if only a microphone and a sound-level meter had been used. In this instance the meter on the sound-level meter would be acting as the read-out device.

A slightly more involved instrumentation system is shown in Fig. 4.1*b*. In this instance a data-storage element in the form of a magnetic tape recorder has been employed. The output of the tape recorder is connected to an octave-band analyzer which acts as both a frequency analyzer and a read-out device. Another system is shown in Fig. 4.1*c*. Again, data storage is employed, and the read-out device is now a graphic level recorder.

An instrumentation system has two basic functions: (1) to obtain noise or vibration data in the field, and (2) to reduce measured data to a usable form, e.g., values of sound-pressure level or velocity or acceleration which can be tabulated or plotted. In some systems the elements serving these two functions cannot easily be separated. With the basic measuring system, which consists of a combination of microphone, sound-level meter, and octave-band analyzer, sound-pressure levels are both measured and read directly in the field. Generally, where a magnetic tape recorder is employed, these two functions can be separated. The measuring system is composed of the transducer and the magnetic tape recorder. The data-reduction system includes the magnetic-tape playback and the associated frequency analyzer and read-out device.

Basic Considerations. The degree of success of a noise survey depends upon one's understanding of the problem at hand and the manner in which the survey is planned. Therefore, before selecting instrumentation we should ask ourselves the following three important questions: (1) What *kind of noise* is *involved?* (2) How much *information* about the noise is *desired?* (3) How much *time* is *available?*

Various kinds of noises may be encountered in practice. For purposes of this discussion, they can be classified into the following general categories:

1. *Steady Wide-band Noise.* Jet noise, air moving through ducts or orifices, ambient noise
2. *Steady Narrow-band Noise.* Circular saws, planers, transformer noise, jet-engine compressor whine
3. *Impulse (Impact) Noise.* Drop forge hammer, dog barking, pistol shot, door slamming
4. *Repeated Impulse (Impact) Noise.* Riveting, pneumatic hammers, machine guns
5. *Intermittent Noise.* Aircraft flyovers, automobile traffic, trains

The type of noise influences the selection of instrumentation. For example, use of a system employing a tape recorder for data storage is almost a necessity for obtaining data on aircraft taking off. On the other hand, the noise output of an aircraft operating on the ground can be measured by simply employing a sound-level meter and an octave-band analyzer. Yet either of these instrumentation systems would be a poor choice for the measurement of impulse noise.

The amount of time available and the amount of information desired are interrelated factors. Usually more information requires more time, either in making measurements in the field or reducing data in the laboratory or both. The amount of data taken will depend on the number of different measurements to be made, how often certain measurements are to be repeated (to determine statistical variation), number of noise-source conditions to be tested and how much detail is required in analyzing the measured data.

The amount of time available for measurements may be dictated by several factors. First, and often most important, is the economic consideration. If only a half-day noise survey can be afforded, then the planning of this survey will certainly be different from the planning of a week-long noise survey. Sometimes the operating schedule of the device to be measured clearly defines the time available. Take for instance a rocket motor that fires approximately 45 sec once a week or a blow-down wind tunnel that operates for perhaps 2 min every 2 days. These time limitations impose more severe restrictions on the planning of a noise survey than would continuously running devices, such as air conditioners or office equipment.

The amount of information desired depends on the use to which the data are to be put. For example, assume that it is known that there are no outstanding pure-tone components in a continuous spectrum noise, at least above 300 cps. Assume also that the data are to be used to determine potential damage to hearing, interference with speech communication, or annoyance. Then octave-band analyses of the noise using the basic measuring system of this chapter are usually adequate. On the other hand, if the purpose of taking the data is to redesign a noisy machine so that it is quieter, narrow-band analyses of the noise are often necessary.

So we see that the type of noise involved, the amount of information about this noise that is desired, and the amount of time available to obtain this information all influence the choice of instrumentation for a noise survey.

4.2 Selection of Transducers

Selection of Microphones. No one microphone is suitable for all types of noise measurements or for all environmental conditions encountered. The choice of a microphone or microphones for a particular noise survey depends on several factors. In this section the most important of these factors are discussed briefly. Throughout this discussion the reader may wish to refer to Table 6.1 (and Fig. 6.1), wherein the salient characteristics of several of the commercially available microphones are tabulated.

1. *Environmental Conditions.* One should first consider the environmental conditions to which a microphone will be exposed. The most important environmental conditions are temperature and humidity. The maximum temperature at which most microphones will continue to operate satisfactorily ranges from 110°F to about 500°F. For measurements at normal room temperature, all microphones will satisfy the temperature requirement. On the other hand, at temperatures above about 300°F, only the condenser-type microphone should be employed. It is important to note that at the higher temperatures the weakest link in the microphone system is likely not to be the microphone but the associated preamplifier and cabling.

Humidity also affects the operation of many microphones. Condenser microphones are particularly susceptible because of internal electrical leakage caused by moisture on the insulators employed. This condition can be avoided by keeping the microphone at a temperature higher than the ambient temperature. Sometimes the heat generated by the microphone preamplifier is utilized for this purpose.

2. *Range of Pressure Levels.* Next in importance is a consideration of either the maximum or minimum sound-pressure levels that will be encountered. When very low noise levels, e.g., nighttime ambient noise in a residential area, are to be measured, then the minimum sound-pressure level that a microphone can measure should be the determining factor. On the other hand, the choice of a microphone for measuring extremely intense noise levels, such as produced by jet or rocket engines, is influenced by the maximum sound-pressure level that a microphone will measure without excessive distortion or failure. No one microphone serves satisfactorily to measure both extremely high and extremely low sound levels. A microphone that can measure very low noise levels is usually limited in the maximum sound-pressure level that it can measure. Conversely, high-intensity microphones are not suitable for measuring low noise levels.

3. *Frequency Response.* The final important factor to be considered in selecting a microphone is its frequency response. Assuming that considerations of environment and range of sound-pressure levels to be measured have narrowed the choice of microphones to just a few, the microphone that should be selected from this group

is the one having the smoothest frequency response over the frequency range of interest.

Selection of Vibration Pickups. As in the case of microphones, there are several factors that influence the choice of a vibration pickup (or accelerometer) for a particular application. In this section, the most important of these factors are discussed briefly. For detailed characteristics of several of the commercially available vibration pickups, the reader is referred to Table 6.2 (and Fig. 6.3).

1. *Frequency Response.* Perhaps the most important factor to consider in selecting a vibration pickup is its frequency response. The reason that this characteristic is so much more important than it is in the selection of microphones is that accelerometers, as a group, cover a much wider frequency range. It extends from dc (0 cps) to more than 30,000 cps—covering four to five decades! Also of importance is the fact that, although the frequency response ($\pm$ 1 db) of most pickups extends down to about 1 cps, the upper frequency limit differs widely among pickups. So, from the point of view of selection, one can probably eliminate from consideration several of the available pickups by first considering the frequency range to be measured and comparing it with the frequency response of the available pickups.

2. *Range of Acceleration Levels.* Next in importance is the amplitude range to be covered, more specifically, the estimated maximum acceleration to be measured. Here again the range covered by the available accelerometers varies by at least a factor of 1,000, from a maximum of 10 *g* to an allowable maximum acceleration of about 10,000 *g*. Note that "*g*" equals the acceleration due to gravity.

3. *Weight of Accelerometer.* The weight of the accelerometer is also important. The combined weight of a pickup and its mounting must be small enough so that it will not materially affect or "load" the motion of the device being measured. The weights of commercially available pickups vary from about 0.1 oz to almost 10 oz, i.e., by a factor of 100.

4. *Environmental Conditions.* Environmental conditions also influence the choice of an accelerometer but, comparatively speaking, they are a minor consideration. The maximum allowable temperature varies over a rather narrow range from about 110°F to 230°F, indicating that, for the most part, temperature would not appreciably influence the choice of an accelerometer. The output of a pickup varies with temperature, but for most accelerometers the variation is less than 1 db over about the same temperature range, from about −40°F to the maximum allowable temperature.

4.3 Selection of Analyzers

Frequency Analysis. The simplest measure of noise that one can obtain is the over-all sound-pressure level (SPL) in decibels. This is a single number read from a sound-level meter using the C (or flat) network. It generally represents the SPL in the frequency range from about 20 to 10,000 cps. The over-all reading can be, and often is, misleading for the majority of noises encountered in practice.

Some of the ways in which the over-all SPL reading can be misleading are demonstrated in Fig. 4.2. These data were obtained by use of a sound-level meter and octave-band analyzer. Except for Fig. 4.2*d*, the data are plotted in terms of measured sound-pressure levels vs. octave bands of frequency. Also shown at the left of the graph are the over-all SPL's one would have obtained using only a sound-level meter.

In Fig. 4.2*a* two octave-band spectra are shown that have very different characteristics. One spectrum peaks in the lower-frequency range, the other in the higher-frequency range, yet both have the same over-all SPL. Clearly, the over-all reading gives no indication of the distribution of noise as a function of frequency. The shape of a noise spectrum can be extremely important, particularly in problems involving noise control, since low-frequency noise is much more costly to reduce than high-frequency noise.

A second way in which the over-all SPL can be misleading is illustrated in Fig. 4.2*b*. The curve represents a design criterion, i.e., a plot of acceptable noise levels as a function of frequency. Although in the example the criterion is associated with a particular speech-communication environment, the shape of the curve is similar to other criterion curves having to do with annoyance, damage to hearing, etc., in that the acceptable levels are appreciably higher at the low frequencies than at high frequencies. What is the design objective for this example? In terms of the over-all SPL it is 80 db. This number is meaningless since it gives no indication of the criterion levels in the seven octave bands above the lowest. As can be seen, the acceptable level in the 150- to 300-cps octave band, for example, is 68 db, which is 12 db less than the over-all. In the 1,200- to 2,400-cps band, it is even lower—60 db. The range in levels over the octave bands underlines the necessity of stating more than the over-all level when specifying a noise criterion.

Finally, another example is the one shown in Fig. 4.2*c* and *d*. The noise levels on each side of a noise-reduction treatment have

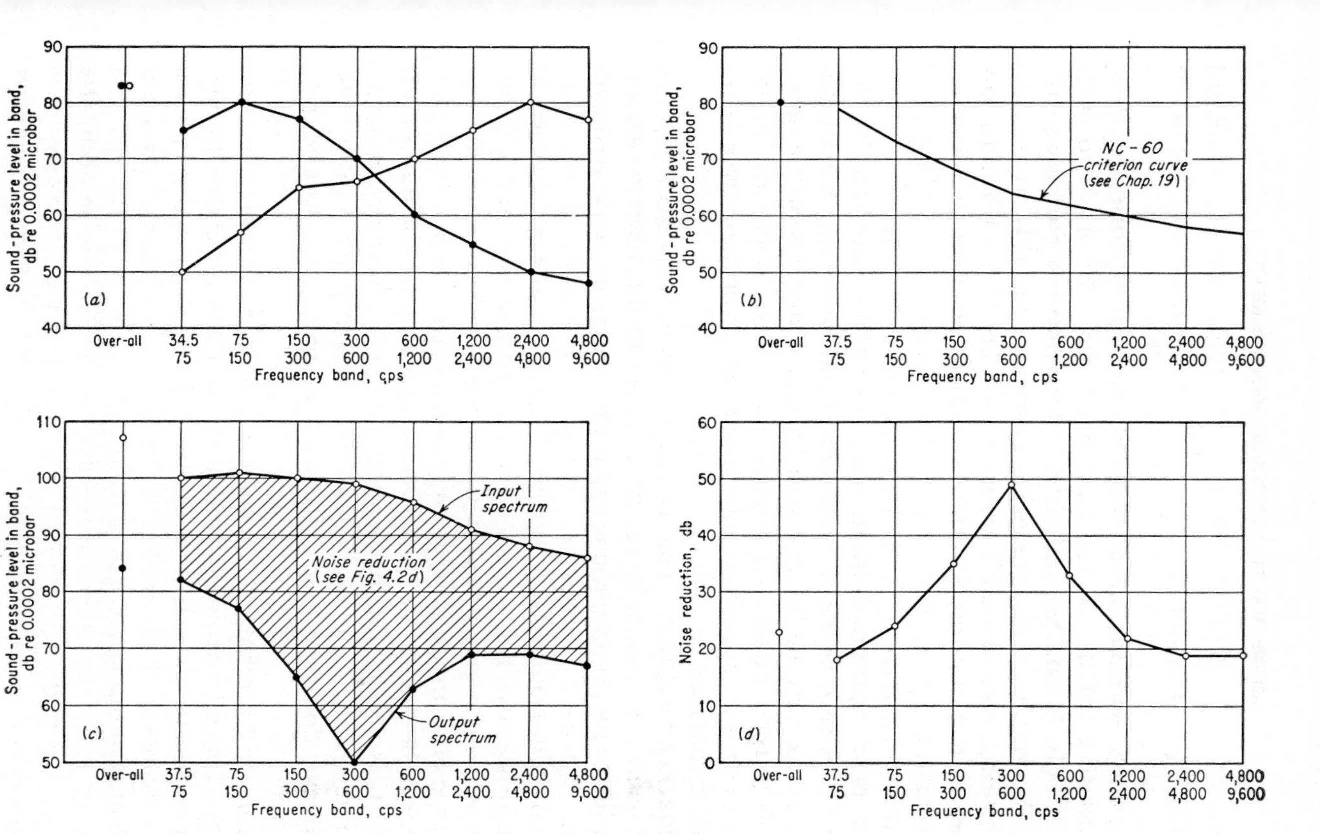

FIG. 4.2 Examples illustrating the limitations of the over-all sound-pressure-level reading: (*a*) two octave-band noise spectra with the same over-all sound-pressure level; (*b*) example of a criterion curve describing acceptable noise levels in octave bands; (*c*) noise spectra at input and output of noise-reduction treatment; (*d*) noise reduction in octave bands for measurement shown in (*c*).

been measured and plotted in Fig. 4.2*c*, along with the associated over-all SPL's. The noise reduction of this treatment, or the difference in SPL's on either side, is plotted in Fig. 4.2*d*. For this example, the noise reduction varies appreciably with frequency, ranging from 18 db to 49 db. On the other hand, the difference in over-all levels is only 23 db. Not only is this difference misleadingly small, but it also gives no indication of the effectiveness of this noise-reduction device in meeting, say, a criterion like that of Fig. 4.2*b*. In particular, the fact that this acoustical treatment provides almost 50 db of noise reduction in the 300- to 600-cps octave band is completely hidden in a statement of the over-all difference.

The above examples illustrate only a few of the ways in which the over-all SPL reading can be misleading and deceptive. This problem can be overcome by making a frequency analysis of the noise. For this purpose one uses a frequency analyzer. There are several types of analyzers available, differing primarily in the degree of information that they provide.

Frequency Analyzers. For our purpose, we may classify frequency analyzers as (1) constant bandwidth, (2) constant-percentage narrow bandwidth, and (3) octave, one-half-octave, or one-third-octave bandwidth.

1. *Constant-bandwidth Analyzer.* The first of these, the constant-bandwidth analyzer, has a filter or passband that is a fixed number of cps wide. Common bandwidths range between 2 and 200 cps. This type of analyzer, particularly the ones with the narrower passbands, can be used to define the shape of a noise spectrum in great detail. It is also useful for singling out the harmonic components of the noise, when the frequencies are sufficiently stable so that the components do not shift in and out of the narrow passband during measurements.

2. *Constant-percentage Narrow-bandwidth Analyzer.* The constant-percentage narrow-bandwidth analyzer has a bandwidth that varies automatically with the tuning of the analyzer and remains a fixed percentage of the midband frequency. For example, a 4 per cent bandwidth analyzer has a bandwidth of 4 cps at 100 cps, and 40 cps at 1,000 cps. Because this type of analyzer is continuously tunable, it is particularly useful in searching for discrete frequency components of a sound whose fundamental frequency fluctuates slowly. If the passband is wide enough to contain the varying fundamental, it will also be wide enough at the higher frequencies to contain any of the harmonics.

3. *Octave-, One-half-, and One-third-octave-band Analyzers.* Analyzers that are very commonly used but that provide less detail than

the two types described above, are the octave-, one-half-octave-, and one-third-octave-bandwidth analyzers. Although more commonly known by these designations, these analyzers are also constant-percentage bandwidth analyzers. For the octave-band analyzer the upper cutoff frequency of each passband is twice the lower cutoff frequency. For the one-half-octave-band analyzer the upper cutoff frequency is $\sqrt{2}$ times the lower, and for a one-third-octave-band analyzer the ratio is $\sqrt[3]{2}$.

Table 4.1 illustrates more clearly how the types of frequency analyzers discussed above compare with one another. The bandwidths shown for the octave, one-half-octave and one-third-octave types are not necessarily commercially available but are given as shown to provide a common basis of comparison.

Table 4.1

Comparison of Frequency Analyzers at Two Center Frequencies

Type of Analyzer	Center Frequency of 100 cps		Center Frequency of 1,000 cps	
	Passband, cps	Bandwidth, cps	Passband, cps	Bandwidth, cps
Constant-bandwidth (4 cps)	98–102	4	998–1,002	4
Constant-percentage—narrow-band (2%)	99–101	2	990–1,010	20
Octave	71–141	70	707–1,414	707
½-Octave	84–119	35	841–1,189	348
⅓-Octave	89–112	23	891–1,122	231

For center frequencies of 100 and 1,000 cps the upper and lower cutoff frequencies have been tabulated as well as the bandwidths. For all analyzers except the constant-bandwidth analyzer, the bandwidth increases in proportion to the center frequency. For example, the one-third-octave bandwidth centered around 100 cps is about 23 cps, whereas at 1,000 cps, it is 231 cps wide. Also in evidence is the variation in size of the bandwidth for these particular analyzers at any one center frequency. At a center frequency of 1,000 cps, for example, the bandwidth for the analyzers shown ranges from 4 cps to 707 cps.

Analyzer Comparisons. Practically speaking, one might ask how much these analyzers differ from one another in measuring a noise spectrum. How much more detail does one provide than another? Let us consider the illustrative example described below.

For purposes of discussion let us assume that the noise output of a particular noise source, measured at some distance, is given by the spectrum shown in Fig. 4.3. This curve is plotted in terms of "true" spectrum level as a function of frequency. By "true" spectrum level

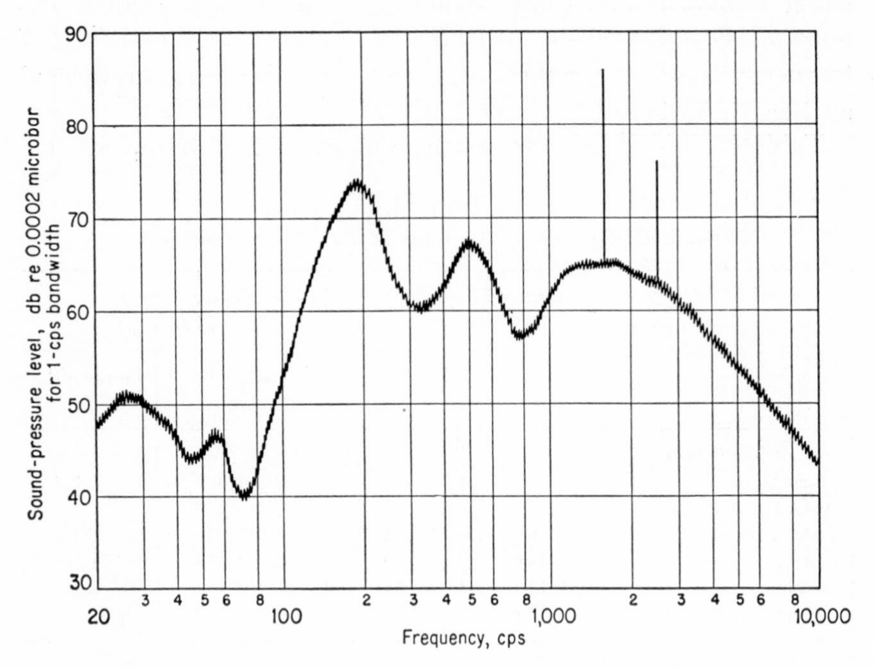

Fig. 4.3 Spectrum levels vs. frequency of spectrum used in illustrative example in text (see Fig. 4.4).

we mean that it was measured by an analyzer actually having a bandwidth of 1 cps at all frequencies. Examination of that spectrum reveals that the noise appears to be continuous with frequency and has two broad peaks at 200 and 500 cps. Furthermore, two pure-tone components are in evidence, one at 1,600 cps and the other at 2,500 cps. These components are 21 db and 13 db, respectively, above the level of the continuous noise spectrum.

How would the noise spectrum of Fig. 4.3 appear as measured on several different frequency analyzers? In Fig. 4.4 are shown the noise spectra measured with an octave-band, one-half-octave-band, one-third-octave-band, and a 4-cps constant-bandwidth analyzer. Except in the bands where the two pure tones occur, these curves have been computed from Fig. 4.3 by use of the following expression:

$$\text{SPL}_{\text{band}} = S(f) + 10 \log_{10} \Delta f \qquad \text{db}$$

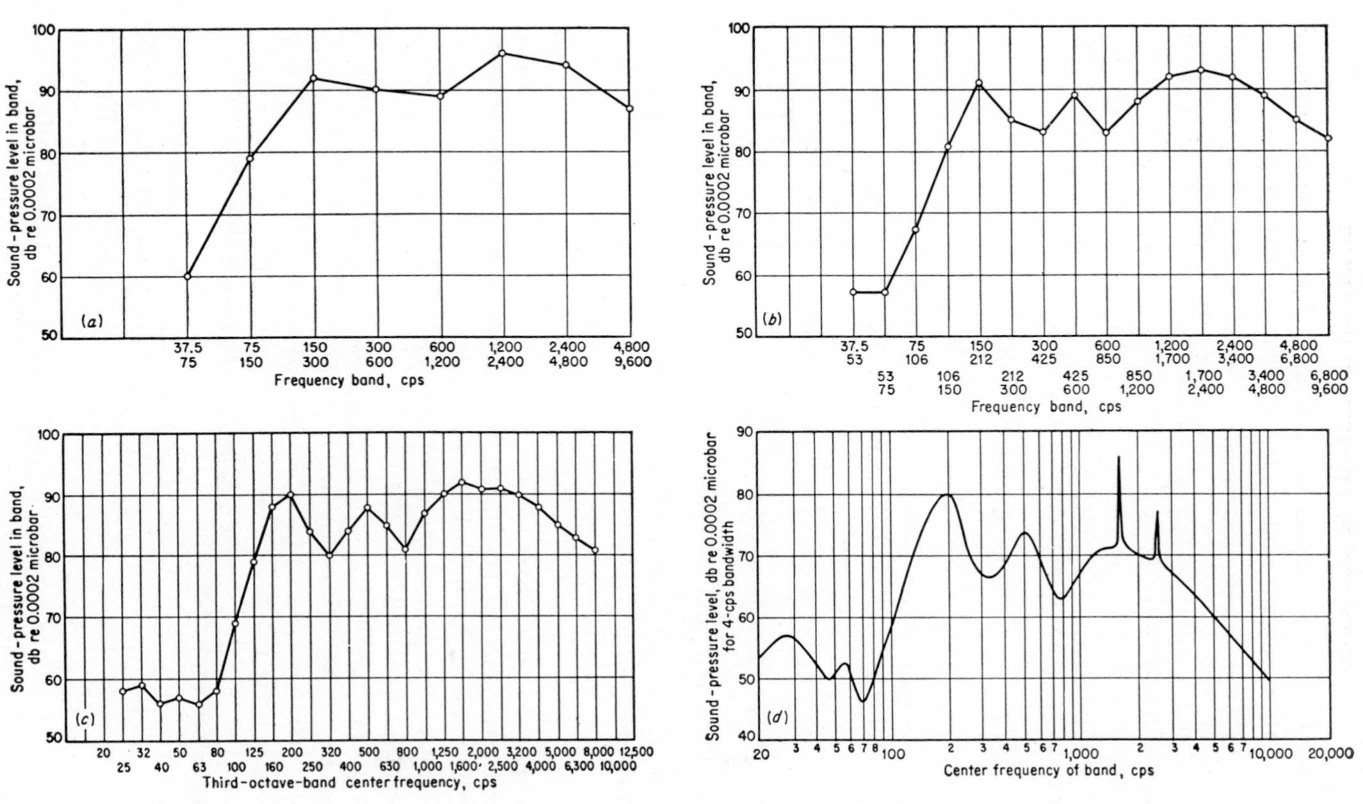

FIG. 4.4 Plots of noise spectrum of Fig. 4.3 as measured on four different frequency analyzers: (*a*) octave-band analyzer; (*b*) one-half-octave-band analyzer; (*c*) one-third-octave-band analyzer; (*d*) constant-bandwidth (4-cps) analyzer.

where SPL_{band} = sound-pressure level measured with bandwidth Δf
Δf = bandwidth, cps
$S(f)$ = spectrum level, db

In converting spectrum level to a band level by the equation given above, the spectrum level in decibels that is used for a given band is the one that is associated with the average sound pressure in that band. For example, the one-third-octave-band level for the band centered at 100 cps was computed in the following way:

$$SPL_{band} \text{ (continuous noise)} = 55 \text{ db} + 10 \log (23) \text{ db}$$
$$= 69 \text{ db}$$

If a pure tone occurs within the band in question, the level of the pure tone must be added separately. For example, the continuous-noise part of the one-third-octave-band level for the center frequency of 1,600 cps was calculated in the following manner:

$$SPL_{band} = 65 \text{ db} + 10 \log (370) \text{ db} = 91 \text{ db}$$

$$\text{Pure-tone SPL} = 86 \text{ db}$$

From Fig. 3.7 we combine the continuous part of the spectrum and the pure-tone part on an incoherent basis to obtain

$$\text{Total } SPL_{band} = 92 \text{ db}$$

Comparison of the four spectra in Fig. 4.4 reveals that the wider the passband, the less resolution is obtained. Only measurements with the constant-bandwidth (4-cps) analyzer retain the fine detail and shape of the original spectrum. Some of the basic characteristics of the noise spectrum, e.g., the two broad peaks in the continuous-spectrum part of the noise occurring at 200 and 500 cps, begin to emerge in the one-half-octave-band spectrum. These peaks are more pronounced in the one-third-octave-band spectrum, which also begins to show a little more detail at the very low end of the spectrum. None of the broader bands reveals the presence of the two pure-tone peaks at 1,600 and 2,500 cps.

Associated with loss of resolution resulting from use of a frequency analyzer with a comparatively wide passband is loss of information about a steeply sloping spectrum. The noise spectrum in Fig. 4.3 changes by 30 db from 75 cps to 150 cps. An octave-band analyzer would provide only one reading in this band and, hence, little or no information about the spectrum slope. (With some experience it may be possible, by considering the adjacent octave-band levels, to salvage some information about the slope of the spectrum.) On the other hand, with a one-third-octave-band analyzer, three readings could be obtained in the frequency region between 75 and 150 cps. These readings would provide enough information to adequately describe the slope of the noise spectrum in this region for practical purposes.

A further illustration of why octave-band readings are poor indicators of steep spectrum slopes is given in Fig. 4.5. For each of the

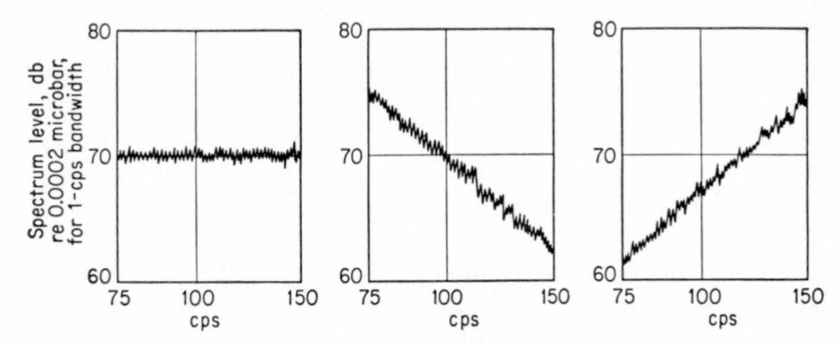

FIG. 4.5 Illustration of effect of spectrum slope on band reading. The 75- to 150-cps octave-band reading is identical for each of these examples.

spectrum samples shown—a flat spectrum, a positively sloping spectrum (+13 db/octave), and a negatively sloping spectrum (−13 db/octave)—the 75 to 150 octave-band readings are identical!

Clearly, the narrower the bandwidth of the frequency analyzer used, the more accurately one can measure the true noise spectrum. Very narrow band analyzers are necessary to single out pure-tone components or narrow-band peaks in the noise spectrum. If it is important to know the slope of the noise spectrum, the type of frequency analyzer used should be chosen carefully.

One must realize that the narrower the passbands of the frequency analyzer used, the greater the amount of time required to take data in the field. The simplest and quickest reading is the over-all sound-pressure level as read on a sound-level meter. A single set of octave-band levels can also be taken quickly. With experience one can satisfactorily read a complete octave-band spectrum in less than a minute. A one-half-octave-band spectrum takes at least twice as long, and a one-third-octave-band spectrum, of course, even longer. It is difficult to estimate the amount of time to make a complete narrow-band analysis; but generally speaking, such an analysis is impractical in the field without automatic equipment because of the amount of time required just to measure one spectrum. In many problems, noises requiring detailed narrow-band analysis are tape-recorded for later study in the laboratory.

A logical question to ask at this point is: Is the additional detail obtained, for example, by use of a one-third-octave-band analyzer or even a narrow-band analyzer really necessary? The answer, of course, depends upon the problem at hand and the questions to be answered. In many problems, an octave-band analysis will do very well. In particular, where the noise is fairly continuous as a function of frequency and does not exhibit any sharp peaks and dips in sound-pressure level or contain any pure-tone components, octave-band measurements are completely satisfactory. In cases where the noise is being measured to learn if there is possible damage to hearing, one only needs octave-band levels and a knowledge of whether or not there probably are pure tones present and whether these lie above or below 300 cps. Usually an experienced listener armed with some knowledge of the machine can ascertain this much supplementary detail by ear.

There are many noise problems, however, where more detail is required than provided by an octave-band analysis. For example, in attempting to redesign a complicated piece of equipment to make it quieter, one can make very good use of detailed information about the noise spectrum, particularly in isolating the various components of the noise source that contribute to its total noise output.

A window air-conditioning unit is a good example of such a complex noise source. The total noise produced is composed of noise generated by the compressor, by the fan, and by air escaping through openings. Also, noise from one or more of these elements may be radiated from the case itself. Separation of sources can be accomplished in various ways: by turning off the compressor and the fan individually and noting the change in the noise spectrum; by making measurements on the fan and the compressor as separate units; by varying the operation of the fan and the compressor, if possible, and noting changes in noise output; by adding vibration mounts or damping material or absorbing treatment and noting the resultant change in noise spectrum. There are other possible approaches, of course, but the results in any case are changes in the noise spectrum.

These changes are most easily observed and measured by means of narrow-band analyzers. Gross changes in the noise spectrum can, of course, be obtained by use of an octave-band analysis. However, octave-band readings can sometimes obscure the fact that a reduction in noise has been effected. For example, a pure-tone component can be reduced by 10 to 20 db without changing the octave-band reading (note that a reduction of the 1,600-cps component of Fig. 4.3 by 15 db does not alter the octave-band reading). Also, changes in octave-band levels can be very misleading. Assume that, in one octave band, noise is contributed approximately equally by the compressor and the fan. Turning either one or the other off results in reduction in the octave-band level of about 2 to 3 db. Assuming, for the sake of discussion, that both units cannot be turned off together, the experimenter might conclude incorrectly that little noise reduction could be achieved in this octave band by treating the fan and the compressor. A narrow-band analysis would clearly indicate that the octave-band level dropped only 2 to 3 db when one unit was turned off because noise from the other unit still was a major contributor, and that when both units were off, the noise level would drop by 10 to 15 db. Hence, the noise level in this octave band could be reduced by some attention to both the compressor and the fan.

Another example that illustrates the way in which narrow-band measurements can be used to great advantage has to do with a noise-control study for the Convair 340 commercial airliner.[1,2] In the first phase of the program the problem was to determine how much of the noise in various parts of the airplane cabin was due to the propellers and how much was due to the engine exhausts. Both propeller noise and engine-exhaust noise contain discrete pure-tone components. Knowing the operating characteristics of the engines and the propellers, one could calculate fairly well the frequencies at which these discrete tones will occur. Then narrow-band measurements of sound-pressure level were made at these frequencies at various points within the aircraft. In this way the relative contributions of propeller and exhaust noise could be determined throughout the length of the cabin.

In the Convair 340, the measurements showed that propeller noise predominated in the forward part of the cabin and exhaust noise in the rear of the cabin. From the relative sound-pressure levels of these various components, one could then calculate how much total noise reduction would be achieved within the cabin by certain noise-control measures, such as a muffler for the engine exhausts, additional weight in the cabin sidewalls at various points, interior sound absorption in various parts of the cabin, etc. Octave-band or even one-third-octave-band measurements would have been essentially useless in this particular problem.

An example that illustrates where the octave-band-measuring system is adequate to the need is the measurement of jet aircraft takeoff noise. Here, the purpose of the measurements is to determine the noise levels as perceived by residents in the communities surrounding an airport. It is known that on takeoff the jet exhaust noise predominates over the whine of the inlet compressor. It is also known that the exhaust noise is of the continuous-spectrum type. In measurement, the noise is usually recorded on a magnetic tape recorder and analyzed later into octave-band levels. These levels are then inserted into a formula for conversion into perceived-noise levels (PNdb). Psychological-engineering tests have shown that perceived-noise levels (PNdb) are a good indication of the disturbing effects of aircraft noise in neighborhoods.

4.4 Read-out Devices

Introduction. After the noise or vibration levels have been detected, amplified, and perhaps filtered, there remains one more operation to be performed by the instrumentation system. The noise or vibration information must be presented on a read-out instrument. This presentation of information usually takes the form of a meter reading, a paper record of the signal, or an oscilloscope pattern. Commonly used read-out instruments are vacuum-tube voltmeters, indicating meters on the sound-level meter and octave-band analyzer, graphic level recorders, oscilloscopes, and impact-noise analyzers.

The choice of a read-out instrument will be influenced primarily by the type of noise or vibration in question and, in some cases, by whether or not an automatic record of the signal is desired. For so-called "steady" wide- and narrow-band noise which does not fluctuate markedly with time, vacuum-tube voltmeters or the meters on the sound-level meter and the octave-band analyzer can be used satisfactorily. For intermittent noises, where the level varies rapidly with time, or for impulse (impact) noises, where the peak-to-rms ratio is quite high, these indicating devices are not suitable. One must resort to the use of a cathode-ray oscilloscope or an impact-noise analyzer.

Meters. The reason that certain indicating instruments cannot be used for intermittent and impulse noises is that their meter response is not fast enough. Even on the "fast" scale of a sound-level meter or octave-band analyzer, the meter requires approximately 0.2 sec to attain a final reading after the application of a signal. In other words, a transient signal that reaches its peak and decays in a time shorter than 0.2 sec would not be correctly indicated on even the fast scale. The "slow" scale requires that the signal be on for even a longer period of time. The primary purpose of the slow scale is to perform an "averaging" of fluctuating noise levels.

Impulse Noise Read-out. Impulse and repeated-impulse noise can be satisfactorily read on an oscilloscope or one of the available impact-noise analyzers. An oscilloscope has a very short response time which makes it extremely suitable for the measurement of transient noises. It has a further advantage in that a picture of the waveform of the noise or vibration signal can be displayed on the face of the scope and can either be examined visually or photographed for later study.

1. *Cathode-ray Oscilloscope.* Study of impulse noises by means of a cathode-ray oscilloscope permits a determination of the peak level of the noise, the manner in which the noise decays, the amount of time necessary for decay, etc. The oscilloscope can be calibrated rather easily so that one can read directly the magnitude of signal as well as the amount of time any part of the signal is on. This calibration is accomplished simply by displaying on the scope face a sinusoidal signal of known amplitude and frequency—perhaps from the output of a sound-level meter being fed by an acoustic calibrator. The number of cycles of the signal displayed on the scope and the frequency of the signal set the time base. The height of the signal and the sound-level meter reading can be used to calibrate the vertical scale.*

2. *Impact-noise Analyzer.* Although the oscilloscope provides a useful means for observing the complete waveform of an impulse or transient signal, it is a bulky instrument and is suitable primarily for laboratory use. Of more practical use in the field is an "impact-noise analyzer," several of which are commercially available. These analyzers simplify the procedure of measuring peak sound-pressure level and are especially designed for this purpose. The response time is of the order of 0.0001 sec. These analyzers are simple, compact, and very satisfactory for use in the field. Their disadvantage in comparison with the oscilloscope is, of course, that they are a direct-reading device and do not permit examination of the complete time pattern of the noise. One commercially available impact-noise analyzer, however, permits a measure of the duration of the impact noise, as well as the peak amplitude. In addition, this analyzer is useful for measuring repeated impact noises.

Graphic Level Recorder. A read-out device that is very useful in certain situations is the graphic level recorder. It can be connected to the output of a sound-level meter, a tape recorder, or a

* One must keep in mind that the face of the scope reads linearly in voltage (which in this instance is proportional to sound pressure or vibration amplitude). Therefore, a point halfway down from the peak deflection is equal to a sound-pressure level that is 6 db less than the peak. One-third the peak amplitude is equal to a sound-pressure level that is 10 db lower than the peak, etc. Also one should remember that the reading on the sound-level meter of the calibrating sinusoidal signal is an rms reading. To obtain the peak level of the calibrating pure-tone signal on the scope, 3 db must be added to this reading.

frequency analyzer. It produces a permanent paper record of the variation of sound-pressure level as a function of time.

The graphic level recorder can be used in many ways. Once it is properly set up and calibrated, it can be left unattended to obtain a long-time recording of the variation of sound-pressure level with time. A disadvantage in this particular application is that it can record the output of only one frequency band at a time. One of its most common uses is to measure the reverberation time of a room. It can also be used to plot in detail the frequency spectrum of a particular noise. For example, it can be set up to operate with a General Radio 736A frequency analyzer (constant bandwidth of 4 cps) to obtain automatically a plot of a noise spectrum similar to that shown in Fig. 4.4*d*.

Although the response of most graphic level recorders is not fast enough to properly record short-duration impulse noises, it is an ideal instrument for recording intermittent noise such as from aircraft flyovers, passing traffic, etc. Use of the instrument is generally limited to the laboratory; however, with careful handling the instrument can be satisfactorily employed in the field.

An important factor to bear in mind when using graphic level recorders is that as of this time (1959) they are usually peak-indicating instruments, although when one calibrates one of them he adjusts it to read the rms value of the sine wave (pure tone) used in the calibration. Because it is a pseudo-peak instrument it will read differently on different kinds of noises depending on the peak-to-rms ratio of the noise in question. For example, a sinusoidal wave used in calibration has a difference in peak and rms values of 3 db. A repeating impulse noise might have a peak-rms difference of 10 db. Whereas the recorder was calibrated to read the rms value of the sine wave correctly, it could read up to 7 db above the rms value of the impulse noise.

One way to circumvent this problem is to calibrate the graphic level recorder with a sample of the noise to be measured. This is accomplished by first calibrating the graphic level recorder (GLR) in the usual way with a sinusoidal signal whose amplitude is read on an octave-band analyzer (OBA). In other words the GLR and OBA are made to read alike. Next, a sample of the noise to be analyzed is played through the octave-band analyzer into the graphic level recorder. Simultaneously the octave-band levels on the read-out device and on the analyzer are written down. Now, since both the OBA readings and the GLR readings were alike when using a sinusoidal signal, the difference in their readings in each band is a correction number to be used to convert recorded data to the equivalent of direct OBA readings. One should find that the GLR reads higher than the OBA on noise if they read alike when calibrated with a sinusoidal tone. Obviously, if the calibration were required in other than octave bands, the appropriate frequency analyzer must be employed.

The GLR is extremely useful when used in conjunction with a magnetic tape recorder. The tape recorder records noise in the field that cannot easily be read directly or analyzed in the field. This information is then played back successively through appropriate filter bands onto a GLR to obtain a permanent record which can be more easily analyzed and studied.

4.5 Data Storage

The most practical and widely used data-storage device is the magnetic tape recorder. Noise or vibration information fed to the tape recorder by a microphone or vibration pickup is permanently recorded on a magnetic tape. At a later time the magnetic tape can be replayed, and the output fed through any kind of a frequency analyzer and displayed on any one of the read-out devices mentioned above. All types of noises can be recorded on a tape recorder including steady-state wide-band noise, impact noise, intermittent noise, etc.

Implicit, of course, in the above discussion is the fact that the tape recorder employed must be of high quality and dependable. It should have a fairly flat frequency response extending over the range of frequency of interest. It should have a fairly high signal-to-noise ratio, a minimum of flutter and "wow," low hum and internal noise level, constant speed drive, good mechanical construction, etc. (More information on the characteristics of a good, high-quality tape recorder is given in Chap. 6.)

The use of a tape recorder saves time in the field. Rather than read directly the noise output as a function of frequency on one of the various frequency analyzers, one need only record the noise in question. In the laboratory this noise sample can be recorded on an endless tape loop and played over and over again. By this technique a detailed frequency analysis can be performed even on very short samples of noise, a practice that would be very time-consuming and often impossible in the field. With a tape recorder one can record data at many locations within the time that would have been required to take direct readings at only a few of these locations.

For measuring intermittent noises, the tape recorder is ideal. The noise from a single firing of a missile, for example, lasting less than a minute, can be completely recorded and analyzed in any detail desired in the laboratory. Without a tape recorder it might be possible to obtain only a few octave-band readings of this type of noise directly in the field. Also, these readings must necessarily be taken in haste, increasing the possibility of error. Further, they are not strictly comparable because of the time-varying nature of the noise. Using a tape recorder in conjunction with a GLR in the laboratory, one can obtain a complete picture of the time variation of these and similar intermittent noises as a function of frequency.

The use of a tape recorder also simplifies the gathering of impact-noise data. For example, suppose that an octave-band analysis of the noise output of a drop hammer were desired at a particular position. With only an OBA and the impact-noise meter in the field, at least eight operations of the hammer would be required in order to obtain readings in each of the eight octave bands. In practice, because of setup problems, trial attenuator settings, etc., many more than eight operations would be required to obtain one octave-band spectrum. With a tape recorder one needs to obtain only one valid recording* of the hammer operation. This recorded signal can then be analyzed through the OBA at some later time.

Further, if a picture of a transient signal is desired, a tape recording made in the field can be played back onto an oscilloscope. If the signal is put on an endless tape loop, the impulse noise can be played over and over again and either studied on the oscilloscope or photographed.

The tape recorder has also been used to great advantage to obtain directivity patterns of a noise radiator. For example, it is suited to the measurement of the directivity pattern of the noise field around an airplane. By moving the microphone on a circle of fixed radius around an aircraft, say, using an automobile, one can obtain very quickly a continuous plot of sound-pressure level as a function of angle. This technique reduces engine operating time and field time for the measuring crew.

Although the tape recorder is an extremely useful instrument, it is not the panacea for all noise-measurement problems. The performance of a complete noise survey using only a tape recorder could lead to disastrous results. A poorly functioning tape recorder might render all data invalid. More often the problem is one of taking a large amount of data that may have little or no bearing on the problem at hand. One does not learn this fact until the survey

* Because of the usual high peak-to-rms ratio of impulse noise, one should adjust the attenuator on the tape recorder so that the peak VU meter indication is 10 to 20 db below full-scale reading. Otherwise there is danger of the recorder overloading on the peak signal.

is over and the data have been reduced and analyzed, and by that time it may be too late to make additional measurements. In this respect, taking at least part of the data in the field with direct-reading equipment represents a distinct advantage in that the data can be examined on the spot with a view toward planning additional measurements. If the noise problem is a familiar one, then of course the tape recorder can be used almost exclusively and to great advantage. If not, and if a large amount of data must be collected, then some of the recorded data should be reduced in the field, or direct readings should be taken to serve as a check on the data already collected as well as a guide indicating what additional information is required.

The ease with which one can gather data in the field with a tape recorder can also serve as a disadvantage. Oftentimes too much data are recorded. Since these data must be reduced and converted to some usable form in the laboratory, unnecessary time (and money) is expended, both in the laboratory and in the field. Therefore, when using a tape recorder, it is well to keep in mind the amount of information required and the amount of time available.

4.6 Summary

The selection of instrumentation for a noise survey is not always a simple task. A number of factors influence the choice of instrumentation. A basic understanding of the nature of the problem at hand is an obvious primary requirement. One must have a fairly good idea of the type of noise or vibration that is involved, the amount of information about the noise that is required and the amount of time that one can spend taking the data in the field and reducing it either in the field or in the laboratory.

Further, the choice of instrumentation is based on a knowledge of the limitations and applications of the various types of instrumentation available. If all of the instrumentation discussed here and in Chap. 6 are available, then the problem is one of selecting the proper equipment for the job at hand. If, as is usually the case, only a limited amount of instrumentation is available, then the problem is one of recognizing the limitations of the instrumentation at hand with reference to the particular noise survey in question. For example, if an impact-noise (impulse-noise) analyzer or a cathode-ray oscilloscope is not available, then measurements of the noise from drop hammers in a forge shop should simply not be undertaken. Likewise, measurements of the noise of passing automobile and truck traffic can be obtained with a combination of sound-level meter and octave-band analyzer, but one should recognize that little or no information about the transient nature of the noise can be obtained in this manner.

As a rule of thumb, if the purpose of the measurement is to determine potential damage to hearing, interference with speech communication, or annoyance and if there are no prominent pure-tone components above 300 cps, octave-band analyses are usually adequate. If the purpose of the measurement is to yield data to aid in the redesign of a machine or if there are prominent components in the spectrum above 300 cps, narrower-band analyses are usually necessary.

Finally, there is one other instrument that merits some discussion. Although we have available a number of different microphones and vibration pickups, and we can store any type of noise for an indefinite period of time on a tape recorder, and we can analyze this noise in various ways as a function of frequency, and we can measure impact noises with relative ease, there is one instrument available to all of us which we tend to overlook, but which can be very useful. That instrument is the human ear. It is true that in several respects it may not compare favorably with some of the instrumentation discussed in this chapter, but with regard to portability, availability, and freedom from maintenance, it is an instrument that has no peer. We should all use our ears more often. A wise acoustical engineer with an inquisitive ear may learn in an hour what an unobserving assistant with perfect equipment may never learn.

REFERENCES

1. Beranek, L. L., and E. M. Kerwin, Jr.: Noise Reduction in a Passenger Airplane, *J. Acoust. Soc. Am.,* vol. 27, p. 1014(A), 1955.

2. Hunter, G. S.: Sound Reduction Program for Convair-Liner 340, *Noise Control,* vol. 2, pp. 27–32, January, 1956.

SUGGESTIONS FOR FURTHER READING

Beranek, L. L.: "Acoustics," Chap. 12, McGraw-Hill Book Company, Inc., New York, 1954.

Harris, C. M. (ed.): "Handbook of Noise Control," Chaps. 15, 16, and 17, McGraw-Hill Book Company, Inc., New York, 1957.

Peterson, A. P. G., and L. L. Beranek: "Handbook of Noise Measurement," 3d ed., General Radio Company, Cambridge, Mass., 1956.

Pomper, V. H.: "Noise Simplified—Instrumentation, Measurement, Analysis, Control," Herman Hosmer Scott, Inc., Cambridge, Mass., 1954.

Williams, C. R., and J. R. Cox, Jr.: Industrial Noise Measurement—Science or Art? *Proc. 3d Annual NNAS,* vol. 3, Oct. 10, 1953.

Young, R. W.: A Brief Guide to Noise Measurement and Analysis, *U.S. Naval Electronics Lab. Research and Develop. Rept.* 609, May, 1955.

Chapter 5
THE BASIC SOUND-MEASURING SYSTEM

David N. Keast

5.1 Introduction

The simplest complete system that is generally used for making precise quantitative sound measurements consists of four elements: a microphone, a sound-level meter, an octave-band analyzer, and an acoustic calibrator. Chapter 6 discusses the characteristics of various microphones. The present chapter contains a general description of the functions and features of the other three elements. In addition, detailed performance information on these and on more complicated instruments is given in Chaps. 6 and 7.

The most common instrument available to measure sound pressures in air is the *sound-level meter.* This instrument contains an amplifier with a calibrated attenuator, a set of frequency response networks (weighting networks), and an indicating meter.

The sound-level meter was originally standardized [1] to indicate *sound level* in decibels, a single number giving the total sound-pressure level *weighted* by an approximation to the loudness-level sensitivity of the human ear for pure tones. This approximation was made in three steps: the A scale for sound levels below 55 db, the B scale for sound levels between 55 and 85 db, and the C scale for sound levels above 85 db. Unfortunately, this single number is not very closely related to the loudness level of a complex noise, and complex noises are encountered much more frequently in noise-control problems than individual pure tones. As a result, the sound-level meter is often used primarily as a calibrated amplifier and attenuator between the microphone and some other analyzing instrument. As such it is operated with the C (or flat) scale only and its meter readings are unweighted (true) *over-all sound-pressure levels* in decibels *re* the standard reference of 0.0002 μbar. Much of the value of the sound-level meter is derived from its portability, its simplicity of use, its stability under battery operation, and its reliability in the field.

The *octave-band analyzer,* possessing attributes similar to those just mentioned for the sound-level meter, is commonly used in the field to obtain a frequency analysis of the output signal from the sound-level meter. This instrument contains an amplifier, a set of filters, and a meter which indicates the signal strength in each filter passband. The standard [2] instrument has the filter bands listed in Table 5.1, in addition to an "over-all" band which passes all frequen-

Table 5.1

Standard Octave-band-analyzer Filters

Low pass to 75 cps*
75–150 cps
150–300 cps
300–600 cps
600–1,200 cps
1,200–2,400 cps
2,400–4,800 cps
High pass from 4,800 cps*

* The lower and upper cutoff points are controlled by the amplifier bandwidth, the microphone response, or both. Assuming typical values for these limits, the first band is sometimes called the 20- to 75-cps band, and the last the 4,800- to 10,000-cps band.

cies from 20 to 10,000 cps. It can be seen that six of the bands span an octave in frequency, hence the name *octave-band analyzer.* Some of the commercially available versions of this instrument also permit an analysis in narrower-frequency bands.

For use in the field, the sound-level meter and octave-band analyzer may be connected as indicated on Fig. 5.1. The microphone may be mounted on the sound-level meter (the instruments are usually sup-

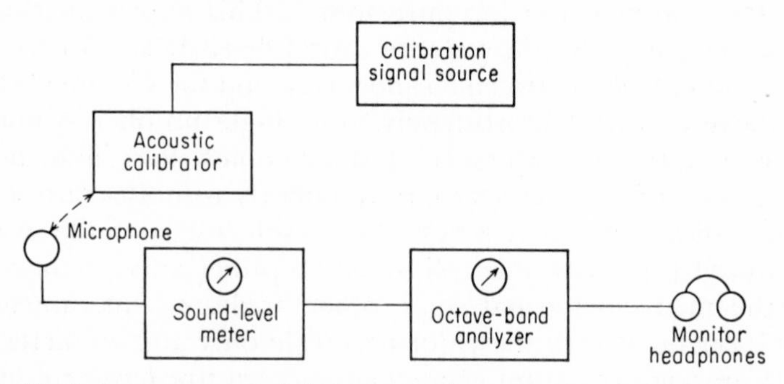

FIG. 5.1 Typical arrangement of instruments for measuring sound-pressure level in octave frequency bands.

plied this way by the manufacturer), but for reasons which will be pointed out later, it is often preferable to separate them and to provide a cable between the two. Another short cable connects the output jack on the sound-level meter to the input terminals on the octave-band analyzer. In addition, it is often wise to monitor the output of the octave-band analyzer with headphones in order to detect spurious signals or possible malfunction of the equipment.

After checking the batteries in the instruments and carrying out their individual calibrations as described in the manufacturer's manual, it is necessary to obtain an adequate calibration of the system before proceeding to make measurements. This is often done with an *acoustic calibrator*. The acoustic calibrator consists of a small transducer in a housing designed to fit over a particular type of microphone, leaving a coupling cavity. When a known voltage is applied to the terminals of the transducer, a corresponding known sound-pressure level is produced at the diaphragm of the microphone. The meters on the sound-level meter and the octave-band analyzer can then be set to read this sound-pressure level when the calibrator is on the microphone. Some commercially available calibrators will fit several types of microphones, producing, however, a different sound-pressure level with each one. The actual value of sound-pressure level produced is supplied by the manufacturer. The acoustic calibrator, combined with a small portable signal source, is a most convenient way to calibrate a measurement system in the field. It is evident, however, that the accuracy of the system then hinges upon that of the calibrator and upon the voltmeter that measures the voltage supplied to the terminals of the calibrator.

When the system of Fig. 5.1 is properly calibrated, the readings of the meter on the octave-band analyzer are *octave-band sound-pressure levels* in decibels *re* the standard reference of 0.0002 μbar.

Some of the commercially available versions of the sound-level meter, octave-band analyzer, and acoustic calibrator are pictured on Figs. 5.2, 5.3, and 5.4. Equipment manufactured by the General Radio Company is illustrated in Fig. 5.2. The sound-level meter, normally supplied with a detachable Shure Brothers type 9899 microphone, is pictured on the top of the figure. This may be connected directly to the octave-band analyzer pictured below. A small battery-powered transistor oscillator (400 and 1,000 cps) for the calibrator is to the left of the sound-level meter.

Equipment manufactured by Herman Hosmer Scott, Inc., is illustrated in Fig. 5.3. The unusually shaped sound-level meter, designed to be hand held, has a detachable Shure Brothers 9898 microphone at one end and an indicating meter at the other. (The latter

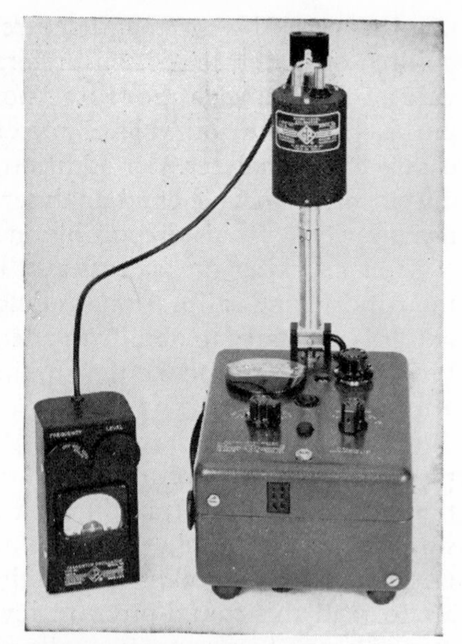

FIG. 5.2 Sound-level meter, acoustic calibrator, calibration oscillator, and octave-band analyzer manufactured by the General Radio Company, Cambridge, Mass. (*Courtesy of General Radio Company.*)

is not visible in the photograph.) This instrument can be connected to the frequency analyzer which permits analysis in half-octave as well as in octave steps up to 20,000 cps. The Scott acoustic calibrator is designed to operate from a random noise generator (not pictured).

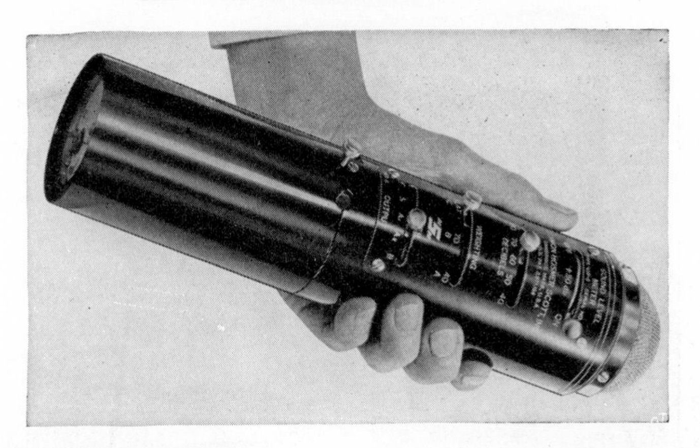

Fig. 5.3 Sound-level meter, acoustic calibrator, and frequency analyzer manufactured by Herman Hosmer Scott, Inc., Maynard, Mass. (*Courtesy of H. H. Scott, Inc.*)

The Soundscope, manufactured by Mine Safety Appliances Company and illustrated in Fig. 5.4, combines some of the features of the sound-level meter with a frequency analyzer and acoustic calibrator, all in the same instrument. The Soundscope is normally supplied with a Shure Brothers type 98B99 microphone on a detachable gooseneck mounting and with a matching acoustic calibrator (both

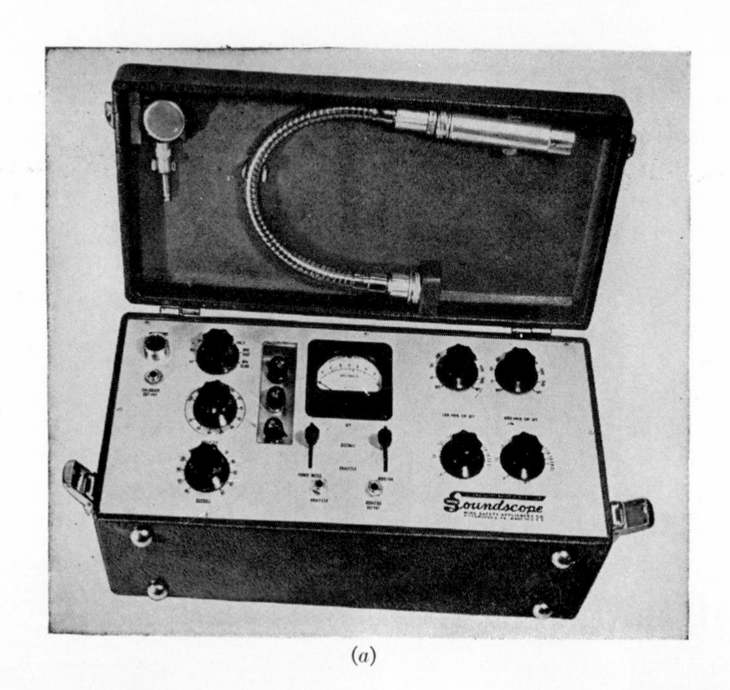

(*a*)

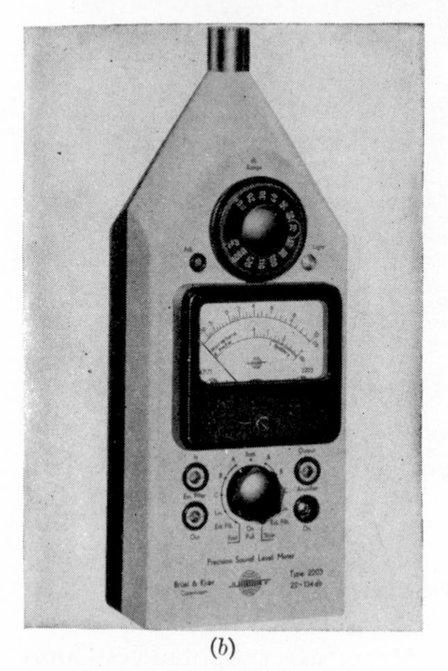

(*b*)

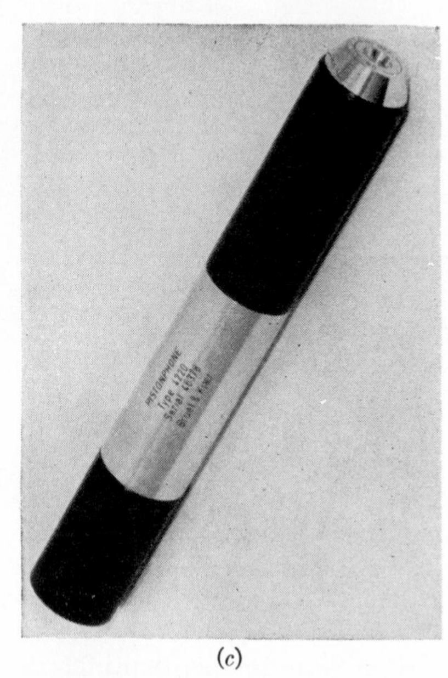

(*c*)

FIG. 5.4 Instruments of Bruel and Kjaer of Naeram, Denmark, and of Mine Safety Appliances Company of Pittsburgh, Pa. (*a*) Soundscope, combining the functions of a sound-level meter, frequency analyzer, and acoustic calibrator in a single instrument. (*Courtesy of Mine Safety Appliances Company.*) (*b*) and (*c*) Precision sound-level meter and pistonphone (acoustic calibrator), both battery operated. (*Courtesy of Bruel and Kjaer.*)

mounted in their carrying brackets in the lid of the instrument in the photograph). The calibrator obtains its signal from a 900-cps oscillator built into the instrument. The frequency analyzer has filter bands that are adjustable in width and permit analysis in steps as small as one-tenth octave from 67.5 to 20,160 cps, as well as analysis in the eight standard octave bands of Table 5.1.

Equipment manufactured by Bruel and Kjaer is also illustrated in Fig. 5.4*b* and *c*. The precision sound-level meter (length 12.5 in., width 4.7 in., and depth 3.5 in.) uses a 0.95-in.-diameter condenser microphone. The case is shaped to minimize acoustic diffraction effects and has an accuracy of ± 1 db between 20 and 15,000 cps. The instrument uses 17 transistors and has a battery life of 100 hr. The pistonphone produces a 250 cps ($\pm 1\%$) pure tone. It is battery-operated (length 9 in. and diameter 1.4 in.) and can be used on a number of different types of microphones.

5.2 The Sound-level Meter

A block diagram of a typical sound-level meter is indicated on Fig. 5.5. The signal from the microphone goes through the input preamplifier stage to a 10-db step attenuator.* From the attenuator, it enters a second feedback-stabilized amplifier stage whose gain may be set during calibration. In some instruments, calibration gain control is obtained by adjusting the B+ battery voltage.

The signal then goes through the spectrum-weighting networks to separate output amplifiers for a meter and for an output jack. The separate amplifiers are used so that the meter reading is independent of loading on the output jack and so that the signal at the output jack is not distorted by the nonlinear meter element. In some instruments, a single output amplifier is used, and the meter is disconnected when a plug is inserted into the output jack.

The meter is provided with "fast" and "slow" response speeds chosen by a switch. The "fast" response is specified by standard [1] so that the meter gives a true indication within 0.2 to 0.25 sec after a steady 1,000-cps tone is applied and does not overshoot more than 1 db. The "slow" response has not been standardized and may vary between instruments of different manufacture. The sound-level reading of the sound-level meter is the sum of the meter reading and the attenuator setting. The meter can be switched to read the battery voltages as well.

* In some instances, the attenuator may be split into sections between various amplifying stages. This is done to improve the signal-to-noise ratio of the instrument and to limit the dynamic range required of the individual amplifier stages. However, a single dial varying in 10-db steps is still used.

The weighting networks, of which there are three, adjust the response of the instrument to fall within the limits specified by ASA Standard Z24.3–1944.[1] These responses are indicated in Fig. 5.6. The A scale permits the instrument (including its microphone) to have a response approximating the 40-phon equal-loudness contour, the B scale the 70-phon contour, and the C scale provides a flat response to about 8,000 cps. These networks are used principally to make sound-level measurements. The range of deviations permitted by the standard are principally to accommodate commercially available microphones. In addition to the weighting networks, some in-

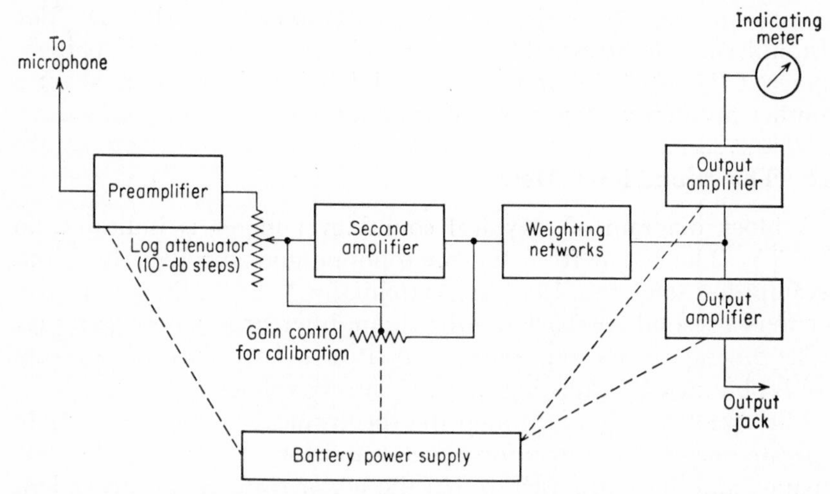

FIG. 5.5 Block diagram indicating typical internal arrangement of the sound-level meter.

struments have a 20-kc scale that provides a nominally flat amplifier response (not including the microphone) up to 20,000 cps. When making sound-pressure-level measurements or using the sound-level meter as an amplifier to feed a frequency analyzer, the C scale (for the microphone supplied with the instrument) or the 20-kc scale (for other microphones) is used.

Some of the other requirements for this instrument have already been touched upon in Sec. 5.1. The sound-level meter is basically a device for field use and as such should be reliable, portable, reasonably stable under battery operation, and light in weight. The input impedance should be sufficiently high to provide minimal loading for high-impedance microphones. (Because of this high input impedance, capacitive loading of the microphone by coaxial cables between it and the sound-level meter is often a problem, and cables must be chosen carefully.) The output impedance should be low and the output level should be high enough to drive the associated analyzer. Within limits, it can be said that all the commercially available instruments meet these requirements.

It is perhaps of even greater importance to discuss some of the limitations of the commercially available sound-level meters. These limitations are seldom evident to the inexperienced user of the instruments and they indicate the directions in which improvements of the instruments might be made.

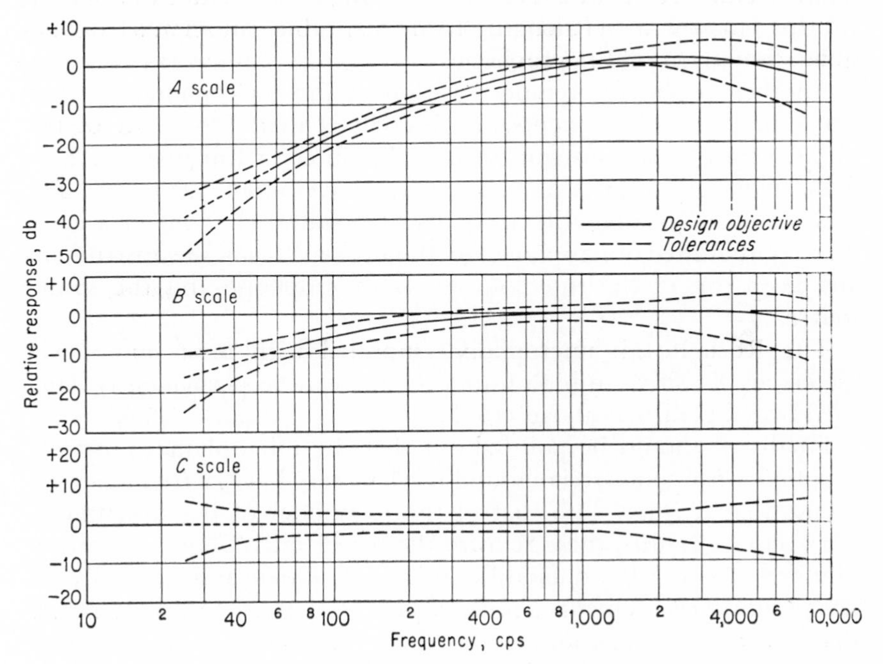

FIG. 5.6 Responses of standard sound-level-meter weighting networks (ASA Standard Z24.3–1944).

The electrical noise floor of most instruments is sufficiently high so that measurements of sound-pressure level below about 30 db are, particularly at high frequencies, impossible. When the sound-level meter is used in conjunction with a frequency analyzer, it is often possible to obtain erroneous readings which are simply readings of the noise floor of the sound-level meter rather than the sound field being studied. At the other end of its dynamic range, in the presence of sound-pressure levels above 110 to 120 db, the sound-level meter is generally limited by microphonic excitation of its vacuum tubes. Here we find one of the reasons why it is desirable to be able to mount the microphone at a location remote from the sound-level meter: both the experimenter and the instrument are unhappy in the presence of high sound levels.*

Fortunately, electrical noise or microphonics can be detected by replacing the measurement microphone with a "dummy" microphone having the same electrical impedance but no electrical output. Readings obtained on the sound-level meter with the dummy microphone should be at least 6 to 10 db below the acoustic data. Often, for this purpose, a 10-ft length of shielded cable can serve as an adequate dummy crystal microphone.

Another reason for remote location of the microphone, at least when precise measurements are desired, is that diffraction of the sound waves by the instrument and the observer can produce 2- to 3-db errors in readings at frequencies above about 1,000 cps. For this reason, it is advisable when using a sound-level meter with a directly mounted microphone to stand away from the instrument and keep out of the line joining the microphone and the sound source.

The Rochelle-salt microphones supplied with most sound-level meters are sensitive to high humidity and can be permanently damaged by temperatures above 115°F.

Finally, it should be pointed out that the microphones normally supplied with commercial sound-level meters have a frequency response that is inferior to that of the remainder of the instrument. For accurate results, these require the use of microphone correction numbers above a few thousand cycles per second and generally do not permit measurements above 8,000 cps. Better (condenser rather than crystal) microphones are available as accessories from the manufacturer and are generally worth the additional expense.

In discussing these problems, it should be pointed out that the sound-level-meter standard is presently undergoing review. It is possible that in a few years sound-level meters will be manufactured to conform to more stringent requirements, thus alleviating some of the present difficulties. Nevertheless, the presently available instruments represent remarkable achievements in design and utility for a reasonable price.

At least two manufacturers presently market a pocket-sized instrument similar to the sound-level meter (see Fig. 5.7). Complete with an integral crystal microphone, transistor amplifier, and batteries, this instrument is convenient for initial examinations in noise studies and for acoustical comparisons of different noise sources. The

* One must be wary, however, of microphonic excitation of the cables linking the transducer and the sound-level meter. This is discussed in greater detail in Chap. 6.

microphone is not removable, and no portable acoustic calibrator is available. The gain of the amplifier is so set that the sensitivity of the instrument is correct at 1,000 cps within ±1 db. The *B* and *C* weighting positions are in accordance with Fig. 5.6 when the tolerances on Fig. 5.6 are increased by 1 db.

FIG. 5.7 Sound-survey meter. (*Courtesy of General Radio Company.*)

5.3 The Octave-band Analyzer

It has been pointed out that in present-day practice it is common to use the sound-level meter solely as a preamplifier (set to the *C* weighting scale or to the 20-kc position) between the microphone and an octave-band analyzer. The latter instrument permits an examination of the spectrum of the sound being measured.

A typical octave-band analyzer might have the internal arrangement indicated in Fig. 5.8. The input signal is fed directly to the "over-all" position or to one of the filters and thence to a 10-db step attenuator. The output of the attenuator goes to a feedback-stabilized amplifier stage whose gain may be set during calibration. Finally, a meter and an output jack similar to those on the sound-level meter are fed through parallel output amplifiers. It is usually possible to vary the damping in the meter with a switch marked "fast" and "slow," as on the sound-level meter.

Although some of the commercially available octave-band analyzers permit measurements in bands smaller or larger than the standard octaves, they can all be set to read conveniently in the standard bands of Table 5.1. In addition, it is always possible to obtain a broad-band (over-all) reading. The latter permits the octave-band analyzer to be "calibrated" by adjusting its meter to the same reading as that on the sound-level meter preceding it.

The shapes of the octave-band filters should fall within the range specified by ASA Standard Z24.10–1953.[2] This range is plotted in Fig. 5.9, along with the measured responses of the octave filters in three commercially available octave-band analyzers. It can be seen

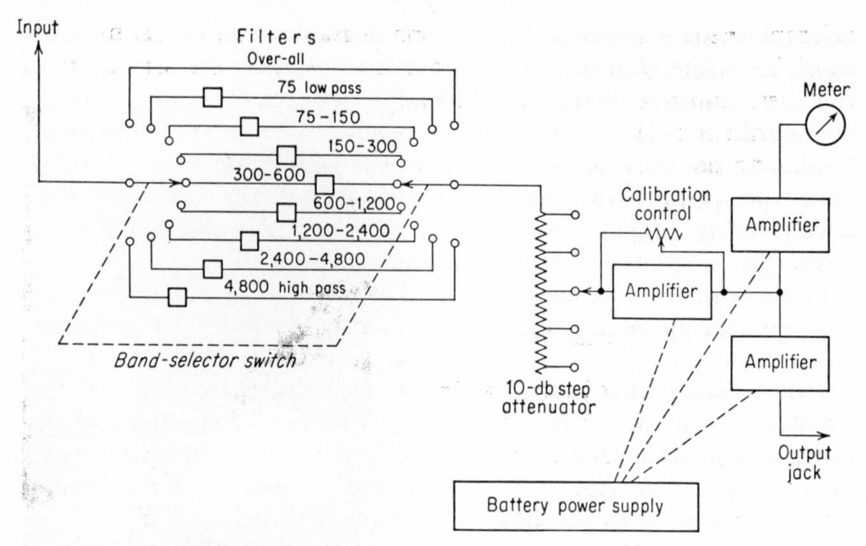

FIG. 5.8 Block diagram indicating typical internal arrangement of the octave-band analyzer.

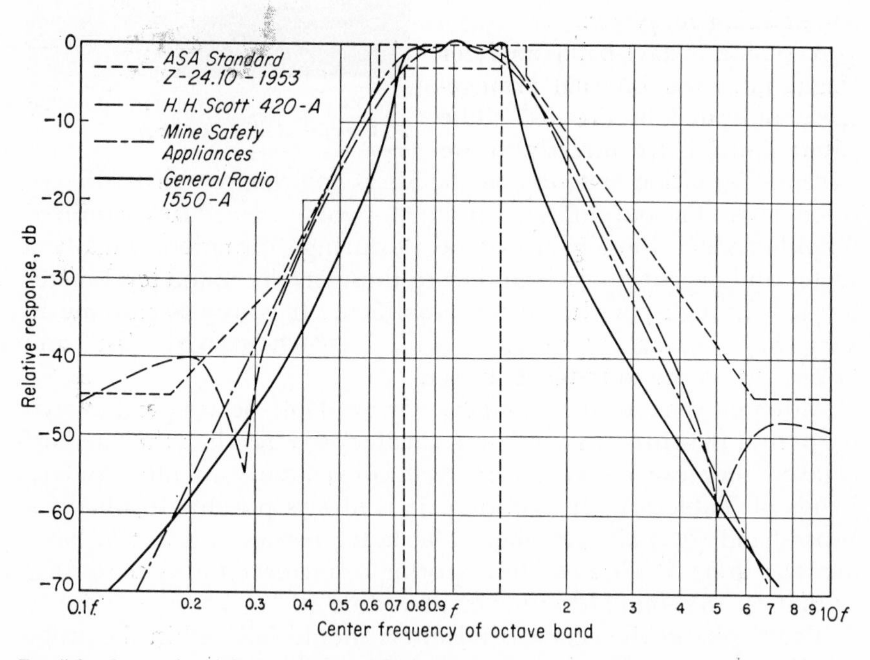

FIG. 5.9 Octave-band filter characteristics of several commercially available instruments. (*Courtesy of G. W. Kamperman, Bolt Beranek and Newman Inc.*)

that, for most practical purposes, the commercially available instruments do meet the standard.

The advantages of the octave-band analyzer are similar to those of the sound-level meter. Stability under battery operation is of primary importance, as are portability and simplicity of use. Because of the susceptibility of inductors (components of the filter networks) to stray electromagnetic fields, care is generally given to electrostatic and magnetic shielding in the instrument.

In many ways the design of a "good" octave-band analyzer is more difficult than the design of a "good" sound-level meter. Microphonics and susceptibility to hum pickup are often a problem. The advantage of steep filter characteristics beyond the nominal cutoff frequencies is evident in that this prevents erroneous readings due to high levels in adjacent bands. However, such filters tend to be more complex and thus heavier to carry around in the field. When using those analyzers which permit measurements above 10,000 cps, one must remember that the normal crystal sound-level meter microphone is inadequate in this range.

Finally, before purchasing an octave-band analyzer or a sound-level meter, check on the ease of calibration and maintenance and the availability of batteries and any accessories which may be required. Most manufacturers supply accessory devices such as carrying cases, microphone extension cables, special-purpose microphones and vibration pickups, tripods, peak-reading meters, a-c operated power supplies, and acoustic calibrators.

5.4 The Acoustic Calibrator

All sound-level meter microphones should be periodically calibrated in the laboratory throughout their frequency range. When this is done, and the microphone is known to be stable, the sound-level meter can then be used in the field with only an electrical calibration of its amplifiers. However, the use of an acoustic calibrator in the field * not only permits a calibration of the electronics, but also insures that the sensitivity of the microphone, at least at one frequency, is included in the calibration.

When the sound-level meter is used with a magnetic tape recorder,

* Acoustic calibrators are occasionally used for laboratory calibration of microphones, but this is generally considered to be inferior to free-field comparison or reciprocity calibrations[3,4] because the latter permit determination of the effects of angle of incidence and diffraction around the microphone and can extend to higher frequencies. For these reasons, acoustic calibrators are considered here primarily as devices for use in the field.

it is mandatory for subsequent analysis of the tape that a signal of known level be recorded. Although this could be provided by an insert voltage, an acoustic calibrator is more often used. Even when only a sound-level meter and an octave-band analyzer are used, it is often more accurate to calibrate the analyzer with a signal of known level from an acoustic calibrator rather than to adjust the over-all reading of the analyzer to that of the sound-level meter in the noise field present at the time.

Some commercially available acoustic calibrators have a reasonably flat frequency response in the range from approximately 40 to 1,000 cps. They may be used either with a pure-tone oscillator or with a random-noise generator (one manufacturer recommends the latter). In the field, calibration with a single pure tone is often easier because small, stable, battery-operated, single-frequency oscillators are available.* However, calibration with broad-band noise has the advantage of "ironing out" small variations, if any, in the frequency response of the microphone. Because the crystal microphone has a uniform (flat) response below about 1,000 cps, the pure-tone check is adequate in the measurement of low-frequency noises. Because of the more irregular response of the crystal microphone at high frequencies, calibration with random noise is advantageous in the measurement of high-frequency noises.

There are some drawbacks associated with the use of acoustic calibrators in the field. Obviously, one must have access to the microphone. If the microphone happens to be inaccessible, as is often the case, it may be hazardous as well as inconvenient to use an acoustic calibrator. Also, an acoustic calibrator cannot be used if the noise field in the vicinity is comparable to or exceeds the level of the calibration signal. For this reason, it is best to have an acoustic calibrator which produces high sound levels.

Occasionally, the microphone will be operating at elevated or reduced temperatures, and thus may have a different sensitivity than at ambient conditions. When a metal acoustic calibrator which is at ambient temperature is placed on such a microphone, it may quickly bring the microphone to ambient temperature and thus result in an erroneous indication of the microphone sensitivity during the measurements.

Finally, fitting some acoustic calibrators on their respective microphones can be a tricky job, and an improper fit can result in a change in the sound-pressure level at the diaphragm of the microphone. In the press of a field measurement survey, a poorly fitted calibrator is easy to overlook.

* Frequencies of 400, 900, or 1,000 cps are commonly used.

5.5 Summary

We have described the designs, the uses, and the limitations of the sound-level meter, the octave-band analyzer, and the acoustic calibrator and have introduced the reader to some commercially available versions of these instruments. It is evident that much of the value of these instruments stems from their ability to obtain reliable data under field conditions with a minimum of transportation, power supply, and calibration difficulties. However, being electronic devices of moderate cost, these instruments are subject to certain difficulties, some of them obvious, and some subtle. Electrical noise, microphonics, amplifier drift, and temperature effects may possibly interfere with measurements. Frequency-response limitations may render the instrument useless when very low or very high frequency acoustic signals are encountered.

With the considerable present interest in noise problems, these basic field measurement instruments will be widely used, and some familiarity with their operation will be essential for workers in the field. It is hoped that the preceding discussions will enable users to choose the instruments best suited to their needs and to understand the operation of these instruments.

More detailed information on the presently available instruments can, of course, be obtained from the manufacturers, and some readers may wish to study the pertinent ASA standards and other publications listed under "Suggestions for Further Reading."

REFERENCES

1. ASA Standard Z24.3–1944, "Sound Level Meter for Measurement of Noise and Other Sounds," American Standards Association, 70 E. 45th St., N.Y. 17, N.Y. (This standard has been revised, but as of press time has not been accepted by all interested bodies.)

2. ASA Standard Z24.10–1953, "Octave Band Filter Set for the Analysis of Noise and Other Sounds." (See Ref. 1.)

3. Beranek, L. L.: "Acoustic Measurements," John Wiley & Sons, Inc., New York, 1949.

4. ASA Standard Z24.4–1949, "Method for the Pressure Calibration of Laboratory Standard Pressure Microphones." (See Ref. 1.)

SUGGESTIONS FOR FURTHER READING

Peterson, A. P. G., and L. L. Beranek: "Handbook of Noise Measurement," 3d ed., and E. E. Gross, Jr.: "Measurement of Vibration," The General Radio Company, Cambridge 39, Mass., 1956.

Pomper, V. H.: Noise Simplified, *Safety Maintenance and Production,* March,

1954, and April, 1954; also available from Herman Hosmer Scott, Inc., 111 Powder Hill Road, Maynard, Mass.

Noise—Pertinent Questions and Answers, Mine Safety Appliances Company, 201 N. Braddock Ave., Pittsburgh 8, Pa.

Beranek, L. L.: "Acoustics," McGraw-Hill Book Company, Inc., New York, 1954.

Beranek, L. L.: "Acoustic Measurements," John Wiley & Sons, Inc., New York, 1949.

Scott, H. H., and D. von Recklinghausen: "A Compact Versatile Filter-type Sound Analyzer," *J. Acoust. Soc. Am.*, vol. 25, p. 727, 1953.

Bruel and Kjaer Company *Tech. Rev.*, nos. 3–1952, 4–1952, 2–1953, 4–1953, 1–1954, 2–1954, 4–1954, 1–1955, 2–1955, 3–1955, 1–1956, 3–1956, 2–1957, 3–1957; copies of these may be obtained from Bruel and Kjaer Company, Naeram, Denmark.

Symposium on Sound Level Meters, *J. Acoust. Soc. Am.*, vol. 29, pp. 1330–1342, 1957.

Chapter 6

PERFORMANCE OF SOUND AND VIBRATION INSTRUMENTATION

G. W. Kamperman

6.1 Introduction

In the previous chapter we have described the basic sound-measuring system, namely, the sound-level meter, the octave-band analyzer, the associated microphone, and the acoustic calibrator. That system is particularly useful where studies are to be made of noise environments in relation to their effect on man. In this application it can be used satisfactorily to obtain data on continuous noises in problems involving the possibility of hearing loss, speech interference, and office and neighborhood annoyance. With the basic system, the acoustical engineer can accomplish many of his assignments.

When detailed engineering information is desired, more specialized equipment is frequently needed. The bases for selection of such instrumentation are covered in Chap. 4. In the present chapter we shall describe the performance characteristics of acoustical instrumentation.

6.2 Transducers

There are many types of transducers available for the measurement of sound and vibration. In this section we shall concern ourselves primarily with general-purpose microphones and vibration pickups. There is no one transducer that can adequately solve all acoustical measuring problems.

Microphones. Microphones that respond to sound pressure are generally used in sound measurements. In some special cases it may be desirable to measure some quantity other than sound pressure. There are relatively low-cost microphones available for measuring pressure gradient, which is directly related to particle velocity. No satisfactory general-purpose microphone is available for measuring sound intensity or particle displacement.

Desirable microphone characteristics are reliability, smooth frequency response, minimum phase distortion, high sensitivity, small size (to minimize disturbance of the sound field), simplicity, and reasonable price. There are few commercially available microphones that have a completely uniform frequency response in the frequency range of major interest (from 20 to 10,000 cps). Some microphones are available that can be electrically compensated to approach this ideal frequency response.

The physical size of the microphone also determines its frequency response. When the wavelength of sound being measured is less than 10 times the physical diameter of the microphone, the microphone performance will not be the same for various angles of incidence of the sound. The microphones described here can be considered nondirectional for all frequencies below 1,000 cps because their size is less than about 0.1 wavelength at 1,000 cps (λ at 1,000 cps equals 1.1 ft). With increasing frequency above 1,000 cps, the sensitivity becomes strongly dependent on the angle of incidence. Microphones that are as much as 2 in. in diameter become 10- to 15-db more sensitive at 10,000 cps to sounds arriving perpendicular to the sensing element than to sounds arriving parallel to it.

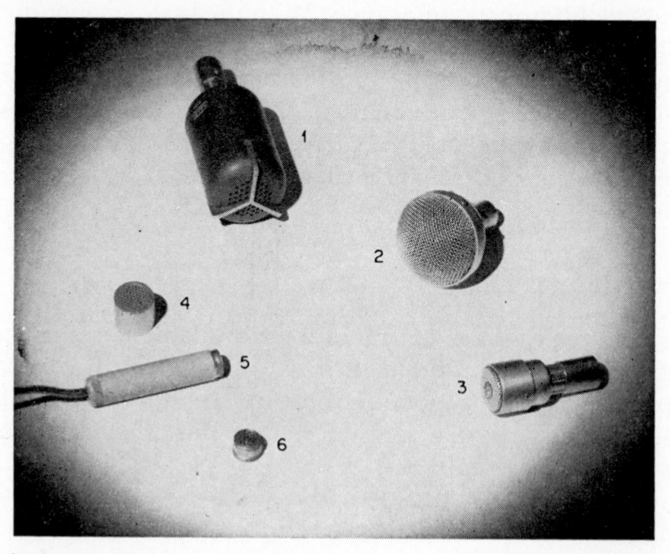

Fig. 6.1 Some general-purpose microphones for measuring sound pressure: (1) Altec 633A dynamic (moving coil); (2) Shure 9898 crystal; (3) Shure 9899 crystal; (4) Western Electric 640AA condenser; (5) Altec 21BR150 condenser with preamplifier; and (6) Altec 21BR200 condenser.

In Fig. 6.1 several types of microphones commonly used for the measurement of sound pressure are shown. Specific data on their characteristics are outlined in Table 6.1. The detailed construction of these microphones and theory of operation have been fully discussed in the literature.[1-5]

Table 6.1

Characteristics of Some General-purpose Microphones

Manufacturer	Altec			Shure			Bruel and Kjaer	Bruel and Kjaer	Western Electric
Model	633A	21BR150	21BR200	9898	9899	98108	4132	4134	640AA
Open-circuit sensitivity, db *re* 1 volt for sound-pressure excitation of 1 μbar	−90	−60	−90	−56	−59	−59	−46	−57	−50
Impedance or capacitance, mmfd	30 ohms	6	4	2,000	700	460	60	20	50
Maximum pressure level for linear response, db *re* 0.0002 μbar	110 150	160	190	155	155	155	155	164	155
Maximum temperature, °F	150	500	500	110	110	210	400	400	400
Stability vs. temperature	Fair	Poor	Good	Fair	Fair	Good	Good	Good	Good
Operation in high humidity	Good	Fair	Fair	Fair	Fair	Good	Fair	Fair	Poor
Frequency range, cps, for ±1 db *re* 400-cps deviation, random-incidence sound field	200–600	20–3,000	20–10,000	20–2,000	20–6,000	20–8,000	10–8,000	20–25,000	10–10,000
Frequency range, cps, for ±3 db *re* 400-cps deviation, random-incidence sound field	50–5,000	10–12,000	10–16,000	20–3,000	10–8,000	10–8,500	1–10,000	10–40,000	1–12,000

1. *Crystal Microphones.*[2-5] Crystal microphones have been used more than any other type for acoustical measurements. They have the advantages of being rugged and inexpensive. However, microphones that use a Rochelle-salt sensing element, as do the Shure Brothers types 9898 and 9899, exhibit properties that can impose serious limitations if the microphones are incorrectly used.

The electrical capacitance of the Rochelle salt crystal varies with temperature. This phenomenon does not present a problem in the over-all behavior of the basic system if the microphone is connected directly to a high-input-impedance amplifier having a low shunt capacity. However, when the microphone is used with an extension cable, a portion of the electrical current from the microphone will flow through the capacitance of the cable. This means that the voltage arriving at the input to the amplifier (following the extension cable) will be reduced by the voltage drop across the internal capacitance of the microphone. This drop depends on the temperature of the crystal in the microphone. In Fig. 6.2 the effect of microphone temperature on the response of the over-all system (like a sound-level meter) is shown for the Shure Brothers 9898 and 9899 microphones

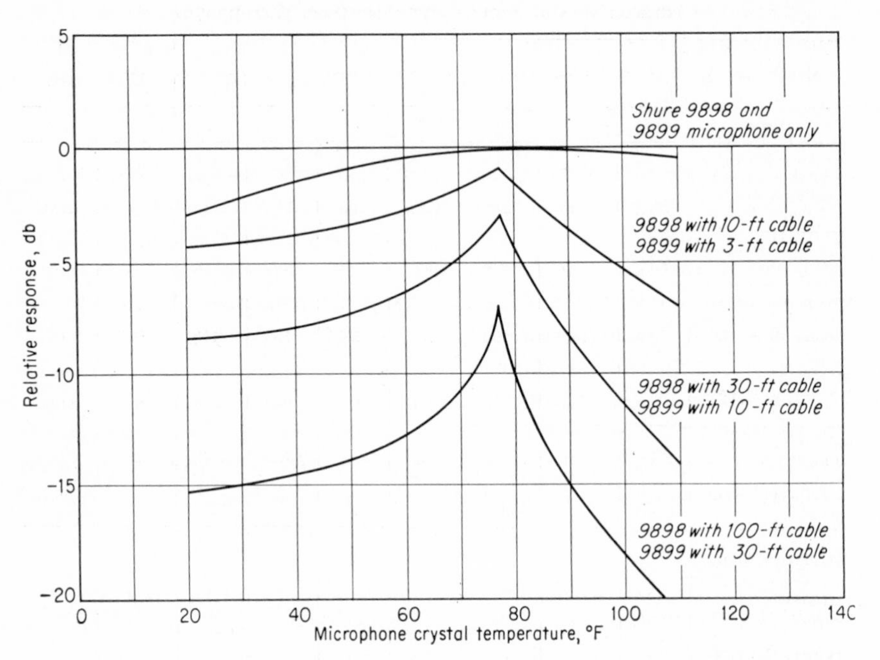

FIG. 6.2 Effect of adding extension cables to a sound-level meter for Shure Brothers 9898 and 9899 Rochelle-salt-crystal microphones as a function of the temperature of the crystal inside the microphone. The extension cables are assumed to have a nominal capacitance of 30 $\mu\mu$f/ft.

with several lengths of cables of the type recommended by the manufacturers. Fortunately, the change in sensitivity is nearly independent of frequency. Therefore, the change in sensitivity can be taken into account during field measurements without a knowledge of the actual temperature by frequent use of an acoustic field calibrator at a single frequency. The maximum safe temperature limitation of about 110°F for the Rochelle salt crystal must also be kept in mind when using this type of microphone.

The Shure 98108 microphone uses a lead zirconate element to obtain superior temperature characteristics. The electrical capacitance and sensitivity of lead zirconate changes very little with temperatures below 200°F. The Shure 9899 and 98108 have identical exterior appearance and similar frequency-response characteristics. Both microphones are adequate for most noise-reduction problems.

2. *Dynamic Microphones.*[1,2,5] The moving-coil or dynamic microphone has been used extensively for sound-pressure measurements. Like the crystal microphone it can tolerate moderate rough handling normally encountered in the field. It does not have as serious (low) a maximum temperature limitation as does the Rochelle-salt-crystal microphone. The maximum safe temperature lies somewhere above 200°F. Due to the low internal impedance of the dynamic microphone, long extension cables are practical. The microphone is very susceptible to electrical pickup from magnetic fields, which often restricts its use in the vicinity of electrical equipment. The dynamic microphone has in it several acoustic resonators to extend its useful frequency range. The stability of these resonant circuits is difficult to control with time and temperature. In this respect the dynamic microphone is less stable than the crystal microphone. The frequency response of a dynamic microphone should be checked frequently to assure its uniformity and stability.

3. *Condenser Microphones.*[1,3,5] The condenser microphone is more stable with time and temperature than either the dynamic or crystal microphone. It also has a more uniform frequency response. On the other hand, condenser microphones are expensive, require a special auxiliary power supply, and need special attention in high humidities.

There are two types of condenser microphones, one using a clamped diaphragm as the sensing element, the other using a stretched diaphragm. Microphones with stretched diaphragms are fragile and are usually seriously damaged if dropped.

Condenser microphones generate more self-noise due to local air turbulence than do the common crystal or dynamic types. If microphone wind noise is a problem, it can be reduced with the aid of a windscreen surrounding the microphone. A small cylindrical windscreen (3 in. in diameter by 5 in. long) of wire mesh covered with light silk will reduce the wind noise 20 to 30 db and will, for all practical purposes, be transparent to sound. More is said about windscreens in Chap. 7.

The most annoying disadvantage of a condenser microphone is its sensitivity to high humidity. Unless special precautions are taken, a condenser microphone may temporarily become noisy and even inoperative due to electrical leakage over the surfaces of the internal insulators.

Vibration Pickups.[6,7] The measurement of vibration plays an important role in the solution of many noise-reduction problems. In the majority of cases, the measurement of velocity or acceleration as a function of frequency is desired.

1. *Velocity Pickups.* Velocity-sensitive pickups are of the electromagnetic type. They have a useful frequency range from about 10 to 1,000 cps.

The primary advantage of the velocity pickup is its low electrical output impedance. Like the dynamic microphone, a disadvantage of the velocity pickup is its sensitivity to stray magnetic fields.

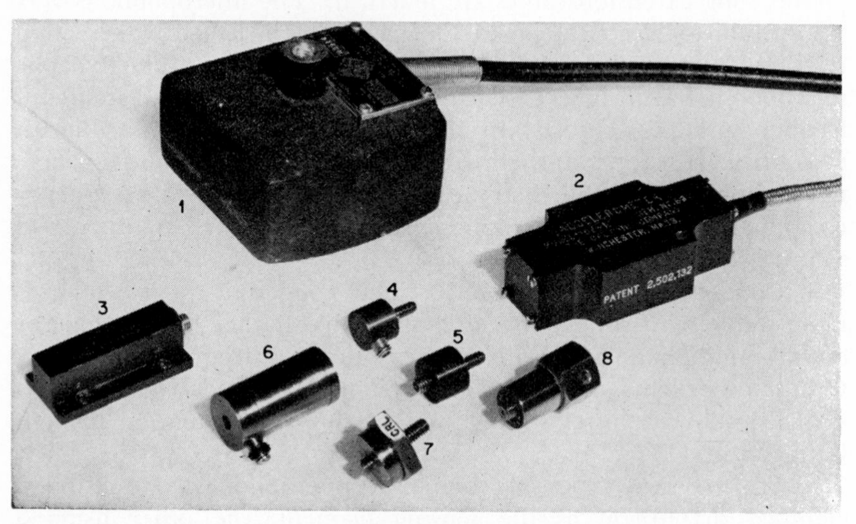

FIG. 6.3 Accelerometers for the measurement of vibration; (1) General Radio 759-P35; (2) Calidyne 18D-10; (3) Gulton A312; (4) Gulton A314; (5) Gulton A321; (6) Massa M-191; (7) Columbia 406; (8) Bruel and Kjaer 4308.

Present-day velocity pickups are best suited for special applications and are not usually considered general-purpose vibration pickups.

2. *Accelerometers.* The problems of the airframe manufacturers and others have created a need for lightweight vibration pickups with a wide frequency range. Many piezo- and ferro-electric accelerometers suitable for acoustic measurements are now available. A few representative units are illustrated in Fig. 6.3. The characteristics of these accelerometers are outlined in Table 6.2.

To ensure reasonable accuracy of measurement, it is frequently necessary to calibrate these accelerometers with respect to frequency response and sensitivity. Variations in time and temperature may alter the sensitivity several decibels.

Many companies have designed and manufactured piezo- and ferro-

Table 6.2

Some General-purpose Accelerometers for the Measurement of Vibration

Manufacturer	Model	Type of Transducer	Sensitivity, mv/g	Impedance, mmfd	Weight, oz.	Maximum Acceleration, g*	Frequency Range ±1 db, cps	Resonant Frequency, kc	Resonance Peak, db	Lateral Rejection, db	Temperature Range, °F, for Output, Error of 1 db*	Temperature Maximum, °F*	Response to Air-borne Sound
Bruel and Kjaer	4308	Barium titanate	20	700	0.9	1,000	1–10,000	30	26	20–30	−40–+194	194	Low
Bruel and Kjaer	4309	Barium titanate	6	700	0.6	2,000	1–10,000	45	20	20–30	−40–+194	194	Low
Calidyne	18D-10	Vacuum tube	1,000	100,000 ohms	2.5	10	0–100	0.2	0	30–40	<0–>120	150	Moderate
Columbia	406	Barium titanate	30	500	0.5	>10,000	1–20,000	75		20–30	−40–+190	190	Low
Endevco	2202	Ceramic	15	550	1	10,000	1– 8,000	35		>26	−65–+230	230	Low
Endevco	2218	Ceramic	50	500	2	1,000	1– 6,000	20		>26	−65–+230	230	Low
General Radio	1560-P51	Barium titanate	50	8,000	2	100	1– 800	2.3	12	>20	−40–+194	194	Low
General Radio and H. H. Scott	759–P35 and 410-X6	Rochelle salt	1,000	8,000	8	10	1– 500	2	30		+40–+100	110	High
Gulton	A312	Barium titanate	100	1,000	0.7	300	1– 500	1– 3	20	25–35	−40–+175	194	Moderate
Gulton	A314	Barium titanate	1	2,000	0.1	1,000	1–10,000	35		20–30	−40–+194	194	Low
Gulton	A320	Barium titanate	6	2,000	0.25	1,000	1– 3,000	8–10	20	30–40	−40–+194	194	Low
Massa	M-191	ADP	40	110	1	1,000	10–30,000			20–30	−40–+160	160	Low

* Data from manufacturer not verified by the author. Note that "g" equals acceleration due to gravity.

electric accelerometers that are intended to operate over the frequency range from 10 to 10,000 cps. The lower-frequency limit of these accelerometers is determined by their internal electrical leakage and the input impedance of the amplifier to which they are connected. In general, this type of pickup has a half-power point, or is 3 db down in response, in the range between 1 and 10 cps. Specifically, this point occurs where the capacitative reactance of the pickup, including cable, equals the amplifier input resistance.

The resonance frequency of the seismic system in the accelerometer determines the useful upper-frequency limit. This resonance frequency varies from 1 to 100 kc for different types of crystal accelerometers. It is generally acceptable to operate up to a frequency equal to one-third the resonance frequency. The peak in response at the resonance frequency is at least 10 times the nominal response below resonance. If there should be considerable energy in the vibration spectrum at the resonance frequency of the accelerometer, it may be necessary to use a low-pass electrical filter following the accelerometer, to prevent overloading the amplifier following the accelerometer.

An accelerometer with a high resonance frequency usually has a low sensitivity. It is therefore generally necessary to compromise sensitivity with the maximum upper-frequency limit.

Only piezo- and ferro-electric accelerometers are intended for operation over most of the audio-frequency range. However, there are several types of accelerometers, such as the unbonded straingauge accelerometer and the linear variable differential transformer, designed for operation below 100 cps. These accelerometers have low sensitivity and a narrow dynamic range. They are not considered suitable general-purpose accelerometers for acoustical measurements.

Transducer Signal Cables. The signal extension cable connected between the transducer and its amplifier often presents serious problems because of noise generated when the cable is shaken, because of electrical coupling to it, and because it accentuates the temperature effects in the transducer. These difficulties are particularly acute in the case of high-impedance devices like crystal accelerometers. It is frequently desirable to have the accelerometer connected to a high-input-impedance amplifier by means of a lightweight flexible cable. The electrical shielding of the cable and the self-noise generated in it are very important considerations.

In Table 6.3 we show some of the pertinent characteristics of cables. The column on relative microphonic noise is of particular interest. It can be seen that the microphonic or vibration sensitivity of these cables varies over a range of 100,000 to 1, i.e., 100 db.

The noise generated in a cable is proportional to the displacement of one portion of the cable with respect to another. The distance over which the cable is flexed is of secondary importance. The cable noise voltage appears to be relatively independent of the frequency

Table 6.3

Some Signal Cables for Microphones and Accelerometers

Manufacturer	Model	Number of Shielded Conductors	Relative Microphonic Noise, db *re* 1 volt †	Shunt Capacity, mmfd/ft	Outside Diameter of Cable, in.
Amphenol	RG-58A/U	1	0	29	0.195
	RG-62/U	1	−10	13.5	0.242
Belden ‡	8,425	5	−50	80	0.349
Birnback	1,872	1	−30	70	0.245
Bruel and Kjaer*	Mini Noise	1	−40	30	0.101
Federal Cable*	K250	1	−80	44	0.250
Gulton*	C-5 (black line)	1	−100	36.5	0.111
Microdot*	50–3,804	1	−30	27.5	0.088
Tensolite ‡	1,883-H6	6	−40	100	0.205
U.S. Rubber*	908–26	1	−70	40	0.165

* Advertised by manufacturer as special low-noise cable.

† The microphonic noise data are from a nonstandard test used by Bolt Beranek and Newman Inc. The frequency range of the test is from 10 to 10,000 cps.

‡ Noise figure for cable operated with Altec M-14 microphone system.

of excitation. The microphonic data presented in Table 6.3 were obtained by the following method: A 10-ft length of cable of the type to be measured is connected to a high-input-impedance (>100 megohms) amplifier. A section of the cable is excited with a sinusoidal motion, having a peak-to-peak displacement of 1 in. The equivalent peak (not peak-to-peak) noise voltage generated by the cable is that shown in Table 6.3.

Another test that gives comparable results is as follows: A 5-ft loop of the 10-ft cable undergoing test is held in the hand and slapped vigorously against a concrete floor. The peak voltage induced by this excitation leads to data similar to those obtained from the above experiment. The frequency passband of the measuring instruments used in the experiment was 1 to 10,000 cps. Both of these experiments appear to yield the maximum noise voltage that can be expected from these cables. Since the output noise is inversely proportional to the capacitive load on the cable, longer lengths of cable will reduce both the transducer signal level and the cable noise level.

The NBS Report[13] describes another method of measuring cable noise.

Some so-called "low-noise" cables lose their low-noise characteristics after they have been used for a short period of time. It is advisable to check the cables that receive continued use to determine whether the self-noise generated by the cables has changed.

Whenever a ceramic or crystal accelerometer is used for vibration measurement below 100 cps, the noise characteristics of the accompanying signal cable should be carefully considered. When the acceleration of a vibrating surface, measured in narrow frequency bands (e.g., 5 cps), is constant with frequency, the displacement of the surface quadruples, i.e., increases 12 db per halving of frequency. The cable noise increases at this rate whereas the output voltage of the accelerometer remains unchanged. Below some frequency the cable noise will predominate. When low-noise signal cables are used with relatively insensitive accelerometers, the frequency at which the cable noise produces a larger signal than the accelerometer produces may be as high as 20 to 100 cps.

6.3 Transducer Preamplifiers

There are many preamplifiers available for amplifying the low-level signals from microphones and accelerometers. It is important that preamplifiers selected have sufficiently high input impedance and low internal noise. The voltage gain stability, frequency response, and dynamic range must also be satisfactory. Very often microphonics in the preamplifier may present subtle problems. If some or all of the instruments incorporated in the measuring setup are powered from the a-c line, it may be necessary to ground the instrumentation to reduce hum and noise. The most satisfactory point for grounding the measuring system is usually near the point of lowest signal level, in other words, at the preamplifier base.

6.4 Indicating Instruments

Data read-out instruments fall in two general categories. One type displays the complete electrical wave (frequency, phase, and magnitude) of the applied signal. The other type indicates only the magnitude of the applied signal and is called an envelope detector.

The Sanborn level recorder, item 5 in Fig. 6.4, is an example of an indicating instrument which displays the complete electrical signal. This instrument has a useful frequency range from 0 to 60 cps. There are many more elaborate direct-writing recorders (some with many signal channels) that operate over the frequency range from

0 to 3,000 cps or higher. The cathode-ray oscillograph has a very wide frequency range.

The principal advantage of the direct-writing recorder for noise measurements is the ability to compare phase information between signals from different transducers. When a direct-writing recorder is used, the transducer signal is usually sent directly to the indicating instrument (following amplification), without any electrical filtering. The frequency analysis of the signal from the transducer is performed by a graphical Fourier analysis of the waveform of the signal

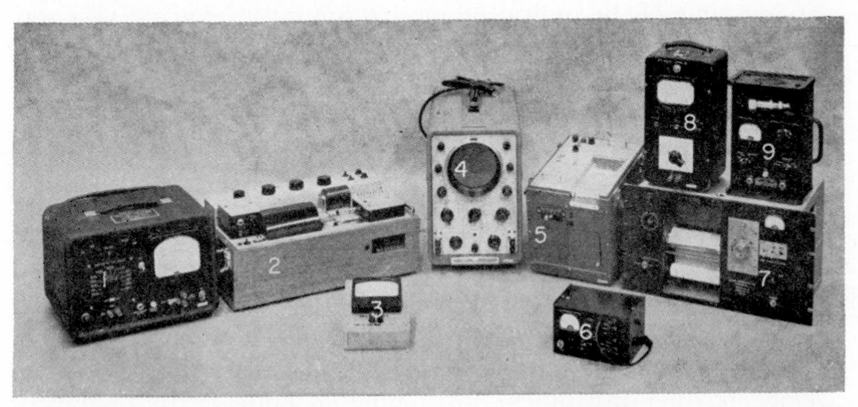

Fig. 6.4 Some indicating instruments for reading out noise and vibration information: (1) Ballantine 320-rms voltmeter; (2) Bruel and Kjaer BL-2304 graphic level recorder; (3) Triplett model 420, type A, VU meter; (4) Hewlett-Packard 130A oscilloscope; (5) Sanborn 151–100A single-channel recorder; (6) General Radio 1556-A impact-noise analyzer; (7) Sound Apparatus SL-4 sliding-coil recorder; (8) Ballantine 302B voltmeter; (9) General Radio 1551-A sound-level meter.

from the recorder or oscillograph display. The amplitude of the signal is displayed linearly. If the highest component in the signal registers full-scale, other components that are 20 db below this maximum will register only 10 per cent of full scale. For acoustical measurements, we are generally interested in levels of at least 40 db, and frequently 60 db or more below the highest component. On the linear scale just illustrated, these would represent readings of 1 per cent and 0.1 per cent of full scale. Obviously, little information could be obtained at these small signal levels.

The magnitude-indicating instrument or envelope detector has widespread application in acoustical measurements. We will consider here only those instruments that have a logarithmic amplitude response. Several of the commonly used instruments (items 1 through

through 9) are illustrated in Fig. 6.4 and their characteristics described in Table 6.4.

The majority of the instruments shown in Fig. 6.4 are designed for operation in the audio-frequency range. However, auxiliary equipment may be used to extend the useful frequency range to the subaudio range for vibration measurements (and to 0 cps if necessary).

Envelope detector instruments can be separated into two categories, indicating meters and graphic level recorders. The type of detector or rectifier used in an indicating instrument and the response time of the indicating mechanism must be given very serious consideration. If the instrument is to be used for indicating the levels of steady pure tones, there is little concern over these factors. For the indication of noise from either a vibration pickup or microphone, the type of detector and the response time of the instrument may greatly influence the reading.

Rectifiers that are sensitive to the rms of the input signal are desirable because most noise and vibration data have been presented in terms of rms values. A true average rectifier is also suitable for the indication of noise signals. Although the difference between the rms value and the rectified average value of a random noise is not the same as for a sine wave, it is small. In fact, the indication of an average rectifier instrument for a random noise having a normal distribution is about 1 db below what it should read if it were an rms instrument. This assumes that the average and rms instruments read alike on sinusoidal tones. The common practice is to ignore the difference.

Indicating instruments that incorporate peak rectifiers should be used with full knowledge of their behavior, because their readings can be much higher than those of the rms and average types. Two of the level recorders described in Table 6.4 have peak-sensitive rectifiers which indicate neither the rms, average, nor the true peak of the noise signal. The indication lies somewhere between rms and peak, depending upon the amplitude distribution of the noise signal. The difference from the rms value is in the order of 1 to 10 db when used to measure noise. The difference can be partly eliminated by calibrating these instruments with a sample of the noise to be analyzed. Discussion of this subject is given in Chap. 4.

The response time of an indicating instrument becomes an important consideration when the noise level varies rapidly with time. The sound-level meter and other instruments using a similar indicating meter have response times that are sufficiently fast for faithfully reproducing the average level of most noises. The sound-level

Table 6.4

Characteristics of Some Indicating Instruments Shown in Figure 6.4

Manufacturer	Model	Full-scale Voltage	Frequency Range, cps	Type of Detector	Input Impedance	Type of Scale	Scale Range	Attenuator Steps	Power Source	Response Time
Ballantine	320 VTVM	0.3 mv–320 volts	5–0.5 mc	Full-wave rms	10 megohms, 8–25 $\mu\mu$fd	Log	0–10 db 1–3.16 volts 3.16–10 volts	10-db steps	A-c line	0.2 sec
	300	10 mv–100 volts	10–150,000	Full-wave average	500,000 ohms, 50 $\mu\mu$fd	Log	20 db	20 db	A-c line	0.2 sec
	302-B VTVM	1 mv–100 volts	2–150,000	Full-wave average	2 megohms, 8–15 $\mu\mu$fd	Log	20 db	20 db	Battery	0.5 sec
Bruel and Kjaer	2304 Level recorder	10 mv–100 volts	20–200,000	Half-wave peak	50,000 ohms	Log	0–10, 25, 50, 75 db	Continuous	A-c line	5–100 cm/sec
	2305 Level recorder	10 mv–100 volts	10–200,000	Full-wave rms, average and peak	18,000 ohms	Log	0–10, 25, 50, 75 db	10-db steps	A-c line	0.4–200 cm/sec
	2604 Microphone amplifier	0.1 mv–1,000 volts	10–200,000	Full-wave rms, average and peak	1.1 megohms, 30 mmfd	Linear	0–20 db 0–3.16 v 0–10 v	10 db	A-c line	0.2 sec 0.8 sec
General Radio	1521-A Level recorder	1 mv–100 volts	20–200,000	Full-wave rms	10,000 ohms	Log	0–10, 20, 40, 80 db	10 db	A-c line	20 in./sec
	1551B Sound-level meter	10 mv–3 volts	20–20,000	Full-wave rms	20 megohms	Log	16 db	10 db	Battery	0.5 sec
	1556A Impact-noise analyzer	1–10 volts	5–20,000	Peak, quasi peak, time average	25,000–100,000 ohms	Log	20 db	Continuous	Battery	50 msec (for peak)
Hewlett-Packard	130A Scope	10 mv–200 volts	0–300,000	None	1 megohm, 50 $\mu\mu$fd	Linear	1 mv–20 v/cm	1:2:5 ratio	A-c line	5 msec
Sanborn	DC 150–1,300 Recorder	5 mv–250 volts	0–100	None	5 megohms	Linear	50 mv–50 v/cm	1:2:5 ratio	A-c line	5 msec
Sound Apparatus	SL-4 Level recorder	1 mv–100 volts	20–100,000	Full-wave peak	10,000 ohms	Log	0–10, 15, 20, 30, 40, 60, 80		A-c line	0.9–0.3 sec
Triplett	VU 420 meter	0.8 volt at 0 VU	dc–100,000	Full-wave average	3,900 ohms	Linear	20 db			0.5 sec

meter can be used to obtain the average sound-pressure level of repetitive sounds (if the peak-to-average signal level is less than 15 db) that have a repetition rate of 2 pulses/sec or greater, e.g., riveting hammers.

Noises that are impulsive in character, like explosions and drop forges, require special instrumentation to obtain an indication of the true peak noise level. For such signals, the peak value of the signal is generally desired rather than the average or rms value. A cathode-ray oscillograph or peak-indicating meter, like items 4 and 6 in Fig. 6.4, may be used to obtain this information. When using an oscillograph, keep in mind that the amplitude scale is linear rather than logarithmic and that the indication is always the peak value of the signal. If the oscillograph is calibrated with a pure tone, it is important to take into consideration the 3-db difference between the rms and peak values of the pure tone. Detailed information on the measurement of impulse (impact) noise is given in Sec. 7.5.

6.5 Frequency Analyzers[8]

In order to understand a noise-reduction problem, some type of frequency analysis of the noise and vibration is generally required. In some applications a wideband analysis is sufficient. Special problems may require a very narrow-band frequency analyzer to obtain

Fig. 6.5 Some general-purpose frequency analyzers for the analysis of noise and vibration data: (1) H. H. Scott 420-A sound analyzer; (2) Hewlett-Packard 300A harmonic-wave analyzer; (3) Allison Type 2A audio-frequency filter; (4) Bruel and Kjaer 1609 one-third-octave-band filter; (5) General Radio 1550-A octave-band analyzer; (6) General Radio 760-B sound analyzer; (7) Spencer Kennedy Laboratory 302 variable electronic filter; (8) Krohn Hite 330-F low-frequency octave-bandpass filter; (9) Krohn Hite 330-A ultra-low-frequency bandpass filter; (10) General Radio 736-A wave analyzer.

Table 6.5

Characteristics of Some Frequency Analyzers for the Analysis of Noise and Vibration Data*

Manufacturer	Model	Frequency Range, cps	Bandwidth Constant	Type of Filter	Type of Tuning	Input Z	Output Z	Load Z	Input noise, db *re* 1 volt	Full-scale voltage	Output Meter
Allison Lab	2A	15–10,000	%	High-pass, low-pass, bandpass, passive	Octave steps and continuous variables	600 ohms	600 ohms	600 ohms	−120	1 volt	No
Bruel and Kjaer	2111	35–32,000	%	⅓-octave, passive	⅓-octave steps	2.2 megohms	50 ohms		−106	0.1 mv–1,000 volts	Yes
	2107	20–20,000	%	6% bandpass, active	Continuous	2.2 megohms	50 ohms		−106	0.1 mv–1,000 volts	Yes
General Radio	736A	20–16,000	Frequency	4 cps bandpass, active	Continuous variable	1 megohm			−100	0.3 mv–300 volts	Yes
	760B	25–7,500	%	2% bandpass, active	Continuous variable	20,000–30,000 ohms	20,000 ohms	10,000 ohms	−80	1 mv–10 volts	Yes
	1550A	20–10,000	%	Octave bandpass, passive	Octave steps	20,000 ohms		0.5 megohm 100 μμf	−120	1 mv–10 volts	Yes
	1554A	2.5–25,000	%	8% plus ⅓ octave active	Continuous	0.1 megohm	5,000 ohms			1 mv–30 volts	Yes
Hewlett Packard	300A	20–16,000	Frequency	7–28 cps bandpass, active	Continuous variable	0.2 megohm			−80	0.1–500 volts	Yes
	302A	20–50,000	Frequency	7 cps bandwidth, active	Continuous	0.1–1 megohm	600 ohms			0.03 mv–300 volts	Yes
Krohn Hite	330A	0.02–2,000	%	Bandpass, active	Continuous variable	22 megohms 200 μμf	High 5,000 Low 500	500 ohms	High −60 Low −80	High 10 volts Low 1 volt	No
	330F	1–4,800	%	Octave bandpass, active	Octave steps	22 megohms	High 5,000 Low 500	500 ohms	High −70 Low −90	High 10 volts Low 1 volt	No
Muirhead	D-489	20–20,000	%	Variable bandpass, active	Narrow- to ⅓-octave steps	0.1 megohm 100 μμf			−86	1 mv–300 volts	Yes
SKL	302	20–200,000	%	High-pass, low-pass, bandpass, active	Continuous variable	1 megohm 30 μμf	300 ohms	10,000 ohms	−80	4 volts	No
H. H. Scott	420-A	20–9,600	%	Bandpass, active	½-octave steps	0.5 megohm	5,000 ohms	10,000 ohms	−80	5 mv–20 volts	Yes
Telefon Fabrika	11203	45–12,000	%	⅓ octave, passive	⅓-octave steps	5,000–20,000 ohms		1 megohm	−100	3 volts	No
Western Electro-Acoustic Lab	600A	10–50,000	%	High-pass, low-pass, bandpass, active	Octave steps and continuous variable	1 megohm	20 ohms	7,500 ohms	−72	3 volts	Yes

* See Fig. 6.6.

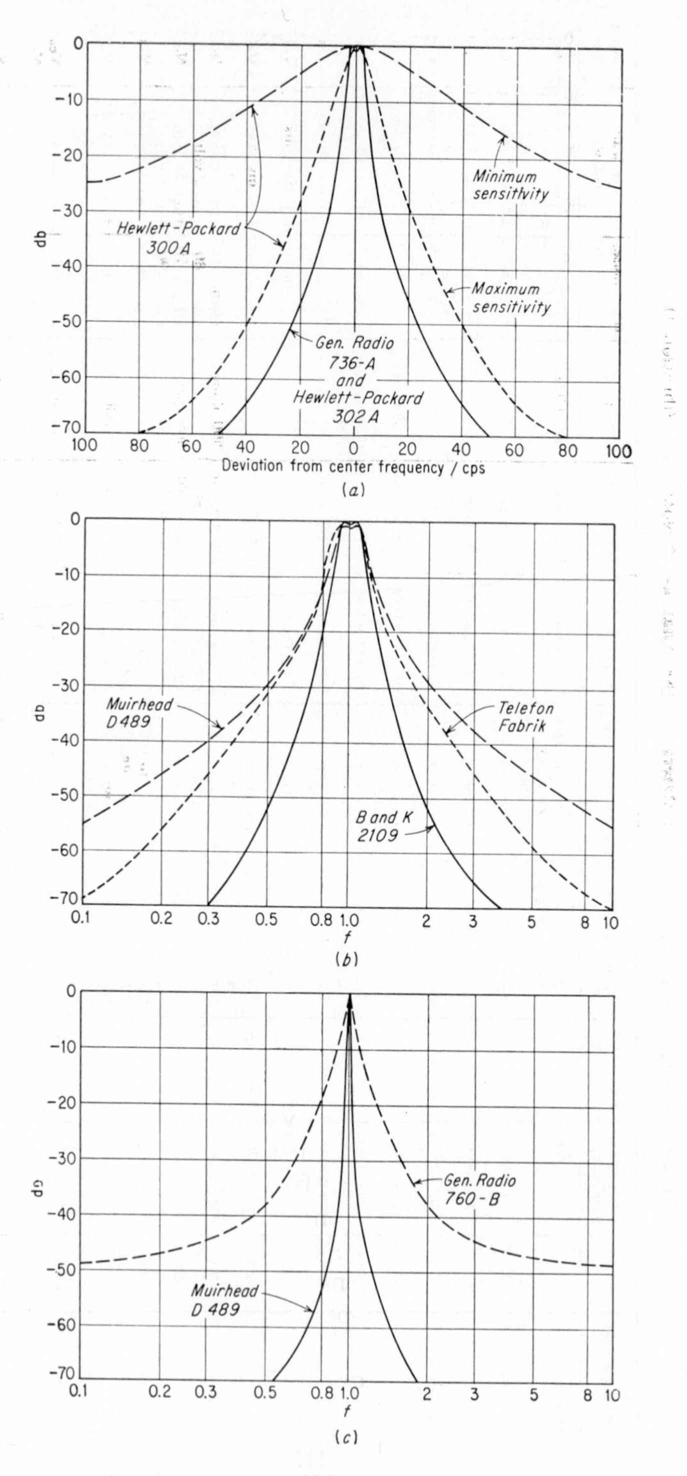

Hewlett-Packard 300A
Minimum sensitivity
Maximum sensitivity
Gen. Radio 736-A and Hewlett-Packard 302A
db
Deviation from center frequency / cps
(a)
Muirhead D489
Telefon Fabrik
B and K 2109
f
(b)
Gen. Radio 760-B
Muirhead D 489
(c)

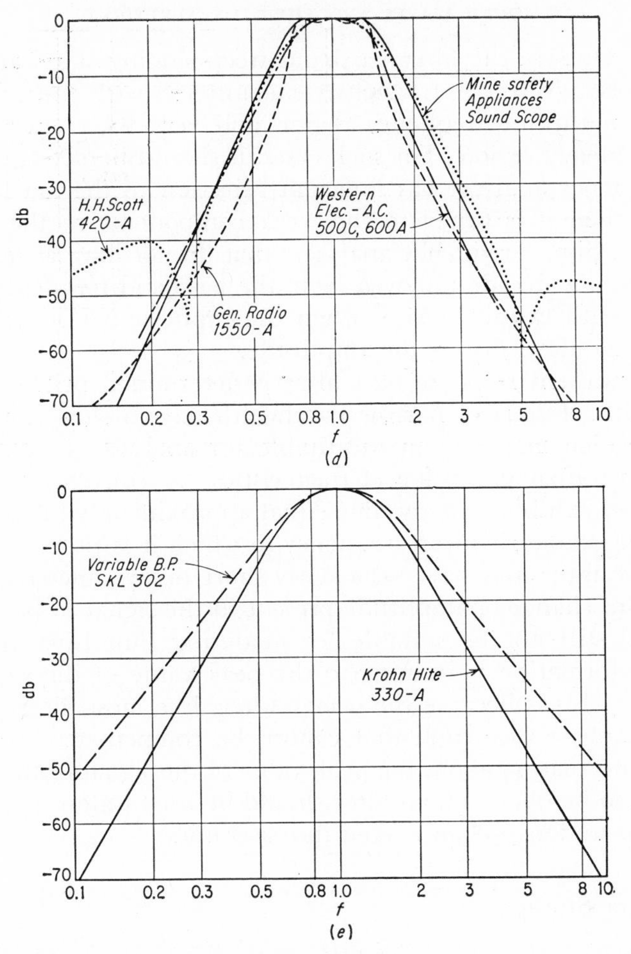

FIG. 6.6 Frequency response of analyzers described in Table 6.5. (*a*) Constant-frequency-bandwidth filters; (*b*) one-third-octave filters; (*c*) constant-percentage-bandwidth filters; (d) octave-band filters; (e) variable bandpass filter, set to yield one octave.

sufficient detail about the frequency spectrum. The selection of the type of analyzer needed for a particular job is discussed in Chap. 4.

There are two general types of frequency-band analyzers available for noise analysis. They are constant-bandwidth and constant-percentage-bandwidth filters. The bandwidth of the latter is a fixed percentage of the center frequency of the filter. In addition to band-pass filters, low-pass and high-pass filters are used for some applications.

Several general-purpose frequency analyzers are illustrated in Fig. 6.5, and their characteristics are described and illustrated in Table 6.5 and Fig. 6.6. Within the filter passband a uniform response is desired. Outside the filter passband steep sloping skirts are important. Noise spectra are frequently encountered with energy predominantly in the low-frequency region and very little energy in the high-frequency region. For such cases, it should be determined that the frequency analyzer has sufficient rejection to the low-frequency energy when it is tuned to the weak components in the high-frequency region. Frequency analyzers that do not have at least 60-db rejection, one decade removed from the passband frequency, should be used with caution. Also, when the input signal is too high in level, some filters distort the amplitude.

The transient response of a filter is determined primarily by its bandwidth. Filters of narrow bandwidth have a slow transient response which makes them unsuitable for analysis of noise having rapidly varying amplitude characteristics. A conventional octave-band filter exhibits a decay time equal approximately to the reciprocal of the geometric mean frequency to which it is tuned. Although an octave filter may have sufficiently short decay time compared to the rate of change of amplitude present in the signal to be analyzed, it would still not be suitable for analyzing impulsive-type noises where information pertaining to the peak value of the signal is of interest. This is because the octave filter, like most filters, exhibits complex phase distortion that cannot be compensated for readily. When information about the peak value of the signal from the transducer is to be obtained, no filter should be used unless its phase distortion is understood and taken into account.

6.6 Data Storage

In recent years, the availability of precision portable magnetic tape recorders has made it convenient for the acoustical engineer to tape-record pertinent data in the field for analysis at a later date in the laboratory with instrumentation that would not have been possible or convenient to take into the field. Another major feature of tape-recording is the ability to record a signal, such as an impulse or an explosion, for repeated analysis.

Three portable tape recorders[9] are shown in Fig. 6.7 for recording data on a single channel, on two channels, and on fourteen channels, simultaneously. Tape recorders suitable for recording noise signals are similar to tape recorders used for speech and music, except that it is desirable for the data recorder to use a constant-current recording process instead of the usual high-frequency preemphasis common in standard audio recorders. With constant-current recording, the signal level for a given distortion on the magnetic tape is relatively uniform over the operating frequency range.

It is also desirable to have available some type of peak signal indicator to avoid peak clipping. Such a peak overload indicator becomes of primary importance for explosive or impulsive types of noises. If not built into the instrument, an oscillograph or other peak-indicating device can be used to indicate overload.

Another important feature desirable in a portable data recorder is a precision decade gain control in the recording amplifier, similar to that on a sound-level meter. The ability to change the record amplifier gain by precise amounts, and preferably in 5- or 10-db in-

FIG. 6.7 Left to right, an Ampex 601-2 for tape-recording two-channel data, an Ampex 800 for recording fourteen-channel data, and an Ampex 600 for single-channel data.

crements, can greatly facilitate the recording of data and the calibration of the instrument.

Where more than one data channel is recorded simultaneously on magnetic tape, it is important to investigate the interchannel cross-talk as a function of frequency. If the recording and playback assemblies have been properly designed, the interchannel cross-talk will usually be of little consequence, except at very low frequencies.

Both direct-recording and frequency-modulation (FM) techniques can be used profitably for the recording of sound and vibration signals. Frequency-modulation techniques have the advantage of very faithfully recording and reproducing the amplitude, frequency, and phase characteristics of a signal from zero to several thousand cycles per second, depending upon the operating tape speed. The principal disadvantage of the frequency-modulation method is limited signal dynamic range caused by flutter and other irregularities in the tape speed. It is also necessary to use a tape speed approximately five times greater than is required for the direct-recording method in order to obtain the same upper-frequency range. On the other hand, with direct-recording methods, the lower-frequency response of the recorder is limited by the tape speed and the physical shape of the reproducing head.

The frequency-response characteristics using direct recording are not as flat as with the frequency-modulation method. The useful frequency range is confined to a bandwidth of three decades. Another problem with direct recording is fluctuation in the reproduced signal due to variations in the magnetic oxide coating on the tape and the contact between the magnetic tape and the record and reproducing heads.

Instrumentation quality tape is not required for the recording of acoustical data; radio broadcast quality tape is sufficient. To minimize the variation in tape sensitivity from reel to reel, it is recommended that the user purchase tape from only one manufacturer. The variation between reels can be reduced even further if the recording tape used on a given series of measurements is from the same production lot.

6.7 Data Reduction

The availability of precision magnetic tape recorders has evolved great changes in the techniques of data reduction. The tape recorder has made it possible to do rapid analysis of sound and vibration data. For those applications where a large amount of data is to be reduced, the data-reduction system may be mechanized to obtain the reduced data in an automatic fashion. For example, very elaborate automatic data-reduction systems have been set up at various locations throughout the country to reduce sound and vibration data transmitted from missiles in flight.

Even for very simple data-reduction problems, it may prove profitable to mechanize some of the basic analysis instrumentation to reduce the data in a semiautomatic way. The system may be designed not only to save data-reduction time, but also to obtain analyses that would not otherwise be possible.

One of the most common forms of data reduction of sound and vibration information is a frequency analysis. For a one-third-octave-band frequency analysis or a narrow-band analysis it is frequently convenient to have the frequency spectrum presented in graphical form. A magnetic tape recorder, a frequency analyzer, and a graphic level recorder may be connected together to perform this basic analysis in a semiautomatic fashion. The tape-recorded signal is usually made into a continuous tape loop, either by cutting the original tape into segments or by dubbing a portion of the recording onto a separate tape loop. The same sample of data can be reproduced continuously for the detailed frequency analysis; for example, it may be reproduced once for each band of the analyzer.

The frequency range of the frequency-analyzing instruments may be extended with the use of a multispeed magnetic tape recorder. This is done by recording the data at one speed and reproducing the recorded data at a different speed.

When planning a data-reduction system, it is essential to consider and compensate for the shortcomings of each instrument used to make up the system. The performance of the entire system should be evaluated. Some of the things that should be considered are frequency response, transient response, dynamic range, distortion, linearity, and stability.

6.8 Phase Measurements

It is frequently desirable in vibration measurements, and occasionally in noise measurements, to measure the phase relations among the signals from two or more microphones or vibration pickups. A simple technique for performing phase measurements using the instrumentation described in this chapter is illustrated in the block diagram of Fig. 6.8. Here phase information is obtained from the

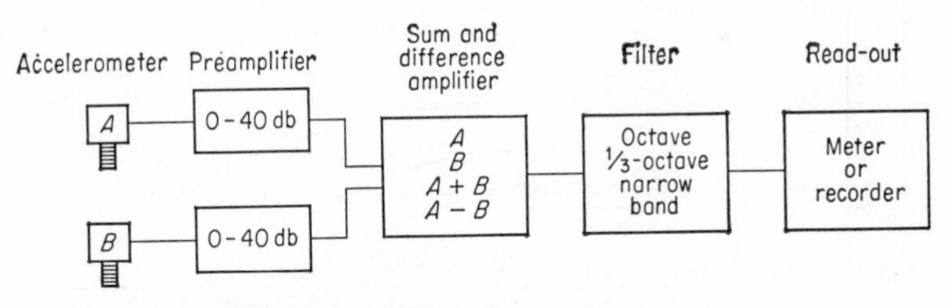

FIG. 6.8 A system for the measurement of phase.

addition and subtraction of the signals from the two transducers. The operation is accomplished with a sum and difference amplifier which consists of a phase inverter and simple resistor summing networks.

The following example illustrates the method of operation: Assume that we desire to obtain the phase between the motions of two accelerometers *A* and *B* mounted on a machine as a function of frequency in one-third-octave filter bands. The read-out meter is read as the sum and difference amplifiers are connected to yield, in order, (1) the output of accelerometer A, (2) the output of B, (3) the linear sum of A and B $(A+B)$, and (4) the linear difference of A and B $(A-B)$. These linear quantities are read in decibels, and the deci-

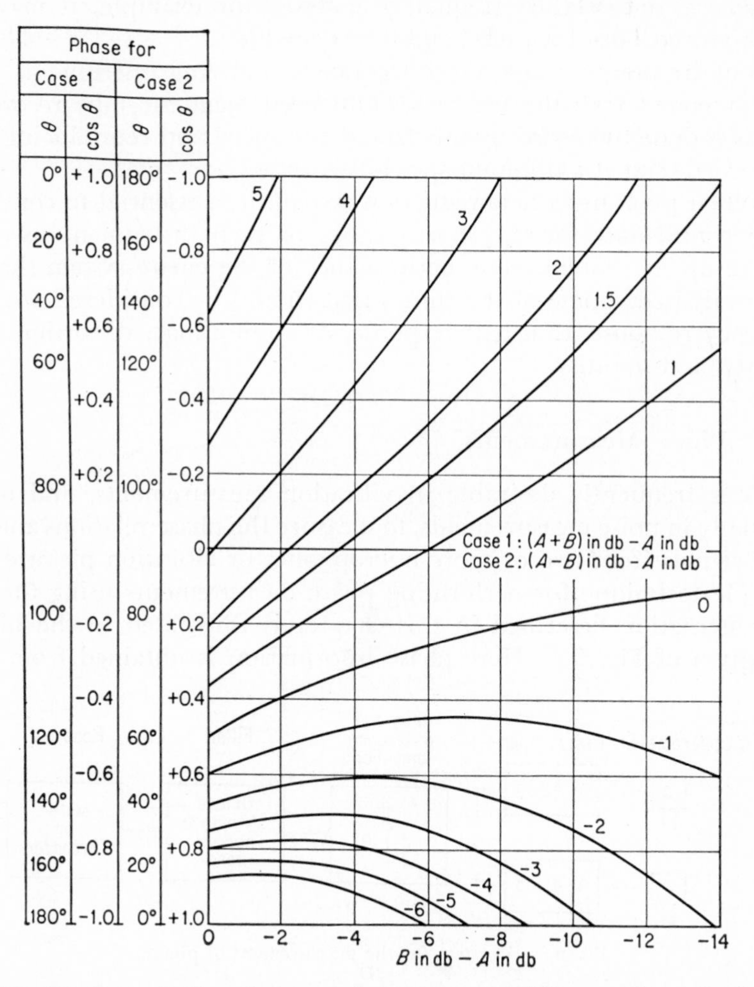

FIG. 6.9 Determination of phase angle from sum and difference of two signals.

bel readings are used, with the help of Fig. 6.9, to obtain the phase relation between the signals from accelerometers A and B.

For our example, let us assume that the read-out meter readings were as follows: $A = 82$ db, $B = 78$ db, $A + B = 85$ db, $A - B = 81$ db. The abscissa of the graph in Fig. 6.9 is the quantity B minus the quantity A, both in decibels. For our example, this difference is -4 db. A phase angle can be obtained using either the decibel reading from $A + B$ (case 1) or $A - B$ (case 2). Let us take $A - B$ less A, both in decibels (case 2). This equals -1 db. Note that the -1-db curve intersects with -4 db on the abscissa at a phase angle of 60° (or $\cos \theta = +0.5$) in the case-2 column on the ordinate of the graph. If we choose to use case 1, then $(A + B) - A$ in decibels, or $85 - 82 = 3$, you will note that the $+3$-curve coincides with a phase angle of 60° in the case-1 column on the ordinate.

There are several advantages to determining phase information by this simple method. Filtering, with its inherent phase distortion, does not affect the phase relation of the signal between the two transducers since it is accomplished in the circuit following the sum and difference amplifier.

6.9 Microphones for the Measurement of High-intensity Noise[10]

The measurement of intense noise is, in general, similar to the measurement of sound and vibration described earlier, except that the problems encountered are more difficult to solve. Part of the reason for this difficulty is that the transducer or microphone used must frequently operate in adverse environments where the temperature, pressure, humidity, and vibration may vary considerably. The electronic instrumentation system following the transducer usually need not be exposed to these environments. When such environments are unavoidable, special equipment may be required.

The microphone or transducer for measuring high-intensity noise must, because of its very purpose, be able to measure faithfully the high-intensity noise and yet not be influenced by the other environmental conditions. A representative group of transducers currently available for the measurement of high sound-pressure levels is illustrated in Fig. 6.10. One or more of these microphones may be found suitable for a particular application. To aid in the selection of the proper microphone for a particular task, certain operating characteristics important to their behavior are shown in Table 6.6.

It can be seen from Table 6.6 that the sound-pressure sensitivity of many high-intensity microphones is very low. Often, the less-sensitive transducers were designed from conventional transducers simply by increasing the stiffness of the moving element, so that the same voltage output could be developed at a much higher sound pressure. When microphones of low sensitivity are used, the extraneous noise produced by the associated cables and amplifiers must be seriously considered at all times. As we have already said, the cable for conducting the electrical signal from the transducer to its pre-

Table 6.6

Some Microphones for the Measurement of High-intensity Sound

Manufacturer	Model	Type of Transducer	Open Circuit Sensitivity, db *re* 1 volt *re* 1 μbar	Maximum Positive Pressure Level, db *re* 0.0002 μbar for Linear Response	Maximum Positive Pressure Level, db *re* 0.0002 μbar Without Damage	Operating Temperature Range, °F
Altec	21BR150	Condenser	−60	160*	175	The microphone
	21BR180	Condenser	−70	170*	185	will operate
	21BR200	Condenser	−90	190*	205	−50°+500°
	21BR220	Condenser	−110	210*	225	maximum
	6117 Probe	Condenser	−75	170*	185	temperature
	21BR180 pin hole	Condenser	−70	170*	185	for preampli-
	21BR180-5 flush mount	Condenser	−70	170*	185	fier, 150°*
Atlantic	BC-10	Barium titanate	−117	230	250	0–+194
	BC-30	cylinder	−111	230	250	0–+194
	LC-60	Zirconate	−112	230	250	−40–+210
	LC-33	Zirconate	−105	225	230	−40–+210
	BD-10	Barium titanate	−111	230	250	0–+194
	BD-30	disks	−111	230	250	0–+194
	LC-5	Lead zirconate probe	−124	224	230	−40–+210
C. E. C.	4-340	Electrokinetic	−107	211	—	−10–+140
Chesapeake	AB-130	Barium titanate	−109	230	230	0–+194
	NM126/T	Ceramic cylinder	−104	200	224	−100–+400
	NM-135/T	Ceramic cylinder	−94	200	224	−100–+400
Endevco	2501	Piezo-electric	−125	224	230	−20–+230
	2503	Piezo-electric	−125	191	230	−30–+230
Gulton	P600	Ceramic sphere	−106	>168	—	<160
	P420-M6	Ceramic	−104	>170	188	−65–+212
	KTP900	Ceramic disk	−101	>180	186	−65–+350
Massa	M-141	ADP crystal	−106	>200	230	−40–+160
	M-125	ADP crystal	−93	>200	230	−40–+160
	M-213	ADP crystal	−102	>200	244	−40–+160
	M-214, M-215	ADP crystal	−98	>200	244	−40–+160
Photocon	320	Condenser	−88	190	220	−50–+400
	374	Microphone	−77	180	200	−50–+400
	364	Plus carrier system	−57	160	180	−50–+400
Statham	PG132TC-15-350	Straingauge	−154	194	200	−65–+400
Western Electric	640AA	Condenser	−50	155*		<−40–>200

* Data verified by author.

† Western Electro-Acoustic Laboratory Type 130A, 120A, 100D.

Table 6.6

Some Microphones for the Measurement of High-intensity Sound (*Continued*)

Frequency Response Range, cps; Free-field Random Incidence Response ±3 db	Frequency Response Range, cps; Pressure Response ±3 db	Acceleration Response in Equivalent Sound-pressure Level for 1 *g* Excitation	Stability of Sensitivity over Specified Temperature Range	Microphone with Flush-mounted Sensing Element	Operation in the Pressure of High Humidity	Internal Impedance, μμf	Model Number of Microphone Preamplifier
10–12,000	10–12,000	100*	±5 db*	No	No	6	M-14 system preamplifier, and power supply
10–14,000	10–14,000	107*		No	No	6	
10–16,000	10–16,000	105*	±1 db*	No	No	6	
10–16,000	10–16,000	109*		No	No	6	
10–10,000	10–10,000	107*		Yes	No	6	
10– 8,000	10– 8,000	107*		Yes	No	6	
10– 8,000	10– 8,000	107*		Yes	No	6	
1– 5,000	1–75,000	130*	±3 db	No	Yes	6,000	104
1– 2,500	1–40,000	145*	±3 db	No	Yes	8,000	104
1–15,000	1–50,000	134*	±0.5 db to	Yes	Yes	2,000	104
1–10,000	1–80,000		175°F	No	Yes	2,500	104
1–20,000	1–20,000	138*	±3 db	Yes	Yes	500	104
1–10,000	1–15,000		±3 db	Yes	Yes	3,000	104
10–20,000	2–50,000			No	Yes	224	104
3– 2,000	3–25,000	130	±0.5 db	No	Yes	100 k	
1– 5,000	1–100,000			No	Yes	2,000	N-120
1–12,000*	1- 12,000	130*	±1 db	No	Yes	1,700	
1–10,000	1–25,000			No	Yes	1,200	N-120
2–10,000	2–10,000	133	±1 db	Yes	Yes	400	2608
2–10,000	2–10,000	125	±1 db	Yes	Yes	400	2608
10– 6,000*		140		No	Yes	400	
1– 8,000*	1–15,000	105*	±1 db	No	Yes	2,000	
30–10,000	30–>10,000	144	±1 db	Yes	Yes	Preamplifier out	KTP 900
20–15,000	20–35,000	138*	±0.25 db	Yes	Yes	500	M-114B
20–15,000	20–35,000		±0.25 db	Yes	Yes	120	M-114B
20–40,000	20–125,000	140*	±0.25 db	Yes	Yes	12	M-114B
20–30,000	20–80,000	140*	±0.25 db	Yes	Yes	18	M-114B
0–12,000	0–12,000	135*	Very stable if water cooling feature used	Yes	Yes	3,500 ohms out of DG-400	DG-400
0–12,000	0–12,000			No	Yes		DG-500
0–10,000	0–10,000	84*		No	Yes		DG-60C Dynagage
0– 6,000	0– 6,000	124	±0.5 db	Yes	Yes	350 ohms	
1–10,000	1–15,000	92*	±1.5 db	No	No	50	†

amplifier is frequently a source of microphonic noise. This type of noise is generally induced by vibrating structures that come in contact with the cable. The self-noise generated by the microphone cable should always be independently evaluated for each situation.

The vibration sensitivity of the microphone warrants consideration whenever it is mounted on or very near the sound source. Vi-

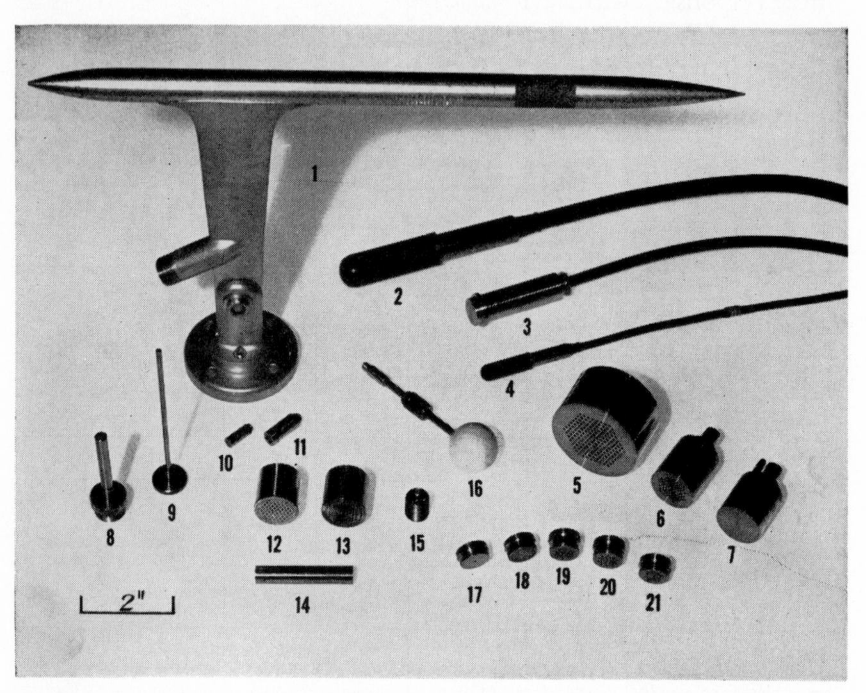

FIG. 6.10 Microphones for the measurement of high-intensity sound: (1) Chesapeake AB-330S; (2) Atlantic BC-30; (3) Massa M-141; (4) Atlantic BC-10; (5) Photocon 354; (6) Photocon 364; (7) Photocon 320; (8) Altec 6117 probe; (9) Altec 159-B; (10) Massa M213; (11) Massa M215; (12) Western Electric 640AA; (13) Tokyo Riko MR-103; (14) Atlantic BD-10; (15) Atlantic BC-60; (16) Gulton P600; (17) Altec 21BR180, pin hole; (18) Altec 21BR180-5, flush mount; (19) Altec 21BR180; (20) Altec 21BR200; (21) Altec 21BR150.

bration levels on missile and aircraft structures, induced by air-borne noise, are frequently in excess of 1 *g* throughout much of the audio-frequency range. Acceleration levels of this order present at the microphone may easily lead to erroneous sound-pressure measurements, as is evidenced from the data in Table 6.6. The vibration sensitivity of microphones is usually reasonably independent of frequency. The vibration data in Table 6.6 are for an acceleration response perpendicular to the microphone sensing element. Microphones are generally more sensitive to vibration perpendicular to the sensing element than to vibration at other angles.

The maximum linear sound-pressure-response data shown in Table 6.6 represent typical values for the microphones considered. If these microphones are to be used in high sound fields approaching sound-pressure levels within 20 db of those given in the table for linear response, the linearity of the particular microphone should be measured. To a first approximation, the microphone linearity is independent of frequency.

Two techniques are commonly used for determining linearity of response. A standing-wave tube, using a loudspeaker or horn driver, can develop sound-pressure levels of 180 db *re* 0.0002 μbar. A piston phone, though less convenient than the standing-wave tube, can develop almost any pressure desired.[11]

Few microphones remain stable in sensitivity over a wide temperature range. It is recommended that the precise calibration for a particular microphone be obtained at the operating temperatures encountered. As can be seen in Table 6.6, there are very few microphones that operate in an environment above 250°F.

High humidity and moisture conditions are detrimental to the operation of high-impedance transducers, such as condenser microphones. The electrical leakage caused by excessive moisture may temporarily deteriorate the low-frequency response of the microphone and at times cause the microphone to be completely inoperative.

6.10 Calibration

There are various techniques in common practice for calibrating a complete measuring system. Ideally, one should apply a known sound pressure or acceleration, whichever applies, to the transducer and note the response of the indicating instrument at the frequencies and amplitudes of interest. To a limited extent, this is feasible.

The complete frequency-response calibration of a microphone in a free sound field is straightforward but time-consuming. The same applies to the calibration of accelerometers and other vibration pickups over their intended operating frequency range. Experience has shown that it is generally sufficient to perform a complete frequency response calibration on the transducers approximately twice a year, under normal conditions of use. For those applications where the transducers are exposed to adverse environments, the performance of the transducers in those environments should be carefully determined. The characteristics of the complete electrical system following the transducer can be conveniently evaluated using insert-voltage calibration techniques.[12]

After the acoustical engineer has become completely familiar with the instrumentation system and understands all of its shortcomings, less elaborate calibration techniques will suffice when the instruments have been assembled in the field for obtaining data. Calibration at one frequency and at one signal level of the complete system, including the transducer, is usually an adequate check on the system during the course of measurements. There are small reliable portable acoustic and vibration calibrators available for this.

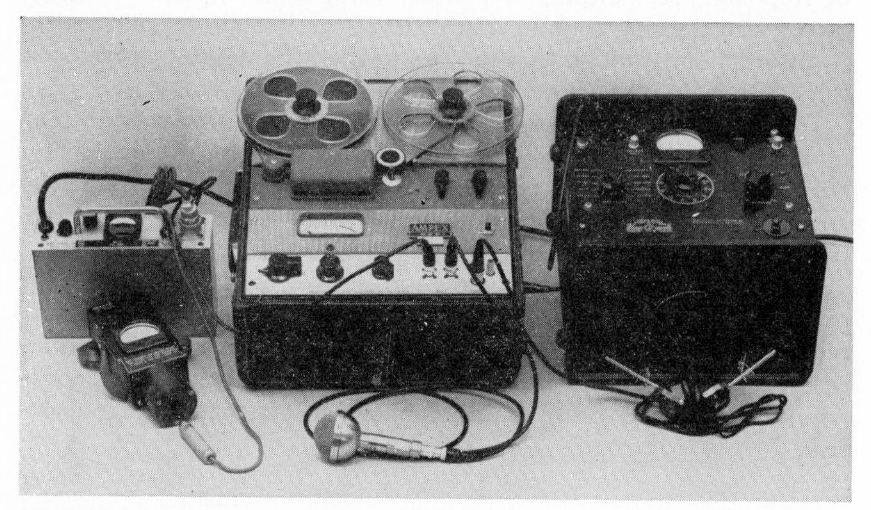

FIG. 6.11 A typical measuring system is shown above. It consists of an acoustic calibrator, Altec condenser microphone system, an Ampex 600 magnetic tape recorder, and a General Radio octave-band analyzer.

To illustrate the calibration process in the field, the typical sound-measuring system of Fig. 6.11 is used. This photograph shows a condenser microphone (the left one of the two microphones shown) with accompanying preamplifier and power supply connected to a portable magnetic tape recorder. The output of the tape recorder is fed to an octave-band analyzer. A small transistorized oscillator and associated acoustic calibrator, located in the lower left-hand corner of the picture, is placed over the microphone for calibrating the complete measuring system. This particular acoustic calibrator produces a 400-cps tone with an amplitude of 130 db sound-pressure level *re* 0.0002 μbar at the microphone. A sample of this calibration tone is recorded on the magnetic tape, and the setting of the precision decade attenuator on the recorder is written down or spoken on a second channel through the crystal (larger) microphone.

The sound-measuring system is now calibrated and ready to use for recording the sound spectrum of interest. It is always necessary to write down or speak into the tape recorder, the position of the precision decade attenuator of the recorder amplifier. At the completion of the measurement, a sample of the acoustic calibration tone is again applied to verify the stability of the measuring system.

If it is convenient to turn off the sound source that is being measured and tape-recorded, this should be done, and a sample of the background noise should be recorded. If the source cannot be turned off, the microphone should either be capped or placed in a small airtight enclosure, and a sample of the resulting signal recorded on the magnetic tape. The residual noise should be lower by 10 db or more at all frequencies than the noise from the source under consideration.

Before the data are analyzed with an octave-band analyzer, the recorder signal should be reproduced and listened to on headphones to determine that no spurious noises exist. The recorded acoustic calibration tone is used to adjust the calibration of the octave-band analyzer. The spectrum of the noise source is then analyzed in octave bands. The recorded section of background noise following this data is also analyzed and compared with the analyzed data from the sound source. This experiment should always be performed before the instruments are disassembled, to be certain that the acoustic background noise and electrical noise within the sound-measuring system are sufficiently below the levels of the sound source being studied in all frequency bands.

6.11 Summary

In this chapter we have outlined the performance characteristics of typical instrumentation for acoustical measurements. The information presented here provides a guide for selecting appropriate instrumentation for a particular measurement situation.

The performance characteristics given in this chapter can be achieved under *optimum* conditions. It is important to evaluate carefully the performance of a specific sound and vibration measurement system and not merely to assume the typical performance information discussed in this chapter. The experimenter should always have a thorough understanding of the shortcomings and advantages of his instrumentation before making acoustical measurements. It is common practice to spend three-quarters of the measurement program effort evaluating the instrumentation and only one-quarter of the effort obtaining the important acoustical data.

REFERENCES

1. Olson, H. F.: "Acoustical Engineering," D. Van Nostrand Company, Inc., Princeton, N.J., 1957.

2. Beranek, L. L.: "Acoustic Measurements," John Wiley & Sons, Inc., New York, 1949.

3. Beranek, L. L.: "Acoustics," McGraw-Hill Book Company, Inc., New York, 1954.

4. Hueter, T. F., and R. H. Bolt: "Sonics," John Wiley & Sons, Inc., New York, 1955.

5. Harris, C. M. (ed.): "Handbook of Noise Control," McGraw-Hill Book Company, Inc., New York, 1957.

6. See Ref. 4, pp. 152–162.

7. See Ref. 5, pp. **15**-1–**15**-28.

8. See Ref. 5, p. **16**-26.

9. Kamperman, G. W.: A Portable Magnetic Tape Recorder for Acoustical Measurements, *Noise Control,* vol. 4, pp. 23–27, January, 1958.

10. Kamperman, G. W.: Measurement of High Intensity Noise, *Noise Control,* vol. 4, pp. 22–27, 56, September, 1958.

11. See Ref. 2, p. 172.

12. See Ref. 2, pp. 601–604.

13. A Simple, Objective Test for Cable Noise Due to Shock, Vibration or Transient Pressures, *Natl. Bur. Standards Rept.* 4094, May, 1955.

Chapter 7

SOME PRACTICAL ACOUSTICAL MEASUREMENTS

Norman Doelling, David L. Klepper, and Leo L. Beranek

7.1 Introduction

In the three chapters preceding this one, you have read about various types of instrumentation for the measurement of sound. In this chapter we shall discuss some practical acoustical measurements using this equipment that are part of the every-day business of noise control. First, we shall discuss simple things that you must remember in using sound-measuring equipment if you are to obtain answers that have meaning.

Every piece of equipment has its limitations and so does every person using it! Because the actual measurements involve the reading of a meter, there is a temptation to do merely this. Readings taken with partially dead batteries or with the instrument standing on a steam radiator may fulfill an assignment, but they do not accomplish a mission.

Intelligent use of sound-measuring equipment calls for a knowledge of how it operates, of its inherent limitations and weaknesses, of the effect of various external influences on its accuracy, and of when it is functioning properly. There is a great risk of obtaining data that are unreliable because a written or mental check list covering the above-named items has not been gone over before writing down meter readings.

7.2 Check Lists

Do You Understand Your Equipment? The experienced operator reads the instruction book. The tyro usually turns to the instruction book only if smoke is seen emerging from the instrument. The experienced operator learns the answers to questions such as the following:

1. What is the principle of operation of the instrument? One should be able to draw a block diagram of the circuit and associate each control knob and jack on the panel of the instrument with the appropriate block in the diagram.

2. Is there danger of damage to the instrument, say from overload, too much heat, twisting something too far or hard, etc.? What are the lowest and highest temperatures and humidities at which the instrument can be operated safely? Is the instrument said by the manufacturer to be delicate or microphonic or limited in application?

3. What are the lowest and highest sound-pressure levels and frequencies that can be read correctly? (See Chap. 6.)

4. If an extension cable is used with the microphone, how are the readings changed? Are the changes dependent on frequency, humidity, temperature, etc.? (See Chap. 6.)

5. What is the inherent accuracy? Is the accuracy a function of frequency? If the meter has a range-multiplying switch, does a reading taken with the upper end of a lower-range scale agree with a reading taken with the lower end of a higher-range scale? If not, which end of the scale is more accurate? Will plugging in external analyzers or recorders affect the accuracy or stability of the instrument? Does this instrument meet the quality standards expected in the industry? (See Chaps. 5 and 6.)

6. How is the accuracy of the calibration maintained and checked? How may its readings be intercompared with those of other instruments? Calibration procedures should be followed meticulously (see Chaps. 5 and 6).

7. Does the meter read the rms, average, or peak levels of a complex sound wave? One must understand how readings on peak meters differ from those on rms meters (see Sec. 6.4).

8. If weighting networks are included, when are they used? How is the ambiguity in readings resulting from switching from one weighting network to another handled when the readings have values that fall between those specified for either network? (See Chap. 5 and Sec. 7.4.)

9. When one connects an analyzer or recorder to the output of a sound-level meter (SLM), for example, does one use a high gain in the SLM and a low gain in the analyzer, or vice versa? What is the best way to set the gains to get maximum signal-to-noise ratio with assurance that the peak levels of the noise are not being clipped?

10. What are the characteristics of the filters in the analyzer? If

the noise slopes at ±6 to ±12 db/octave, are the skirts of the filter response sufficiently steep and extended to eliminate the higher signal levels outside the filter band? (See Secs. 4.3 and 6.5.)

11. If the lowest and highest filter bands are just low-pass and high-pass filters, does the instrument have such extended response with frequency that it will read infrasonic and ultrasonic signals that have no significance to the problem at hand?

12. Does the sum of the squared-pressure readings of the individual bands equal the over-all reading? If this result is not obtained, is it due to faulty design or malfunction of the instrument? (See Sec. 3.3 and farther in this section, i.e., Sec. 7.2.)

13. When do you use the "fast" meter and when do you use the "slow" meter? (See Chap. 5 and farther in this section.)

14. Is the reading of the instrument different when it is hand-held, placed on a stand, or when the microphone is separated and operated with an extension cable on a tripod? Which reading is more nearly correct? (See Chap. 5.)

15. Does the reading depend on the direction from which the sound arrives at the microphone? (See Chap. 6.)

16. Will the use of earphones while taking data help to reduce errors from noise, microphonic hum, overload, etc.? (See Chap. 5 and further in this section.)

Is the Equipment Functioning Properly? Although a highly experienced operator need not employ a written check list as an aid to obtaining correct data, he automatically goes through some routine on arrival at the scene of a measurement such as this:

1. He looks over the equipment to make sure that it shows no sign of damage or evidence of having been dropped.

2. He has worked long enough with the equipment before going out on the job to know whether, when the switch is first tuned on, there should be any visible "flick" of the meter or audible sound during the warm-up period. Any strange behavior must be investigated immediately.

3. He checks the batteries (if any) or the line voltage if he has reason to be uncertain of it. He utters some series of sounds into the microphone that he previously had uttered in the laboratory to make certain that the meter is reading in the right range.

4. He disconnects the microphone and turns up the gain to reassure himself that the electrical background noise is at its normal level. With the microphone disconnected, he slaps the side of the instrument with his hands and observes the meter to make certain that nothing is loose inside and that the instrument has not become microphonic.

5. He inspects the cables (if any) for loose connections, fraying, and moisture or mud in the connectors.

How Do External Influences Affect the Readings? Let us assume now that the instrument itself has been completely understood and that it is functioning properly. Now turn to the problem of obtaining meaningful data. Good data are, in part, the result of taking into account the effects of external influences on the readings. Although more difficult to do, let us form a partial check list of factors to consider. More complete data may be found in Refs. 1 and 2.

1. *Temperature.* What corrections for temperature must be applied to the data? Some types of microphones give readings that vary only slightly for wide ranges in temperature. Other microphones give readings that vary greatly with temperature, particularly when used with extension cables. The outstanding example of a temperature-sensitive transducer is the Rochelle-salt-crystal microphone.

The corrections to the responses of two common types of Rochelle-salt-crystal microphones to be applied as a function of temperature for various cable capacitances are given in Fig. 6.2. The stabilized *internal* temperature of the microphone must be known—not the temperature of the cable. We see that for a long cable the microphones almost become temperature-sensing devices because the variation in response immediately above and below 76°F is more than 5 db per 10°F.

2. *Humidity.* Humidity may affect the behavior of a microphone. The condenser microphone becomes very noisy if the insulators inside collect any moisture. Whenever not in use, condenser microphones should be stored in desiccators. When in use, they should be operated at a temperature (say from the heat of the preamplifier) higher than ambient temperature so as to drive off moisture.

The chemical Rochelle salt gradually dissolves if the humidity is too high (above about 84 per cent). In a modern, well-built, Rochelle-salt microphone, the crystal unit is protected by a coating that is relatively unaffected by high humidity, at least for periods of exposure under a week or so. A Rochelle-salt microphone should not be stored for long periods in a very dry atmosphere, since the crystal can dry out.

3. *Directionality.* All microphones exhibit some directionality at high frequencies. The larger the microphone, the greater the change in response with angle of incidence. The microphone should be oriented to minimize directivity effects during measurement.[1] Because the microphone is usually calibrated for the case of a sound wave traveling in a direction parallel to the surface of the diaphragm, the microphone should be positioned that way in taking data. Also,

the response to randomly incident sound (reverberant sound field) is usually similar to that for parallel incident sound.

4. *Reflections.* The room and nearby objects affect the readings. Sometimes one wants the readings to include the effects of the room or nearby reflecting surfaces, and other times one does not. The effects of surroundings are reduced when the microphone is placed very near to the source of sound, because the direct sound dominates.

If standing waves in a measurement room are expected to influence a result in an unwanted way, the surfaces of the room should be sufficiently well treated with absorbing material so that no appreciable standing wave exists.

Objects in a room reflect the sound waves just as do the walls of the room. Consequently, all unnecessary hard-surface objects should be removed from the measurement room when a source is being measured. In general, *no* objects, including the observer, should be close to the microphone. If it is impractical to follow this principle, the objects should be covered with an appropriate sound-absorbing material, although at middle and lower frequencies this may not be adequate.

5. *Background Noise.* The environmental factor most likely to affect the readings of a sound-measuring instrument is background noise. To determine whether or not background noise is a factor, the sound source being measured should be turned off. Then measure the levels of the remaining noise, using the same filter bands as are being used for the sound-source measurement. Table 7.1 contains information for making an approximate correction for the background noise. The larger the correction, the less accurate this calculation becomes, hence the table is stopped at a correction of 3 db. If it is at all possible to increase the difference between the background-noise level and the sound level under consideration by

Table 7.1

Correction for Background Noise to Be Applied Separately for Each Band

Total Sound Level Less Background-noise Level, db	*Subtract from the Total Sound Level to Get the Sound Level Due to the Source, db*
8–10	0.5
6–8	1.0
4.5–6	1.5
4–4.5	2.0
3.5	2.5
3	3

moving closer to the source, this might be preferable to making the approximate background correction.

6. *Microphonics.* If the equipment is set up in a measurement location in which the levels to be measured are over 90 db or in which there is noticeable vibration, an immediate test should be performed to see if microphonics will influence the measurements. This test should be run with the microphone alternately connected and disconnected from the cable and with the noise source operating at or very near the maximum intensity expected. If readings in any of the bands with the microphone disconnected are less than 10 db below those with the microphone connected, steps should be taken to reduce the microphonic response.

The microphonic response is caused either by vibration transmitted by the floor or supporting structure to the tubes or by sound transmitted through the case to the tubes. If the response is caused by vibration, lifting the equipment will cause a difference in the meter reading. If there is no difference, the trouble is probably airborne microphonics. If the problem is one of vibration, placing the equipment on rubber pads or on a pneumatic-tired truck or both may eliminate the microphonic response.

7. *Internal Electrical Noise.* On some of the older models of sound-level meters, the internal tube noise is high enough that if the noise spectrum slopes downward steeply, say greater than 6 db/octave, the sound being measured will merge with the internal tube noise. If the spectrum slopes down as much as 12 db/octave, this merging may occur in as many as four of the octave bands. This effect, of course, is only observable with an analyzer because the overall level will be set by the high intensities in the lower bands where the signal-to-noise ratio is adequate. One antidote for this trouble is to use a 500-cps high-pass filter in the microphone cable when measuring noise above 600 cps. Another is to use the *A* network on the sound-level meter when analyzing above 600 cps, because the responses of the *A*, *B*, and *C* networks are about the same above 600 cps and the *A* network reduces the signal level below 600 cps. One must be careful to utilize the *A* network for this purpose *only* when analyzing *above* 600 cps.

8. *Wind Noise.* Winds of greater than a few miles per hour may introduce errors in noise measurement. The air passing by the microphone will generate vortices or whirlpools of air because of the microphone's irregular shape. These vortices cause fluctuations in atmospheric pressure which actuate the microphone. Thus the wind generates noise at the microphone which adds to the noise arriving from the source being measured. Bands other than the lowest band

usually contribute very little to the total intensity of wind noise. One can usually detect the presence of wind noise by listening to the output of the sound-level meter with a pair of earphones. One can familiarize himself with the quality of wind noise by blowing gently over the microphone while listening with earphones. Means for reducing wind noise during measurement are discussed later in this chapter.

9. *Instrument Stray Pickup.* Other environmental influences such as electrostatic fields, magnetic fields, and excessive vibration can usually be detected by comparing the noise aurally with that heard in the earphones. Electrostatic and magnetic pickup will appear as components not present in the original noise. In addition, the amount of pickup can be varied by changing the orientation of the microphone or equipment. Dynamic microphones will be influenced by strong magnetic fields.

10. *Vibration.* Excessive vibration may cause unwanted signals to be developed in the microphone, cable, sound-level meter, or octave-band analyzer. Noise produced by vibration of the microphone can be eliminated by a resilient mounting. The microphone should either be held in the hand as described previously or simply suspended by means of its cable. If the latter scheme is chosen, precautions should be taken to prevent the microphone being blown about by a high wind.

11. *Microphone Stray Pickup.* One can be certain that stray pickup and vibration noise originate in the microphone if the symptoms disappear when the microphone is removed from the cable. During this test the location of the cable and equipment should be left unchanged. The open end of the cable should be shielded with a piece of metal foil if electrostatic pickup is suspected.

12. *Meter Dynamics.* The characteristics of the meter mechanism of a sound-level meter or octave-band analyzer require (ASA Z24.3–1944) that a sound be established at a steady level for at least ¼ sec in order that the meter respond correctly on the fast setting. The slow scale requires that the sound be established for a longer time. The exact length of time depends on the individual meter circuit since nothing is specified in the standards about the slow setting. Because of the wide variations in the slow-setting-meter response, the fast setting should be used unless it has been established that the two settings read the same. Proper measurement of sounds which have a shorter duration necessitates the use of more complicated equipment such as a cathode-ray oscilloscope. The measurement techniques involved require considerable discussion which will not be attempted here.

It is possible, however, to obtain a rough idea as to whether the sound is of sufficient duration to be measured by the sound-level meter. If a large difference in the over-all reading between the slow and fast setting (greater than 5 db) is observed, the sound is probably not of sufficiently long duration. When the total intensity of the individual octave-band readings is greater than the over-all reading, the sound is surely too short in duration, if other causes have been ruled out. If a sound of short duration is repeated with great rapidity, such as the noise from some riveting guns, the amplifier may be overloaded during part of the cycle. This may be true even though the slow and fast settings of the meter are the same and are clearly not off scale. This difficulty may cause the total intensity level of all the octave bands to be greater than the over-all level.

13. *Over-all Levels Calculated from Band Levels.* It is also good practice to make a rough addition of the intensities in the individual octave bands to be sure that the over-all level agrees with the band readings. With a little practice this can be done mentally after each run. This frequently serves to detect a band level misread by 10 db and sometimes faulty positioning of band-selection switches. The procedure for adding decibels is given in Sec. 3.3.

If the total intensity level does not check the over-all reading to within ±2 db, the octave-band readings should be rechecked. If no error is found in reading the bands and if the sound has a very short duty cycle, the discrepancy may be a result of the slow mechanical response of the meter or overloading the amplifier.

If neither one of the two above explanations applies and if the over-all level is fairly constant throughout the measurement of all the octave bands, a thorough check of the octave-band analyzer should be undertaken by an electronic engineer or the manufacturer.

7.3 Record of Measurements[1]

One important part of any measurement problem is obtaining sufficient data. The use of data sheets designed specifically for a noise problem helps to make sure that the desired data will be taken and recorded. Sample data sheets are shown in Tables 7.2 and 7.3. The following list of important items may be found helpful in preparing data sheets of this type:

1. Description of the space in which measurements were made
   a. Nature and dimensions of floor, wall, and ceiling
   b. Description and location of nearby objects and personnel

2. Description of device under test (primary noise source)
   a. Dimensions, name-plate data, and other pertinent facts, including speed and power rating
   b. Kinds of operations and operating conditions
   c. Location of device and type of mounting
3. Description of secondary noise sources
   a. Location and types
   b. Kinds of operations
4. Types and serial numbers on all microphones, sound-level meters, and analyzers used—lengths and types of microphone cables

Table 7.2

A Two-page Sound-survey Data Sheet*

Page 1 *Engineer* Date ________________
Organization ________________________________
Address ________________________________
Instruments used:
Sound-level meter—Type ________________ Model No. ________________
Microphone ____________ Preamp ____________ Cable (length) ____________
Analyzer—Type ________________ Model No. ________________
Others ________________________________
____
Industry ________________ Type of machine ________________
Machine model No. ________________ Number of machines ________________
Location of machine in room ________________________________
Environment (Type of building, walls, ceiling, etc.; other operations, any attempts at sound control, particularly note acoustical materials) ________________________________
____
Personnel exposed—Directly ________________ Indirectly ________________
Exposure time pattern ________________________________
Are ear plugs worn? ________________ Type ________________
Are there audiometric examinations? ________________________________
Preplacement ________________ Periodic ________________
Note information as to who makes these examinations, conditions under which they are made, time of day they are made, where records are kept.
____
____

Page 2 Date ________________

Sound-pressure-level Values, db *re* 0.0002 μbar

Note: Record *A*, *B*, and *C* networks on the sound-level meter

Location †	Temper-ature	Time	Flat OA	Frequency bands, cps								Flat OA
				20–75	75–150	150–300	300–600	600–1,200	1,200–2,400	2,400–4,800	4,800–10,000	

* After Loss Prevention Dept., Liberty Mutual Insurance Company, with modifications.

† *Note:* If noise and/or microphone are directional, record distance of the source, microphone position, incidence on microphone (normal, grazing, random).

5. Positions of observers
6. Positions of microphones
   a. Direction of arrival of sound with respect to microphone orientations
   b. Tests of standing-wave patterns and decay of sound level with distance
7. Temperature of microphones
8. Results of maintenance and calibration tests
9. Weighting networks and meter speeds used
10. Measured over-all and band levels at each microphone position —extent of meter fluctuation
11. Background over-all and band levels at each microphone position—device under test not operating

Table 7.3

A Noise-level Field Data Card *

Front side			
	Test no.	Location	Sketch
	Date	Operation	
	Meter		
	Analyzer		
	Microphone		
	Cable		
	Temperature		

Recorded by ______

	Test No.			Rec. by			Second sheet No.		
	Decibel range *re* 0.0002 μbar—*C* network—fast position								
Rear side Time	Over-all	to 75	75–150	150–300	300–600	600–1,200	1,200–2,400	2,400–4,800	4,800 up
1									
2									
3									
	Comments, duration, hours per week, number of workers, etc.								
1									
2									
3									

* Courtesy of Illinois Committee on Noise in Industry, sponsored by the Industrial Hygiene Unit, Factory Inspection Division, Illinois State Dept. of Labor, with modifications.

12. Cable and microphone corrections
13. Date and time
14. Name of observer

When the measurement is being made to determine the extent of noise exposure of personnel, the following items are also of interest:

1. Personnel exposed—directly and indirectly
2. Time pattern of the exposure
3. Attempts at noise control and personnel protection
4. Audiometric examinations
   a. Method of making examinations
   b. Keeping of records

7.4 Measurement of Steady Noise

Factors to Remember. The measurement of a steady noise involves the following few important factors:

1. Choose the proper equipment (see Chap. 4).

2. If a suitable analyzer is not available, take readings on all three scales of the sound-level meter so as to obtain some information on the distribution of the sound intensity with frequency.

3. Decide whether to use the slow or fast scale. If time permits, take readings with both. As a general rule, if when using the fast scale the fluctuations cover a 6-db or greater range, read 3 db down from the frequent peaks. If the fluctuations cover less than a 6-db range, read the average of the range. By this rule, usually the fast- and slow-scale readings on fluctuating noises will be alike.

4. Calibrate the instrument according to the manufacturer's instructions at frequent intervals during each half-day of measurements. Preferably, an acoustical calibrator should be used.

5. Be certain that your microphone is placed (successively) at the locations where you want the noise measured and that you write down or sketch the locations exactly. Use as many measurement positions as are necessary to establish the characteristics of the noise being measured.

6. Orient the microphone properly in relation to the source.

7. Make certain that microphonics or background noises are not influencing the readings.

8. Read the stabilized temperature of the microphone frequently if it is of the crystal type or if the temperatures are very low or high.

9. Make sketches, notes, and tabulations of the space where the measurements are being performed, including sound-reflecting and sound-absorbing surfaces.

Averaging Readings in Space.* Frequently it is necessary to determine the average sound intensity radiated by a source through a hemispherical, spherical, or plane surface. If the assumption is made that the sound-pressure level is equal to the intensity level (as defined in Chap. 3), then the average IL can be determined from the average SPL. The simplest way of doing this is to compute an average sound-pressure level $(\overline{SPL})$ in decibels, by taking the arithmetic average of the SPL's in measuring sections of equal areas. Of course, some error will be present if this is done, except for the case in which all SPL's are equal. The value of $\overline{SPL}$ obtained by arithmetic averaging of decibels is always too low. The maximum error is approximately —2.5 db for the case where there is an even distribution of the data over a 10-db spread for between two and ten measuring stations. Thus, the following rule of thumb may be adopted: If the spread of measured values of SPL at the measuring stations is around 10 db and if the SPL's in decibels are averaged, add 1 db to the average to get the $\overline{SPL}$. The resulting $\overline{SPL}$ usually will be off not more than ± 1 db. If the spread of measured values is 5 db or less, add no correction. Here also, the error will be less than 1 db.

Readings on All Three Networks. In general, when no analyzer is used, it is recommended that readings on noises be taken with *all three* weighting positions. This procedure avoids ambiguities found in switching from one scale to another as mentioned above, and, at the same time, the three readings provide some indication of the frequency distribution of the noise. If the level is essentially the same on all three networks, the sound probably predominates in frequencies above 600 cps. If the level is greater by several decibels on the *C* network than on the *A* and *B* networks, much of the noise is probably below 600 cps.

A more complete statement of this approximate analysis is given by the charts of Fig. 7.1, which can be used to give an approximation of the sound distribution in three frequency bands. These charts should be used only as a guide for determining in a preliminary way what the spectrum might be, and they should not be regarded as obviating a complete octave-band analysis. There are occasions, however, when it is impractical to make more than this preliminary analysis, and then the charts of Fig. 7.1 may help in making a more satisfactory decision about a noise problem than can be done with only one reading of a noise meter.

Certain noises in which the energy is localized at one end or the other of the lower and middle bands of this approximate analysis

* See also Chap. 8.

cannot be analyzed using this method. This type of spectrum will usually give sound-level readings that do not fit on the charts. Simi-

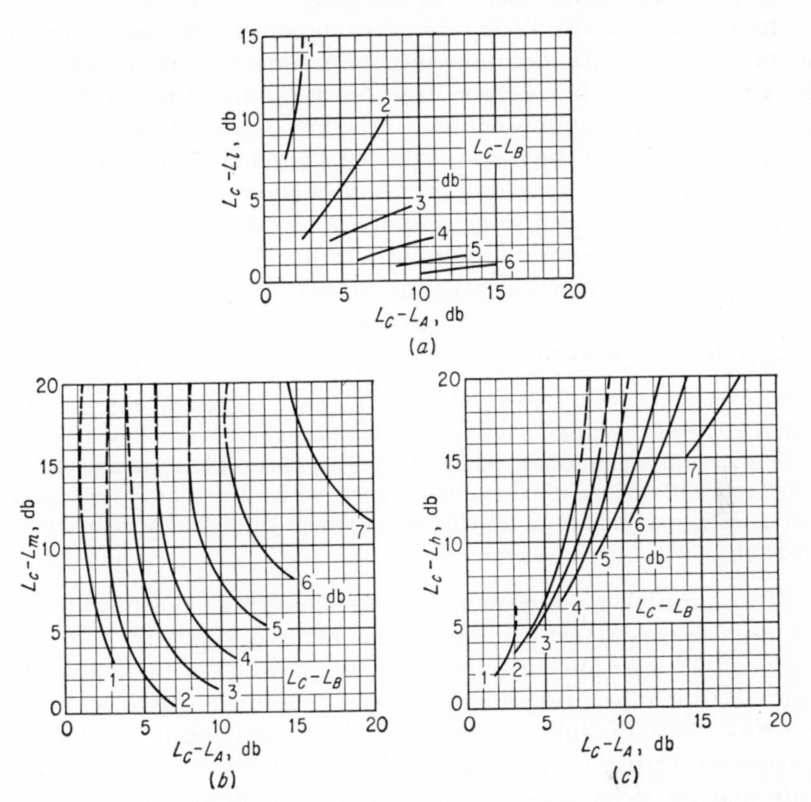

FIG. 7.1 Curves for calculating an approximate frequency analysis in three bands from level readings taken when using the three sound-level meter weighting networks *A, B,* and *C*: (*a*) low band (20 to 150 cps); (*b*) middle band (150 to 600 cps); (*c*) high band (600 to 8,000 cps). The measured value of $L_C - L_A$, as read from the SLM, is entered at the abscissa of each graph, proceeding vertically to the curve labeled with the measured value of $L_C - L_B$, then horizontally to the ordinate value for each of the three bands corresponding to the difference between the individual band levels and the over-all level. (*This method of analysis was developed by J. R. Cox, Central Institute for the Deaf, St. Louis.*)

larly, the dotted portions of the curves are regions of poor accuracy of analysis.

7.5 Measurement of Impulse (Impact) Noise

Physical Characteristics of Impulse Noises. Acoustic noises of the impulse type generally are associated with the impact of two solid bodies or with explosions. Impulse noises may range from the tick of a watch to the impact of large drop forges to explosive sounds such as arise from the firing of a rifle or cannon.

Impulse noises are distinguished from continuous noises in two ways. First, the duration of the impulse noise is very short compared with the duration between occurrences of the impulse. Second, the peak value of the impulse noise is very much greater than the average or rms value of the noise. Thus we say generally that a riveting gun is an impulsive noise source even though it is periodic.

The description of a single impulse noise is quite different from our description of continuous noises. In principle, an impulse noise is completely described by its time history, i.e., a record of the instantaneous amplitude of the signal as a function of time, or by an *energy*-density spectrum.

The energy-density spectrum and a quantity related to the temporal behavior of the instantaneous amplitude of a signal form a Fourier transform pair and are uniquely related to one another. These parameters are relatively difficult to obtain and to work with. In order to assess the effects of impact noises on hearing acuity, more simplified parameters are generally used.

If a simplifying assumption is made concerning the time history of impulse noises, two parameters may be used to provide a description of the physical characteristics of the noise. Let us assume that the rise time of the impulse noise is very rapid compared with typical time constants inherent in the hearing mechanism. Various experiments relative to the growth of loudness level, for example, indicate that the delay action in the middle ear muscles is at least 10 msec. If the rise to the peak is much less than 10 msec, the rise time can be considered as infinitely fast. This assumption seems reasonable for most impact noises.

With this assumption a reasonable and useful description of the total energy in impact noise may be obtained from (1) the peak value of the noise and (2) some measure of its duration such as decay time. The decay time can be defined in the same way as is done in electrical circuits, namely, the time required for the envelope of the (exponentially) decaying wave to drop 8.7 db in level from its initial value. The sum of the peak instantaneous sound pressure (expressed in decibels) and ten times the logarithm of the decay time is proportional to the total energy in the impact noise. Such a sum will rank-order impact noises on the basis of the total energy but does not give an absolute value of the total energy.

As an alternative, one may use an average value of sound pressure over some specified period of time to obtain another measure of the relative energy in impact noises.

Some Measurement Procedures. As we have indicated above, there are several parameters which can be used to describe the physical characteristics of impulse noises. The physiological aspects of the relations of impulse-noise characteristics to loss of hearing acuity are, as yet, not defined well enough to form a basis of judgment concerning which parameters to use. The selection of a set of parameters to describe an impulse noise will therefore depend upon the instrumentation available to the engineer.

The measurement of impact noises is greatly facilitated by the use of a magnetic tape recorder. Use of a tape recorder allows a single measurement in the field to be analyzed in essentially any manner which may be desired at a later time. As noted in Chap. 4, an octave-band analysis of an impact noise in the field will require at least eight operations of the impact source. In practice, one might require ten or twelve operations to obtain one reading in each octave band. Analysis in the field without a tape recorder will require many more operations of the source if the levels of the source vary significantly from operation to operation. It has been found that noise arising from the firing of rifles, machine guns, and cannons is very stable acoustically. However, stability may not be an attribute of other impact noises such as drop forges, punch presses, etc.

If a magnetic tape recorder is used, the over-all peak value should be monitored. First, one obtains an approximate estimation of the stability of the source. This observation is useful for estimating how many tape recordings will be required to obtain sufficiently stable average values of the characteristics of the noise. If, for example, the peak values for four consecutive operations are within 1 db of one another, there is little point in obtaining more than four recordings. If, on the other hand, the spread for four consecutive times is 6 db, then 10 or more operations of the impact source may be required to obtain a stable average value. The number of operations recorded will depend upon tolerable "error" in the final results.*

Second, by monitoring one can prevent overload of the tape recorder by the noise signal. The VU meter on the tape recorder does not respond rapidly enough to indicate a peak overload. The recorder can be instantaneously overloaded and the peak can be "clipped," even though the VU meter indicates a very low reading.

For most industrial noise problems measurements should be made near the position of a machine operator's head, without the operator

* See, for example, Moroney[3] for methods of estimating the required number of measurements for a given confidence level for the resulting average.

at his position. It may also be necessary to make measurements at other positions throughout the manufacturing space to determine the variation of noise intensity in space. Estimates of the variation in intensity should not be made on the basis of the usual room equations such as those given in Chap. 11. The direct noise intensity cannot be accurately estimated from those equations because the propagation of sound for high peak levels is not a simple linear phenomenon. One finds that outdoors the peak levels drop off at a rate of about 9 to 12 db per doubling of distance. The uncertainty in the propagation losses requires that the direct and reverberant fields be determined experimentally.

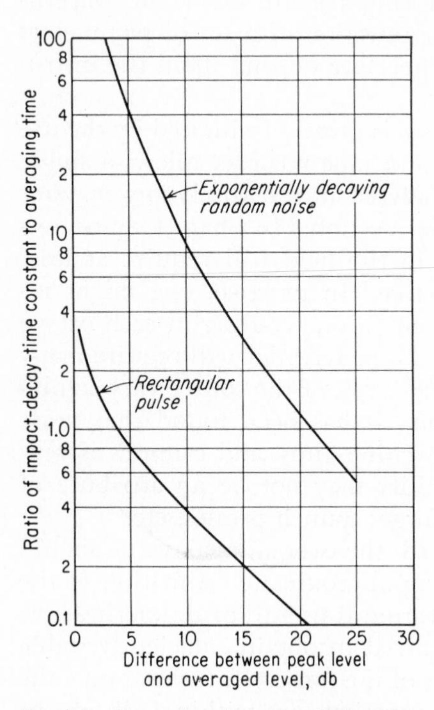

FIG. 7.2 Chart showing the relations between (1) the difference between the peak level (db) and the averaged level (db) and (2) the ratio of the decay time constant of the impact to the averaging time of the circuit. The lower curve is for a rectangular pulse, and the upper one is for usual impact noises.

Magnetic tape recorders may be susceptible to microphonics induced by the high peak levels associated with typical impact noises. Wherever possible the tape recorder should be operated at a point that is distant from the noise source. If this is not possible, the microphone should be removed from the recorder input, and the record attenuator should be set at the approximate level which will be used during the actual measurements. A recording should be made of the microphonic output of the recorder itself in the presence of the impact noise to determine if it is possible to obtain useful data during these conditions.

Analysis of Impact Noise. The time history of an impact noise can be obtained from high-speed oscillographs or by photographing the trace on an appropriately calibrated oscilloscope. Such techniques have been described in Chap. 4. The energy-density spectrum can be obtained by appropriate filtering and integration. These techniques are familiar to the electrical engineer and will not be discussed here. The analysis of an impact noise by use of an impact-noise meter, such as the General Radio type 1556A, is perhaps more interesting since the meter is relatively compact and can be used in the field or in conjunction with a tape recorder.

The impact meter is a flexible device, and a significant amount of information can be obtained by proper use of it. The impact meter can be used to obtain (1) the peak level of the sound pressure and (2) a time-averaged level of the sound pressure. By measuring both the peak level and the averaged level, an estimate can be obtained of the duration (decay time constant) of the impact noise.

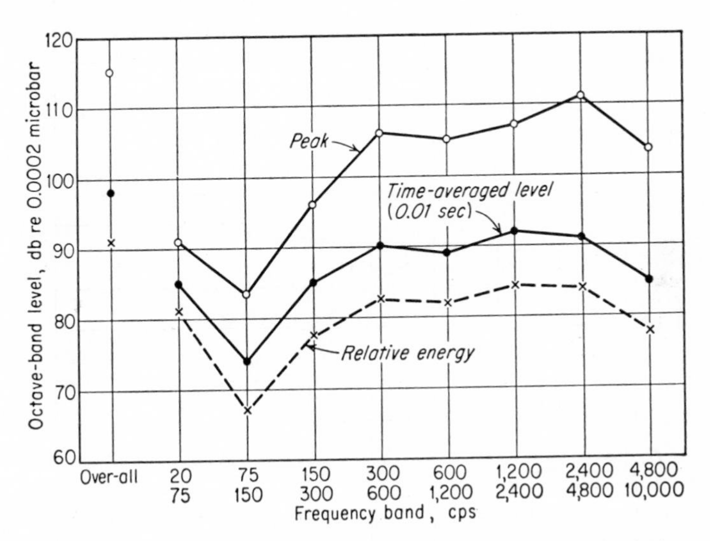

FIG. 7.3 The results of an octave-band analysis of the noise from a single impact of a punch press as measured by the impact-noise analyzer on the output of the octave-band analyzer.

Figure 7.2 shows the ratio of impact decay time to averaging time, as a function of the difference between the peak level and the averaged level, for exponentially decaying noise signals. A particular example[4] will show how this relation can be used.

Measurements were made of a small punch press stamping out blanks. The measurements indicated a peak level of 115 db and a time-averaged level of 98 db when the "time constant" switch was set to 0.01 sec. The difference in level is 17 db. From the upper curve of Fig. 7.2 it is found that this difference corresponds to a time ratio of 2. The equivalent impact decay time is then twice the averaging time or 0.02 sec. The noise from the punch press was also analyzed in octave bands of frequency. Both the peak value and the time-averaged value were obtained for each octave band and are given in Fig. 7.3. It is not clear that the measured peak level in a band has significance by itself, since this peak is not one that actually occurs in the physical sound wave. However, when the peak is taken in conjunction with the time-averaged level, one can get an estimate of how the sound energy is distributed in frequency.

The information in Figs. 7.2 and 7.3 can be used to determine the

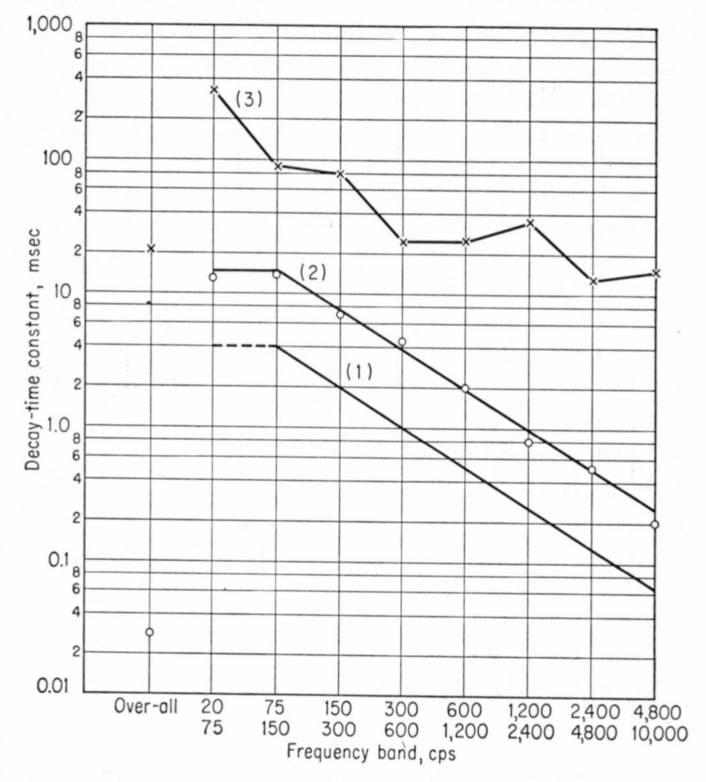

FIG. 7.4 Response time for (1) single-tuned circuit, (2) General Radio 1550-A octave-band analyzer, and (3) decay time of noise from small punch press measured on General Radio 1556-A impact-noise analyzer.

decay time of the noise in each octave band. For example, take the 150- to 300-cps band in Fig. 7.3. The difference between the "peak" and "0.01-sec time-averaged" levels is 11 db. From Fig. 7.2 the ratio of the impact-decay time constant to the averaging time is 8. So, the desired number is 0.08 sec (or 80 msec). The decay time in each octave band is given in Fig. 7.4.

It should be noted that the octave-band filters themselves impose a lower limit on the decay time. The decay time constant for the General Radio type 1550A octave-band noise analyzer is given by the middle curve in Fig. 7.4. If the measured decay times of any noise approach the decay times for the octave-band analyzer, one must conclude that the responses measured are indicative of the response of the analyzer and are not indicative of the decay time of the measured impact noise. For comparison, the decay time of a single-tuned circuit one octave wide is given by the lowest curve of Fig. 7.4.

7.6 Measurement of the Performance of Mufflers

Introduction. Measurements of the noise-control performance of mufflers under field conditions are frequently required to determine if a muffling device has met performance specifications. Such measurements are also made to determine the noise-reduction characteristics of a given muffler installation in order to facilitate extrapolation to the behavior of other mufflers of the same type. For example, one might measure the noise-reduction characteristics of some parallel baffles 10 ft long in order to estimate the performance of 15 ft of the same parallel baffles. As we shall see below, the noise reduction of 15 ft of baffles is not simply 1.5 times the noise reduction of 10 ft of baffles.

Windscreens. The selection of appropriate instrumentation is based on the principles outlined in previous chapters. There is, however, one special problem involved in the measurement of muffler noise reduction which has not been discussed specifically before now. This is the problem of the flow of air through the muffler. In order to avoid measuring noise induced by airflow past the microphone (wind noise) rather than the intended noise, one must use a windscreen over the microphone.

Subtle techniques are required in the field to determine if one measures wind noise or the acoustic signal propagated through a muffler. Wind-noise level depends upon the microphone, the dimensions of the windscreen covering it (if any), and the flow resistance of the windscreen covering.[5] Unfortunately, the magnitude of the wind-noise level depends also upon the structure of the turbulence in the moving-air stream. Therefore, if the magnitudes of wind-noise levels are experimentally determined in a straight air duct, for example, the same noise levels may not be found in the very turbulent regions near a change in cross section typically found near the inlet or outlet of a muffler. Because of the complicated dependence of wind noise on many quantities, "generalized" wind-noise levels (as a function of velocity) may be considered as only rough estimates of the noise levels induced by air flow.

The reduction of wind-noise level caused by adding a windscreen to a microphone has been experimentally found to be almost independent of turbulence level for a given average linear velocity. In addition, the wind-noise reduction does not vary strongly with wind velocity. This conclusion has been drawn from a limited number of experiments using one condenser microphone with a small windscreen. The noise reduction for other windscreens may not be independent of turbulence level.

If it can be experimentally determined for a given windscreen-microphone combination that the reduction in wind noise is independent of turbulence level, then one can estimate by a simple experiment whether the desired noise signal or the windscreen noise is being measured. One measures the noise level at some point in the system with and without the windscreen. If there is no change in noise level when the windscreen is added, the desired noise signal has been measured both before and after the windscreen is attached. If the change in noise level when the windscreen is attached is equal to the experimentally determined noise reduction for the windscreen, then one concludes that wind noise predominates in the measurement both with and without the windscreen attached. If the change in level when the windscreen is added is smaller than the noise reduction, then the measurement without the windscreen indicates wind-noise levels and the measurement with the windscreen indicates the levels of the desired signals.

Source and Filter Characteristics. As will be discussed and illustrated in Chaps. 16 and 17, acoustical characteristics of mufflers generally depend upon the type of noise source which they are associated with. Furthermore, the performance of some mufflers is dependent upon the direction and rate of air flow through the muffler. Whenever possible, one should make acoustical measurements utilizing the source with which the muffler is used. A loudspeaker, pistol, etc., should be used only if it can be demonstrated by independent measurements that the noise reduction obtained using such a substitute source is the same as that obtained using the actual source.

Over-all or octave-band measurements of noise reduction have limited meanings. The measured value of noise reduction will generally depend upon the slope of the input spectrum and the slope of the noise-reduction function. For example, if both the spectrum and the noise reduction slope downward with increasing frequency, the levels will be the greatest at the lower edge of the octave band. Hence, the values of noise reduction obtained will be those at the lowest frequency of the octave band. This dependence on slopes becomes less significant as the bandwidth is decreased. The dependence is usually not very significant for one-third-octave bands, but may be quite significant for octave bands.

Measurement Technique

1. *Measurement of Noise Reduction.* The *noise reduction* of a muffler is defined as the difference between the sound-pressure level measured near the inlet of the muffler and the sound-pressure level measured near the outlet of the muffler. In general, one will have to make measurements at several locations to determine the average value of SPL in the inlet plane. The variation of noise levels in engine test cells has been studied in some detail. It is found from these studies[6] that about five microphone positions at the inlet or outlet of a muffler are needed to obtain an average value which will be within 1 db of the "true" space average value 70 per cent of the time. The location of the microphones in the measurement plane is not critical, but it is generally desirable to use microphone positions which are randomly spaced from the sides rather than a symmetrical array of positions.

Studies of wind-noise levels indicate that the measurement plane on the upstream side of the muffler should be about 0.5 to 1 ft away from the muffler. The wind velocity and the turbulence level there are lower than at the plane of the inlet to the muffler. On the downstream side the measurement plane should be located at the discharge plane of the muffler. The velocity may be somewhat lower downstream of the discharge plane, but the increase in turbulence level will severely increase the wind-noise levels.

2. *Measurement of Insertion Loss.* The *insertion loss* of a muffler is defined as the difference between the sound-pressure level at one point in space before and after the muffler is attached to the noise source. Obviously, the muffler is inserted between the source and the position of measurement. The definition quite adequately describes the measurement technique. If the measurements are made in a duct, a space-averaging technique such as that described in the previous paragraph should be used. In many cases, the insertion loss will be nearly independent of the measurement position selected. Therefore, the measurement position should be chosen so as to minimize wind noise.

3. *Measurement of Attenuation. Attenuation* is defined as the decrease in sound power between two points in a system. Because there is no commercially available acoustic wattmeter, it is possible only to make measurements of sound pressure. Under certain circumstances, it is possible to calculate sound power from sound-pressure measurements. For example, if the direction of travel of a

sound wave is perpendicular to a measurement plane with an area of A ft^2 and if the wave through that plane travels in only one direction (no reflected wave), then the PWL in decibels (*re* 10^{-13} watt) equals the sound-pressure level SPL plus 10 times the logarithm to base 10 of the area in square feet. If the cross-sectional area in a system is constant, then the change in acoustic power in decibels between two locations is equal to the change in sound-pressure level in decibels.

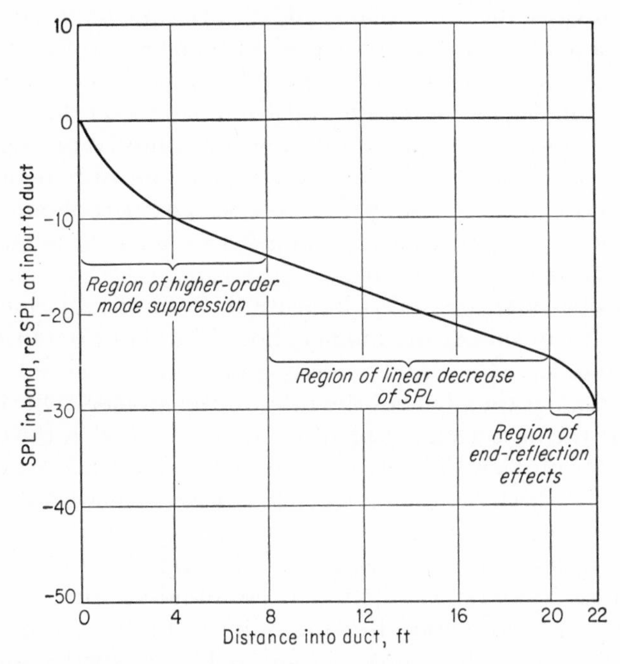

FIG. 7.5 Field measurements of sound-pressure levels in a lined duct.

At the inlet and at the outlet of a muffler there may be reflected sound waves, and the requirement that the sound waves travel in only one direction will be violated. Also, near the inlet there may exist higher-order waves, which are propagated at various oblique angles through the duct. Thus sound-power levels cannot be determined near the inlet from sound-pressure measurements. Higher-order waves usually die out at a distance equal to about 1 to 2 duct widths from the entrance to the muffler. Also, at a distance of 1 to 2 diameters from the outlet of a dissipative muffler, the reflected wave will become small compared to the forward-traveling wave. Therefore, if the muffler is long enough, there will be a region in which the reflected sound wave from the outlet is small and in which the oblique waves near the inlet have died out. In this region, sound-power levels can be determined from sound-pressure measurements. This region is found by making a traverse with a microphone along the axis of the muffler. In this region the *slope* of a curve of sound-pressure level vs. distance will be constant and equal to the attenuation per unit distance. The total attenuation for the muffler then equals this slope multiplied by the length of the muffler.

The results of a traverse through a duct are shown in Fig. 7.5. As indicated in the figure, in the first 6 to 8 ft there is a rapid drop of sound-pressure level with distance. In this range, the sound power in the higher-order waves is rapidly attenuated (see Chap. 17). In the last few feet another abrupt change of slope is seen which arises from the reflection of a sound wave (out of phase with the transmitted wave) back into the duct. In a region between about 8 and 20 ft into the duct, the sound-pressure level decreases linearly with distance. In this region the change in sound-pressure level is approximately equal to the change in sound-power level. The attenuation per foot for this muffler is seen to be about 1 db/ft or a total of 22 db for the entire muffler.

Obviously if the length of this muffler is increased 50 per cent to 33 ft, the noise reduction will increase only 11 db (11 ft $\times$ 1 db/ft). The magnitude of the "end effects" is not increased by the increase in length and thus the increase in noise reduction is less than 50 per cent.

7.7 Measurements of Reverberation Time

In Chaps. 10 and 11, discussing sound in small and large enclosures, we shall see that the sound levels measured in any enclosed space depend on the acoustical characteristics of that space, as well as on the sound-power level of the sound sources. As part of the problem of understanding a particular noise-control situation, the concept of reverberation time in the space being considered is often a necessary quantity. More will be said about this concept in Chaps. 10 and 11. For the present, we shall simply define it as follows:

> The reverberation time in an enclosure is the time required for the sound-energy density in that enclosure to decay 60 db, that is, to 10^{-6} of its original value.[7]

Reverberation time, like other measurable acoustical quantities, is a function of frequency. Therefore, we often speak of the reverberation-time characteristic, which is the reverberation time plotted as a function of frequency. It may be measured in octave bands, one-half-octave bands, one-third-octave bands, narrower bands or at any one particular frequency of interest.

To obtain the reverberation characteristics of a space, we need the following items of equipment:

1. A *sound source,* whose output can readily be cut off instantaneously. Examples are a cannon, or a pistol, or a noise generator (with amplifier and loudspeaker), or an audio-oscillator with amplifier and loudspeaker.

2. A *microphone* or microphone system, such as any of several discussed in Chap. 6.

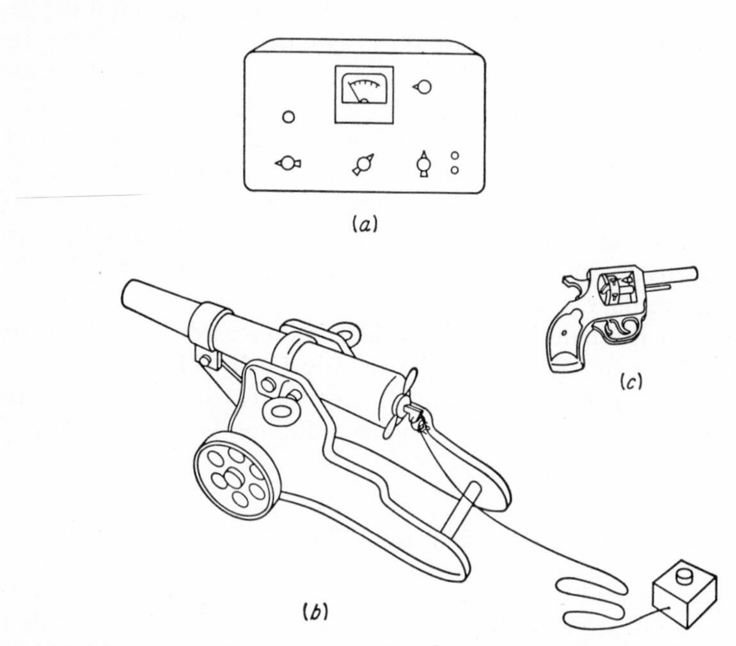

FIG. 7.6 Illustration of alternate sound sources used in reverberation measurements: (*a*) Random noise generator to be used with amplifier and loudspeaker; (*b*) yachting cannon shooting blank shells; (*c*) pistol shooting blank cartridges.

3. A *frequency analyzer*—either an octave-band analyzer or a one-third- or one-half-octave-band analyzer, or a narrow-band analyzer as discussed in Chaps. 4 and 6.

4. A *graphic level recorder* which plots the decay of the sound intensity (see Chap. 6).

5. Additional items that may prove useful are a *tape recorder* to permit recording the sound-intensity decay in the field, making it possible to use the frequency analyzer and graphic level recorder back at the laboratory. Also useful is a *reverberation-time protractor* often furnished by the manufacturer of the graphic level recorder to simplify reading the data.

Figure 7.6 illustrates several sound sources; Fig. 7.7 illustrates a reverberation-time protractor together with a typical decay curve. The remainder of the equipment has been illustrated and discussed in previous chapters.

Selection of Equipment. We may divide the choice of the sound sources into two classes: (1) impulse sources, such as a pistol or a small cannon, and (2) continuous sources, such as audio-oscillators

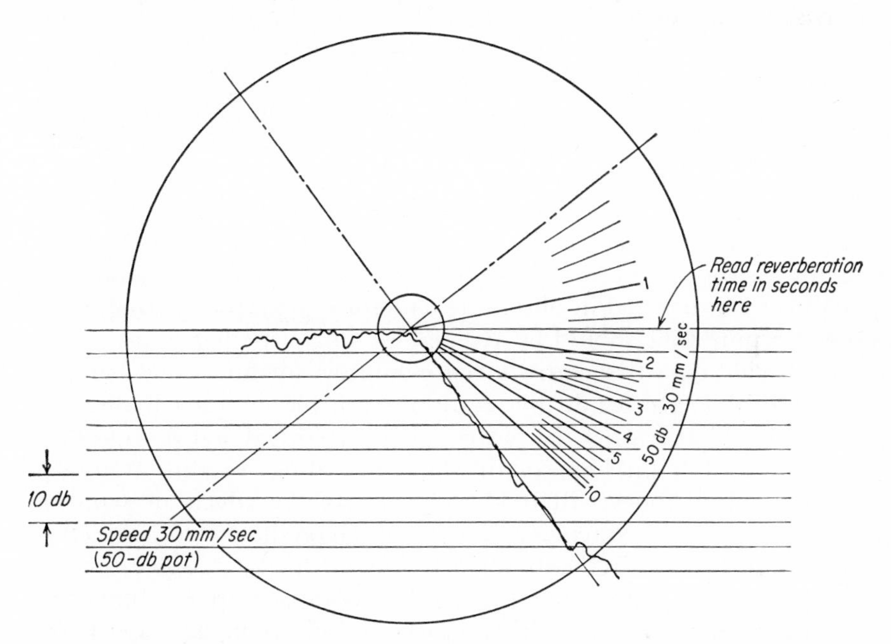

FIG. 7.7 Method for using a reverberation-time protractor. The appropriate sector line on the protractor is laid over the reverberation curve. The reverberation time is read from the scale to the right. The center of the protractor must lie over the intersection of the line through the decay curve and the horizontal line from which the reading is taken.

and noise generators. The results obtained from reverberation-time measurements using either type of source may differ in certain cases; however, this refinement is generally not of great importance in noise-reduction work. The choice may be based on other factors.

The pistol is light in weight. It produces a signal of sufficient intensity to obtain an adequate dynamic range under almost any circumstance. The greatest difficulty in its use is that its sound output falls off very sharply at low frequencies; therefore, it is rather useless for reverberation-time characteristics below 300 cps. A further disadvantage is that its output may annoy and frighten some people.

A small "yachting" cannon can furnish an output of sufficiently high intensity to provide a useful dynamic range under any circumstance. Also, it has sufficient power output throughout the entire frequency range of interest. Its disadvantages are these: its greater weight and complexity of firing with respect to the pistol, the louder output and greater tendency to frighten personnel, the tendency of the sudden noise to vibrate surfaces of the room and ventilation ducts and thereby possibly to create a dust problem.

Noise generators and audio-oscillators used with amplifiers and loudspeakers are bulky and heavy for field use. Their output is usually not enough to cause annoyance or fright. However, unless the amplifiers are sufficiently powerful and a loudspeaker with a large output below 250 cps is used, the useful dynamic range will be limited at low frequencies.

One important factor should be borne in mind when pure-tone oscillators are used as a noise source for reverberation-time measurements (or for almost all other acoustical measurements). Unless the engineer is interested in obtaining data for one particular frequency, use of a pure tone should be avoided. A single frequency may often not provide data that is truly representative for an octave band or a one-third-octave band. Better results can be obtained if the tone is "warbled" at least $\pm$ 10 per cent.[8] Prerecorded warble tones are available either on high-quality tape recordings or disk transcriptions. Warble tones will tend to eliminate the effects of standing waves at discrete frequencies. The use of an oscillator will, of course, eliminate the requirements for a frequency analyzer, since the frequency analysis is provided directly by the sound source. Dual-channel tape recorders, with high cross-talk rejection, are useful for simultaneously providing the warble tone and recording the transmitted sound.

Generally we are not interested in obtaining reverberation times at high frequencies for noise-control work; therefore, the frequency response of Rochelle-salt-crystal microphones is acceptable.

The Bruel and Kjaer model 2305 and the General Radio model 1521A level recorders are acceptable for use in reverberation-time measurements. Experience indicates that a writing speed of 25 to 30 mm per sec, with a logarithmic potentiometer having a range of 50 db, proves adequate for most reverberation-time measurements.

Data Evaluation. The first step in evaluating decay curves obtained from the graphic level recorder is the drawing of an idealized slope through the actual decay curve. The slope chosen should be along the middle of the decay curve; the effects of background noise at the bottom of the slope and transient effects or "rounding off" at the top of the slope should be discounted.

Given the length of the logarithmic vertical scale in millimeters per decibel and the writing speed in millimeters per second, the reverberation time may be calculated from the slope of the trace according to the formula:

$$T \text{ (reverberation time)} = \frac{60 \times \text{(vertical scale)}}{\text{(writing speed)} \times \text{(slope)}}$$

where "vertical scale" is in millimeters per db, and "writing speed" is in millimeters per second. "Slope" is dimensionless, and "60 times the scale" gives the millimeters of vertical scale for a 60-db decay.

If an appropriate protractor is used, the reverberation-time results can be read directly from the scale when it overlays the slope. The instruction booklet contained with the recorder describes the use of the protractor. A sound-decay curve is given in Fig. 7.8 showing how the slope is determined and illustrating T for a 60-sec decay.

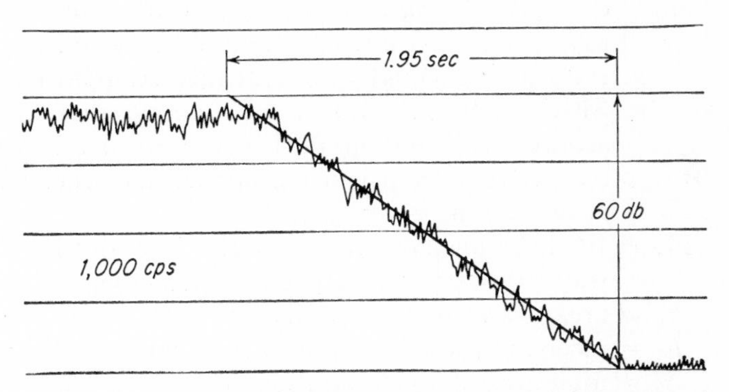

FIG. 7.8 Method for reading the decay time from a sound-decay curve by graphic measurement.

7.8 Measurement of Transmission Loss

In evaluating the solution to a noise problem, it is often necessary to measure the sound transmission loss provided by a wall or any other portion of a structure enclosing the sound source or separating it from a critical area.

Transmission loss (TL) in decibels is defined as 10 $\log_{10}$ of the ratio of the acoustic intensity (acoustic power per unit area) incident upon the wall to the acoustic intensity transmitted through it. This definition is discussed again in Chap. 13.

There are a number of methods for measuring the transmission loss of a structure in a laboratory. These techniques are based on the definition cited above. In the field, however, laboratory techniques are seldom possible. It is usually somewhat simpler to measure another quantity called noise reduction. The *noise reduction* (NR) is the difference between the sound-pressure levels in decibels on the two sides of the structure. It is essential to realize that noise reduction and transmission loss are different quantities. The TL applies to the panel and its mounting, whereas NR is a function of both the TL and the acoustical properties of the source and receiving room. For example, if the receiving room is highly reverberant, the TL is greater than the NR.

The mathematical relations for different receiving-room characteristics and different microphone techniques are*

Case A

$$NR_A = TL - 10 \log_{10}\left(\frac{1}{4} + \frac{S_w}{R_2}\right) \quad \text{db} \tag{7.1}$$

where NR_A = difference in sound-pressure levels, db, on two sides of wall, determined by measuring SPL on primary side with a microphone that moved around in reverberant sound field and then subtracting from it SPL measured with a microphone that is moved around in a region fairly near the surface on the secondary side (it is also assumed that receiving room is not highly reverberant).

TL = 10 times $\log_{10}$ of ratio of sound intensity incident on wall to sound intensity transmitted through wall

S_w = area of transmitting wall

R_2 = receiving-room constant = $S_2\bar{\alpha}/(1 - \bar{\alpha})$

S_2 = total area of receiving-room surfaces, same units as S_w

$\bar{\alpha}$ = average sound-absorption coefficient for surfaces of area S_2. [If there are people or if there is air absorption, these must be included in the calculation of $\bar{\alpha}$ (see Sec. 11.2).]

Case B

$$NR_B = TL - 10 \log \frac{S_w}{S_2\bar{\alpha}} \quad \text{db} \tag{7.2}$$

where NR_B = same as NR_A except that SPL on secondary side is measured at distance that is one-half to one wall width away from wall *and* the receiving room must be quite reverberant ($\bar{\alpha}$ less than about 0.2)

Case C

$$NR_C = TL + 6 \text{ db} \tag{7.3}$$

where NR_C = same as NR_A except that secondary side of panel opens outdoors or into a room that is very dead ($\bar{\alpha}$ greater than about 0.8)

* See Ref. 7, pp. 324–327.

Sources. The noise source used in the measurements may consist of the generator, and amplifier, and a loudspeaker. For greatest convenience in handling, the generator and amplifier should be in one unit. The noise source may be set up and left operating on one side of the wall structure while measurements are taken on the other side. The use of pure-tone sources is not recommended, for reasons given in Sec. 7.7.

Other Equipment. The measurement of the effectiveness of any sound-attenuating device requires that measurements be made on the source side and on the receiving side. If the source room is either small (approximately a normal office size, for example) or reasonably reverberant, the loudspeaker should be located in such a manner that the wall is in the reverberant field, rather than the near field of the sound source (see Chap. 11). This means that the loudspeaker should be located as far as possible from the wall under test. In such a case, a number of different microphone locations should be used, all in the reverberant field. The use of five or more microphone positions will tend to further eliminate effects of standing waves at particular parts of the room.

If the source space is very large or fairly dead, the microphone should be moved around in a plane close to the wall.

At the start of the tests, sound-pressure-level measurements in the receiving room should be made at a number of places. Evaluation of the data obtained will determine whether the receiving room can be considered a "dead" space or a reverberant space. If the sound-pressure levels measured are reasonably uniform throughout the space, it may be considered reverberant, and Eq. (7.2) may be used. If sound-pressure levels immediately adjacent to the transmitting wall are significantly higher than in the center or opposite side of the room, the receiving room may be considered sufficiently large or dead to require the use of Eq. (7.1). If the wall opens outdoors, use Eq. (7.3). For Eqs. (7.1) and (7.3) the locations should be at many positions in a plane close to the wall. For Eq. (7.2) the microphone should be moved around at locations fairly far from the wall.

7.9 Summary

The previous three chapters have described instrumentation for noise measurement. This chapter tells how to use some of this equipment in common applications. First, general tips for successful use of measuring equipment are given to the operator. Second, the measurement of steady and impact noise using the more common commercially available types of equipment is discussed. Third, methods for measuring the performance of mufflers is covered. Fourth, the field measurement of reverberation time is described. This chapter ends with a section on the measurement of transmission loss using simple equipment.

REFERENCES

1. Peterson, A. P. G., and L. L. Beranek: "Handbook of Noise Measurement," 3d ed., and E. E. Gross, Jr., "Measurement of Vibration," General Radio Company, Cambridge, Mass., 1956. (Both books are bound in one volume.)

2. Williams, C. R., and J. R. Cox, Jr.: Industrial Noise Measurement—Science or Art? *Proc. 3d Annual NNAS,* pp. 1–18, 1952. (Copies may be obtained from the authors by writing to Dr. Williams at Liberty Mutual Insurance Company, Boston, Mass.)

3. Moroney, M. J.: "Facts from Figures," Penguin Books, Inc., Baltimore, 1951.

4. *Gen. Radio Exptr.,* vol. 30, no. 9, General Radio Company, Cambridge, Mass.

5. Dyer, I.: Design of Cylindrical Microphone Windscreens, *J. Acoust. Soc. Am.,* vol. 27, p. 206(A), 1955.

6. Doelling, Norman: Noise Control for Aircraft Engine Test Cells and Ground Run-up Suppressors, vol. 3, An Engineering Analysis of Measurement Procedures and of Design Data, *WACD Tech. Rept.* 58–202(3), August, 1958.

7. Beranek, L. L.: "Acoustics," p. 306. McGraw-Hill Book Company, Inc., New York, 1954.

8. Beranek, L. L.: "Acoustic Measurements," pp. 799–806, John Wiley & Sons, Inc., New York, 1949.

Chapter 8

THE MEASUREMENT OF POWER LEVELS AND DIRECTIVITY PATTERNS OF NOISE SOURCES

Francis M. Wiener

8.1 General

In the preceding chapters, the reader has familiarized himself with the more common instrumentation necessary to perform acoustical measurements. It was shown that among the physical quantities by which sound waves are quantitatively described, sound-pressure level in decibels stands out as the quantity which is usually measured. To measure it, a calibrated standard sound-level meter and an octave-band or one-third-octave-band analyzer are usually used, because we are generally interested in how the sound-pressure level varies over the audible frequency range. For some noise-control problems it is sufficient to measure the sound-pressure levels in bands at a specified location. This location is given in advance, such as the position where a person's ear or a piece of sensitive equipment is most likely to be. While single-position data are adequate in some cases, an analytical approach to a noise-control problem usually calls for a more complete description of the noise source.

The acoustic power in the frequency bands of interest is part of the fundamental information describing the acoustic performance of a source. Another part of the fundamental data is the directive behavior of the source in those same frequency bands. Some sources of sound radiate more or less uniformly in all directions (see Fig. 8.1); the majority do not. A sound-level meter placed successively at several points in space a fixed distance away from a source will generally read different sound-pressure levels (see Fig. 8.2). This happens when the source is large compared with the wavelength of the sound radiated or when parts of it vibrate out of phase with each other and produce acoustic cancellation in some directions.

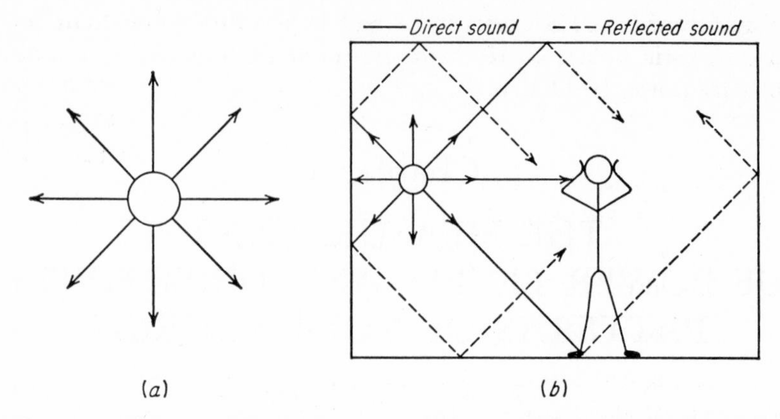

FIG. 8.1 A sound source which radiates uniformly in all directions: (*a*) Sound source in free space radiating uniformly in all directions; (*b*) the same source in an enclosure showing reflections from interior surfaces.

The concepts of acoustic power and directivity with which we shall deal in this chapter are extremely useful ones. For example, given the acoustic power and directivity of a source of sound, it is possible to make an engineering estimate of the sound-pressure levels generated by that source in almost any acoustic environment. Moreover, the acoustic power alone can frequently be used as a noise rating which is useful in the comparison of different noise sources. The rating of a sound source is not unique because both the radiated power and the directivity are influenced by the method of source mounting in general and by the proximity of reflecting surfaces in particular. The measurement of acoustic power should therefore be carried out with the source mounted in its typical way; any mounting surfaces, corners, etc., are to be considered a part of the noisy device to be evaluated.

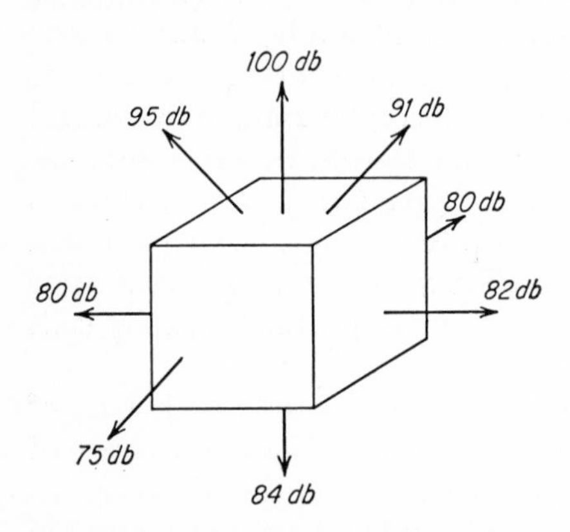

FIG. 8.2 A sound source with directional radiation into free space. This behavior is typical of equipment noise.

8.2 Sound Power and Directivity

Sound-power Level. The sound powers radiated in air by sources of practical interest range from less than a microwatt to megawatts.

A logarithmic scale suggests itself and it becomes convenient to express acoustic power in terms of power level referred to a suitable reference power (see Chap. 3).

$$\text{PWL} = 10 \log_{10} W - 10 \log_{10} W_0 \qquad \text{db} \tag{8.1}$$

where PWL = sound-power level, db, *re* W_0 watts
W = acoustic power, watts
W_0 = acoustic reference power, watts*

Consequently

$$\text{PWL} = 10 \log_{10} W + 130 \qquad \text{db } re \ 10^{-13} \text{ watt} \tag{8.2}$$

When stating a sound-power level, it is important to state the reference level also so that no confusion will result.

Directivity Pattern. Most sources of sound of practical interest are directive. If one measures the sound-pressure level in a given frequency band a fixed distance away from the source, different levels will generally be found for different directions. A plot of these levels in polar fashion along the direction for which they were obtained is called the *directivity pattern* of the source. This directivity pattern forms a three-dimensional surface, a hypothetical example of which is sketched in Fig. 8.3. The particular pattern shown exhibits rotational symmetry about the direction of maximum radiation, which is typical of many sources of noise of practical interest. It is also in the nature of things that many sources of noise are nondirectional, or nearly so, at low frequencies. As the frequency increases, directionality also increases. The directivity pattern is usually defined as evaluated in the far radiation field (see below) and in the absence of obstacles and reflecting surfaces other than those associated with the source itself.

Directivity Index and Directivity Factor. Numerical measures of the directionality of a source are the *directivity index* (DI) and *directivity factor* (Q):

$$\text{DI}_\theta = \text{SPL}_\theta - \overline{\text{SPL}_S} \qquad \text{db} \tag{8.3}$$

and

$$Q_\theta = \text{antilog}_{10} \frac{\text{DI}_\theta}{10} \qquad \text{dimensionless} \tag{8.4}$$

where the subscript θ is added to denote the fact that DI_θ, SPL_θ, and Q_θ are measured in the direction θ. We have also assumed that DI_θ

* The literature in general uses 10^{-13} watt for W_0 because the sound-pressure level and the sound-power level will be nearly alike if the sound power passes uniformly through 1 ft² of area. Some writers use 10^{-12} watt, for which pressure and power levels are nearly alike for an area of 1 m². Because the engineers in the United States use the English system of units in their daily work, $W_0 = 10^{-13}$ watt is selected for this text.

and Q_θ are independent of the distance r from the source. This is true in the far free field as will be discussed shortly. The source is assumed to radiate a power of W watts and for this power produces a sound-pressure level of SPL_θ in the direction θ at distance r from the source. The quantity $\overline{SPL}_S$ is the sound-pressure level that would be produced by a nondirectional source radiating the same power W at

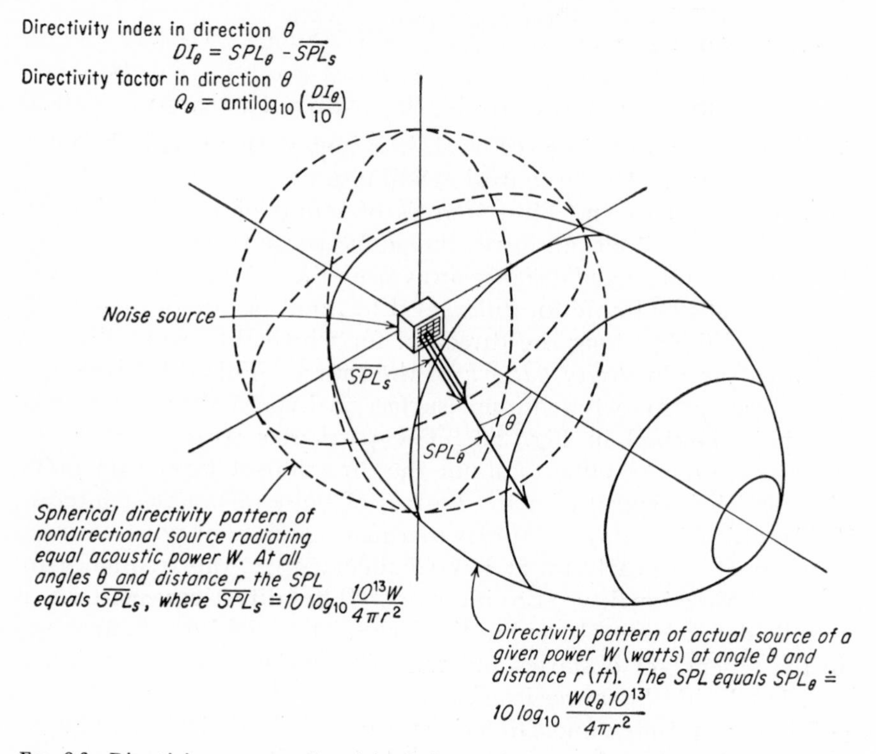

Directivity index in direction θ
DI_θ = SPL_θ - SPL_S
Directivity factor in direction θ
Q_θ = antilog10 (DI_θ/10)
Noise source
SPL_S
θ
SPL_θ
Spherical directivity pattern of nondirectional source radiating equal acoustic power W. At all angles θ and distance r the SPL equals SPL_S, where SPL_S ≐ 10 log10 10^13 W / 4πr^2
Directivity pattern of actual source of a given power W (watts) at angle θ and distance r (ft). The SPL equals SPL_θ ≐ 10 log10 WQ_θ 10^13 / 4πr^2

FIG. 8.3 Directivity pattern, directivity index, and directivity factor of a noise source radiating in free space.

the same distance r (see Fig. 8.3). In the case of nondirectional radiation, $Q_\theta = 1$ and $DI_\theta = 0$ at all angles θ. Because we assume that the acoustic intensity (in watts per unit area) is proportional to the sound pressure squared (see Chap. 2), we may write for a nondirectional source that

$$\overline{SPL}_S \doteq 10 \log_{10} \frac{W\,10^{13}}{4\pi r^2} \quad \text{db} \tag{8.5}$$

where W is in watts, and r is in feet. It is obvious from Eqs (8.3) and (8.4) that

$$\text{SPL}_\theta \doteq 10 \log_{10} \frac{WQ_\theta\, 10^{13}}{4\pi r^2} \quad \text{db} \tag{8.6}$$

At normal room temperatures and pressures these equations are accurate to within 0.5 db.

8.3 A Sound Source and Its Radiation Field

Near Field, Far Field, and Reverberant Field. The character of the radiation field of a typical noise source may vary with distance from the source. In the vicinity of the source, the particle velocity is not necessarily in the direction of travel of the wave, and an appreciable tangential velocity component may exist at any point. We speak of this as the *near field*. It is frequently characterized by appreciable variations of the sound pressure with position even when the source is in free space (anechoic chamber). Moreover, the acoustic intensity is not simply related to the sound pressure squared.

The extent of the near field of a noise source depends on the frequency, on a characteristic source dimension, and on the phases of the radiating parts of the surface. The characteristic dimension may vary with frequency and angular orientation. It is difficult therefore to establish general limits for the near field of an arbitrary source with accuracy. It is often necessary to explore the sound field experimentally.

In the *far field* the sound-pressure level decreases 6 db for each doubling of distance* if the sound source is in free space (anechoic chamber) or if the absorption in an enclosure is great enough that the *reverberant field* has not yet been reached (see Fig. 8.4). In this *free-field part of the far field,* the particle velocity is primarily in the direction of propagation of the sound wave and the acoustic intensity is proportional to the sound pressure squared.

If the source is radiating within an enclosure, one also finds considerable fluctuations of sound pressure with position in the reverberant part of the far field, i.e., in the region where the waves reflected from the boundaries of the enclosure are superimposed upon the incident field (see Fig. 8.4). Many reflected wave trains are generally present and the sound pressure, averaged over several feet, reaches a level which is essentially independent of distance from the source. The reverberant field is called a *diffuse field* if a great many reflected wave trains cross from all possible directions and the sound-energy density is very nearly uniform throughout the field.

* These distances are measured from the acoustic center which is located somewhere within the region occupied by the source.

Anechoic Chamber and Reverberation Chamber. Two special types of acoustic environment in which sound sources are studied are of particular interest. In an *anechoic chamber,* where all boundaries are highly absorbent, the free-field region extends very nearly to the

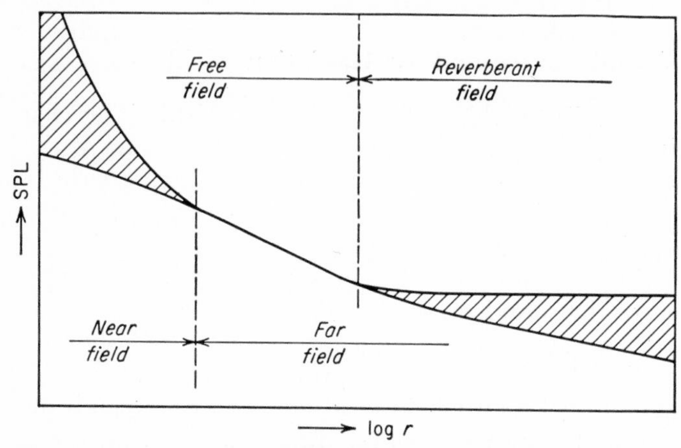

FIG. 8.4 The variation of sound-pressure level in an enclosure along a radius r from a typical noise source. Shaded areas indicate regions of level fluctuating with distance. In between, the *SPL* decreases at the rate of 6 db for each doubling of distance from the acoustic center of the source (see the text).

boundaries. In a *reverberation chamber* all boundaries are hard and the reverberant field extends over nearly the entire room volume with the exception of a small region around the source.

8.4 The Experimental Determination of the Acoustic Power and Directivity of a Noise Source

The acoustic power could be found by adding up the products of the areas times the acoustic intensities for the areas on any closed surface which contains the source. Unfortunately, equipment for the direct measurement of acoustic intensity is not available at this time in commercial form. The acoustic intensity and, therefore, the acoustic power must be computed from sound-pressure measurements. This can be accurately and simply accomplished in the free field beyond the near field. There, as we have said before, the intensity is directly related to the (mean-squared) sound pressure and the sound power is obtained by integrating over the closed surface.

In addition to the above case of a free progressive wave, the computation of sound power from sound-pressure measurements can also be made in the diffuse sound field. In a diffuse sound field the sound-energy density is constant and the net intensity is zero. But the sound-energy density is directly related to the (mean-squared) sound pressure and the acoustic power of the source. Hence it is again possible to compute the acoustic power from sound-pressure measurements. In all cases, the indicating meter of the measuring apparatus should read the rms sound pressure at the microphone location.

In practice, sufficiently close engineering approximations to free-field conditions can be achieved in properly designed anechoic rooms, or outdoors. Approximately diffuse sound fields can be obtained in large, hard-walled, sometimes irregularly shaped reverberation chambers. Where these environments are not available or where it is not possible to move the noise source under test, the semireverberant-room method (see below) is available for the determination of the acoustic power. While not as accurate as the first two methods, it nevertheless permits useful engineering measurements to be made in ordinary spaces.

The three methods for determining the sound power radiated by a noise source from SPL measurements will be discussed in some detail below.

All procedures apply to the determination of sound power in octave bands. To obtain greater frequency resolution, one-third-octave bands are sometimes used. The total sound power is obtained by adding the power in all frequency bands which contain appreciable contributions from the source. Pure-tone components must be detected by a narrow-band analyzer.

Where the directivity factor or directivity index of the source is to be determined, an essentially free-field environment must be used.

Our principal concern here is with sources that radiate wideband noise of a relatively sustained character. The indicating meter of a standard sound-level meter will then give valid indications. This would not be the case for such devices as drop hammers and other machines that produce impulsive sounds.

Most practical noise sources radiate sound as a byproduct to their primary function. For such sources, changes in acoustic environment have very little influence on their primary power input and on their mechanical mode of vibration. Such devices, provided they are mounted and supported in the same way, will radiate very nearly the same power in a free field, a reverberant field, or a semireverberant field. Mounting conditions and the proximity of reflecting surfaces can and do influence the radiated power. It is therefore important to choose typical mounting conditions for the source under test in all cases and to regard any mounting surfaces, corners, etc., as part of the source.

Sound-power Measurements in the Free Field. The sound power radiated by many types of practical noise sources is difficult to measure under true free-field conditions. Some devices are too heavy to be suspended in the center of an anechoic room, some are too large to fit in the room available, and a great many sources are normally supported by, or associated with, one or more hard surfaces.

A simple acoustic environment that closely approximates actual operating conditions for many types of noise sources is an otherwise free field above a smooth, hard reflecting plane. This environment is of great practical importance and can be readily approximated even for radiators of large size by placing them on a flat, paved strip outdoors. Alternatively, the source may be placed on the hard floor of a room whose five other boundaries are anechoic.

The determination of the sound power radiated by a noise source in a free field above a reflecting plane is based on the premise that the reverberant field is negligible at the positions of measurement and that the total radiated sound power is obtained by a space integration of the sound intensity over a hypothetical hemisphere (test hemisphere) centered on the source of noise. The surface of the test hemisphere must, in addition, be in the far field of the source. Then, in lieu of the average intensity, the space average of the sound pressure squared over the hemisphere is determined. This may be done by choosing an array of microphone locations over the hemisphere. How many points are needed will depend on the accuracy required and how directive the source is. A number of convenient point arrays are given in Ref. 1, p. 78, and Ref. 2, p. 110. Symmetry in the radiation pattern may be utilized to reduce the number of necessary measurement positions.

To obtain the average mean-square sound-pressure level over the test hemisphere ($\overline{\text{SPL}}_H$), the following procedure is followed: (1) The sound-pressure level at each microphone position is converted into sound pressure squared and is multiplied by the area associated with that position; (2) these products are added, and the result is divided by the total area of the test hemisphere, i.e., $2\pi r^2$; (3) the result is converted into average sound-pressure level in decibels by taking the logarithm (to the base 10) and multiplying by 10.

The point arrays given in Refs. 1 and 2 have the advantage that the fraction of the hemispherical area associated with each test point is constant. Hence, the computations are greatly simplified.

Alternatively, a single microphone may be used which moves in a circular arc about the rotating source (Ref. 3). The output is squared and averaged by electronic means, giving suitable weight to the surface areas of the sphere.

Example 8.1. Sound-pressure levels were measured at 12 points on a 12-segment hemispherical array[1,2] around a small source. Column

2 of Table 8.1 shows the levels measured in the 300- to 600-cps band, in decibels *re* 0.0002 μbar (i.e., 2×10^{-5} newton/m²). Find $\overline{\text{SPL}}_H$, the sound-pressure level that would be produced by a nondirectional source radiating the same power at the same radius.

Solution. This array of points is special in that the areas associated with eight of the particular points are equal. Four of the points lie in the reflecting plane and have only one-half of their share of area associated with them.[1,2] These points are identified by boldface type in the calculations of Table 8.1. For these points, therefore, only

Table 8.1

Point No.	SPL, db	p^2, newtons ²/m⁴
1	60	4×10^{-4}
2	60	4×10^{-4}
3	61 (58)	2.5×10^{-4}
4	65	1.2×10^{-3}
5	60 (57)	2×10^{-4}
6	55	1.2×10^{-4}
7	55	1.2×10^{-4}
8	55	1.2×10^{-4}
9	61 (58)	2.5×10^{-4}
10	64	10^{-3}
11	62 (59)	3×10^{-4}
12	61	5×10^{-4}

one-half of the calculated value of pressure squared has been inserted in the table. Instead of computing then the product of sound pressure squared for each segment times the area of that segment, the sound-pressure levels for the segments are added on a pressure-squared (intensity) basis, and the sum is divided by 12 to get the average. Addition of column 3 gives:

$$\Sigma p^2 = 4.9 \times 10^{-3} \text{ newton}^2/\text{m}^4$$

$$\overline{\text{SPL}}_H = 10 \log_{10} \frac{\Sigma p^2}{12 p^2_{\text{ref}}} = 10 \log_{10} \frac{\Sigma p^2}{p^2_{\text{ref}}} - 10 \log_{10} 12$$

Since 10 log 12 = 10.8 db, and $p^2_{\text{ref}} = 4 \times 10^{-10}$ newton²/m⁴, the reference pressure squared, the result is 70.8 − 10.8 = 60 db = $\overline{\text{SPL}}_H$. Note that this is the average sound-pressure level in decibels over the surface of the hemisphere.

Alternatively, since the areas are equal here, one can add the sound-pressure levels directly, on an intensity basis, taking two levels at a time. The line chart in Fig. 8.5 will be helpful. Before adding, subtract 3 db from the sound-pressure levels pertaining to the four boldface points (see the numbers in parentheses in Table 8.1). After subtracting 10 log 12 = 10.8 db, $\overline{\text{SPL}}_H = 60$ db is obtained.

For rough guesses, one can simply average the levels in decibels (if they are not different by more than 10 db) and add 1 db to the

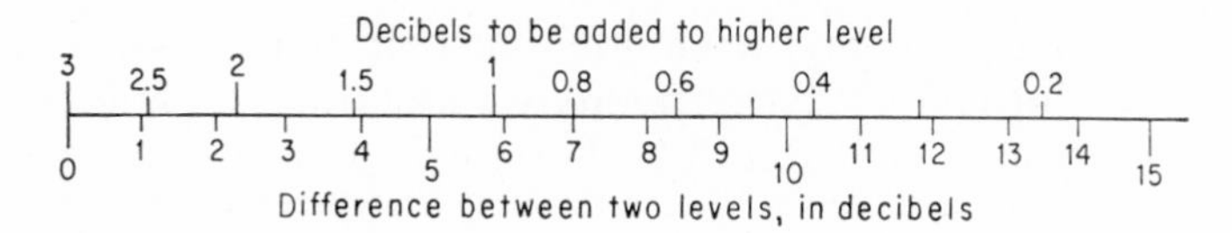

FIG. 8.5 Line chart for the combination of sound-pressure levels on a pressure squared (intensity) basis. *Note:* Combine two levels at a time.

result. Before adding, subtract 3 db from the sound-pressure levels pertaining to the four boldface points (see the numbers in parentheses in Table 8.1). This procedure gives very nearly 60 db as the answer also.

To realize the proper acoustic conditions, a test room of adequate volume is required to accommodate the "test hemisphere" with some space to spare. This room must have anechoic boundaries except for a hard floor and must be free from obstacles which may give rise to disturbing reflections. Outdoors, the proper environment calls for the source to be mounted on a hard smooth plane extending continuously from the noise source beyond the farthest microphone position. The area must be free from obstacles. A homogeneous atmosphere, free from appreciable temperature and wind gradients, is a prerequisite.

The test hemisphere which determines the several microphone positions has its center on the reflecting plane beneath the acoustic center of the sound source. In order that the measurements be carried out in the far field, the hemisphere radius should be equal to at least two major source dimensions or four times the average source height above the reflecting plane, whichever is the larger. It is good engineering practice for the radius of the test hemisphere to be always greater than about 2 ft. No microphone position should be closer to the room boundaries than $\lambda/4$ ft, where λ is the wavelength of sound at the center frequency of the lowest frequency band of interest. Outdoors, atmospheric effects are likely to influence the measurements if the radius of the test hemisphere is much greater than about 50 ft, even in favorable weather.

After the sound-pressure-level measurements have been carried out in hemispherical space as described above, the sound-power level in each frequency band is computed as follows:

$$\text{PWL} \doteq \overline{\text{SPL}}_H + 20 \log_{10} r + 10 \log_{10} 2\pi \text{ db} \qquad (8.7)^*$$

where PWL = sound-power level, db *re* 10^{-13} watt
$\overline{\text{SPL}}_H$ = mean-square sound-pressure level, db *re* 0.0002 μbar over test hemisphere
r = radius of test hemisphere, ft

PWL should be computed for the frequency bands of interest.

The radiated acoustic power W follows from Eq. (8.2):

$$W = 10^{-13} \text{ antilog}_{10} \left(\frac{\text{PWL}}{10} \right) \quad \text{watts} \qquad (8.8)$$

In view of the accuracy of measurements attainable at present and the limitations of standard instrumentation, PWL should be rounded off to the nearest decibel, and W should be rounded off accordingly.

Example 8.2. In the example previously discussed, assume that the sound-pressure levels were measured over a test hemisphere of 10-ft radius. Determine the radiated acoustic power W.

Solution. The power level according to Eq. (8.7) is PWL $\doteq 60 + 20 \log 10 + 8 \doteq 88$ db *re* 10^{-13} watt. The corresponding acoustic power is

$$W = 10^{-13} \text{ antilog } 8.8 \doteq 6 \times 10^8 \times 10^{-13} = 6 \times 10^{-5} \text{ watt}$$

The directivity index of the source evaluated in a given direction θ and for a given frequency band follows very simply:

$$\text{DI}_\theta = \text{SPL}_\theta - \overline{\text{SPL}}_H + 3 \text{ db} \qquad (8.9)$$

where DI_θ = directivity index in direction θ, db
SPL_θ = sound-pressure level, db *re* 0.0002 μbar, in direction θ, r ft from source
r = radius of test hemisphere, ft
$\overline{\text{SPL}}_H$ = mean-square sound-pressure level, db *re* 0.0002 μbar, over test hemisphere

* In Eqs. (8.7), (8.10), (8.16), and (8.17) two additive constants were neglected. One equals —0.5 db. The other depends on the barometric pressure B and the ambient temperature °F at the time of the measurement. For example, for $B = 30$ in. Hg and °F = 70, Eq. (8.13), as written, yields sound-power levels that are too large by 0.5 db. The second constant is determined from Fig. 3.4 on p. 54 and is *subtracted* from the power level. For example, for $B = 20$ in. Hg and °F = 140, one finds a correction of —2 db. Hence, due to this second constant, the PWL must be *increased* by 2 db in order to yield the correct answer.

Note that 3 db must be added in this equation because the $\overline{\text{SPL}}$ was measured over a hemisphere instead of a full sphere, as defined in Eq. (8.3). From Eq. (8.3) it follows that $\text{DI}_\theta = \text{DI} = 3$ db if a source were to radiate uniformly into hemispherical space.

The directivity factor in the direction θ is given by Eq. (8.4).

Some pieces of noisy equipment are normally associated with more than one reflecting surface. An air conditioner standing on the floor against a wall is a case in point. The power level of noise sources of this type should be measured with those surfaces in place. This is done best in a test room with anechoic walls but with one hard wall forming an "edge" with the hard floor. The general considerations of the preceding paragraphs apply here as well. For such a source, radiating into a quadrant of a sphere, the additive constant of Eq. (8.7) is 5 db instead of 8 db. The directivity index is calculated according to Eq. (8.9), except that the additive constant is 6 db instead of 3 db.

Sometimes the noise source under test is not associated with a hard surface and perhaps is small enough to be placed near the center of an anechoic chamber. In this true free-field case, it is necessary to determine the average mean-squared sound-pressure level over a test sphere ($\overline{\text{SPL}}_S$). The general principles stated earlier for the hemispherical case apply. References 1 and 2 contain point arrays suitable for measurements over a sphere (pp. 78 and 110, respectively).

After the sound-pressure-level measurements have been carried out in spherical space, the sound-power level in each frequency band is computed as follows:

$$\text{PWL} \doteq \overline{\text{SPL}}_S + 20 \log_{10} r + 11 \text{ db} \qquad (8.10)$$

where PWL = sound-power level, db *re* 10^{-13} watt

$\overline{\text{SPL}}_S$ = mean-square sound-pressure level, db *re* 0.0002 μbar, over test sphere

r = radius of test sphere, ft

PWL should be computed for all frequency bands of interest.

The radiated acoustic power W follows from Eq. (8.8). The directivity index and the directivity factor of the source evaluated in a given direction θ and for a given frequency band are computed from Eqs. (8.3) and (8.4).

Figure 8.6 shows the results of power-level measurements of a small electric motor in the free field.* Measurements are shown for various

* Unpublished data taken by R. J. Wells of the General Electric Company, reproduced here by permission.

values of the radius of the test hemisphere. The measured power levels are in good agreement down to values of $r/a = 2$, where r is the radius of the test hemisphere and a the maximum motor dimension. Evidently, for $r/a = 1$ the microphone positions are in the near-radiation field, which results in the deviations of the readings shown.*

Sound-power Measurements in the Reverberant Field. The determination of the sound power radiated by a noise source in the reverberant field is based on the premise that the measurements are

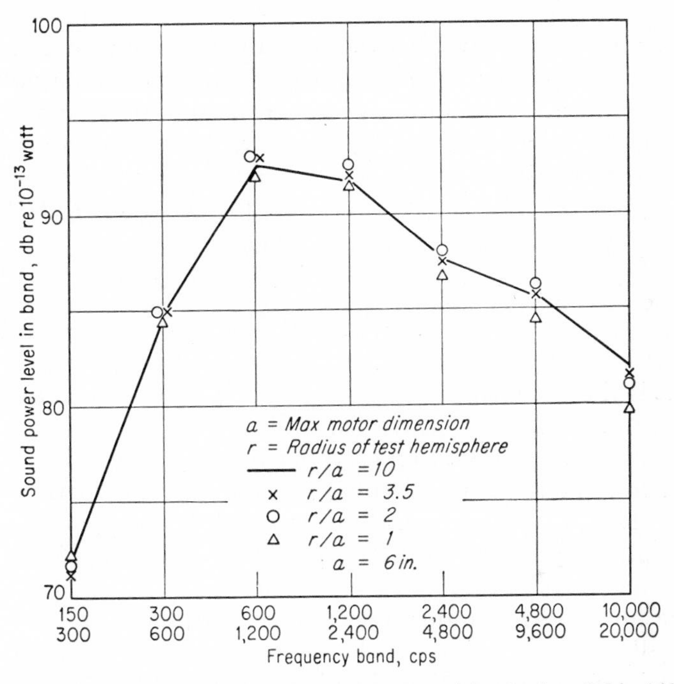

FIG. 8.6 Power level of a small electric motor measured in the free field. (*After R. J. Wells.*)

carried out entirely in a diffuse sound field. In such a field the sound-pressure level is essentially uniform and is simply related to the sound power radiated by the source. It is clear that information about the directivity of the source cannot be obtained.

To realize the proper acoustic conditions, a test room of adequate volume and suitable shape is required, whose boundaries, over the frequency range of interest, can be considered hard. The volume of

* One would ordinarily expect that the apparent power levels measured very near the source would be larger than the true values. In the present case the opposite is true, especially at the high frequencies; this is probably due to local shielding of the noise emanating from the motor interior.

this reverberation chamber should be large enough so that at least 20 modes of vibration of different frequency (see Chaps. 10 and 11) are excited in the lowest frequency band of interest. The approximate number n of modes in a frequency band Δf cps wide and of center frequency f cps is given by[4]

$$n = (f^2 V \times 10^{-8} + 1.3 f A \times 10^{-6})\,\Delta f \qquad (8.11)$$

where V = total volume of reverberation chamber, ft^3
A = total surface area of reverberation chamber, ft^2

It is clear from the above that the sound power radiated by a source which emits a line or narrow-band spectrum cannot be accurately determined in a reverberation chamber. Assuming a source radiating wideband noise and assuming an analyzing filter one octave wide, Eq. (8.11) can be simplified. For practical chamber shapes, $A > 6V^{2/3}$. Because Δf is one octave wide, $\Delta f/f_m = 1/\sqrt{2}$, where f_m is the geometric-mean frequency of the band. Then

$$V \geq \frac{4}{3}\lambda^3 \qquad (8.12)\,*$$

where λ is the wavelength of sound in feet, evaluated at the center frequency of the lowest octave band used.

A rectangular reverberation chamber should be such that the ratio of any two dimensions is not equal or close to an integer. The proportions $1:2^{1/3}:4^{1/3}$ are frequently used. Hard objects, such as polycylindrical diffusers or rotating paddles, are sometimes placed in the room in an effort to improve the diffusion.

If the noise source under test is normally associated with a hard floor, wall, edge, or corner, it should be placed in a corresponding position in the reverberation chamber.[5] It is not advantageous generally to place the source near the center of the chamber since this limits the available test radius (see below) and an appreciable portion of the resonant modes in the chamber may not be excited.

In selecting a microphone position one should bear in mind that near the room boundaries and near the source the sound field departs from the ideal state of diffusion.[6] Assuming a source radiating wideband noise and assuming an analyzing filter one octave wide, it is good engineering practice to place the microphone at least $\lambda/4$ ft away from all room surfaces.[5,6] The wavelength λ is evaluated at the center of the lowest frequency band of interest.

* This simplified equation is the work of R. J. Wells of the General Electric Company.

In addition to the requirements of the preceding paragraph, the microphone should be placed far enough away from the source to be in the reverberant field (see Sec. 8.3). This is realized if the microphone is placed at least one major source dimension but not less than a distance $\frac{1}{6}Q^{1/2}V^{1/3}$ away from the source,* where Q is the source directivity. If the directivity is not known, a practical upper limit of $Q = 16$ yields $\frac{2}{3}V^{1/3}$, which is very nearly equal to the mean-free path $4V/A$ (Ref. 2) and should be used as an alternative expression for the minimum source-microphone distance.

Within this permitted region of microphone placement, several microphone positions should be chosen and the results averaged into a single value of sound-pressure level SPL. Alternatively, a space average may be obtained by moving the microphone about in the permitted region.

After carrying out the measurements over the frequency range of interest, the sound-power level in each frequency band is computed as follows:

$$\text{PWL} \doteq \overline{\text{SPL}} + 10 \log_{10} V - 10 \log_{10} T - 19 \text{ db} \qquad (8.13)$$

where PWL = sound-power level, db *re* 10^{-13} watt
$\overline{\text{SPL}}$ = space average of sound-pressure level, db *re* 0.0002 μbar in reverberation chamber
V = total volume, ft^3, of reverberation chamber with source in place
T = reverberation time of chamber with source in place, sec

The reverberation time T can either be measured by the use of a high-speed level recorder, or it can be estimated from the total number of absorption units in the room (see Chap. 11).

Alternatively, there may be available a source of known sound-power level PWL′ whose noise spectrum encompasses the frequency range of the noise source under test. This source is placed in the reverberation chamber and the space average of its sound-pressure level $\overline{\text{SPL}}'$ determined as outlined above. The power level of the unknown source PWL is then simply:

$$\text{PWL} = \text{PWL}' + (\overline{\text{SPL}} - \overline{\text{SPL}}') \qquad (8.14)$$

where PWL and $\overline{\text{SPL}}$ are defined above.

* This estimate assumes that the direct sound field is no more than one quarter of the reverberant field, that $A \doteq 6V^{2/3}$ and the average sound absorption coefficient in the chamber is less than 0.06.

Sound-power Measurements in the Semireverberant Field. High-quality anechoic or reverberation chambers are test facilities which frequently cannot be used for economic reasons, and outdoor tests are often impractical. In some cases, the noise source of interest cannot be moved from its location into a special test environment. Under such conditions measurements must be carried out in factory or laboratory areas, where the walls and ceiling are neither completely absorbing nor completely hard. A hard floor is almost always present. On the other hand, it is sometimes desirable to test wall-mounted or floor-mounted devices in a room designed to represent the "average living room." In such rooms the sound field is neither predominantly diffuse nor free. The following discussion deals with the measurement of power level in such a *"semireverberant field."*[3] Although limited in accuracy, the practical importance of this method warrants a short description here. It is not advisable to attempt directivity measurements in the semireverberant field except for rough orientation purposes.

No specific assumptions are made concerning the test room, except that it should be large enough so that the microphone can be placed in the far field of the source without being too close to the room boundaries. The noise source under test should be mounted as it is normally used. This mounting condition will typically include a hard floor. Unusually long, narrow rooms or rooms with sound-absorbing material localized in patches are best avoided.

In order to compute the power level, the room constant (Ref. 2 and Chap. 11) of the test room must be known. While it could be computed from the amount of absorption present in the room or determined from a measurement of the reverberation time, it is frequently advantageous to "calibrate" the test room and thus obtain a direct measure of the room constant. For this purpose a standard sound source is selected, whose noise spectrum encompasses the frequency range of the noise source under test. The standard source is placed on a hard surface in a free field and the mean-squared sound-pressure level $\overline{\text{SPL}}_{H_1}$ over a test hemisphere is measured as in Example 8.1. The standard source is now transferred into the semireverberant-test environment and placed on the floor in the position of the unknown source.* Keeping the operating conditions of the standard source constant, the mean-squared sound-pressure level $\overline{\text{SPL}}_{H_2}$ over a like test hemisphere is determined. A convenient expression containing the room constant of the semireverberant test room can be computed from the following relation:[3]

* In the event that the unknown source cannot be moved, an acoustically equivalent location in the room may be found for the standard source.

$$\frac{4}{R} = \frac{1}{S}\left[-1 + \text{antilog}_{10}\left(\frac{\overline{\text{SPL}}_{H_2} - \overline{\text{SPL}}_{H_1}}{10}\right)\right] \qquad \text{sq ft}^{-1} \qquad (8.15)$$

where $S = 2\pi r^2$, which is the area of the test hemisphere in square feet.

The noise source under test is now operated in the test room and the mean-square sound-pressure level $\overline{\text{SPL}}_H$ over a test hemisphere is determined, proceeding as if the walls of the test room were anechoic. The power level is computed as follows:

$$\text{PWL} \doteq \overline{\text{SPL}}_H - 10 \log_{10}\left(\frac{1}{S} + \frac{4}{R}\right) \qquad \text{db} \qquad (8.16)$$

where PWL = sound-power level of source under test, db *re* 10^{-13} watt

$\overline{\text{SPL}}_H$ = mean-square sound-pressure level of source under test, db *re* 0.0002 μbar, determined in semireverberant test room over test hemisphere of surface $S = 2\pi r^2$ ft^2

$4/R$ = computed from Eq. (8.15)

If r is the same as that used for the standard source and the sound fields of the two sources are not greatly different, the power level of the unknown source can be computed directly from the mean-square sound-pressure levels of the two sources measured in the semireverberant test room.

PWL should be computed for all frequency bands of interest.

It is of interest to compare the power levels of a source of noise as measured in the free field and in a semireverberant field.* Figure 8.7 shows the results obtained for the same small electric motor discussed in connection with Fig. 8.6. The power levels measured in the reverberant field are within 2 db of the free-field values for a radius of the test hemisphere equal to ten times the maximum motor dimension.

Sound-power Measurements in a Duct. The sound-power level in a duct can be computed easily from sound-pressure-level measurements, provided one deals essentially with a plane progressive wave.[7]

$$\text{PWL} \doteq \text{SPL}_D + 10 \log_{10} S_D \qquad \text{db} \qquad (8.17)$$

where PWL = sound-power level, db *re* 10^{-13} watt

SPL_D = mean-square sound-pressure level, db *re* 0.0002 μbar, obtained over cross-sectional area S_D of duct, ft^2

* Unpublished data taken by R. J. Wells of the General Electric Company, reproduced here by permission.

The above relation assumes not only a nonreflecting termination but also essentially uniform sound intensity across the duct. At frequencies beyond the first cross mode of the duct the latter assumption is no longer satisfied. But Eq. (8.17) can still be used provided SPL_D is replaced by a suitable space average $\overline{SPL}_D$ over the cross-sectional area S_D obtained from sound-pressure level measurements at several

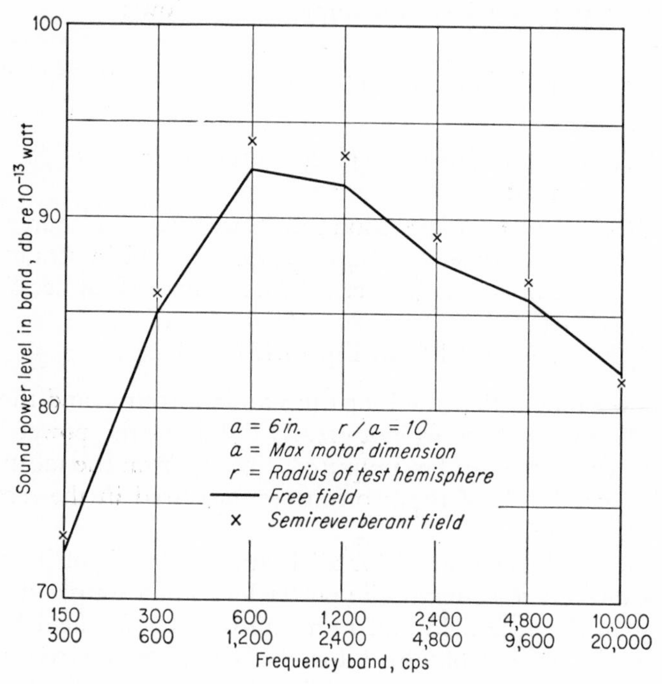

FIG. 8.7 Power level of a small electric motor measured in the free and semi-reverberant fields. (*After R. J. Wells.*)

points across the duct. The number of measurement positions across the cross-section used to determine $\overline{SPL}_D$ will depend on the accuracy desired and the frequency. The necessary number of positions increases with both. A properly chosen single microphone position can be used in the duct with minimum error, as Dyer[8] has shown.

In practical situations, reflections occur at the open end of the duct, and the effect of branches and bends must be considered.[9] It is also necessary to surround the microphone by a suitable windscreen[7] to reduce the aerodynamic noise that may interfere with the measurements.

8.5 Summary

For some noise-control problems it is sufficient to measure the sound-pressure levels at a specified location over the frequency range of interest. Other problems require an analytical approach which calls for a knowledge of the acoustic power and directivity of the source or sources of sound involved, with the aid of which it is possible to make an engineering estimate of the sound-pressure levels generated by that source in almost any acoustic environment. The acoustic power alone can frequently be used as an acoustic rating of sources.

In the absence of a practical acoustic intensity meter, sound-power determinations must be carried out indirectly via sound-pressure measurements. Suitable environments for such measurements are anechoic rooms or suitable locations outdoors, and reverberation chambers. These constitute good engineering approximations to free-field and diffuse-field conditions, respectively. In such fields, the acoustic power is directly related to the sound pressure squared. By necessity, the practicing acoustician must frequently carry out sound-power determinations in ordinary spaces such as factory or office areas. Sound-power measurements in such semireverberant spaces can be carried out with accuracies adequate for many practical problems.

In ducts sound-power determinations can be made within certain limitations knowing the sound-pressure-level distribution in the duct.

REFERENCES

1. Peterson, A. P. G., and L. L. Beranek: "Handbook of Noise Measurement," 3d ed., and E. E. Gross, Jr., "Measurement of Vibration," General Radio Company, Cambridge, Mass., 1956.

2. Beranek, L. L.: "Acoustics," McGraw-Hill Book Company, Inc., New York, 1954.

3. Wells, R. J.: Apparatus Noise Measurement, *AIEE, Power Apparatus and Systems,* December, 1955.

4. Morse, P. M.: "Vibration and Sound," 2d ed., McGraw-Hill Book Company, Inc., New York, 1948, p. 395.

5. Waterhouse, R. V.: Output of a Sound Source in a Reverberation Chamber and Other Reflecting Environments, *J. Acoust. Soc. Am.,* vol. 30, p. 4, 1958.

6. Waterhouse, R. V.: Interference Patterns in Reverberant Sound Fields, *J. Acoust. Soc. Am.,* vol. 27, p. 247, 1955.

7. Beranek, L. L., J. L. Reynolds, and K. E. Wilson: Apparatus and Proce-

dures for Predicting Ventilation System Noise, *J. Acoust. Soc. Am.*, vol. 25, p. 313, 1953.

8. Dyer, I.: Measurement of Noise Sources in Ducts, *J. Acoust. Soc. Am.*, vol. 30, p. 833, 1958.

9. "Heating, Ventilating, Air Conditioning Guide," chap. 25, American Society of Heating, Refrigerating and Air Conditioning Engineers, Inc., New York, 1960, published annually.

Part Two

FUNDAMENTALS UNDERLYING NOISE CONTROL

Chapter 9

SOUND PROPAGATION OUTDOORS

Francis M. Wiener

9.1 General

The open atmosphere is by its nature in constant motion and fluctuation. Density and temperature, wind and humidity are never uniform in a given volume of air under observation, nor are they constant in time. Sound waves traveling through the atmosphere are affected by these nonuniformities in two ways: the sound-pressure level at the receiving end fluctuates, and its mean value depends materially on the mean wind, temperature, and humidity conditions along the path of propagation. The longer the transmission path through the atmosphere, the more important is the effect of these factors on the received sound-pressure level.

Typically, in a noise-control problem, the receiving point is near the ground, and very frequently, the noise source is near the ground also. The discussion in this chapter is therefore concerned primarily with the propagation of sound in the audible frequency range through the lower atmosphere along the ground (see Fig. 9.1). We are interested in the ground wave and the conditions in the atmospheric boundary layer. Acoustic sky-wave transmission will not be discussed here, although enormous distances are occasionally bridged by it under favorable conditions. A short discussion of air-to-ground propagation of sound is given in Sec. 9.8, because of its obvious practical importance in connection with aircraft in flight.

The strong interdependence between sound transmission along the ground and the "weather" is a matter of everyday knowledge. It has been a matter of scientific investigation for centuries. However, only recently have adequate micrometeorological instrumentation techniques become available and been brought to bear on the problem. On the other hand, for air-to-ground transmission of the type discussed here, the "weather" seems to play a much smaller role.[1] Significant strides have been made toward an engineering solution to the problem by means of a combination of theory and empirical design curves. The material presented here is an attempt to reduce to engineering practice what has been learned from theory and experiment to date.

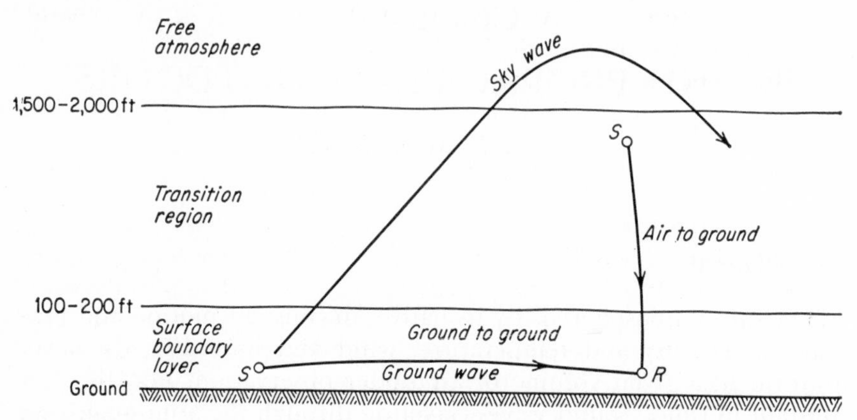

FIG. 9.1 Two important types of sound propagation in the atmosphere. (*Adapted from Ingard.*[20])

9.2 Spherical Divergence and Excess Attenuation

In an ideal, homogeneous, loss-free atmosphere the sound pressure decreases inversely with distance when one is in the far free field (see Chap. 8) of the source. In other words, there is a 6-db decrease in sound-pressure level SPL for each doubling of distance. Assuming a source of power level PWL located several feet above an ideal hard-ground plane in this ideal atmosphere, we have

$$\mathrm{SPL}_\theta \doteq \mathrm{PWL} + \mathrm{DI}_\theta - 20 \log_{10} r - 11 \text{ db} \tag{9.1}$$

where SPL_θ = sound-pressure level at receiver, db *re* 0.0002 μbar in direction θ, r ft from source
PWL = sound-power level of source, db *re* 10^{-13} watt
DI_θ = directivity index of source in direction θ, db
r = distance of receiver from source, ft

Equation (9.1) follows directly* from Eqs. (8.7) and (8.9). In evaluating the directivity index of the source it should be noted that

* The additive constant in Eq. (9.1) depends on the ambient temperature and barometric pressure. At 67°F and 30 in. Hg the additive constant in Eq. (9.1) is precisely −10.5. In addition, if these quantities are widely different from 67°F and 30 in. of mercury, a further correction should be applied to the right-hand side of Eq. (9.1) as given in Fig. 3.4 on p. 54.

if a source is nondirectional in *hemispherical* space, $DI_\theta = 3$ db, by definition.

Example 9.1. A jet aircraft of PWL = 170 db is run up on the ground. What is the sound-pressure level in an ideal atmosphere 2 miles away measured 120° in front of the direction of the exhaust?

Solution. Assuming a directivity index of -10 db for $\theta = 120°$, one has from Eq. (9.1)

$$SPL_{120°} = 170 - 10 - 20 \log_{10} (10{,}560) - 11$$
$$\doteq 69 \text{ db}$$

Due to atmospheric conditions and the presence of obstacles, the sound-pressure levels measured outdoors are almost always lower, sometimes drastically so, than those predicted from spherical spreading alone. The important factors that affect sound propagation along the ground are: (1) sound absorption in the air; (2) presence of fog, rain, or snow; (3) presence of walls and trees; and (4) effect of wind and temperature gradients, atmospheric turbulence, and the acoustic effect of the presence of the ground. All these factors are to some extent interrelated in that the effect of one will be dependent on the presence of the others. In the case of sound propagation from air to ground, the effect of obstacles on the ground and the effects of ground-created wind and temperature gradients are usually not present. It is useful to lump the net effect of all of the above enumerated factors into a single quantity, the excess *attenuation* A_e, in decibels, over and above the effect of spherical spreading.

It is clear that if we know the sound-pressure level near the source (but in its far field) at one distance and the sound-pressure level in the same direction at a greater distance, the total attenuation A is given by:

$$A \equiv SPL_0 - SPL_1 = 20 \log \frac{r}{r_0} + A_e \quad \text{db} \tag{9.2}$$

or

$$SPL_1 = SPL_0 - 20 \log \frac{r}{r_0} - A_e \quad \text{db} \tag{9.3}$$

where SPL_1 = sound-pressure level, db *re* 0.0002 μbar, at angle θ and distance r ft from source

SPL_0 = sound-pressure level, db *re* 0.0002 μbar, at angle θ and distance r_0 ft from source. (Note that r_0 is a reference distance and that the distance r is always greater than r_0. The point r_0 must be in the far field of the source. See Fig. 8.4.)

A_e = excess attenuation, db

SPL_0 is obtained experimentally or it can be estimated from Eq. (9.1) provided the distance r_0 is small enough (of the order of 50 ft or less) so that A_e is unimportant. Since A_e generally exhibits a frequency dependence, it must be evaluated over the entire frequency range of interest.

To obtain an engineering estimate of A_e, the excess attenuation can be split into its several contributions assumed to be independent of each other, as follows:

$$A_e = A_{e_1} + A_{e_2} + A_{e_3} + A_{e_4} \qquad \text{db} \tag{9.4}$$

where A_e = excess attenuation
A_{e_1} = attenuation due to absorption in air
A_{e_2} = attenuation due to fog, rain, or snow
A_{e_3} = attenuation due to walls and trees
A_{e_4} = attenuation due to wind and temperature gradients, atmospheric turbulence, and ground effect

In the following, we shall discuss these various contributions to the excess attenuation quantitatively and from a practical point of view. They are evaluated primarily in terms of their average values, ignoring fluctuations in the received sound-pressure levels for the moment. These fluctuations will be discussed in general terms in Sec. 9.9. Sound propagation over hilly terrain is taken up briefly in Sec. 9.7, and propagation from a source in the air directly to the ground is discussed in Sec. 9.8. It is assumed throughout that the sound source is not so intense that the air is "overdriven." Therefore, nonlinear effects (air distortion) need not be considered.

9.3 Attenuation Due to Absorption in the Air, A_{e_1}

A sound wave traveling through still homogeneous air loses energy by two processes. Energy is extracted from the wave through the effects of heat conduction and radiation, viscosity, and diffusion. This so-called "classical absorption" is proportional to the frequency squared and is independent of the humidity of the air. Energy is also extracted from the sound wave by a second process, which has to do with the molecular relaxation behavior of the oxygen molecules in the air. This so-called "molecular absorption" depends not only on frequency but also on humidity; it typically exceeds the classical absorption in most noise-control problems. For example, for a 4,000-cps sound wave, heat conduction and viscosity effects account for less than 1 db excess attenuation in 1,000 ft, whereas at 40 per cent relative humidity and a temperature of 70°F, molecular absorption causes an excess attenuation of more than 10 db in 1,000 ft.

As is indicated by the above example, we need to be concerned here only with molecular absorption; classical absorption is unimportant for most noise-control problems in the audible frequency range. Moreover, any error caused by omitting classical absorption is overshadowed by the considerable uncertainties of the numerical values of molecular absorption published in the literature.

Molecular absorption of sound in still homogeneous air is an exponential effect. The corresponding excess attenuation can therefore be expressed in terms of an attenuation coefficient α_1, conveniently defined in terms of db/1,000 ft.

Consequently

$$A_{e_1} = \alpha_1 \frac{r - r_0}{1{,}000} \doteq \alpha_1 \frac{r}{1{,}000} \qquad \text{db} \tag{9.5}$$

where α_1 is the attenuation coefficient due to molecular absorption in db/1,000 ft, and A_{e_1}, r, and r_0 are defined in Eqs. (9.3) and (9.4).

The functional dependence of the attenuation coefficient α_1 on frequency, temperature, and humidity has been determined theoretically by Kneser.[2,3] According to the theory, the maximum value of α_1 depends only on frequency and the temperature. Measurements by Knudsen,[4] Delsasso and Leonard,[5] and Harris[6] have essentially confirmed $\alpha_{1,\max}$ as calculated from Kneser's theory. Measurements of α_1 by the above investigators have disagreed, however, with Kneser's theory at low and high humidities. A recent comprehensive

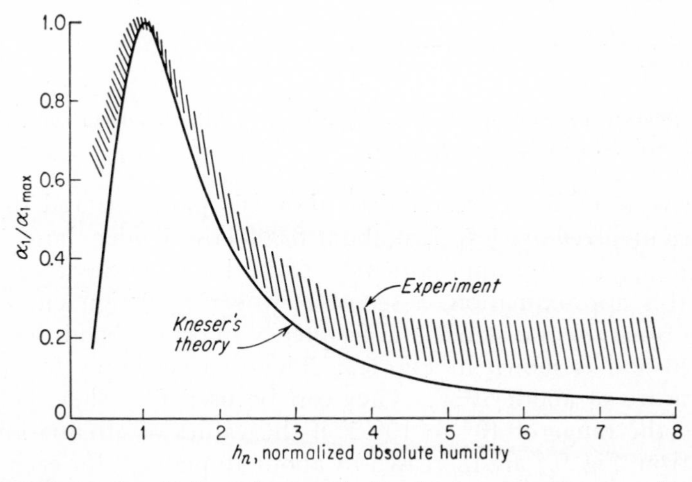

FIG. 9.2 Calculated and observed normalized molecular absorption as a function of normalized absolute humidity. (*Adapted from Nyborg and Mintzer.*[3])

investigation by Evans and Bazley[7] has produced results at 68°F which are consistently higher than Kneser's values of α_1. The nature of the discrepancies between theory and experiment is illustrated in Fig. 9.2 where $\alpha_1/\alpha_{1,\max}$ is plotted as a function of absolute humidity h_n, normalized to the absolute humidity at $\alpha_{1,\max}$.

Design Chart for A_{e_1}. In many practical problems the absolute humidity h is greater than about 8 gm/m³, which corresponds at 60°F

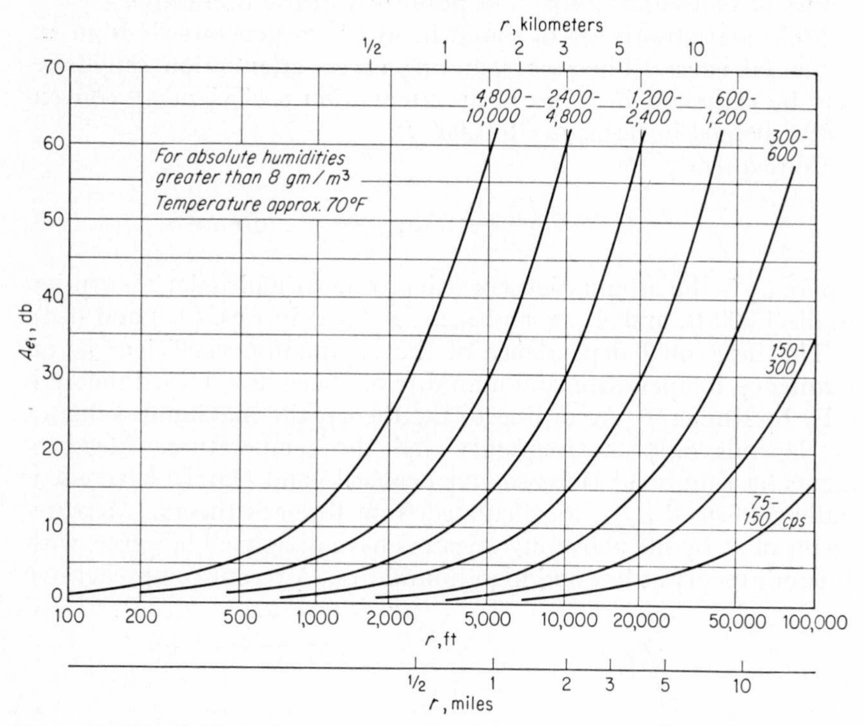

FIG. 9.3 Engineering estimate of excess attenuation in still air due to molecular absorption. (See text for other temperatures.)

to relative humidities greater than about 60 per cent, and the frequencies involved are less than about 8,000 cps. Under those conditions, $\alpha_1/\alpha_{1,\max} = 0.2$, approximately, from the experimental results. Using this approximation, a set of engineering design curves for A_{e_1} as a function of distance and frequency in octave bands has been prepared and is shown in Fig. 9.3. These curves apply for an air temperature of about 70°F. They can be used for other temperatures in the range of 40° to 100°F if the values of attenuation obtained from Fig. 9.3 are increased by about 10 per cent for each 10°F in excess of 70°F and decreased by about 10 per cent for each 10°F below 70°F,[3] provided the absolute humidity exceeds about 8 gm/

m³.* It has been assumed, furthermore, that the geometric mean frequency of each octave band is representative of the band as a whole. Recent work by the author indicates that a frequency of 1.2 times the lower-cutoff frequency of the octave bands yields results which correlate better with experiment in some instances. For conditions where the absolute humidity is less than about 8 gm/m³, Rudnick's[8] nomogram can be used to estimate α_1 for single frequencies.†

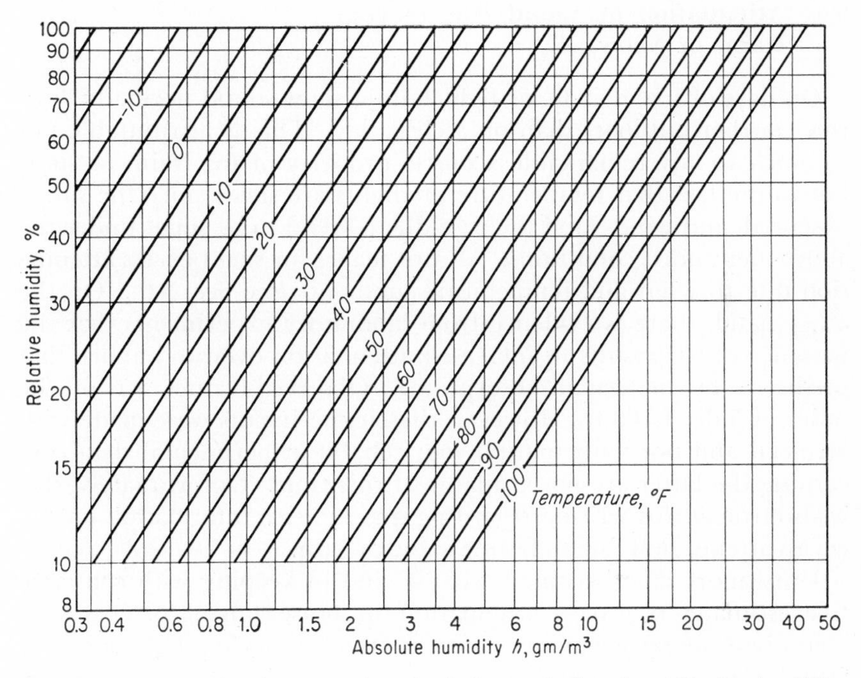

FIG. 9.4 Nomogram for the conversion of relative to absolute humidity in air. (*After Nyborg and Mintzer.*[3])

Example 9.2. A jet aircraft of PWL = 171 db and a directivity index of −10 db in the 1,200- to 2,400-cps band is run up on the ground. What is the sound-pressure level in a still homogeneous atmosphere 2,000 ft away? The air temperature is 70°F and the relative humidity is 75 per cent.

Solution. From Eq. (9.1) one finds the sound-pressure level near the source, say, at $r_0 = 100$ ft

* For convenience, a nomogram relating temperature, relative and absolute humidity is given in Fig. 9.4.[3]

† Note that in Rudnick's graph (Fig. 3.1 in Ref. 8) the absolute humidity scale is labeled incorrectly above 10 gm/m³.

$$\text{SPL}_0 = 171 - 10 - 20 \log (100) - 11 = 110 \text{ db}$$

At $r = 2{,}000$ ft, $\text{SPL}_1 = 110 - 20 \log (20) - A_e = 84 - A_e$ db. To find the excess attenuation A_{e_1} due to molecular absorption one first finds the absolute humidity from Fig. 9.4, i.e., $h \doteq 14$ gm/m^3. Since $h > 8$ gm/m^3, the temperature is 70°F, and $f < 8{,}000$ cps, Fig. 9.3 applies. With $(r - r_0) = 1{,}900$ ft, one finds $A_e = A_{e_1} = 5$ db. *Ans.* 79 db.

9.4 Attenuation of Sound Due to Fog, Rain, or Snow, A_{e_2}

On days of fog, drizzle, or light falling snow sound normally "carries" farther outdoors than on a clear day. This is attributable not so much to any remarkable acoustic property of fog, rain, or snow but to the fact that temperature gradients are small since the sun is obscured, and winds (and their gradients) tend to be small under the above-mentioned conditions. All this makes for small excess attenuation due to wind and temperature gradients (see Sec. 9.6). On the other hand, there is evidence from laboratory experiments that the presence of fog results in increased attenuation over and above that predicted by molecular absorption.[9] Observed values are of the order of 5 db/1,000 ft. However, these experiments were made with artificial and not with natural fog; reliable experimental data concerning the latter are not available at this time. Unfortunately, this is also true at this writing regarding quantitative information on the excess attenuation due to falling snow or rain.

Until more data become available, and in keeping with conservative estimates of excess attenuation in noise-control problems, it seems best not to assign any excess attenuation to fog, light snow, or falling rain, i.e., assume

$$A_{e_2} \doteq 0 \tag{9.6}$$

When sound is propagated through the atmosphere with precipitation present, some thought may have to be given to the effect of changes, due to that precipitation, in background-noise levels at a receiving station near the ground. When it rains, these background levels may increase appreciably; when snow is on the ground, the background levels normally present are frequently effectively muffled.

9.5 Attenuation Due to Walls and Trees, A_{e_3}

Rigid Barriers. It is well known that walls, buildings, and other large rigid barriers, if interposed between source and receiver, result in appreciable excess attenuation. This is due primarily to the acoustic shielding effect of such structures. Since sound waves are diffracted around an obstacle, one would expect the excess attenuation to increase with frequency. This has, indeed, been confirmed experimentally and theoretically. Fehr[10] has invoked optical-diffraction theory in calculating the expected noise reduction by a thin rigid wall of infinite lateral extent when source and receiver are on the ground and the air is still and homogeneous. (Again, unfortunately, no large body of data is available to verify the theoretical computations for a wide variety of geometrics and a wide range of

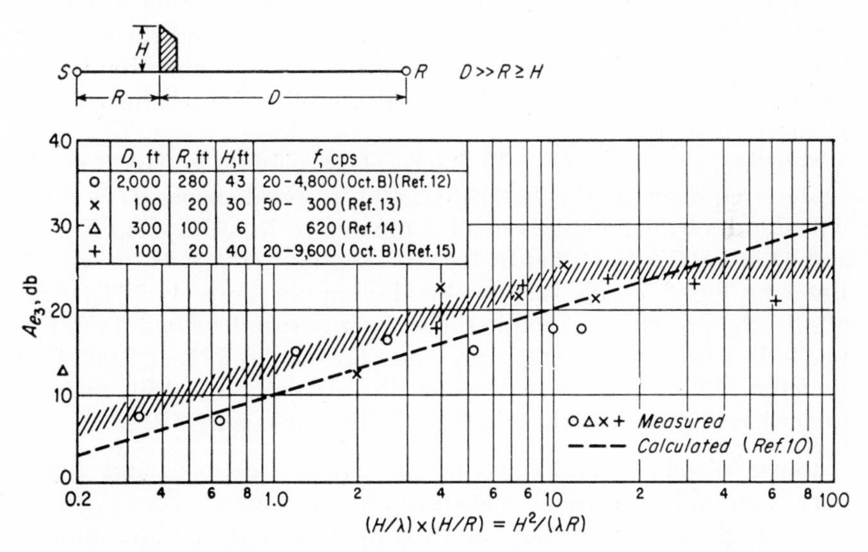

FIG. 9.5 Chart for estimating excess attenuation due to a rigid barrier.

frequencies.) The situation is further complicated by the finite lateral size of practical obstacles, by the fact that most sources are located somewhat above ground, that wind and temperature gradients are present, and that many barriers, such as buildings, are not "thin."

To estimate the attenuation due to shielding by barriers, Fig. 9.5 can be used. Experimental points are presented from Refs. 11 through 15. The experimental data show that a practical limit (due to transmission through the barrier or bypassing around it) of the excess attenuation due to shielding is somewhere near 25 db as indicated on the graph. The theoretical relation,[10] indicated by a dashed line, shows no such limit.

Despite its limitations, the graph permits useful engineering estimates to be made of the excess attenuation A_{e_3} to be expected from buildings and walls for the important case where the receiver is far away from the barrier and the source is not too close to the barrier. In extrapolating from this chart, common sense must be used: for obstacles of finite thickness the height on the source side is significant; the source must not be too close laterally to the edge of an obstacle of finite lateral extent; and when the source is located appreciably aboveground, interference effects between the field of the source and its image in the ground plane are to be expected. Wind is also of some importance,[11] as it tends to reduce the maximum attenuation obtainable.

Example 9.3. A hangar 40 ft high, 50 ft deep, and 300 ft wide is interposed between the jet aircraft of Example 9.2 and the receiver near the ground 2,000 ft away. What is the excess attenuation resulting from the presence of the barrier if the side of the hangar is 200 ft distant from the source?

Solution. $R = 200$, $H = 40$, hence $H/R = 0.2$. $\lambda \doteq 0.65$ ft at the center frequency of the 1,200- to 2,400-cps band, hence $H/\lambda \doteq 62$. The depth of the hangar is small compared to R and D, also $R > H$ and $D = 1{,}800 >> R$, hence Fig. 9.5 applies. $H/\lambda \times H/R = 12.4$. Entering Fig. 9.5 with 12.4 on the abscissa one finds $A_e = A_{e3} \doteq 23$ db. In the preceding example the sound-pressure level was found to be 79 db without the hangar present. The actual SPL at 2,000 ft is therefore $79 - A_{e3} = 79 - 23 = 56$ db. Actually, the SPL may be somewhat higher since the width of the hangar is only 300 ft.

Dense Woods. Sound traveling through dense woods and shrubbery is attenuated through absorption by the leaves on trees and on the ground and by multiple scattering by the tree trunks and limbs.[16] Measurements by Eyring in dense jungles show that the excess attenuation can be described in terms of an attenuation coefficient α_3 conveniently expressed in db/1,000 ft; the lower the visibility, the higher α_3. Recent measurements by Keast and Wiener[18,19] and Pietrasanta[12] have led to the estimates of the excess attenuation in dense evergreen woods where absorption predominates (shown in Fig. 9.6). For comparable visibility, these values are not greatly different from the values of excess attenuation found in dense jungles.[17]

Fences of living trees and hedges have frequently been suggested as a means of noise control. This seldom turns out to be practicable.[15] Not only is great density of planting required with branches reaching to the ground but also an appreciable height of planting is necessary; otherwise the sound is propagated over rather than through the tree barrier. To estimate the attenuation of planting in depth, Fig. 9.6 can be used as a guide, although it applies primarily to the case where source and receiver are both in the woods. The graph shows that the excess attenuation is rather small unless the depth of planting is quite large.

Example 9.4. A jet-engine test stand is surrounded by dense woods, 1,000 ft in depth. What is the excess attenuation provided by the woods in the 300- to 600-cps band?

Solution. Assuming that Fig. 9.6 applies, we find $A_{e_3} \doteq 18$ db. [It should be noted that this example does not take into account the excess attenuation produced by temperature and wind gradients; however, in dense woods these effects are negligible. They are very important in open terrain (see Sec. 9.6).]

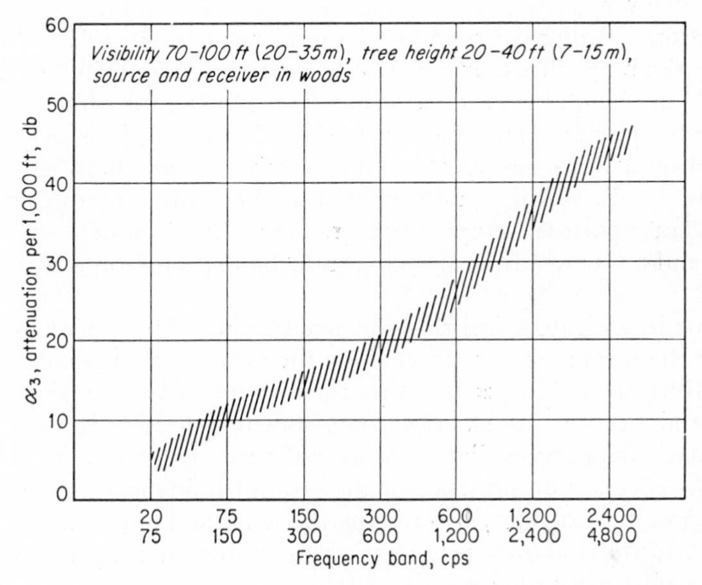

FIG. 9.6 Estimated excess attenuation coefficient in dense evergreen woods where absorption predominates.

9.6 Attenuation of Sound Due to Wind and Temperature Gradients, Atmospheric Turbulence, and Ground Effect, A_{e_4}

Over open level ground appreciable vertical temperature and wind gradients almost always exist; the former because of the heat exchange between the ground and the atmosphere, the latter because of the friction between the moving air and the ground. Because of these gradients, the speed of sound varies with height above the ground, and sound waves are refracted, that is to say, bent upward or downward. Under such conditions, it is possible to have a "shadow zone" into which no direct sound can penetrate.[18–21] These shadow zones are never sharp in the sense of light propagation because sound energy is diffracted into the shadow zone as well as scattered into it by turbulence.

A shadow zone is most commonly encountered upwind from a source, where the wind gradient* bends the sound rays upward. Downwind, the wind gradient bends the sound rays downward, and no shadow zone is produced. Crosswind, there is a zone of transition. This asymmetric behavior is characteristic of wind-induced sound refraction. Temperature-induced sound refraction tends to be symmetrical about the source. A shadow may completely encircle a source in the presence of a strong negative temperature gradient (large temperature lapse) and a low windspeed, such as may be expected on a calm, sunny day. On the other hand, there will be no shadow at all, within a mile or two of the source, in the presence of a strong positive temperature gradient (large temperature inversion) and a low windspeed, such as may be expected on a clear, calm night.

Some investigators feel that the acoustic impedance of the ground affects the sound-pressure level along the ground in the shadow zone. The finite ground impedance presumably also affects the propagation of sound outside the shadow zone, downwind, for example. Few systematic data are available for an estimate of this type of acoustic ground effect. The presence of the ground influences the excess attenuation most strongly by creating the wind and temperature gradients mentioned above in general terms, rather than by virtue of its finite acoustic impedance.

In the present state of the art it seems best to approach the problem of providing engineering estimates for the excess attenuation of sound propagated along the ground from an experimental point of view. Not only does a comprehensive theory not exist at this time but also the measurement or estimate of the micrometeorological parameters which must be used in any computation is probably beyond the capabilities of the average acoustician interested in noise control. On the other hand, useful engineering estimates of at least the maximum of the excess attenuation to be expected from temperature and wind gradients over open level terrain can be had from considering recent experimental data.

Consider Fig. 9.7. Source and receiver are shown a distance r

* Wind gradients near the ground are, on the average, always positive, i.e., the windspeed increases with height.

apart. The average direction from which the wind is blowing is indicated by a wind vane. The angle between the direction of the wind vane and the line connecting the source and receiver is called ϕ. There will generally be a shadow zone (the shaded region of Fig. 9.7) produced on the upwind side of the source because sound waves traveling upwind tend to be bent upward by the wind. Oftentimes the air near the ground is warmer than that farther up, so that there is a tendency for the sound waves to be refracted upward on all sides

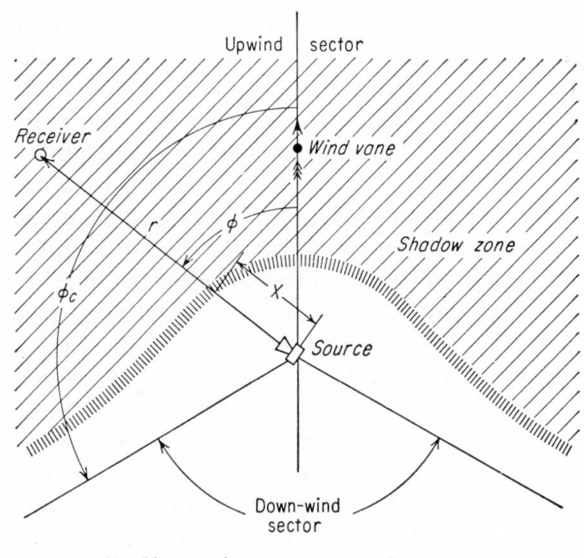

FIG. 9.7 Geometry of sound propagation over open level terrain (plan view). Average daytime conditions are shown.

of the sound source, in addition. However, any wind present tends to bend the sound waves downward in the downwind direction. At some critical angle ϕ_c the effects of wind and temperature gradients may cancel each other and the shadow zone vanishes. As a result the plane is divided into an upwind sector $2\phi_c$ and a downwind sector $360° - 2\phi_c$. Experiments have shown that the excess attenuation is frequently radically different upwind and downwind, with a gradual transition at the boundaries $\phi = \pm \phi_c$. On a sunny day, with moderate winds, the excess attenuation upwind inside the shadow zone is typically 20 to 30 db higher than that for the same distance downwind.

The results of extensive measurements in the frequency range of from about 300 to 5,000 cps taken under a large variety of micrometeorological conditions are available in the literature.[18,19] These measurements have been made over open level terrain with sparse low ground cover (1 ft high), a source height of 12 ft and a receiver height of 5 ft using octave bands of random noise as the transmitted signal. Windspeeds encountered ranged from 2 or 3 to about 15 mph. From the extensive series of experiments cited above, empirical charts were prepared [19] with the aid of which the excess attenuation can be estimated for any angle ϕ, for distances r up to about 1 mile and the terrain conditions noted above, provided that the temperature and wind gradients are known from measurements made at approximately half the average source and receiver heights. Since the experiments did not include tests at the very low and very high audio frequencies, the design curves are subject to confirmation there.

Ingard [22] has performed similar measurements over open level terrain with sparse low ground cover and over a concrete runway. The source of sound was an aircraft on the ground and the receiver height was about 8 ft. The signal frequencies covered the entire audible frequency range but the distances covered were less than 2,000 ft. Detailed meteorological data are not given. In the frequency range of about 300 to 5,000 cps, the results agree generally with those reported in Refs. 18 and 19. In particular, the excess attenuation upwind into the shadow zone was found to be essentially independent of frequency. At frequencies below about 300 to 400 cps the excess attenuation inside the shadow zone was found to increase with distance at a rate which decreased with decreasing frequency. Excess attenuations measured for propagation over concrete showed a tendency to be smaller than when a ground cover was present.

In a typical noise-control situation the problem is usually not that of estimating the excess attenuation A_{e_4} for a given set of conditions, but of estimating A_{e_4} on a year-round basis perhaps, or for many values of ϕ, since the wind direction is generally subject to diurnal and seasonal changes. Diurnal and seasonal variations occur also in the temperature and wind gradients.[23]

Except for the case of the very low frequencies, where it can be assumed that $A_{e_4} \doteq 0$ upwind, crosswind, and downwind up to distances of about 1 mile from the source, it seems best to restrict ourselves to two conditions: (1) straight downwind propagation ($\phi = 180°$) and (2) straight upwind propagation ($\phi = 0$). These conditions bracket the extreme values of A_{e_4} for any given set of conditions. The engineer must then make his choice appropriate to the problem at hand.

Figure 9.8 shows the empirical design chart appropriate for the determination of A_{e_4} for straight downwind propagation over open level terrain with low ground cover and for frequencies in the range of about 300 to 5,000 cps. The abscissa is plotted in terms of the product $f_m \times r$, where r is the distance from the source to the receiver in feet and f_m is the center frequency of the band in question, in cycles per second. There is no theoretical explanation of the change in slope shown in the graph available at this time.

Figure 9.9 shows the companion design chart, giving values of

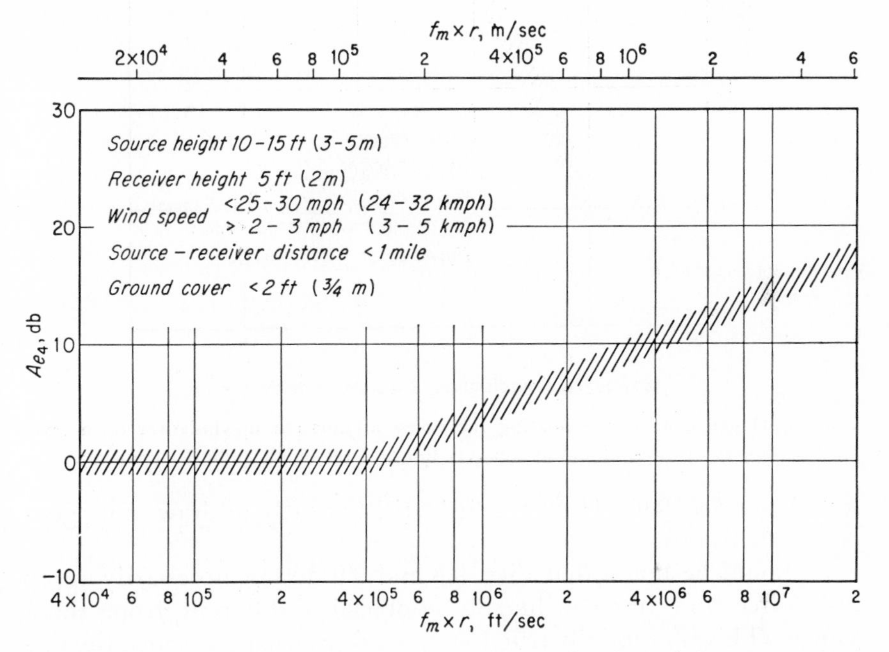

FIG. 9.8 Design chart for estimating the excess attenuation downwind over open level terrain (subject to confirmation at the very low and very high audio frequencies).

A_{e_4} for straight upwind propagation. The abscissa is plotted in terms of source-receiver distance, normalized to the minimum distance to the shadow zone X_0. The distance X_0 can either be obtained by direct measurement in the field or estimated from Table 9.1.

It should be noted that the above design charts contain implicitly the effect of atmospheric turbulence near the ground on the average received sound-pressure level. Work is in progress by several research groups to determine the effect of atmospheric turbulence itself as distinct from other effects. Preliminary results indicate that turbulence causes an increase in the average attenuation, as well as the superposition of fluctuations on the mean level of the received

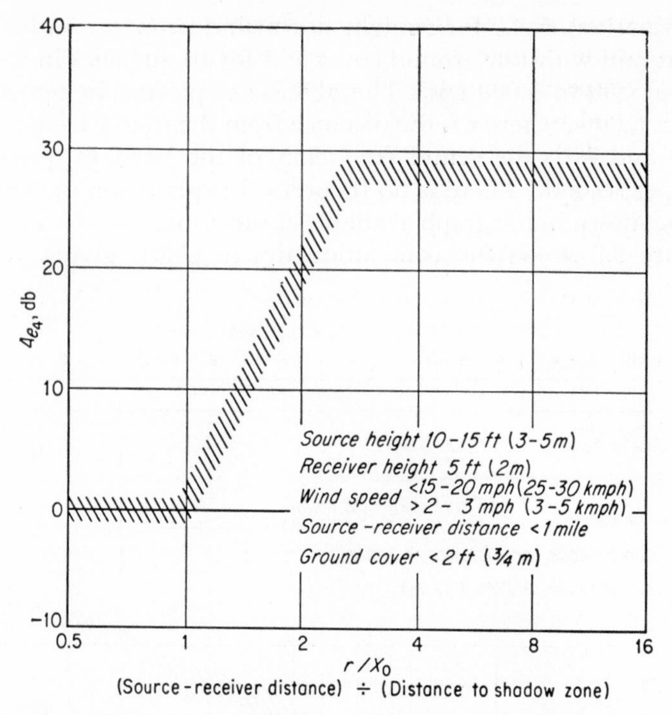

FIG. 9.9 Design chart for estimating the excess attenuation upwind over open level terrain (subject to confirmation at the very low and very high audio frequencies).

sound (see Sec. 9.9). However, it is still too early to draw more general conclusions.

It should be noted that Figs. 9.8 and 9.9 do not necessarily apply for source and receiver heights appreciably different from those shown. This is especially true for upwind propagation, since the distance X_0 to the shadow zone increases with both source and receiver height. Hayhurst[24] has found that for propagation along a concrete runway with source height fixed the excess attenuation decreases

Table 9.1

Estimates of X_0 Upwind, 300 to 5,000 cps, $\phi = 0$
(Source height 10–15 ft, receiver height 5 ft)

Time		Sky		Temperature Profile			Wind, mph	X_0 ft
Day	Night	Clear	Overcast	Lapse	Neutral	Inversion		
	x	x				x	2–4	2,000
x			x		x		10–15	400
x		x		x			10–18	250

with increasing receiver height for upwind propagation. He found no appreciable change downwind for receiver heights up to 30 ft. As the windspeed drops below about 2 to 3 mph, the temperature gradients tend to control and the wind-to-sound angle loses its importance; under these conditions the design charts do not apply.

Example 9.5. A jet aircraft of PWL = 171 db in the 1,200- to 2,400-cps band and directivity index −10 db is run up on the ground near an airfield. The terrain is flat and covered with grass of low height. It is a clear, sunny day, and a 10-mph wind is blowing from the west. The air temperature is 70°F and the relative humidity is 75 per cent. Estimate the sound-pressure level about 5 ft above the ground 2,000 ft away, (*a*) to the east, and (*b*) to the west.

Solution. In the absence of temperature and wind gradients we have SPL = 79 db at 2,000 ft from a previous example. It remains here to estimate A_{e_4}.

(*a*) To the east—downwind: the mean band frequency $f_m = 1{,}700$ cps; $f_m \times r = 3.4 \times 10^6$ ft-sec^{-1}. From Fig. 9.8, $A_{e_4} \doteq 9$ db. *Ans.* SPL = 79 − 9 = 70 db at 2,000 ft.

(b) To the west—upwind: estimate $X_0 = 250$ ft from Table 9.1; $r/X_0 = 8$. From Fig. 9.9, $A_{e_4} \doteq 29$ db. *Ans.* SPL = 79 − 29 = 50 db at 2,000 ft.

9.7 Propagation of Sound over Hilly Terrain

Experiments where sound is propagated from hilltop to hilltop across a valley indicate that wind direction, temperature, and wind gradients play a much smaller role than if sound propagation had taken place over level terrain. After accounting for the excess attenuation resulting from molecular absorption (see Sec. 9.3) and possibly fog, rain, or snow (see Sec. 9.4), the experimental data show a residual excess attenuation which is roughly independent of frequency.[18,19] Residual attenuations of 10 to 20 db have been measured in the frequency range of about 400 to 2,000 cps for distances of 1 to 1½ miles. This residual attenuation is probably due to turbulence. For conservative engineering estimates and until more systematic data are available, it seems best to disregard such attenuation due to turbulence in sound-control problems involving propagation over hilly terrain.

9.8 Propagation of Sound from Air to Ground

This case is of considerable practical importance for the acoustician who needs to estimate the sound-pressure level near the ground due to aircraft overhead. In many ways this situation is not unlike the case of hill-to-hill propagation discussed in the preceding section. The excess attenuation of sound has been measured by several investigators using propeller aircraft, helicopters, and jet aircraft flying at moderate altitudes (½ mile or less) under various atmospheric conditions.[1,25,26,27] It was assumed in every case that the excess attenuation measured for various distances to the airplane (slant range) can be represented in terms of an attenuation coefficient α_e which is independent of the slant range. Figure 9.10 shows, as a function of fre-

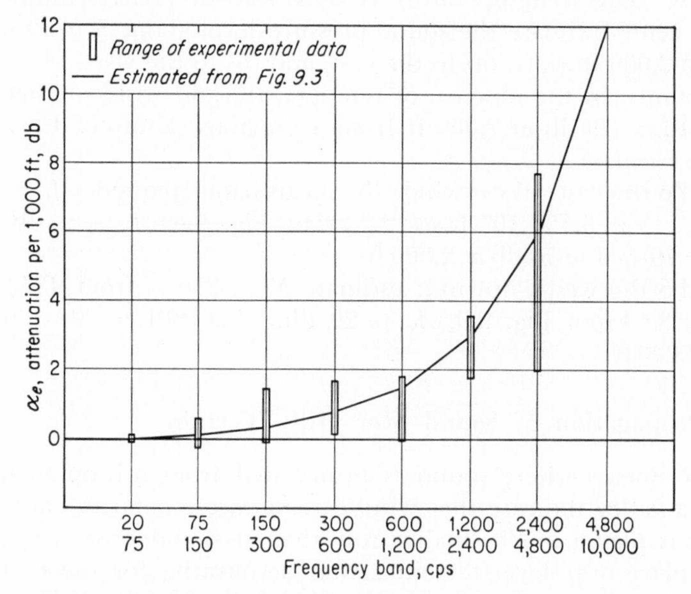

FIG. 9.10 Estimated excess attenuation coefficient for air-to-ground propagation. The range of available experimental data is also shown.

quency, the attenuation coefficients so obtained from the various studies. It can be inferred from the experiments that the excess attenuation can be accounted for by that resulting for molecular absorption (see Sec. 9.3), possibly fog, rain, or snow (see Sec. 9.4) and atmospheric turbulence. The latter is best disregarded for conservative estimates until more systematic data are available. These calculations are to be regarded as an estimate at best, since it is difficult to infer the average atmospheric conditions along the transmission path from meteorological measurements at ground level.

Nonetheless, the values of attenuation due to molecular absorption taken from Fig. 9.3 and plotted on Fig. 9.10 account reasonably well for the results of the measurements of air-to-ground attenuation, assuming precipitation was either absent or had no appreciable effect on the results.

As the aircraft passes overhead in level flight, the sound-pressure level at a fixed point on the ground will rise, go through a maximum, and fall again. The position of the aircraft for maximum sound-pressure level depends on its acoustic directivity pattern. As a first approximation, a 45° position can be assumed for jet aircraft. Consequently, this slant distance is the effective length of the transmission path and must be used not only to compute the excess attenuation but also the spherical divergence. For propeller aircraft the minimum distance to the flight path is relevant. Note that in using Eq. (9.1) the power level in flight should be used.

9.9 Fluctuations of the Received Sound-pressure Level

Fluctuations in level are characteristic of sound that has traveled through the atmosphere. These fluctuations typically encompass a rather wide frequency spectrum and peak-to-peak fluctuations of appreciable magnitude occur. The peak-to-peak fluctuations of the received sound-pressure level at various distances over level ground have been investigated by Baron,[28] Ingard,[22] and by Wiener and Keast.[19] The following general conclusions can be drawn: (1) For downwind propagation the magnitude of the fluctuations increases with the frequency of the signal and with distance; (2) for upwind propagation the magnitude of the fluctuations is greatest near the shadow boundary; (3) in a stable atmosphere (clear night, weak winds) the peak-to-peak fluctuations are typically about 5 db; (4) in an unstable atmosphere (clear sunny day, strong winds) the peak-to-peak fluctuations are typically 15 to 20 db; (5) the spectrum of the fluctuations measured over open level ground encompasses components from fractions of a cycle to several cycles per second; and (6) sound propagation from hilltop to hilltop and from air to ground is frequently characterized by large low-frequency fluctuations in the received sound-pressure level in addition to the faster fluctuations observed over level terrain.

9.10 Summary

Sound waves traveling through the atmosphere over appreciable distances are affected by the inhomogeneities in the medium in essentially two ways: First, the sound-pressure level at the receiver fluctuates in time; and second, the mean value of the received sound-pressure level is generally below that expected from inverse-square-law attenuation alone. This excess attenuation of sound propagated along the ground is profoundly affected by the presence of temperature and wind gradients. Attenuations measured upwind may exceed those measured downwind by as much as 25 to 30 db. The proposition that the excess attenuation of sound propagated along the ground can basically be accounted for by an attenuation coefficient which is independent of distance is not borne out by the facts.

In hilltop-to-hilltop propagation, wind direction is of secondary importance. In dense woods, sound absorption and scattering by the trees and ground cover are important. Considerable depths of planting are required to yield the noise reductions called for in many practical noise-control situations. The excess attenuation of sound radiated by low-flying aircraft can be estimated in many cases by considering molecular absorption alone.

The fluctuations of the received sound propagated along the ground depend on the thermal stratification of the atmosphere through which propagation has taken place. In an unstable turbulent atmosphere (daytime) the fluctuations exceed those experienced under stable (nighttime) conditions. Fluctuations also generally tend to increase with the source-receiver distance and signal frequency.

There are many problems as yet unresolved in the field of atmospheric acoustics, and further work is greatly needed.

REFERENCES

1. Parkin, P. H., and W. E. Scholes: Oblique Air-to-Ground Sound Propagation over Buildings, *Acustica,* vol. 8, pp. 99–102, 1958.

2. Kneser, H. O.: Interpretation of the Anomalous Sound Absorption in Air and Oxygen in Terms of Molecular Collisions, *J. Acoust. Soc. Am.,* vol. 5, p. 122, 1933.

3. Nyborg, W. L., and D. Mintzer: Review of Sound Propagation in the Lower Atmosphere, *WADC Tech. Rept.* 54–602, ASTIA AD–67880, 1955.

4. Knudsen, V. O.: The Absorption of Sound in Air, in Oxygen, and in Nitrogen—Effects of Humidity and Temperature, *J. Acoust. Soc. Am.,* vol. 5, p. 112, 1933.

5. Delsasso, L. P., and R. W. Leonard: Field Measurements of the Attenuation of Sound in the Low Audio Range, *J. Acoust. Soc. Am.* (*Abstr.*), vol. 25, p. 835, 1953.

6. Harris, C. M.: Columbia University, private communication.

7. Evans, E. J., and E. N. Bazley: The Absorption of Sound in Air at Audio Frequencies, *Acustica,* vol. 6, p. 238, 1956.

8. Rudnick, I.: Propagation of Sound in the Open Air, chap. 3 of C. M. Harris (ed.), "Handbook of Noise Control," McGraw-Hill Book Company, Inc., New York, 1957.

9. Knudsen, V. O., J. V. Wilson, and N. S. Anderson: The Attenuation of Audible Sound in Fog and Smoke, *J. Acoust. Soc. Am.,* vol. 20, p. 849, 1948.

10. Fehr, R. O.: The Reduction of Industrial Machine Noise, *Proc. 2d Annual NNAS,* vol. 2, p. 93, 1951.

11. Stevens, K. N., and R. H. Bolt: On the Shielding of Noise Outdoors, *J. Acoust. Soc. Am.* (*Abstr.*), vol. 26, p. 938, 1954.

12. Pietrasanta, A. C.: private communication.

13. Beranek, L. L.: "Acoustics," McGraw-Hill Book Company, Inc., New York, 1954.

14. Rettinger, M.: Noise Level Reductions of Barriers, *Noise Control,* vol. 3, p. 50, September, 1957.

15. Hayhurst, J. D.: Acoustic Screening by an Experimental Running-up Pen, *J. Roy. Aeronaut. Soc.,* vol. 57, p. 1, 1953.

16. Givens, M. P., W. L. Nyborg, and H. K. Schilling: Theory of the Propagation of Sound in Scattering and Absorbing Media, *J. Acoust. Soc. Am.,* vol. 18, p. 284, 1946.

17. Eyring, C. F.: Jungle Acoustics, *J. Acoust. Soc. Am.,* vol. 18, p. 257, 1946.

18. Keast, D. N., and F. M. Wiener: Some Experimental Results Concerning the Propagation of Sound over Ground, *J. Acoust. Soc. Am.* (*Abstr.*), vol. 29, p. 772, 1957.

19. Wiener, F. M., and D. N. Keast: An Experimental Study of the Propagation of Sound over Ground, *J. Acoust. Soc. Am.,* vol. 31, p. 724, 1959.

20. Ingard, U.: The Physics of Outdoor Sound, *Proc. 4th Annual NNAS,* vol. 4, p. 11, 1953.

21. Ingard, U.: A Review of the Influence of Meteorological Conditions on Sound Propagation, *J. Acoust. Soc. Am.,* vol. 25, p. 405, 1953.

22. Ingard, U.: private communication; also Field Studies of Sound Propagation over Ground, *Report to NACA, Contract NAw*–6341, M.I.T., Oct. 29, 1954.

23. Sutton, O. G.: "Micrometeorology," McGraw-Hill Book Company, Inc., New York, 1953.

24. Hayhurst, J. D.: The Attenuation of Sound, *Acustica,* vol. 3, p. 225, 1953.

25. Parkin, P. H., and W. E. Scholes: Air-to-Ground Sound Propagation, *J. Acoust. Soc. Am.,* vol. 26, p. 1021, 1954.

26. Benson, R. W., and H. B. Karplus: Sound Propagation near the Earth's Surface as Influenced by Weather Conditions, *WADC Tech. Rept.* 57–353, 1958.

27. Clark, W. E., A. C. Pietrasanta, et al.: Intrusion of Aircraft Noise into Communities near Two USAF Air Bases, *WADC Tech. Note* 58–213, 1958.

28. Baron, P.: Propagation du son dans l'atmosphère et audibilité des signaux avertisseurs dans le bruit ambiant, *Cahiers d'Acoustique, Ann. Télécomm.,* vol. 9, p. 258, 1954.

29. Sieg, H.: Über die Schallausbreitung im Freien und ihre Abhängigkeit von den Wetterbedingungen, *El. Nachr. Technik,* vol. 17, p. 193, 1940.

Chapter 10

SOUND IN SMALL ENCLOSURES

Leo L. Beranek and R. H. Bolt

10.1 Introduction

In the previous chapters we discussed sound propagation under two simplifying conditions. The first condition was the assumption of one-dimensional waves. By this we mean either waves that are required to travel in one direction only, say down a tube whose cross-sectional dimensions are small compared to a wavelength, or waves that travel outward from a uniformly pulsating sphere or point source. The second condition was that the waves must be free progressive. By this we mean that there are no boundaries in the path of the outward-traveling wave to reflect the sound back toward the source.

In this chapter we deviate from both of these restraining conditions. We first deal with a reflecting surface in the path of a wave traveling along a one-dimensional tube. Then, as a second step, we expand the cross-sectional dimensions of the tube so that waves may travel transversely as well as along the length. In other words, we change our problem from that of a speaking tube to that of a room.

In bringing the story to you, we introduce some mathematics. In order to make the material as easy as possible to read we have, to the best of our ability, described every symbol and every step in detail. You should be able to follow the development if you have had some trigonometry and calculus and if you exercise some imagination and patience. For those with a good college foundation in physics and mathematics, we recommend that you study the texts by Rayleigh[1] and Morse,[2] or their easier counterparts by Beranek,[3] Hunter,[4] Kinsler and Frey,[5] Swenson,[6] and Wood.[7] This chapter is intended for the engineer who has been assigned a noise problem and who probably is encountering room acoustics for the first time.

10.2 Standing Waves in a Tube

Wave Equation. In Chap. 2 we discussed a standing wave as being a combination of two oppositely traveling free waves produced by separate sound sources. Now, let us produce a backward-traveling wave in a tube as a result of reflecting the forward-traveling wave off a rigid wall at the right side of the tube, say at $x = +l_x$. Also, let us put another rigid end at $x = 0$. In other words we now have a tube closed at both ends by rigid walls (see Fig. 10.1).

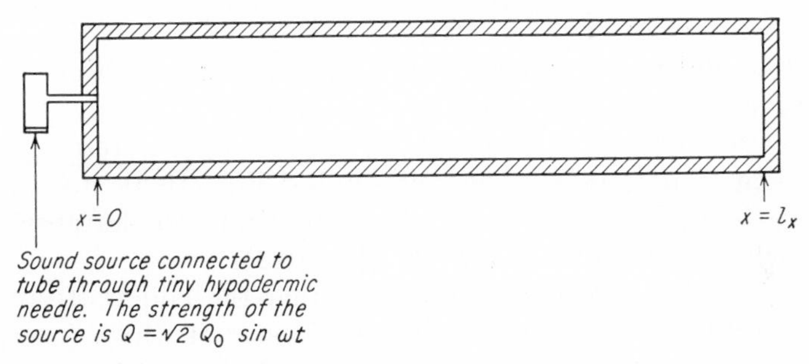

Fig. 10.1 A rigidly closed tube with lateral dimension small compared to a wavelength so that plane standing waves will occur along its length. The sound source is coupled to the main tube through a tube so small that the wall at $x = 0$ appears to be acoustically rigid (no leak).

We shall specify the source of sound as being pinpoint in size at the center of the wall at $x = 0$. Its strength will be

$$Q = \sqrt{2}\, Q_0 \sin \omega t \tag{10.1}$$

where[3] Q_0 is the rms magnitude (in cubic meters per second) of the volume of air per unit time squeezed in and out of the tiny hole at $x = 0$.

The steady-state wave equation with a source present is given by

$$\frac{\partial^2 p}{\partial x^2} - \frac{1}{c^2}\frac{\partial^2 p}{\partial t^2} = \frac{\rho_0}{V}\frac{\partial Q}{\partial t} \tag{10.2}$$

where ρ_0 = density of air, kg/m^3
V = volume of tube = Sl_x, m^3
S = cross-sectional area of tube, m^2
l_x = length of tube, m
p = sound pressure at plane x in tube at time t, newtons/m^2 (this quantity varies both with x and t)

Let us set

$$Q = \sqrt{2}\, Q_0(x) \sin \omega t \tag{10.3}$$

$$p = \sqrt{2}\, P(x) \cos \omega t \tag{10.4}$$

In other words, we are assuming that in general the source may be located at any point x with a sinusoidal output whose rms amplitude Q_0 remains constant with time. This is called the steady-state case. From the nature of Eq. (10.2) we know that if Q varies in this manner, p must also. Then

$$\frac{\partial^2 P(x)}{\partial x^2} + \frac{\omega^2}{c^2} P(x) = \frac{\rho_0 \omega}{V} Q_0(x) \tag{10.5}$$

where $P(x)$ and $Q_0(x)$ are now rms values and are functions only of x because we canceled out $\cos \omega t$.

Sound Pressure in the Tube. It would be interesting to go through the mathematical steps of solving Eq. (10.5), but there is not space in this chapter. Instead, let us jump to the answer for the case where the source is at $x = 0$ (Eq. 10.1).

The solution to Eq. (10.5), assuming that the end walls are only slightly absorbing, is a *sum of terms, each term* of which (except for a phase angle that is dependent on ω) is given by[2]

$$|P_m(x)| = \frac{2KQ_0\rho_0\omega\,[\cos\,(m\pi x/l_x)]}{V[4\omega_m{}^2k_m{}^2 + (\omega^2 - \omega_m{}^2)^2]^{1/2}} \tag{10.6}$$

The vertical bars around $P_m(x)$ indicate that we are not including the phase angle. Now let us look at each symbol in Eq. (10.6) carefully. [Q_0, ρ_0, l_x, and V are defined near Eqs. (10.1) and (10.2).]

$|P_m(x)|$ = rms sound pressure associated with one of the terms (a particular value of m) at a point x along tube, newtons/m²
m = integer, i.e., 1, 2, 3, . . .
$\omega = 2\pi f$ = angular frequency of source, radians/sec
f = frequency of source, cps
K = constant that depends on speed of sound, etc.
k_m = damping (dissipation) constant of ends of tube, radians/sec
$\omega_m = 2\pi f_m$ = normal (natural) angular frequency of tube, radians/sec

The pressure at any point for each mode of vibration also varies sinusoidally, so that

$$p_m(x,t) = \sqrt{2}\, P_m(x) \cos \omega t \tag{10.7}$$

The factor $\sqrt{2}$ is in the equation because $P_m(x)$ is defined as an rms quantity.

The *normal frequencies* (i.e., the "natural" frequencies) of the tube are given by

$$f_m = \frac{\omega_m}{2\pi} = \frac{cm}{2l_x} \tag{10.8}$$

where m is an integer, 1, 2, 3, . . . (10.9)

The *resonance frequencies* of the tube f_{0m} are found by maximizing Eq. (10.6) as a function of frequency. In other words, as the frequency ω is varied and approaches ω_m, the denominator of Eq. (10.6) becomes smaller so that $P_m(x)$ becomes larger. Because ω also appears in the numerator, the exact frequency ω_0 at which $P(x)$ becomes a maximum is given by

$$\omega_{0m} = \sqrt{\omega_m{}^2 + k_m{}^2} \tag{10.10}$$

In the case where $k_m{}^2 << \omega_m$, such as in this tube where the ends are not covered with an efficient sound-absorbing material,

$$\omega_{0m} \doteq \omega_m \tag{10.11}$$

The full meaning of Eqs. (10.6) to (10.11) can probably best be learned from a numerical example.

Example 10.1. Given a tube 0.688 m long, the speed of sound equal to 344 m/sec, and a source of sound whose frequency f can be varied. (*a*) Find the first four normal frequencies of the tube; (*b*) find the first four resonance frequencies of the tube, assuming that the damping constant for the tube varies as follows: $k_1 = 8\pi$, $k_2 = 40\pi$, $k_3 = 100\pi$, and $k_4 = 200\pi$ sec^{-1}; (*c*) at $x = l_x$ plot a resonance curve for the second resonance frequency; and (*d*) plot the sound-pressure distributions in the tube for the first and fourth normal modes of vibration.

Solution. (*a*) The first four normal frequencies, ω_m, occur for $m = 1$, 2, 3, and 4 in Eq. (10.8). So

$$f_1 = \frac{344}{2 \times 0.688} = 250 \text{ cps}$$

$$f_2 = 2 \times 250 = 500 \text{ cps}$$

$$f_3 = 3 \times 250 = 750 \text{ cps}$$

$$f_4 = 4 \times 250 = 1{,}000 \text{ cps}$$

(*b*) The first four resonance frequencies f_{0m} are given by Eq. (10.10)

$$f_{01} = \sqrt{250^2 + 4^2} = 250.032 \doteq 250 \text{ cps}$$

$$f_{02} = \sqrt{500^2 + 20^2} = 500.40 \doteq 500 \text{ cps}$$

$$f_{03} = \sqrt{750^2 + 50^2} = 751.67 \doteq 752 \text{ cps}$$

$$f_{04} = \sqrt{1{,}000^2 + 100^2} = 1{,}005.0 \doteq 1{,}005 \text{ cps}$$

(*c*) The pressure vs. frequency curve for the $m = 2$ mode of vibration at $x = l_x$ is computed from the equation below and is plotted in Fig. 10.2. We get $\omega_2 = 2\pi \times 500 \doteq 3{,}140$ radians/sec.

$$\left|\frac{P_2(\omega)}{P_2(\omega_{02})}\right|_{x=l_x} \doteq \frac{2\omega k_2}{[(4\omega_2{}^2k_2{}^2) + (\omega^2 - \omega_2{}^2)^2]^{1/2}}$$

$$= \frac{2 \times 20(\omega/3{,}140)}{\left[4 \times 400 + \left(1 - \frac{\omega^2}{3{,}140^2}\right)^2\right]^{1/2}}$$

$$= \frac{\omega/3{,}140}{\left[1 + \frac{1}{1{,}600}\left(1 - \frac{\omega^2}{3{,}140^2}\right)^2\right]^{1/2}}$$

(*d*) The curves of pressure distribution as a function of x for $m = 1$ and $m = 4$ are computed from

$$P_1(x) = [P_1(x = 0)] \cos \frac{\pi x}{0.688}$$

and

$$P_4(x) = [P_4(x = 0)] \cos \frac{4\pi x}{0.688}$$

The results are given in Fig. 10.3.

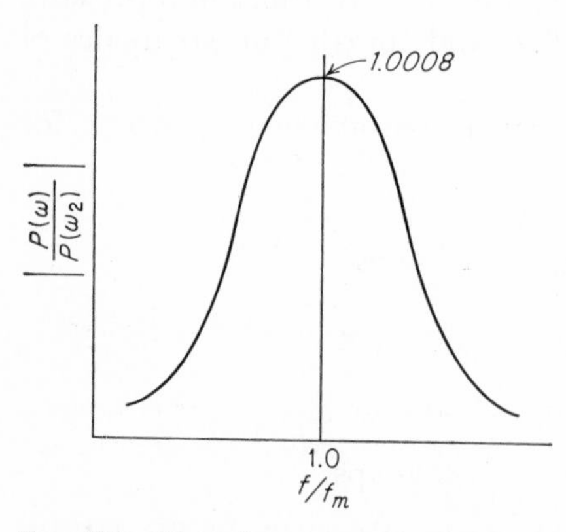

FIG. 10.2 Plot of a resonance curve as computed in Example 10.1 for $x = l_x$ and $m = 2$.

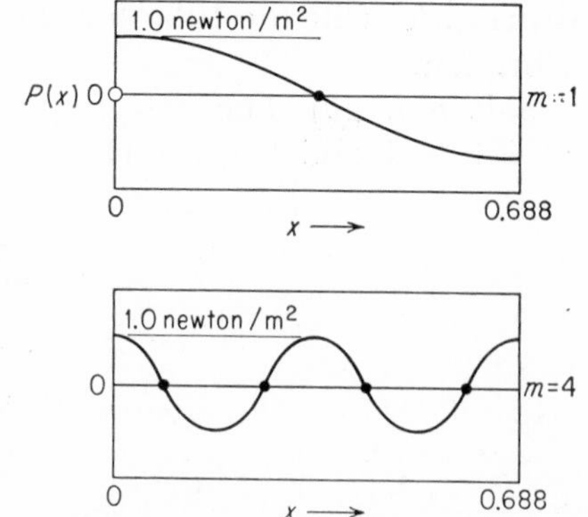

FIG. 10.3 Plot of the sound-pressure distribution in the tube of Example 10.1 for $m = 1$ and $m = 4$.

In both plots we have arbitrarily assumed that at $x = 0$, the pressures $P_1(x = 0)$ and $P_4(x = 0)$ are equal to 1 newton/m^2. The actual values would come from Eq. (10.6), provided we knew the value of K. Because K is difficult to determine, in practice the constant is generally measured with a microphone.

Summary. We have shown that a tube closed at both ends will resonate at an infinite number of frequencies. The normal frequencies f_m are given by Eq. (10.8) and are harmonically related to each other, as we see from Eq. (10.9).

For each normal mode of vibration, the magnitude of the sound pressure is distributed in the tube according to a cosine function (Eq. 10.6) with maximum values at the two ends, $x = 0$ and $x = l$, and at every half-wavelength point in between (Fig. 10.3).

If the frequency of the source of sound f is about equal to one of the normal frequencies f_m, resonance is said to occur, and $P_m(x)$ becomes very large [Eq. (10.6) and Fig. 10.2]. If the source has many frequencies, the tube will resonate at every frequency where the source and normal frequencies are nearly alike. As we vary frequency, the value of $P_m(x)$ increases along one side of the resonance curve (Fig. 10.2), goes through a maximum at resonance (f_0), and then decreases along the other side. However, $P_m(x)$ will soon begin to increase along the left side of the resonance curve for the next higher normal frequency. So, for a varying frequency f, the magnitude of the rms sound pressure passes through a maximum each time f about equals one of the normal frequencies f_m. Remember that the instantaneous sound pressure varies sinusoidally with time above and below atmospheric pressure all of the time, at the rate of f times a second, according to Eq. (10.7).

The maximum value to which the pressure rises at resonance is determined by the dissipation, i.e., the losses at the ends of the tube. That is to say, from Eq. (10.6),

$$P_m(x = 0) \text{ at resonance} = \frac{KQ_0\rho_0/V}{k_m} \tag{10.12}$$

where k_m is the damping constant. If the damping constant is halved, the sound pressure at resonance doubles.

Just as we saw for a simple standing wave (Chap. 2), the sound pressure is zero at points in the tube halfway between the maximum points (see Fig. 10.3). Zero pressure points are called *nodal pressure* points. Actually, when there is dissipation in the tube, i.e., when k_m is not equal to 0, the sound pressures at the nodal points do not equal 0.

Particle Velocity in the Tube. The particle velocity at any point is the instantaneous velocity of the molecules of the air at that point (first averaging out random motion). The particle velocity u is always back and forth along a line parallel to the direction of travel of the sound wave. Of basic importance is the fact that it is always related to the *sound-pressure gradient* at that point through the equation

$$u = -\frac{1}{\rho}\int \frac{\partial p}{\partial x}\,dt \tag{10.13}$$

If we again assume the steady state, then p is given by Eq. (10.4), and u is given by

$$u = \sqrt{2}\,U(x)\cos(\omega t + \theta) \tag{10.14}$$

where θ is a time phase angle relative to the sound pressure. The phase angle is determined when the operations of Eq. (10.13) are executed.

In the steady state (and assuming a particular value of m) the insertion of Eqs. (10.7) and (10.14) in Eq. (10.13) yields

$$U_m(x) = \frac{-1}{\omega\rho_0}\frac{\partial P_m(x)}{\partial x} \qquad \underline{/-90^\circ} \tag{10.15}$$

The symbol $\underline{/-90^\circ}$ means that the phase angle introduced into $U_m(x)$ by the time integral is $-90°$. The total angle θ (see Eq. 10.14) is the sum of $-90°$ and the phase angle introduced by the derivative of $P_m(x)$ with respect to x.

To obtain the magnitude of the particle velocity $U(x)$ in the tube of Fig. 10.1, we substitute Eq. (10.6) into Eq. (10.15). Remembering Eq. (10.7), this operation yields

$$U_m(x) = \frac{|P_m(x = 0)|}{\rho_0 c}\sin\frac{m\pi x}{l_x} \qquad \underline{/-90^\circ} \tag{10.16}$$

where, as we can see from Eq. (10.6),

$$|P_m(x = 0)| = \frac{2KQ_0\rho_0\omega}{V[4\omega_m{}^2k_m{}^2 + (\omega^2 - \omega_m{}^2)^2]^{1/2}} \tag{10.17}$$

Equation (10.16) contains four pieces of information that we should know: (1) The magnitude of the particle velocity is directly proportional to the $x = 0$ values of the sound pressure. That is to say, when ω is varied, the particle velocity will follow the same type of resonance curve as that shown in Fig. 10.2. (2) The particle velocity is at a maximum at those points in the tube where the sound pressure is near-zero and vice versa. This can be seen by the presence of the sin $(m\pi x/l_x)$ in Eq. (10.16) as compared to the cos $(m\pi x/l_x)$ in Eq. (10.6). In particular, at $x = 0$ and $x = l_x$, the particle velocity is near-zero because the walls are near-rigid there (not very absorbing. (3) The maximum value of $U_m(x)$, i.e., when sin $(m\pi x/l_x) = 1$, is related to the maximum value of $P_m(x)$, i.e., when cos $(m\pi x/l_x) = 1$, by the factor $1/\rho_0 c$. In this respect there is similarity with a free progressive plane wave, wherein $p/u = \rho_0 c$. (4) The particle velocity lags in time behind the sound pressure by $T/4 = 1/4f$ sec, that is to say, by a phase angle of $-90°$. This means that the instantaneous sound pressure at any point in the tube rises to its maximum one-fourth of a period before the particle velocity.

Example 10.2. Assume the same tube and speed of sound as for Example 10.1. The normal frequencies, resonance frequencies, and the resonance curve are, of course, the same as for that example. Plot the rms particle velocity distributions in the tube for the first and fourth normal modes of vibration.

Solution. The curves of rms particle velocity distribution as a function of x for $m = 1$ and $m = 4$ are computed from

$$U_1(x) = \frac{P_1(x = 0)}{410} \sin \frac{\pi x}{0.688}$$

$$U_4(x) = \frac{P_4(x = 0)}{410} \sin \frac{4\pi x}{0.688}$$

Assuming as we did in the previous examples that $P_1(x = 0) = P_4(x = 0) = 1$ newton/m², we obtain the results of Fig. 10.4.

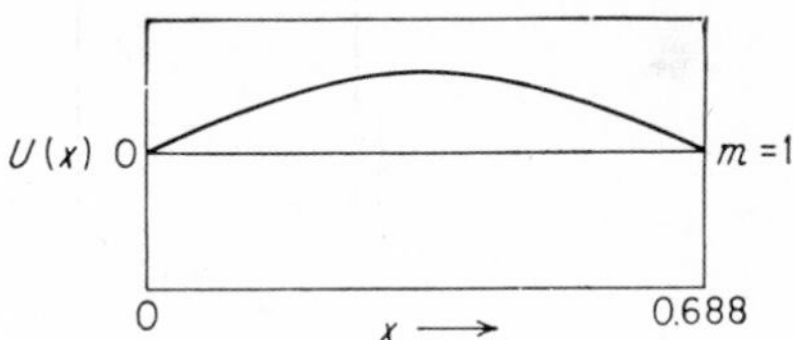

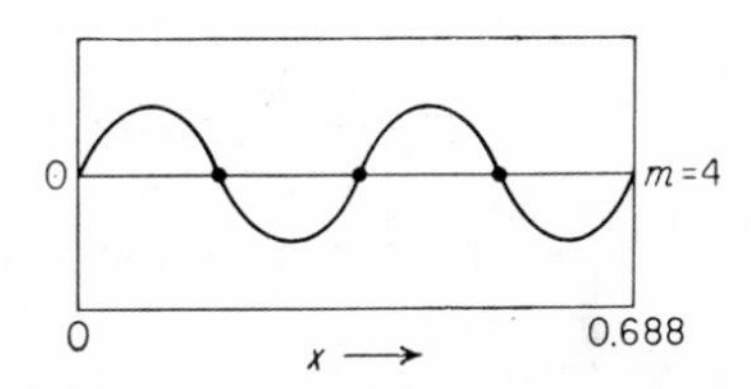

FIG. 10.4 Plot of the particle velocity distribution in the tube of Example 10.1 for $m = 1$ and $m = 4$.

10.3 Waves in Rectangular Enclosures

Having already studied the acoustical conditions in a tube, it is easy to extrapolate to the case of a three-dimensional enclosure, using the concepts introduced so far. Let us start with a perfectly rectangular enclosure with smooth parallel walls. Let us assume that there is only a very little damping on the walls. Furthermore, let us put our tiny source at a corner of the room, say at $x = y = z = 0$ (see Fig. 10.5).

When the source of sound is started, a sound wave spreads out in all directions because the wave is no longer confined by the small cross section of the tube. It can reflect obliquely off the walls at $x = l_x$, $y = l_y$, and $z = l_z$ and travel around the room in every direction. Now, if each path that the wave takes is traced around the room, there will be certain paths of travel that will repeat on themselves just as was the case for the tube, and resonances will occur. In other words, the room has its *normal modes of vibration* (resonances) and its *normal frequencies*.

There are three types of normal modes of vibration in the enclosure:

1. *Axial modes*—in which the component waves move parallel to an axis (one-dimensional)

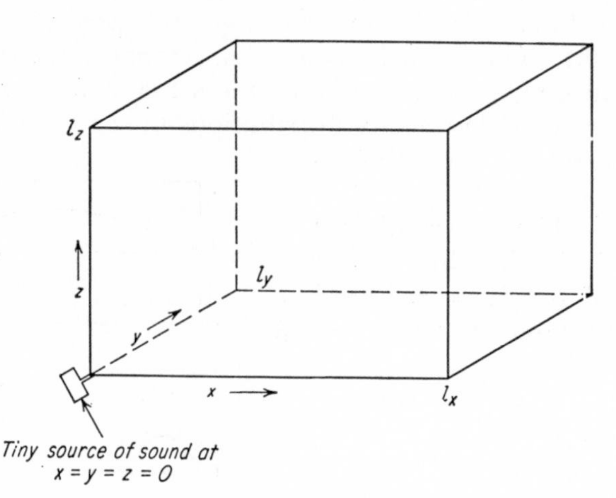

FIG. 10.5 Rectangular enclosure with dimensions l_x, l_y, and l_z.

2. *Tangential modes*—in which the component waves are tangential to one pair of surfaces, but are oblique to the other two pairs (two-dimensional)

3. *Oblique modes*—in which the component waves are oblique to all three pairs of the walls (three-dimensional)

Steady-state Sound Pressure in Rectangular Enclosures. For each normal mode of vibration, the rms sound pressure in the rectangular room of Fig. 10.5 at the point x,y,z is given by an equation almost exactly like Eq. (10.6), namely[2]

$$|P_n(x,y,z)| = \frac{2KQ_0\rho_0\omega\psi_n(x,y,z)}{V[4\omega_n{}^2k_n{}^2 + (\omega^2 - \omega_n{}^2)^2]^{1/2}} \tag{10.18}$$

where K = constant that depends on location of source, speed of sound, etc.

Q_0 = rms strength of source, m^3/sec
ρ_0 = density of air, kg/m^3
ω = angular frequency of source, radians/sec
V = volume of room, m^3
ω_n = normal angular frequency for nth mode, radians/sec
k_n = damping constant, units of inverse time, for nth mode

The factor $\psi(x,y,z)$ gives the sound-pressure distribution in the room, namely:

$$\psi(x,y,z) = \cos\frac{n_x \pi x}{l_x} \cos\frac{n_y \pi y}{l_y} \cos\frac{n_z \pi z}{l_z} \tag{10.19}$$

The normal frequencies of the room, in cycles per second, are given by

$$f_n = \frac{\omega_n}{2\pi} = \frac{c}{2}\sqrt{\left(\frac{n_x}{l_x}\right)^2 + \left(\frac{n_y}{l_y}\right)^2 + \left(\frac{n_z}{l_z}\right)^2} \tag{10.20}$$

where n_x, n_y, and n_z can independently take on all integral values 0, 1, 2, 3, . . . , ∞; c is the speed of sound in meters per second; and l_x, l_y, and l_z are the dimensions of the room in meters.

Example 10.3. Assume that the dimension l_z is less than 0.1 of all wavelengths being considered. This corresponds to n_z being zero at all times. Hence,

$$f_{n_x,n_y} = \frac{c}{2}\sqrt{\left(\frac{n_x}{l_x}\right)^2 + \left(\frac{n_y}{l_y}\right)^2} \tag{10.21}$$

Let $l_x = 1.5$ m and $l_y = 1.0$ m. (*a*) Find the normal frequencies of the $n_x = 2$, $n_y = 0$; the $n_x = 1$, $n_y = 1$; and the $n_x = 2$, $n_y = 1$ normal modes of vibration. (*b*) Plot the sound-pressure distribution for these three normal modes of vibration.

Solution. (*a*) From Eq. (10.20) we have

$$f_{2,0,0} = \frac{344}{2}\sqrt{\left(\frac{2}{1.5}\right)^2} = 229 \text{ cps}$$

$$f_{1,1,0} = \frac{344}{2}\sqrt{\left(\frac{1}{1.5}\right)^2 + \left(\frac{1}{1}\right)^2} = 207 \text{ cps}$$

$$f_{2,1,0} = \frac{344}{2}\sqrt{\left(\frac{2}{1.5}\right)^2 + \left(\frac{1}{1}\right)^2} = 287 \text{ cps}$$

(*b*) The pressure distribution is calculated from Eq. (10.19), with

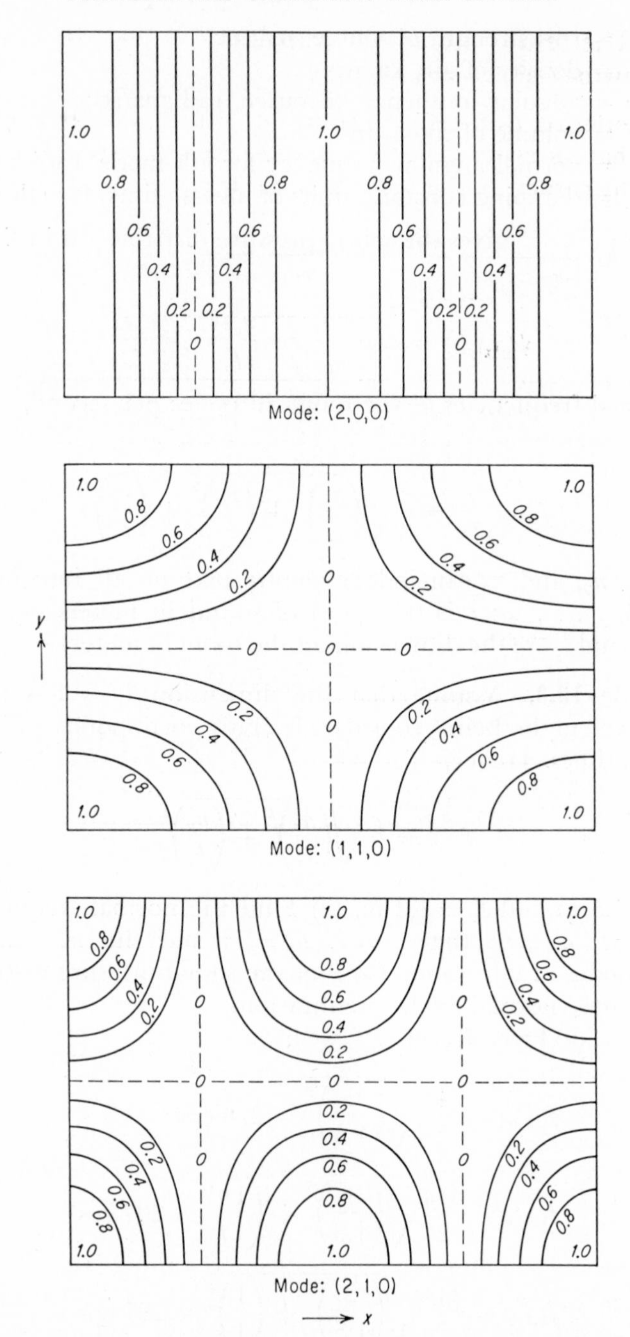

FIG. 10.6 Sound-pressure contour plots on a section through a rectangular room for three different modes of vibration. The numbers on the plots indicate the relative sound pressure.

$n_z = 0$. The results are shown graphically in Fig. 10.6, assuming $P_n = 1$ newton/m^2 at $x = y = z = 0$.

An interesting thing is seen from Eq. (10.19) and from Fig. 10.6, namely, that for every mode of vibration the sound pressure is maximum at the corners of the box. Also, for every mode of vibration for

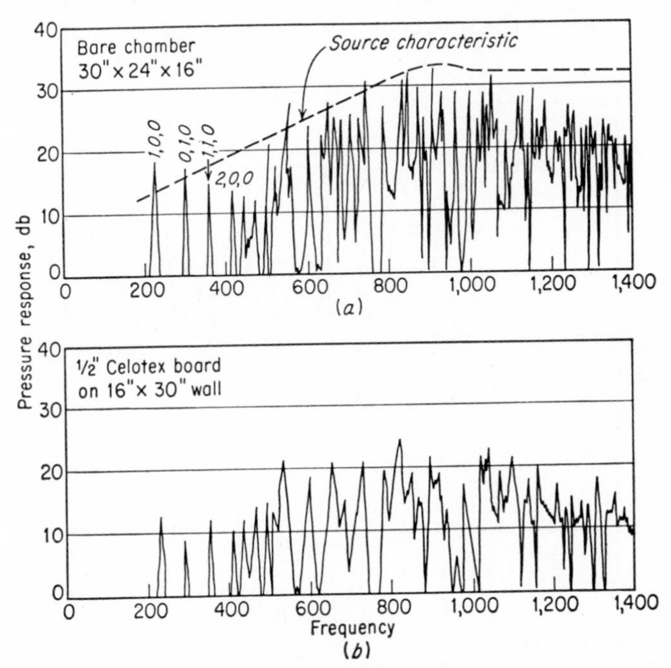

FIG. 10.7 Comparison of two transmission curves recorded with and without a sound-absorbing sample on a 16- by 30-in. wall of a model chamber with dimensions 16 by 24 by 30 in. The microphone was in one corner, and the source was diagonally opposite. The dashed line shows the relative response of the small loudspeaker (3/8 in. diameter) measured at 2 in. in free space. Zero decibel for the source curve is about 50 db *re* 0.0002 μbar and for the transmission curve is about 78 db *re* 0.0002 μbar. (*From Hunt, Beranek, and Maa, Analysis of Sound Decay in Rectangular Rooms, J. Acoust. Soc. Amer., vol.* 11; *pp.* 80–94, 1939.)

which one of the numbers n_x, n_y, or n_z is *odd,* the sound pressure is zero at the *center* of the room, hence at the geometrical center of the room only one-eighth of the modes of vibration have a finite sound pressure. Extending this further, at the center of any one wall the modes for which two of the n's are odd will have zero pressure, so that only one-fourth of them are detectable. Finally, at the center of one edge of the room, the modes for which one of the n's is odd will have zero pressure, so that only one-half of them are detectable there. This latter case would also occur for a microphone at $x = l_x/2$ in the tube of Fig. 10.1. Note from Eq. (10.6) that at the $l_x/2$ plane in the middle of the tube all modes of vibration for which m is odd produce zero sound pressure because $\cos(m\pi/2) = 0$.

Examples of the sound pressure as a function of frequency produced in a small rectangular enclosure measured with the microphone at one corner and either (1) the source at the diagonally opposite corner or (2) the source in the center of the room are given re-

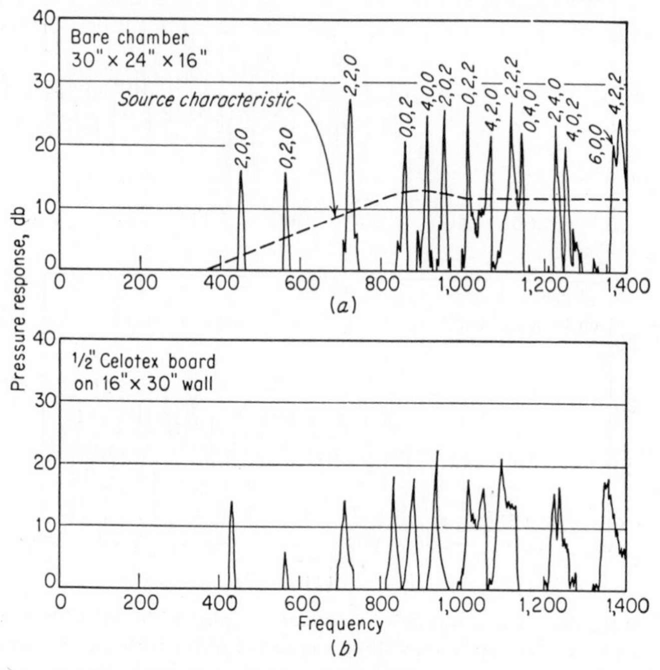

FIG. 10.8 Same as Fig. 10.7, except that the source was in the center of the room and the 0-db reference for the source characteristic is about 71 db *re* 0.0002 μbar.

spectively in Figs. 10.7 and 10.8. The curves were obtained by slowly varying the frequency (a pure tone) of the loudspeaker and simultaneously recording the output of the microphone. The eightfold increase in the number of modes of vibration that were excited with the sources at the corner over that with the source at the center is apparent. It is apparent also that the addition of sound-absorbing material on the 16- by 30-in. wall decreases the height of resonance peaks, particularly at the higher frequencies where it is most effective.

Sound Decay in Small Rectangular Enclosures. Thus far in this chapter, we have dealt only with the case of a steady source—the so-called "steady-state" case. When the source of sound is turned off, the sound in the room dies out, or decays, at a rate that depends on the losses, or damping, in the room. In fact, each mode of vibration behaves in a manner independent of the others. The total process of sound decay is the rms summation of the sound pressures associated with the individual modes.

At the time $t = 0$, when the sound source is turned off, the sound pressure associated with a particular mode is given by Eq. (10.18). A particular mode of vibration has its angular normal frequency ω_n, and its damping constant k_n. Let us call the sound pressure at $t = 0$, $p_n(t = 0)$.

The sound pressure for each normal mode of vibration, after $t = 0$, decays at its own normal frequency at an exponential rate of k_n. That is to say,

$$p_n(t) = p_n(t = 0)e^{-k_n t} \cos \omega_n t \qquad (10.22)$$

Let us take the rms time average of $\cos \omega_n t$ and designate $p_n(t)$ as $\overline{p}_n(t)$. Then Eq. (10.22) says that on a log $\overline{p}_n$ scale vs. time, the *envelope* of the sound pressure decays linearly with time (see Fig. 10.9*a*).

When many normal modes of vibration (each with its own amplitude, normal frequency, and damping constant) decay simultaneously, the total rms sound pressure is given by

$$\overline{p}(t) = \sqrt{\overline{p}_1^2 + \overline{p}_2^2 + \cdots + \overline{p}_n^2} \qquad (10.23)$$

where n represents any combination of the indices n_x, n_y, n_z describing the modes involved in the decay. The logarithm of $\overline{p}(t)$ as a function of time for two such multiple cases is shown in Fig. 10.9*b* and *c*. We see that the decay is irregular because the different modes of vibration have different frequencies and beat with each other during decay.

10.4 Summary

The main points we have tried to cover are:

1. In a narrow tube of length l_x a normal mode of vibration will occur at every frequency for which l_x is an integral multiple of a half wavelength, i.e., for $f_m = mc/2l_x$, where c is the speed of sound and m is an integer.

2. The extent to which a normal mode of vibration is excited depends on its damping constant k_m and the nearness of the frequency of the source f to the normal frequency f_m. If the two frequencies are equal, then the extent of the excitation is propor-

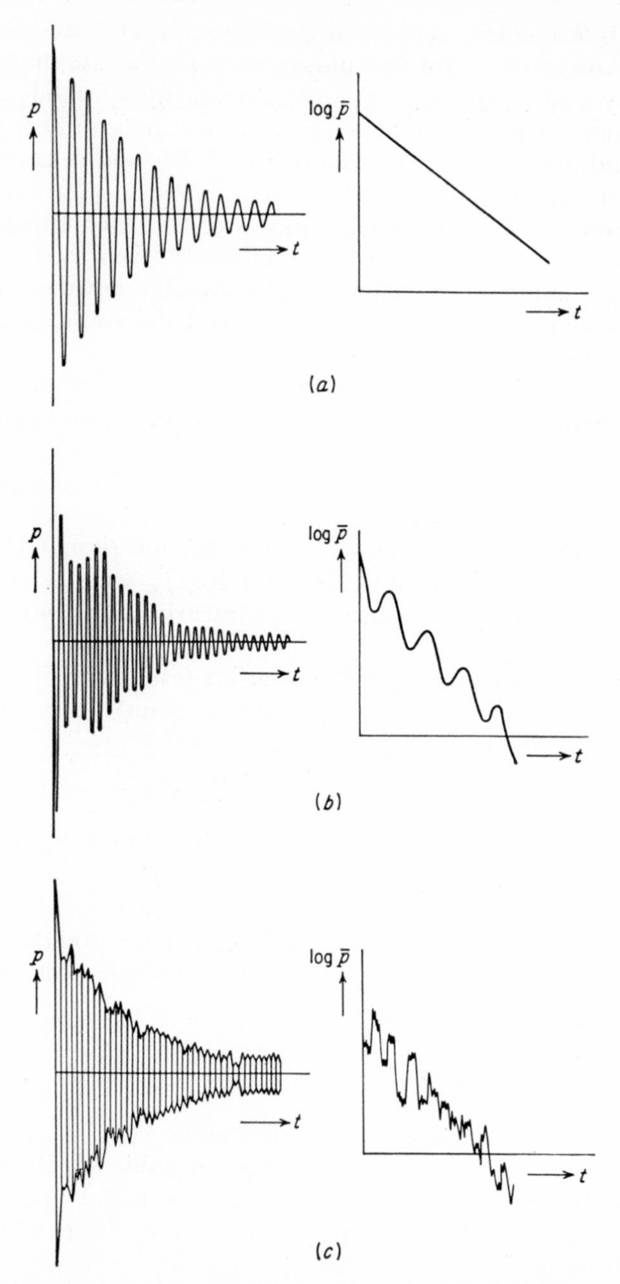

FIG. 10.9 Sound-pressure decay curves: (*a*) for a single mode of vibration; (*b*) for two closely spaced modes of vibration with the same decay constant; (*c*) for a number of closely spaced modes of vibration with the same decay constant. The graphs on the left show the course of the instantaneous sound pressure, and those on the right show the curve of the envelope of the left graphs plotted in a log $\bar{p}$ vs. t coordinate system.

tional to Q_0/k_m, where Q_0 is the strength of the source (see Eq. 10.12).

3. The particle velocity in a standing wave lags behind the sound pressure by a 90° phase angle, i.e., a time lag of one-fourth a period ($T/4$). Also, the particle velocity in such a wave is zero at pressure maxima and is maximum at pressure zeros. In a plane free progressive wave (Chap. 2) the particle velocity and the sound pressure are in phase and are related to each other at all points in space by the factor $\rho_0 c$.

4. In three-dimensional enclosures, a normal mode of vibration occurs at a frequency for every permutation of n_x, n_y, and n_z in Eq. (10.20), where these indices are confined to integral values: 0, 1, 2, 3, ...

5. Statements 2 and 3 above also apply to the three-dimensional case.

6. When sound decays in a room, the total sound pressure equals the rms summation of the rms sound pressures for the individual modes of vibration. Each mode has its own normal frequency ω_n, damping constant k_n, and initial amplitude $p_n(t = 0)$. The summation is given by Eq. (10.23) and is illustrated in Fig. 10.9.

REFERENCES

1. Lord Rayleigh: "The Theory of Sound," vols. I and II, Dover Publications, New York, 1945. (Vol. I was first published in 1877, revised in 1894; vol. II was first published in 1878, revised in 1896. Both volumes are available since 1945 in the United States as two volumes bound in one.)
2. Morse, P. M.: "Vibration and Sound," 2d ed., McGraw-Hill Book Company, Inc., New York, 1948.
3. Beranek, L. L.: "Acoustics," McGraw-Hill Book Company, Inc., New York, 1954.
4. Hunter, J. L.: "Acoustics," Prentice-Hall, Inc., Englewood Cliffs, N.J., 1957.
5. Kinsler, L. E., and A. R. Frey: "Fundamentals of Acoustics," John Wiley & Sons, Inc., New York, 1950.
6. Swenson, G. W., Jr.: "Principles of Modern Acoustics," D. Van Nostrand Company, Inc., Princeton, N.J., 1953.
7. Wood, A. B.: "A Textbook of Sound," The Macmillan Company, New York, 1955.

Chapter 11
SOUND IN LARGE ENCLOSURES

*R. F. Lambert**

11.1 Introduction

The acoustics of enclosures is one of the most important aspects of the science of sound. It is of interest to the architect, scientist, engineer, musician, and for that matter to everyone who can speak and hear. From a technical point of view, almost every problem involving large, irregular enclosures is a new one because almost no two are exactly alike. Thus, the attack on each problem must be formulated from judgments based upon technical knowledge, experience, and ability. Strictly speaking, there is no exact solution to any acoustics problem involving large enclosures. A satisfactory solution generally comes about by methods involving some empiricism.

However, in the past decade our knowledge and understanding of acoustics of large enclosures have increased to the point where satisfactory solutions to many problems can be systematized and put into the hands of individuals who are not necessarily acoustical experts or specialists. Some of the most rapid advances have been in the field of absorbing materials which play a major role in room acoustics and the technology of noise reduction. We now feel confident that in many applications acoustical materials will perform as specified, thereby eliminating the necessity for costly cut-and-try procedures.

For the large, irregularly shaped enclosures which are discussed in this chapter, no precise mathematical description can be given for the sound field. Instead, we formulate a statistically reliable statement about the average conditions of the sound field in the room. Very frequently, in noise-reduction problems this is all that is necessary. We shall see that statistical studies lead to simpler mathematical statements about acoustical conditions than the detailed study of

* Professor of Electrical Engineering, Institute of Technology, University of Minnesota, Minneapolis.

a particular room geometry. This is fortunate, because there are only a small handful of geometries that lend themselves to detailed study. Progress is being made in the mathematical treatment of geometries which do not lend themselves to an exact calculation by employing various perturbation techniques,[1] but this science still has far to go before practical solutions are forthcoming. Problems involving such considerations are best left to the theoretician for the time being.

11.2 Sound-absorption Coefficients

Definition of Absorption and Noise-reduction Coefficients. Probably the one aspect of noise control of greatest interest in applied problems is that of sound-absorption coefficients of materials and objects. When a sound wave falls on a surface, its energy is partially reflected and partially absorbed. The sound-absorbing ability of the surface is given in terms of an absorption coefficient designated by the symbol α. This absorption coefficient is defined as the ratio of the energy absorbed by the surface to the energy incident upon the surface. Accordingly, α may take on all numerical values between 0 and 1. The values[2] of absorption coefficients usually range from about 0.01 for marble slate up to about 1.0 for long absorbing Fiberglas wedges such as are used in anechoic chambers.

We expect that the absorption of a given material will differ depending upon how it is used, for example, for various types of mounting. In addition, all materials have absorption coefficients which are frequency dependent. Materials containing resonant cavities have a frequency region of high absorption surrounded on either side by regions of low absorption. Thus, to specify the absorbing qualities of a material or object, a curve of α as a function of frequency is needed. If, for simplicity, such a curve for a commercial material is not given, its performance may be specified at 500 cps or by a noise-reduction coefficient (NRC) obtained by averaging (to the nearest multiple of 0.05) the absorption coefficients at 250, 500, 1,000, and 2,000 cps. Typical absorption characteristics of porous commercial materials used in noise-reduction applications are shown in Figs. 11.1 and 11.2. It is well to note that some spread in absorption coefficients for commercial materials is usually found.[2,3] Further data are given in Chap. 15.

There is still another added complication which enters the picture. As a sound wave travels around inside an enclosure it suffers many reflections from walls and other objects. The wave will strike these surfaces at various angles of incidence. It is known that the absorption coefficient varies with angle of incidence. In order to obtain a more realistic index for practical use, a statistical average absorption coefficient at each frequency is usually measured and given in the literature.

In a very large and irregular reverberant enclosure excited by noise, the number of sound waves striking a given surface per second is so large that at each surface sound waves from all directions are equally probable. The sound field in such a room is said to be

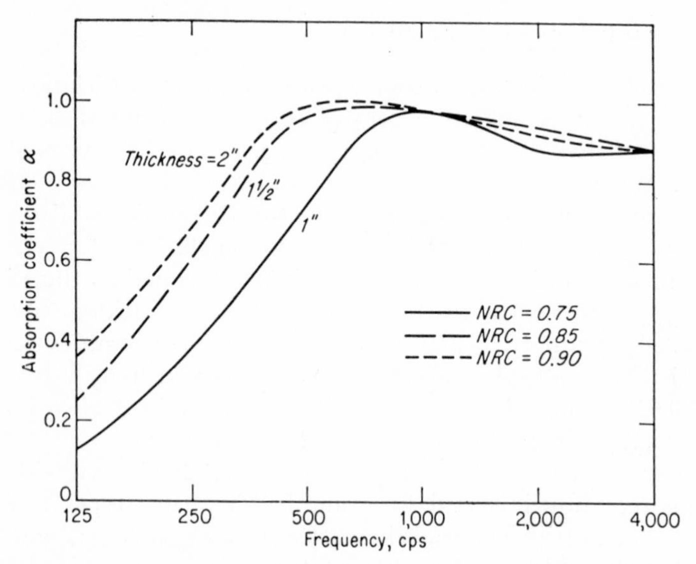

Fig. 11.1 Absorption coefficients and noise-reduction coefficients of porous rigid glass-fiber formboard mounted directly on hard backing. The results show that when the thickness of the board is less than $\frac{1}{16}$th wavelength the absorption halves for each halving of frequency, approximately.

diffuse. The sound-absorption coefficient of importance in this type of room is obtained by averaging the ratio of absorbed to incident energy over *all* angles of incidence between 0° and 90°. The absorption coefficients of commercial materials given in published tables are usually measured in a reverberation chamber where a diffuse field is not fully achieved. However, the degree of diffusion is much greater than that existing in most inhabited rooms and offices. The use of a reverberation chamber is to be contrasted with measuring the absorption coefficient in a tube where the wave impinges at only one angle of incidence, namely, 0°. Hence, the values of α obtained from published tables such as those of Ref. 2 are more directly applicable to noise-reduction and reverberation calculations than those from laboratories where only tube measurements are made. As a word of caution, however, the published values of α may differ greatly from the values of α measured in some particular situation, because of the particular angles of incidence in the wave field at the acoustical material. This is one of many reasons why in acoustical engineering one must use judgment in addition to published formulas and data.

Average of Absorption Coefficients in Rooms. There is still another type of averaging of absorption coefficients which is necessary for simple calculations of sound decay in rooms. To simplify calculation of sound-pressure level and reverberation time for a room, an average absorption coefficient for the room as a whole must be defined. This average absorption coefficient for the room as a whole is obtained by weighting the absorption coefficients of the individual surfaces according to their respective areas as follows:

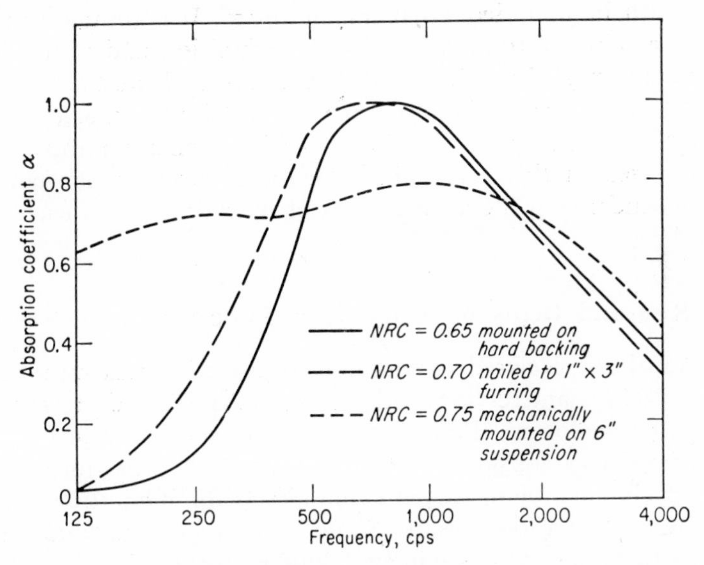

FIG. 11.2 Effect of mounting conditions on absorption coefficient and noise-reduction coefficient of ¾-in. textured acoustic tile. The decrease in absorption at high frequencies is due to the nonopen nature of the textured surface.

$$\bar{\alpha} = \frac{S_1\alpha_1 + S_2\alpha_2 + \cdots + S_q\alpha_q}{S_1 + S_2 + \cdots + S_q} \tag{11.1}$$

where $S_1, S_2, \ldots, S_q$ are the areas of particular sound-absorbing surfaces in question and $\alpha_1, \alpha_2, \ldots, \alpha_q$ are the statistical average absorption coefficients of the materials associated with these areas.

We shall see later that there are certain conditions which must be fulfilled before this calculation of $\bar{\alpha}$ is meaningful. Equation (11.1) is accurate only if $\bar{\alpha}$ is small and all surfaces have nearly the same absorption coefficient. In the following statistical analysis of sound in large enclosures we assume that sound waves of nearly equal energy density are traveling in all possible directions with equal probability. Obviously, if these conditions are to exist in practice, it is necessary that no one part of the room be heavily absorbing. A diffuse sound field cannot exist in an enclosure if $\bar{\alpha}$ is large because the traveling waves die out rapidly.

Absorbing objects such as chairs, seats, tables, desks, and even people must be included when calculating $\bar{\alpha}$. It is common practice to assign a value[2] of total absorption to each object and to add it into the numerator of Eq. (11.1). No modification of the total area S is made. In other words, the total area S is usually taken as that of the boundaries of the room. In the case of audience seating, caution must be used in this regard because the absorption per person is quite dependent on the average area of floor devoted to each seated person.[4]

11.3 Rise and Decay of Sound in a Tube

Assumptions. Related to sound absorption in enclosures are the rise and decay characteristics and the steady-state sound-pressure level (SPL) in an enclosure. Before taking up these calculations for large enclosures, let us consider the somewhat simpler situation of sound in a long closed tube of length l (see Fig. 11.3). This tube is driven near one end by a loudspeaker S and is terminated at *both* ends by an absorbing material T having an absorption coefficient α.

The diameter of the tube is assumed small compared with a wavelength at the highest frequency considered, and the length l is assumed long compared with a wavelength at the lowest frequency considered. Thus, we may conceive of this tube as being a sort of large, one-dimensional enclosure as far as its acoustic characteristics are concerned. The assumptions of great length compared to a wavelength means that in the frequency range being considered, a very large number of normal modes of vibration exists. Because of this fact, the sound field can be dealt with on a statistical basis, rather than by the mean-square sound pressures for discrete modes of vibration as was suggested in the previous chapter.

The sound signal exciting the tube consists of a large number of components having equal amplitudes and different but closely spaced frequencies. The absorption coefficient of the terminations α is assumed to be constant for the range of frequencies considered. The existence of a different standing-wave pattern for each of the many frequencies involved produces an approximately uniform steady-state sound-energy density throughout the tube. It is important to note that while only two directions of travel are possible, namely, backward and forward in the tube, the time phases of the various components are randomly related to each other.

The mean-square sound pressure is detected by the microphone

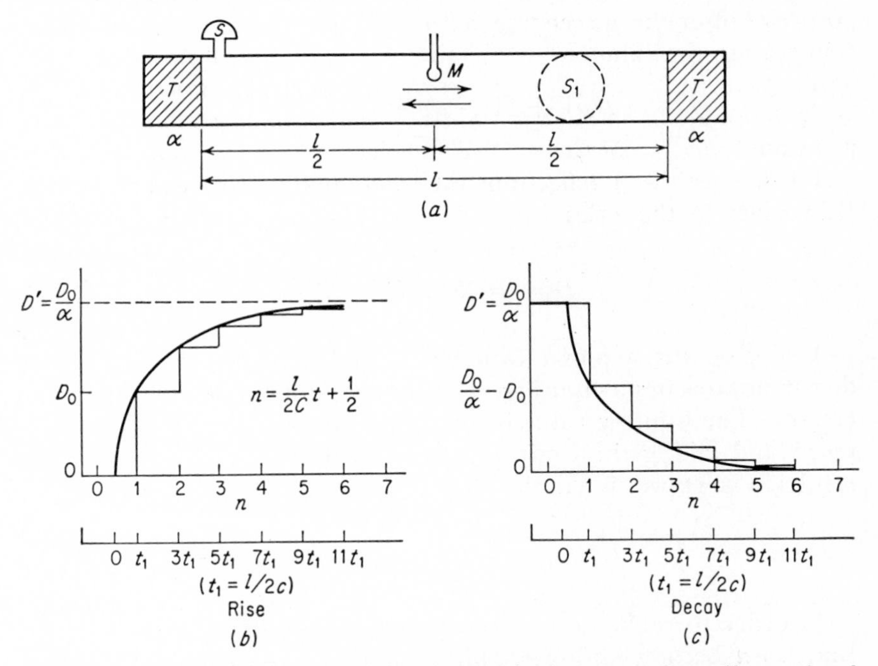

FIG. 11.3 Rise and decay characteristics of sound in the center of a uniform tube of length l terminated at both ends by absorbing materials having an absorption coefficient α. [(b) and (c) *after Swenson,*[1] *p.* 156, *fig.* 7.3.]

M at the center of the tube $l/2$, and a space average is obtained by moving the microphone back and forth over a small distance (about a half wavelength at the lowest frequency) about this point and averaging the fluctuating microphone output voltage in an rms rectifier.

Growth of the Sound Field. Initially the energy in the tube is zero. At time $t = 0$ the source begins to emit sound energy at a rate of W watts. The energy density (see Chap. 2) at the center remains zero until $t = t_1 = l/2c$ at which time the wavefront traveling at the speed c reaches the center of the tube. The energy density then rises abruptly to a value $D_0 = \dfrac{W}{cS_1}$ where S_1 is the cross-sectional area of the tube. As the wave progresses to the right, the energy density remains constant at the value D_0 until the wavefront is reflected from the right-hand termination and returns at a time $t = 3t_1 = 3l/2c$. The energy density at the point $l/2$ then rises to a value

$$D_0 + D_0(1 - \alpha) \tag{11.2}$$

since the wave returning from the right has now been diminished in energy by the factor $(1 - \alpha)$. The total energy density at the point $l/2$ after the succeeding reflection from the left-hand termination rises to the value

$$D_0[1 + (1 - \alpha) + (1 - \alpha)^2] \tag{11.3}$$

at a time $t = 5t_1 = 5l/2c$.

Thus, after $n - 1$ reflections the total energy density at the point $l/2$ is given by the series

$$D(n) = D_0 \sum_{k=0}^{n-1} (1 - \alpha)^k \tag{11.4}$$

A plot of this step-wise function is shown in Fig. 11.3*b*. It evidently approaches a limiting value as the number of reflections increases. The limiting value is found by summing the series of $n - 1$ terms and taking the limit as n goes to infinity. The summation results in an expression of the form

$$D(n) = \frac{D_0}{\alpha}[1 - (1 - \alpha)^n] \tag{11.5}$$

In order to calculate the steady-state energy density we take the limit as n becomes infinite. The quantity $(1 - \alpha)$ has a value less than 1 by definition, and the limit is easily calculated to be

$$D' = \frac{D_0}{\alpha} = \frac{W}{cS_1\alpha} = \frac{2W}{cS\alpha} \tag{11.6}$$

where $S = 2S_1$ is the total absorbing area in the tube. D' is defined as the steady-state (ultimate) energy density in the tube.

We note that the final energy density varies inversely with the absorption $S\alpha$, a result which is important in noise-control problems.

It is convenient in many calculations to divide the total steady-state energy density, Eq. (11.6), into two components: one associated with the direct field, $D_0 = 2W/cS$, and the other associated with the field set up by all reflections, namely, the reverberant field D_R. In the case of the tube we obtain for the steady-state energy density

$$D' = D_0 + D_R = \frac{2W}{cS} + \frac{2W}{c}\frac{1-\alpha}{S\alpha} \tag{11.7}$$

The reverberant component is frequently written in the form

$$D_R = \frac{2W}{cR} \tag{11.8}$$

where $R = \dfrac{S\alpha}{1-\alpha}$ is analogous to the room constant[3] in the acoustics of enclosures which is discussed in more detail in Sec. 11.5.

We see from Fig. 11.3 that for large values of time the number of reflections n is nearly equal to ct/l, where statistically the length of the tube now may be interpreted as the *mean free path*. By mean free path we mean the average distance which the sound wave travels between reflections. After this value is substituted into Eq. (11.5) for n, the result is a time function

$$D(t) = \frac{2W}{c}\,\frac{1-(1-\alpha)^{ct/l}}{S\alpha} \tag{11.9}$$

This function is plotted as a smooth curve in Fig. 11.3*b*. Strictly speaking the sound field within any enclosure rises and decays in a series of steps as shown in the figure. However, this discontinuous function can be closely approximated by a smooth curve if the number of reflections is large. This observation is of fundamental importance in the understanding of the decay of sound in enclosures.

Decay of the Sound Field. In reverberation measurements the sound-energy density is allowed to reach a steady-state value and then the source is suddenly turned off. Call this instant the time $t = 0$. Returning again to the tube, the energy density $D' = D_0/\alpha$ persists until $t = t_1 = l/2c$, at which time the trailing edge of the direct wave reaches $l/2$. At this instant the energy density decreases to a value $D_0/\alpha - D_0$ and remains at this value until the once-reflected wave arrives at a time $t = 3t_1 = 3l/2c$. At this instant the energy density drops to a value $D_0[1/\alpha - 1 - (1-\alpha)]$. After sufficient time has elapsed for the edge of the $n-1$ reflected wave to reach the point $l/2$, the energy density has been reduced to a value of

$$D(n) = D_0\left[\frac{1}{\alpha} - \sum_{k=0}^{n-1}(1-\alpha)^k\right] \tag{11.10}$$

which may be expressed in the convenient form

$$D(n) = \frac{D_0}{\alpha}(1 - \alpha)^n = D'(1 - \alpha)^n \tag{11.11}$$

through the use of Eqs. (11.5) and (11.6).

This function is plotted vs. n as a stepwise function in Fig. 11.3*c*. Clearly, the energy density approaches zero as n approaches infinity. Just as before, when t is large we replace n by $(c/l)\ t$ and obtain a smooth exponential decay curve

$$D(t) = D'(1 - \alpha)^{ct/l} = D'e^{(c/l)[-\ln(1-\alpha)]t} \tag{11.12}$$

where $\ln(1 - \alpha)$ is the logarithm to the base e of $(1 - \alpha)$.

This function is also plotted as a smooth curve in Fig. 11.3*c*. It will be observed in Fig. 11.3 that approximating the actual step response by a smooth rise or decay curve is grossly in error for the first few reflections. However, as the number of reflections increases, the fit becomes better, and finally, after many reflections, the fit is almost perfect.

This is about as far as it is profitable to continue the discussion of sound buildup and decay in a tube. We will now apply the above ideas to the buildup and decay of sound in a large enclosure. An exact calculation of the buildup and decay of sound in a large enclosure is much more complicated than that of the tube. Sound in an enclosure strikes the bounding surfaces with varying angles of incidence, and the absorption coefficients differ from one location to another. Furthermore, the path length between successive reflections varies widely with frequency and source location. Thus, an exact calculation of the rise and decay of sound in a large enclosure would be complicated, to say the least, and recourse is taken to purely statistical methods. Some of the important considerations in this analysis are discussed in the following section.

11.4 Mean Free Path and Sound Decay in a Large Enclosure

Basic Matters. If a sound source having components extending over a band of frequencies radiates energy into a large, irregular enclosure, we observe fluctuations in the sound pressure as a microphone is moved about the enclosure. However, in this case the maxima and minima of sound pressure lie much closer together in position than for either a small or a regular enclosure. One finds that the wider the band of frequencies the smaller the fluctuations, other factors remaining the same.

In the low-frequency range there will be a few room resonances while at high frequencies there will be many resonances in any given band of frequencies. Thus, high-frequency sound waves whose average wavelength is small compared with the dimensions of the enclosure will excite a large number of standing waves. As in the case of the tube, we assume that the source produces a uniform sound-energy density throughout the enclosure. The mean-square pressure is determined by moving the microphone back and forth over a short distance in order to obtain a satisfactory average.

Whereas in the case of the tube there were only two possible directions of wave travel (forward and backward), in an enclosure all directions are possible. We now imagine that the sound field in the enclosure consists of a superposition of plane waves traveling in all directions with equal probability. As we said before, this hypothetical situation leads to what is commonly called a *diffuse sound field*. We are neglecting for the moment the influence of the source itself. Strictly speaking, the diffuse field exists only at a distance from the source. This topic will be discussed in Sec. 11.6.

The number of reflections which a sound wave makes in 1 sec is equal to the ratio c/d, where d is called the *mean free path of the wave*. We visualize the situation for a large enclosure by imagining a wave bouncing from one wall to another inside the room. This wave travels in a straight line until it strikes a wall surface or other object. It is reflected off the surface at an angle which is equal to the angle of incidence. It then travels in a new direction until it strikes another surface, and so on. Because sound travels at a speed $c = 344$ m/sec (1,130 ft/sec), many reflections will occur during each second of travel. Thus, from a statistical standpoint the mean free path d is defined as the average distance which a sound wave travels in an enclosure between reflections from bounding surfaces. This condition is shown in Fig. 11.4.

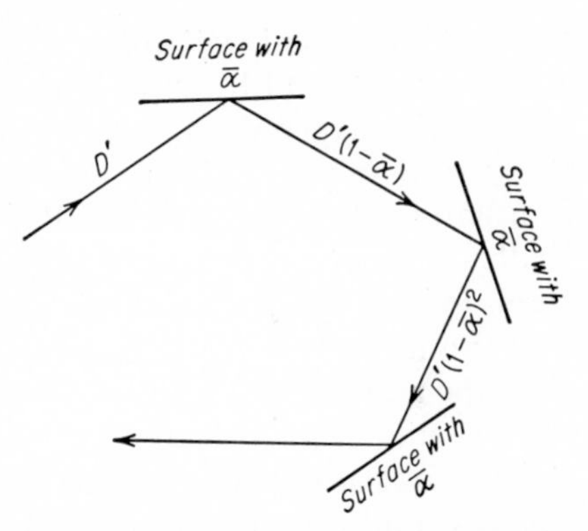

FIG. 11.4 Path of sound waves after successive reflections from surfaces with average absorption coefficient $\bar{\alpha}$ and spaced a mean free path d apart. (*After Beranek*,[3] *p.* 305, *fig.* 10.14.)

Under the above assumptions the total energy which strikes an element of wall area ΔS during 1 sec can be shown[5] to equal

$$\frac{Dc}{4}\,\Delta S \tag{11.13}$$

The quantity $\frac{Dc}{4}$ is the energy flux through a unit of area per second which is by definition *sound intensity*. The total energy per second (power) absorbed by the walls therefore will be

$$\frac{Dc}{4}(\alpha_1 S_1 + \alpha_2 S_2 + \cdots + \alpha_q S_q) = \frac{Dc}{4} S\bar{\alpha} \tag{11.14}$$

where $\bar{\alpha}$ is defined in Eq. (11.1).

On the other hand, the total energy per second removed from the enclosure will be

$$DV\frac{c}{d}\bar{\alpha} \tag{11.15}$$

where V is the volume of the enclosure and d is the *mean free path*. By equating the results of Eqs. (11.14) and (11.15) one obtains a statistical formula for the mean free path d of a large irregular enclosure which depends only upon geometry, namely

$$d = \frac{4V}{S} \tag{11.16}$$

where S is the total wall area.

Knudsen[5] found by experiment that the mean free path of various irregularly shaped enclosures was nearly equal to this value. Recent theoretical studies indicate that the mean free path in actual rooms is usually a few per cent smaller than given by Eq. (11.16).[6] We shall use the formulation of Eq. (11.16) in this chapter. It is of interest to note in the case of the one-dimensionl enclosure of Sec. 11.3 that the mean free path was exactly $2V/S$ where V is the volume of the tube and $S = 2S_1$ is the total absorbing area. Thus, in extending the one-dimensional formulation of diffuse sound fields in large enclosures to a three-dimensional one, the mean free path is altered by roughly doubling the ratio of volume to total absorbing-surface area within the enclosure.

We are now in a position to describe what is meant by a large enclosure. First, the ratio of the mean free path to the wavelength must be large. Second, the enclosure must be irregular in shape or contain reflecting objects which set up many reflections per second. Since wavelength varies inversely with frequency, an enclosure which is large at comparatively high frequencies may not necessarily be considered large at low frequencies. Thus, one must know what frequency range is of importance and base the discussions on this knowledge. This information must be determined from characteristics of the source.

Rate of Decay of the Sound Field. Before discussing the manner in which the source influences sound distribution in an enclosure, let us discuss how sound decays and what important factors govern the decay rate. We have shown that the mean free path for an enclosure of volume V and total wall area S having an average absorption coefficient $\bar{\alpha}$ is approximately $4V/S$.

After the source is turned off, the sound energy stored within the enclosure will decrease with each reflection according to the relation

$$D(n) = D'(1 - \bar{\alpha})^n \tag{11.17}$$

in the same manner as in the case of the tube (see Eq. 11.11) where D' is the steady-state energy density before the source was turned off and n is the number of reflections. By replacing n by $ct/d = (cS/4V)t$ we obtain the decay formula

$$D(t) = D'(1 - \bar{\alpha})^{\frac{cS}{4V}t} = D'e^{-(cS/4V)[-\ln(1-\bar{\alpha})]t} \tag{11.18}$$

where as before $\ln (1 - \bar{\alpha})$ is the logarithm to the base e of $(1 - \bar{\alpha})$.

In Chap. 2 we stated that in a reverberant sound field, the energy density is proportional to the mean-square sound pressure. Therefore, we shall rewrite Eq. (11.18) as follows,

$$\left|p_{av}\right|_t^2 = \left|p_{av}\right|_0^2 e^{-(cS/4V)[-\ln(1-\bar{\alpha})]t} \tag{11.19}$$

where $\left|p_{av}\right|_t^2$ and $\left|p_{av}\right|_0^2$ are the mean-square pressures in the enclosure at the time t and time zero, respectively. From the definition of sound-pressure level (SPL) the above equation may be rewritten in the form

$$(\text{SPL})_t = (\text{SPL})_0 + \frac{4.34cS}{4V}[2.30 \log_{10}(1 - \bar{\alpha})]t \tag{11.20}$$

since $\log_{10} e^{-(cS/4V)[-\ln(1-\bar{\alpha})]t} = 4.34 \dfrac{cS}{4V}[2.30 \log_{10}(1 - \bar{\alpha})]t$

Hence the SPL decays at a rate of

$$1.085 \frac{cS}{V}[-2.30 \log_{10}(1 - \bar{\alpha})] \quad \text{db/sec} \tag{11.21}$$

The Reverberation Equation. The reverberation time of the enclosure is defined as the time required for the SPL to fall 60 db. This time corresponds to about that required for the SPL to diminish from a normal value to the lower level of audibility.

The well-known Norris-Eyring equation for reverberation time T

is obtained from Eq. (11.20) by replacing $(SPL)_0 - (SPL)_t$ by 60 db and calculating the required time in seconds. This calculation yields

$$T = \frac{60V}{1.085cS\,[-2.30 \log_{10}(1 - \bar{\alpha})]} \quad \text{sec} \qquad (11.22)$$

A chart giving the values of $[-2.30 \log_{10}(1 - \bar{\alpha})]$ for typical values of $\bar{\alpha}$ is shown in Fig. 11.5. The values of $\bar{\alpha}$ for most enclosures range from about 0.05 at the low end to about 0.4 at the high end.

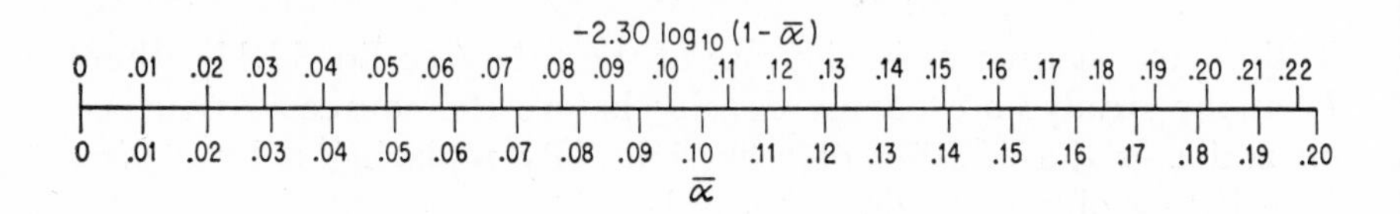

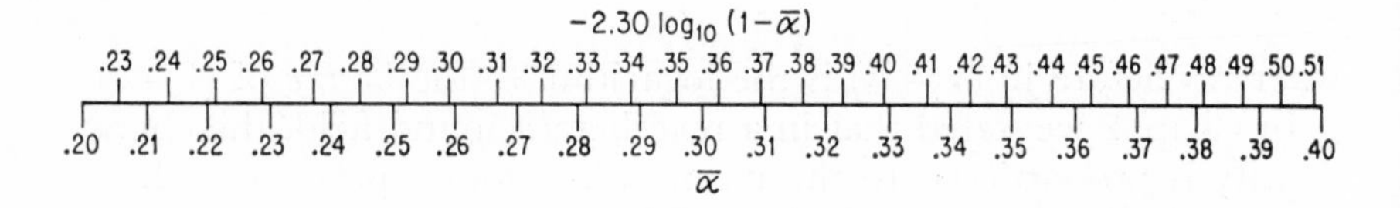

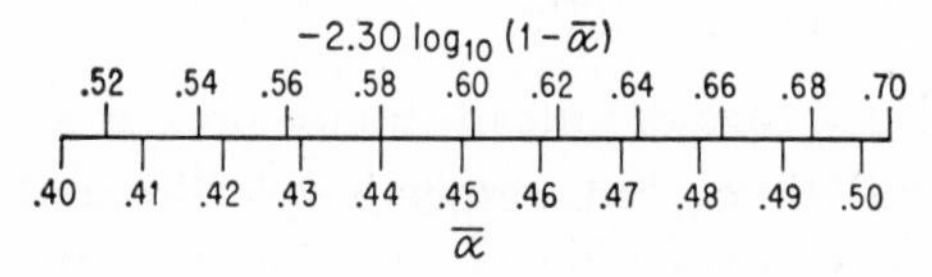

FIG. 11.5 Chart for computing absorption factor, $-2.30 \log_{10}(1 - \bar{\alpha})$, from a knowledge of $\bar{\alpha}$. (*After Beranek.*[3])

Norris-Eyring Formulas. Metric Units: In mks units, the reverberation equation is

$$T = \frac{0.0161V}{S[-2.30 \log_{10}(1 - \bar{\alpha})]} \quad \text{sec} \qquad (11.23)$$

where V = volume of room, m^3
S = area of bounding surface, m^2

English Units: The reverberation equation in English units reads

$$T = 0.049\,\frac{V}{a} \quad \text{sec} \qquad (11.24a)$$

where

$$a = S[-2.30 \log_{10}(1 - \bar{\alpha})] \qquad (11.24b)$$

The units of a are *sabins* in ft² and

$$V = \text{room volume, ft}^3$$
$$S = \text{area of bounding surfaces, ft}^2$$

Reverberation time vs. volume of the enclosure is shown in Fig. 11.6, using the total absorption $S\bar{\alpha}$ as a parameter.

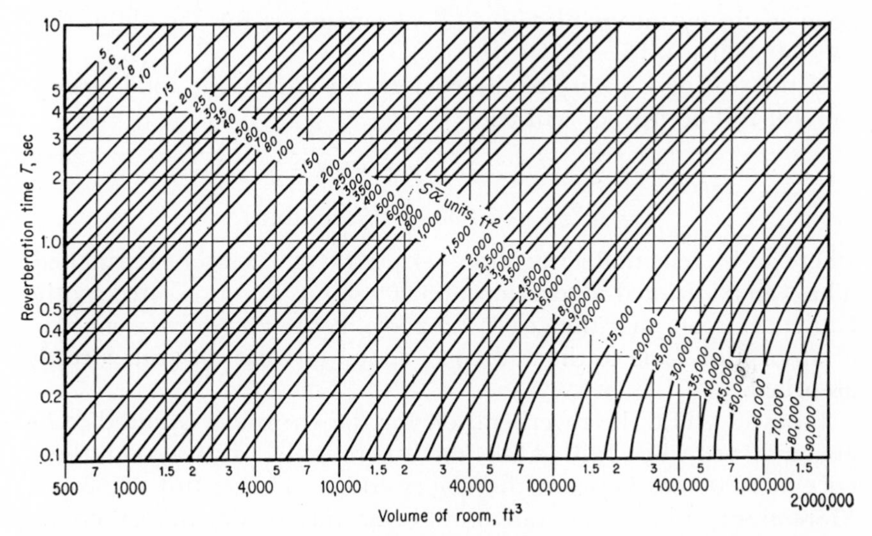

FIG. 11.6 Chart showing relation between reverberation time, volume, and the absorption $S\bar{\alpha}$ in English units. The $-2.30 \log_{10} (1 - \bar{\alpha})$ correction factor is incorporated. Maximum error in reverberation time is less than 5 per cent for enclosures with maximum dimensions less than three times minimum dimension and for $\bar{\alpha} < 0.6$. (*After Beranek,*[3] *p.* 307, *fig.* 10.16.)

Sabine (Approximate) Formulas. The original Sabine reverberation formula is obtained from Eq. (11.20) for small α by noting that $[-2.30 \log_{10} (1 - \bar{\alpha})] \doteq \bar{\alpha}$ if $\bar{\alpha} << 1$. This is also evident by examining Fig. 11.5. In this case we have in place of Eq. (11.24) the approximate formulas

$$T \doteq \frac{0.161V}{S\bar{\alpha}} \quad \text{sec} \quad \text{(metric units)} \tag{11.25}$$

$$T \doteq \frac{0.049V}{S\bar{\alpha}} \quad \text{sec} \quad \text{(English units)} \tag{11.26}$$

Example 11.1. Consider a large irregular room having a volume of 400,000 ft³ and a total wall area of about 45,000 ft². Let the average absorption coefficient for the room as a whole be $\bar{\alpha} = 0.15$ at a frequency of 2,000 cps including the seats but without occupants.

From Fig. 11.5 we find that $[-2.30 \log_{10} (1 - 0.15)] \doteq 0.16$. The absorption a (Eq. 11.24*b*) is about 7,200 ft². The reverberation time from Eq. (11.24*a*) is

$$T = \frac{0.049 \times 400{,}000}{7{,}200} = 2.7 \text{ sec}$$

for the unoccupied room. This calculation might alternatively have been made by means of Fig. 11.6. Remember, for Fig. 11.6, that $S\bar{\alpha} = 45{,}000 \times 0.15 = 6{,}750$ ft².

If the seating capacity of the enclosure is 2,000, the total absorption for full occupancy will be

$$S\bar{\alpha} = (45{,}000 \times 0.15) + (2{,}000 \times 1.5)$$
$$= 6{,}750 + 3{,}000 = 9{,}750 \text{ sabins}$$

where an incremental absorption per person of 1.5 sabins is assigned. Thus, an additional absorption of 3,000 sabins is contributed by the audience. Note that the factor 1.5 above is the *incremental* amount of absorption (due to the audience) added to what the seats already absorbed.

From Fig. 11.6 the reverberation for the enclosure is found to be about 1.7 sec when full. Thus, there is roughly a full second difference at 2,000 cps between the empty room and one full of people. More absorption on the walls would certainly reduce this difference as well as the reverberation time itself.

In order for the reverberation formula to be generally useful, a correction for air absorption is sometimes included. We shall discuss how this correction comes about and its use in the following section.

11.5 Boundary and Air Absorption

In noise-control applications we are interested not only in the decay rate but also in the steady-state sound-pressure level. This calculation not only involves $\bar{\alpha}$ but also the power level (PWL) of the source and its location within the enclosure. One of the items of interest will be the room constant of the enclosure.

Let us start by writing down the power supplied by the source to the reverberant sound field [see Eqs. (11.15) and (11.16)]. We have

$$D'V \frac{cS}{4V} \bar{\alpha} = W(1 - \bar{\alpha}) \qquad (11.27)$$

The sound absorbed by the first reflection, namely $W\bar{\alpha}$, has been subtracted from the total power supplied by the source W. As before. D' is the steady-state energy density in the room. By definition, the reverberant field energy is the energy in the enclosure after the first reflection. The energy that has not suffered a reflection is called the "direct field energy."

Rearranging Eq. (11.27) gives us the steady-state sound-energy density in the reverberant field

$$D' = \frac{4W}{c}\left(\frac{1-\bar{\alpha}}{S\bar{\alpha}}\right) = \frac{4W}{cR} \tag{11.28}$$

where R is the *room constant* defined as

$$R = \frac{S\bar{\alpha}}{1-\bar{\alpha}} \tag{11.29}$$

The room constant R of an enclosure having proportions 1:1.5:2 is plotted vs. volume of the enclosure in Fig. 11.7 for a range of average absorption coefficients $\bar{\alpha}$.

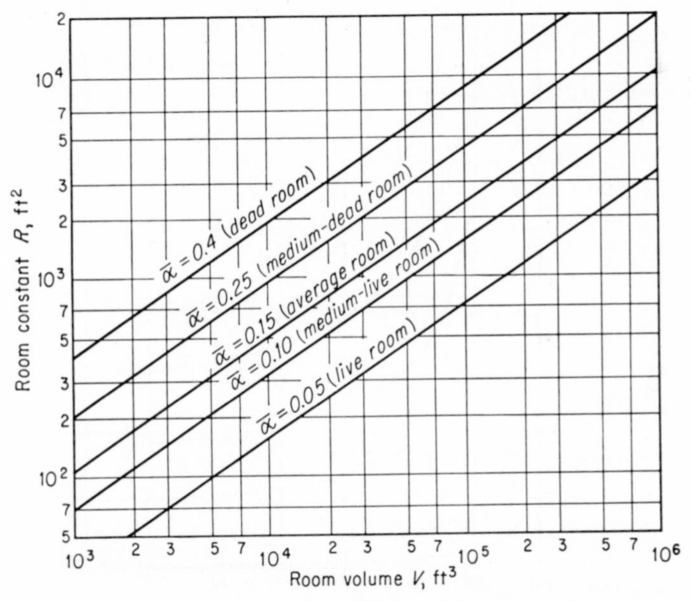

Fig. 11.7 Calculated values of room constant R in the mid-frequency range of 500 to 1,000 cps as a function of room volume for rooms with proportions of about 1:1.5:2. These proportions give $S = 6.25\ V^{2/3}$. (*After Beranek*,[3] *p.* 316, *fig.* 10.20.)

Sound energy is also absorbed into the air itself to a slight degree. This fact has not been taken into account in previous discussions of sound decay in a tube and other enclosures. One clearly sees that if the enclosure is small, the number of reflections is large and the mean free path is also small. In this case air absorption is generally not as important as boundary absorption. In a very large enclosure, however, the mean free path is large, and the energy loss in air cannot be ignored, particularly for frequencies above 1,000 cps.

During the period of time that a wave is traveling one mean free path, the energy density will be decreased by air absorption according to the decay formula

$$D(md) = D'e^{-md} \tag{11.30}$$

where m is the energy attenuation constant in units of reciprocal length. Measured values of m under some typical atmospheric conditions using frequency as a parameter are shown in Fig. 11.8. It

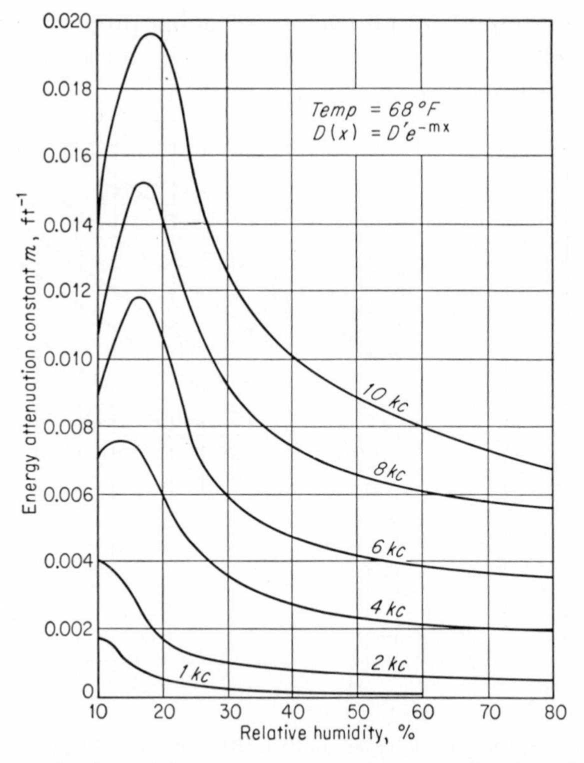

Fig. 11.8 Measured values of the energy attenuation constant for air m (ft^{-1}) as a function of relative humidity for different frequencies and at a temperature of 68°F. (*After Knudsen and Harris,* "Acoustical Designing in Architecture," *p.* 160, *fig.* 8.10, *John Wiley & Sons, Inc., New York,* 1950.)

will be observed that the value of m depends strongly upon frequency and relative humidity. To a somewhat lesser degree it is also temperature dependent.

Without going into the derivation here, it is known that the room constant must be modified to take into account the effects of air absorption as follows:[3]

$$R_T = \frac{S\bar{\alpha}_T}{1 - \bar{\alpha}_T} \tag{11.31}$$

where

$$\bar{\alpha}_T = \bar{\alpha} + \frac{4mV}{S} \tag{11.32}$$

and R_T is defined as the total room constant. In deriving Eq. (11.32) it is assumed that $4mV/S$ is less than about 0.4.

Obviously, when the total boundary absorption $S\bar{\alpha}$ becomes large in comparison with $4mV$, i.e., the absorption per reflection is large compared with air absorption over a mean free path, air absorption may be ignored.

When the reverberation time formula, Eq. (11.24), is corrected to account for air absorption it is modified [3] to read

$$T = 0.049 \frac{V}{a + 4mV} \quad \text{sec} \qquad \text{(English units)} \tag{11.33}$$

The above relations can also be used to determine $\bar{\alpha}_T$ once T has been measured, whence

$$\bar{\alpha}_T = 1 - \left\{\log_{10}{}^{-1}\left[\frac{V}{2.30ST}(0.049 - 4mT)\right]\right\}^{-1} \tag{11.34}$$

Example 11.2. If in Example 11.1 the room has a relative humidity of 30 per cent at a temperature of 68°F, the attenuation constant $m = 0.001$ at 2,000 cps. The reverberation time for the empty room now becomes

$$T = \frac{0.049 \times 400{,}000}{7{,}200 + 4 \times 0.001 \times 400{,}000} = \frac{19{,}600}{7{,}200 + 1{,}600} = 2.2 \text{ sec}$$

Thus, appreciable air absorption will decrease the reverberation time of the enclosure from the value based upon boundary absorption alone.

11.6 Total Steady-state Sound-pressure Level

We are now in a position to incorporate the direct sound field from a source into the energy equations and calculate the total steady-state sound-pressure level.

The space-average sound-energy density at a distance from a small directional source radiating W watts is

$$D_r = \frac{|p_r|^2}{\rho_0 c^2} = \frac{W}{4c\pi r^2} Q \tag{11.35}$$

where $|p_r|^2$ is the mean-square sound pressure, Q is the directivity factor[7] along any axis, and r is the distance from the acoustical center of the source. (See Chap. 8 for the definition of Q.)

From Eq. (11.35) the mean-square pressure at a distance r along any axis is calculated to be

$$|p_r|^2 = \frac{\rho_0 c W Q}{4\pi r^2} \tag{11.36}$$

For the reverberant field contribution we calculate the mean-square pressure from Eq. (11.28), namely,

$$|p_{\text{av}}|^2 = \frac{4\rho_0 c W}{R_T} \tag{11.37}$$

By adding the mean-square-pressure contributions at a point r from the source along any axis, one obtains the total mean-square pressure in a large enclosure:

$$|p|^2 = |p_r|^2 + |p_{\text{av}}|^2 \tag{11.38a}$$

$$|p|^2 = W\rho_0 c\left(\frac{Q}{4\pi r^2} + \frac{4}{R_T}\right) \tag{11.38b}$$

Expressing the above relation on a logarithmic scale, one obtains the relations between a given directional source and the SPL existing within the enclosure:

MKS Units

$$\text{SPL} \doteq \text{PWL} + 10 \log_{10}\left(\frac{Q}{4\pi r^2} + \frac{4}{R_T}\right) - 10 \text{ db} \tag{11.39}$$

where PWL $= 10 \log_{10} W + 130$ db *re* 10^{-13} watt
W = power radiated, watts
R_T = total room constant, m^2
r^2 = distance from source squared, m^2

English Units

$$\text{SPL} \doteq \text{PWL} + 10 \log_{10}\left(\frac{Q}{4\pi r^2} + \frac{4}{R_T}\right) \quad \text{db} \tag{11.40}$$

where PWL and W are as above and R_T and r^2 equal the total room constant and distance from the source squared in *square feet*.

The relative importance of the direct and reverberant field contribution at a given point is readily evident by comparison calculations of the magnitudes of the quantities $Q/4\pi r^2$ and $4/R_T$. If the former prevails, then the SPL is largely due to direct radiation, and small changes in R_T will have little effect upon the existing SPL. On the other hand, if $4/R_T$ is large compared with $Q/4\pi r^2$, the room absorption is an important factor in determining the SPL because the point lies in the reverberant field. A generalized graph plotting SPL–PWL vs. $r/\sqrt{Q}$ is shown in Fig. 11.9.

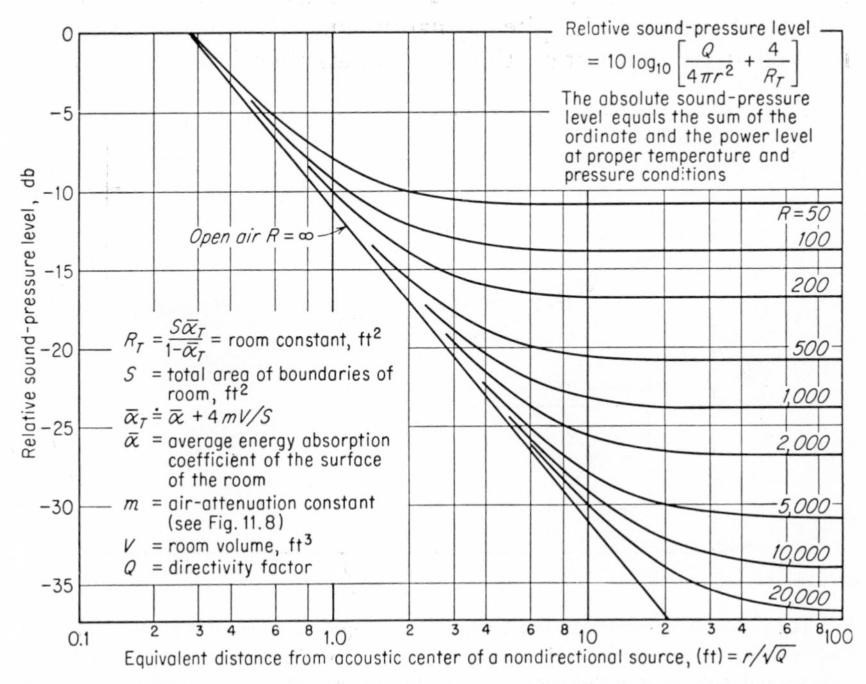

FIG. 11.9 Chart for determining the sound-pressure level in a large irregular enclosure at a distance r from the center of a source with directivity factor Q. The ordinate is SPL — PWL db. The parameter is R_T.

We should be aware that in many rooms it is not possible to get down to the flat part of the curve. Also, it should be remembered that the assumption of a diffuse sound field was made in the derivation of Eqs. (11.39) and (11.40) and that the results shown in Fig. 11.9 are valid, strictly speaking, only under this ideal condition. There are, moreover, important examples such as offices where nearly all the absorption is on one surface and consequently the sound field is not diffuse.[8] Another kind of such an enclosure is one where all of the absorbing material is found on the walls.[9] One would expect some deviations in the measured curves from those calculated using the results of Eqs. (11.39) and (11.40), and indeed this seems to be the case.

Figure 11.10 shows an experimental plot of the relative SPL vs. distance for a classroom area approximately 150 ft by 75 ft and with a 15-foot-high ceiling treated with acoustical tile. The floor and walls are relatively hard, yet the room is adequately treated from a noise-reduction point of view. The room constant R was estimated to be no larger than about 3,500, and the expected curve is shown by the top line. However, the measured curve falls somewhat below the calculated curve and in spite of small fluctuations does not appear to level off at say 20 ft but continues to fall another 6 db at 100 ft.

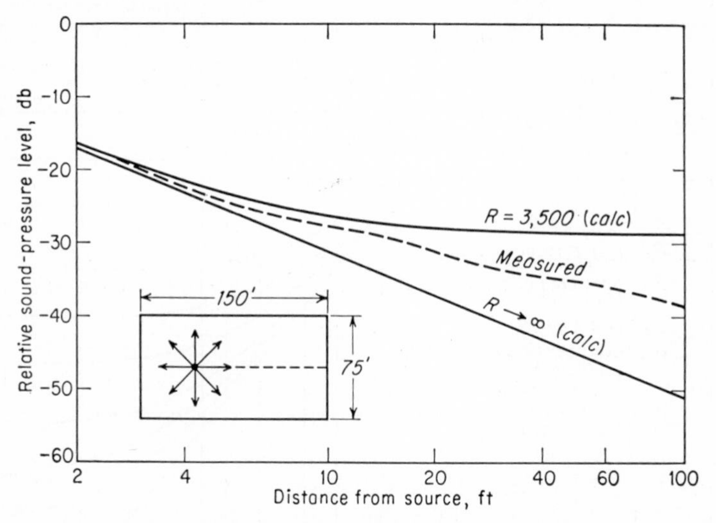

FIG. 11.10 Measured and calculated relative sound-pressure levels in a room approximately 150 by 75 ft with a 15-ft ceiling, plastered walls, wooden floor, acoustic-tile ceiling, and numerous scattering objects in the room. Source and receiver both at a height of 6 ft from floor. The comparison here is with a room having acoustic characteristics which do not satisfy the simple diffuse-room theory.

This same observation has been made by others[9] who have encountered rooms which do not qualify for the diffuse theory. In the situation shown in Fig. 11.10 there were large drawing desks along the outside walls, large light fixtures hanging from the ceiling, and a number of other objects around the room which could scatter sound. This may mean that as the waves progress to the extreme boundaries of the room they are scattered back toward the source by such objects. What we would commonly call the diffuse part of the field then also decreases with distance. This may be due more to the scattering of sound than to absorption.

Example 11.3. From Example 11.1 the average absorption coefficient for the room filled with 2,000 people is

$$\bar{\alpha} = \frac{9,750}{45,000} = 0.22$$

Considering air absorption, the total absorption coefficient at 68°F, relative humidity = 30 per cent, and a frequency of 2,000 cps is

$$\bar{\alpha}_T = \bar{\alpha} + m\frac{4V}{S} = 0.22 + \frac{1,600}{45,000} = 0.22 + 0.04 = 0.26$$

Hence, the room constant has a value of

$$R_T = \frac{S\bar{\alpha}_T}{1 - \bar{\alpha}_T} = \frac{45,000 \times 0.26}{1 - 0.26} = 16,000 \text{ ft}^2$$

For a nondirectional source, the difference between the PWL of the source and the SPL in the reverberant field ($r > 60$ ft), Fig. 11.9, is about 35 db. Thus, if the source has a PWL of 95 db at 2,000 cps, the SPL in this part of the enclosure is 95 − 35 = 60 db.

If a given source has a Q value of 1 in free space, the actual Q value within a room may increase due to reinforcing reflections from nearby surfaces. If the room is large and does not react on the source its directivity will still be 1 when located at the center of the room. Near the center of a wall or floor it will have a value of about 2 since the power is being radiated into a hemisphere. Near an edge of the enclosure it will have a value of 4, and at a corner it will go as high as 8.

It is clear from Eqs. (11.39) and (11.40) that if the SPL within an enclosure is to be estimated, some knowledge of the directivity factor Q is necessary. It is important to keep in mind that the PWL of the source as well as Q may change with changes in source position.[10] Hence, detailed information about source characteristics is vital to satisfactory engineering solution of noise-reduction problems. These matters were discussed in Chap. 8.

The relations of Eqs. (11.39) and (11.40) hold either when the sound source is very small or when the observation points lie outside the region of the near field of the source. Practical noise sources are frequently distributed over regions of significant size. Near-field effects are avoided only when the distance from source to observation point is several times the largest dimension of the source. When studies involving a sound source in an enclosure are made, the variation of SPL with distance should be studied by means of direct measurements. By this means one can determine the regions of inverse distance (6 db per doubling of distance) and the reverberant region or region of uniformity. This is an important consideration when taking reverberation measurements because data obtained in the vicinity of the source will not be characteristic of the enclosure alone.

11.7 Summary

Sound-absorbing materials play an important role in the abatement of noise in an enclosure. Sound-absorption coefficients of materials depend upon frequency and angle of incidence of the sound as well as upon other parameters. For meaningful and simple calculation of sound characteristics, several absorption coefficients are defined and measured. The statistical average absorption coefficient of a given surface, α, is obtained by averaging the ratio of absorbed to incident energy over all possible angles of incidence. An average absorption coefficient for the room as a whole $\bar{\alpha}$ is obtained by averaging the statistical coefficient over the total absorbing area of the enclosure (see Eq. 11.1).

Sound energy in an enclosure rises and decays by reflection from surfaces in a series of steps. However, if the average number of reflections per second is sufficiently large, the rise and decay become exponential. For diffuse sound fields the time rate of decay and the reverberation time, $T = 0.049V/a$ (English units), are controlled by the mean free path, average absorption coefficient, and energy absorption in the air. Air absorption becomes important when the energy reduction over a mean free path becomes comparable with the energy reduction per reflection.

The acoustic field within an enclosure excited by a sound source of finite size may be divided satisfactorily into three regions for noise-reduction calculations. These are the near field which lies within a few diameters of the source, the direct radiation or free field which is characterized for simple sources by a reduction in sound-pressure level of 6 db per doubling of distance from the center of the source, and the reverberant field existing at distances where multiple reflections from bounding surfaces become important. See also Chap. 8.

The significant parameters which control the distribution of sound within an enclosure are the source-power level (PWL), source directivity factor Q, radial distance from source r, and the room-absorption constant R_T—all related by the formula

$$\text{SPL} \doteq \text{PWL} + 10 \log_{10} \left(\frac{Q}{4\pi r^2} + \frac{4}{R_T} \right)$$

for a diffuse field (English units). The source-power level is important when calculating the sound-pressure level in all regions. The directivity factor and distance from the source are important in the free field, while the room constant becomes important in the reverberant field. Deviations from these ideal conditions due to directional effects probably caused by scattering of sound by objects in an enclosure are important. These deviations result in a somewhat greater difference between the source PWL and the field SPL than expected on the basis of the diffuse-field theory.

REFERENCES

1. Swenson, G. W., Jr.: "Principles of Modern Acoustics," chap. IX, D. Van Nostrand Company, Inc., Princeton, N.J., 1953.

2. Sound Absorption Coefficients of Architectural Acoustical Materials, *Acoust. Materials Assoc., Bull.* XIX, New York, 1959.

3. Beranek, L. L.: "Acoustics," chap. 10, McGraw-Hill Book Company, Inc., New York, 1954.

4. Beranek, L. L.: "Acoustics of Halls for Music," John Wiley & Sons, Inc., New York, in press; and Audience and Seat Absorption in Large Halls, *J. Acoust. Soc. Am.*, vol. 32, pp. 661–670, June, 1960.

5. Knudsen, V. O.: "Architectural Acoustics," chap. V, John Wiley & Sons, Inc., New York, 1932.

6. Allred, J. C., and A. Newhouse: Applications of the Monte Carlo Method to Architectural Acoustics, *J. Acoust. Soc. Am.*, vol. 30, pp. 1–3, January, 1958, and pp. 903–909, October, 1958.

7. See Ref. 3, chap. 4.

8. Beranek, L. L.: Performance of Rectangular Rooms with One Treated Surface, *J. Acoust. Soc. Am.*, vol. 12, pp. 14–23, 1940.

9. Sabine, H.: The Use of Acoustical Materials in the Control of Industrial Noise, *Proc. 3d Annual NNAS*, 1952.

10. Waterhouse, R. V.: Output of a Sound Source in a Reverberation Chamber and Other Reflecting Environments, *J. Acoust. Soc. Am.*, vol. 30, pp. 4–13, January, 1958.

Chapter 12

PROPERTIES OF POROUS ACOUSTICAL MATERIALS

Leo L. Beranek and S. Labate

12.1 Basic Physical Properties of Porous Acoustical Materials

Acoustical materials generally serve one of two purposes. Either they are used to reduce standing waves in an enclosure and hence reduce the reverberation time inside the enclosure, or they are used as a barrier to reduce the intensity of sound in its travel from one point to another. Some materials can serve both of these functions at once, although in many noise-reduction problems, separate materials are used to perform the two tasks more efficiently.

Because porous materials are essential for efficient noise reduction, we shall have to learn something about their basic properties. These properties have not been extensively used in acoustical design, probably because very little has been published other than highly mathematical treatises.[1–5] However, the acoustical engineer must learn to apply analysis techniques in the design of noise-control structures, so he should have a working knowledge of the properties of common porous materials. Applications of this information are given in Chaps. 14 and 17.

Now, what are these basic physical properties? Of prime importance is the resistance that the material offers to the flow of air through it—it is the friction of the moving air at the surface of the fibers of the material that causes loss of energy. Of course, the density of a material is of importance, and the porosity—i.e., the percentage of a given volume of the material in the voids—affects its performance. The bulk elasticity also evidences itself under some circumstances. Finally, the orientation of the fibers or passageways in the material must be taken into account in any exact analysis, although here we shall consider materials only in a grosser manner.

Adiabatic vs. Isothermal. When one draws air through a cigarette, he encounters *flow resistance.* This resistance is said to be due to friction between the air and the fibers, but actually the energy loss takes place in a very narrow layer of air adjacent to each fiber in the material. When air is blown through a small tube (see Fig. 12.1) this layer forms next to the walls (a) of the tube. At the walls the air particles are at rest. In the center of the tube they move back and forth freely with the sound field. In between, in regions (a), there is a gradient in velocity, a shear gradient, which causes the air particles to slide over each other. Due to the viscosity of air, energy is drawn from the sound wave, and a slight warming of the air takes place which is almost instantly absorbed by the sidewalls.

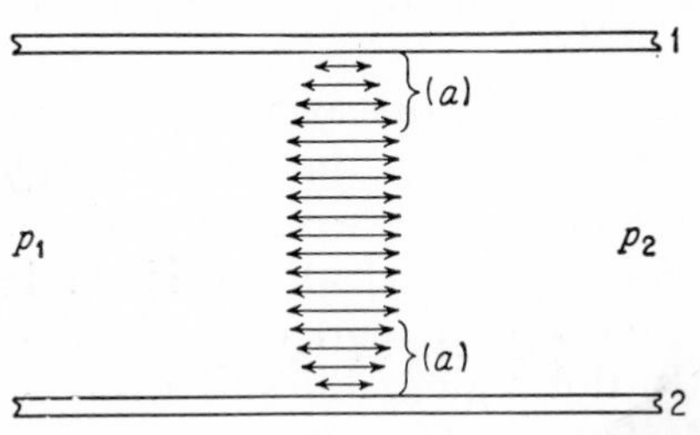

FIG. 12.1 Sketch showing the variation in amplitude of vibration of air particles in a sound wave near surfaces 1 and 2. (a) Regions in which viscous losses occur. The variations cause adjacent layers of air in regions (a) to slide on each other, thereby causing loss of energy in the viscosity of the air.

At the center of the tube of Fig. 12.1, the expansion and rarefactions of the air take place *adiabatically,* i.e., there is no loss of heat during compression. An incremental change in pressure in the air (due to the sound wave) causes a change in volume of the air as given by

$$\Delta P = -\gamma C \,\Delta V \tag{12.1}$$

where γ = ratio of specific heats for air—for diatomic gases like air, $\gamma = 1.4$

C = dimensional constant—negative sign indicates that increase in pressure causes decrease in volume

On the other hand, adjacent to the walls, the heat that would normally be built up during compression is absorbed by the walls so that the gaseous expansions and rarefactions take place *isothermally.* The relation between incremental pressure and volume is now

$$\Delta P = -C \,\Delta V \tag{12.2}$$

where C is the same constant as for Eq. (12.1).

The implication of Eq. (12.2) is that a gas that is behaving isothermally is 40 per cent more compressible than one that is adiabatic.

In an acoustical material where the spaces (voids) between fibers are very small, the compressions will be isothermal at very low frequencies, because there is time for heat exchange with the solid fibers from all parts of the gas. At high frequencies regions (*a*) of Fig. 12.1 become very thin because of the rapid reversals in sound pressure and the resulting small time available for heat exchange. Hence, on the average at high frequencies the compressions and rarefactions in the voids are adiabatic. For a narrow range of frequencies in between high and low frequencies the compressions are neither isothermal nor adiabatic. This transition region usually occurs in some part of the frequency range between 50 and 800 cps. In this region, γ lies between 1.0 and 1.4.

Specific Flow Resistance R_1. One of the most important quantities needed to determine the sound-absorbing characteristics of a porous material is its specific flow resistance per unit thickness of material. In the mks system* this is defined as

$$R_1 = \frac{\Delta p}{\Delta T u} \qquad \text{mks rayls/m} \tag{12.3}$$

where R_1 = specific flow resistance, mks rayls/m
Δp = sound-pressure differential across thickness ΔT of sample, measured in direction of particle velocity, newtons/m^2
u = particle velocity through sample, m/sec
ΔT = incremental thickness, m

The specific flow resistance R_1 may be measured using an actual sound wave. However, for engineering calculations it is usually accurate enough to measure it with a steady airflow. An apparatus like that in Fig. 12.2 is commonly used. Other apparatuses are described in Ref. 6.

In using the flow-resistance apparatus, the sample is carefully and securely mounted in a sample holder *A* with an inside area *S*. Two coarse screens *C* and *D* accurately define the upper and lower surfaces. The thickness ΔT of the sample is read from a scale. Air is drawn through the sample, and its volume velocity is determined, so that

$$u = \frac{U}{S} \qquad \text{m/sec} \tag{12.4}$$

where u = linear velocity, m/sec
U = volume velocity, m^3/sec
S = area of sample, m^2

The pressure Δp across the two faces of the sample is determined by a gauge, for example, a slant manometer with a range of 0- to 1-cm

* Conversion tables are given in Appendix C at the end of this book.

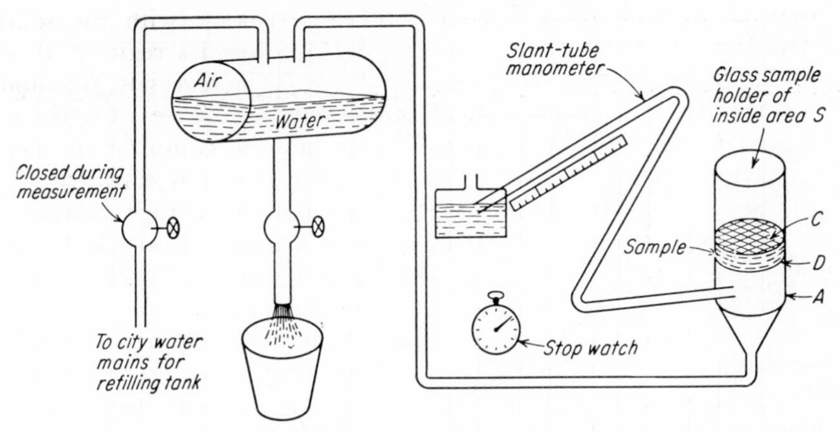

FIG. 12.2 Apparatus for measuring the flow resistance of porous materials. The sample of area S is held in cylinder A between two wire screens C and D. When water is removed from the tank steadily, air is sucked through the sample at a rate equal to the rate of water removed. The pressure drop across the sample is measured with a slant manometer. Note that to convert from inches of water on the gauge to newtons/m² multiply by 249, to convert from sample area in square inches to square meters multiply by 6.45×10^{-4}, and to convert from inches to meters divide by 39.37. Also to convert from rayls to mks rayls multiply by 10, and to convert from dynes/cm² to newtons/m² multiply by 10^{-1}. (See also Appendix C.)

H_2O. The manometer reading is converted into newtons per square meter. Then, the *flow resistance* R_f equals

$$R_f = \frac{\Delta p}{U/S} \qquad \text{mks rayls} \tag{12.5}$$

The *specific flow resistance* R_1 equals

$$R_1 = \frac{R_f}{\Delta T} \qquad \text{mks rayls/m} \tag{12.6}$$

The value of ΔT is in meters. Values of R_f and R_1 for a particular flexible blanket are shown in Fig. 12.3. Flow resistance is sometimes a function of velocity u. It is desirable to make measurements at three or so different rates of air flow and to extrapolate to zero airflow if necessary.

Porosity Y. The porosity of a porous material is defined as the ratio of the volume of the voids in a material to its total volume:

$$Y = \frac{V_a}{V_m} \tag{12.7}$$

where V_a is the volume of the air in the voids and V_m is the total volume of the sample of acoustical material being tested.

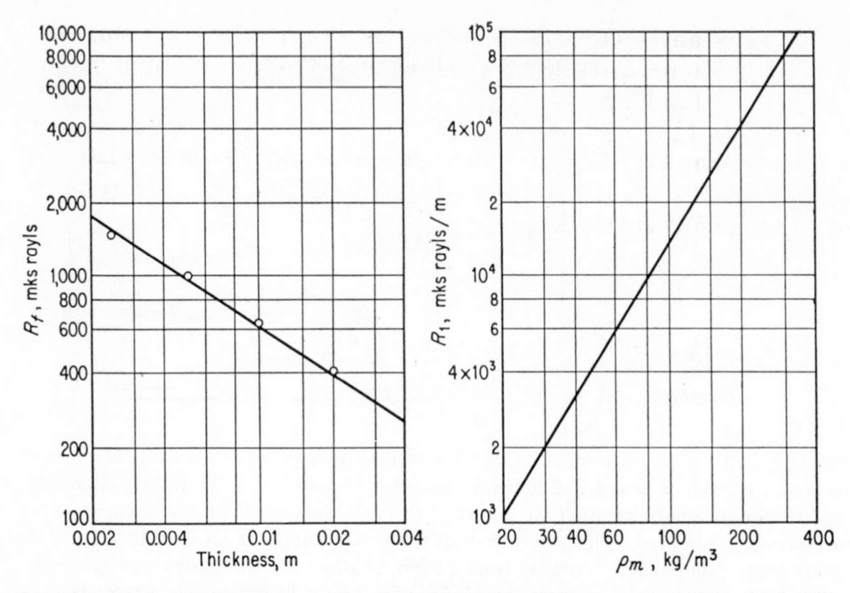

FIG. 12.3 Typical flow-resistance data obtained with the apparatus of Fig. 12.2. The sample used here had a surface density of 2.5 kg/m². Hence, at a thickness of 0.0415 m, ρ_m was about 60 kg/m³, R_f was about 250 mks rayls, and R_1 was about 6,000 mks rayls/m.

The simple apparatus[6] of Fig. 12.4 may be used for the measurement of Y. The acoustical material is contained in a rigid chamber of volume V whose temperature is held very constant. Initially, the stopcock is opened and the height of the water in the two sides of the U manometer observed. Let us call this height h. Then the stopcock is closed, and one side of the manometer is elevated until the levels have changed from h to h_1 and h_2, respectively. The pressure change ΔP_0, in meters of water, equals $h_2 - h_1$, and the reduction of volume of air $\Delta V_a'$ in the chamber equals $(h_1 - h)S$, where S is the area of cross section of the connecting tube. The volume V_a' equals V_a plus the volume of air not contained in the material. The heights h and h_1 are observed by means of a cathetometer, and $h_2 - h$ may be read with sufficient accuracy from the graduated scale. The porosity is given by

$$Y = 1 - \frac{V}{V_m} - \frac{P_0\,\Delta V_a'}{V_m\,\Delta P_0} \tag{12.8}$$

where P_0 = atmospheric pressure = 1.013×10^5 newtons/m² (10.35 m of water)

V_m = volume of acoustical material, m³

$V = V_m + V_0$, m³

V_0 = any volume in chamber not occupied by V_m plus volume in tube of cross section S at time liquid has a level h m

$V_a' = V_a + V_0$, m^3

In interpreting Eq. (12.8) we must remember that by Boyle's law the ratio of $\Delta V_a'$ to ΔP_0 is a negative quantity, so that the third term of Eq. (12.8) finally is positive.

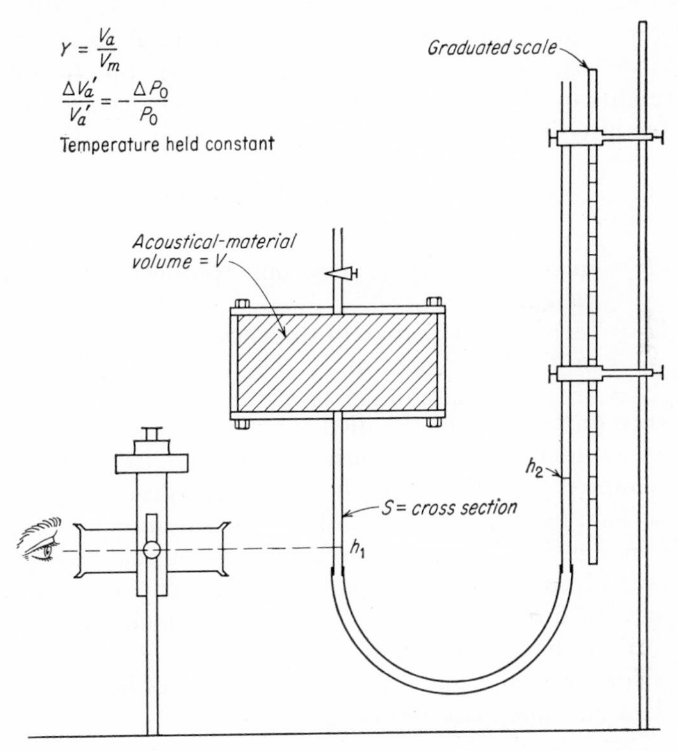

FIG. 12.4 Apparatus for measuring the porosity Y of an acoustical material. The volume of air V_a in the voids of the material plus the additional air in the cavity and tube equals V_a'.

Volume Coefficient of Elasticity of Air K. The volume coefficient of elasticity really comes from Eqs. (12.1) and (12.2). Let us rewrite them as follows:

$$\Delta P = -K \frac{\Delta V}{V} \qquad \text{newtons/m}^2 \tag{12.9}$$

where V = total volume of air being considered, m^3

ΔV = incremental change in volume, m^3, of air produced by incremental change in pressure ΔP, newtons/m^2

K = volume coefficient of elasticity, newtons/m^2. [For isothermal conditions K equals barometric pressure P_0 (about 10^5 newtons/m^2) and for adiabatic conditions $K = \gamma P_0$, where $\gamma = 1.4$ for air. K may vary between these two limits.]

Volume Coefficient of Elasticity of the Skeleton Q. The volume coefficient of elasticity of the skeleton Q is determined as follows: Place a small sample of the material on a smooth, flat surface. Lay a flat, lightweight sheet of metal or plastic on top. Be sure it is parallel to the surface. Measure the thickness of the sample T. Then add small weights to the plate and determine the changes in thickness. The value of Q in newtons per square meter is given by

$$dF = -Q \frac{dT}{T} S \tag{12.10}$$

where dF = incremental force, newtons, produced by adding a weight

dT = incremental change in thickness, m

T = original thickness, m

S = area of sample under test, m^2

Structure Factor k. The detailed inner structure of a porous material has some effect on its acoustical behavior in addition to the effect accounted for by the specific flow resistance. This effect has been described by Zwikker and Kosten[1] and amounts to an apparent increase in density of the air in the voids of material. The increase results from any tortuous paths that the air particles must take in the material during the excursion of their velocities.

The authors have found that for flexible blankets the structure factor varies between 1 and 1.2 and for rigid tiles between 1 and 3. Actual data on some materials are given below.

12.2 Measured Values of Specific Flow Resistances, Structure Factors and Elasticities of Porous Materials

Having defined the basic physical properties of porous acoustical materials, it may be well to provide the practicing acoustical engineer with some measured values of these quantities for a number of commercial acoustical materials which are commonly used for many noise-control applications.

Specific Flow Resistance. Figures 12.5 to 12.8 give the specific flow resistances for a wide variety of homogeneous porous materials currently available commercially. The unit-thickness flow resistance (in cgs rayls per inch of material thickness) is plotted as a function of volume density (pounds per cubic foot) of the material. Alternatively, a scale of mks rayls per meter of thickness is given in the right-hand ordinates, and a scale of kilograms per cubic meter is given along the upper edge of the graphs. The data in these figures

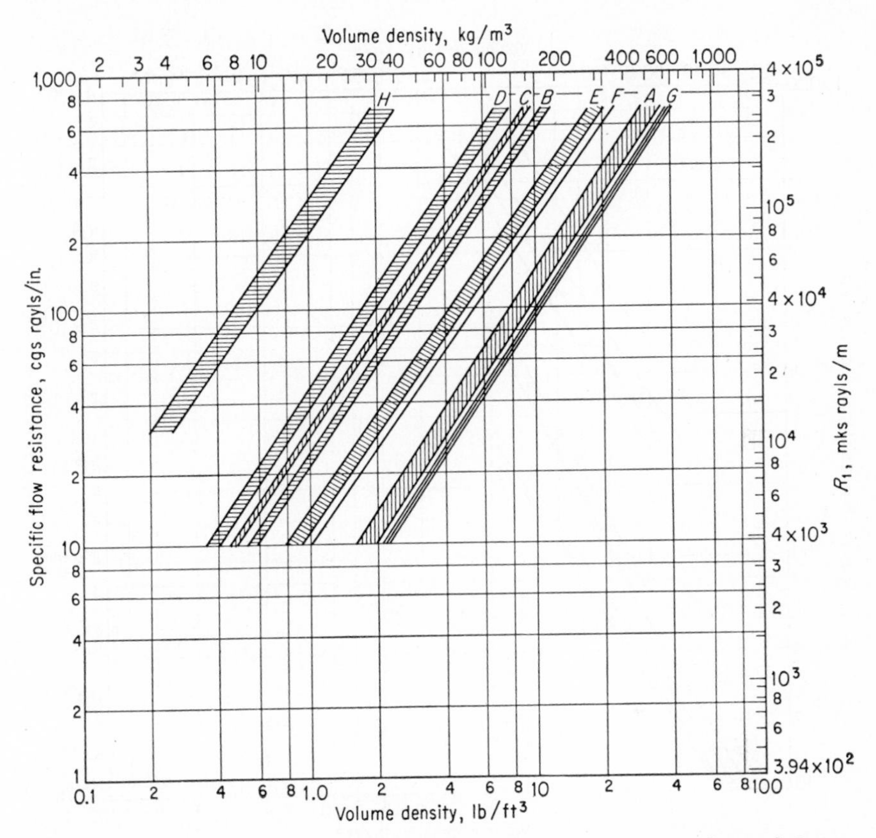

FIG. 12.5 Specific flow resistance versus volume density for some products of Owens-Corning Fiberglas Corporation: (*A*) PF 610 series Fiberglas (1954); (*B*) 318 Aerocor Fiberglas (1954); (*C*) 336 Aerocor Fiberglas (1954); (*D*) 334–335 Aerocor Fiberglas (1954); (*E*) PF 450 Aerocor Fiberglas (1953); (*F*) TWL white wool Fiberglas (1955); (*G*) TWF white wool Fiberglas (1953); (*H*) PF-105, XAA Fiberglas (1950). (*Measurements courtesy of Bolt Beranek and Newman Inc.*)

represent flow resistance values obtained by taking a particular sample of porous material, compressing it to different volume densities, and measuring its corresponding value of flow resistance.

The dates listed in the captions of each figure indicate the years in which the measurements were made.

It may be well to discuss a few practical considerations associated with the use of flow-resistance data. In the first place, the acoustical engineer must be on the alert for small changes in materials which manufacturers sometimes make. Since the flow resistance depends on the diameter of the fibers, porosity of the material, and orientation of the fibers, small manufacturing changes can sometimes ap-

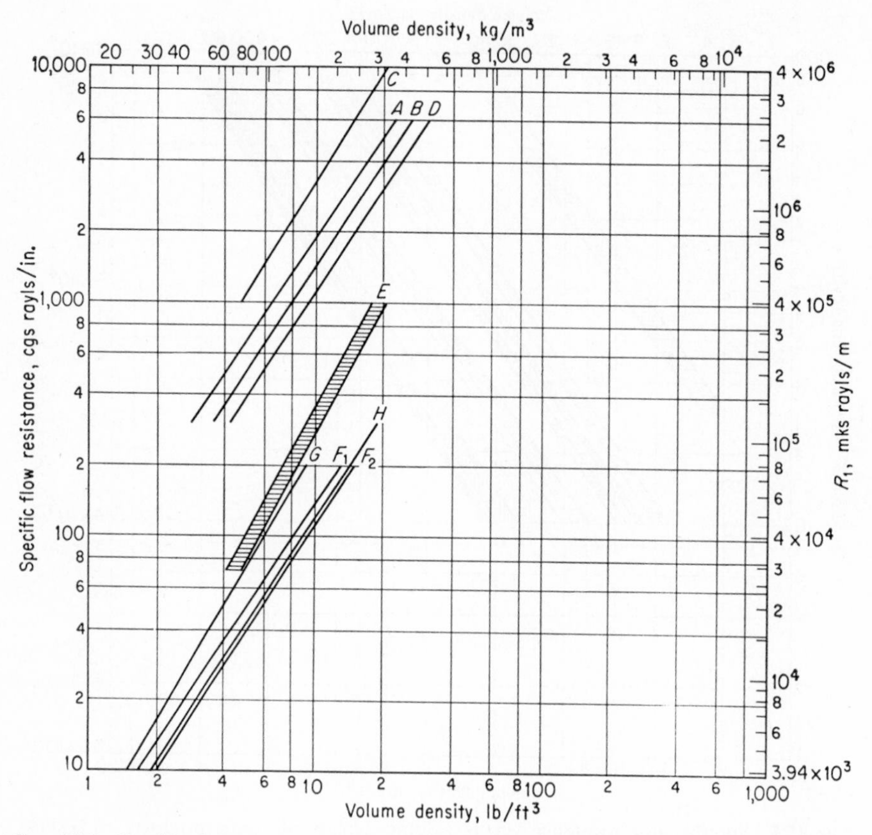

FIG. 12.6 Specific flow resistance versus volume density for some products of Johns-Manville Sales Corporation: (*A*) Thermoflex RF 300 (1954); (*B*) Thermoflex RF 400 (1954); (*C*) Thermoflex RF 600 (1954); (*D*) Thermoflex RF 800 (1954); (*E*) Spincoustic (1954); (F_1, F_2) Spintex 305—2½ lb/ft³ nominal density (1954); (*G*) Spintex 305—3½ lb/ft³ (1954); (*H*) Spintex 305—4½ lb/ft³ (1954). (*Measurements courtesy of Bolt Beranek and Newman Inc.*)

preciably affect the value of flow resistance. For example, a 2:1 change in fiber diameter may result in a 16:1 change in flow resistance.

Many manufacturers produce several different densities in a particular line of porous materials. In some cases an increase in density involves a change in fiber diameter (e.g., Spintex), and in others it may involve a change in porosity owing to an increase in binder material (e.g., Ultrafine and Ultralite). In addition, the measured densities of some materials encountered in actual field installations do not always agree with the nominal or rated densities given by the manufacturer. Such changes in densities will necessarily result in a differ-

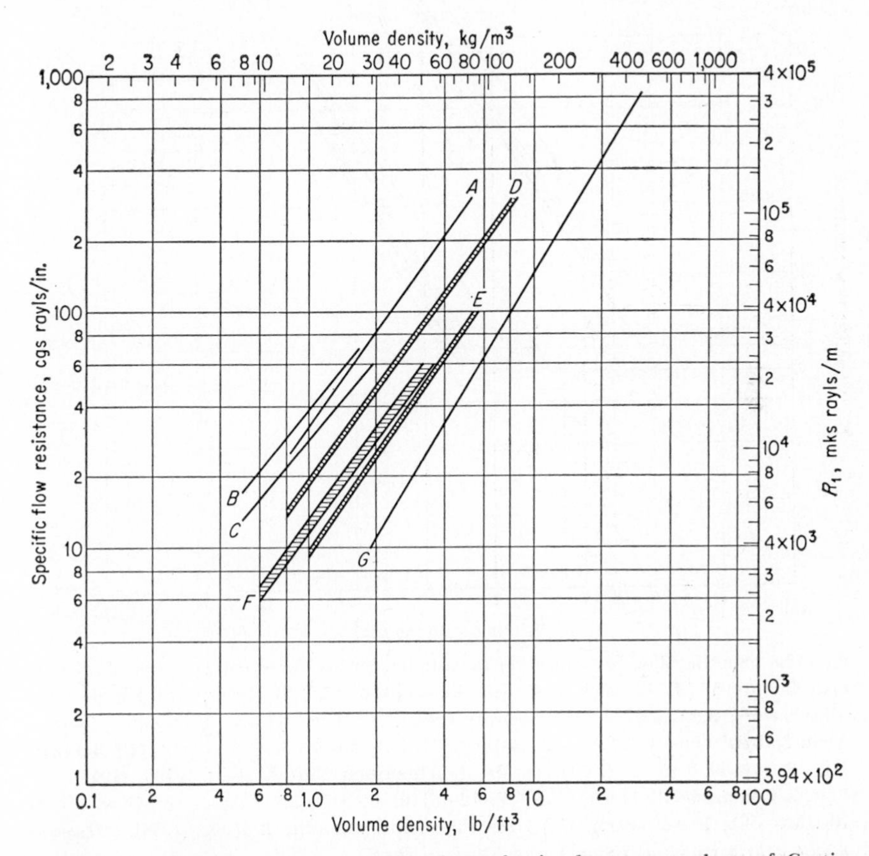

Fig. 12.7 Specific flow resistance versus volume density for some products of Gustin Bacon Company. (*A*) Ultrafine No. 1001 (1954); (*B*) Ultrafine No. 751 (1954); (*C*) Ultrafine No. 501 (1954); (*D*) Ultralite No. 200 (1954); (*E*) Ultralite No. 100 (1954); (*F*) Ultralite No. 50 and No. 75 (1954); (*G*) Nylabond (1955). (*Measurements courtesy of Bolt Beranek and Newman Inc.*)

ence in the flow resistance of a given material from that taken from these charts.

Structure Factors. There are few data in the literature on structure factors. For homogeneous materials made of fibers or granules, with interconnecting pores and few "blind alleys," the structure factor is related to the porosity according to Fig. 12.9.

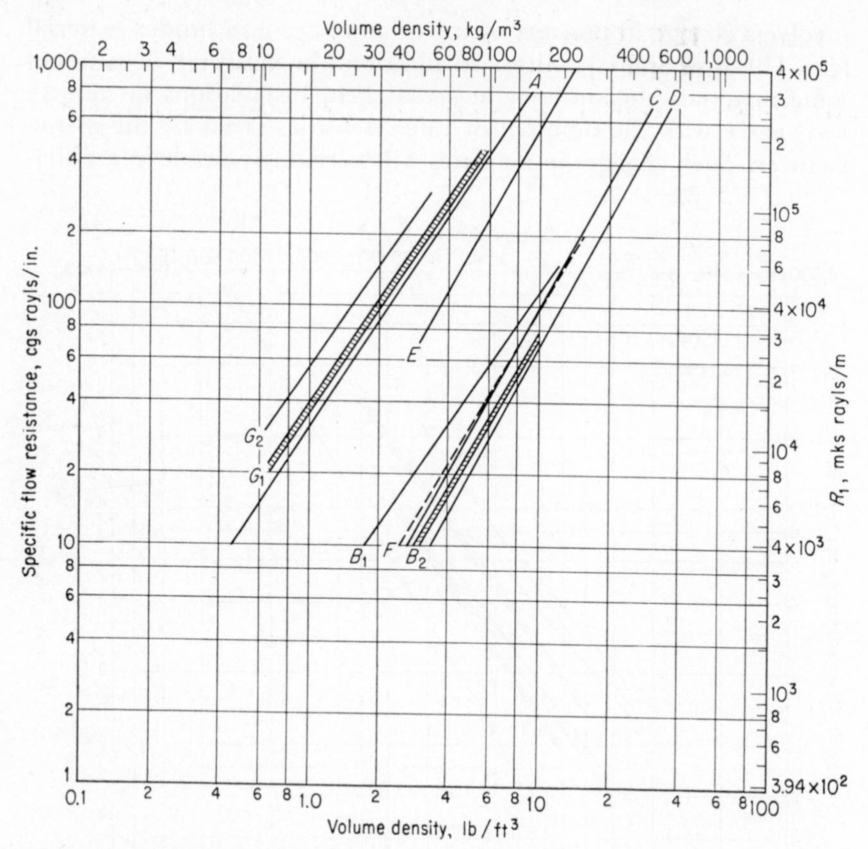

FIG. 12.8 Specific flow resistance versus volume density for some products of various manufacturers. (*A*) Babcock & Wilcox Kaowool Blanket B (1956); (B_1) Kittell Lacy (distributors) Basaltwool, 6 lb/ft³ density (1955); (B_2) Kittell Lacy (distributors) Basaltwool, 3 lb/ft³ density (1956); (*C*) Baldwin Hill Rockwool Style 1 (1951); (*D*) Baldwin Hill Rockwool Style 2 (1951); (*E*) H. I. Thompson Refrasil Batt B-100 (1954); (*F*) Felters Company All Fab (4–3080–72, 12–45110, 16–2076–72) (1955); (G_1) Glass Fibers Inc. Microlite B-305 (1954); (G_2) Glass Fibers Inc. Microlite B-310/0 (1954). (*Measurements courtesy of Bolt Beranek and Newman Inc.*)

Porosities. The porosity of a material may be calculated where the fibers are solid, provided the weight and volume of the binder, if any, can also be estimated. Assume a material of solid fibers with no binder; we obtain

$$Y = 1 - \frac{M_s}{V_s \sigma_F}$$

where V_s = total volume of sample, m³
σ_F = density of fibers, kg/m³
M_s = total mass of sample, kg

Example 12.1. Glass fibers have a density of about 2.5×10^3 kg/m³. Assume a sample that weighs 25 kg/m³. What is the porosity?

Solution

$$Y = 1 - \frac{25}{2.5 \times 10^3} = 0.99$$

For materials of complex composition, the porosity must be measured.

Volume Coefficients of Elasticity for Materials. For soft, cotlike materials the volume coefficient of elasticity Q is within the range of $0.01\ K$ to $0.002K$, where K is the volume coefficient of elasticity for air and equals $\gamma \times 10^5$ newtons/m². The quantity γ* lies between 1.0 and 1.4 as we said earlier in this chapter.

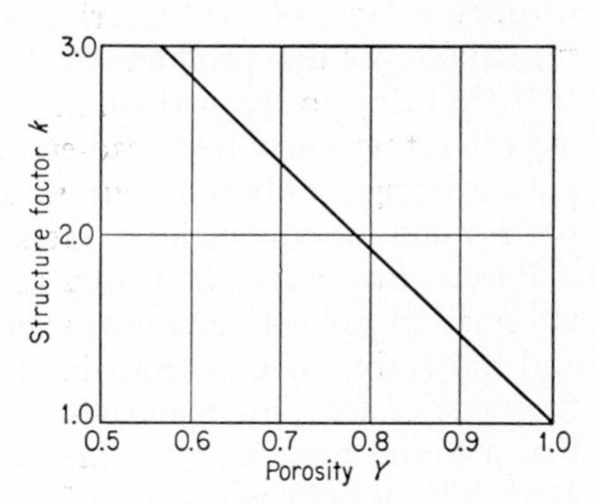

Fig. 12.9 Approximate relation between porosity Y and structure factor k for homogeneous materials made of fibers or granules with interconnecting pores and few "blind alleys."

12.3 Propagation Constants and Characteristic Impedance of Homogeneous Porous Materials

Many noise-reduction problems require the use of light-weight structures in which the sound must both be attenuated and absorbed by porous materials. In Chap. 13 we deal with solid structures—structures that generally weigh of the order of 10 to 100 lb/ft². Such weights are prohibitive in aircraft and in most types of surface vehicles. Furthermore, they offer no aid in reducing reverberant sound because their surfaces are highly reflecting.

Where weight is a consideration and where appreciable reduction of noise only at frequencies above, say, 500 cps is desired, structures containing porous blankets are frequently used. Structures containing porous materials will be discussed in Chap. 14. Before studying that chapter, we must first know the propagation constants of the materials. This is the purpose of this section.

A porous sound-absorbing material is characterized by a skeleton that may range from very soft to quite rigid. This skeleton contains interconnecting pores or channels through which air particles may

* Strictly speaking γ is used as a constant in most of the literature because the problem of variability of volume coefficient of elasticity only rises in acoustical materials. Rather than introduce still another term in the equations we shall arbitrarily permit γ to vary between 1.0 and 1.4.

move. An air particle in passing from one side of a porous layer to the other under steady-flow conditions usually has to follow a tortuous path. If the flow is nonturbulent, the volume of air that passes through a layer of the material is directly proportional to the pressure difference that produces the flow.

In addition to the paths that go more-or-less straight through the material, there may be "dead-end streets" in which no motion takes place during steady flow but which may participate in alternating particle motion under the influence of a sound wave.

Porous materials when used as sound-attenuating sheets produce attenuation partly by acting as a reflecting surface—as do solid walls—and partly by viscous losses in the interstices. So effective at high frequencies are some materials, such as Fiberglas PF–105* blankets, that a sound wave at 1,000 cps is attenuated by 60 db in traveling 1 ft! The weight per square foot of a wall 1 ft thick made of this material would be only 0.6 lb/ft^2. The weight of a solid partition necessary to provide this much reduction at 1,000 cps would be about 40 lb/ft^2. At low frequencies the story is much different. For example, at 250 cps the attenuation for a Fiberglas PF–105 wall would be only about 10 db, while that for a 40-lb/ft^2 solid wall would be about 45 db.

One might invent an axiom for sound-attenuating barriers, namely, that barrier thickness may be traded for barrier weight, provided the thickness exceeds ¼ the sound wavelength.

To determine the acoustical behavior of a porous, homogeneous material we need to know its complex propagation constant and its characteristic impedance. The *complex propagation constant* b has two parts, the real part α called the attenuation constant, in nepers (or in decibels) per meter, and the imaginary part β called the phase constant, in radians (or degrees) per meter. (Note that the number of nepers must be multiplied by 8.69 to convert to decibels and the number of radians is multiplied by 57.3 to convert to degrees.) That is to say

$$b = \alpha + j\beta \qquad m^{-1} \tag{12.11}$$

where $j = \sqrt{-1}$.

The propagation constant is a property of the material itself and is not dependent on mounting conditions, when large areas of the material are being considered.

The attenuation constant α tells how much a sound wave will be reduced as it travels through the material. The phase constant β

* This is the designation of a type of material commonly used in aircraft.

is a measure of the speed of propagation of the sound wave through the material.

The relation for determining the speed of sound in the material is given by the formula

$$c_m = \frac{\omega}{\beta} \qquad \text{m/sec} \tag{12.12}$$

where $\omega = 2\pi f$ = angular frequency, radians/sec
β = phase constant, radians/m
f = frequency, cps
c_m = speed of sound in material, m/sec

The *characteristic impedance* Z_0 is the ratio of the sound pressure to the particle velocity at the entrance surface of a material of infinite depth on which a plane sound wave is falling perpendicular to the surface.

In the general case, the skeleton and the air in the interstices may both participate in the transmission of the sound wave through the material. In this chapter we shall deal only with materials that may be called *rigid* or with those that we may classify as *flexible*. In these two extremes, the stiffness of the skeleton may be neglected in the calculations with attendant simplicity. The important acoustical properties of a sound-absorbing material that determine the propagation constant are gathered together here.

R_1 = alternating flow resistance for unit thickness of material due to difference between velocity of skeleton and velocity of air in interstices, mks rayls/m.

R_{f1} = unit-thickness flow resistance measured by steady-airflow method, mks rayls/m. At lower frequencies R_1 is equal to this quantity. (To convert from in. to m, multiply number of in. by 0.0254.) At frequencies above about 1,000 cps, R_1 is often found to be greater than R_{f1}.

ρ_m = density of acoustical material, kg/m^3. (To convert from lb/ft^3 to kg/m^3, multiply by 16.)

ρ_0 = density of air, kg/m^3 (equal to 1.18 kg/m^3 at atmospheric pressure of 10^5 newtons/m^2 and temperature of 22°C or 71.6°F).

Y = porosity equal to ratio of volume of voids in material to total volume; porosity equals total volume minus fiber volume, all divided by total volume.

$K = \gamma \times 10^5$ newtons/m^2 = volume coefficient of elasticity of air. For isothermal conditions, γ = 1.0. For adiabatic conditions, it equals 1.4. Generally, isothermal conditions prevail for frequencies below 100 cps, and adiabatic conditions prevail above 1,000 cps. In between, γ varies between 1.0 and 1.4.

Q = volume coefficient of elasticity of an acoustical material, newtons/m^2.
k = structure factor that introduces into equations nature of interstices in skeleton. Its value is always greater than unity (see Fig. 12.9).
c = speed of sound, m/sec (about 344 m/sec or 1,128 ft/sec in air).
Z_0 = characteristic impedance of acoustical material, mks rayls; equals ratio of sound pressure to particle velocity just outside surface of semi-infinite sample with plane wave striking surface perpendicularly.

Zwikker and Kosten[2] and Beranek[3] have developed equations for the solution of the propagation constants in soft blankets and rigid tiles. Using the results of Beranek (which include porosity), the propagation constant b for a rigid tile is

$$b = j\omega \sqrt{\frac{\rho_0 k Y}{K}} \sqrt{1 - j\frac{R_1}{\rho_0 k \omega}} \tag{12.13}$$

The characteristic impedance is

$$Z_0 = \frac{-jKb}{\omega Y} \tag{12.14}$$

For a soft blanket where $K > 20Q$

$$b = j\omega \sqrt{\frac{Y}{K}} \sqrt{\langle\rho_1\rangle - j\frac{\langle R_1\rangle}{\omega}} \tag{12.15}$$

where

$$\langle R_1\rangle = \frac{R_1[1 - \rho_0(1 - Y)/\rho_m]}{\left[1 + \frac{\rho_0(k-1)}{\rho_m}\right]^2 \left[1 + \frac{R_1^2}{\rho_m^2\omega^2[1 + \rho_0(k-1)/\rho_m]^2}\right]} \tag{12.16}$$

and

$$\langle\rho_1\rangle = \rho_0 k - \frac{\dfrac{R_1^2(Y/k + \rho_m/\rho_0 k)}{\rho_m^2\omega^2[1 + \rho_0(k-1)/\rho_m]^2} + \dfrac{1 + \rho_0 Y(k-1)/\rho_m k}{1 + \rho_0(k-1)/\rho_m}}{1 + \dfrac{R_1^2}{\rho_m^2\omega^2[1 + \rho_0(k-1)/\rho_m]^2}} \tag{12.17}$$

The quantity $\langle\rho_1\rangle$ is called the *effective density* of the gas particles. It is seen that at low frequencies it approaches $(\rho_m + \rho_0 kY)$ and at high frequencies it approaches $\rho_0 k$. The physical reasons for these limits are as follows: At low frequencies the mass reactance of

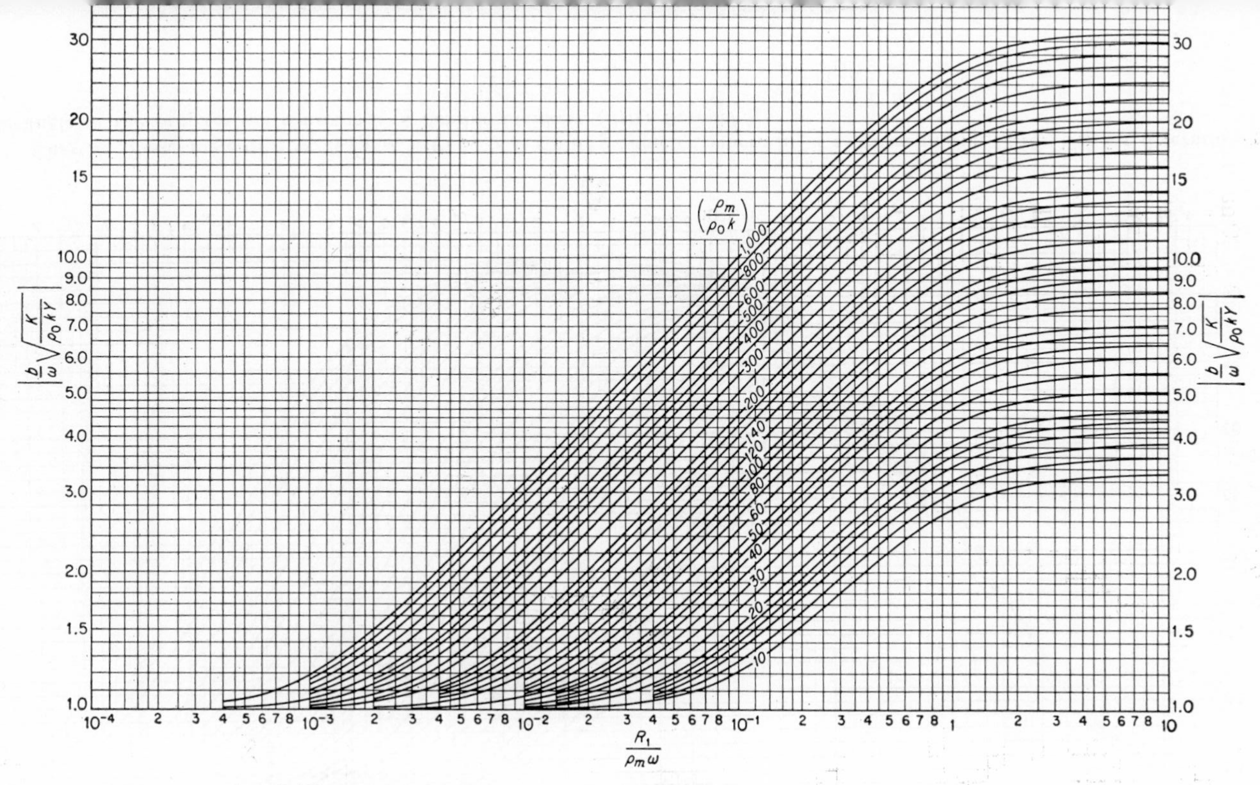

FIG. 12.10 Chart of $\left|b\sqrt{\frac{K}{\rho_0 kY}}\frac{1}{\omega}\right|$ vs. $\frac{R_1}{\rho_m\omega}$, with $\frac{\rho_m}{\rho_0 k}$ as the parameter, where

$$b = j\omega\sqrt{\frac{\rho_0 kY}{K}}\sqrt{\frac{(R_1/\rho_m\omega)^2(1 + \rho_m/\rho_0 k) + 1 - j(R_1/\rho_m\omega)(\rho_m/\rho_0 k)}{1 + (R_1/\rho_m\omega)^2}}$$

Use for determining the magnitude of the propagation constant b of flexible blankets. Under the large radical the porosity Y has been assumed to be near unity. The abscissa, the ordinate, and the parameter are all dimensionless quantities. (*After Beranek.*[3])

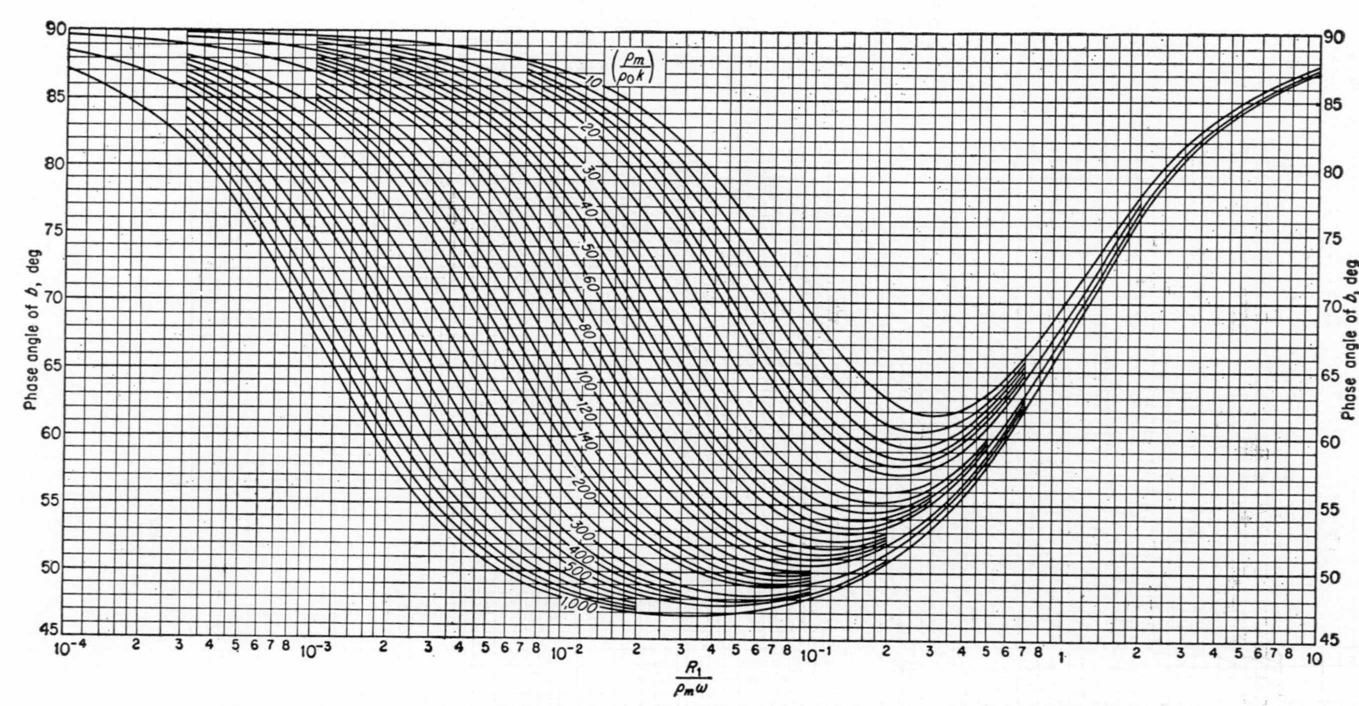

FIG. 12.11 Chart of phase angle of b vs. $R_1/\rho_m\omega$, with ρ_m/ρ_0k as the parameter, where $|b|$ is as given in Fig. 12.10. Use for determining the phase angle of the propagation constant b of flexible blankets. (*After Beranek.*[3])

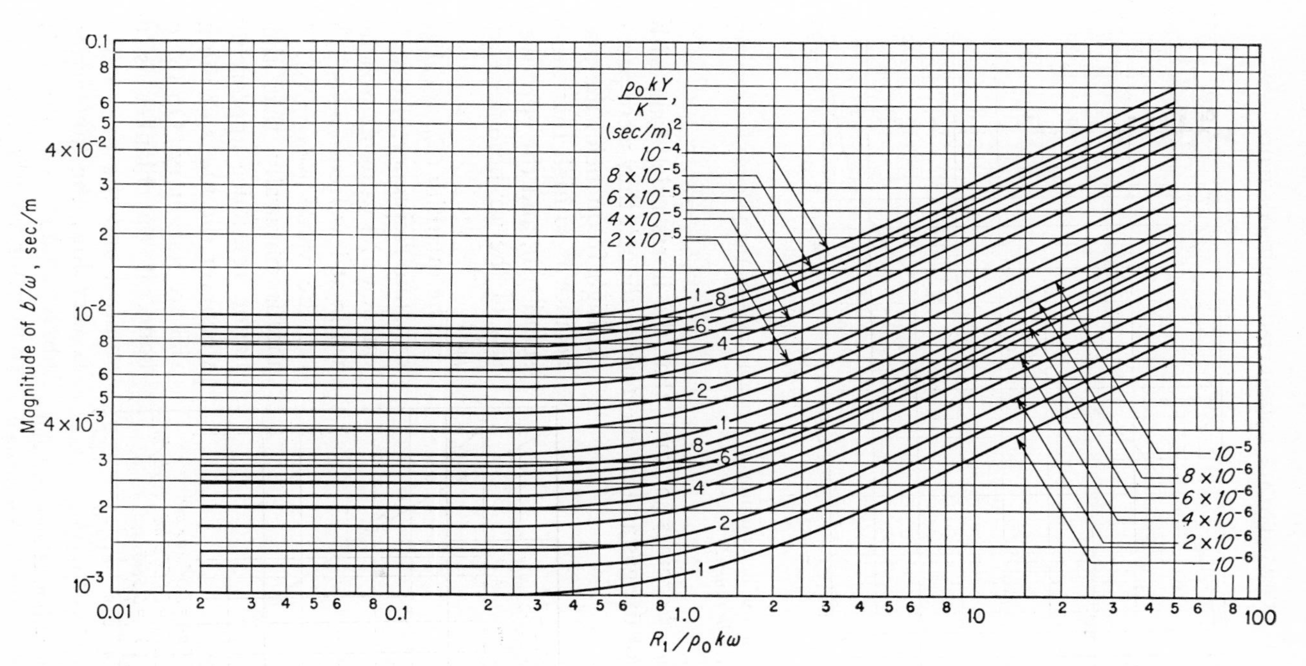

FIG. 12.12 Chart of $|b/\omega|$ vs. $R_1/\rho_0 k\omega$, with $\rho_0 kY/K$ as the parameter, where $b = j\omega\sqrt{\rho_0 kY/K}\sqrt{1 - j(R_1/\rho_0 k\omega)}$. Use for determining the magnitude of the propagation constant b of rigid tiles. The abscissa is dimensionless; the ordinate is in dimensions of sec/m and the parameter in (sec/m)^2. (*After Beranek.*[6])

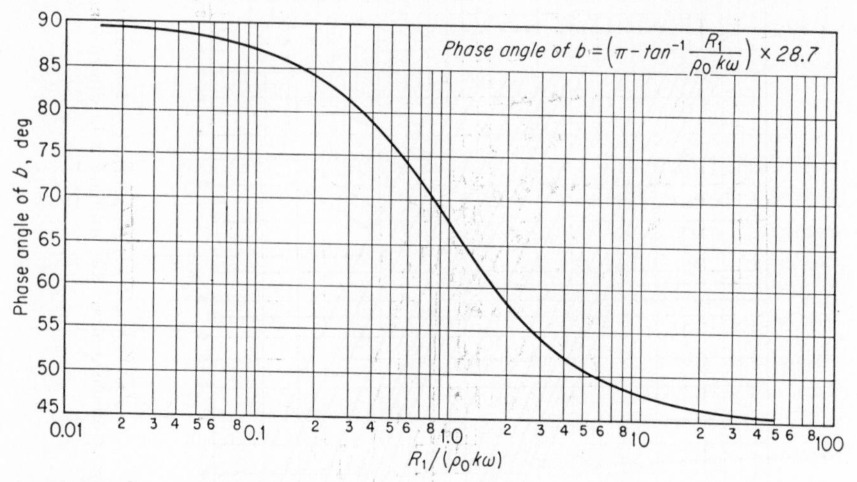

FIG. 12.13 Chart for determining the phase angle of the propagation constant b of rigid tiles. (*After Beranek.*[6])

the fibers $\omega\rho_m$ becomes so small that the fibers ride along with the oscillations of the gas and in effect act as though they were part of it. At high frequencies the mass reactance is so high that the fibers stand still, and the effective density of the air becomes equal to $\rho_0 k$.

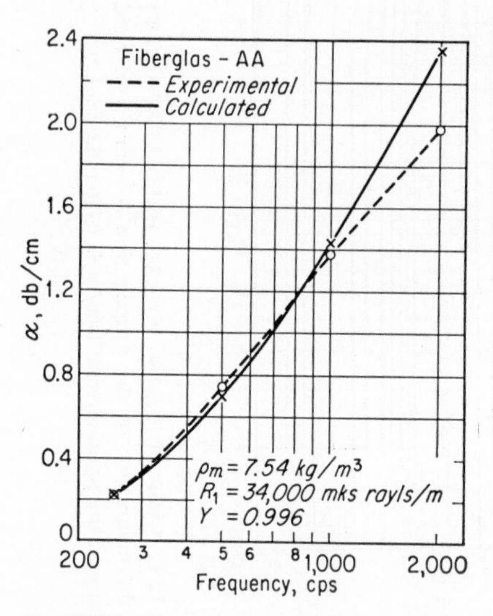

FIG. 12.14 Comparison of measured and calculated values of attenuation constant for Fiberglas AA. The values of α, Y, ρ_m, and R_1 were measured.

The quantity $\langle R_1 \rangle$ is called the dynamic resistance. At high frequencies its value approaches R_1 while at low frequencies it takes on the value $(\omega^2 \rho_m^2/R_1)$. This means physically that the dynamic resistance approaches R_1 whenever the fibers of the material are stationary. On the other hand, if the blanket is unrestrained and the mass reactance is very low, the air carries the fibers along, and there is no chance for viscous dissipation at the surface of the fibers; so, from Eq. (12.16) $\langle R_1 \rangle$ approaches zero at low frequencies.

Graphical solutions of Eqs. (12.13) and (12.14) are given in Figs. 12.10 through 12.13. The assumption made in converting Eq. (12.15) to Figs. 12.11 and 12.12 was that

$$\frac{\rho_0(k-1)}{\rho_m} \ll 1$$

To determine the propagation constants for various kinds of materials, values of the specific flow resistance R_1 in mks rayls per meter, structure factor k and volume coefficient of elasticity Q were given in Sec. 12.2.

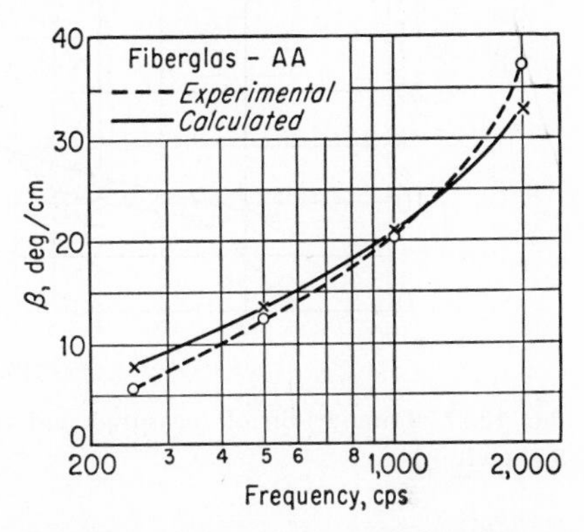

FIG. 12.15 Comparison of measured and calculated values of phase constant for Fiberglas AA.

Calculations of the real and imaginary parts of the propagation constant for four materials are compared in Figs. 12.14 through 12.21. It is seen that the agreement between measured data and calculations is satisfactory in most cases. The deviations at low frequencies in the case of Aerocor and J. M. Stonefelt are known to be due to artifacts in the measuring setup. At high frequencies, the measured attenuation constants for several of the materials are higher than calculated, which may indicate that the dynamic flow resistance R_1 increases with frequency. In the case of TWF Fiberglas significant discrepancies exist for frequencies between 100 and 400 cps. It appears that some sort of structural resonance occurred that could be partly eliminated by placing heavy expanded-metal screens every inch in the material parallel to the wavefront.

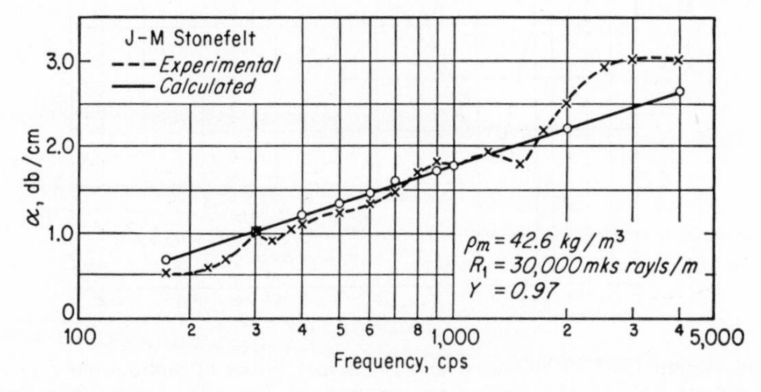

FIG. 12.16 Comparison of measured and calculated values of attenuation constant for J-M Stonefelt. The values of α, Y, ρ_m, and R_1 were measured.

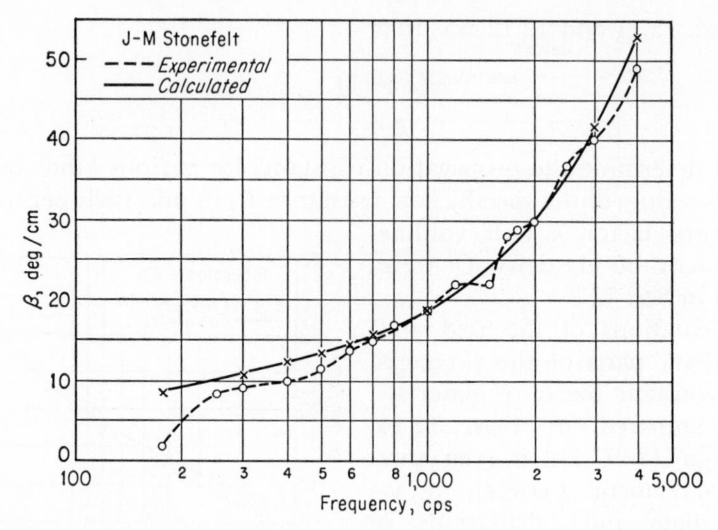

FIG. 12.17 Comparison of measured and calculated values of phase constant for J-M Stonefelt.

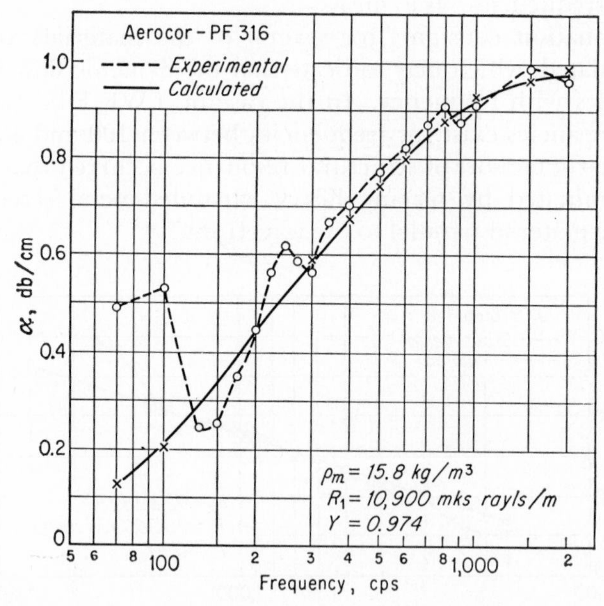

FIG. 12.18 Comparison of measured and calculated values of attenuation constant for Aerocor PF-316. The values of α, Y, ρ_m, and R_1 were measured.

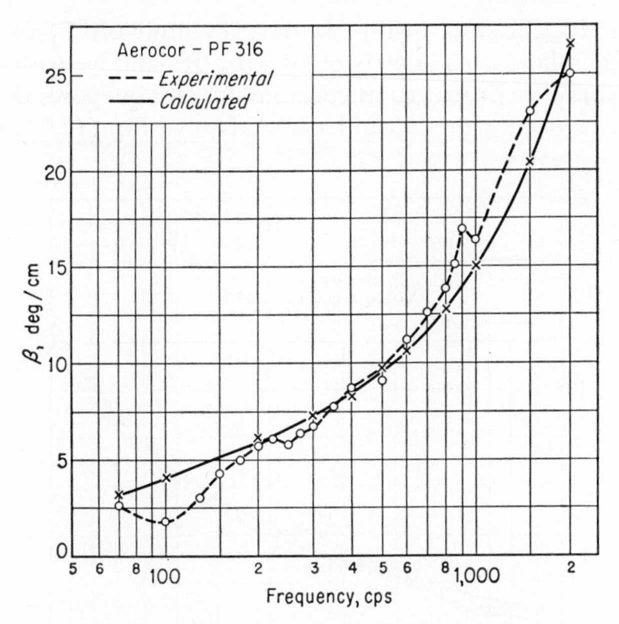

FIG. 12.19 Comparison of measured and calculated values of phase constant for Aerocor PF-316.

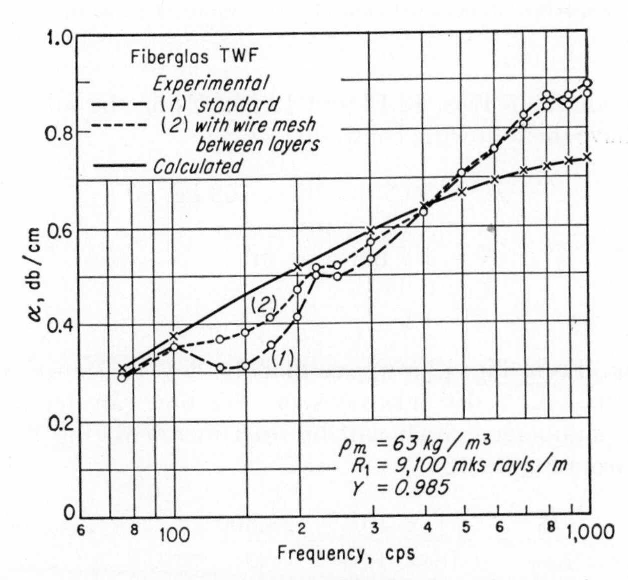

FIG. 12.20 Comparison of measured and calculated values of attenuation constant for Fiberglas (TWF). The values of α, Y, ρ_m, and R_1 were measured.

Example 12.2. Let us assume we have a sample of Owens-Corning PF–610 Fiberglas with a density of 10.5 lb_m/ft^3, and we wish to determine the complex propagation constant b. For purposes of this calculation we consider this a rigid tile and solve Eq. (12.13) with the

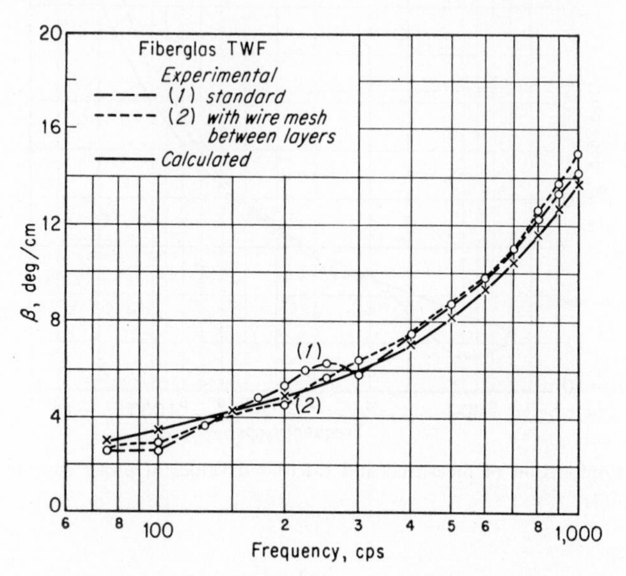

FIG. 12.21 Comparison of measured and calculated values of phase constant for Fiberglas (TWF).

use of the charts in Figs. 12.12 and 12.13. Using the mks system of units we have the following constants:

$$\rho_m = 10.5 \text{ lb}_m/\text{ft}^3 = 168 \text{ kg/m}^3$$
$$\rho_0 = 1.18 \text{ kg/m}^3$$
$$K = 10^5 \text{ newtons/m}^2$$
$$Y = 0.95$$
$$k = 1.5$$

Solution. From Fig. 12.5 we see that the flow resistance $R_1 = 130$ cgs rayls/in. $= 5.1 \times 10^4$ mks rayls/m. We now can determine the following parameters which will be used in calculating the magnitude of b from Fig. 12.12:

$$\frac{R_1}{\rho_0 k\omega} = \frac{5.1 \times 10^4}{1.18 \times 1.5\omega} = \frac{28{,}800}{\omega}$$

$$\frac{\rho_0 k Y}{K} = \frac{1.18 \times 1.5 \times 0.95}{10^5} = 1.68 \times 10^{-5}$$

We can set up the values as shown in Table 12.1.

Table 12.1

f	ω	$\frac{R_1}{\rho_0 k\omega}$	$\left\lvert\frac{b}{\omega}\right\rvert$ (Fig. 12.12)	$\lvert b\rvert$, m^{-1}	ϕ (Fig. 12.13)	$b = \alpha + j\beta$,* nepers/m	α†, db/in.
(1)	(2)	(3)	(4)	(5)	(6)	(7)	(8)
100	628	46	2.7×10^{-2}	17	45.5	$11.9 + j\,12.1$	2.6
200	1,256	23	1.9×10^{-2}	23.8	46.3	$16.4 + j\,17.2$	3.6
400	2,512	11.5	1.35×10^{-2}	34	47.5	$23 + j\,25$	5.0
800	5,024	5.8	9.5×10^{-3}	47.7	50	$30.6 + j\,36.6$	6.7
1,600	10,048	2.9	7.0×10^{-3}	70	54.5	$40.6 + j\,57$	8.9
3,200	20,096	1.45	5.2×10^{-3}	105	63.0	$47.6 + j\,93.5$	10.5

* By usual vector theory, $|b| = \sqrt{\alpha^2 + \beta^2}$ and $\phi = \tan^{-1}(\beta/\alpha)$. The use of a trigonometric slide rule is recommended for easy conversions. Note that ϕ in column 6 must be divided by 57.3 to obtain radians.

† The attenuation constant α in column 7 is given in nepers per meter. In order to convert to decibels per inch, we multiply by $8.686/39.37 = 0.22$. The resulting values are given in column 8.

If, in addition, we also wish to obtain the specific acoustic impedance (Z_0), we solve it from Eq. (12.14).

$$Z_0 = -j\frac{Kb}{\omega Y} = -j\frac{10^5}{0.95}\frac{b}{\omega} = -j\,1.05 \times 10^5\,\frac{b}{\omega}$$

$$Z_0 = |Z_0| \;\underline{/\theta}$$

$$= 1.05 \times 10^5 \left|\frac{b}{\omega}\right| \;\underline{/-90 + \phi}$$

Table 12.2

f	$\left\lvert\frac{b}{\omega}\right\rvert$, m^{-1}	$\phi°$	$\lvert Z_0\rvert$	$\theta°$	$\frac{\lvert Z_0\rvert}{\rho_0 c}$
(1)	(2)	(3)	(4)	(5)	(6)*
100	2.7×10^{-2}	45.5	2,840	−44.5	6.9
200	1.9×10^{-2}	46.3	1,990	−43.7	4.9
400	1.35×10^{-2}	47.5	1,420	−42.5	3.5
800	9.5×10^{-3}	50.0	1,000	−40.0	2.4
1,600	7.0×10^{-3}	54.5	735	−35.5	1.8
3,200	5.2×10^{-3}	63.0	545	−27.0	1.33

* The units of impedance Z_0 are mks rayls. It is customary to express the impedance in $\rho_0 c$ units where $\rho_0 c$ is the characteristic impedance of air and is equal to 410 mks rayls. Column 6 gives the impedance in $\rho_0 c$ units (dimensionless).

12.4 Measured Values of Attenuation Constants and Wavelengths for Some Commercial Porous Materials

The method of calculation of propagation constants given in the previous section is useful when no measured data are available, but it is tedious. Hence, it is desirable to have available measured propagation constants on a number of materials. Also, above about

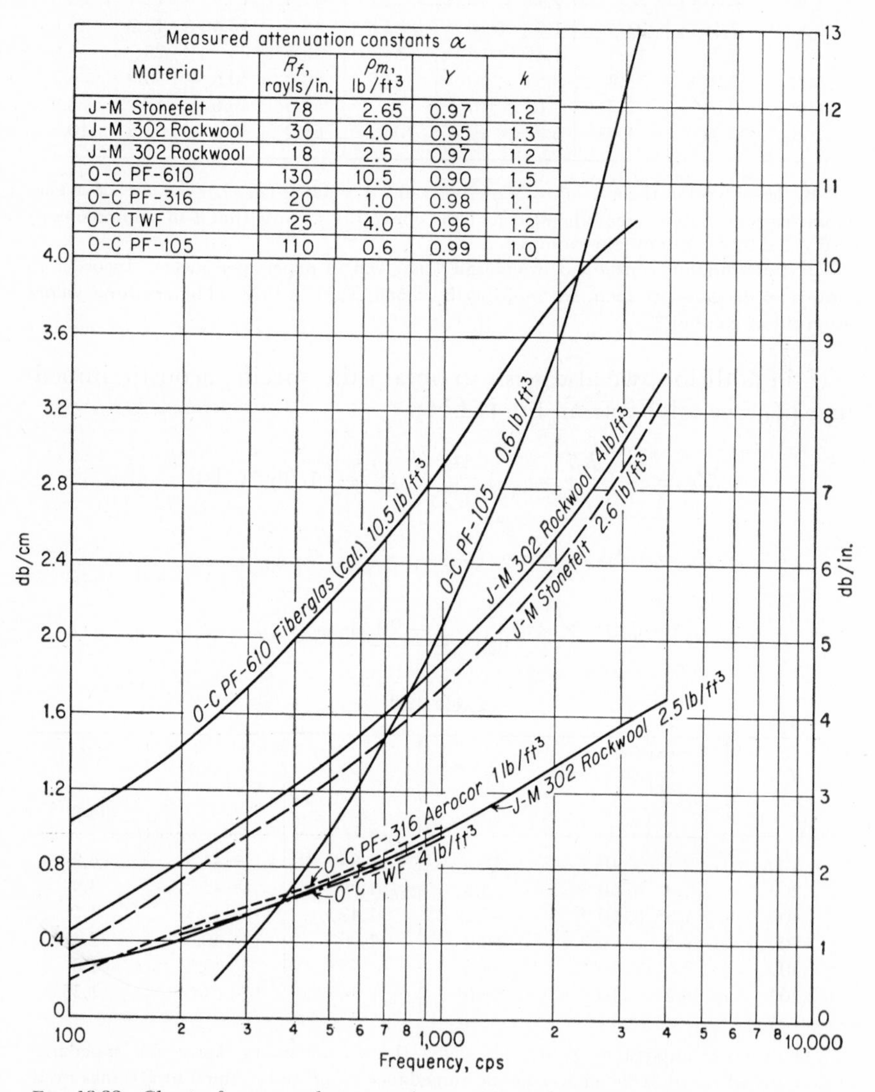

Measured attenuation constants α				
Material	R_f, rayls/in.	ρ_m, lb/ft³	Y	k
J-M Stonefelt	78	2.65	0.97	1.2
J-M 302 Rockwool	30	4.0	0.95	1.3
J-M 302 Rockwool	18	2.5	0.97	1.2
O-C PF-610	130	10.5	0.90	1.5
O-C PF-316	20	1.0	0.98	1.1
O-C TWF	25	4.0	0.96	1.2
O-C PF-105	110	0.6	0.99	1.0

FIG. 12.22 Chart of measured attenuation constants for seven common materials.

1,000 cps, the attenuation constant for some materials seems to rise more rapidly with increasing frequency than one would calculate. This is presumably due to an increase in R_1 with increasing frequency, probably induced by the particular internal structures of the material.

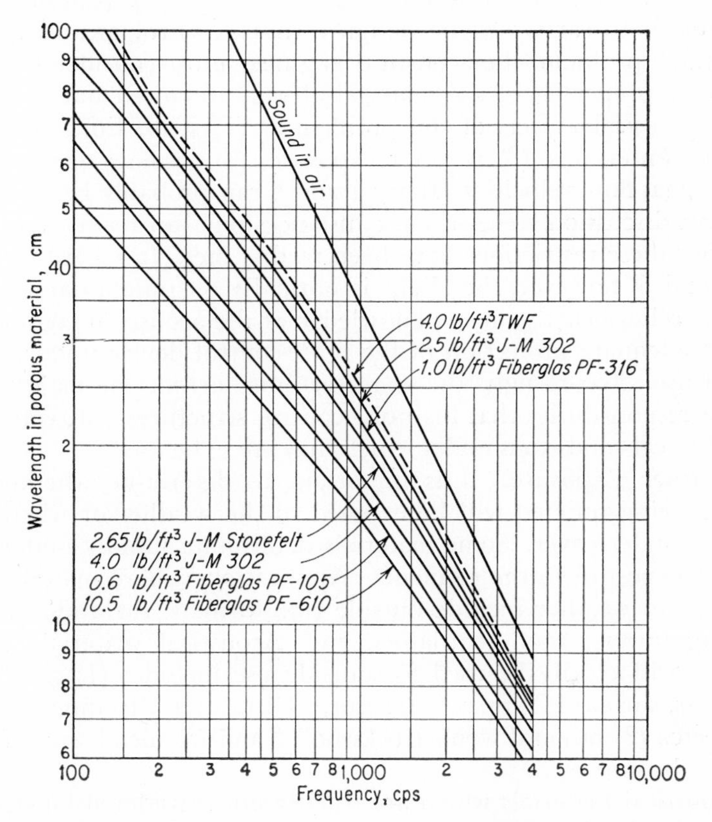

FIG. 12.23 Chart of measured wavelengths (inside the material) for seven common materials.

It is also useful to know the wavelength of the sound wave that is propagated inside the material so that the thickness of the blanket being considered may be compared with it.

The attenuation constants in decibels per inch and decibels per centimeter for seven materials are given in Fig. 12.22. The wavelength of the wave propagated inside the same were computed from Eq. (12.18) below and are plotted in Fig. 12.23:

$$\lambda_M = \frac{2\pi}{\beta} \tag{12.18}$$

In the case of some materials, such as Johns-Manville type 302 rockwool, the propagation constants are different in different directions in the material. An average value is shown in the figures.

12.5 General Considerations

In order to carry out the design of noise-control measures for a particular problem, the acoustical engineer must consider not only the basic properties of the material already discussed but also such other practical aspects of the problem as (1) space limitations, (2) weight limitations, (3) weather exposure, (4) gas-flow temperature, and (5) gas-flow velocities. Items 1 and 2 can probably be dismissed without discussion since in most instances one can readily ascertain whether these restrictions have been readily met. In regard to items 3, 4, and 5, however, the client is normally on unfamiliar ground, and he relies heavily on the knowledge of the acoustical engineer to insure adequate protection of the noise-control device under operational use. Accordingly, the engineer must assume the major share of the responsibility that his noise-control structures will withstand the physical conditions under which they are to be used.

Weather Exposure. It is important to determine whether the device being quieted will be exposed to the weather or whether it will be under cover. Some materials deteriorate rapidly under continued exposure to the elements. Applications where materials are exposed to weather include outside transformers, turbines, motors, cooling towers, roof ventilators, etc. Acoustical materials which might satisfactorily be used for outside use include: (1) glass fiber blankets, such as Fiberglas,* Aerocor,* Ultralite,† Ultrafine,† Microlite,‡ etc., (2) mineral wools (rockwool),‡ and (3) metal wools (steel, copper, bronze, etc.).

Acoustical materials which are usually not considered satisfactory for outside use include: (1) natural fibers (kapok, paper, cotton, etc.), (2) wood fibers, (3) cellulose fibers, and (4) hair felt and wool felt.

In addition to weather exposure, some noise-control problems involve spaces where combustible liquids or fumes are prevalent. Entrainment in the fibers may create a fire hazard. Such conditions would require some protection of the fibrous acoustical material. This protection might be achieved by enclosing the material in a bag of plastic or other impervious material. This wrapping usually reduces the absorptive properties of the material, particularly at high

* Owens-Corning Fiberglas Company, Newark, Ohio.

† Gustin-Bacon Company, Owens, Ill.

‡ Johns-Manville Corporation, Manville, N.J.

frequencies, and this fact has to be considered in the over-all results.

A very common situation is for an acoustical material to be located where it will be damaged by contact with people or moving objects, or where it may need frequent cleaning or painting. In these cases it is customary to cover the material with perforated metal or asbestos sheets, expanded metal grilles, rigidized metal sheets, or woven plastic (Saran) cloth. Unless the percentage open area of the covering sheet is large, the reduction of sound absorption at high frequencies may be considerable.

Gas-flow Temperature. In many noise-control problems temperature is a very important factor. Sometimes high-temperature ducts or pipes that are radiating noise, such as are associated with compressors, diesel engines, wind tunnels, etc., must be wrapped. With proper selection, thermal and acoustical insulation can often be combined in one material. Another common need is for the installation of an acoustical material inside fluid-flow devices, such as the exhausts of diesel engines, jet-engine test cells, ventilating systems, wind tunnels, compressors, etc. In such cases the extremely high temperatures tend to decrease the tensile strength of the materials and to subject them to thermal shock.

Examples of some of the materials that are currently available for use where temperature must be considered are tabulated in Tables 12.3 and 12.4. The approximate maximum allowable temperature to which the material should be exposed is indicated.

Table 12.3

Porous Materials for Pipe and Duct Wrappings or Fillers in Wall Structures

Material	*Maximum allowable temperature, °F*
Fibrous Materials:	
Rockwool fibers (e.g., Baldwin-Hill Koldboard); nylon fibers (e.g., Gustin Bacon Nylabond)	150–300
Asbestos felts	300
Bonded glass fibers (e.g., Microlite, Aerocor, PF Fiberglas board)	350–400
Wool felts	600
Asbestos fibers, rockwool (e.g., Johns-Manville Spintex)	700–900
Unbonded glass fibers (e.g., TWF and TWL Fiberglas, felted asbestos fiber)	1,000
Mineral wool (e.g., Baldwin-Hill rockwool) and asbestos and rockwool combinations	1,200
Mineral wool, felted block (e.g., Baldwin-Hill Monoblock)	1,400
Vitreous fiber-silica (e.g., H. I. Thompson Refrasil)	1,800
Refractory fiber (e.g., Johns-Manville Thermoflex)	2,000
Alumina-silica fiber (e.g., Babcock and Wilcox Kaowool)	2,000

In Table 12.3, materials are given that can be used to wrap high-temperature pipes and ducts or as inner fillers in wall panels for acoustical enclosures. Not all these materials are satisfactory for use in noise-control problems where the acoustical material is to be used as an absorbing medium in the path of gas streams—mainly because the flow resistance of some of them is much too high.

Table 12.4

Porous Materials for Use in Hot Gas Streams

Material	*Maximum allowable temperature, °F*
Fibrous Materials:	
Some mineral wools (e.g., Johns-Manville Airacoustic)	125–150
Wool felts	150–200
Hair felts and nylon fibers (e.g., Gustin Bacon Nylabond)	200–250
Bonded glass fibers* (e.g., Microlite, PF Aerocor, Ultralite, Ultrafine)	350–400
Asbestos fibers (Johns-Manville Spintex and Spincoustic)	800
Unbonded glass fibers (TWF and TWL Fiberglas)	1,000–1,100
Mineral wool, felted block (e.g., Baldwin-Hill Rockwool)	1,200
Basaltwool	1,400
Vitreous fiber-silica (e.g., H. I. Thompson Refrasil)	1,800
Alumina-silica (Babcock and Wilcox Kaowool)	2,000
Other Materials:	
Haydite block (cracks under high transient temperatures)	900
Porous firebricks (including ceramic)	1,600–3,000
Gravel	3,000

* In these materials the temperature limits generally apply to the binder; the glass fibers themselves are good to about 1,000 to 1,100°F. After the binder melts, the glass fibers may have a tendency to sift under vibration.

In Table 12.4 some porous materials that can be used satisfactorily in the presence of hot gases and that have desirable flow resistances are listed. The fibrous materials are normally encased behind perforated metal sheets forming the exterior of parallel baffles or acoustical liners for tuned ducts. Perforated bricks or blocks are sometimes used to line exhaust mufflers or to form parallel baffles. These bricks usually do not have as good acoustical properties as the fibrous materials and in many instances are very expensive.

The temperature limits presented in Tables 12.3 and 12.4 do not in all cases agree with values published in advertising literature. Rather, the values represent the results of field data accumulated from a variety of installations throughout the country and are generally more conservative.

One other important aspect is the problem of material shrinkage owing to temperature effects. Some of the materials mentioned in this section have relatively little shrinkage loss (about 2 per cent), while others have shown substantially higher values. Not enough data are available at this time to present them in any organized manner. The acoustical engineer, however, should be alerted to this problem. In many designs provisions might be made, for example, to use somewhat larger sizes of material to allow for a certain amount of shrinkage.

Gas-flow Velocities. Noise-control problems often involve the use of acoustical materials in high-velocity gas streams; for example, in the exhausts of jet engines, diesel engines, compressors, high-velocity ventilating ducts, etc. The problem generally is the deterioration of the acoustical material due to the high-velocity gases flowing past the material. This leads to a tearing off of particles of the material and is a cumulative process which eventually may lead to complete deterioration. In addition, turbulence in the high-velocity streams subjects the materials to vibration which may lead to further deterioration. One solution to this problem is to install the acoustical material behind some type of protective facing. Protective facings vary in complexity depending upon the velocity of the gases.

In many cases, high temperatures and high-velocity gas streams are present simultaneously so that the acoustical engineer must take into account the effects due to their interaction.

There is virtually no quantitative information on the allowable limits of structures for noise control in the presence of high-speed gas flows. A limited amount of information is available on the basis of field experience, and some of this information is tabulated in Table 12.5. The values given in this table (in feet per second) are tentative and may change as more field experience is accumulated. We believe that these values are useful in the design of acoustical structures and in most instances are believed to be conservative.

The gas velocities given in Table 12.5 represent average values for smooth, diffuse gas flow (no flow separation) at grazing incidence only. If high-velocity gradients (turbulence) exist near the surfaces of the protective facings, such as might be encountered in 90° bends or in the vicinity of sharp edges, or sharp constrictions, local gas velocities might be expected to increase to values several times the calculated average velocities. It is generally wiser not to place acoustical structures where the gas turbulence is high, because erosion is highly probable. If acoustical structures are used in turbulent gas streams, extreme care should be exercised to protect the porous materials inside as much as possible.

The thickness of the perforated protective facing material shown in Table 12.5 is governed by both gas temperature and gas velocities. For ventilating systems the thickness of the perforated facing is generally 20 gauge. For large panels, such as might be used in engine test-cell installations, the thickness ranges from about 20 gauge at

Table 12.5

Protective Facings for Acoustical Linings Subjected to High-velocity Gas Streams

Materials	Maximum allowable velocity - fps (In straight runs)	Materials	Maximum allowable velocity - fps (In straight runs)
A Ventilating duct systems		*B* Large panels (e.g. test cell applications)	
1. Uncoated linings Examples of suitable materials *a*) Baldwin Hill Sound Liner *b*) Q-T Duct Liner *c*) Johns Manville Airacoustic	25	1. Perforated metal Glass fiber cloth Blanket	≤ 75
2. Coated linings Examples of suitable acoustical blankets *a*) Ultralite *b*) Microlite *c*) Fiberglas	25 - 50	2. Perforated metal Wire screen Glass fiber cloth Blanket	75 - 150
3. 20 gauge perforated metal minimum 25% open area maximum hole size 1/8" Blanket	50 - 75	3. Perforated metal 2 corrugated steel sheets Wire screen Glass fiber cloth Blanket	150-225
4. Perforated facing as in 3 Glass fiber cloth Blanket	75-100	4. Perforated metal 2 corrugated steel sheets Perforated metal Wire screen Glass fiber cloth Blanket	225-250
5. 20 gauge perforated metal Wire screen 10 - 20 mean Glass fiber cloth Blanket	100-125		

Note: Selection of an acoustical blanket will depend also on gas temperature (see Table 12.4). In general, PF Fiberglas board should not be used in cases where the velocities exceed 75 fps because the binder has a tendency to shift, owing to the effects of vibration.

normal room temperature and a maximum velocity of 75 fps to about 14 gauge for temperatures of 400°F at 225-fps velocities. The increased thickness in the metal is usually required because of the expected higher degree of turbulence associated with many problems involving high temperatures. In these large panels the perforated facings should be at least 20 per cent open. In the case of items *B*3 and *B*4 in Table 12.5, the facing consists of two orthogonal corrugated sheets, each about 30 per cent open. In item *B*4 the innermost perforated metal should be at least 40 per cent open. The information given in Table 12.5 applies to acoustical panels which are installed in sections about 3 ft in length. If smaller sections of perhaps half this length are used, the velocity limits given may be somewhat increased.

The glass-fiber cloth indicated in Table 12.5 should also have a very low flow resistance (see next section) and it should always be securely clamped at all edges, e.g., sewn to form a bag around the porous material.

In general, the acoustical engineer should give serious consideration to those problems in which a high-velocity air stream is to be encountered (above about 200 fps). It is not unusual in many instances to have the air stream generate so much noise that the stream itself may act as a flanking noise source to the air-borne noise treatment. The importance of this self-generated aerodynamic noise will of course depend on how quiet a noise environment one is seeking in a particular noise problem. If the noise-reduction requirements are not too stringent, the flanking noise source may not be a problem.

Flow Resistance of Protective Facings. We see from this table that in some instances a complex sandwich structure is required to provide adequate protective cover in the high-velocity gas applications. In the selection of the wire screen and protective cloth, care must be exercised to insure that these materials do not have too high a value of flow resistance. Careless selection may nullify the acoustical performance of the installation. In general, it is safe to assume that the wire screen and cloth should have a combined flow resistance not in excess of approximately 100 mks rayls in order to insure that their presence negligibly affects the total flow resistance of the particular acoustical lining.

Table 12.6 lists the flow-resistance values of some wire-mesh

Table 12.6

Flow Resistance of Some Wire-mesh Screens

Number of Wires/lin. in.	Wire Diameter, mils (in. $\times 10^{-3}$)	Flow Resistance, mks rayls (newton-sec/m^3)
30	13	5.7
50	8.7	5.9
100	4.5	9.0
120	3.6	13.5
200	2.25	24.6

screens that are considered adequate for use with acoustic fillers for panels.

A cloth commonly used as a protective material covering acoustical blankets is woven from glass fibers. This type of cloth withstands temperatures up to approximately 1100°F. Table 12.7 gives flow-resistance values of a number of currently available glass-fiber cloths. For some of the cloths, the flow resistance of a small sample was found to vary by as much as ±20 per cent from the average value. Variations of this magnitude may be expected for cloths where the resistances of the paths through the threads and the paths between the threads are nearly equal in magnitude. In general, a large number of samples of each material were measured and the results averaged.

Table 12.7

Flow Resistance of Glass-fiber Cloths

Manufacturer*	Cloth No.	Surface Density, oz/yd	Construction, Ends × Picks	Flow Resistance, mks rayls (newton-sec/m^3)†
(1)	(2)	(3)	(4)	(5)
1, 2, 3	120	3.16	60 × 58	300
1, 2, 3	126	5.37	34 × 32	45
1, 2, 3	138	6.70	64 × 60	2,200
1, 2, 3	181	8.90	57 × 54	380
3	1032		14 × 16	5
3	1044	19.2	14 × 14	36
2	1544	17.7	14 × 14	19
2	1550-24	3.4	24 × 32	42
3	3862	12.3	20 × 38	≈350
1	HG19	1.87	24 × 24	≈10
1	HG32	1.94	30 × 16	<5
1	HG63	9.60	16 × 14	13
1	HG82	14.5	60 × 56	400
1	HG84	24.6	42 × 36	220
1	HG89	12.0	13 × 12	≃11
1	X91		14 × 18	<5

* Code numbers for manufacturers are as follows: (1) Hess Goldsmith and Company, (2) J. P. Schwebel and Company, and (3) United Merchants Industrial Fabrics.

† To convert to cgs rayls, divide column 5 by 10.

12.6 Summary

This chapter is devoted to a discussion of the basic properties of acoustical materials: propagation constants, flow resistance, porosity, volume coefficients of elasticity, and structure factor. Measured data are given for many kinds of materials. Charts for calculating the propagation constants from the other properties of the materials are presented. These charts are shown to agree well with measured data. Finally, general considerations in the application of acoustical materials are discussed, such as weather exposure, gas-flow temperature, gas-flow velocities, and protective facings.

REFERENCES

1. Zwikker, C., C. W. Kosten, and J. v. den Eijk: Absorption of Sound by Porous Materials, *Physica,* vol. 8, pp. 149, 469, 1094, and 1102, 1941; C. W. Kosten and C. Zwikker, Extended Theory of the Absorption of Sound by Compressible Wall Coverings, *Physica,* vol. 8, p. 968, 1941.

2. Zwikker, C., and C. W. Kosten: "Sound Absorbing Materials," Elsevier Press, Inc., New York, 1949.

3. Beranek, L. L.: Acoustical Properties of Homogeneous Isotropic Rigid Tiles and Flexible Blankets, *J. Acoust. Soc. Am.,* vol. 19, no. 4, pp. 556–568, July, 1947.

4. Kosten, C. W., and J. H. Janssen: Acoustic Properties of Flexible and Porous Materials, *Acustica,* vol. 7, no. 6, pp. 372–378, 1957.

5. Scott, R. A.: The Absorption of Sound in a Homogeneous Porous Medium, *Proc. Phys. Soc. London,* vol. 58, p. 165, 1946.

6. Beranek, L. L.: "Acoustic Measurements," pp. 836–869, John Wiley & Sons, Inc., 1949.

Chapter 13

THE TRANSMISSION AND RADIATION OF ACOUSTIC WAVES BY SOLID STRUCTURES*

Leo L. Beranek

13.1 Introduction

Two kinds of sound waves are found in elastic media, namely, compressional and shear waves.

In gases, sound waves are compressional. Only in second order, because of viscosity, do shear waves exist. In perfect gases, the air particles would always move back and forth in a direction parallel to the direction in which a sound wave is traveling.

In liquids, compressional waves are also of primary importance. Shear waves occur in second order but are of greater importance than in gases.

In structures, both compressional and shear waves are of primary importance. When a bar is excited at its end by motion parallel to its axis, a near-compressional wave travels along the bar. So little air-borne sound is radiated by a compressional wave in a bar that the ear is not likely to hear it. The wave usually may be perceived only by touch or by an electromechanical transducer. Distortional (torsional) waves in a circular rod are a form of shear wave, but, owing to the shape of the rod, produce very little sound in a surrounding medium. The waves of most interest in structures are bending (flexural) waves, a combination of compressional and shear waves. They have associated with them large transverse displacements that may readily couple to compressional waves in surrounding fluids. Bending waves are easily excited in a bar or plate by air-

* Presented as the 45th Thomas Hawksley Lecture before the (British) Institution of Mechanical Engineers, London, Nov. 21, 1958.

borne or water-borne sound waves. In turn, bending waves readily radiate sound energy into fluid media. Thus, a vibrating engine in a ship excites bending waves in the ship's hull. These waves radiate sound into the surrounding water.

Some of the ways in which sound may travel from a source in one room to an observer in a second room are shown in Fig. 13.1. Air-

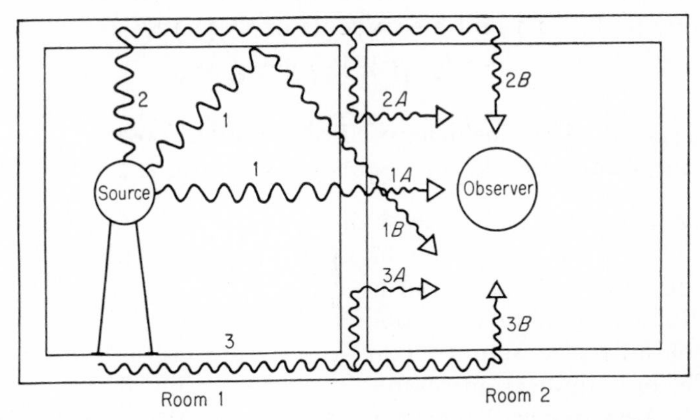

FIG. 13.1 Diagrammatic representation of some of the ways in which a machine (source of sound) may create sound in an adjoining room.

borne sound waves, 1, may impinge, either directly or after reflection, on the common wall between rooms 1 and 2 and excite bending waves that cause air-borne sound waves 1*A* and 1*B* to be radiated into room 2. Air-borne sound waves, 2, may impinge on other walls of room 1 and the resulting bending waves may be transmitted by the walls to room 2 and radiated as air-borne sound waves 2*A* and 2*B*. The mechanical motion of the machine may induce bending waves in the floor (or walls) that may travel through the floor and walls to produce air-borne waves 3*A* and 3*B*.

In airplanes or missiles, the fuselage may be set into vibration (1) by sound waves produced by the engines or propellers, (2) by direct excitation from the vibration of the engines, or (3) by traveling turbulence vortices over the exterior surfaces due to the forward motion of the body through the air. Structure-borne waves in a fuselage radiate air-borne waves to the passengers' ears or cause vibrations of equipment.

In automobiles, vibrations of the engine or road surfaces are transmitted through the structure to the sides and roof of the car. Air-borne sound is then radiated to the interior. Air turbulence and engine intake and exhaust noise also create structure-borne sound waves.

Much practical information and data on structure-borne sound have appeared in publications of the British Building Research Station;[1] Parkin and Humphreys;[2] The Netherlands Institute for Public Health Engineering, TNO, The Hague;[3] in the U.S. Air Force research reports;[4] and in the Proceedings of the Goettingen Symposium.[5] Some American information on sound transmission in buildings is summarized by Cook and Chrzanowski.[6] Extensive work has been done by Cremer and his students in Germany[7] and recently in the United States by Watters, Kurtze, and Dyer.[8-10]

13.2 Transmission Coefficients, Transmission Losses, and Room Levels

Throughout this paper we shall be concerned with the type of wall* that is used as a barrier between a region in which there is a source of air-borne sound and a region in which there is an observer of air-borne sound. For example, one application of the results is the party wall between two flats in a multifamily dwelling. The terminology and formulas in common use for specifying how well a wall transmits sound are now described.

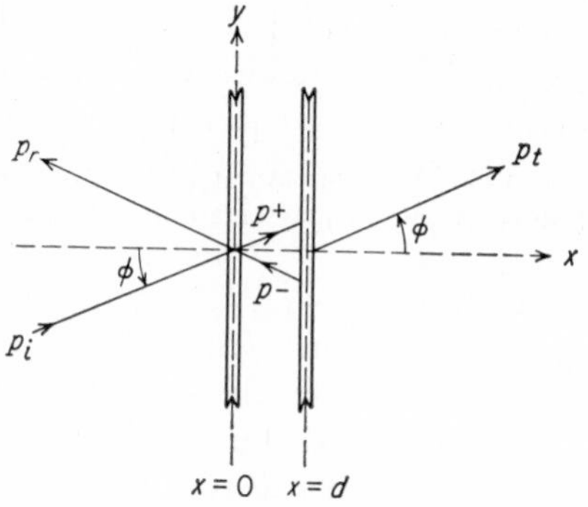

The "sound transmission coefficient" τ of a wall is the fraction of sound power in the incident air-borne wave that appears in the transmitted air-borne wave on the secondary side of the wall.

The "sound transmission loss" TL of a wall in decibels is

$$\mathrm{TL} = 10 \log_{10} \frac{1}{\tau} \quad \text{db} \quad (13.1)$$

A situation of common interest, namely, a wall with two parallel panels, is shown in Fig. 13.2. Here are illustrated (1) the incident and reflected sound waves on the source side of a double wall,

FIG. 13.2 Geometrical situation illustrating the transmission of sound through a double wall with parallel panels. The incident sound wave p_i impinges on the first panel at angle ϕ. Part of the sound energy is reflected (wave p_r) and part passes into the air space between the walls. In the interwall air space, a standing wave is set up with components traveling to the right $p+$ and components traveling to the left $p-$. Part of the sound power passes through the double wall and produces the transmitted sound wave p_t at an angle ϕ.

* We shall speak of a wall as being made up of one or more panels, perhaps separated by air spaces and perhaps containing sound-absorbing materials.

(2) the standing sound wave between the two panels of a double wall, and (3) the wave transmitted through the double wall. For a single wall, item 2 above vanishes. Various authors (Schoch,[11] London,[12,16] Beranek and Work,[13] Cremer,[14] and Schoch and Fehér[15]) have derived for both double and single walls the desired ratio of the acoustic intensities of the incident and transmitted waves for air medium on both sides defined as

$$\frac{1}{\tau} = \left|\frac{p_i}{p_t}\right|^2 \tag{13.2}$$

The vertical bars indicate that the rms of the ratio of the sound pressures is desired.

A *single panel* is one whose two sides move together. Hence, if v_1 is the velocity of the panel surface on the source side taken perpendicular to the surface and v_2 is the same on the opposite side, then $v_1 = v_2 = v_n$. For a homogeneous panel, this relation is generally satisfied provided its thickness is less than one-sixth of the wavelength of the bending wave λ_B in the panel at the frequency being considered. We shall discuss λ_B shortly.

Secondly, let us define the *specific transmission impedance of a panel* as the complex ratio of the pressure difference on the two sides of the wall to the velocity of the wall perpendicular to its surface. Thus,

$$Z_s = \frac{\Delta p}{v_n}$$

where Z_s = complex number, $\text{lb}_f\text{-sec/ft}^3$ (or newton-sec/m^3)
Δp = lb_f/ft^2 (or newtons/m^2) (see Appendix B)
v_n = ft/sec (or m/sec)

Both p and v_n are complex quantities. The complex aspects of these quantities indicate the time phases relative to a reference time.

These quantities are point quantities and are equal in magnitude at all positions on an infinite plate driven by a wave of infinite extent. Finite plates are handled in practice as discussed later.

Room Levels.[42] A source of sound in Room 1, separated from Room 2 by a partition with a transmission loss TL (for reverberant sound on the Room 1 side) produces a sound pressure level SPL_2 in Room 2 of

$$\text{SPL}_2 = \text{PWL} - \text{TL} + 10 \log_{10}\left(\frac{S_W}{R_1}\right) + 10 \log_{10}\left(\frac{1}{S_W} + \frac{4}{R_2}\right) \text{db} \tag{13.3a}$$

or alternatively,

$$\mathrm{SPL}_2 = \mathrm{SPL}_1 - \mathrm{TL} + 10 \log_{10}\left(\frac{S_W}{4}\right) + 10 \log_{10}\left(\frac{1}{S_W} + \frac{4}{R_2}\right) \text{ db} \tag{13.3b}$$

where PWL is the sound-power level of the source in Room 1 *re* 10^{-13} watt (see Eq. 3.6); SPL_1 is the sound-pressure level in the reverberant sound field of Room 1; S_W is the area of the interroom partition in square feet; R_1 is the room constant for Room 1, in sq ft [see Eqs. (11.29) and (11.31)]; and R_2 is the room constant for Room 2, in sq ft. It is assumed that the sound-pressure level in Room 2 (SPL_2) is measured near the surface of the partition.

13.3 Some Properties of Elastic Waves in Plates

It is beyond the scope of this paper to cover thoroughly the various types of elastic waves. Excellent papers have been presented recently which review experimental and theoretical work in this field (Davies,[17] Schmidt,[18] Exner, Gueth, and Immer,[19] Cremer,[7] Kurtze,[20] Junger,[21] Deresiewicz,[22] and Watters and Kurtze[8,9]).

The structure that we are most concerned with is the vibrating plate. Several of the formulas needed in the latter part of this chapter are presented here.

Longitudinal-wave Velocities. In a thin bar, a longitudinal wave can be propagated along its axis whose low-frequency phase velocity is the "longitudinal-bar velocity"

$$c_L = \left(\frac{E}{\rho_p}\right)^{1/2} \quad \text{ft/sec (m/sec)} \tag{13.4}$$

where E = Young's modulus, $\mathrm{lb}_f/\mathrm{ft}^2$ (or newtons/m^2)
ρ_p = density of plate, slugs/ft^3 (or kg/m^3)*

* To get slugs per cubic foot, pounds mass (lb_m) per cubic foot are divided by 32.2.

In a plate, owing to the constraint on the sides, the low-frequency phase velocity is slightly different. It is known as the "longitudinal-plate velocity," and is given by

$$c_L' = \sqrt{\frac{E}{\rho_p(1 - \sigma^2)}} \doteq c_L \tag{13.5}$$

In calculations we generally make the approximation that c_L' is given by Eq. (13.4). The quantity σ is the Poisson's ratio, equal to 0.3 in most cases.

The Wave Equation. The classical equation of motion for a bending wave on a thin plate is

$$\nabla^4\eta + \frac{12}{(hc_L')^2}\frac{\partial^2\eta}{\partial t^2} = 0 \tag{13.6}$$

where η is the displacement of the plate perpendicular to its surface in feet (or meters) and h is the thickness of the plate in feet (or meters).

Bending-wave Velocity and Wavelength. From a solution to Eq. (13.6) (Morse[23]) we find that the velocity of propagation of the bending wave is

$$c_B = (1.8hfc_L')^{1/2} \qquad \text{ft/sec (or m/sec)} \tag{13.7}$$

where h is the thickness of plate in feet (or meters) and f is the frequency of the wave in cycles per second.

The wavelength of the bending wave in the plate is given by

$$\lambda_B = \frac{c_B}{f} = \sqrt{\frac{1.8\,hc_L'}{f}} \qquad \text{ft (or m)} \tag{13.8}$$

We see immediately that there is an essential difference between a longitudinal wave in a thin bar (or in air) and a bending wave in a plate. In the case of a longitudinal wave, the wavelength is inversely proportional to frequency because the speed of travel is constant. In the case of a bending wave, the wavelength is inversely proportional to the square root of the frequency. Because bending waves of different frequencies travel at different velocities, the waveform of a complex wave is not preserved, and the medium when excited into bending waves is said to be "dispersive."

Supported Rectangular Plate. For any plate of finite size there is a galaxy of frequencies at which resonance occurs. For each resonance there is a different arrangement of nodal lines and of maximum lines of vibration over the surface of the plate. These vibrating regions radiate air-borne sound waves. The strength of the radiation depends on how hard the plate is driven and on the particular configuration of the pattern of vibrations in the plate.

Assume that the wall between a room and out of doors is made from a solid, single sheet of metal supported at the edges. This sheet of metal has mass and stiffness distributed throughout its surface. If we were to strike the center of this wall with a hammer, it would ring or resonate at a number of frequencies, and it would radiate sound. The lowest of these resonance frequencies would occur when the panel moves everywhere in phase. For example, if the center of the panel were displaced and then released, the panel would vibrate back and forth perpendicular to the plane of the sheet. The largest amplitude of vibration would be at the center, and the amplitude would decrease gradually to zero at the corners. The manner in which the panel vibrates at the higher resonance frequencies may be visualized by assuming that bending waves, excited by the hammer blow, travel out to the boundaries and are reflected backward across the panel. Whenever the wavelength of the sound in the panel is such that reflected waves reinforce other traveling waves of the same wavelength, a resonance condition is created, and the panel vibrates vigorously at some parts and exhibits lines of zero motion at others (Fig. 13.3). After the initial hammer blow, the vibration persists for a length of time that depends on the internal and external damping of the panel. The panel could also be set into vibration by an incident sound wave.

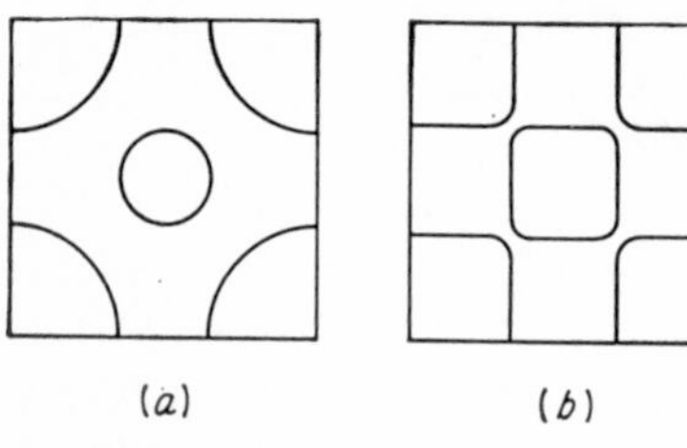

FIG. 13.3 Vibration patterns of flat plates mounted horizontally on a single support at the center with no clamping at the edges. The plates were set into vibration with a violin bow, and sand was sprinkled on top. The sand collected at the lines of zero motion as shown above. The two patterns shown are produced by pulling the bow across the edge at two different places, thereby exciting two different resonant conditions.

A rectangular thin plate, supported (but not clamped in frames) at the four edges, has resonance frequencies (called "normal frequencies") as computed by

$$f_{n_x n_y} = 0.45\, c_L h \left[\left(\frac{n_x}{l_x}\right)^2 + \left(\frac{n_y}{l_y}\right)^2\right] \quad \text{cps} \tag{13.9}$$

where c_L = longitudinal-wave velocity, ft (or m) / sec
h = plate thickness, ft (or m)
l_x, l_y = lateral dimensions, ft (or m)

By a "supported plate" we mean that at the boundaries, no transverse motion is possible but the slope of the plate is not constrained.

In Eq. (13.9) the quantities n_x and n_y can have any integral values, independent of each other, 1, 2, 3, ... The lowest normal frequency is $f_{1,1}$ with $n_x = 1$ and $n_y = 1$.

Let us consider now excitation of the panel with a sound wave. If the panel were rigidly clamped at the boundaries, thereby restricting both the motion and slope to zero, the lowest resonance frequency would be twice as high and all others would be somewhat higher than for the "supported" case.

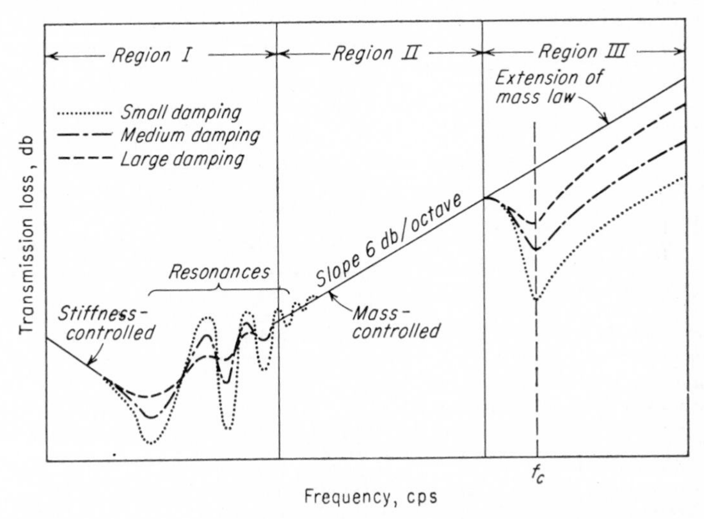

FIG. 13.4 Separation of the behavior of a panel into three frequency regions: (I) stiffness controlled and resonances; (II) mass controlled; (III) wave-coincidence controlled. The critical frequency is called f_c.

Below the lowest resonance frequency stiffness alone generally controls the movement of the panel. That is to say, mass and damping are unimportant (see Fig. 13.4). Above the first few resonance frequencies the mass generally becomes most important. We must not conclude that for all frequencies above these lowest resonance frequencies the panel is mass controlled. Equation (13.9) shows that hundreds of resonances are possible which require the existence of both mass and stiffness. Indeed, for undamped systems, like a bell, a multitude of resonances can be perceived. However, if there are significant energy losses either in the plate or at the boundaries, the higher resonances are less pronounced and the surface of the plate on the average (averaged both in space and over a band of frequencies) moves nearly as though it were mass controlled, that is to say, as if it were made of a large number of little masses free to slide by each other without intermass constraining forces (see Fig. 13.4). The mass-controlled region may extend from two or three times the lowest

Table 13.1

Units Used in Connection with Transmission Loss Formulas*

a. Consistent Systems of Units

System of Units	Units				k_n [See (Eq. 13.37)]	ρc †	Approximate P_0 at Sea Level
	p or P_0	v	M_s †	d			
mks	newton/m²	m/sec	kg/m²	m	1	410 newton-sec/m³	10^5
cgs	dyne/cm²	cm/sec	gm/cm²	cm	1	41 dyne-sec/cm³	10^6
Foot-slug-sec	lb_f/ft^2	ft/sec	slug/ft²	ft	1	2.61 lb_f-sec/ft³	2.12×10^3

b. Mixed English System of Units

System of Units	Units				k_n [See Eq. (13.37)]	(ρc)† $\rho c \times k_n$	Approximate P_0 at Sea Level
	p or P_0	v	w†	d			
ft-lb-sec	lb_f/ft^2	ft/sec	lb_m/ft^2	ft	32.2	84 lb_f-sec/ft³	2.12×10^3
in.-lb-sec	lb_f/in.²	in. sec	lb_m/in.²	in.	386	0.584 lb_f-sec/in.³	14.7

* See also Appendix B.

† $M_s/\rho c$ in Table 13.1*a* equals $w/(\rho c)$ in Table 13.1*b*.

Note: To convert from lb_f/ft^2 to newton/m² multiply by 47.88.
To convert from lb_m/ft^2 to kg/m² multiply by 4.882.
To convert from lb/ft² to lb/in.² multiply by 0.00695.
To convert from slug/ft² to kg/m² multiply by 157.2.
To convert from newton/m² to dyne/cm² multiply by 10.
To convert from kg/m² to gm/cm² multiply by 0.1.

resonance frequency up to the "critical frequency." By critical frequency we mean the frequency at which the bending wavelength λ_B in the panel equals the wavelength of the radiated wave λ in the second medium. As we shall see shortly, above the critical frequency the stiffness of the panel again becomes very important.

Summary. In this section it was stated that stiffness in a panel is important in determining the amount of radiated sound (1) below the lowest resonance frequency and (2) above the critical frequency. Between areas 1 and 2, on the average, practical panels excited by sound waves often may be treated as though they were mass controlled. The TL associated with these three regions is illustrated in Fig. 13.4. In the stiffness-controlled and the basic-resonance region it is very difficult to calculate the TL. In the mass-controlled region, the TL is easy to calculate. In the wave-coincidence region calculations are more difficult, but good approximations can be made, as will be shown later in this paper.

13.4 Transmission Losses of Double and Single Walls

Double Wall. Assume that both panels in the double wall of Fig. 13.2 are identical. Then, from the above references, the complex ratio of the incident and transmitted sound pressures for a double wall (London[16]) is

$$\frac{p_i}{p_t} = 1 + \frac{Z_s \cos\phi}{\rho c} + \left(\frac{Z_s \cos\phi}{2\rho c}\right)^2 (1 - e^{-2j\beta}) \qquad (13.10)$$

where, by definition, $$\beta \equiv \frac{\omega d \cos\phi}{c} \equiv \frac{2\pi d \cos\phi}{\lambda} \qquad (13.11)$$

ρc = characteristic resistance of air
ρ = density
c = velocity of propagation
$\omega = 2\pi f$
f = frequency, cps
j = operator, $\sqrt{-1}$
d = separation of panels

Under the special case where the specific transmission impedance of each panel in the wall is reactive, that is to say, the real part is near-zero

$$Z_s \equiv R_s + jX_s \doteq 0 + jX_s \qquad (13.12)$$

The quantity X_s is called the "specific transmission reactance." We may now, with the help of Eqs. (13.1) and (13.2), write for the TL of a double wall

$$\mathrm{TL} = 10 \log_{10} \left\{1 + \left(\frac{X_s}{\rho c}\right)^2 \cos^2\phi \left[\cos\beta - \frac{1}{2}\left(\frac{X_s}{\rho c}\right)\cos\phi \sin\beta\right]^2\right\} \qquad (13.13)$$

Table 13.1 enables conversion from metric to English to mixed English systems of units.

The specific transmission impedance Z_s of a wall is near-reactive and equal approximately to jX_s only when the frequency of the incident wave is below the "critical frequency." Above the critical frequency, any damping of the bending wave, either internal or applied to the surface, introduces a significant dissipation term R_S.

Single Wall. For a single wall of specific transmission impedance Z_s, Eq. (13.10), with $d = 0$, reduces to

$$\frac{p_i}{p_t} = 1 + \frac{Z_s \cos \phi}{2\rho c} \tag{13.14}$$

The factor of 2 appears in the denominator because (Fig. 13.2) not only the separation d has been reduced to zero, but one of the walls of the impedance Z_s has been eliminated. That is to say, Z_s for a single wall is taken to be half that for the wall in Fig. 13.2.

If a purely reactive specific transmission impedance, Eq. (13.12), is assumed, the TL for a single wall, using Eqs. (13.1) and (13.2), becomes

$$\text{TL} = 10 \log_{10} \left[1 + \left(\frac{X_s \cos \phi}{2\rho c} \right)^2 \right] \quad \text{db} \tag{13.15}$$

Reverberant-source Room. Frequently a wall is located between two rooms, with the source room being fairly reverberant. In a fairly reverberant room, the sound impinges on a wall from all angles below a limiting angle ϕ_{limit} that is usually not much greater than 80°. To obtain the "average transmission coefficient" $\bar{\tau}$ we need to perform the operation given by

$$\bar{\tau} = \frac{\int_0^{\phi_{\text{limit}}} \tau \cos \phi \sin \phi \, d\phi}{\int_0^{\phi_{\text{limit}}} \cos \phi \sin \phi \, d\phi} \tag{13.16}$$

Summary. In this section formulas were presented for the TL's of double and single walls at any angle of wave incidence ϕ in terms of the specific transmission impedance Z_s of the wall. Also presented was the equation for determining the average transmission coefficient for a wall with the source side exposed to a highly reverberant sound field in which many angles of incidence below a limiting angle ϕ_{limit} exist simultaneously. These formulas are of particular importance in determining the TL's of double and single walls in the frequency region where the component panels are mass controlled.

13.5 Wave Coincidence

It is not surprising that an intense wave is established by a panel when it is driven at a frequency equal to one of its resonance frequencies. Less expected is the fact that for every frequency above a certain critical frequency there is a particular angle ϕ_0 at which even an infinitely large vibrating panel vibrates when excited by a sound wave as though it were at resonance. The reason for this behavior is that above this critical frequency the wavelength of the bending wave in the panel λ_B can become equal to the wavelength in air λ projected on the panel, and a high degree of coupling between the panel and the air is achieved. This condition of equal wavelengths in the panel and in the air is designated *wave coincidence*. Similarly, if an air-borne sound wave is incident on a panel at such a frequency and angle that wave coincidence occurs, the panel is set into motion readily, just as though it were at resonance.

A condition of wave coincidence is illustrated in Fig. 13.5. The

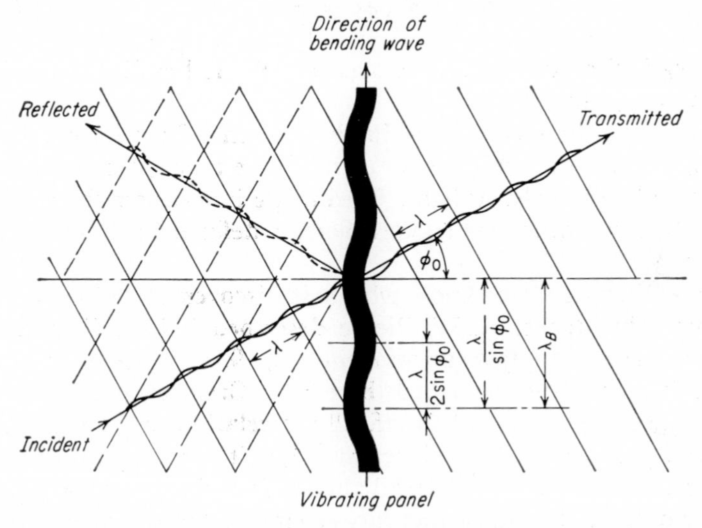

FIG. 13.5 Wave coincidence. The wavelength of the bending wave in the panel is λ_B. A sound wave in air, whose wavelength is λ, impinges on a plate at the angle ϕ_0. When $\lambda/\sin \phi_0$ is equal to λ_B, the intensity of the transmitted wave approaches the intensity of the incident wave. The frequency for which $\lambda = \lambda_B$ is called the critical frequency.

incident wave with wavelength λ travels from the lower left and impinges on the panel, which is set into vibration. The resulting vibration appears as a bending wave with wavelength $\lambda/\sin \phi_0$ that travels as shown (vertical on the page). If λ is less than λ_B, then, at a particular angle of incidence ϕ_0, it is possible for the projected wavelength ($\lambda/\sin \phi_0$) to equal the bending wavelength λ_B, as shown. Under this condition, the panel vibrates at an amplitude almost equal to the amplitude of the air particles in the incident wave. In turn, the panel radiates a transmitted wave with almost this same amplitude also at the coincidence angle ϕ_0. In other words, the panel radiates a wave (the transmitted wave) that is almost as intense as the exciting wave (the incident wave). Hence, the TL at that frequency and angle is very small.

The condition for wave coincidence is

$$\sin \phi_0 = \frac{\lambda}{\lambda_B} \tag{13.17}$$

Obviously, if the wavelength of the sound in air is greater than the wavelength of the sound in the plate, no wave coincidence can occur, because the sine cannot be greater than 1.0.

For a given frequency f, $c = \lambda f$, and $c_B = \lambda_B f$, so that

$$\sin \phi_0 = \frac{c}{c_B} \tag{13.18}$$

where c_B is the velocity of propagation of the bending wave in the plate and c is that in air.

To develop the terminology further, when a fixed frequency is assumed, the angle at which wave coincidence takes place is defined as the *coincidence angle*. When a fixed angle is assumed, the frequency at which wave coincidence takes place is defined as the *coincidence frequency*.

The author does not know who first discovered the phenomenon of wave coincidence. G. W. Pierce described it in a United States patent[24] that was applied for on August 2, 1933. In the field of ultrasonics, Sanders[25] also described the effect. Cremer[7] of Germany presented the concept in 1942 specifically in relation to the transmission of sound through walls, and deserves credit for explaining the difference between the simple wall theory (assuming normal wave incidence only) and actual panel measurements that involve radiations of sound at angles oblique to the panel.

Critical Frequency. The *critical frequency* f_c, defined as the lowest frequency at which wave coincidence occurs, is that frequency for which $\lambda_B = \lambda$ or $c_B = c$. In other words, the critical frequency is the lowest possible coincidence frequency and occurs for grazing incidence sound, that is, $\phi_0 = 90°$. From Eq. (13.8) and setting $c_B = c$, we get in a consistent set of units

$$f_c = \frac{c^2}{1.8hc_L'} \doteq \frac{c^2}{1.8h}\sqrt{\frac{\rho_p}{E}} \quad \text{cps} \tag{13.19}$$

In mixed English units, we have

$$f_c = \frac{6.6c^2}{hc_L'} = \frac{1.16c^2}{h}\sqrt{\frac{\rho_p}{E}} \quad \text{cps} \tag{13.20}$$

where h is in inches, c and c_L' are feet per second, ρ_p is in pounds mass (lb_m) per cubic foot, and E is in pounds force (lb_f) per square foot.

Equations (13.19) and (13.20) are strictly valid only for cases where the wavelength of the flexural wave is greater than about six times the thickness h of the panel ($\lambda_B > 6h$).

Coincidence Frequency. Another way of looking at the phenomenon of wave coincidence is to say that the air-borne wave divides the panel into segments, each $\lambda/(2 \sin \phi)$ in length. Each segment behaves like a supported bar. Its effective mass varies in direct proportion to its length and its thickness, and its effective stiffness varies in inverse proportion to the cube of the length and in direct proportion to the cube of the thickness. The bar is naturally expected to vibrate very easily when the mass reactance equals the stiffness reactance.

Table 13.2

Surface Densities and Weights of Common Building Materials per Unit Thickness*

Material	w/h, lb_m/ft^2 per inch of thickness	M_s/h, slug/ft² per foot of thickness	M_s/h, kg/m² per meter of thickness
Aluminum	14	5.2	2,700
Asbestos board (Transite)	9	3.4	1,700
Brick	10–12	3.7–4.5	1,900–2,300
Concrete:			
Dense	12	4.5	2,300
Cinder	8	3.0	1,500
Cinder fill	5	1.9	1,000
Glass	13	4.8	2,500
Lead	59	22.0	11,000
Plaster:			
Lightweight aggregate (Perlite or Vermiculite binder)	5	1.9	1,000
Sand aggregate	9	3.4	1,700
Plexiglas or Lucite	6	2.2	1,150
Sand:			
Dry loose	7–8	2.6–3	1,300–1,500
Dry packed	9–10	3.4–3.7	1,700–1,900
Wet	10	3.7	2,000
Steel	40	15.0	7,700
Wood:			
Timber	3–5	1.1–1.86	580–1,000
Fir plywood	3	1.1	580

* To avoid confusion in this chapter we shall speak of "surface densities" in slugs/ft² or kg/m² and of "surface weights" in lb_m/ft^2 or lb/ft² (see Appendix B)

Because the stiffness reactance of the segmented length of panel varies directly as the square of the frequency and because the mass reactance is independent of frequency (due to the fact that the segmented length is inversely proportional to frequency), it is found that the critical frequency has the form shown in Eq. (13.19). In addition we observe by this logic that the coincidence frequency varies in inverse proportion to $\sin^2 \phi$. We can also show this relation by combining Eqs. (13.8) and (13.17) to yield.

$$f_{\text{coincidence}} = \frac{c^2}{1.8c_L'h \sin^2 \phi_0} \tag{13.21}$$

13.6 Some Physical Properties of Common Materials

It is intended that this chapter may be used for numerical calculations. Accordingly the properties of several common structural materials are presented in Tables 13.2, 13.3, and 13.4.

Surface Density and Mass. Surface densities, weights, and masses in mks, English, and mixed English units for a number of materials are given in Tables 13.2 and 13.3. These quantities are used in the calculation of transmission loss at frequencies below the critical frequency.

Table 13.3

Surface Densities and Weights of Building Blocks

Material	Weight/ft² w, lb_m/ft^2	Mass/ft² M_s, slug/ft²	Mass/m² M_s, kg/m²
Hollow cinder block, 6 in. thick*	25	0.78	120
Hollow cinder block, 6 in. thick, ⅝ in. plaster on each side	35	1.09	170
Hollow dense concrete block, 6 in. thick	35	1.09	170
Hollow dense concrete block, 6 in. thick, sand filled	54	1.68	260
Solid dense concrete block, 4 in. thick	37	1.15	180

* Nominal thickness. Approximately ⅜ in. smaller to allow for mortar joint.

$M_s f_c$ Product and η. Two quantities are needed for the calculation of TL at frequencies above the critical frequency. These are the internal damping factor η and the product of the surface density M_s (or, in mixed English units, the surface weight w) and the critical frequency f_c. The numerical values of these quantities for a number of common building materials are given in Table 13.4.

Table 13.4

Internal Damping Factors and Products of Surface Density and Critical Frequency for Common Building Materials

Material	Product of Surface Density and Critical Frequency			Internal Damping Factor at 1,000 cps
	($lb_m/ft^2 \times$ cps) wf_c	($slug/ft^2 \times$ cps) $M_s f_c$	($kg/m^2 \times$ cps) $M_s f_c$	η*
Aluminum	7,000	217	34,700	10^{-4}
Brick	7,000–12,000	217–373	34,700–58,600	0.01
Concrete, dense poured	9,000	279	43,900	0.005
Concrete (clinker) slab plastered on both sides, 2 in. thick	10,000	310	48,800	0.005
Masonry block:				
Hollow cinder (nominal 6 in. thick)	4,750	148	23,200	0.005
Hollow cinder, ⅝ in. sand plaster each side (nominal 6 in. thick)	5,220	162	25,500	0.005
Hollow dense concrete (nominal 6 in. thick)	4,720	147	23,000	0.007
Hollow dense concrete, sand-filled voids (nominal 6 in. thick)	8,650	269	42,200	Varies with frequency
Solid dense concrete (nominal 4 in. thick)	11,100	345	54,100	0.012
Fir timber	1,000	31	4,880	0.04
Glass	7,800	242	38,800	0.002
Lead:				
Chemical or tellurium	124,000 (approx.)	3,850	605,000	0.015
Antimonial (hard)	104,000	3,240	508,000	0.002
Plaster, solid, on metal or gypsum lath	5,000	165	24,500	0.005
Plexiglas or Lucite	7,250	225	35,400	0.002
Steel	20,000	621	97,500	10^{-4}
Gypsum board (½ in. to 2 in.)	7,000	217	34,200	0.03
Plywood (¼ in. to 1¼ in.)	2,600	81	12,700	0.01
Wood waste materials bonded with plastic, 5 lb/ft²	15,000	466	73,200	

* These values for η are approximate and in most cases are based on very limited data.

Critical Frequencies. A chart of critical frequencies for several common structural materials as a function of plate thickness is given in Fig. 13.6.

13.7 Transmission Loss of Single Panels

Mass-controlled Limp Panels. Laboratory measurements of the transmission loss of panels that are either damped internally or at the boundaries reveal that at some frequencies such panels behave nearly as though they were a lot of little masses free to slide on each other. This frequency range lies below the critical frequency and above about twice the frequency of the first panel resonance. Because the stiffness of the panel can be neglected in predicting the TL, the panel is described as "limp" in this frequency region. Then, the specific transmission impedance of Eq. (13.12) is simply a mass reactance

$$Z_s = j\omega M_s \tag{13.22}$$

where M_s is the mass per unit area in slugs per square foot (or kilograms per square meter).

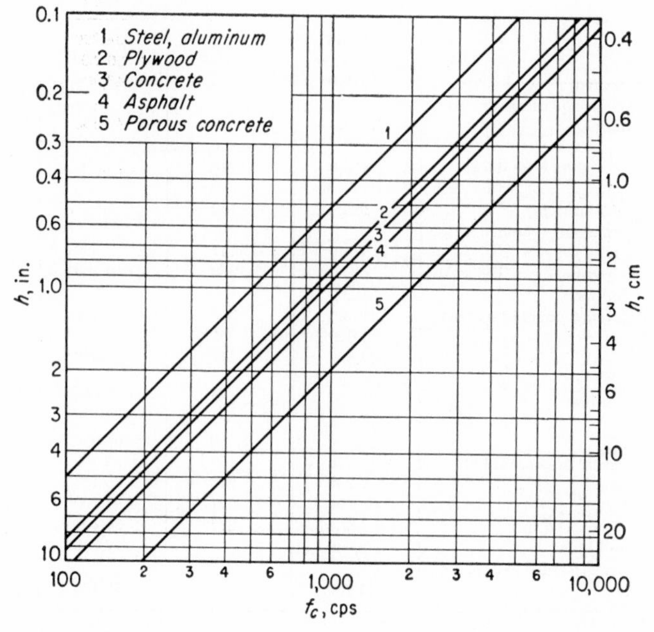

FIG. 13.6 Critical frequency f_c plotted as a function of plate thickness h. This is the lowest frequency at which the coincidence effect is possible. At this frequency, the TL is quite small.

Substitution of Eq. (13.22) into Eqs. (13.14) and (13.2) yields*

$$\frac{1}{\tau} = 1 + \left(\frac{\omega M_s \cos \phi}{2\rho c}\right)^2 \tag{13.23}$$

Limp-wall Transmission Losses. If the incident sound wave impinges on the limp, mass-controlled panel at normal incidence so that $\phi = 0$ (Fig. 13.2) we obtain, using Eq. (13.1), the so-called "normal-incidence, limp-wall mass law,"

* See Table 13.1 for values of ρc and units for M_s. Note that if ρc is taken as 84, the quantity w in pounds per square foot may be substituted for M_s.

$$(\text{TL})_0 = 10 \log_{10} \left[1 + \left(\frac{\omega M_s}{2\rho c} \right)^2 \right] \quad \text{db} \qquad (13.24)$$

Further, insertion of Eq. (13.23) in Eq. (13.16) and integration up to $\phi_{\text{limit}} = 90°$ and then inserting the result in Eq. (13.1) yields the so-called "random-incidence, limp-wall mass law"

$$(\text{TL})_{\text{random}} \doteq (\text{TL})_0 - 10 \log_{10} [0.23(\text{TL})_0] \quad \text{db} \qquad (13.25)$$

Equation (13.25), as written, is an approximation valid for $[\text{TL}]_0$ greater than about 15 db (see Fig. 13.7 for values below 15 db). Equa-

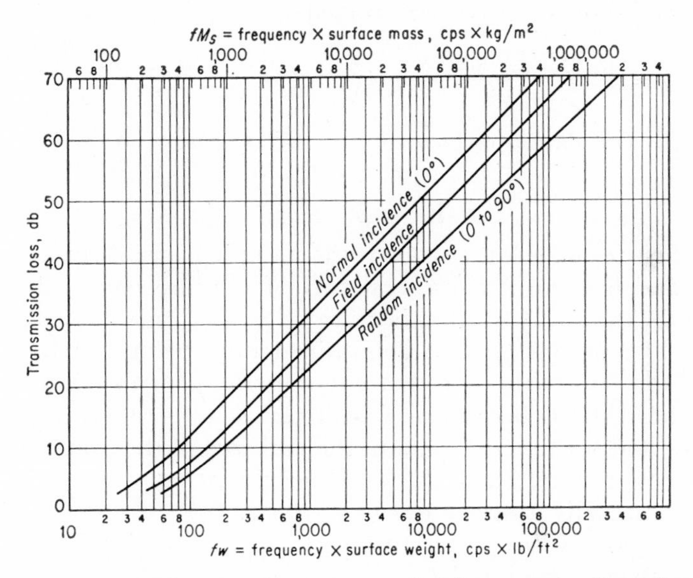

FIG. 13.7 Theoretical TL curves for mass-controlled limp panels. By surface weight is meant lb_m. Field data indicate that in practical situations the field incidence curve gives satisfactory predictions below the coincidence frequency, and above about twice the lowest panel resonance frequency.

tion (13.25) says that if the sound on the primary side comes from a highly reverberant room where all angles of incidence from waves of equal energy impinge on the wall, the transmission loss is significantly reduced compared to its normal incidence value.

Equations (13.24) and (13.25) are plotted as the upper and lower lines, respectively, in Fig. 13.7. These are idealized curves because they assume that all resonances are suppressed. For a finite-sized panel that is not sufficiently damped, the transmission loss curve will vary above and below these curves as the panel goes through antiresonances and resonances.

Measurements made in the field in buildings on walls between flats and offices reveal that TL's generally lie about 5 db below the normal incidence curve. We thereby draw such a curve in Fig. 13.7 and call it the *field incidence curve.* A transmission-loss curve of $[\mathrm{TL}]_0 - 5$ db approximates a diffuse sound field with a limiting upper angle of about 78° [Eq. (13.16)].

Solid Panel above Critical Frequency. In the region near and above the critical frequency f_c, the mass-law curves in Fig. 13.7 are no longer useful because the transmission loss is decreased due to the coincidence effects. In other words, because wave motion in the wall is involved, the bending stiffness of the panel must be considered.

Cremer[26] shows that the wall impedance is given by

$$Z_T = \frac{\eta c_L{}^2 M_s h^2 \omega^3 \sin^4 \phi}{12c^4} + j\left(\omega M_s - \frac{c_L{}^2 M_s h^2 \omega^3 \sin^4 \phi}{12c^4}\right) \tag{13.26}$$

where η is the damping factor from the complex Young's modulus given by

$$E' = E(1 + j\eta) \tag{13.27}$$

The first term in Eq. (13.26) is the damping term because it contains η and because it is the real part of the impedance. It also is affected by the panel stiffness because $c_L{}^2 = E/\rho_p$, as we saw from Eq. (13.4). The second term is determined only by the surface mass. The third term is the one that contains the principal information about the stiffness because it would persist even if there were no damping and because $c_L{}^2$ contains the real part of the Young's modulus E. Note that although the panel density M_s appears in the third term it is actually canceled out by ρ_p in $c_L{}^2$.

Substitution of Eq. (13.26) in Eq. (13.10) and solving for $1/\tau$ from Eq. (13.2) yields (Feshbach[27])

$$\frac{1}{\tau} = \left[1 + \frac{\omega M_s}{2\rho c}\,\eta\left(\frac{f}{f_c}\right)^2 \cos\phi \sin^4\phi\right]^2 + \frac{\omega^2 M_s{}^2}{4\rho^2 c^2}\cos^2\phi\left[1 - \left(\frac{f}{f_c}\right)^2 \sin^4\phi\right]^2 \tag{13.28}$$

We see that if ϕ is set equal to zero, which is the case for a wave incident on the panel normally, we obtain for $1/\tau$ exactly the same expression as we had in Eq. (13.23). In other words, when the wave is normally incident and is plane, it cannot set the bending wave into existence.

In the practical case we must know how to determine the TL for field incidence sound. Without going into the mathematical ramifications, the approximate results are shown in Fig. 13.8. To use this figure the values of w or M_s, f_c, and η for common materials are needed. These values are found in Table 13.4.

Practical Design Chart for a Solid Panel. The theoretical chart in Fig. 13.8 reveals several interesting facts: first, below about 0.6 of the critical frequency the TL is independent of damping and de-

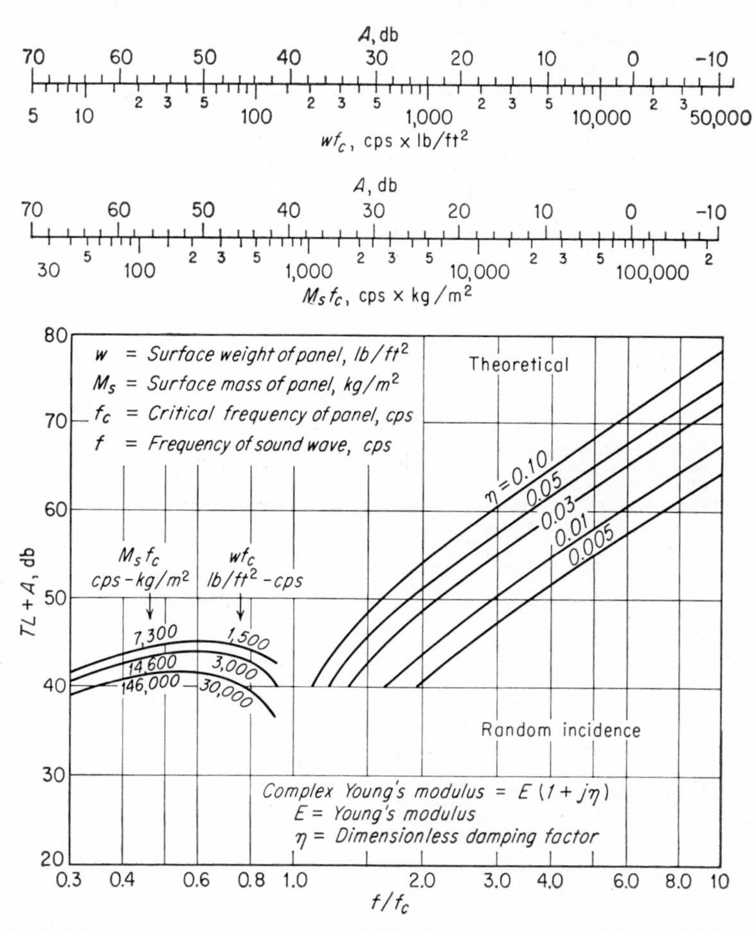

FIG. 13.8 Transmission loss for field incidence sound on a single homogeneous panel of surface weight w lb_m/ft^2 (or surface mass in kg/m^2) and damping constant η. The frequency range near and above the critical frequency f_c only is considered. The curves are not continued below 40 db in the f/f_c region between 1 and 2 because there they become a function of both wf_c and η. Note that 1 kg equals 2.2 lb_m.

pends almost entirely on A (given by the scale at the top). Secondly, we note that at $f/f_c = 0.3$, A is related to wf (or $M_s f$) according to the "field incidence" curve in Fig. 13.7. As an example, assume $wf_c =$ 5,000 cps·lb/ft² and $f/f_c = 0.3$. Then $A = 10$ db and $(\text{TL} + A) - A$ $= \text{TL} = 40 - 10 = 30$ db. From Fig. 13.7 for $wf = 5{,}000 \times 0.3 =$ 1,500 cps·lb/ft², we obtain from the field incidence curve, TL = 30 db. Thirdly, all of the lines above $f/f_c = 1$ are nearly parallel to each other, and increase at the rate of about 10 db per doubling of frequency. Finally, from field observation with panels as mounted in buildings, the dip in the TL between $f/f_c = 1$ and $f/f_c = 2$, generally does not drop much below $(\text{TL} + A) \doteq 36$ or above 44 db.

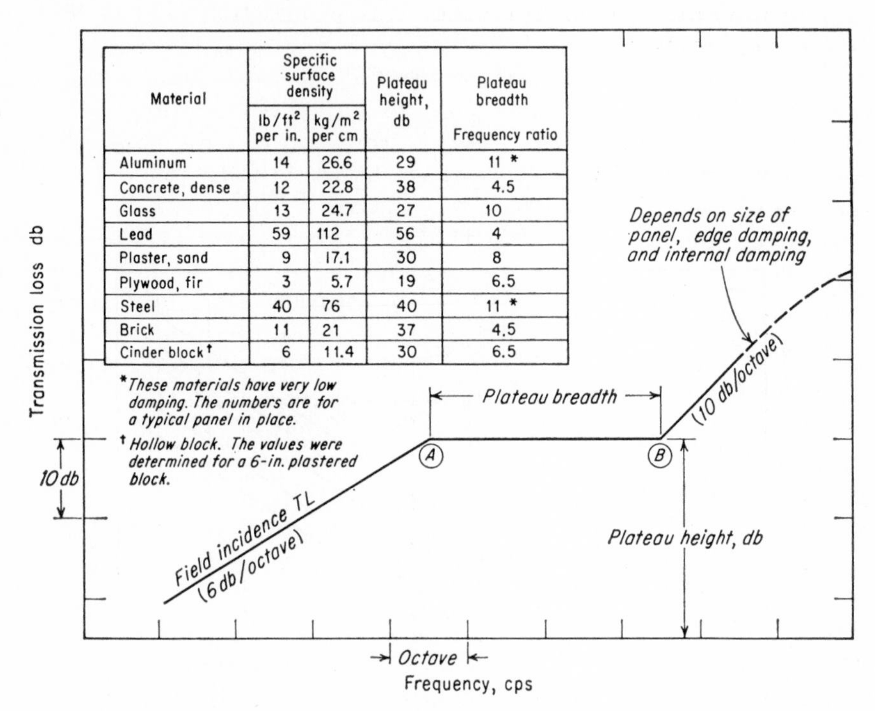

Material	Specific surface density, lb/ft² per in.	Specific surface density, kg/m² per cm	Plateau height, db	Plateau breadth, Frequency ratio
Aluminum	14	26.6	29	11 *
Concrete, dense	12	22.8	38	4.5
Glass	13	24.7	27	10
Lead	59	112	56	4
Plaster, sand	9	17.1	30	8
Plywood, fir	3	5.7	19	6.5
Steel	40	76	40	11 *
Brick	11	21	37	4.5
Cinder block †	6	11.4	30	6.5

FIG. 13.9 A practical design chart for single panels. A reverberant sound field on the source side of the wall is assumed. The part of the curve left of A is determined from the field-incidence curve of Fig. 13.7. The plateau height and the length of the line from A to B are determined from the table. The part above B is determined by extrapolation. This chart is fairly accurate for large panels. For example, masonry panels must be greater than 6 by 8 ft; thin panels, like those in airplanes, must be greater than 2 by 3 ft; etc. Note that 1 kg equals 2.2 lb_m. (*After Watters.*[9])

Taking into account the observations of the previous paragraph, one can, without much loss in generality, make up a single design curve for a wide variety of panels. The desired curve is given in Fig. 13.9.

To estimate the TL curve for a single solid panel (above the lowest panel resonance frequency) the following procedure is followed:

1. Select a piece of graph paper like that used to draw Fig. 13.9. Label the abscissa "Frequency, cps." The abscissa should be a loga-

rithmic scale so that each octave (doubling) of frequency has the same extent along the abscissa. Label the ordinate "transmission loss, db."

2. From the data in the table in Fig. 13.9 and from the panel thickness determine the surface density of the panel. For example, suppose the panel is a 4-in.-thick slab of dense concrete. The surface density (weight) is $4 \times 12 = 48$ lb/ft^2.

3. From the field-incidence curve in Fig. 13.7, determine the TL at one frequency. For example, suppose, for the panel of step 2, that we want the TL at 100 cps. The TL from Fig. 13.7 for $wf = 4{,}800$ cps·lb/ft^2 is 41 db. Plot this point on the graph paper, and draw a line through it with a slope of 6 db/octave (Fig. 13.9).

4. Determine the plateau height in db from the table in Fig. 13.9 and draw a horizontal line at that value. This line intersects the field-incidence TL line at point A. In our example, the plateau height is 38 db. The 100-cps point that we determined in step 3 was 41 db. A difference of 3 db is a frequency change of ½ octave. Hence, point A will fall ½ octave below 100 cps, i.e., at 70 cps.

5. Starting from point A, mark off the length of the line AB according to the plateau breadth shown in the table on Fig. 13.9. In our example, the plateau breadth for dense concrete is 2.2 octaves, or a ratio of 4.5 in frequency. Hence, point B falls at $70 \times 4.5 = 315$ cps.

6. Extend the line above point B at a rate of about 10 db/octave for the first octave, gradually decreasing the slope to about 6 db/octave (Fig. 13.9).

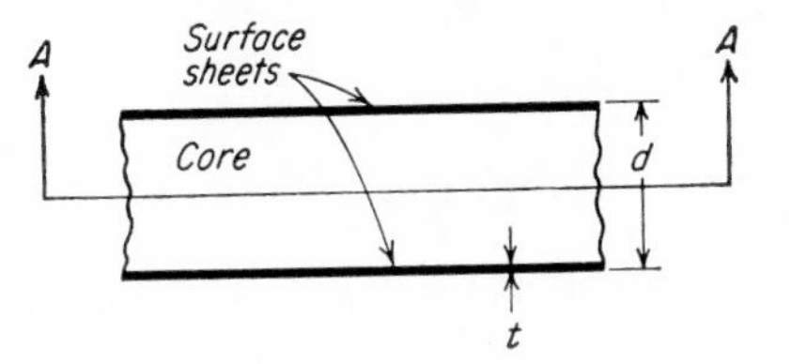

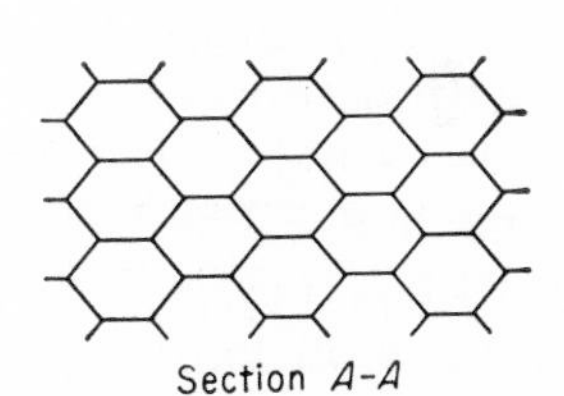

FIG. 13.10 Construction of a sandwich (honeycomb) panel.

Sandwich (Honeycomb) Panels. Today, the demand in building is for stronger, lighter-weight materials. One such structure, shown in Fig. 13.10, is a "sandwich" panel. It has surface sheets, such as aluminum, bonded to both sides of a lightweight honeycomb core.

The sandwich panel structure combines high stiffness with low weight, with the result that bending waves are propagated with higher velocities than in a heavier, solid panel of equal thickness.

If we assume that the stiffness of the panel is concentrated in the surface sheets, with the core merely serving to hold these sheets apart, we find the following approximate equation for the critical frequency:

$$f_c = \frac{c^2}{2\pi d}\sqrt{\frac{2M_s}{tE}} \tag{13.29}$$

where M_s = total surface mass of panel including core, kg/m² in mks units (or slugs/ft² in English units)

d = center-to-center spacing of surface sheets, m (or ft) (Fig. 13.10)

t = thickness of one surface sheet, same units as d (Fig. 13.10)

E = Young's modulus for surface-sheet material, newtons/m² in mks units (or lb/ft² in English units)

c = speed of sound in air, m/sec (or ft/sec)

As we have indicated, it is important to use a consistent system of units throughout. Appropriate constants must be introduced when units other than those specified are used.

Once the coincidence frequency and the surface mass are known, the honeycomb panel is treated as though it were a homogeneous isotropic panel.

13.8 Laboratory Tests for Wall Constants

In order to utilize the formulas and design charts to determine TL of panels, one must have available the relevant physical constants of any particular wall. These constants are the surface mass (or surface weight), the critical frequency, and the damping factor. The test procedures described here are from Cremer[26] and Watters.[9]

The mass and critical frequency may sometimes be determined by static methods. The bar or strip is weighed to yield the mass. The critical frequency is determined by supporting a thin bar or strip horizontally on knife-edge supports located at the ends of the bar. The distance between the supports l and the sag of the bar at the mid-point ξ_{max} are measured. The critical frequency of a homogeneous, isotropic bar or plate is given by

$$f_c = \frac{9 \times 10^4 \xi_{max}^{1/2}}{l^2} \tag{13.30}$$

where ξ_{max} is in inches and l is in feet.

Very stiff materials, such as masonry and honeycomb panels, are too stiff to give an easily observable deflection in such a test. Also, many materials would fail in tension if supported in such a manner.

A simple dynamic test, which gives both the velocity of propagation of bending waves in the material and the internal damping of these waves, is as follows: First mount the bar so that it can freely vibrate (free-free end conditions) and then measure the resonance frequencies and the Q values (bandwidth) of the resonances. The width of the bar must be small compared to the shortest bending wavelength. Care must be taken to avoid torsional modes of vibration by symmetrically locating the exciting force and the mounting supports about the longitudinal axis of the bar.

Standing waves will be set up in this bar, and at resonance the vibrations may be intense. At each resonance, the standing wave will consist of some multiple of a ½ bending wavelength plus an end correction at each end of about ⅛ wavelength (Morse[23]). The bending velocity of propagation is to a close approximation

$$c_B \doteq \frac{lf_0}{n/2 + \frac{1}{4}} \qquad n = 1, 2, 3, \ldots \tag{13.31}$$

where l = actual length of bar
f_0 = frequency of resonance
n = number of standing half waves in bar aside from end corrections

The damping factor η may be determined from either a measurement of the bandwidth of the resonance peaks or of the decay time of the vibration amplitude at the resonance frequency. Also, the resonance frequency must be measured. Thus,

$$\eta = \frac{\Delta f_0}{f_0} \tag{13.32}$$

or

$$\eta = \frac{2.2}{Tf_0} \tag{13.33}$$

where Δf_0 = bandwidth cps, between frequencies on either side of resonance curve where amplitude of vibration has dropped to 0.707 of maximum value
f_0 = frequency of resonance, cps
T = time, sec, required for amplitudes of free-bending vibrations on bar to drop to 10^{-3} of initial value

Equation (13.32) is valid for $\eta < 0.1$, while Eq. (13.33) is exact.

An alternative method for measuring the bending wavelength (and thus c_B) is to scan the standing-wave pattern in the resonant bar with a movable vibration pickup. This technique is recommended if the

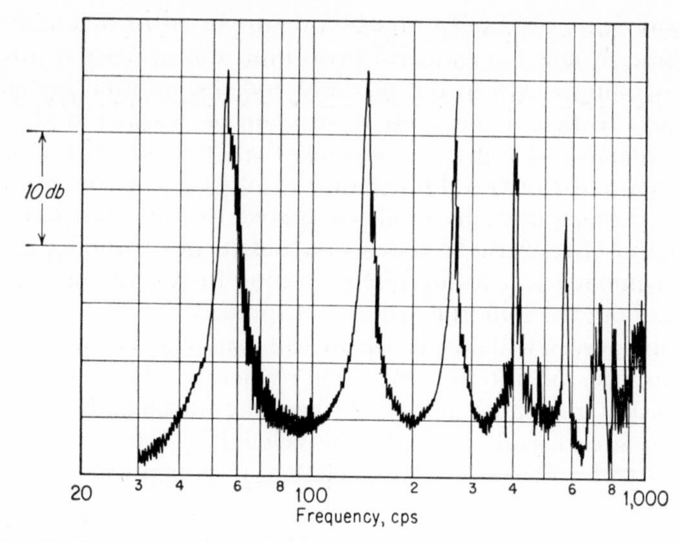

FIG. 13.11 Acceleration at top end of a 6-in.-thick plastered (both sides) cinder block strip, 9.9 ft high and 16 in. wide. The driving force at the bottom end was held constant. (*After Watters.*[9])

bar is relatively thick and if there is any doubt as to the mode of vibration of the bar.

To support a bar (or strip) so that its motion is not significantly altered by the mounts, it may be placed in a horizontal position and

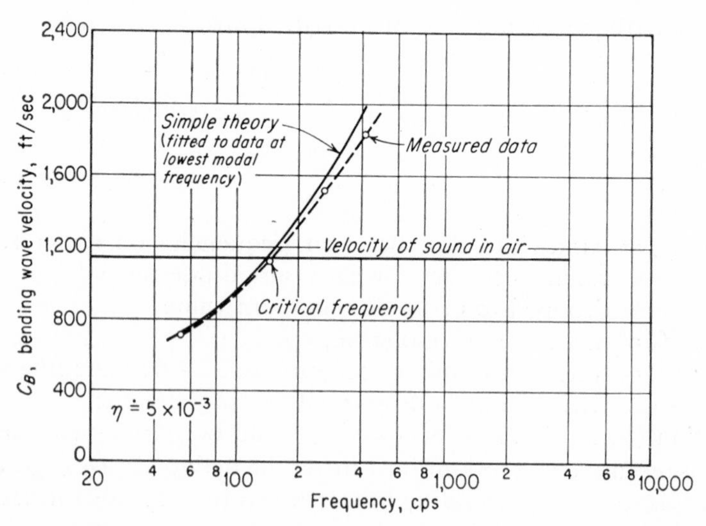

FIG. 13.12 Velocity of propagation of bending waves c_B in plastered 6-in. cinder block. (*After Watters.*[9])

be continuously supported on soft sponges or, if of masonry, seated on a "swing." A swing consists of two thin rods in tension supported from the ceiling with a wooden board spanning the bottom ends. The board is slightly larger than the strip of masonry wall that sits on it, and the steel rods are far enough apart to avoid touching the sides of the strip. For a 12-in.-wide, 4-in.-thick concrete-block wall, a strip one block-width wide and 8 ft high is built on a 5- by 14-in. board hung from ¼-in.-diameter rods that are 13 in. apart. The loading effect of the swing and the vibration shaker appear to be negligible above about 100 cps.

In conducting the experiment, the vibration shaker is tightly con-

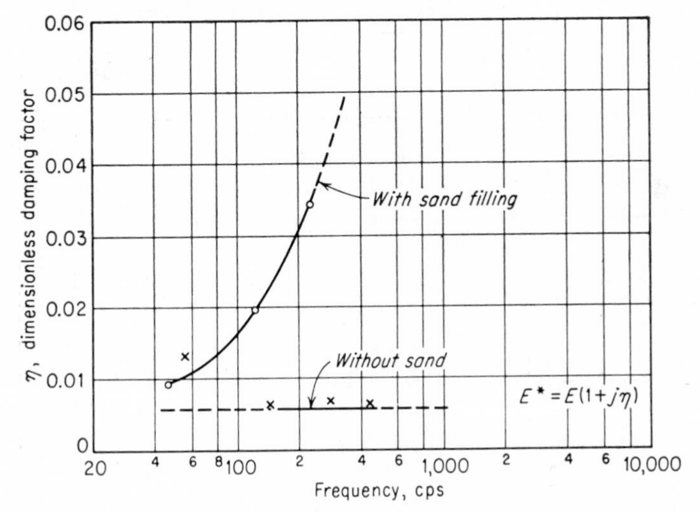

Fig. 13.13 Damping factor η for 6-in. hollow, dense concrete block. (*After Watters.*[9])

nected to the bottom of a vertical strip. A lightweight accelerometer is mounted near the top of the strip. The shaker is excited by an audio-oscillator whose frequency is slowly swept so as to search out the frequency and bandwidth of the various resonances.

The nature of the resonance of a typical strip is shown in Fig. 13.11. The results of the measurements are given in Figs. 13.12 and 13.13. Finally, in Fig. 13.14 there are shown two comparisons between calculated data (using Figs. 13.7, 13.8, and 13.9) and measured field data. The TL at higher frequencies is lower than that calculated. This may be due to the nature of the damping at the edge of the panel and to the size of panel. The effect of adding sand in the cavities of a masonry wall has been reported by Kuhl and Kaiser[28] and is shown by the data of Watters in Fig. 13.13.

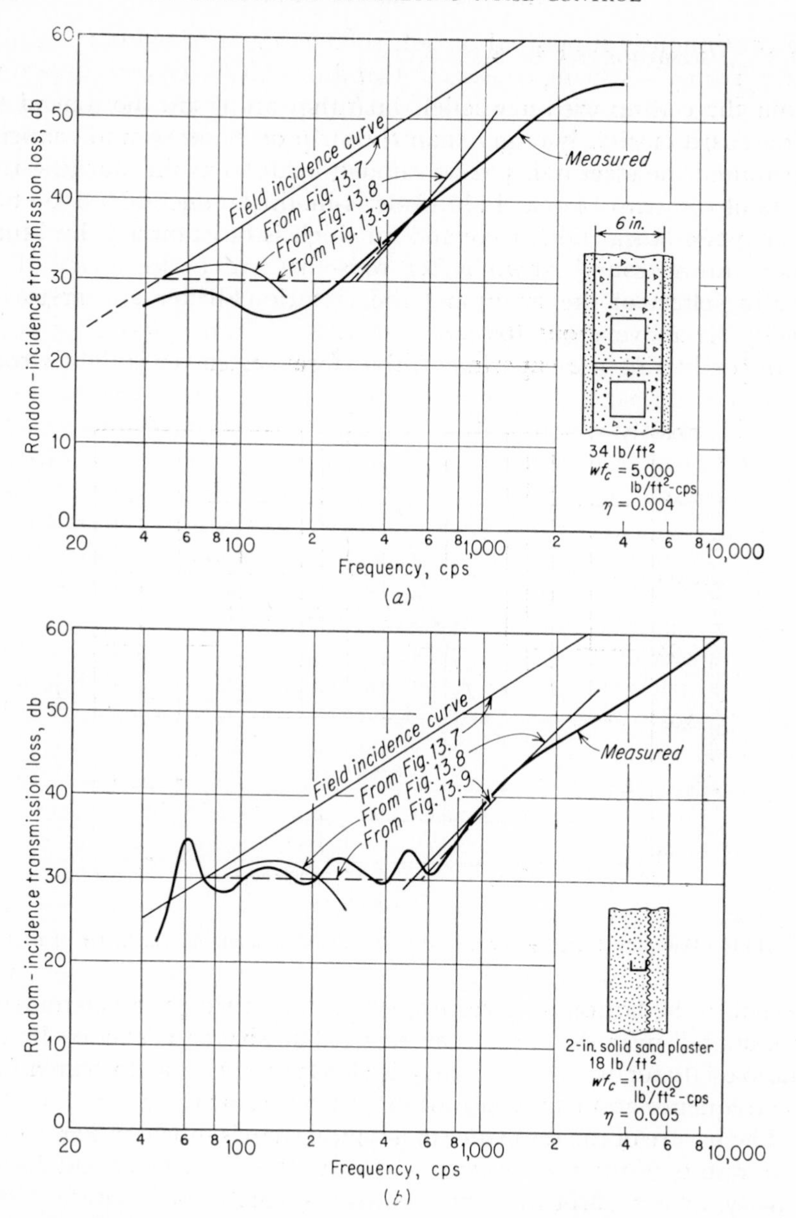

FIG. 13.14 Measured and calculated TL: (*a*) for 6-in.-hollow cinder block plastered; (*b*) for 2 in. of solid sand plaster. Note that w is in units of lb_m/ft^2. (*After Watters.*[9])

13.9 Discontinuities in Structures

Bending waves are attenuated when they travel around a bend or corner (Cremer[29]). Kurtze, Tamm, and Vogel[30] determined experimentally over a frequency range of 200 to 2,000 cps the transmission of bending waves around a 90° bend made by joining two 1.2- by 1.2-in. Plexiglas bars. Qualitatively they show that at a corner the bending waves behave as shown in Fig. 13.15.

The solid and dashed lines in Fig. 13.15 illustrate the displacements (greatly exaggerated) of the bars at two instants of time separated by ¼ period. The source was located about 6 ft from the corner on bar 1. It is seen that when the bending stiffness of the corner is greatly reduced, either by inserting a rubber layer or by a "rolling" joint, the transmitted bending wave is greatly attenuated.

In traveling around a rigid corner, the bending wave was attenuated by 3 to 5 db. With 0.08 in. of rubber, installed as shown in Fig. 13.15, the attenuation was about 20 db; with 0.2 of rubber, about 25 db, and with 1.2 in. of rubber, about 30 db. With the "rolling" joint the attenuation was about 12 db.

The importance of these findings is that sound from a vibrating machine may be produced in an adjoining room by a common wall not only due to air-borne excitation of that wall but also due to structure-borne excitation of that wall (and other walls).

In practice, joints in the form of T's or +'s are important. For a rigid T the attenuation below 1,000 cps is 6 to 7 db between the vertical member and each of the members at right angles. For a rigid + the attenuation is 9 to 10 db (Westphal[31]). These attenuations are exceeded in walls of rooms at frequencies above 1,000 cps.

In double walls of the type shown in Fig. 13.2, a "sound bridge" is often intentionally or unintentionally built in. The "bridge," for example, might be a rod conducting compressional waves, and thereby exciting bending waves in the radiating wall. Cremer[32] has analyzed the effect of such a tie. He considers the particular case of a concrete sheet 1.5 in. thick floated 0.4 in. over a concrete slab 4.5 in. thick. Cremer finds that a sound bridge in the form of a rod with a cross-sectional area of between 0.2 and 0.6 in.^{2} increases the transmission of impact sound over no sound bridge by 10 to 30 db. The exciting impact on the floated slab was applied 6 ft from the sound bridge. The deleterious effect is less pronounced when the point of excitation is removed a greater distance from the bridge.

Masonry walls are frequently held together for structural reasons by resilient wire clips. Their deleterious effects are not great if the walls are very heavy. However, between two flexible leaves, they may

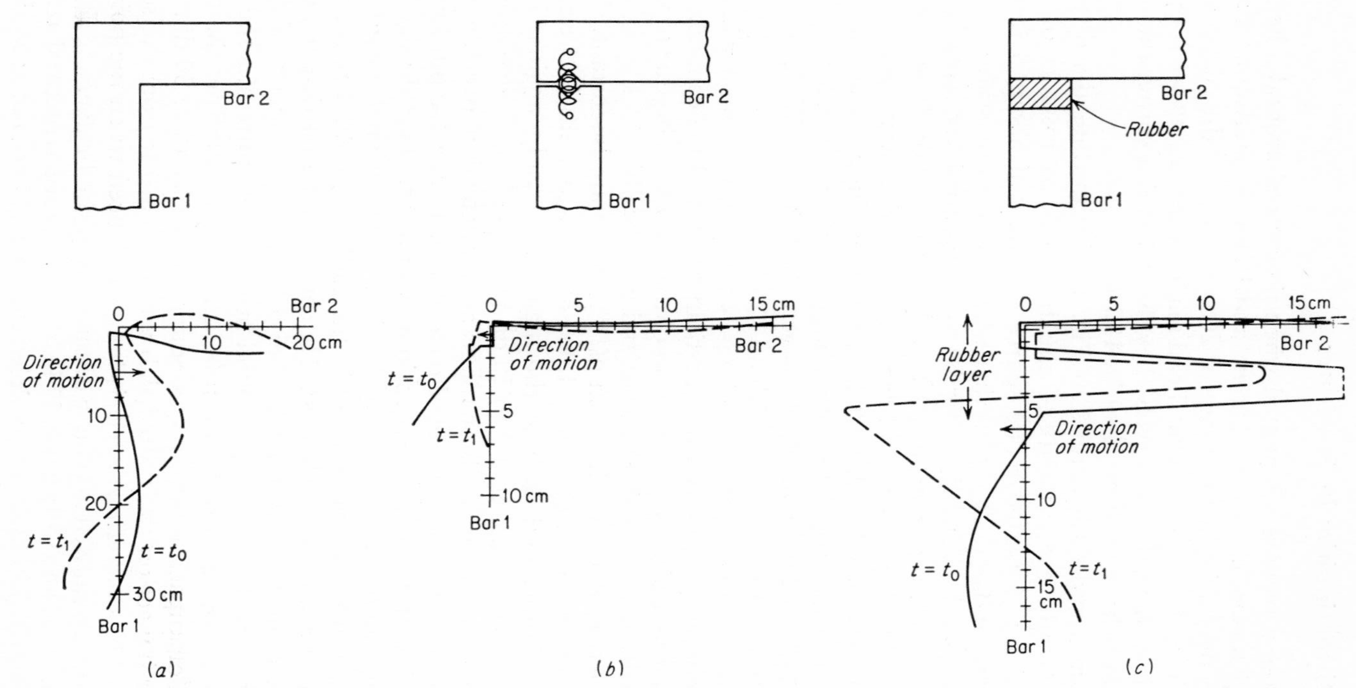

FIG. 13.15 Qualitative representation of the transmission of bending waves around a 90° corner for three types of corner construction: (*a*) rigid joint; (*b*) rolling joint; (*c*) elastic joint. (*After Kurtze, Tamm, and Vogel.*[30])

reduce the TL of a double wall by a significant amount. In the case of a lightweight wall, the coupling will be less for a massive bridge than for a lightweight bridge.

Two additional cases of interest have been studied recently by Kurtze.[33] One of these cases is shown in Fig. 13.16*a*. A bending wave is established in member *A*. The transmitted bending wave is measured in *B*. If the thicknesses h_1 and h_2 are equal, the effect of adding the L-shaped member on top, with a vertical length of leg L_z, is to cause a 4-db TL. This TL will reduce to about 2 db at values of $L_z = \lambda/2$, λ, $3\lambda/2$, 2λ, etc. It will increase to 7 db at $L_z = \lambda/4$, $3\lambda/4$, $5\lambda/4$, etc. We remember that for a simple T, that is, with L_z infinitely long, the TL from *A* to *B* is 4 db and from *A* to *C* is 6.5 db.

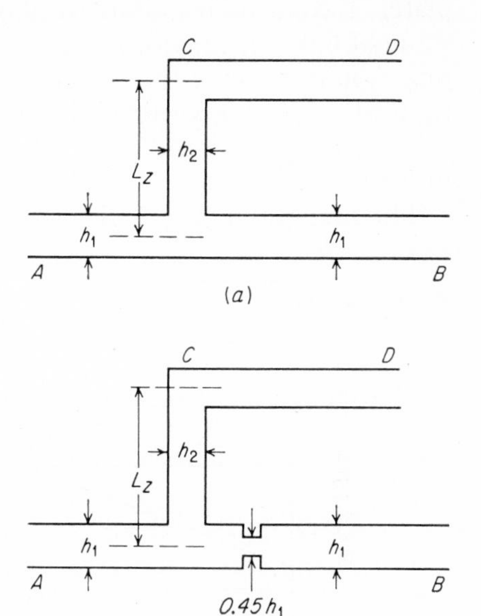

FIG. 13.16 Structures for which bending wave transmission has been computed. (*After Kurtze.*[33])

Now, if (Fig. 13.16*a*) the thickness h_2 is made equal to $1.75h_1$, the TL averages 11 db, with maxima and minima of 16 and 6 db respectively at the same values of L_z/λ as given above.

In Fig. 13.16*b*, a case which embodies a double slot is shown. The material in member *B* is reduced to $0.45h$, as shown. The TL from *A* to *B* averages 12 db with maxima and minima at 16 and 8 db respectively and at the same values of L_z/λ as previously. Finally, if both a double slot and the value of $h_2 = 1.75h_1$ are chosen, the TL from *A* to *B* averages 19 db with maxima and minima of 25 and 13 db, respectively.

13.10 Coupling of Radiating Surface to a Medium

Below the critical frequency, a solid panel does not couple well to an air medium. At the critical frequency, a panel radiates a sound wave with a particle displacement nearly equal to the surface displacement of the panel itself. This is also true at each frequency above the critical frequency—the air-borne sound wave being radiated at the coincidence angle. The radiation from panels above and below the critical frequency has been discussed by Westphal [31] and Goesele.[34] Simply summarized, they show that below the critical frequency f_c the radiated sound may be up to 20 db less than above the critical frequency for the same strength of bending wave in the plate. Two of their graphs are shown in Figs. 13.17 and 13.18. The value of 0 db on the ordinates corresponds to the radiation of sound that would occur if the panel were vibrating as an infinitely large rigid plate, back and forth perpendicular to its surface, with a peak velocity v equal to the peak velocity of the traveling bending wave in the plate.

The first graph (Fig. 13.17) shows the effect of damping (expressed

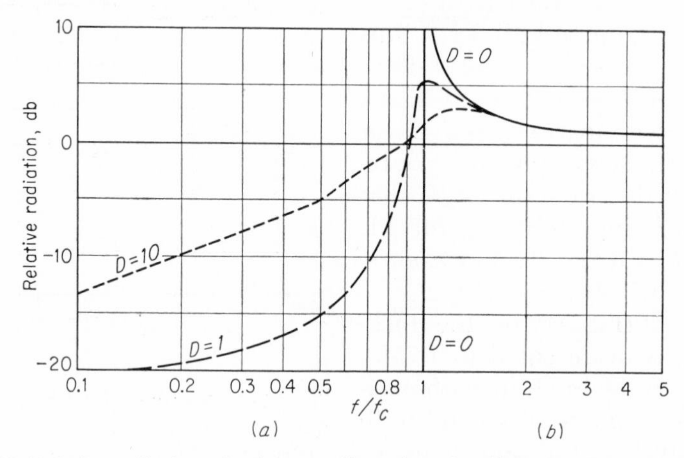

FIG. 13.17 Relative radiation of a free-traveling plane bending wave from an infinitely large plate. The parameter D is the damping of the bending wave (in decibels per wavelength). (*After Westphal.*[31])

as $D =$ number of db the bending wave is attenuated in traveling a distance in the plate equal to a wavelength of the bending wave) on the intensity of the radiated wave. In Fig. 13.18, they show the effect on the intensity of the radiated wave of finite plate size, b/λ_B [where b is the length of a side of a square plate and λ_B is the bending wavelength given by Eq. (13.8)].

It is seen that, below coincidence, for a given intensity of bending wave in the plate, increased damping or making the plate small increases the intensity of the radiated wave. However, for a given excitation of the panel, damping usually decreases the intensity of the bending wave set up in the panel, so that the effect of Fig. 13.17 is counterbalanced.

The practical significance of these findings lies in their application to the transmission of noise (vibrations) via the structure from one room to another or in flanking a double wall. For example, in Fig. 13.1 a sound wave from room 1 may arrive in room 2 as wave 1*A* or 2*A* or 2*B*. If there were no loss in travel of wave 2 through the structure, the waves 2*A* and 2*B* combined would be as strong as wave 1*A*. In other words, each would be 3 db weaker than wave 1*A*. However, the presence of the common wall reduces wave 2*A* by 6.5 db and 2*B* by 4 db. Damping in the structure or a vibration break in the path of the wave will cause greater reductions.

To get the most noise reduction out of a wall, lightweight stiff materials should not be used. Heavy nonstiff materials, a thin sheet of

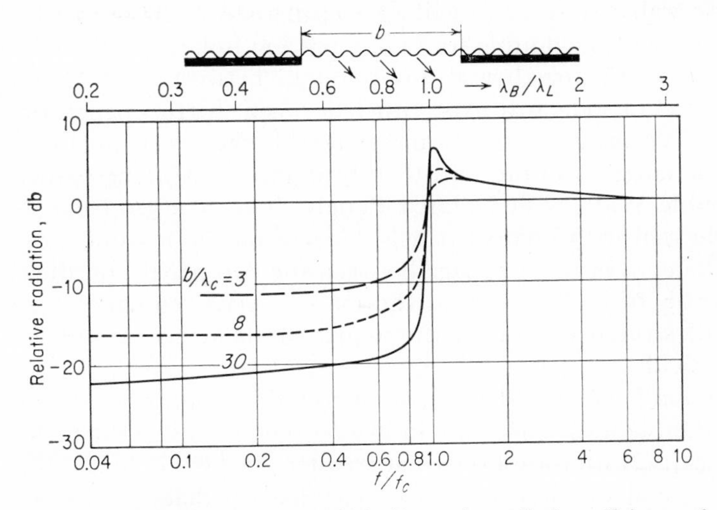

FIG. 13.18 Effect of finite plate size (width b) on the radiation efficiency of plates. Note both the upper and lower ordinates. λ_c = wavelength in the plate at the critical frequency; λ_B = the bending wavelength in the plate at frequency f; λ_L = the wavelength in the air at frequency f. (*After Goesele.*[34])

metal, for example, loaded with ceramic tiles such as are used in bathrooms, would be better.

13.11 Today's Trends

A glance at Fig. 13.7 reveals that, at a given frequency, the field TL for each doubling in mass of a single solid panel increases by only 6 db. In modern building structures where lightweight movable partitions are becoming common practice, and in aircraft and lightweight train structures it is not possible to achieve desired TL values with ordinary single panels. This leaves us with the following possible approaches:

1. Proper choice of critical frequency, i.e., usually high

2. Multiple-leaved walls, perhaps with porous sound-absorbing materials between the leaves
3. Structural discontinuities
4. Optimized damping
5. Specialized designs that possess bending-wave velocities that do not increase with frequency

13.12 Solid Double Walls

In this section we shall treat sound transmission through solid double walls. Two cases will be considered: (1) those in which the two panels of the double wall are vibration isolated at the edges and are not bound together at any point in between, and (2) those in which ties of some sort exist either between the two panels or at the edges. We shall call the former a double wall with isolated panels and the latter a double wall with bridging (or flanking paths).

Double Walls with Isolated Panels. The acoustic behavior of a double wall with isolated panels depends on the mass of the panels, on the depth of the air space between the two panels, on the critical frequency of each panel, on the mass-air-mass resonance f_{res} of the over-all structure, and on the angles of incidence in the exciting sound field.

London[16] has solved the case of sound transmission through isolated double walls under the following restrictions: (1) the panels are identical; (2) they are excited at frequencies below their critical frequency; and (3) they are mass controlled so that panel resonances need not be considered.

With these assumptions and Eqs. (13.13) and (13.2) we may write the reciprocal of the transmission factor as

$$\frac{1}{\tau} = \left|\frac{p_i}{p_t}\right|^2 = 1 + \frac{\omega M_s}{\rho c} \cos^2 \phi \left(\cos \beta - \frac{1}{2} \frac{\omega M_s}{\rho c} \cos \phi \sin \beta \right)^2 \qquad (13.34)$$

where $\beta \equiv (\omega d \cos \phi)/c = (2\pi d \cos \phi)/\lambda$ [see Eq. (13.11)]
$\omega = 2\pi f$ = angular frequency, rad/sec
M_s = mass of *each* panel, slugs/ft^2 (or kg/m^2)
ρc = characteristic resistance of air in suitable units (see Table 13.1)
ϕ = angle of incidence of sound wave (see Fig. 13.2)
d = spacing of panels, ft (or m) (see Fig. 13.2)
c = speed of sound, ft/sec (or m/sec)
ρ = density of air, slugs/ft^3 (kg/m^3)
λ = wavelength of sound in air, ft (or m)

Perfect transmission occurs, that is, $\tau = 1$, when the bracketed portion of the equation vanishes. That is to say when

$$\cot\beta = \frac{1}{2}\frac{\omega M_s}{\rho c}\cos\phi \tag{13.35}$$

In the special but frequently encountered case that $2\pi d/\lambda$ is small, we can replace the cotangent by the reciprocal of its argument, so that

$$\omega_{\text{res}} = \sqrt{\frac{2.8\,P_0}{dM_s\cos^2\phi}k_u} \tag{13.36}$$

where k_u is the correction factor introduced if nonconsistent systems of units for P_0 and M_s are used (see Table 13.1), and P_0 is the atmospheric pressure in pounds per square foot (or newtons per square meter). Note that from the wave equation for gases, γP_0 always equals ρc^2. For air, $\gamma = 1.4$.

We call ω_{res} the "mass-air-mass resonance condition" because the masses of the walls and the depth of the air space are involved. If we define a *basic resonance* ω_0 as that mass-air-mass resonance existing for a wave at normal incidence ($\phi = 0$), then

$$\omega_0 = \sqrt{\frac{2.8\,P_0}{dM_s}k_u} \tag{13.37}$$

and

$$\omega_{\text{res}} = \frac{\omega_0}{\cos\phi} \tag{13.38}$$

Finally, we substitute Eq. (13.37) into Eq. (13.34) and, in turn, into Eq. (13.1) to obtain the TL in decibels.

$$\text{TL} = 10\log_{10}\left[1 + \left(\frac{\omega}{\omega_0}\right)^2\left(\frac{\omega_0 M_s}{\rho c}\right)^2\cos^2\phi \times\left(\cos\beta - \frac{1}{2}\frac{\omega}{\omega_0}\frac{\omega_0 M_s}{\rho c}\cos\phi\sin\beta\right)^2\right] \quad \text{db} \tag{13.39}$$

where

$$\beta = 2\frac{\omega}{\omega_0}\frac{\rho c}{\omega_0 M_s}\cos\phi \tag{13.40}$$

We have in Eqs. (13.39) and (13.40), three dimensionless variables, ω/ω_0, $\omega_0 M_s/\rho c$, and ϕ.

For four values of $\omega_0 M_s/\rho c$, namely 3, 10, 30, and 100, the TL in decibels is plotted in Fig. 13.19 for $\phi = 60°$. For three cases, namely $\omega_0 M_s/\rho c = 3$, 10, and 100, the TL is plotted in Figs. 13.20 to 13.22

for $\phi = 0°$, $30°$, $45°$, $60°$, and $80°$. No assumptions except the three of London given above are incorporated in the graphs.

We note from the graphs and from Eq. (13.38) that the frequency of the mass-air-mass resonance shifts upward as the angle of incidence increases, in proportion to $1/\cos\phi$.

It appears from Eq. (13.39) that when the incident sound wave impinges on the wall at grazing incidence, the TL approaches zero. In practice, zero TL at grazing incidence does not occur for at least

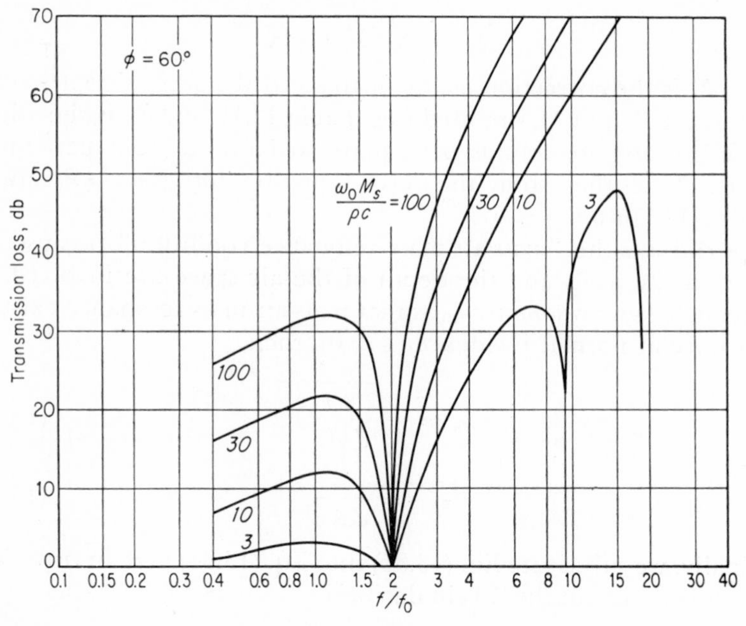

Fig. 13.19 Theoretical TL for limp double wall with completely isolated panels. The mass-air-mass resonance is f_0, the driving frequency is f, and M_s is the mass of each panel. The angle of incidence $\phi = 60°$. (See Table 13.1 for units.)

three reasons. First, grazing incidence sound waves cannot exist if the wall yields. Secondly, absorption of sound in the faces or at the edges of the panels tends to dampen the wave that is traveling in the air space parallel to the panels. Thirdly, zero TL can occur at grazing incidence only for infinitely broad waves traveling along infinitely large walls.

Let us now consider the region of the graphs (Figs. 13.19 to 13.22) above $f/f_0 = 2$ where the curves rise steeply. Assume also that the wall spacing d is less than $\frac{1}{6}$ the wavelength in air. In this region, the reciprocal of the transmission factor $(1/\tau)$ varies nearly as $\cos^6\phi$. Using Eqs. (13.6) and (13.1) we find that if the reverberant incident sound field is composed of waves impinging on the wall from all angles below a limiting angle ϕ_L, the average transmission loss is given by

$$\overline{\mathrm{TL}} = 10 \log_{10} \frac{1}{\tau}$$

$$= (\mathrm{TL})_0 - 10 \log_{10} \left(\frac{1}{2 \cos^4 \phi_L} - \frac{1}{2} \right) \frac{1}{\sin^2 \phi_L} \quad \text{db} \qquad (13.41)$$

where $(\mathrm{TL})_0$ is the transmission loss at normal wave incidence

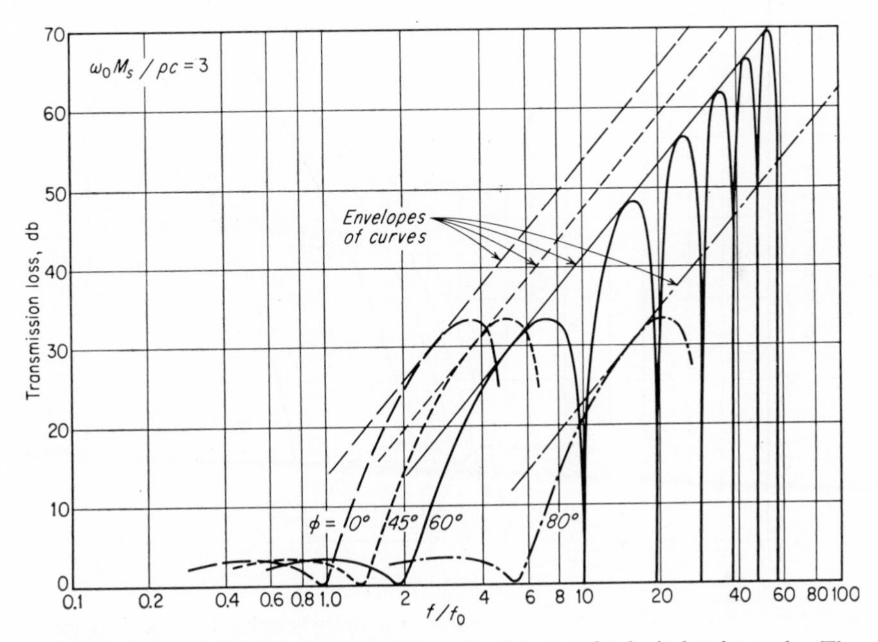

FIG. 13.20 Theoretical TL for a double wall with completely isolated panels. The mass-air-mass resonance is f_0, the driving frequency is f, and $\omega_0 M_s/\rho c = 3$, where M_s is the mass of each panel. (See Table 13.1 for units.)

($\phi = 0$). For example, if we say that the sound field is diffuse, but contains no angles of incidence greater than 80°, then

$$\overline{\mathrm{TL}} = (\mathrm{TL})_0 - 28 \text{ db} \qquad (13.42)$$

In other words, the TL actually obtained is highly dependent on the composition of the incident sound field.

For the frequency region above f/f_0 equal to about 12, resonance due to standing waves between the panels occurs whenever the equality of Eq. (13.35) exists. For $\phi = 0$ and $\omega M_s/2\rho c$ large, these resonances occur when

$$\beta = \frac{2\pi d}{\lambda} \doteq \pi \tag{13.43}$$

or

$$d \doteq \frac{\lambda}{2} \tag{13.44}$$

The behavior of the TL curve for $\phi = 0$ and for frequencies up to $100f_0$ is illustrated in Fig. 13.23. Because the resonance dips are

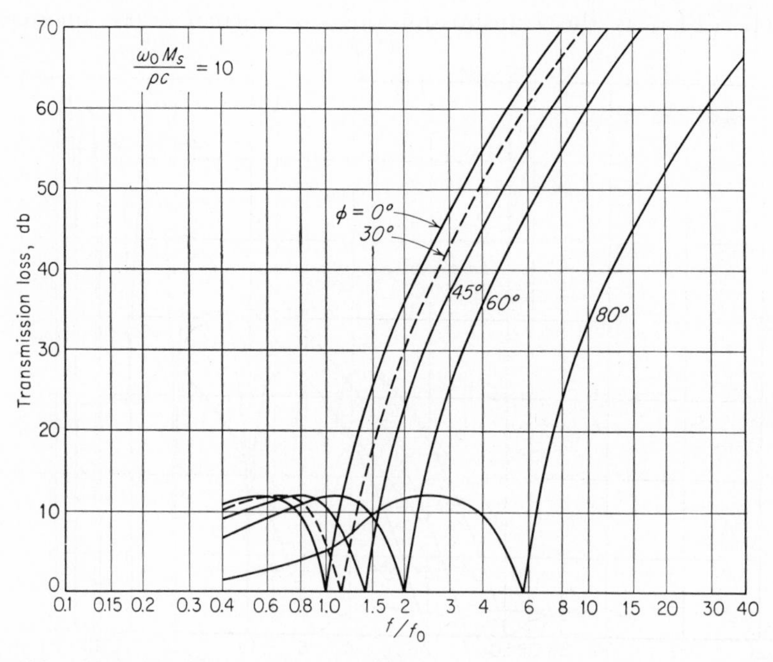

Fig. 13.21 Theoretical TL for double wall with completely isolated panels. The mass-air-mass resonance is f_0, the driving frequency is f, and the angle of incidence is ϕ. The value of $\omega_0 M_s/\rho c = 10$, where M_s is the mass of each panel. In practice, for TL's above 70 db, assume that the curves rise about 6 db/octave unless absorbing materials are introduced into the air space. (See Table 13.1 for units.)

quite narrow, it is customary to assume for simplicity an average transmission loss above $f = 10f_0$ that increases at a rate of about 6 db per doubling of frequency.

For other angles of incidence and for $(\omega M_s \cos \phi/2\rho c)$ large the interpanel resonances occur at

$$d \doteq \frac{\lambda}{2 \cos \phi} \tag{13.45}$$

The addition of sound-absorbing materials in the air space will

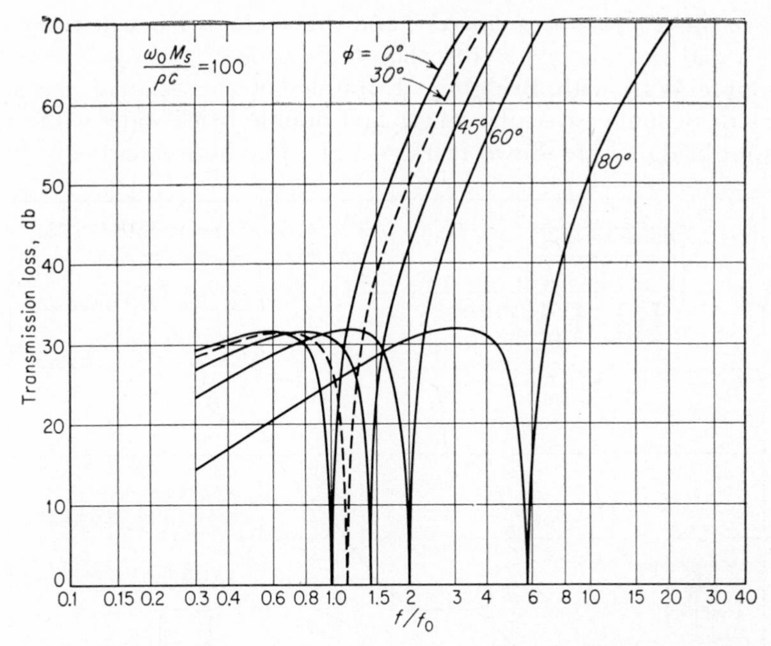

FIG. 13.22 Theoretical TL for double wall with completely isolated panels. The mass air-mass resonance is f_0, the driving frequency is f, and the angle of incidence is ϕ The value of $\omega_0 M_s/\rho c = 100$, where M_s is the mass of each panel. In practice, for TL's above 70 db, assume that the curves rise about 6 db/octave unless absorbing materials are introduced into the air space. (See Table 13.1 for units.)

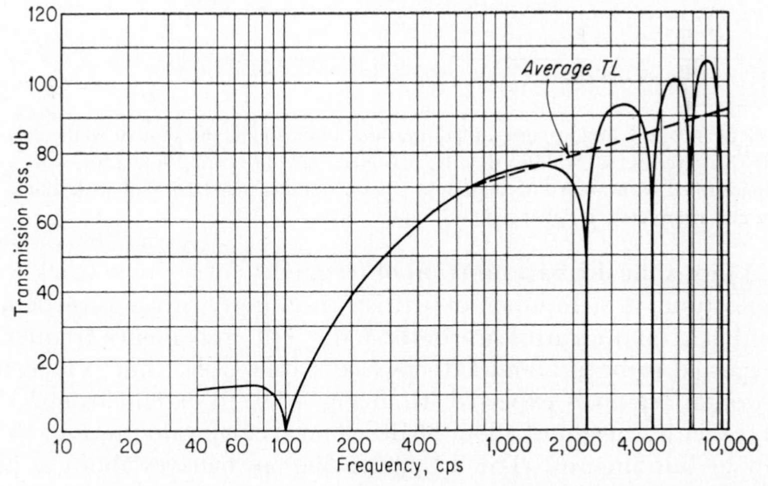

FIG. 13.23 Theoretical TL for an isolated double wall for $\phi = 0°$ and with no restriction on the separation between the walls. Each side = 2 lb/ft²; air space = 3 in.; normal sound incidence; $\omega_0 = 628$; $\omega_0 M_s/\rho c = 15$.

remove these resonances entirely, and the average TL will rise at a more rapid rate above $f = 10f_0$ than 6 db/octave.

Double Wall with Bridging. Examples of the relative acoustic behavior of some particular single and double brick walls with and without bridging are shown in Fig. 13.24. The bottom curve in Fig.

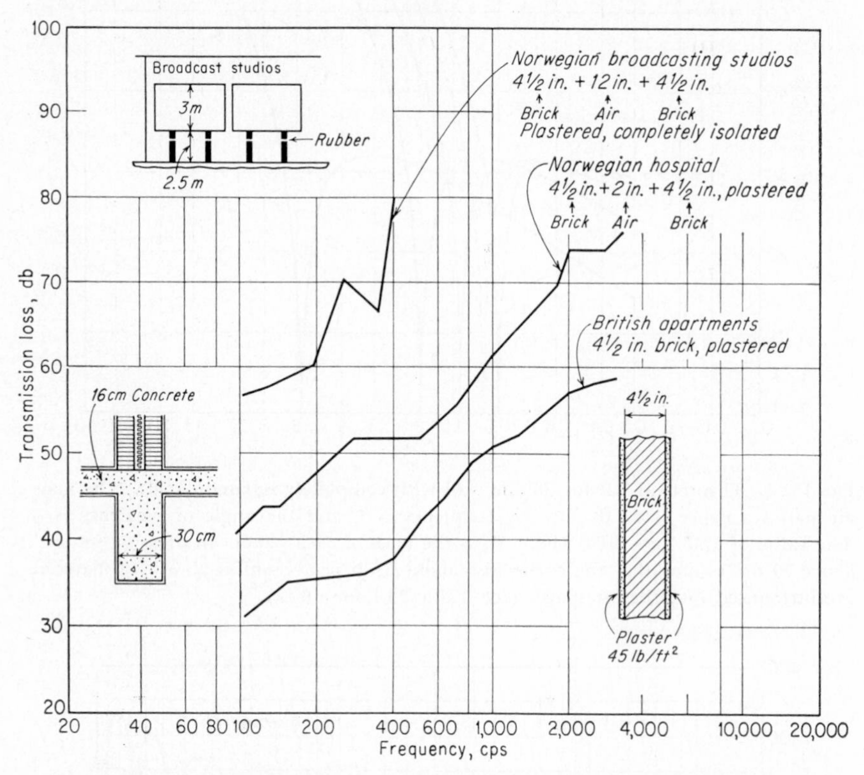

FIG. 13.24 Acoustic performance of 4½-in.-thick brick single and double walls. Lower: single, plastered wall. Middle: double, plastered wall on single foundation with 2-in. air space filled with absorbing material. Upper: double, plastered wall with 12-in. air space and completely isolated panels.

13.24 shows the TL as a function of frequency for a single brick wall as measured in a number of British dwellings with a reverberant sound field on the source side of the wall. The coincidence frequency is expected to be at about 220 cps (see Table 13.4). The TL in the horizontal region is expected, from Fig. 13.9, to extend from 55 to 300 cps with a height of about 33 db. The average value in Fig. 13.24 is about this amount. The TL above 300 cps behaves about as predicted from Fig. 13.9, namely, it rises 10 db/octave and then gradually tapers off to 6 db/octave.

When two plastered brick walls 5 in. thick are combined to produce an isolated double wall with a 12-in.-deep air space between the panels, the results at low frequencies may be predicted from Eqs. (13.39) and (13.41). For such a wall, the mass-air-mass resonance frequency f_0 is about 10 cps and $\omega_0 M_s/\rho c$ is 35. At $f = 100$ cps and $\phi = 0°$, a TL of about 82 db is calculated. For a diffuse sound field with a limiting angle of $\phi_L = 80°$ it is found from Eq. (13.42) that $\overline{\text{TL}} = [\text{TL}]_0 - 28 \text{ db} \doteq 82 - 28 \doteq 54$ db. Above 100 cps one would expect the TL to increase first at a rate of 12 db/octave and later at 6 db/octave. One example of such a completely isolated double wall is found in the literature (Moeller[35]). A double wall is formed by the two facing walls of two adjacent well-isolated broadcast studios (see Fig. 13.24). As is shown by the upper curve, the TL at 100 cps, with a reverberant sound field, was measured as 57 db. For the first two octaves above $f = 100$ cps, the TL rises at a rate of only about 7 db/octave. This lower rate than the predicted 12 db is due to the effects of wave coincidence. In the next octave the TL should rise at a rate of 6 to 10 db/octave, but the data are incomplete and do not show this frequency region.

A double wall of similar construction but with flanking is now to be compared with the preceding wall (see the middle curve in Fig. 13.24). We see that the measured TL values are reduced by 15 to 30 db below those of the upper curve over the entire audible frequency range. The presence of the 30-cm-wide foundation beneath the double wall reduced the flanking transmission because the foundation has a high moment of inertia to bending waves. If the 16-cm slab carried straight through with no 30-cm-wide foundation beneath, the improvement of the double wall over a single wall would be only 2 to 4 db.

In several examples taken from the British Building Research Station tests, almost no improvement over single-wall performance was obtained by use of double masonry walls with butterfly wire ties between.

13.13 Impact-sound Transmission

Noise in a structure can be produced by the impact of a rigid body against one of its surfaces. The chief difference between the excitation of vibration in a panel by air-borne sound waves and that by impact is the extent of the area over which the driving force is applied. Theoretically, one can handle this problem by assuming this area of contact to be a point.

Cremer[26] has made a theoretical study of "the reaction of a driven continuum to a point source." He derives the mechanical point impedance (ratio of an alternating driving force to the resultant alternating velocity) for plates of steel, aluminum, asphalt, and plywood as a function of their thickness. He then derives the spectrum of the impact sound as a result of dropping a mass on the infinite plate, assuming several values of an added elastic layer at the surface of the plate. An added elastic layer could, in practice, be the result of adding a resilient covering to the surface of the plate. Cremer next discusses theoretically the case of adding a floating floor over a floor-ceiling. As indicated earlier in this paper, Cremer also considers the case of "sound bridges," that is, solid ties, between the floating floor and the ceiling below.

Because of the lack of space in this text, the reader is referred to Cremer[26] for further study. Measured data in comparison with predictions of the sound insulation of floating floors with sound bridges have been published by Heckl.[36] He reports good agreement.

13.14 Damping of Panels

It is apparent by now that above the critical frequency of a panel, bending waves are readily excited by air-borne or water-borne waves and, having been excited, they readily produce sound waves on the other side of the panel in an air or water medium. Furthermore, bending waves travel easily along a structure and around corners. Undesired sound (noise) is produced by vibrating structures in buildings, aircraft, ships, automobiles, and so forth. If the induced bending waves are intense enough, the material of the structure may fail, or electronic or mechanical equipment attached to it may malfunction. Failures of these types are often found in jet aircraft and in missiles.

To alleviate these undesired phenomena, methods for controlling or reducing the amplitudes of the bending waves must be found. In structures where high tensile strength with low weight is a primary factor, such as in aircraft, it is not possible to interrupt the continuity of the structure to insert elastic layers. Furthermore, the introduction of heavy strips, with high moments of rotational inertia perpendicular to the direction of travel of the bending wave, may not be possible owing to weight restrictions. Therefore, highly efficient, lightweight damping materials are needed to control the excitation and transmission of bending waves in structures.

Historically, damping has often been treated as a somewhat mysterious phenomenon, best approached by trial-and-error methods.

In recent years, Oberst,[37] Oberst and Becker,[38] and Liénard [39] have successfully analyzed the damping produced by single homogeneous layers of viscoelastic damping materials. They have found the damping mechanism there to be associated with the stretching of the

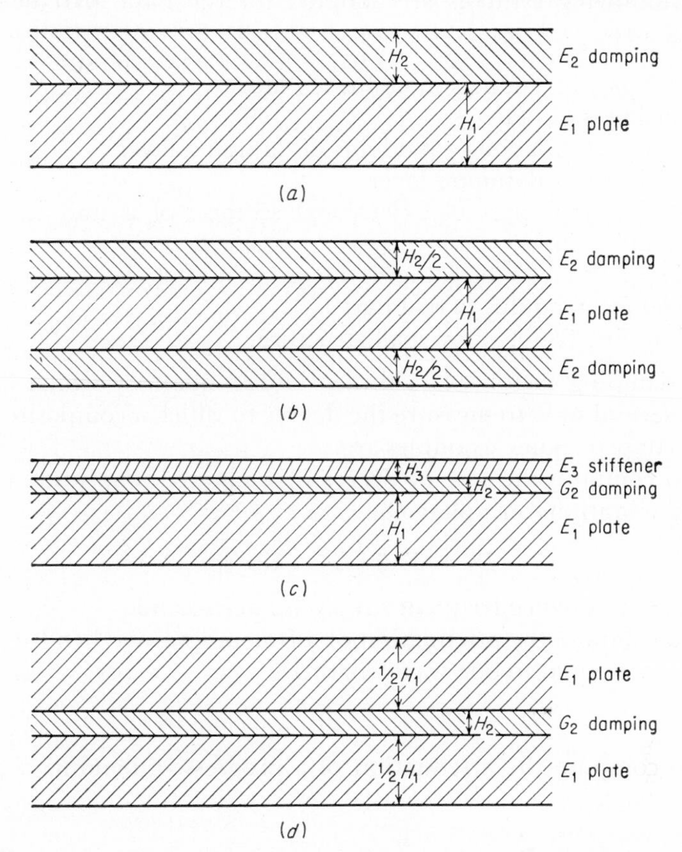

FIG. 13.25 Four types of damped plates: (a) single plate with single layer of damping material; (b) single plate with equal damping layers on both sides; (c) single plate with thin damping layer and thin constraining sheet; (d) two plates with single damping layer between.

damping layer. Kerwin[40] and Ross, Kerwin, and Dyer[41] have shown that the damping of a constrained layer is caused by the shear motion of the damping layer between the plate being damped and the outer constraining foil.

Following Ross, Kerwin, and Dyer, four types of damped structure are shown in Fig. 13.25. As the caption to the figure indicates, single or double plates are damped by one or two nonconstrained or constrained layers of damping material.

Nonconstrained Damping Layers. Two common nonconstrained damping structures for application to a single plate are shown in Fig. 13.25*a* and *b*. Both structures are assumed to have the same weight.

The following symbols are adopted for the basic parameters of the structures:

E_1 = real part of Young's modulus of plate
H_1 = thickness of plate
E_2 = real part of Young's modulus of damping layer
H_2 = thickness of damping layer
$K_2 = E_2H_2$ = real part of extensional stiffness of a unit length of damping layer
η_2 = damping factor of damping layer = ratio of imaginary part of Young's modulus to real part
η = damping factor of over-all structure

The damping factor η is a dimensionless quantity which can be used in several ways to measure the degree to which a composite structure is damped. Some examples are:

1. To compute the rate of decrease of squared amplitude of free bending vibrations on a plate D_t

$$D_t = 27.3\,\eta f_0 \qquad \text{db/sec} \tag{13.46}$$

where f_0 = resonance frequency in cycles per second.

2. To compute the attenuation of a free progressive bending wave in a distance equal to one wavelength D_λ

$$D_\lambda = 13.6\,\eta \qquad \text{db/wavelength} \tag{13.47}$$

3. To compute the attenuation of a progressive bending wave in a unit distance

$$D_1 = 13.6\,\frac{\eta}{\lambda} \qquad \text{db/unit length} \tag{13.48}$$

where λ is the wavelength in the units of distance desired.

4. To compute the sharpness of a resonance of the plate at the 0.7 of maximum amplitude points on either side of the resonance frequency

$$\Delta f = f_0\eta \tag{13.49}$$

where f_0 is the resonance frequency in cycles per second.

It is assumed now, following Oberst, that the product of the damping factor η_2 times the extensional stiffness of the damping layer K_2 is less than one-tenth the extensional stiffness of the original bar K_1.

The relative damping factor for a bar with a single homogeneous layer is a function of the relative elastic (Young's) modulus (E_2/E_1) as well as the relative thickness of the damping layer (H_2/H_1). In Fig. 13.26, the damping is plotted as a function of the relative thickness with the relative Young's modulus as the parameter. The curves show an approximate square-law relation for the damping as a func-

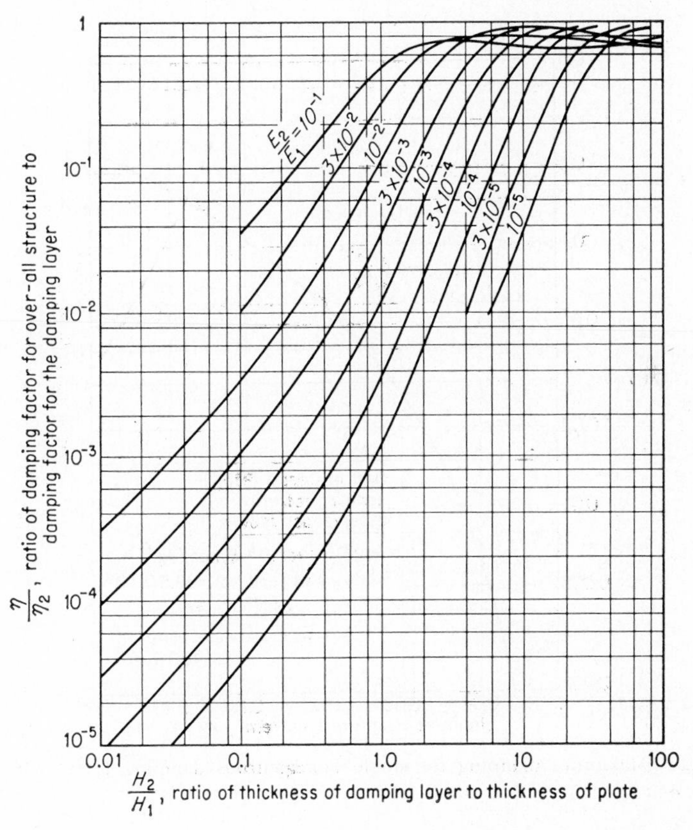

FIG. 13.26 Relative damping factor η/η_2 for homogeneous damping layer. (*After Oberst.*[37])

tion of thickness. They also show that for thick damping layers the factor for the over-all structure becomes about 0.8 that for the damping layer itself.

To illustrate the best performance to expect from a nonconstrained damping layer, calculations were made for the best one of the damping materials reported by Oberst and Becker.[38] The properties of the materials and the results of the calculations are given in Fig. 13.27. The chart is a plot of η_{max} (for best type of damping material) as a function of the ratio of the treatment weight to the plate weight. The damping is seen to rise as the square of the weight ratio.

Ross, Kerwin, and Dyer find that dividing the damping layer into two layers (Fig. 13.25*b*) reduces the resulting damping by about one half. This result is to be expected because the effectiveness of a damping layer is proportional to the square of the thickness of the layer.

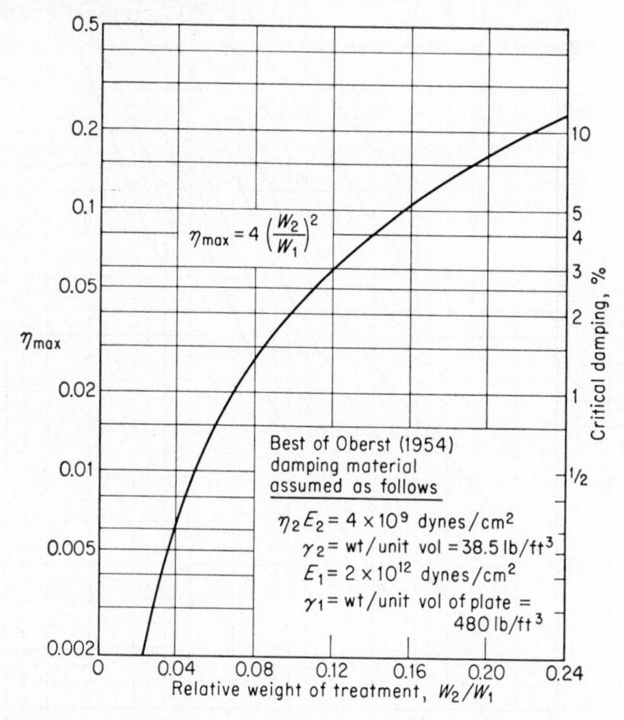

FIG. 13.27 Maximum damping for single homogeneous damping layer. (*After Ross, Kerwin, and Dyer.*[41])

Constrained Damping Layers. When the viscoelastic layer is constrained by a stiff foil (Fig. 13.25*c*), the mechanism responsible for the major component of the damping is the shear motion of the damping layer. A relatively thin layer may be quite effective, and the requirements for optimum damping materials are quite different from those for the single layer.

The symbols of the preceding paragraphs are used here, augmented as follows:

G_1 = real part of complex shear modulus of plate
G_2 = real part of complex shear modulus of damping layer

K_1 = extensional stiffness of plate
K_3 = extensional stiffness of constraining foil
β = material damping factor = ratio of imaginary to real part of complex shear modulus of damping layer
H_3 = thickness of constraining layer
Γ = modified shear parameter = $G_2/(E_1H_1H_2k_p{}^2)$
k_p = bending-wave number in plate = $2\pi/\lambda_p$

Now assume, with Ross, Kerwin, and Dyer, that the thickness of the damping material is smaller than that of the foil. The foil is

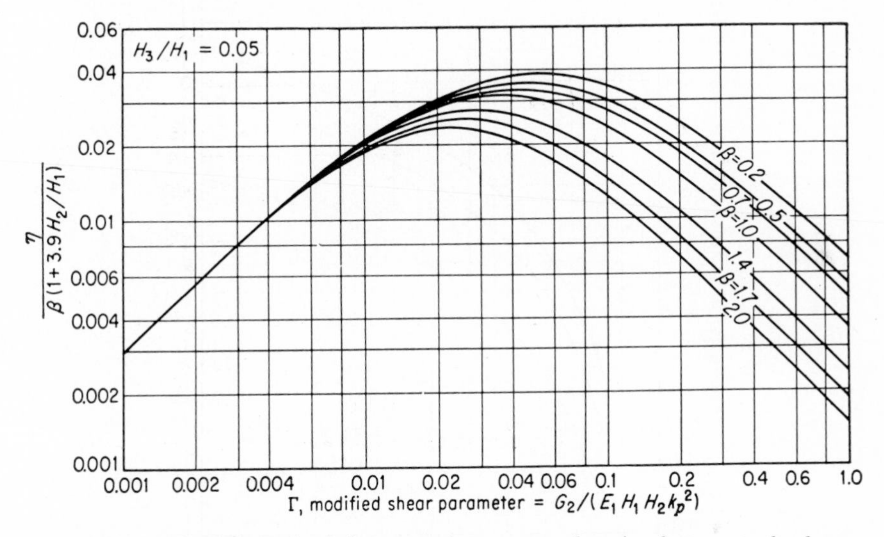

FIG. 13.28 Graph of the ratio of the composite structure damping factor η to the damping factor β of the damping material plotted as a function of the modified shear parameter Γ. The graph is based on calculations assuming a constrained damping layer with $\frac{1}{20}$ the thickness of the plate. (*After Ross, Kerwin, and Dyer.*[41])

assumed to be of the same material as the plate. The results of their computations are shown in Fig. 13.28 for H_3/H_1 equal to 0.05. The ordinate of the graph is the ratio of the composite damping factor to the damping factor for the damping layer, modified by the factor $(1 + 4H_2/H_1)$.

It is interesting to note that for low values of the material damping factor β, the peak of the relative damping curve occurs for Γ in the order of 0.06. For increasing values of β, the relative damping of the composite is unchanged at low values of Γ but is appreciably lower at high values of Γ. Note, however, that because β is in the denominator of the ordinate scale, the structure damping factor η increases with the material damping factor for Γ less than about 0.06. For values of β greater than 0.5, η is nearly independent of the material damping factor above $\Gamma = 0.1$.

Figure 13.28 also shows that for a damping layer with little resistance to shear (low value of shear parameter Γ) the damping is relatively low. As the damping layer becomes stiffer, and Γ increases, the damping increases. As Γ increases further, a maximum in the damping occurs. Finally, as the layer becomes "too stiff" and

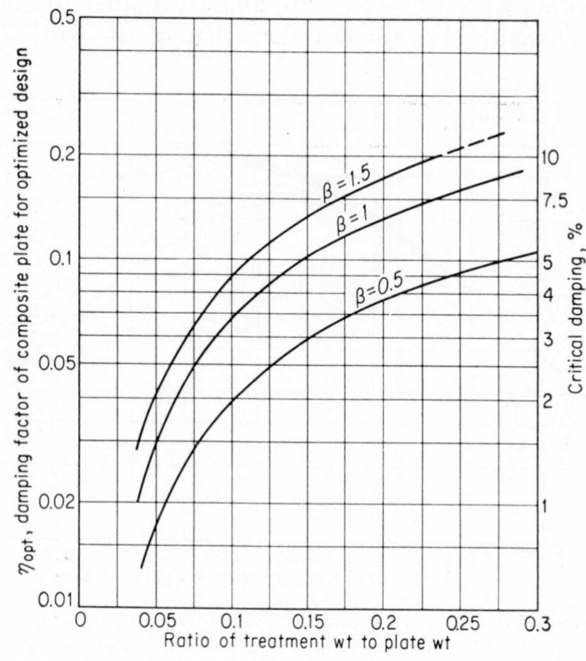

FIG. 13.29 Calculated optimum-composite-structure damping factor of a plate with a constrained damping layer for three values of the damping factor β of the damping material. (*After Ross, Kerwin, and Dyer.*[41])

the foil stretches appreciably as the plate bends, the damping decreases.

The shear parameter is a complicated function of frequency and temperature, depending as it does on the shear modulus of the damping material, which is a function of frequency and temperature. Also, the shear parameter is a function of the thickness of the damping layer. For a given operating temperature and frequency it is usually possible to select a damping material or layer thickness, or both, that will yield close to maximum damping. When this is done, the damping achieved is primarily a function of the constraining layer's relative stiffness and of the loss factor, β.

Assume that a proper choice of damping material has been made. In order to achieve optimum damping the value of Γ will lie somewhere between 0.02 and 0.06. For this case, the optimum damping factor η_{opt} is plotted in Fig. 13.29 as a function of the ratio of the treatment weight to the plate weight. The range of values of β (0.5, 1, and 1.5) covers the range from "easy to achieve" to "possible to achieve."

Comparison of Constrained and Nonconstrained Damping Layers. The damping achievable by nonconstrained damping layers depends

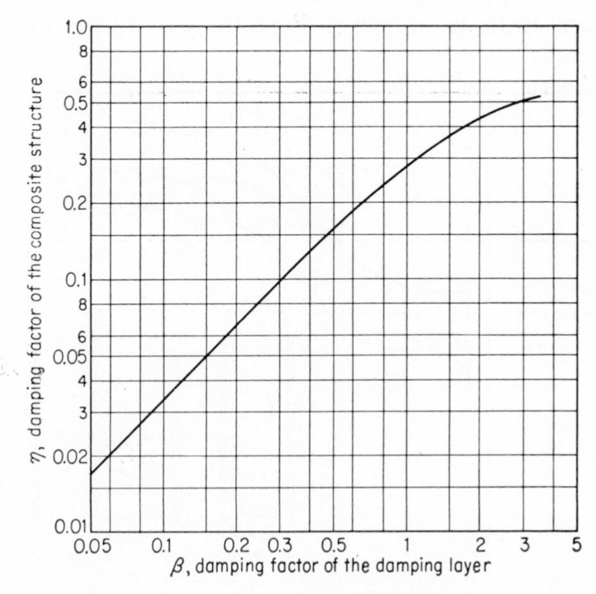

FIG. 13.30 Damping of sandwich structures embodying a thin damping layer. (*After Ross, Kerwin, and Dyer.*[41])

roughly on the square of the weight of the material, while that for constrained layers depends approximately linearly on weight. At this writing, it appears that if we choose the best type of damping material and optimize it for temperature and frequency, the two types are likely to give about equal values of damping for weights between 10 and 20 per cent of the base-plate weight. Below 10 per cent, the constrained layer damping is likely to be more effective. Above 20 per cent the nonconstrained layer damping should be more effective.

Sandwich Plate. A sandwich plate (Fig. 13.25*d*) is a special type of constrained damping layer consisting of a thin damping layer between two identical stiff plates. The maximum damping is obtainable with a value of modified shear parameter, Γ, equal to about 0.2.

The result of calculations assuming this value of Γ is shown in Fig. 13.30 (Ross, Kerwin, and Dyer[41]).

For low values of the material loss factor, the maximum loss factor of the composite sandwich is one-third that of the damping material, while at higher values of β, this ratio decreases.

13.15 Field Measurements of Transmission Loss of Practical Structures

Single and Double Walls. Transmission-loss (TL) measurements on 20 practical walls are shown in Figs. 13.31 to 13.45. Figures 13.31

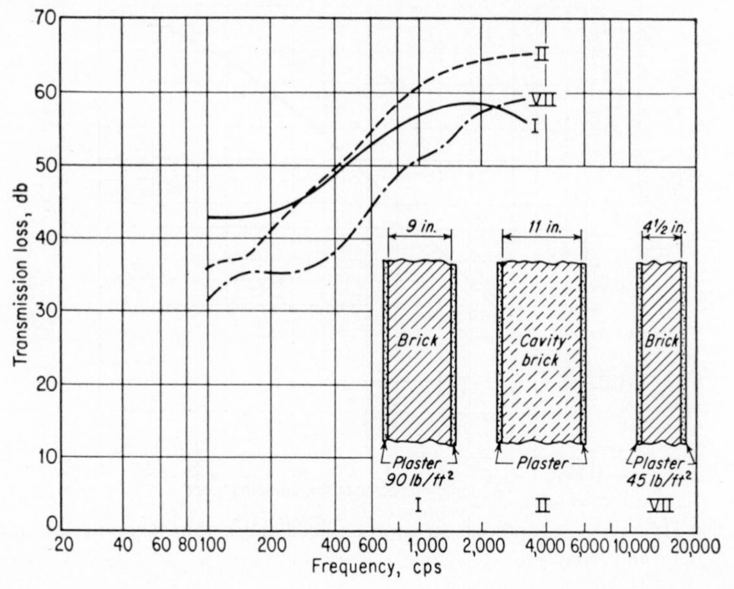

FIG. 13.31 TL's of three types of masonry walls measured in the field at a number of installations. Note that 1 kg equals 2.2 lb. (*Courtesy of British Building Research Station.*)

and 13.32 were reported by the Building Research Station in *Digest* 88.[1] The next data were measured by Watters and Farrell of Bolt Beranek and Newman Inc., for the United States Gypsum Company and are reported here with the express permission of USG. Because of flanking transmission, the double walls are generally very little better than the single ones, except where one or both of the panels are vibration isolated, e.g., see Fig. 13.24. The graphs and their captions should be self-explanatory.

Floating Floors and Hung Ceilings. In a building, sound does not only pass from one room to another acoustically. The sound may be impact created due to objects striking the primary side. In dwellings, air-borne sound arises from talking, the radio, TV, and the like, while impact sound is created by shoes, by movement of furniture, by pounding, or by dropping objects on the floor.

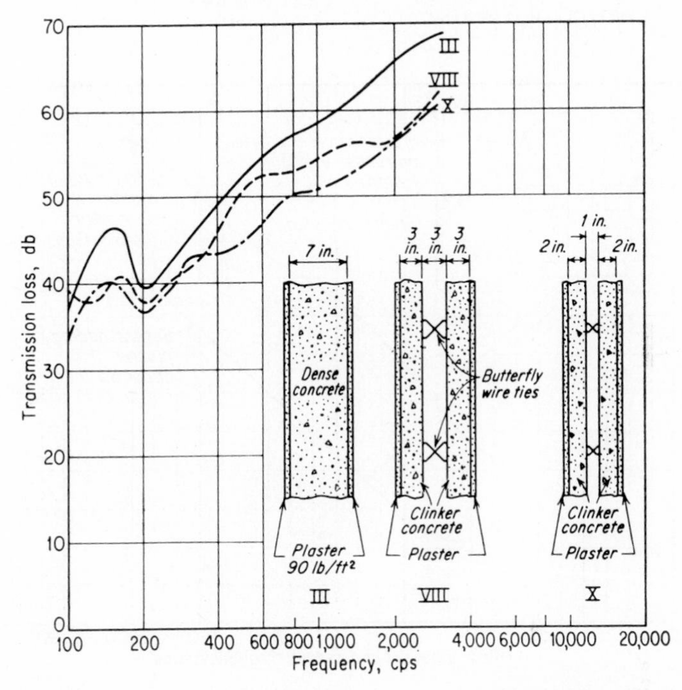

Fig. 13.32 TL's of three types of concrete walls measured in the field on a number of installations. Note that 1 kg equals 2.2 lb. (*Courtesy of British Building Research Station.*)

The control of sounds that travel from one room to another via the common floor-ceiling path can be accomplished by a floating floor, a hung ceiling, a suitable floor covering, or by any two or three combinations of these.

Informative studies on the practical behavior of such floor-ceiling structures have come from the Research Institute for Public Health Engineering TNO (12 Koningskade, The Hague) and the Technical Physics Department TNO (11 Mijnbouwplein, Delft) in Holland. Following World War II, the Dutch Government erected a set of experimental dwellings containing 48 self-contained apartments built in three stories. Some apartments had four rooms and others three rooms (exclusive of kitchen and bath). The rooms were built with different floors and walls for experimental reasons. A total of 38 floors of 6 different categories were tested, namely (1) wood joist, (2) steel joist, (3) concrete joist, (4) reinforced concrete, (5) reinforced brickwork, and (6) prefabricated. Also, a number of floor coverings were tested. Their work has been presented in a series of reports extending over a period of 5 years.[3]

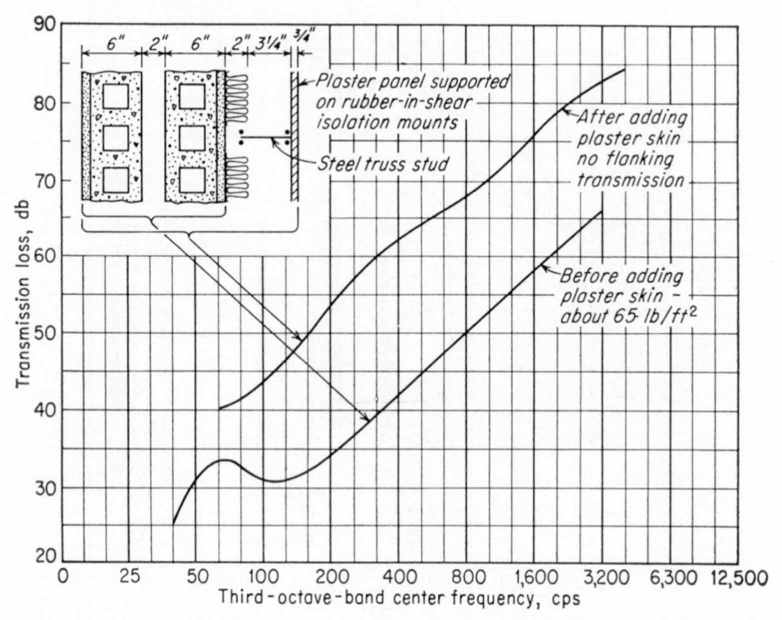

FIG. 13.33 TL of a double concrete block wall before and after adding an isolated plaster panel (field measurement).

Extensive investigations have been carried out at the Building Research Station, Garston, Herts, England. The work of that organization has been presented in numerous journals and government reports.[1]

Floors are tested in much the same way as walls, namely by creating sound in the room above the floor and measuring the amount that is transmitted through into the room below. In addition, an apparatus for determining the ease with which impact sound is transmitted from the upper surface to the room below has been developed. The apparatus is called an "impact machine."[2] This machine contains five striking hammers in a line, each having a mass of 0.5 kg, which are spaced at regular intervals over a straight line, the two outer hammers being 40 cm apart. The cam shaft, driven by an electric motor, raises each hammer in turn to a height of 1 cm, the hammers being afterwards impelled by a spring with such a force that they reach the floor at the same speed as they would have had if allowed to fall freely from a height of 4 cm. The machine produces 10 impacts/sec. Each hammer head is of brass with a diameter of 3 cm and with a convex striking surface that has a radius of curvature of about 50 cm.

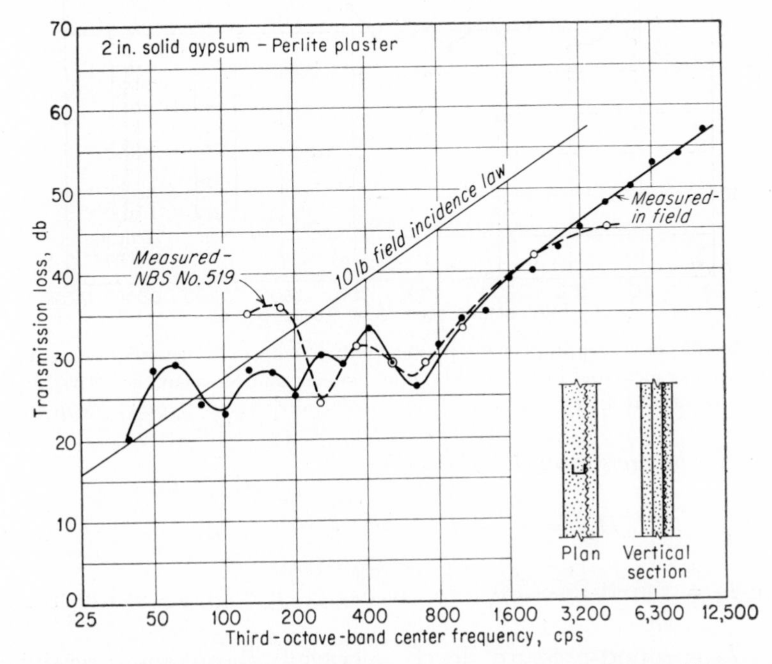

FIG. 13.34 TL of a gypsum-sand plaster wall, between two student residence rooms, measured in the field by Bolt Beranek and Newman Inc. NBS laboratory data are shown for comparison. (*Courtesy of United States Gypsum Company.*)

In testing a floor in the field for air-borne sound transmission, the quantity determined is the sound insulation in decibels—that is to say, the difference in noise levels in two rooms, in one of which there is sound. In determining the effectiveness of a floor for impact-noise transmission, actual noise radiated into the room below is measured.

In the case of the British and Dutch data, all measured results are corrected to a receiving room reverberation of 0.5 sec using the formulas:[2]

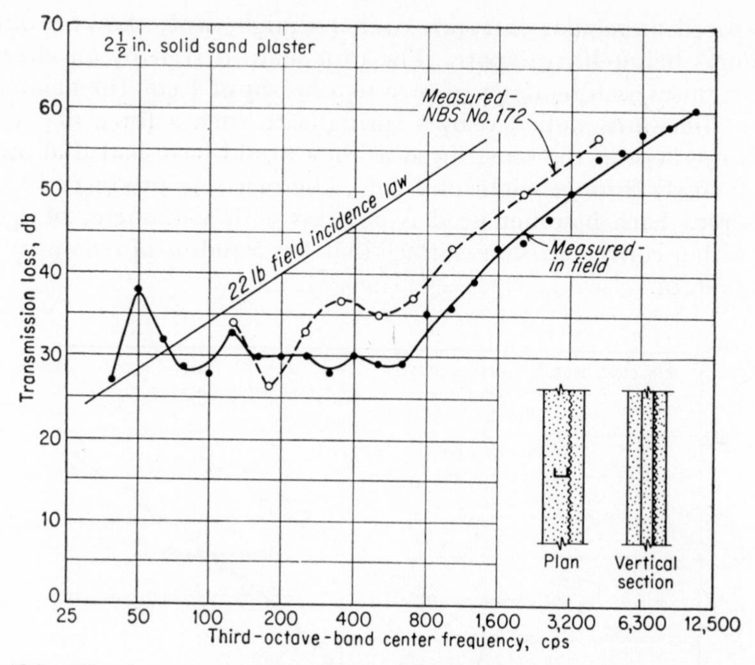

FIG. 13.35 TL of a gypsum-perlite plaster (mill-mixed) wall, between two student residence rooms measured in the field by Bolt Beranek and Newman Inc. NBS laboratory data are shown for comparison. (*Courtesy of United States Gypsum Company.*)

*For Air-borne-sound Insulation**

$$D_{0.5} = L_1 - L_2 + 10 \log \frac{T}{0.5} \qquad \text{db} \tag{13.50}$$

where L_1 = sound-pressure level averaged throughout source room, db

L_2 = sound-pressure level averaged throughout receiving room, db

T = reverberation time in receiving room, sec

The values of $D_{0.5}$ differ only slightly from the TL defined in Sec. 13.2.

*For Impact-sound Transmission**

$$L_{0.5} = L_2 - 10 \log \frac{T}{0.5} \qquad \text{db} \tag{13.51}$$

* At a recent meeting of the International Standards Association, Technical Committee 43, it was decided to specify the D and L equations in terms of 10 m² of absorbing material in the receiving room rather than in terms of 0.5 sec.

where L_2 = average sound-pressure level measured in receiving room using standard tapping machine on floor above

T = reverberation time in receiving room, sec

1. *Holland Data.* In presenting the results of the Holland tests, all data are given in terms of a reference floor structure, designated as floor 1 in Fig. 13.46. This floor is essentially a 4.5-in.-thick concrete floor with plaster finish on top and lime plaster on the bottom.

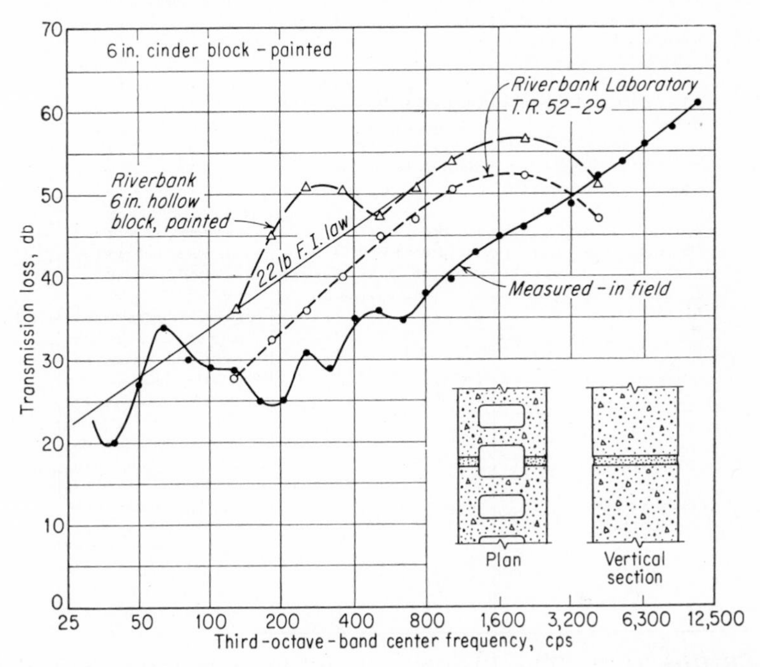

FIG. 13.36 TL of a painted cinder-block wall, between two student residence rooms, measured in the field by Bolt Beranek and Newman Inc. Riverbank Laboratories' data are shown for comparison. (*Courtesy of United States Gypsum Company.*)

The noise insulation and the impact noise for this floor are given in Figs. 13.47 and 13.48, respectively. The noise-insulation values shown are in agreement with measurements made in the laboratory at the National Bureau of Standards, except that below 480 cps the NBS tests yielded results about 4 db lower.

The behaviors of the other 10 Dutch structures are given in Figs. 13.49 through 13.52, and the details of their construction are given in Table 13.5. It is seen that the suspended ceilings give better insulation of both impact and air-borne sources of sound than do the other structures shown.

Table 13.5

Description of Holland Structures of Figs. 13.46 to 13.51

Structure No.	Floor	Floor Finish	Ceiling	Weight, lb_m/ft^2
1	4.3-in. reinforced concrete slab	0.8-in. hard plaster	0.4-in. plaster on soffit	50
2	2-in. reinforced concrete with 4.3-in. concrete joists	0.8-in. hard plaster	1-in. nailing strip; lath, reeds, and mortar	
3	5.1-in. Perfora hollow brick with reinforcement	0.8-in. hard plaster	0.4-in. plaster on soffit	40
4	4.3-in. reinforced concrete slab	0.8-in. hard plaster	Separate wooden ceiling joists on the walls; lath, reeds, and mortar	65
5	4.3-in. reinforced concrete slab	Magnesite tiles on layer of bitumen paper and Tree-tex	0.4-in. plaster on soffit	76
6	Prefabricated floor of 5.1-in. pumice concrete roof slabs with cassettes	Hard plaster		27
7	7-in. wood joists	Boarded	Lath, reeds, and mortar	12
8	7-in. wood joists	Boarded on battens, floating on rockwool sheet (not tied in at walls)	Lath, reeds, and mortar	13
9	7-in. wood joists	Boarded	Separate ceiling joists supported on the walls; lath, reeds, and mortar	13
10	7-in. wood joists	Boarded	Separate ceiling joists not bearing on masonry, suspended from the joists by metal hooks near the joist ends; lath, reeds, and mortar	13
11	8-in. wood joists	Boarded	1.2-in. Wood-fiber board (impregnated with cement binder); 2.4-in. sand fill (pugging); lath and plastered brick, wire-mesh finish	33

2. *British Data.* The British wall and floor data are particularly interesting because the measurements were carried out on a variety of structures in a large number of dwellings after construction according to laboratory specifications. Some of the data were privately supplied by Mr. P. H. Parkin of the Building Research Station[1] and other material was taken directly from *Digests* 88 and 89.

(In the discussion below, two grades of floor are named: *Top Grade* and *Lower Grade.* Top Grade is defined as suitable for all except very critical interapartment requirements. The lowest quality of floor believed acceptable in apartments is the Lower Grade. Seven examples of the Top Grade are shown in Fig. 13.53.)

Descriptions of the British structures are given in Table 13.6. In presenting the results of the British tests, all data are given in terms of a reference floor structure, designated as floor *A* in Table 13.6. This floor is reinforced concrete slab, weighing about 45 lbs/ft^2 with

Table 13.6

British Floors and Ceilings of Figs. 13.53 to 13.57

Structure	Floor	Floor Finish	Ceiling	Weight, lb_m/ft^2
A	Reinforced concrete and hollow pot slab	Hard plaster	0.5-in. plaster on soffit	Not less than 45
B	Concrete	Wood boards or ¼-in.-thick linoleum or cork tiles	Same as *A*	
*C**	Concrete with floating concrete screed	Any surface	Same as *A*	
*D**	Concrete with floating wood raft	Boarded	Same as *A*	
E	Concrete	Hard plaster	Suspended (see *G**)	
F	Concrete	Wood board	Suspended (see *G**)	
*G**	Concrete	Thick cork tiles or rubber on sponge-rubber underlay	Suspended	
H	Concrete with 2-in. lightweight concrete screed	Hard plaster	Same as *A*	
*J**	Concrete with 2-in. lightweight concrete screed	Thick cork tiles or rubber on sponge-rubber underlay	Same as *A*	
K	Reinforced concrete slab 6–8 in. thick	Hard plaster	Same as *A*	Not less than 75
*L**	Concrete	Thick cork tiles or rubber on sponge-rubber underlay	Same as *A*	Not less than 75
M†	Plain joist	Boarded	Plasterboard and single-coat plaster ceiling; no pugging	
*M*1‡	Same as *M*	Same as *M*	Same as *M*	
N†	Plain joist	Boarded	Plasterboard and single-coat plaster; 3 lb/ft^2 pugging	
*O*1‡	Plain joist	Boarded	Heavy lath and plaster; no pugging	
P†	Plain joist	Boarded	Lath and plaster; 17 lb/ft^2 pugging	
Q†	Floating (see *U** or *T*1*)	Boarded	Single-coat plaster plasterboard; 3 lb/ft^2 pugging	
R†	Floating (see *U** or *T*1*)	Boarded	Single-coat plaster plasterboard; 3 lb/ft^2 pugging	
S†	Floating (see *U** or *T*1*)	Boarded	Heavy lath and plaster; no pugging	
*S*1‡	Same as *S*	Same as *S*	Same as *S*	
*T*1*‡	Floating	Boarded	Lath and plaster; 3 lb/ft^2 pugging (sand fill)	
*U**†	Floating	Boarded	Lath and plaster; 17 lb/ft^2 pugging (sand fill)	

* These floors are illustrated in Fig. 13.53.
† Thin walls.
‡ Thick walls.

a hard finish on the floor. The noise insulation and impact noise for this floor are given in Figs. 13.54 and 13.55.

The behavior of the other 13 English structures are given in Figs. 13.56 and 13.57. In regard to the details of design of these floors and ceilings, the following material is extracted from *British Building Research Station Digest* 89.

13.16 Building Research Station—*Digest* 89 (Modified)

Floating Floors on Concrete. *Wood-raft Floating Floors (on Concrete).* The wood-raft floating floor consists simply of floor board-

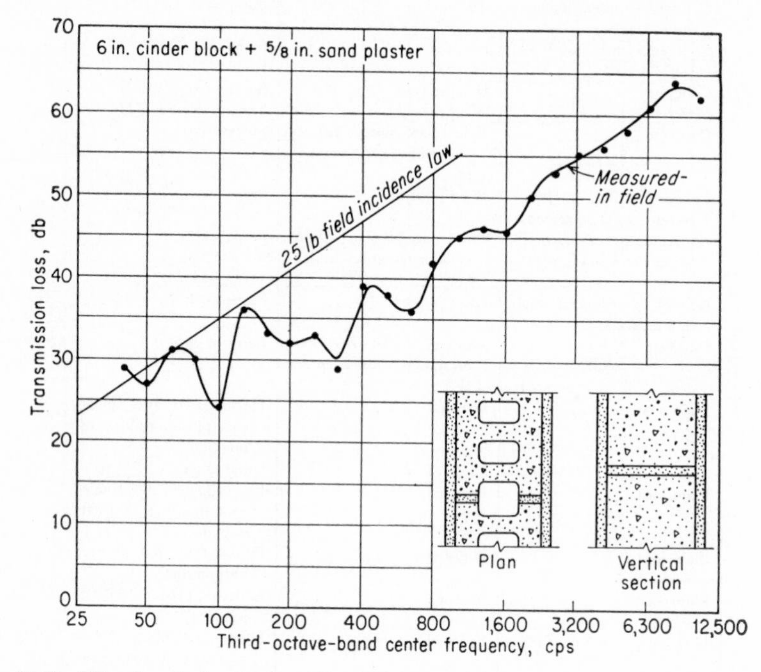

Fig. 13.37 TL of a cinder-block wall, with gypsum-sand plaster surfaces, between two student residence rooms, measured in the field by Bolt Beranek and Newman Inc. (*Courtesy of United States Gypsum Company.*)

ing nailed to battens to form a raft that rests on a resilient quilt laid over the structural floor slab. The battens must not be fixed in any way to the slab. For structural reasons the floor boards should preferably be tongued and grooved and not less than ⅞ in. thick. The battens, which are usually spaced at 16-in. centers, should be 1½ or 2 in. deep and not less than 2 in. wide. The resilient layer is a very important feature of the floating floor and only the most suitable materials should be used. Glass wool and mineral wool are in most common use for resilient layers; quilts of the long-fiber types have been found the most satisfactory. Other materials are sometimes suggested and are occasionally used, but at present none is known of comparable cost that appears to have such generally suitable properties as the glass-wool and mineral-wool quilts. A nominal quilt thickness of 1 in. at a density of 5 to 6 lb/ft^3 is recommended; it compresses to about $\frac{3}{8}$ in. under the battens of a wood-raft floor. Thicker quilts give better insulation but they may allow too much movement of the floating floor. Bitumen-bonded mats are the cheapest form of glass-wool or mineral-wool quilt and they are sometimes

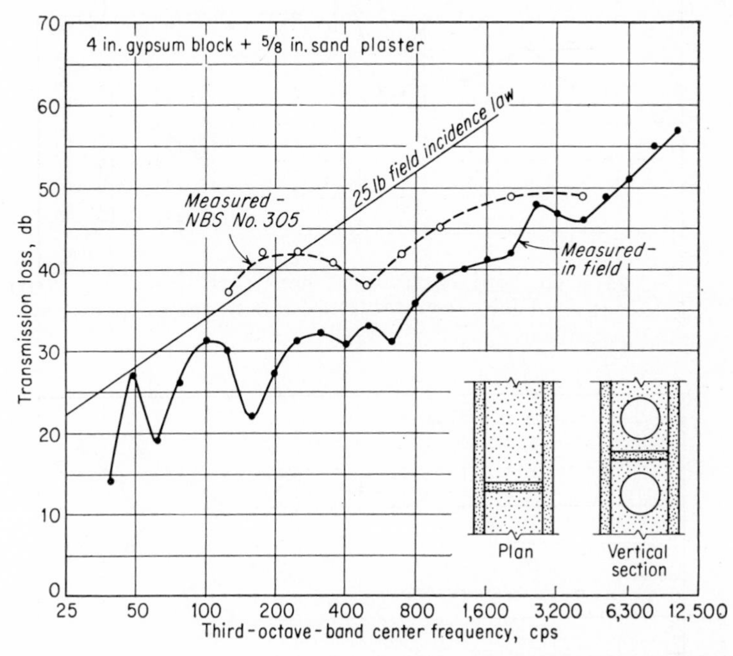

FIG. 13.38 TL of a gypsum-block wall, with gypsum-sand plaster surfaces, between two student residence rooms, measured in the field by Bolt Beranek and Newman Inc. NBS laboratory data are shown for comparison (*Courtesy of United States Gypsum Company.*)

used on this account, but paper-covered plain wool quilts are preferred because of their better performance for impact-sound insulation and because they are less susceptible to damage from the slight rocking movement of the raft battens when the floor is walked on. Most resilient quilts are obtained in rolls 3 ft wide. They should be laid with the edges closely butted but not overlapping. The quilt should normally be turned up a little against the surrounding walls as a means of separating the floating floor from the walls; this separation is important. The skirting should not be fixed to the floor but only to the walls (see Fig. 13.58).

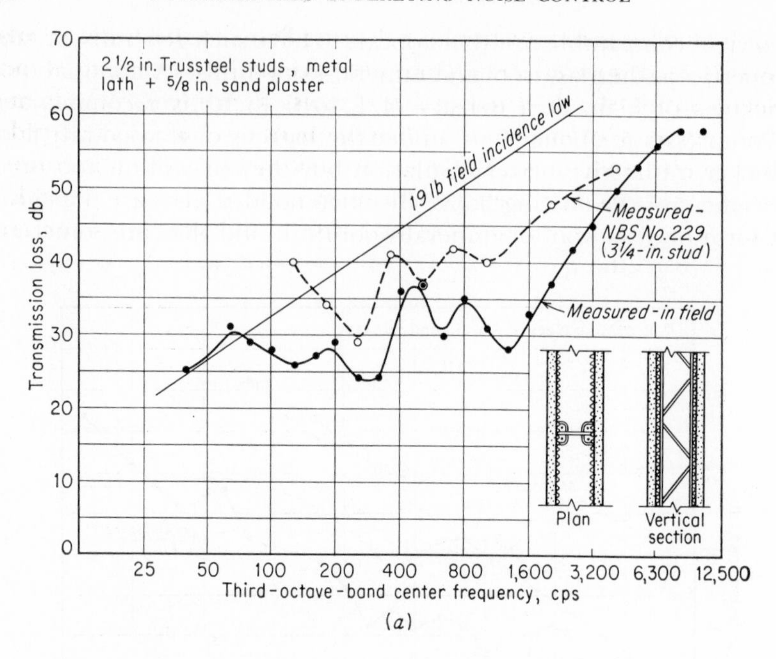

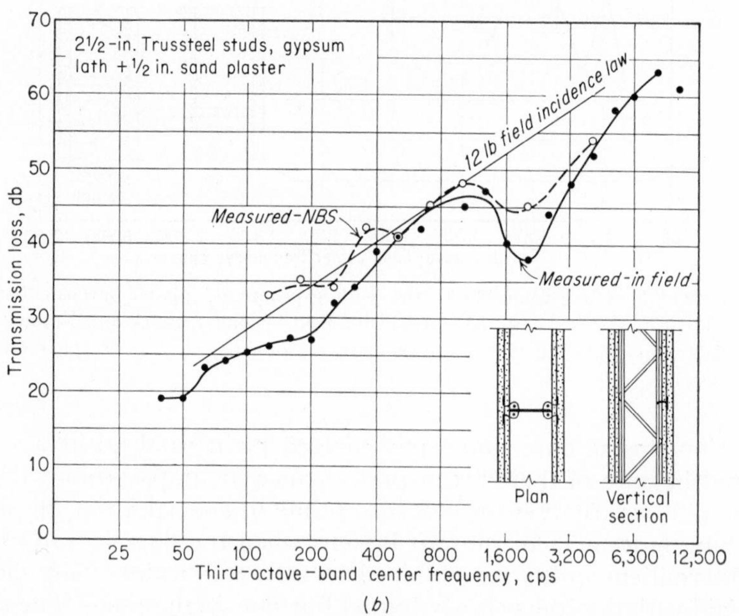

FIG. 13.39 TL of double-leaf gypsum-sand plaster walls on Trussteel (registered trademark) studs, between two student residence rooms, measured in the field by Bolt Beranek and Newman Inc.: (*a*) with metal lath; (*b*) with gypsum lath. NBS laboratory data are shown for comparison. (*Courtesy of United States Gypsum Company.*)

Soft wood is generally used for floating floors in dwellings; if hardwood is used, special precautions are necessary because hardwood is usually supplied kiln-dried to a low moisture content. If this moisture content is lower than it will subsequently be when the building is completed and in use, the wood will swell as it takes up moisture and the floating raft, being unrestrained, will buckle. Therefore, when using hardwood for floating rafts, it is desirable to construct the raft with the hardwood at a higher moisture content than that

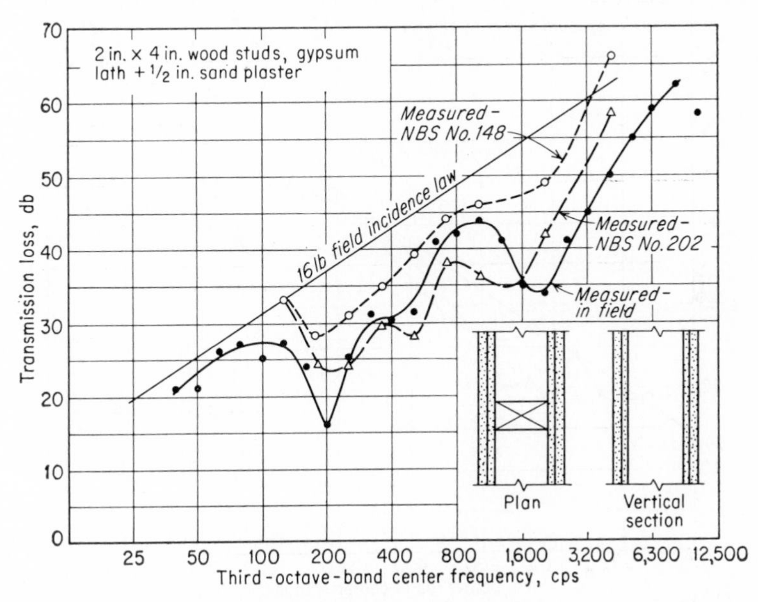

FIG. 13.40 TL of a double-leaf gypsum-sand plaster wall on wood studs, between two student residence rooms, measured in the field by Bolt Beranek and Newman Inc. NBS laboratory data are shown for comparison. (*Courtesy of United States Gypsum Company.*)

at which it is usually supplied, and, if possible, types of hardwood with a comparatively low moisture movement should be used.

Concrete-screed Floating Floors. This type of floating floor consists of a layer of concrete not less than 1½ in. thick (2 in. thick or more is better) resting on a resilient quilt laid over the structural floor slab and turned up against the surrounding walls at all edges. As with wood-raft floating floors the resilient quilt is usually of glass wool or mineral wool of 1 in. nominal thickness at a density of 5 to 6 lb/ft³. The quilt compresses to ⅜ to ½ in. under the load of a floating screed. Paper-covered quilts of plain wool give better insulation than bitumen-bonded mats; nevertheless, the latter have been extensively used under floating concrete screeds because they are cheaper, and in many instances they have proved satisfactory. It is necessary to provide a layer of waterproof building paper over the quilt to prevent wet concrete running through it. If the quilt is supplied with its own covering of waterproof paper, it may be sufficient to provide extra waterproof paper in narrow strips (say 6 in. wide) covering the joints in the quilting only. It is usual to

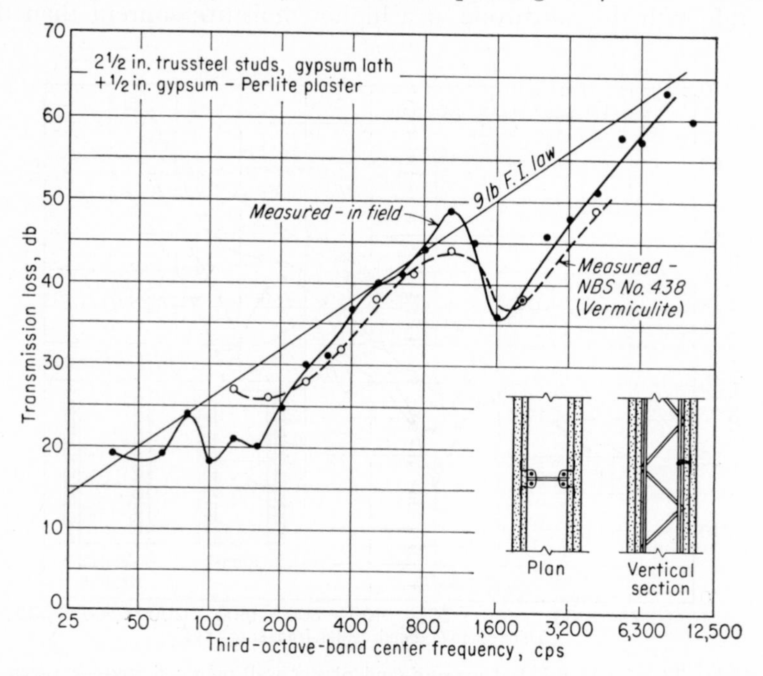

Fig. 13.41 TL of a double-leaf gypsum-perlite plaster (mill-mixed) wall on Trussteel (registered trademark) studs, between two student residence rooms, measured in the field by Bolt Beranek and Newman Inc. NBS laboratory data are shown for comparison. (*Courtesy of United States Gypsum Company.*)

provide wire-mesh reinforcement (e.g., ¾- to 2-in. mesh chicken wire) for the floating screed, and this is normally laid directly onto the waterproof paper. Apart from reducing the risk of cracking of the floating screed, the wire netting protects the building paper and the resilient quilt from mechanical damage during the operation of placing the concrete; any such damage might well result in the concrete making direct contact with the structural slab, which would spoil the insulation of the floating floor. A suitable mix for the concrete screed is 1:2:4, cement-sand-gravel aggregate, with an aggregate size of not more than ⅜ in.

Concrete-screed floating floors may not be suitable for rooms larger than about 150 ft² or 200 ft² at most, or with a greater dimension than 15 to 20 ft. This is because a slight "dishing" or curling up at the corners invariably occurs with floating floors, owing to quicker drying out at the top surface than at the bottom and the lack of bond with the sub floor. This curling is not noticeable in small rooms, but in rooms of larger dimensions than those given, the risk of cracking from drying shrinkage makes it necessary to subdivide

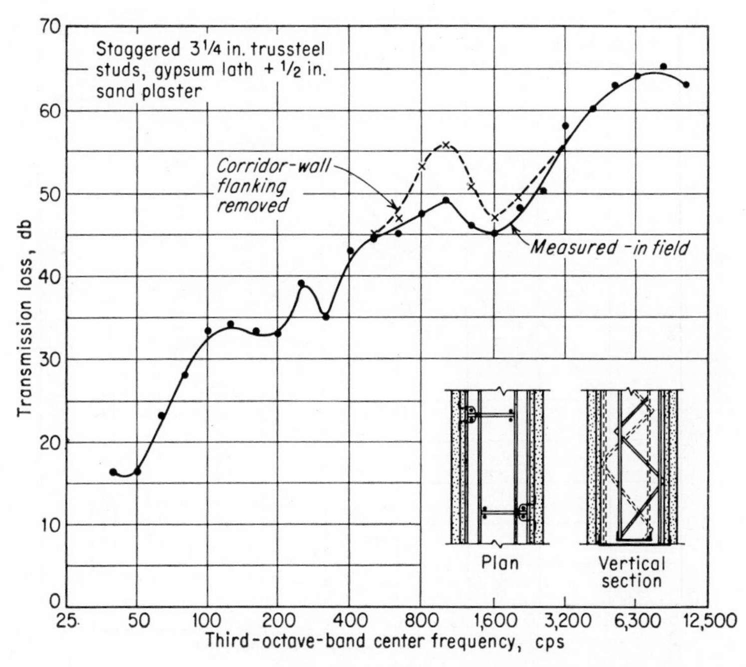

FIG. 13.42 TL of a double-leaf gypsum-sand plaster wall on separated Trussteel (registered trademark) studs, between two student residence rooms, measured in the field by Bolt Beranek and Newman Inc. (*Courtesy of United States Gypsum Company.*)

the floating screed into bays, and the curling of these separate bays can lead to difficulty with many types of floor finish.

Floor Finishes and Coverings. In present-day apartments (flats), hard floor finishes, such as thermoplastic tiles, Pitchmastic, etc., are commonly used. Finishes of this type make virtually no contribution to the sound insulation of a floor. Ordinarily, no floor finish adds anything to the airborne sound insulation, but the impact sound insulation can be considerably improved by means of a finish that is sufficiently soft or resilient. Thus it is possible to raise most floors up to Top Grade for *impact* insulation only simply by adding a finish or covering that is soft enough. Thick fitted carpet or underfelt, for instance, will nearly always give Top Grade impact insulation. If, therefore, a plain concrete floor that is heavy enough to have an air-borne sound insulation of Top Grade, say 6- to 7-in. reinforced concrete, is fitted with a carpet, an over-all insulation of Top Grade is achieved. Other floor finishes giving Top Grade impact insulation when applied to a plain concrete floor include rubber flooring on a sponge-rubber underlay or cork tiles 5/16 to 3/8 in. thick. A floor

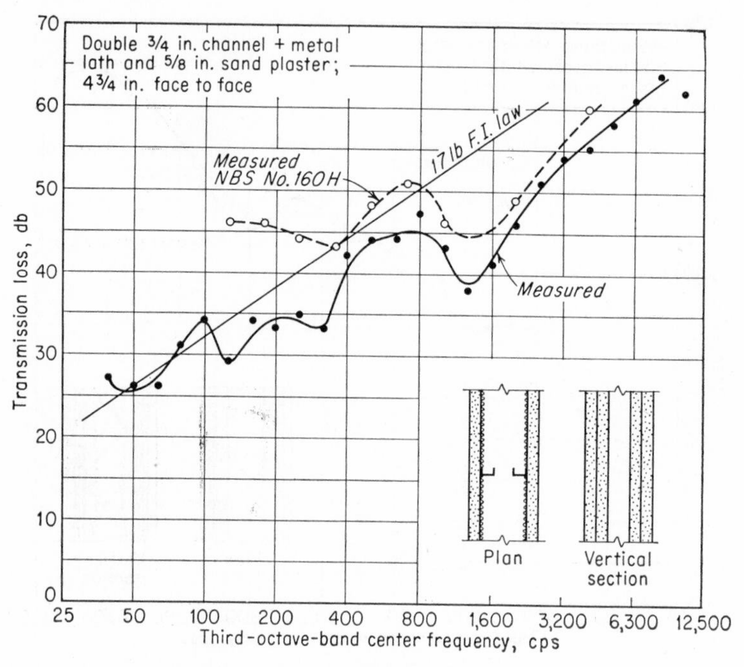

FIG. 13.43 TL of a double-leaf gypsum-sand plaster wall on separated steel channels, between two student residence rooms, measured in the field by Bolt Beranek and Newman Inc. NBS laboratory data are shown for comparison. (*Courtesy of United States Gypsum Company.*)

finish on concrete of 3/16- to 1/4-in. cork tiles or 1/4 in. linoleum or soft rubber would be expected to give *Lower Grade* impact insulation, so that any normal concrete floor with a floor finish of this type would have an over-all sound insulation of Lower Grade.

Partitions on Floating Floors. Partitions (apart from such short lengths as the sides of cupboards) should not be built on top of floating floors, which should be self-contained with each room. Building the partitions on floating floors may overload the resilient quilt and reduce its insulating properties; moreover the movement of the floating floor may cause cracking of the partitions.

Pipes and Conduits in Floating Floors. It is often necessary for services such as electric conduits, gas and water pipes, etc., to traverse a concrete floor. Whenever possible these pipes should be accommodated within the thickness of the floor slab or the leveling screed, but sometimes they have to be laid on top of the slab and contained within the depth of a floating floor. This is more likely to occur with a floating concrete screed. A floating wood raft requires a very

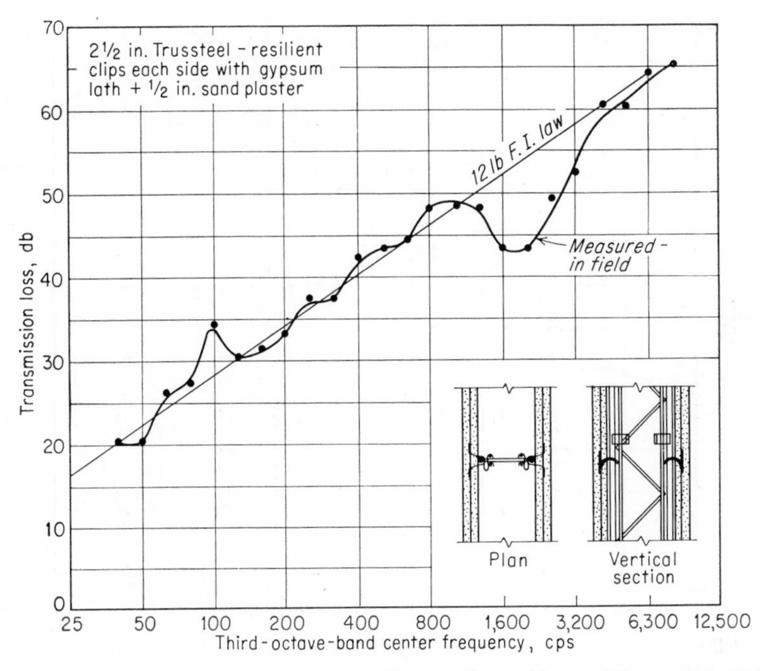

Fig. 13.44 TL of a gypsum-sand plaster wall on resilient clips on Trussteel (registered trademark) studs, between two student residence rooms, measured in the field by Bolt Beranek and Newman Inc. (*Courtesy of United States Gypsum Company.*)

level surface as a base and therefore, as a rule, calls for a leveling screed in which the pipes can be embedded. However, pipes or conduits need not cause trouble with a floating screed provided that (1) they do not extend more than about 1 in. above the base, (2) they are properly fixed so as not to move while the floating floor is being laid, and (3) they are haunched up with mortar on each side to give continuous support to the resilient quilt. When two pipes cross, one of them should be sunk into the base slab. The resilient quilt should be carried right over the pipes. If a wood-raft floating floor is being used and the pipes have to be laid above the slab, the pipes can, of course, be readily accommodated parallel to and between the raft battens, but in the other direction the battens will have to be notched over them; the battens must be thick enough to allow for notching.

Suspended Ceilings. Suspended ceilings are chiefly of benefit against air-borne sound and are comparable with lightweight screeds in that they can be used to raise the sound insulation of a normal concrete floor up to Top Grade, provided a soft floor finish is also

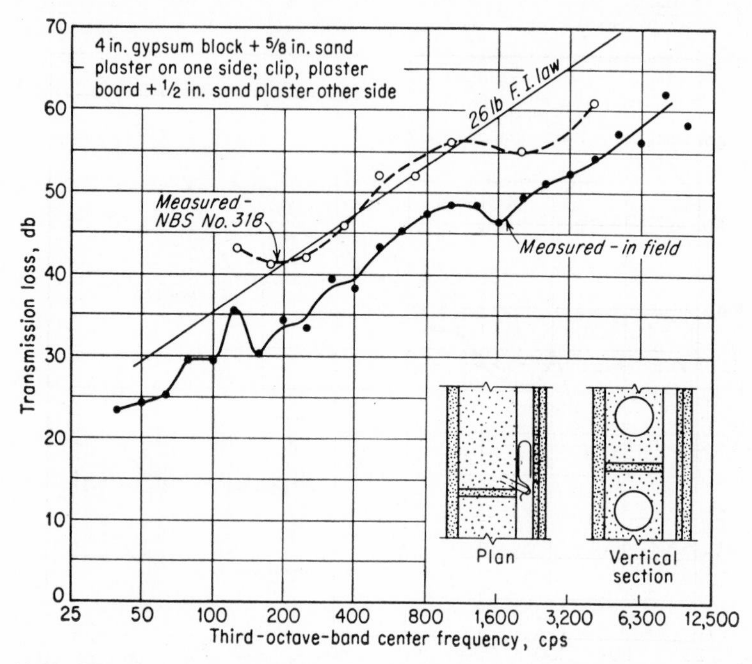

Fig. 13.45 TL of a gypsum-block wall with gypsum-sand plaster on one side and gypsum lath with gypsum-sand plaster on the other side, isolated from the block with resilient clips, between two student residence rooms, measured in the field by Bolt Beranek and Newman Inc. NBS laboratory data are shown for comparison. (*Courtesy of United States Gypsum Company.*)

used to give the necessary improvement in impact insulation. Not all the suspended ceilings that have been measured have given a satisfactory improvement of sound insulation, and the requirements for a successful system of construction are not all known precisely; but the following features would appear to be of importance:

1. The ceiling membrane should be moderately heavy, say not less than 5 lb/ft.^{2} If there is mineral or glass wool in the air space, ceilings as light as 2 lb/ft^2 may be used.
2. The membrane should not be too rigid.
3. The ceiling should be essentially airtight so as to eliminate

direct sound penetration via air paths, such as would occur with open-textured materials or with open joints.

4. The points of suspension from the floor structure should be as few and as flexible as possible, preferably by using specially designed hangars.

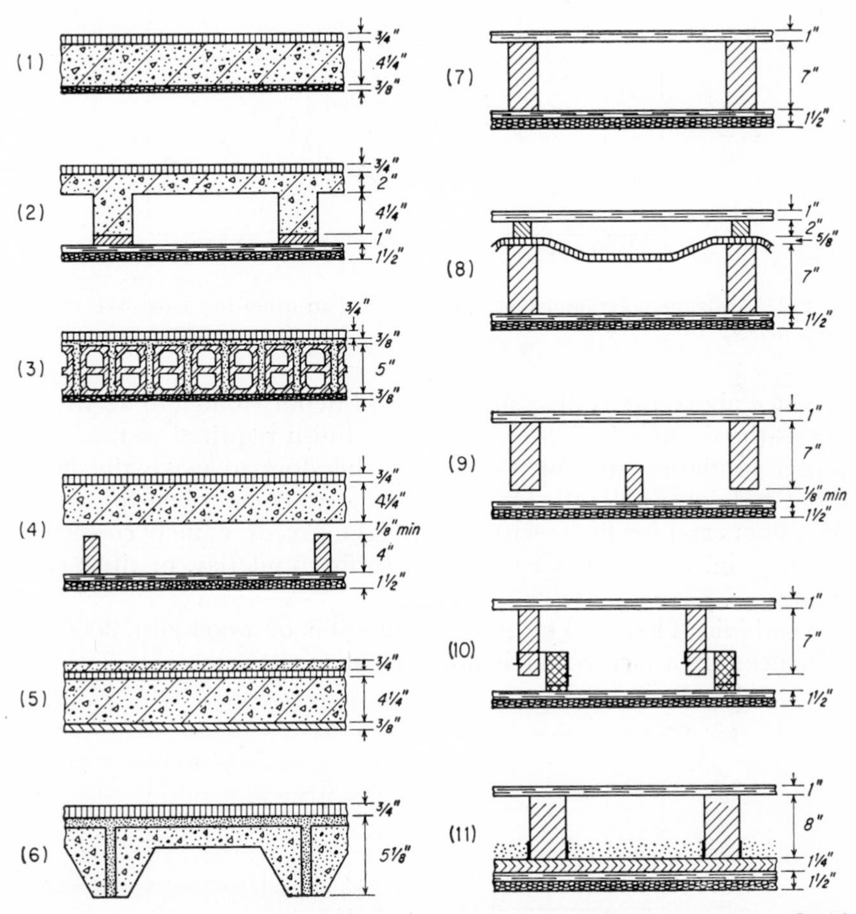

FIG. 13.46 Series of 11 floors measured in Rotterdam, Holland, apartments (flats).[3] Details are given in Table 13.5.

In spite of their usefulness for sound absorption, ceilings of soft insulating fibreboard alone are not recommended for sound insulation because of their lightweight and porous nature. Plastering on expanded metal or on ceiling boards, provided the total weight is not less than the 5 lb/ft^2 mentioned above and provided the whole membrane is supported by light hangers, is usually satisfactory. The

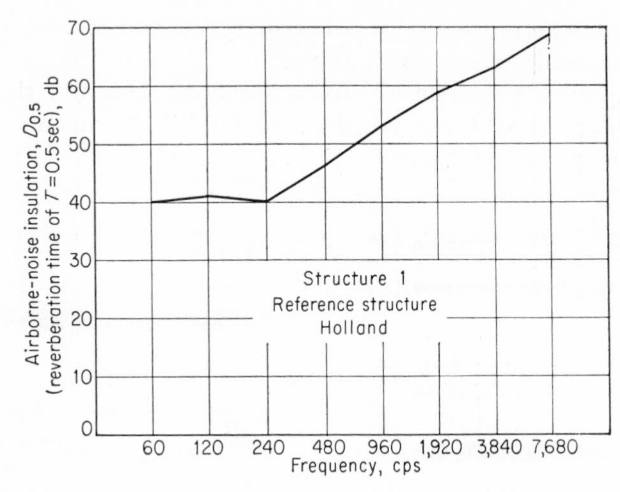

FIG. 13.47 Air-borne noise insulation $D_{0.5}$ of Holland structure No. 1 (see Fig. 13.46). (*From Ref.* 3.)

air space above the ceiling may range in depth from 1 to 12 in. or more—the deeper the better. The floor finish required to raise the impact insulation to Top Grade and therefore to make the floor Top Grade over-all is the same as those already described for heavy plain floors and for floors with lightweight screeds, namely cork tiles $\frac{5}{16}$ to $\frac{3}{8}$ in. thick, rubber or sponge-rubber underlay, or thick carpeting.

Wood-joist Floors. The sound insulation of wood-joist floors is influenced by a factor that is not present to the same extent when

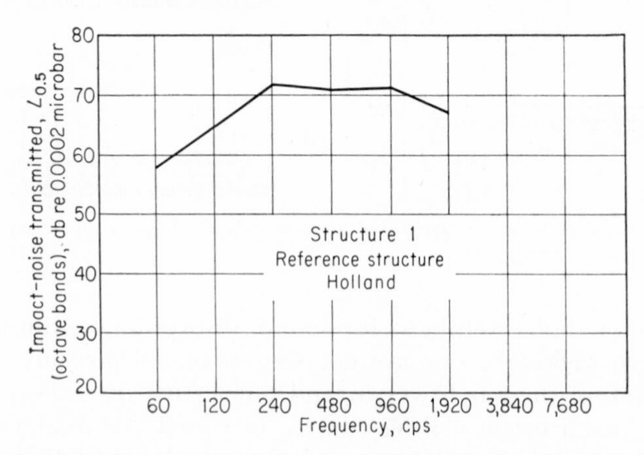

FIG. 13.48 Impact noise transmitted, $L_{0.5}$, for Holland structure No. 1 (see Fig. 13.46). (*From Ref.* 3.)

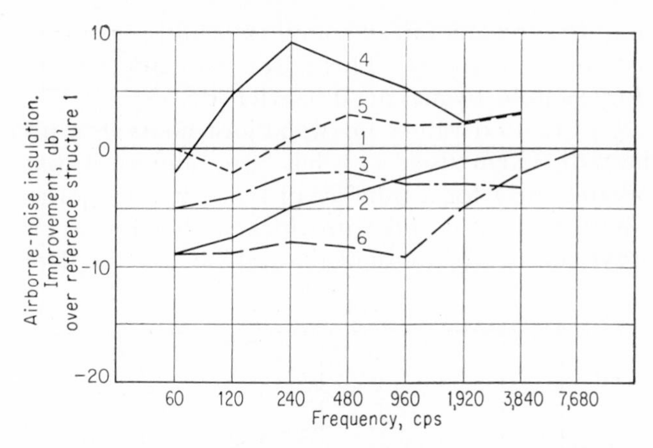

Fig. 13.49 Relative air-borne noise behavior of five concrete structures of Fig. 13.46 relative to structure No. 1. Positive numbers mean improvement in TL. (*Holland field studies.*[3])

concrete floors are used, namely, a variation of the amount of indirect or flanking sound transmitted via the walls. Flanking transmission always has some effect on the over-all insulation, but if the sound energy passing up or down the walls is greater than the energy passing through the floor, then the walls and not the floor will actually control the sound insulation between the rooms. Once the floor has been given enough insulation to ensure that the amount of sound transmitted through it is no more than that going down the walls, then further treatment of the floor will be of little value for sound

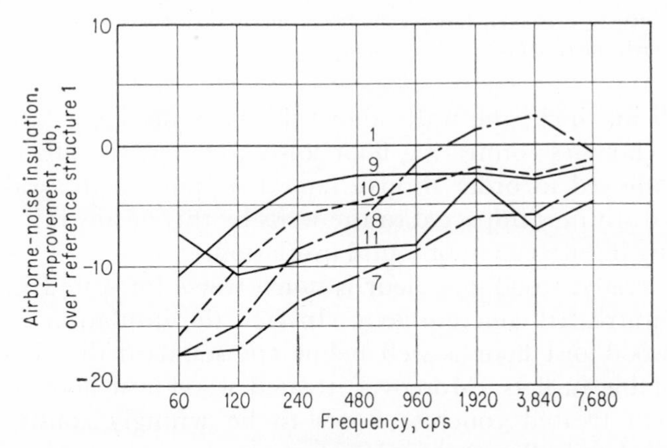

Fig. 13.50 Relative air-borne noise behavior of five wood structures of Fig. 13.46 relative to structure No. 1. Positive numbers mean improvement in TL. (*Holland field studies.*[3])

insulation unless the amount of sound transmitted via the walls can be reduced correspondingly. This means reducing the vibration of the wall, for example, by making it heavier (thicker).

In giving insulation values for wood-joist floors it is necessary to specify the wall system also. This has been done in Table 13.6 and Figs. 13.56 and 13.57 for a number of types of wood-joist floor construction both with thin and with thick walls. To be classed as a thick-wall system, two or more of the walls below the floor must be

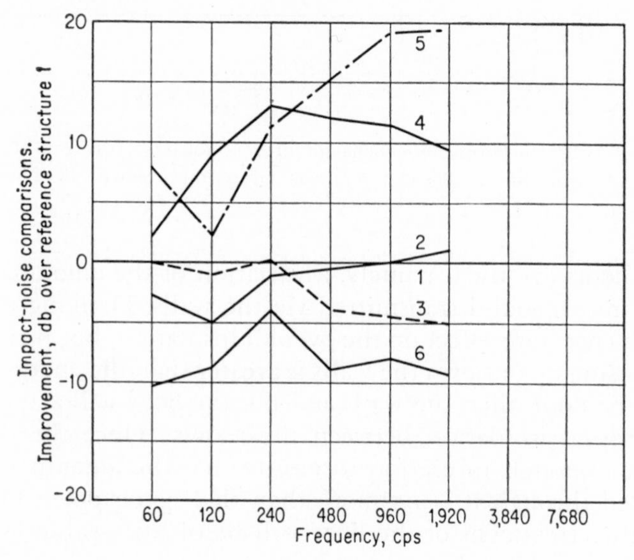

Fig. 13.51 Relative impact-noise behavior of five concrete structures of Fig. 13.46 relative to structure No. 1. Positive numbers mean that less impact noise is transmitted through floor. (*Holland field studies.*[3])

at least 9 in. thick; the walls above the floor need not be as thick. Metal anchorages connecting floor joists to external walls are sometimes employed in order to give lateral support to the walls. The additional stiffness impacted to the walls by this means is insufficient to give any improvement of sound insulation.

An untreated wood-joist floor is much worse for sound insulation than an untreated concrete floor. In fact, the insulation of the untreated wood-joist floor is well below any standard that is likely to be acceptable in flats. Moreover, treated wood-joist floors are more liable than treated concrete floors to be wrongly constructed in matters of detail. But there may, of course, be other reasons for preferring wood-joist floors to concrete floors in some apartments.

Wood-joist Floors with Thin Walls. When the walls are thin (i.e., 4 ½-in. brick or less) most wood-joist floors, even if designed for sound insulation, will fall short of Lower Grade by at least 2 db, because of transmission via the walls. In order to get more insulation than this the wood-joist floor must also be heavy enough to hold the walls effectively laterally—in the manner of a concrete floor. The only satisfactory method known at present is to provide a ceiling of expanded-metal lath and three-coat plaster, loaded directly with a

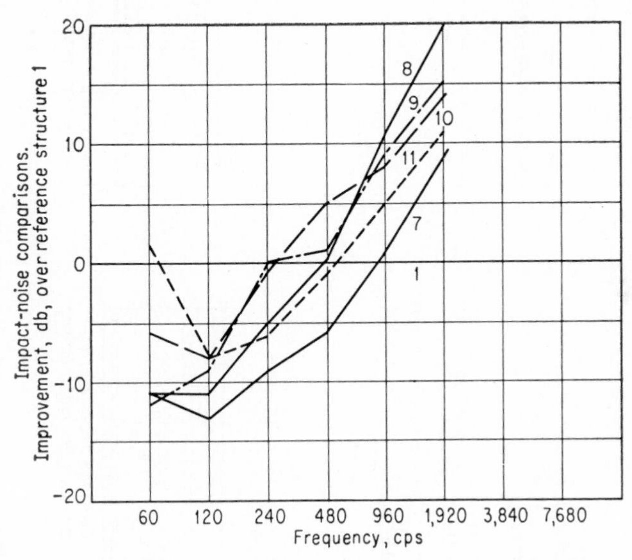

FIG. 13.52 Relative impact-noise behavior of five wood structures of Fig. 13.46 relative to structure No. 1. Positive numbers mean that less impact noise is transmitted through floor. (*Holland field studies.*[3])

pugging of 2 in. of dry sand (or other loose pugging material having at least the same weight, namely, 17 lb/ft²) together with a properly constructed floating floor. It is important that the pugging should be supported by the ceiling; if it is supported independently, the insulation will be less. The metal lathing must, of course, be securely fixed to the joists as it has to support the weight of the sand pugging as well as the plastering. The sand must not be omitted from the narrow spaces between the end joists and the walls. The sand should be as dry as possible when it is placed in the floor, and sand containing deliquescent salts must, of course, be avoided.

Wood-joist Floors with Thick Walls. When the walls are thick (i.e., at least two of the walls below the floor are not less than 9-in.

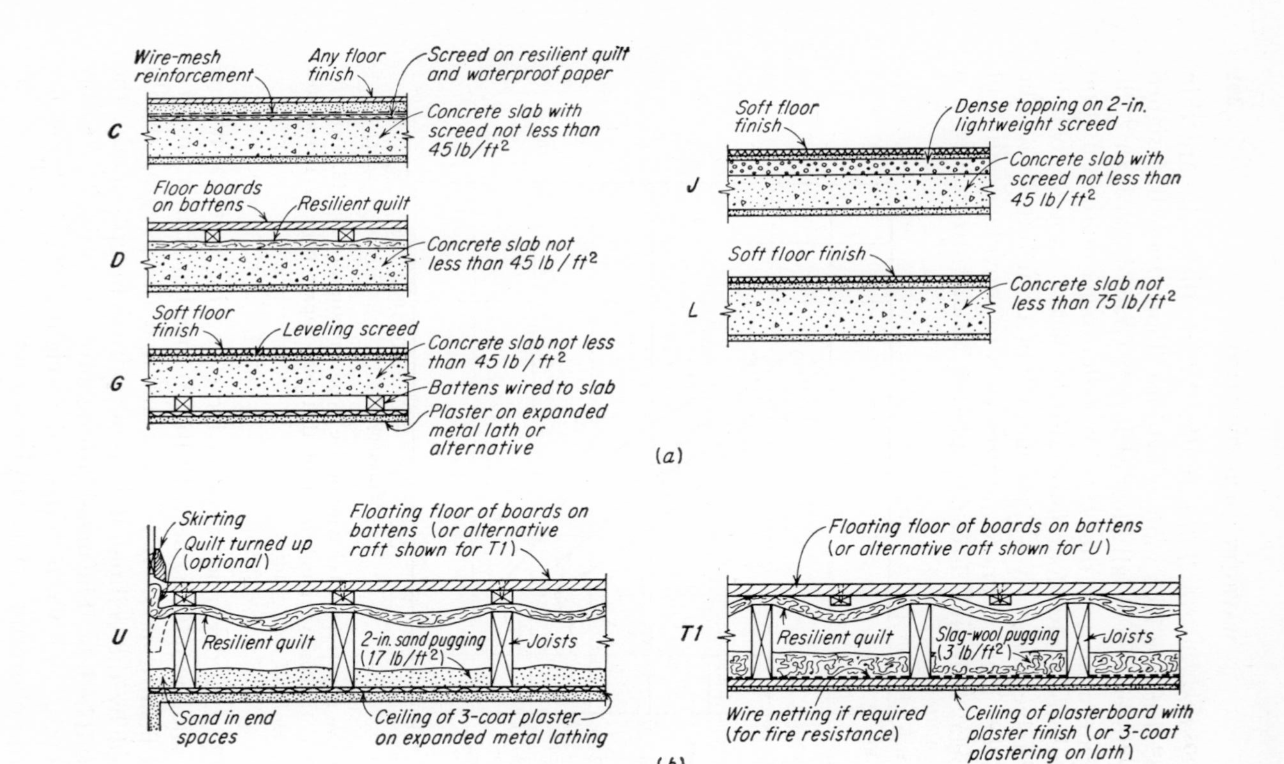

FIG. 13.53 Five Top Grade concrete floors and two Top Grade wood floors. "Top Grade" means "satisfactory for apartment houses (flats)" as distinguished from "Lower Grade," which means "minimum recommended for apartment houses." As alternatives, gypsum lath may be substituted for metal lath and laminated gypsum wallboard may be substituted for plaster on lath. (*From Ref.* 1.)

brick) it is possible to use a lighter form of wood-joist floor construction than the heavy one just described without falling below Lower Grade insulation—though the heavier construction is still to be preferred because with thick walls it can be relied on to give Top Grade insulation whereas the lighter floor cannot. In the lighter form of construction the floating floor remains an essential feature, but the pugging is reduced to a weight of not less than 3 lb/ft²; the ceiling supporting it can therefore be of plasterboard with a single-coat plaster finish. It is even more important than before that the pugging should be supported directly on the ceiling and not independently. Wire netting separately stapled to the joists is sometimes

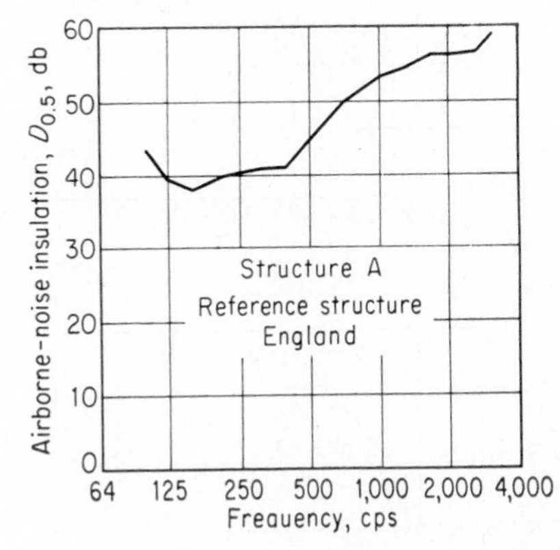

FIG. 13.54 Air-borne noise insulation $D_{0.5}$ of British structure *A* (about same construction as Holland structure No. 1; see Table 13.6).[1]

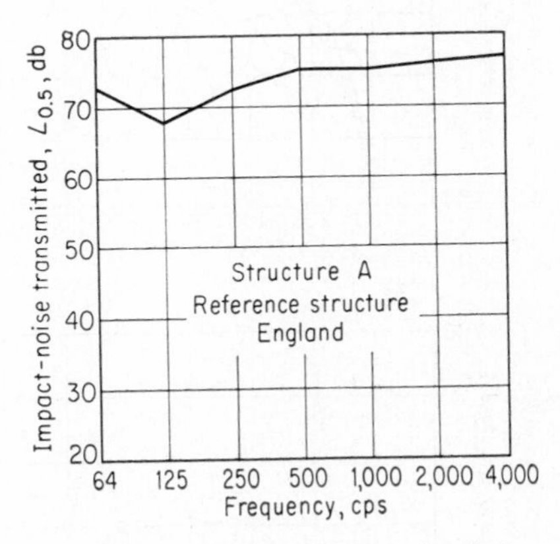

FIG. 13.55 Impact noise transmitted, $L_{0.5}$, for British structure *A*.[1]

inserted above the ceiling to retain the pugging in position in order to ensure that the floor attains a full half-hour fire resistance. This netting must not prevent the pugging from bearing fully on the ceiling. Alternatively, the same fire resistance can be achieved by increasing the plaster finish on the plasterboard ceiling to ½-in. thickness. The pugging material normally recommended is high-density slag wool (12 to 14 lb/ft³), 3-in. thickness being required to give the minimum weight of 3 lb/ft². Other pugging materials can be employed instead, provided they are of loose wool or granular type and the thickness used is sufficient to give the stipulated weight of 3 lb/ft²; very lightweight materials such as glass wool or exfoliated vermiculite are not suitable.

Floating Floors for Wood-joist Floors. The floating floor consists of floor boarding nailed to battens to form a raft, which rests on a resilient quilt draped over the joists. The raft must not be nailed down to the joists at any point and it must be isolated from the surrounding walls, either by turning up the resilient quilt at the edges (which is the better practice) or by leaving a gap round the edges to be covered by the skirting (see Fig. 13.58). The floor boards should not be less than 7/8 in. thick, preferably tongued and grooved. The

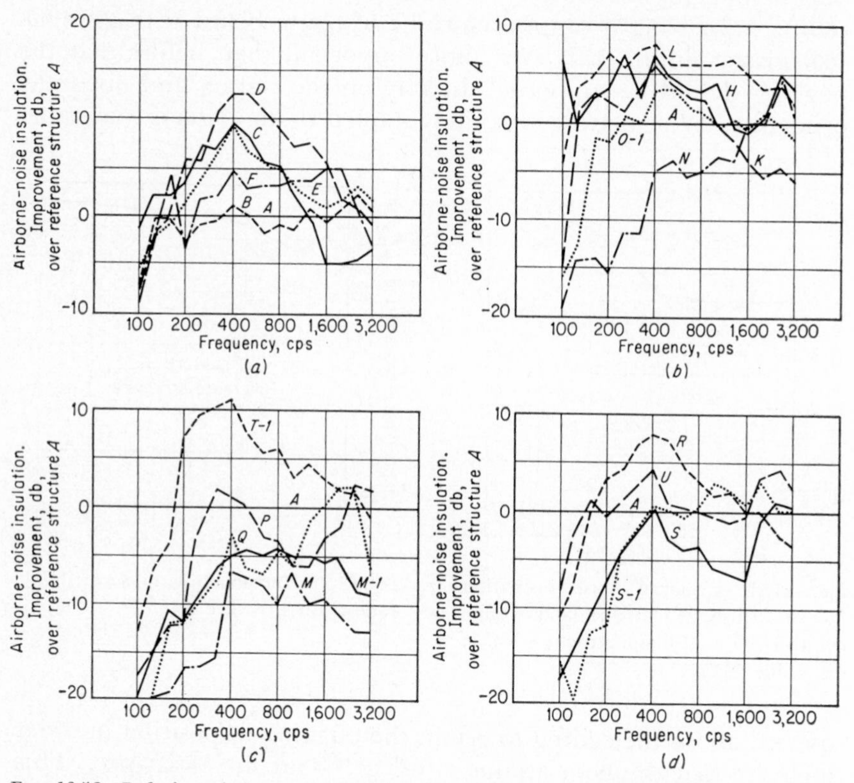

FIG. 13.56 Relative air-borne noise behavior of 19 structures of Table 13.6 relative to British structure *A*. Positive numbers mean improvement in TL. (*British field studies.*[1])

battens should be 2 in. wide and at least 1 in. deep, preferably 1½ or 2 in. deep. They should be parallel with the joists because battens that cross the joists provide too small a bearing area and overload the resilient quilt; the whole area of the top edges of the joists should share the loads transmitted from the floating raft.

There are two common methods of constructing the raft; one is to place the battens on the quilt along the top of each joist and to nail down the boards to the battens in the normal way, as shown in Fig. 13.53-*U*; the other method is to prefabricate the raft in separate

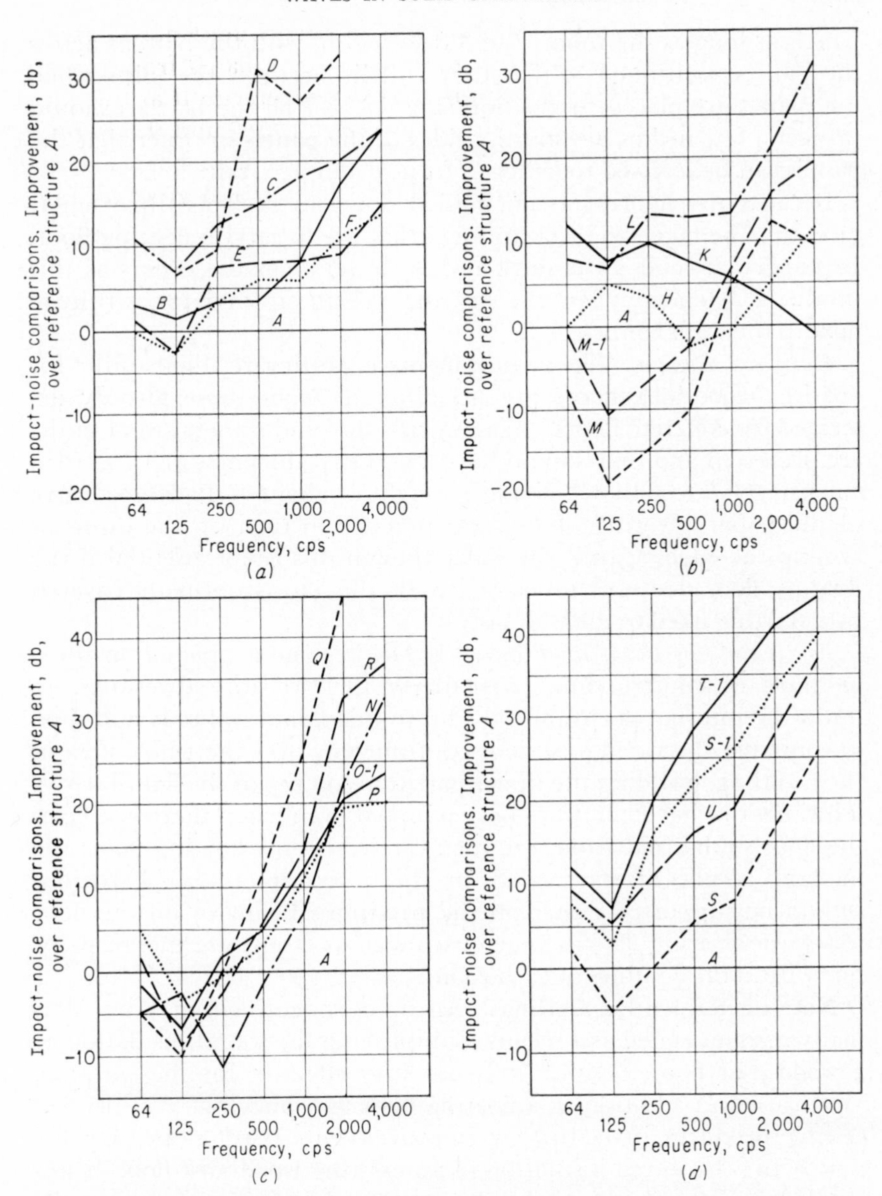

FIG. 13.57 Relative impact-noise behavior of 19 structures of Table 13.6 relative to British structure *A*. Positive numbers mean that less impact noise is transmitted. (*British field studies.*[1])

panels as long as the room, 2 to 3 ft in width, with the battens across the panels positioned so that they will lie between the joists when the panels are placed in position (Fig. 13.53-*T*1); the battens should project a few inches beyond the sides of the panels in order that the panels can be screwed together to form a complete raft.

Because the flooring is not nailed down to the joists, particular care must be taken to level up joists that are to carry a floating floor. In particular, end joists must not be lower than the others as this produces a tendency for the floating raft to tip and for furniture next to the walls to rock.

Resilient Layers. The requirements regarding resilient quilts for floating wood-joist floors are no different from those already described for concrete floors. Again, glass-wool and mineral-wool quilts are preferred and they should have a nominal thickness of 1 in., with a density of 5 to 6 lb/ft^3, giving a weight of about ½ lb/ft^2 exclusive of the paper covering. It is good practice to turn up the quilt all around the edges against the walls, though this is not essential if the floating floor does not touch the wall; the gap is normally covered by a skirting fixed to the wall only.

Partitions on Floating Floors. It has become a practice in some methods of construction, particularly for two-story dwellings, to build the internal partition walls up to first-floor level only and then to construct the wood-joist floor continuously over the whole area of the dwelling, building the upper partitions on top of the floor boards. This practice has sometimes been adopted even when there is a floating floor so that partitions are then carried on the floating floor; the method is not recommended in these circumstances. Partitions should be supported either on the partitions below or on the floor joists, the floating floor being constructed as a separate independent raft within the confines of each room.

Note on Suspended Ceilings. An independent ceiling is not normally recommended as a means of improving the sound insulation of a wood-joist floor, because it is not very effective for the purpose. When used in addition to a floating floor and pugging, a suspended ceiling produces little further improvement. Used alone, say for improving the sound insulation of an existing wood-joist floor, a suspended ceiling to be of much value needs to be so heavy that it might not be practicable to construct it. It is true that the air-borne insulation at high frequencies could be improved by a comparatively light suspended ceiling, but wood-joist floors are mainly deficient in sound insulation at the lower frequencies, and more weight is generally required to remedy this deficiency. Suspended ceilings are not of much benefit for impact insulation. If a floating floor is being built, it is usually a simple matter to pug the ceiling, and nothing worthwhile is then gained by adding a suspended ceiling. If, however, an existing floor cannot be disturbed and carpeting or some other soft floor covering is being added to give improved impact insulation, then the addition of a heavy suspended ceiling also may give a moderate improvement of the air-borne sound insulation; but such a floor is unlikely to approach Lower Grade for air-borne sound insulation unless the walls are thick.

Details at Edges of Floating Floors. In Fig. 13.58 constructional details at the edges of floating wood and concrete floors are shown.

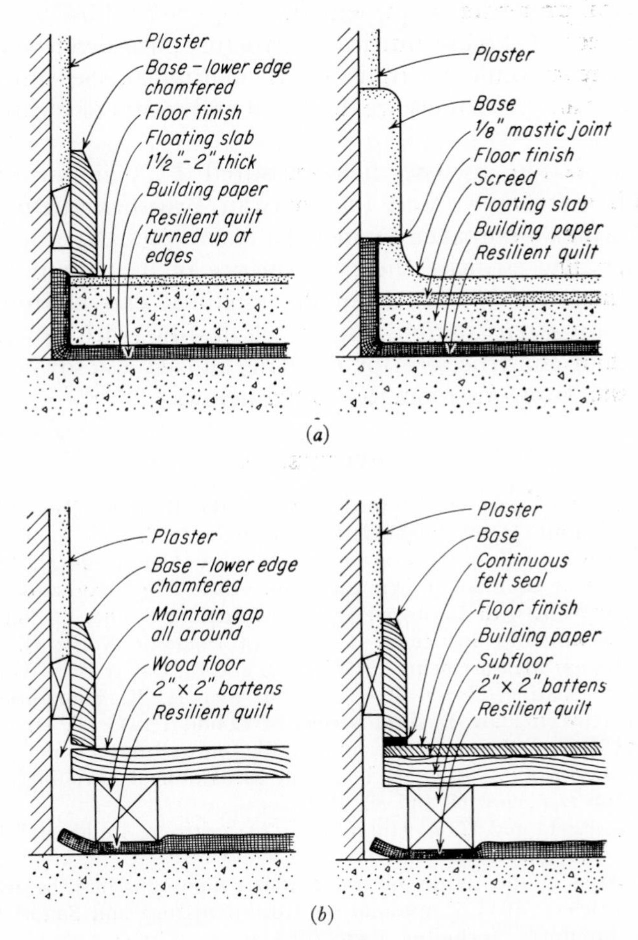

FIG. 13.58 Details of construction of floating floors at walls: (*a*) Top Grade concrete floors; (*b*) Top Grade wood floors. (*Courtesy of British Building Research Station.*)

13.17 Summary

This chapter covers a wide variety of subjects, starting with definitions of sound transmission loss and specific transmission impedance of a wall. A review is made of some properties of elastic waves in walls.

The TL of double and single walls is discussed in detail taking into account wave coincidence and angle of incidence effects. Critical and coincidence frequencies are defined and discussed. Some physical properties of common materials are given and practical design charts for single walls are presented. Laboratory procedures for obtaining the physical properties are presented.

The effects of discontinuities in structures are described qualitatively. Impact sound transmission is mentioned. Several different means for damping panels are discussed and charts of measured data given.

A large section discusses field-measured TL's of practical structures. These structures were in American, British, and Dutch buildings. The data were taken by modern techniques. By interpolation, the reader may estimate the performance of nearly any structure.

The chapter ends with a lengthy modified extract from *British Building Research Station Digest* 89 dealing with concrete and wooden floor constructions for reducing impact and air-borne noise transmission from one room to another below.

REFERENCES

1. Summaries of part of the work done at the Building Research Station, Garston, Watsford, Herts, England, from 1948 to 1956, have appeared in the following publications (articles arranged in chronological order): "1948 Symposium on Noise and Sound Transmission," including these subjects: W. A. Allen: Party Walls with Improved Sound Reduction; Studies of Sound Insulation by Discontinuous Structures; A Study of Domestic Noise; G. H. Aston: Sound Insulation Measurements on Windows and on Cavity Brick Walls; H. R. Humphreys: Floating Floors; P. H. Parkin and H. R. Humphreys: Sound Insulation between Flats, Physical Society of London, 1949. P. H. Parkin and E. F. Stacey, Recent Research on Sound Insulation in Houses and Flats, *J. Roy. Inst. Brit. Architecture,* July, 1954. Sound Insulation of Dwellings, *Bldg. Research Station Dig.,* nos. 88 and 89, 1956.

2. Parkin, P. H., and H. R. Humphreys: "Acoustics of Buildings," Faber and Faber, Ltd., London, 1958.

3. Summaries of the results of two Dutch symposiums may be read in the following articles: "1952 Symposium on Noise Nuisance and Sound Insulation in House Building," including these subjects: C. Bitter: Inquiry about the Noise Nuisance in Flats; M. L. Kasteleyn and J. van den Eijk: The Former

and Present Method of Measuring the Airborne Sound Insulation in the Field; Sound Transmission through Windows; J. van den Eijk, M. L. Kasteleyn, C. Bitter, and A. H. M. Basart: The Reduction of Noise in Staircase Halls; *Ingénieur Utrecht,* nos. 33, 34, and 36, 1952. "1954 Symposium on Noise Nuisance and Sound Isolation," including these subjects: M. L. Kasteleyn: Design of Floors and Sound Insulation; C. W. Kosten: Lightweight Double Walls Having a Large Transmission Loss; C. Bitter and P. van Weeren: Sound Insulation in Floors and Its Bearing upon the Problem of Sound Nuisance; W. P. van Leening: The Tentative Dutch Standard V-1070: Sound Insulation in Dwellings; *Ingénieur Utrecht,* nos. 38, 41, 44, and 47, 1954. C. Bitter and P. van Weeren: Sound Nuisance and Sound Insulation of Dwellings—I, *Research Inst. Public Health Eng. TNO Rept.* 24, The Hague, September, 1955.

4. Bolt Beranek and Newman Inc.: Physical Acoustics, vol. 1 and suppl. 1 to vol. 1 of "Handbook of Acoustic Noise Control," *WADC Tech. Rept.* 52–204, Wright-Patterson Air Force Base, April, 1955.

5. "1955–56 Symposium on Sound and Vibrations in Solid Bodies," Goettingen, Germany, April 19–22, 1955; papers appear in German, French, and English in *Acustica (Akust. Beih.),* vol. 6, p. 49, 1956.

6. Cook, R. K., and P. Chrzanowski: Transmission of Air-borne Noise through Walls and Floors, chap. 20 of C. M. Harris (ed.), "Handbook of Noise Control," McGraw-Hill Book Company, Inc., New York, 1957.

7. Cremer, L.: *Akust. Z.,* vol. 7, p. 81, 1942.

8. Kurtze, G., and B. G. Watters: New Wall Design for High Transmission Loss or High Damping, *J. Acoust. Soc. Am.,* vol. 31, pp. 739–748, June, 1959.

9. Watters, B. G.: The Transmission Loss of Some Masonry Walls, *J. Acoust. Soc. Am.,* pp. 898–911, July, 1959.

10. Dyer, I., and B. G. Watters: Excitation of Finite Plates by Variable Incidence Sound, *J. Acoust. Soc. Am.,* vol. 31, p. 840A, June, 1959.

11. Schoch, A.: "Die physikalischen und technischen Grundlagen der Schalldammung im Bauwesen," S. Hirzel Verlag, Leipzig, 1937.

12. London, A.: Transmission of Reverberant Sound through Single Walls, *J. Research Natl. Bur. Standards, Research Rept.* 1998, vol. 42, p. 605, 1949.

13. Beranek, L. L., and G. A. Work: Sound Transmission through Multiple Structures, *J. Acoust. Soc. Am.,* vol. 21, p. 419, 1949.

14. Cremer, L.: "Die Wissenschaftlichen Grundlagen der Raumakustik," S. Hirzel Verlag, Leipzig, 1950; The Propagation of Structure-borne Sound, German Research Rept. 1, Ser. B, undated, sponsored by *Gt. Brit. Dept. Sci. Ind. Research.* Calculation of Sound Propagation in Structures, *Acustica,* vol. 3, p. 317, 1953.

15. Schoch, A., and K. Fehér: The Mechanism of Sound Transmission through Single Leaf Partitions, *Acustica,* vol. 2, p. 289, 1952.

16. London, A.: Transmission of Reverberant Sound through Double Walls, *J. Acoust. Soc. Am.,* vol. 22, p. 270, 1950.

17. Davies, R. M.: Stress Waves in Solids, *Appl. Mechanics Revs.,* vol. 6, p. 1, 1953.

18. Schmidt, H.: Die Schallausbreitung in koernigen Substanzen, *Acustica,* vol. 4, p. 639, 1954.

19. Exner, M. L., W. Gueth, and F. Immer: Untersuchung des akustichen Verhaltens koerniger Substanzen bei Angregung zu Schubschwingungen, *Acustica,* vol. 4, p. 350, 1954.

20. Kurtze, G.: Koerperschalldaempfung durch koernige Medien, *Acustica* (*Akust. Beih.*), vol. 6, p. 154, 1956.

21. Junger, M. C.: Structure-borne Noise, review paper presented at "First Symposium on Naval Structural Engineering," sponsored by Office of Naval Research and Stanford University, August, 1958. To be published in the proceedings of that symposium.

22. Deresiewicz, H.: A Review of Some Recent Studies of the Mechanical Behavior of Granular Media, *Appl. Mechanics Revs.*, vol. 11, p. 259, 1958.

23. Morse, P. M.: "Vibration and Sound," 2d ed., McGraw-Hill Book Company, Inc., New York, 1948.

24. Pierce, G. W.: "Transmission and Reception of Sound Waves," *U.S. Patent No.* 2,063,945, filed Aug. 2, 1933.

25. Sanders, F. H.: *Can. J. Research*, vol. 1, 1939.

26. Cremer, L.: Insulation of Air-borne Sound by Rigid Partitions, and Insulation of Impact Sound, secs. 11.2 and 11.3 respectively of Bolt Beranek and Newman Inc., *WADC Tech. Rept.* 52–204, vol. 1, suppl. 1, Wright-Patterson Air Force Base, Ohio, 1955.

27. Feshbach, H.: Transmission Loss of Infinite Single Plates for Random Incidence, unpublished report prepared for Bolt Beranek and Newman Inc., 1953 (See Ref. 26).

28. Kuhl, W., and H. Kaiser: Absorption of Structure-borne Sound in Building Materials without and with Sand-filled Cavities, *Acustica,* vol. 2, p. 179, 1952.

29. Cremer, L.: Berechnung von Körperschallvorgängen, *Acustica,* vol. 6, no. 1, p. 59, 1956.

30. Kurtze, G., K. Tamm, and S. Vogel: Modellversuche zur Biegewellendämmung an Ecken, *Acustica,* vol. 5, p. 223, 1955.

31. Westphal, W., Zur Schallabstrahlung einer zu Biegeschwingungen Angeregten Wand, *Acustica,* vol. 4, pp. 603–610, 1954, and Ausbreitung von Koerperschall in Gebauden, *Acustica,* vol. 7, pp. 335–348, 1957.

32. Cremer, L.: Berechnung der Wirkung von Schallbruecken, *Acustica,* vol. 4, no. 1, p. 273, 1954.

33. Kurtze, G.: Noise Transmission by Boundary Layer Turbulence, *Quart. Progr. Rept.,* April–June, 1958; submitted to Office of Naval Research by Bolt Beranek and Newman Inc., July 30, 1958.

34. Goesele, K.: Radiation Behavior of Plates, *Acustica* (*Akust. Beih.,* no. 1), vol. 6, p. 94, 1956.

35. Moeller, F.: Sound Isolation with Structural Engineering, *Tek. Ukeblad,* vol. 93, no. 45, 1946.

36. Heckl, M.: Messungen an Schallbruecken zwischen Estrich und Rohdecke, *Acustica* (*Akust. Beih.*), vol. 6, p. 91, 1956; Untersuchungen über die Luftschalldämmung von Doppelwänden mit Schallbrücken (to be published in the proceedings of the 3d International Congress on Acoustics, Stuttgart, Germany, September, 1959).

37. Oberst, H.: Über die Dämpfung der Biegeschwingungen dünner Bleche durch fest haftende Beläge, *Acustica* (*Akust. Beih.,* vol. 4), vol. 2, p. 181, 1952; Werkstoffe mit extrem hoher innerer Dämpfung, *Acustica* (*Akust. Beih.,* no. 1), vol. 6, p. 144, 1956; Akustische Anwendung von Schaumstoffen, *Kunststoffe,* vol. 46, no. 5, p. 190, 1956.

38. Oberst, H., and G. W. Becker: Über die Dämpfung der Biegeschwingungen dünner Bleche durch fest haftende Beläge II, *Acustica* (*Akust. Beih.*, no. 1), vol. 4, p. 433, 1954.

39. Liénard, P.: Étude d'une Méthode de Mesure du Frottement Intérieur de Revêtements Plastiques Travaillant en Flexion, *Recherche aéronaut.*, vol. 20, p. 11, 1951. Les Mesures d'Amortissement dans les Matériaux Plastiques ou Fibreux, *Ann. Télécomm.*, vol. 12, no. 10, p. 359, 1957.

40. Kerwin, E. M.: Vibration Damping by a Constrained Damping Layer, Rept. 547, submitted to Convair, San Diego, Calif., by Bolt Beranek and Newman Inc., April 29, 1958. Vibration Damping by a Stiffened Damping Layer, Paper U-8, 55th Meeting of the Acoustical Society of America, May 10, 1958.

41. Ross, D., E. M. Kerwin, and I. Dyer: Flexural Vibration Damping of Multi-layer Plates, Rept. 565, prepared for the Office of Naval Research by Bolt Beranek and Newman Inc., June 26, 1958.

42. Beranek, L. L., "Acoustics," McGraw-Hill Book Company, Inc., New York (1954), pp. 324–331.

Chapter 14

SOUND TRANSMISSION THROUGH STRUCTURES CONTAINING POROUS MATERIALS

Leo L. Beranek

14.1 Introduction

Porous acoustical materials are used chiefly for controlling reverberation in rooms. That is to say, they are utilized most often for their *sound-absorbing* properties. In certain applications, they are also effective in the reduction of sound transmitted from one space to another. Such applications are generally those in which either (1) weight is a primary factor or (2) flexibility is desired or (3) control of noise at frequencies above, say, 500 cps is the essential requirement or (4) some combination of these factors must be satisfied. For example, for the control of noise in passenger aircraft, both low weight and the improvement of speech intelligibility (400 to 5,000 cps) are dominant factors in the choice of porous materials as sound barriers at the cabin sidewalls.

Surprisingly little has been written in recent times on the subject of sound transmission through structures containing porous materials. Beranek and Work[1] treat the transmission of sound through six varieties of aircraft structures, embodying two thicknesses of materials, four values of flow resistances, and three depths of air space within the structure. The studies are limited to cases of no flanking transmission, to particular weights of the elements of the structure, and to sound at normal incidence. Cook and Chrzanowski[2] publish TL's measured in the laboratory with random incidence sound for a few structures containing porous materials. This chapter includes selected information from these references.

Basic data for use in this chapter are given in Chap. 12. For example, flow-resistance data are to be found in Figs. 12.5 through

12.8. Measured attenuation constants and wavelengths for some common materials are found in Figs. 12.22 and 12.23.

14.2 Transmission Loss through a Single Porous Blanket

The problem that we consider first is shown in Fig. 14.1. We assume a sound wave to impinge on the left side of a porous blanket at normal incidence and to emerge with reduced amplitude from the right side. Let us specify the following parameters for the blanket (see Chap. 12 for units).

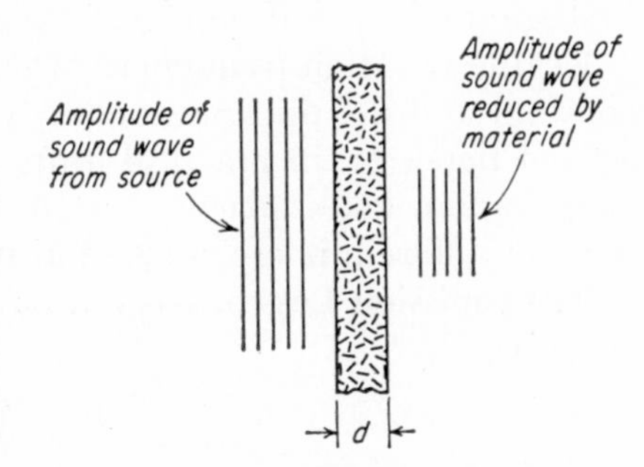

FIG. 14.1 Reduction of a plane sound wave as it passes at normal incidence through a porous acoustic blanket.

d = thickness of blanket
R_f = flow resistance of blanket of thickness d
R_1 = specific flow resistance = R_f/d
M_s = surface mass of blanket
w_s = surface weight of blanket, lb_m/ft^2
b = complex propagation constant = $\alpha + j\beta$ (see Chap. 12)
α = number of nepers sound wave is attenuated in traveling unit distance in very large sample of material (to convert from nepers to decibels, multiply former by 8.69)
β = number of radians that phase shifts when sound wave travels unit distance in very large sample of material (to convert from radians to degrees, multiply former by 57.3)
c = speed of sound in air (about 345 m/sec or 1,128 ft/sec at normal temperature and pressure conditions)
c_2 = speed of sound inside very large sample of material = ω/β
ω = $2\pi f$
λ_M = wavelength of sound inside blanket = $c_2/f = 2\pi/\beta$
ρc = characteristic impedance of air
= 410 mks rayls or 41 cgs rayls or 84 $\text{lb}_f\text{-sec/ft}^3$ for normal temperature and pressure
Z_0 = $|Z_0| \angle\theta$ = characteristic impedance of acoustic blanket (see Chap. 12) ($|Z_0|$ and θ are magnitude and phase angle of Z_0, respectively)

We shall divide the analysis of the problem of Fig. 14.1 into three regions of frequency as follows:

Region A. Low frequencies where the thickness of the blanket is less than 0.1 of the wavelength of sound within the blanket, that is, $d < 0.1\ \lambda_M$. In this region the two sides of the blanket move together, i.e., the blanket moves as a whole.

Region B. Transition region between regions A and C.

Region C. High frequencies where the product of the attenuation constant and the thickness of the blanket are greater than 1, that is, $\alpha d > 1$ neper or, in practical units, $\alpha d > 9$ db. In this region, wave propagation takes place, and at different distances through the blanket different phases and amplitudes occur.

Transmission Loss in Region A. The transmission loss for region A is given by*

$$\mathrm{TL} = 10 \log_{10} \left\{ 1 + \frac{\dfrac{R_f}{\rho c}\left(\dfrac{\omega M_s}{\rho c}\right)^2 \left(4 + \dfrac{R_f}{\rho c}\right)}{4\left[\left(\dfrac{R_f}{\rho c}\right)^2 + \left(\dfrac{\omega M_s}{\rho c}\right)^2\right]} \right\} \quad \text{db} \qquad (14.1)$$

This equation is plotted in Fig. 14.2. The lower abscissa is plotted

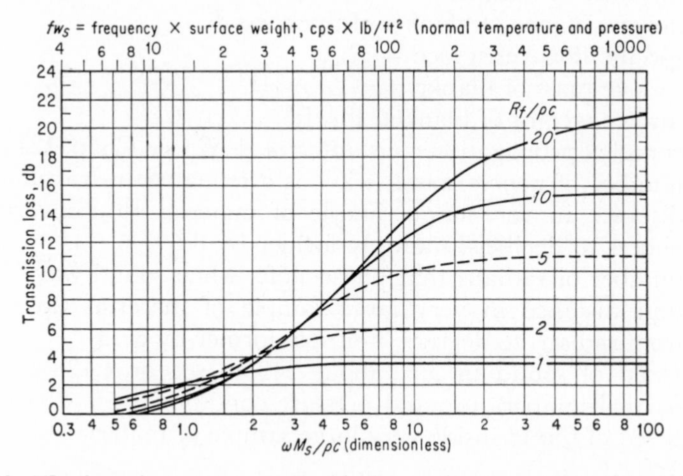

FIG. 14.2 TL through porous acoustic blanket under assumption that wavelength of sound in blanket is greater than ten times blanket thickness. Surface weight is measured in $\mathrm{lb}_m/\mathrm{ft}^2$.

in dimensionless units, while the upper abscissa gives the values in frequency (f) times surface weight ($\mathrm{lb}_m/\mathrm{ft}^2$),† assuming room tempera-

* This result is obtained from elementary electrical circuit theory by finding the decrease in voltage across a load of resistance ρc due to adding in series with it a parallel combination of inductance equal to M_s and resistance equal to R_f. The generator is assumed to have an internal resistance ρc.

† See Appendixes B and C for discussions of units.

ture and standard barometric pressure. The parameter $(R_f/\rho c)$ is dimensionless also.

Transmisson Loss in Region *C* $(\alpha d > 1)$. In region *C*, three different losses contribute to the over-all TL. The first contribution $(\mathrm{TL})_1$ is due to the attenuation in the blanket itself, given by

$$\mathrm{TL}_1 = 8.69\ \alpha d \tag{14.2}$$

Of course, α and d must be in the same system of units.

The second and third contributions $(\mathrm{TL})_2$ and $(\mathrm{TL})_3$ are alike. These contributions occur because part of the energy in a sound wave is reflected backwards when either the wave travels from air into the blanket or from the blanket into the air. A simple chart for converting from the magnitude of the specific characteristic impedance, $|Z_0|$, and the phase angle θ of the blanket to $(\mathrm{TL})_2$ or $(\mathrm{TL})_3$ is given by Fig. 14.3.

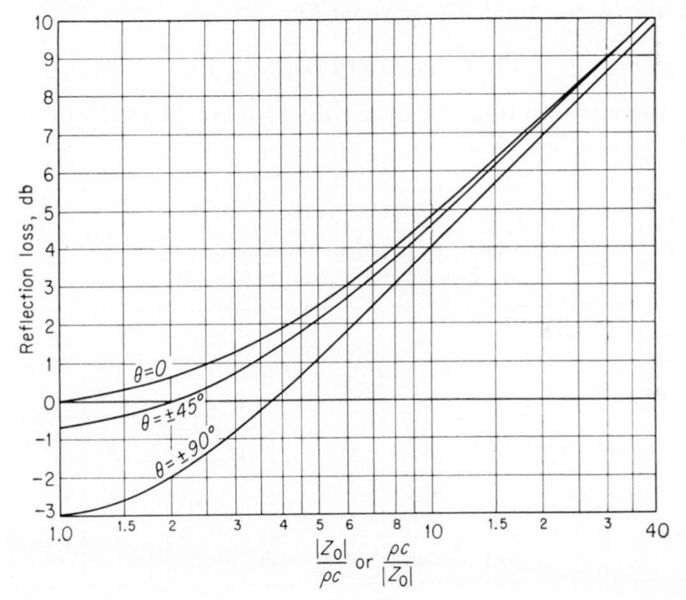

FIG. 14.3 Reflection loss at air-blanket or blanket-air interface. Note that either $|Z_0|/\rho c$ or the inverse may be used, whichever is greater than unity.

Transmission Loss in Region *B*. No detailed method for calculating the TL in Region *B* is given here. It is generally adequate to obtain TL in regions *A* and *C*, to plot the results, and to estimate the curve in the transition region *B*. Detailed calculations for many structures are given in Ref. 1 and Sec. 14.5.

Example 14.1. Determine the TL of a 2-in.-thick layer of Owens-Corning PF Fiberglas with a density of 10.5 lb_m/ft^3. Calculations of the propagation constants and characteristic impedances for this material were performed in Example 12.2.

Solution. Some of the data on the material are as follows:

$$R_f = 260 \text{ cgs rayls} = 2{,}600 \text{ mks rayls}$$
$$w_s = 1.75 \text{ lb}_m/\text{ft}^2$$
$$M_s = 8.52 \text{ kg/m}^2$$
$$d = 2 \text{ in.} = 0.0508 \text{ m}$$
$$\frac{R_f}{\rho c} = 6.35$$

The other data needed are given in Table 14.1. It is noted that region A includes only the points at 50 and 100 cps, because at 200 cps $\lambda_M/d \doteq 6$, which is a little less than the 10 desired. Region C includes all frequencies upward from 400 cps. At frequencies of 400 cps and above, the total TL is given by

$$\text{TL} = \alpha d + 2(\text{TL})_{2,3} \quad \text{db} \tag{14.3}$$

The results from column 11 of Table 14.1 are plotted as the upper curve in Fig. 14.4.

Table 14.1

Data and Computations for Example 14.1

(2-in.-thick Fiberglas sheet, 10.5 lb_m/ft^3)

f, cps	ω, radians/sec	αd,* db	β,* radians/m	λ_M, m	Region	Z_0,* mks rayls	$\theta°$	$(\text{TL})_{2,3}$, db, Fig. 14.3	$(\text{TL})_1$, db, αd	TL, db
(1)	(2)	(3)	(4)	(5)	(6)	(7)	(8)	(9)	(10)	(11)
50	314				A					10.0†
100	628			0.52	A					12.0†
200	1,257	7.2		0.365	B					12.0‡
400	2,510	10.0	25.0	0.25	C	1,420	−42.5	1.1	10	12.2
800	5,030	13.4	36.6	0.172	C	1,000	−40	0.4	13.4	14.2
1,600	10,050	17.8	57.0	0.110	C	735	−35.3	0.2	17.8	18.2
3,200	20,110	21.0	93.5	0.067	C	545	−27	0	21.0	21.0

* Values of α, β, and Z_0 are from Example 12.2.
† From Fig. 14.2.
‡ Interpolated from Fig. 14.4.

Oblique Angles of Incidence. When the sound wave impinges on the blanket at an angle other than normal to the surface, the wave must travel a greater distance inside the blanket, and the attenuation will be increased. By Snell's law (see Fig. 14.5)

$$\sin \phi_2 = \frac{c_2}{c} \sin \phi_1 \qquad (14.4)$$

where ϕ_1 = angle of incidence on blanket, relative to normal incidence
ϕ_2 = angle of travel inside blanket
c = speed of sound in air
c_2 = speed of sound in blanket = ω/β (see Chap. 12)

We observe, as an example, from Table 14.1, that c_2 at 1,600 cps $= 2\pi \times 1{,}600/57 = 176$ m/sec. So that $c_2/c = 176/344 = 0.51$.

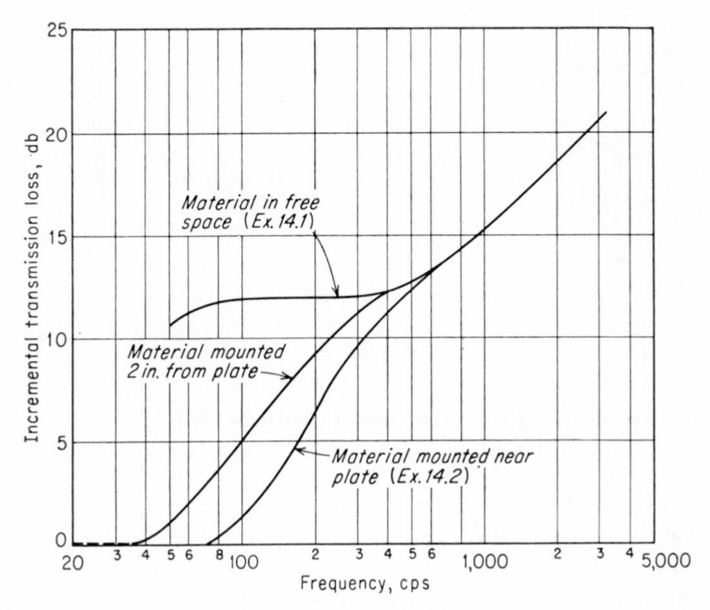

FIG. 14.4 TL of 2 in. of 10.5 lb_m/ft^3 Owens-Corning PF Fiberglas sheet, calculated as per Examples 14.1 and 14.2.

Now, as an example, let us assume that the wave impinges at an angle of $\phi_1 = 45°$. Then

$$\sin \phi_2 = 0.48 \sin 45° = 0.34$$

$$\phi_2 = 20°$$

We see that the angle ϕ_2 is substantially less than the angle ϕ_1 so that the distance that the wave travels in the blanket is not much greater than the blanket thickness. In this case, the travel distance d_2 is

$$d_2 = \frac{d}{\cos 20°}$$

$$= 1.06\, d$$

Generally, in practice, to take account of random incidence, the αd product (in nepers or decibels) is arbitrarily increased by about 10 per cent. However, the reflection losses $(\text{TL})_{2,3}$ remain nearly unchanged.

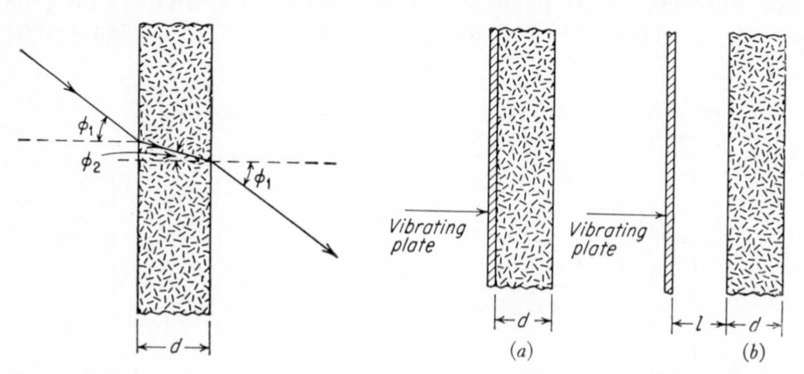

FIG. 14.5 Sound transmission through a blanket, showing angles ϕ_1 in air and ϕ_2 in blanket.

FIG. 14.6 Examples of blankets (*a*) adjacent to and (*b*) spaced from a vibrating plate.

14.3 Transmission Loss through a Porous Blanket Added to a Plate

Frequently, a porous blanket is added to a plate as shown in Fig. 14.6*a*. Because the motion of the plate is unchanged by mounting the blanket against it, the blanket is less effective in reducing the noise radiated than is predicted from Fig. 14.2. In fact, if the thickness of the blanket is less than one-tenth the wavelength, almost no noise reduction will be achieved.

Transmission Loss in Region *C* ($\alpha d > 9$ *db*). In region *C*, the blanket thickness is great enough that the wave propagated is attenuated in the blanket itself. The transmission loss in this region due to the addition of the blanket is given by the equation

$$\text{TL} = 8.69\, \alpha d + (\text{TL})_2 \tag{14.5}$$

The αd product is found in the same way as for Eq. 14.2. The reflection loss is applied only *once* for this structure and is determined from Fig. 14.3.

Transmission Loss in Region *A*. Assume that the TL in Region *A* is substantially zero below the frequency for which $d < \lambda_m/10$.

Transmission Loss in Region B. Starting from the frequency whose wavelength (in the blanket) is equal to 10 times the blanket thickness, gradually increase the TL from zero db up to the number of decibels computed for the point where $\alpha d = 1$ (that is, $\alpha d = 9$ db).

Transmission Loss for Spaced Blankets. In case that there is an air space between the blanket and the plate (Fig. 14.6*b*), the noise reduction is nearly 0 db up to the frequency where the *combined* thickness of material and depth of air space is one-tenth the average of the wavelengths in the air and blanket. Above this frequency, the attenuation is handled as for regions B and C above, if the material is reasonably porous.

Example 14.2. Assuming the blanket given in Example 14.1, calculate TL when this blanket is added to a vibrating plate as shown in Fig. 14.6*a*.

Solution. From Table 14. 1, we can estimate that at about 80 cps, the wavelength in the blanket will be about 10 times the blanket thickness of 5 cm. Below this frequency, the TL is nearly zero.

In region C, the transmission loss at 400 cps will be

$$\text{TL} = 10 + 1.1 = 11.1 \text{ db}$$

In between 80 and 400 cps, we estimate the increase in TL to be about as shown by the lowest curve of Fig. 14.4.

If the blanket is spaced out 2 in. from the plate, the attenuation will be about zero up to about 40 cps and will increase in a regular manner up to 400 cps. Above 400 cps, the TL will be about the same as for the blanket in free space.

As we stated in Sec. 14.2, for *random wave incidence* we should add about 10 per cent to the αd products.

In case the blanket has a rigid skeleton and is placed in direct contact with the vibrating plate, the TL will be substantially decreased due to the vibration of the skeleton. Either a flexible blanket or a flexible separating layer should be used to avoid direct excitation of the skeleton of the attenuating layer.

14.4 Transmission Loss through a Combined Plate, Porous Blanket, and Impervious Sheet

A very common form of problem in noise control is the need to reduce the sound radiated from a pipe or other free-standing object that cannot be enclosed by a wall. One procedure for handling this type of situation is to wrap the pipe or object with several inches of porous material and to cover the porous material with an impervious sheet of some type, say, asphalted paper, thin aluminum sheet, flexible impregnated glass-fiber mat, etc. The impervious sheet not only protects the porous blanket from the weather and from damage, but it also contributes materially to the noise reduction. See Fig. 14.7.

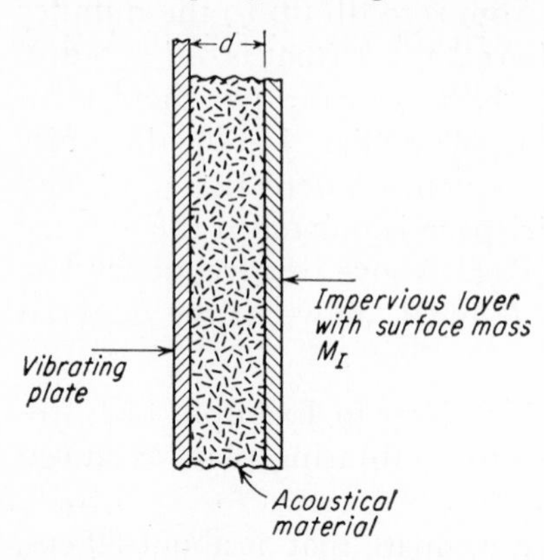

FIG. 14.7 Example of covered blanket adjacent to a vibrating plate.

At the outset, it must be emphasized that if the impervious layer is supported directly from the vibrating plate, the porous blanket in between will be bypassed by the wave transmitted through the supporting members. An example of flanking transmission around a layer of porous material is given in Sec. 14.7.

As was done previously in this chapter we shall treat the calculation of TL in three frequency regions. These regions are:

Region *A*. The frequency region in the vicinity of the basic resonance.

Region *B*. The transition frequency region between regions *A* and *C*.

Region *C*. High frequencies where the product of the attenuation constant and the thickness of the blanket are greater than 1, that is, $\alpha d > 1$ neper or $\alpha d > 9$ db.

Transmission Loss in Region *A*. The TL in the vicinity of the basic resonance is near-zero. The resonance frequency is computed from:*

$$f_{\text{res}} = \frac{1}{2\pi} \sqrt{\frac{2.8P_0}{M_I d \cos^2 \phi} k_u} \tag{14.6}$$

where M_I is the surface mass of the impervious layer and k_u equals 32.2 if P_0 equals 2.12×10^3 lb_f/ft^2, M_I is in pounds mass (lb_m) per square feet, and d is in feet. If M_I, d, and P_0 are in a consistent set of units, $k_u = 1$.

We see from Eq. (14.6) that the resonance frequency is lowest when $\phi = 0$. Let us call this frequency f_0. For $\phi = 45°$, $\cos^2 \phi = 0.5$, and $f_{\text{res}} = 1.4\, f_0$.

In region *A*, we estimate that the TL is 0 between f_0 and $1.5\, f_0$.

Transmission Loss in Region *C* ($\alpha d > 9$ *db*). In region *C* we may compute the transmission loss from the following formula:

* See Sec. 13.12.

$$\text{TL} = 9.6\alpha d + (\text{TL})_{\text{field}} \tag{14.7}$$

where $(\text{TL})_{\text{field}}$ is determined from Fig. 13.7 of Chap. 13. The first term of Eq. (14.7) gives the loss through the blanket (including the extra 10 per cent for field-incidence sound field). The second term gives the loss through the impervious layer. In case the sound transmission is at normal incidence ($\phi = 0$) then we substitute $(\text{TL})_0$ for $(\text{TL})_{\text{field}}$, where $(\text{TL})_0$ is determined from the upper curve of Fig. 13.7, and the constant in the first term of Eq. (14.7) becomes 8.7 instead of 9.6.

Transmission Loss in Region *B*. In region *B*, we estimate the TL graphically, fairing between the computed values for regions *A* and *C*.

Example 14.3. Assume the conditions of Example 14.2, namely, transmission through a 2-in.-thick layer of 10.5-lb_m/ft^3 PF Fiberglas layer added to a vibrating plate. Assume also that an impervious septum with a density of 1 lb_m/ft^2 is used to cover the exposed side of the layer. Compute the total TL.

The basic resonance frequency is given by the following:

$$f_0 = \frac{1}{2\pi}\sqrt{\frac{2.8 \times 2.12 \times 10^3 \times 32.2}{1 \times 0.167}}$$

$$= 196 \text{ cps (about 200 cps)}$$

If all parts of the plate vibrate in phase, this is the only resonance frequency. If various parts of the plate vibrate out of phase, we will assume a region of low TL that extends from about f_0 to 1.5 f_0, or 200 to 300 cps. Actually, there will be some loss due the blanket and septum which we have estimated at 2 db at 200 cps and 10 db at 300 cps.

The results are given in Table 14.2. The αd data are tabulated in column 3. The required $(\text{TL})_0$ or $(\text{TL})_{\text{field}}$ data are found from Fig. 13.7. The two right-hand columns in Table 14.2 were determined from the previous columns in Table 14.2.

The results are given in Fig. 14.8. Comparison with the results given in Fig. 14.4 reveals a substantial improvement due to the addition of the impervious septum to a blanket near a vibrating plate.

It must be emphasized that the methods given in this section make five important assumptions: (1) that the porous blanket is not rigid enough to carry the plate vibrations through its structure—or, if it is rigid, that it is not in direct contact with the vibrating plate; and (2) that the impervious layer is not tied to the vibrating plate nor is

Table 14.2

Data and Computations for Example 14.3
(2-in.-thick Fiberglas sheet, 10.5 lb_m/ft^2 covered by 1 lb_m/ft^2 impervious layer, all added to a vibrating plate)

f, cps	ω, radians/sec	αd, db*	$(TL)_0$, db	$(TL)_{field}$, db	TL, db ($\phi = 0$)	TL, db (field)
200	628	(Estimate)			(Estimate)	2
300	1,257	(Estimate)			(Estimate)	10
400	2,510	10	23.5	19.0	33.5	29.0
800	5,030	13.4	29.6	24.6	43.0	38.0
1,600	10,050	17.8	35.2	30.2	53.0	48.0
3,200	20,110	21.0	42.0	36.0	63.0	57.0

* The α values were taken from Example 12.2.

it shaken by the structure of the porous blanket; (3) that M_I is less than 0.3 of the mass of the plate to which the porous blanket and impervious sheet are added; (4) that the coincidence frequency of the impervious sheet is high; and (5) that the impervious sheet is highly damped so that it does not resonate.

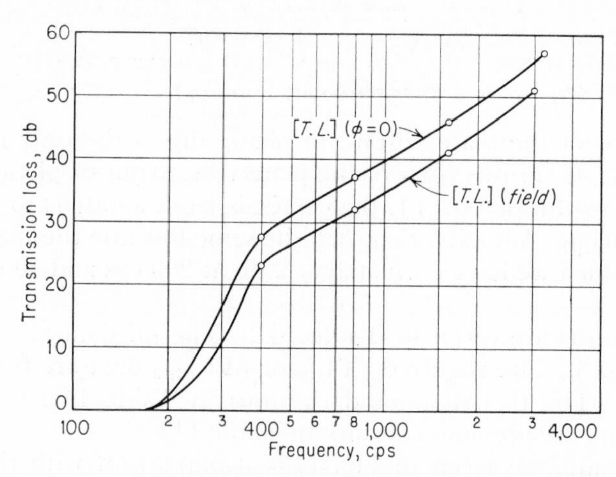

FIG. 14.8 TL of covered 2-in.-thick, 10.5-lb_m/ft^3 Owens-Corning PF Fiberglas sheet, calculated as per Example 14.3. Surface density of covering 1.0 lb_m/ft^2.

14.5 Calculated Transmission Loss through Certain Lightweight Structures with Porous Blankets

Beranek and Work[1] have presented calculated TL data on the six structures shown in Fig. 14.9, using a more exact theory than that given earlier in this chapter. The formulas were converted to charts with the aid of calculating machines. Transmission losses through a representative number of these structures were measured to verify the general results. Recent unpublished data obtained by an aircraft manufacturer (Convair, a division of the General Dynamics Corporation, San Diego, Calif.) also verify the general theory.

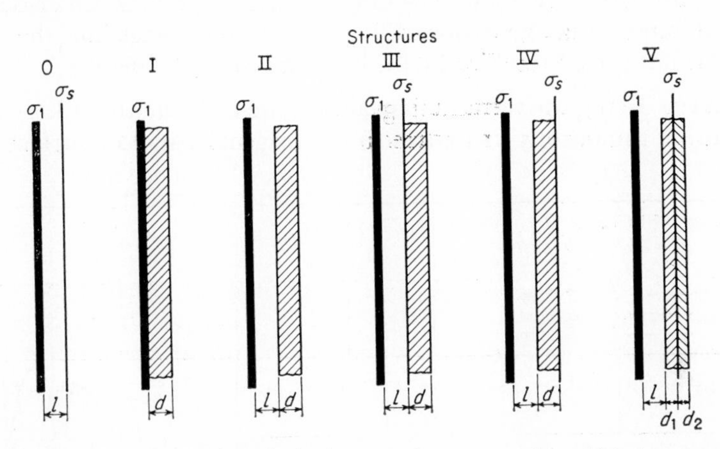

FIG. 14.9 Sketches of the six principal types of structures for which incremental attenuations are given in Figs. 14.10 through 14.15. The density of the septum σ_s was always 0.03 gm/cm²(0.06 lb_m/ft^2). For larger values of σ_s determine the further increase in attenuation from the formula 20 $\log_{10}$ (σ'_s/σ_s) db, where σ'_s is the larger surface density. The source of sound is on the left in each case.

The calculations were for normal incidence sound. However, the results of measurements using semi-random-incidence sound were in good agreement. Therefore, the curves given below can be interpreted as the minimum TL to be expected, *assuming no flanking sound transmission*. Actually when the added TL due to the porous material exceeds 20 to 30 db, the radiation through points of attachment, windows, and so forth, in any practical installation tends to be controlling, and the full value of the porous structure is usually not achieved.

Terminology. The terms used in the charts to follow (Figs. 14.10 through 14.15) are as follows:

d = thickness of acoustical material, cm (divide by about 2.5 to obtain in.)

l = depth of air space, cm

σ_x = surface density of impervious system that appears in structures III, IV, and V. [Only one septum mass was used except for structure IV. The surface density is given in gm/cm² (multiply by about 2 to get lb_m/ft^2). For other densities use the formula in the legend of Fig. 14.9 and see Fig. 14.14, bottom graph.]

R_1 = flow resistance of porous material, cgs rayls/cm (to convert to cgs rayls/in., multiply by 2.5)

ρ_m = density of porous blanket. [In those calculations ρ_m was always assumed to be about 0.04 gm/cm³ (2.5 lb_m/ft^3). Lighter-density blankets may give lower TL's at low frequencies, but the effect is not great; see Fig. 14.15, bottom graph.]

Charts. Using the terminology above, the *incremental* TL's caused by adding the porous structures to the original panel of surface den-

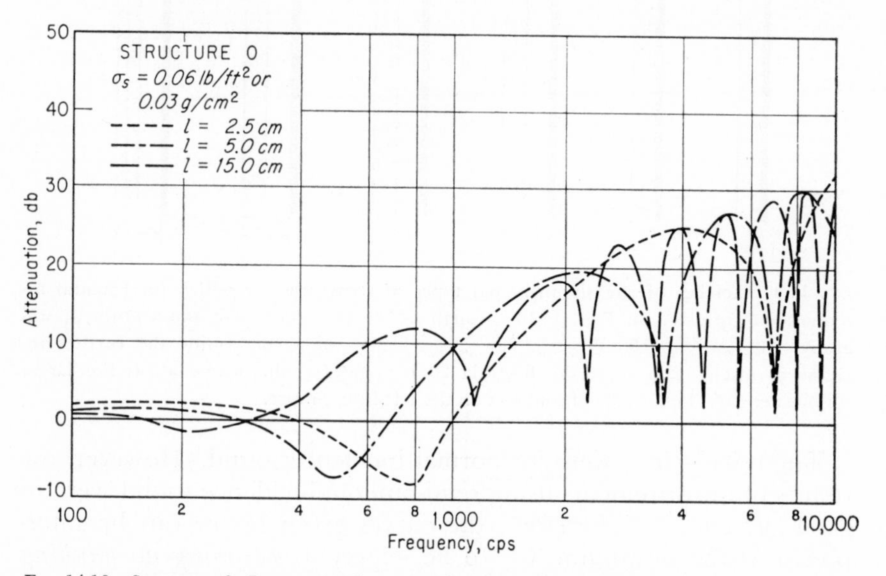

FIG. 14.10 Structure 0: Incremental attenuation in decibels due to adding the septum σ_s to the basic plate, σ_1. The secondary side is assumed to be terminated in the characteristic impedance of air.

sity σ are given in Figs. 14.10 through 14.15. To obtain the total TL for the structures as drawn in Fig. 14.9, add the TL for the panel of surface density σ_1 to the values given on the ordinates of the graphs. This TL is found with the aid of Fig. 13.7.

14.6 Comparison of Measured with Calculated Transmission Losses

The charts of Figs. 14.10 through 14.15 are shown by Beranek and Work[1] to check closely with measured data. A comparison with measured data taken by Convair[3] is made in Fig. 14.16. There, two "averaged" curves are presented, showing the change in TL measured on airplane structures obtained by adding 2 or 3 in. of Owens-Corning Fiberglas PF-105 glass-fiber blanket to a fuselage sidewall panel. The density of the porous material is 0.6 lb/ft³, and the material was spaced out ¾ in. from the panel.

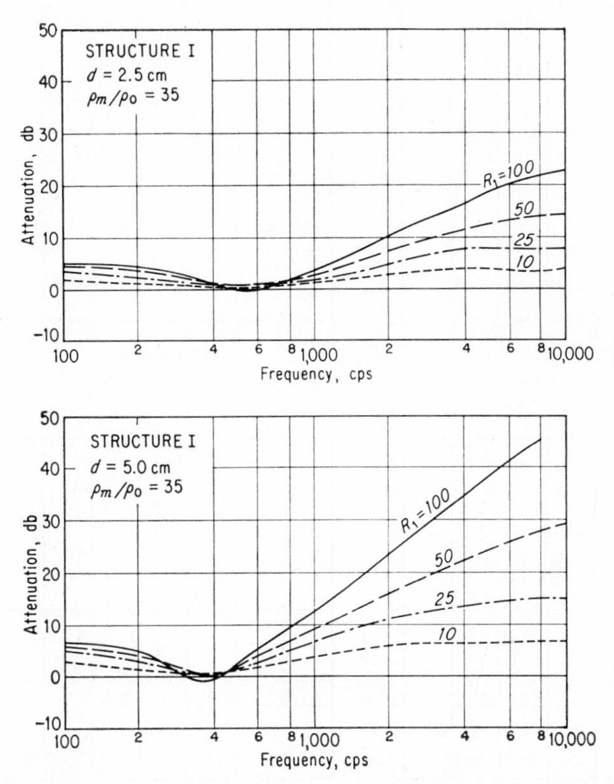

FIG. 14.11 Structure I: Incremental attenuation in decibels due to adding the absorbing structure to the basic plate, σ_1. The secondary side is assumed to be terminated in the characteristic impedance of air.

The x data are from Fig. 14.13, taking into account the specific flow resistance R_1 of the porous material (110 rayls/in.). The "squares" data are the results of using measured propagation constants taken from Fig. 12.22 of Chap. 12. The simple procedures of Sec. 14.3 were used to obtain the curve.

The divergence of the two sets of calculations at 3,000 cps is due to the fact that R_1 increases with frequency above 2,000 cps, while the theory basic to Fig. 14.13 assumes it to be constant with frequency.

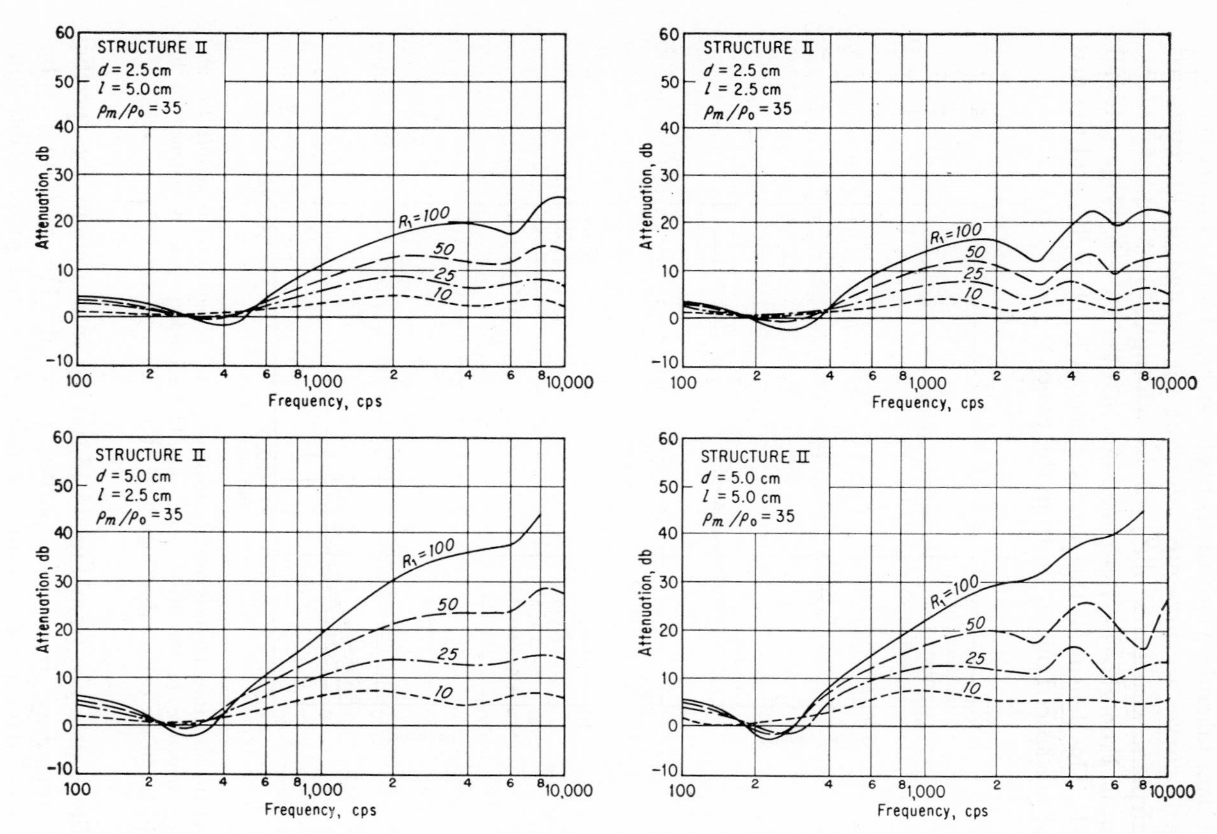

FIG. 14.12 Structure II: Incremental attenuation in decibels due to adding the absorbing structure to the basic plate, σ_1. The secondary side is assumed to be terminated in the characteristic impedance of air.

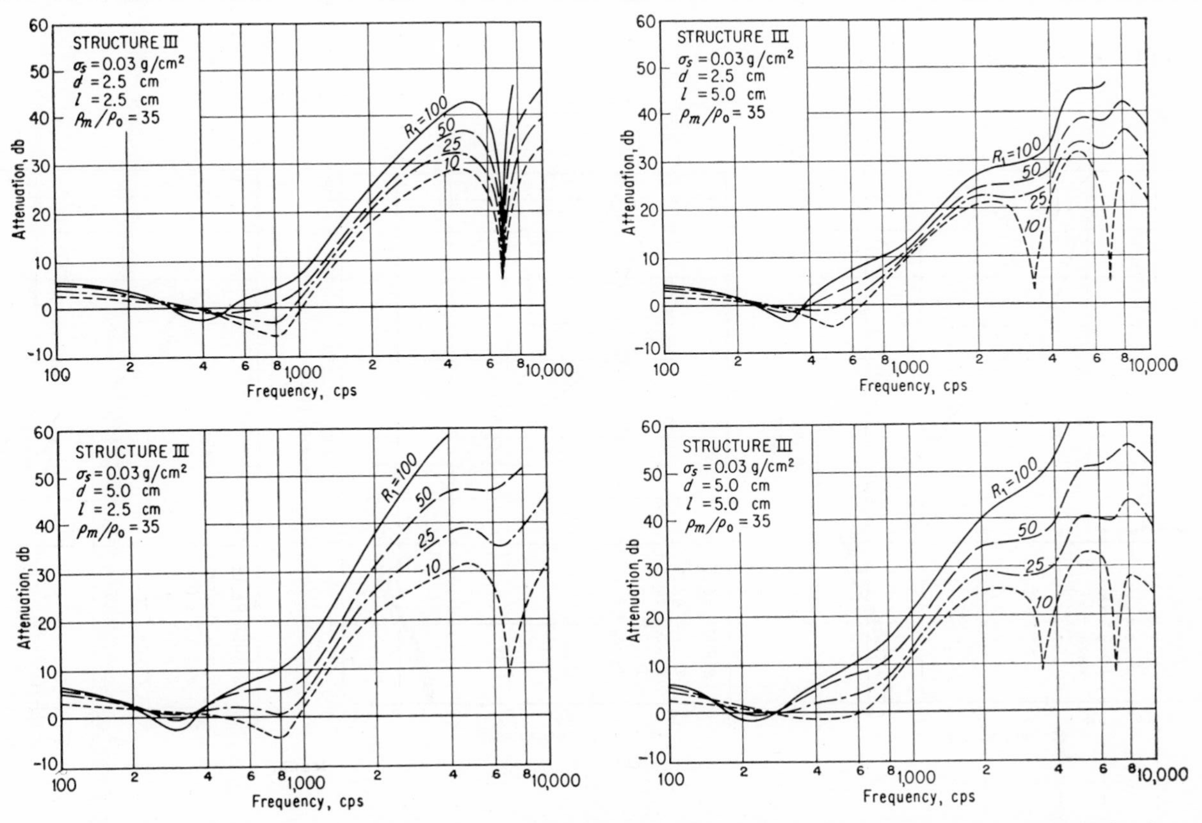

FIG. 14.13 Structure III: Incremental attenuation in decibels due to adding the absorbing structure to the basic plate, σ_1. The secondary side is assumed to be terminated in the characteristic impedance of air.

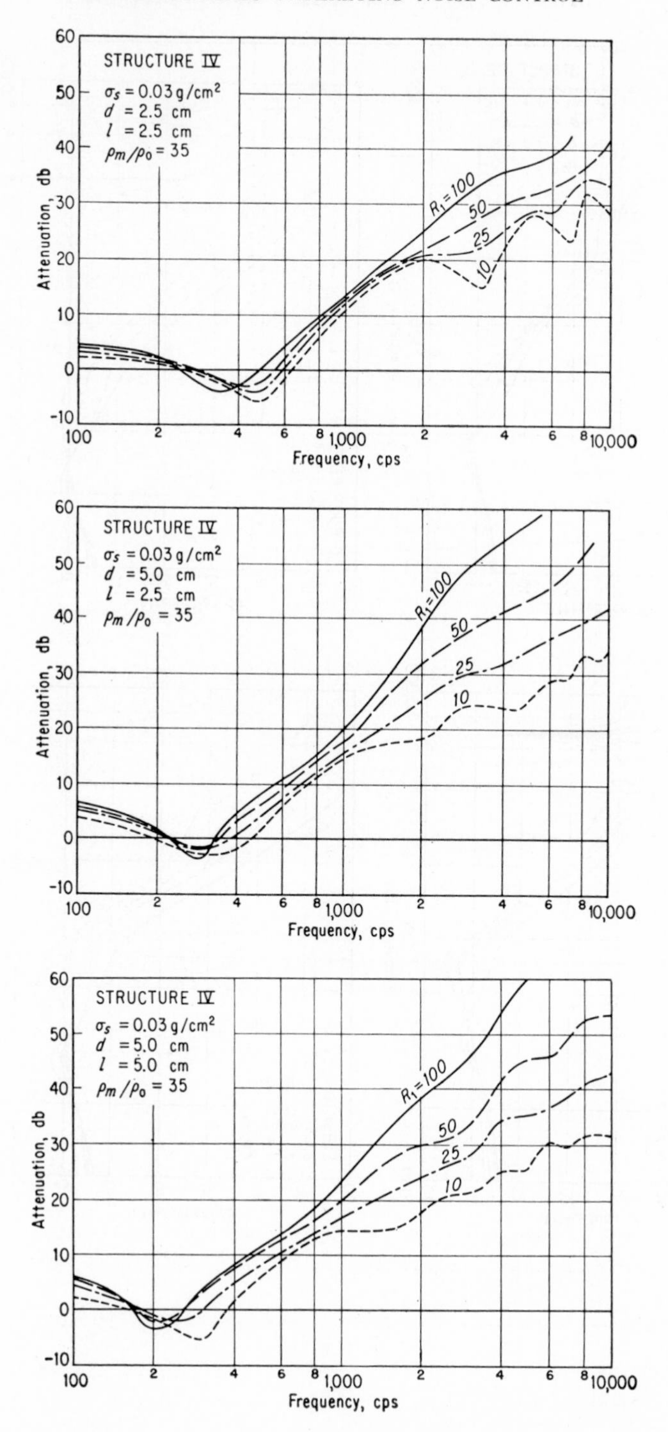
STRUCTURE IV
$\sigma_s = 0.03$ g/cm²
d = 2.5 cm
l = 2.5 cm
$\rho_m/\rho_0 = 35$
$R_1 = 100$
50
25
10
Attenuation, db
Frequency, cps
STRUCTURE IV
$\sigma_s = 0.03$ g/cm²
d = 5.0 cm
l = 2.5 cm
$\rho_m/\rho_0 = 35$
$R_1 = 100$
50
25
10
Attenuation, db
Frequency, cps
STRUCTURE IV
$\sigma_s = 0.03$ g/cm²
d = 5.0 cm
l = 5.0 cm
$\rho_m/\rho_0 = 35$
$R_1 = 100$
50
25
10
Attenuation, db
Frequency, cps

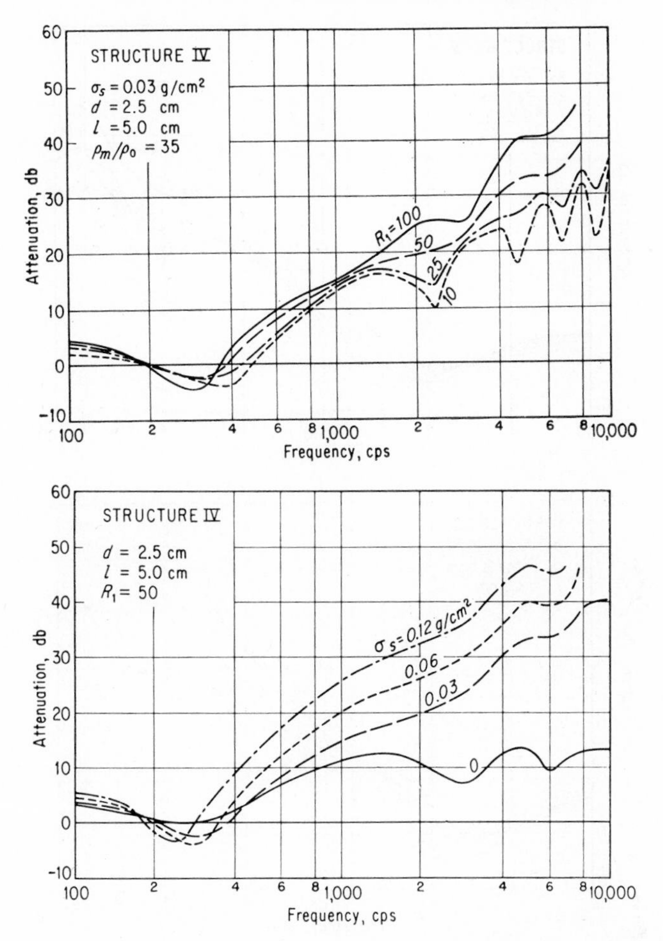

Fig. 14.14 Structure IV: Incremental attenuation in decibels due to adding the absorbing structure to the basic plate, σ_1. The secondary side is assumed to be terminated in the characteristic impedance of air. Note the special case where the surface density of the septum is varied, holding the other parameters constant.

The bending downward of the Convair 3-in. blanket curve above 2,000 cps is probably due to flanking transmission. From this comparison, one may arrive at some feeling for the reliability of measurement and calculations.

Other interesting cases of multiple walls wherein measurements and calculations have been presented were reported by Meyer.[4]

14.7 Porous Materials in Heavy Double-wall Structures

Porous materials are seldom used in the United States in heavy double-wall structures because, as was shown in Chap. 13, flanking

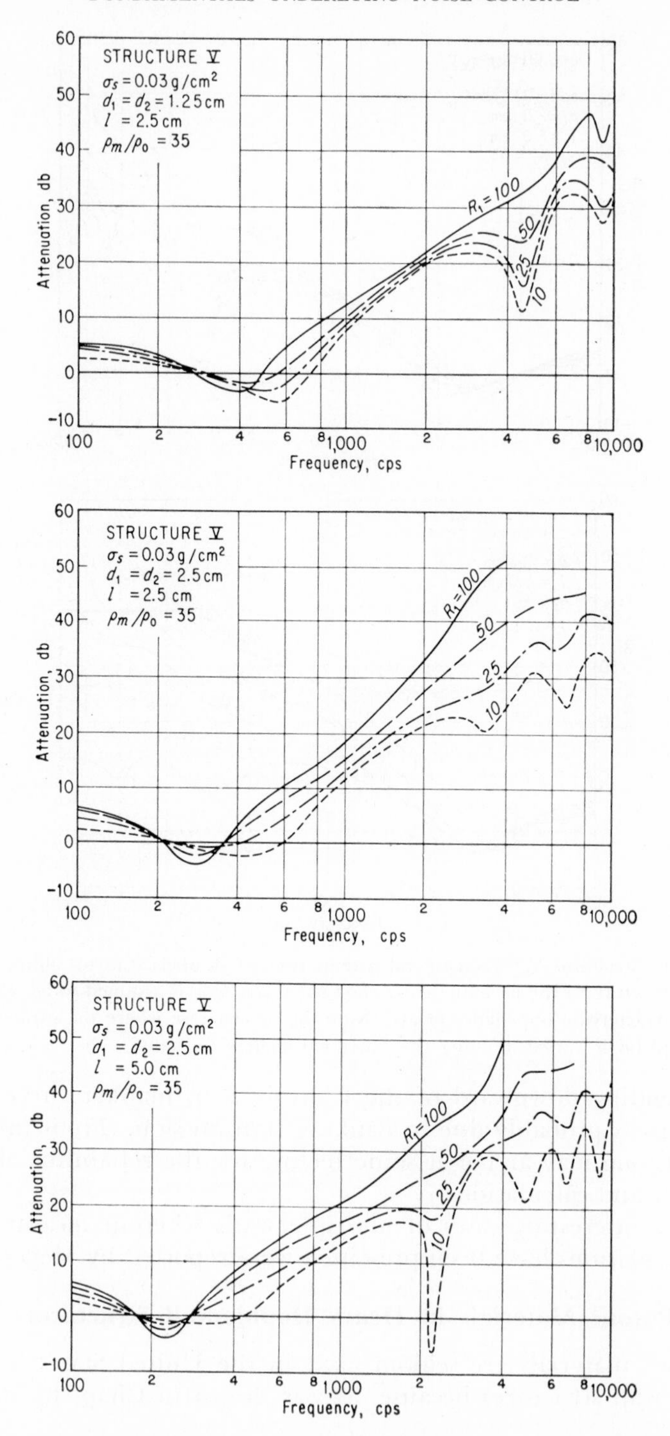
STRUCTURE V
σs = 0.03 g/cm²
d1 = d2 = 1.25 cm
l = 2.5 cm
ρm/ρ0 = 35
R1 = 100
50
25
10
Attenuation, db
Frequency, cps
STRUCTURE V
σs = 0.03 g/cm²
d1 = d2 = 2.5 cm
l = 2.5 cm
ρm/ρ0 = 35
R1 = 100
50
25
10
Attenuation, db
Frequency, cps
STRUCTURE V
σs = 0.03 g/cm²
d1 = d2 = 2.5 cm
l = 5.0 cm
ρm/ρ0 = 35
R1 = 100
50
25
10
Attenuation, db
Frequency, cps

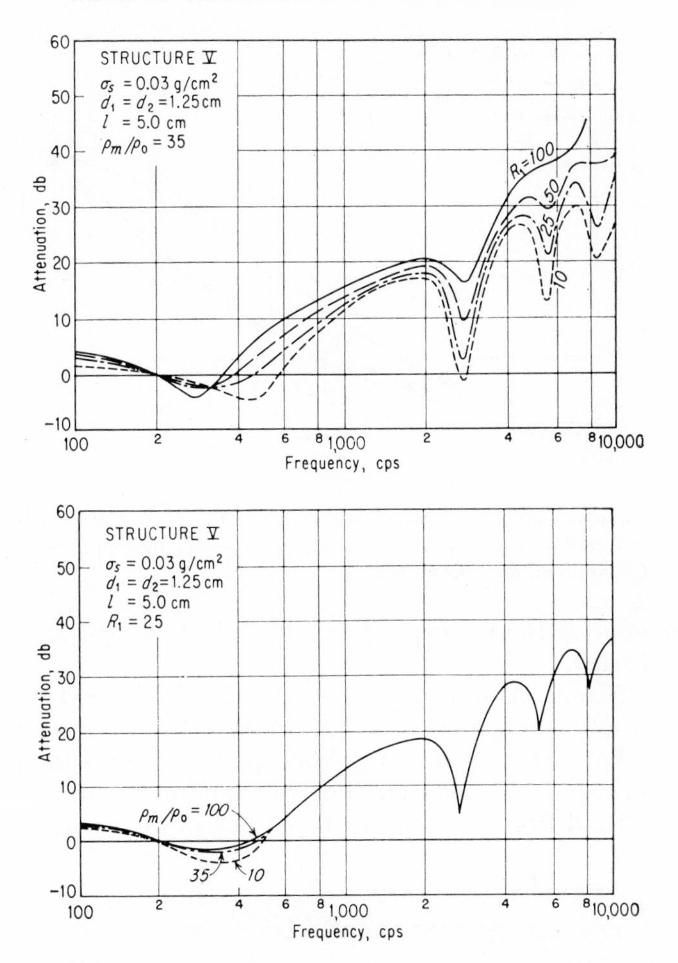

FIG. 14.15 Structure V: Incremental attenuation in decibels due to adding the absorbing structure to the basic plate, σ_1. The secondary side is assumed to be terminated in the characteristic impedance of air. Note the special case where the density of the blanket is varied, holding the other parameters constant.

transmission generally limits the TL. In those cases where a truly isolated pair of walls can be achieved (as, for instance, in the NBS testing laboratory), the effects of added porous materials can be observed.

Figure 14.17 shows two structures measured by the National Bureau of Standards and reported by Cook and Chrzanowski.[2] Panel 166A was a steel-stud partition, with ⅞-in. gypsum plaster on expanded metal lath on both sides of 3¼-in. metal studs, 16 in. OC. The two sides of the panel are tied together by the studs. Panel 166B

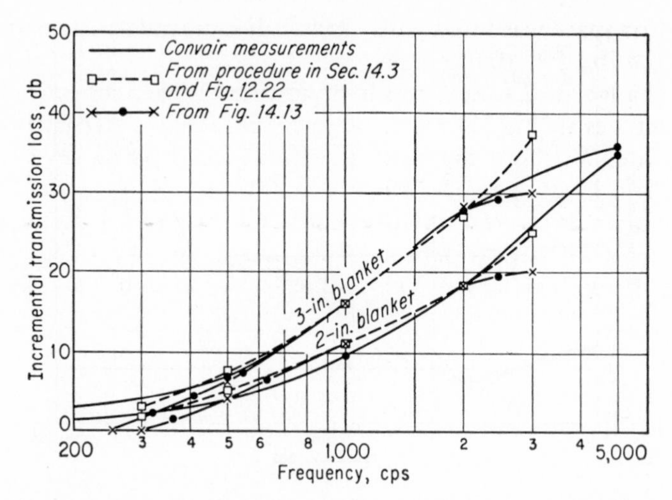

FIG. 14.16 Comparison of calculated incremental TL of a porous blanket with measurements. (*Convair data courtesy of Acoustics Group, Convair Division, General Dynamics Corporation, San Diego, Calif.*)

was the same as 166A except filled with 5.2 lb_m/ft^3 (83 kg/m^3) mineral-wool batts.

Panel 237 was made from staggered 2- by 4-in. wood studs, each set 16 in. OC and spaced 8 in. OC, with a 1½-in. offset from the other set. On each side there was ⅜-in. plain gypsum lath and ½ in. of gypsum vermiculite plaster. The two sides of the panel are completely isolated from each other. Panel 238 was the same as 237 ex-

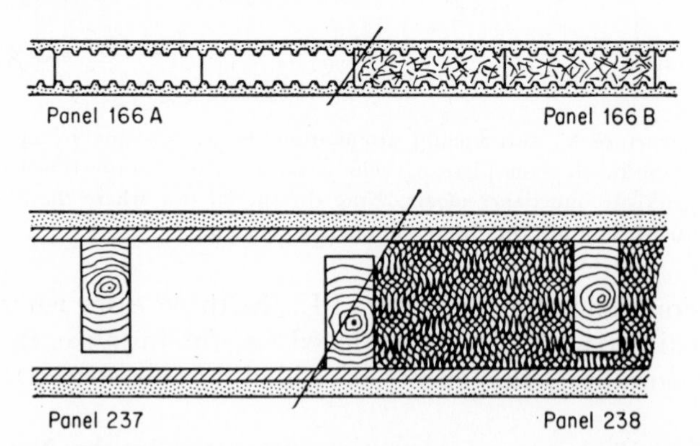

FIG. 14.17 Structures measured at National Bureau of Standards, Washington, D.C. Panel 166A: gypsum plaster, ⅞ in., on expanded metal lath on both sides of 3¼-in. metal studs, 16 in. OC. Panel 166B: Same except with 5.2 lb_m/ft^3 mineral wool filling. Panel 237: Staggered 2- by 4-in. wood studs, each set 16 in. OC and with ½-in. offset. On each side ⅜-in. plain gypsum lath and ½ in. of gypsum vermiculite plaster. Panel 238 same with 6.3 lb_m/ft^3 vermiculite filling.

cept the air space was filled with vermiculite granules. Density of the fill was 6.3 lb_m/ft^3 (100 kg/m^3).

The change in TL as a result of adding the fill is shown for the two structures in Fig. 14.18. It is seen that with flanking, the fill is of no value. Without flanking, however, the increase in TL is between 3 and 15 db between 250 and 4,000 cps.

As a matter of interest, the calculated increase in TL for a particular rockwool fill with a density of 2.5 lb_m/ft^3 is also shown on Fig. 14.18 by the dashed curve. The propagation constant was taken from

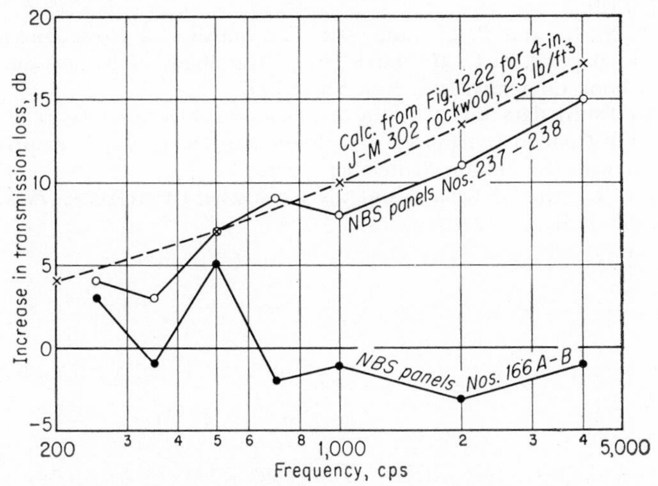

FIG. 14.18 Comparisons of calculations with laboratory measurements of incremental TL's due to porous fillings. The upper solid curve gives the difference between the TL's measured for NBS panels Nos. 237 and 238. The lowest curve is the same for NBS panels Nos. 166A and 166B. No improvement was predicted for the filling of Panel 166A. Acoustical data on vermiculite were not available so calculations were made for Johns-Manville 302 rockwool, which is expected to give higher TL than vermiculite.

Fig. 12.22. Because of the finer structure of the porous rockwool, its acoustical behavior would be expected to be better than that of vermiculite.

As a final note, in both Refs. 1 and 4 the use of sound-absorbing materials around the periphery between double walls is discussed as a means for increasing sound transmission loss.

14.8 Summary

In this chapter practical methods for the estimation of sound transmission loss (TL) through lightweight structures containing porous sound-absorbing blankets are given. Using the basic charts of Chaps. 12 and 13, procedures are given for estimating the TL of

(1) free-hanging blanket, (2) blanket adjacent to a vibrating plate, (3) blanket spaced out from a vibrating plate, (4) blanket adjacent to or spaced out from a vibrating plate but containing an impervious septum. Some comparisons between measurements and calculations are given, and references to the literature are given for others.

REFERENCES

1. Beranek, L. L., and G. A. Work: Sound Transmission through Multiple Structures Containing Flexible Blankets, *J. Acoust. Soc. Am.,* vol. 21, pp. 419–428, July, 1949.

2. Cook, R. K., and P. Chrzanowski: Transmission of Noise through Walls and Floors, chap. 20, in C. M. Harris (ed.), "Handbook of Noise Control," McGraw-Hill Book Company, Inc., New York, 1957.

3. Unpublished data taken by the Acoustics Engineering Group of Convair, A Division of General Dynamics Corporation, San Diego, Calif., in private communication with the author, September, 1958.

4. Meyer, E.: Die Mehrfachwand als akustische Drosselkette, *Elek. Nachr.-Tech.,* p. 393, 1935.

Chapter 15

ACOUSTICAL MATERIALS FOR ARCHITECTURAL USES

Jack B. C. Purcell

15.1 Introduction

In the absence of enclosures or barriers, sound from a small source propagates radially in a spherical pattern similar to that of an ever-expanding balloon. Deviations from this spherical propagation are governed by meteorological and topographical conditions. In an enclosed space, the same sound source produces a sound wave that propagates spherically until it encounters a surface of the enclosure. After the encounter, the progress of the wavefront may be materially altered depending upon the properties, orientation, and configuration of the surface. Specifically, it is possible to accomplish some predetermined acoustical design objective by selecting the enclosure surfaces to absorb, reflect, or transmit the incident wave. How well this objective is accomplished will depend upon the designer's knowledge and skill in the selection and use of materials.

In an auditorium, for example, there is the need to absorb sound at some surfaces for the purpose of controlling undesirable echoes. At the same time, there is the need to reflect sound from other surfaces for the purpose of reinforcing the signal at a listener's position. Each of these objectives is basic to design in room acoustics, yet each requires the use of materials having fundamentally different acoustical characteristics, namely, reflection and absorption. Therefore, since the mechanisms of sound absorption and sound reflection are equally important, we may state that *all* building materials comprising the boundary surfaces of a room may and *should* be thought of as "acoustical materials."

Commonly, the term "acoustical* materials" is applied only to

* ASA terminology states that the adjective as used here should be "acoustic." However, one organization of materials manufacturers has named itself the "Acoustical Materials Association" so that "acoustical" is more commonly used in the United States.

those materials that are capable of absorbing an appreciably high percentage of sound incident on their surfaces. While there is no specified lower limit of percentage of sound absorption which a material must have to qualify as "acoustical," it is generally assumed to be about 20 per cent. By contrast, ordinary building materials, such as glass, plaster, concrete, etc., have a sound absorption generally not in excess of 5 to 10 per cent and are more often in the region of 1 to 5 per cent.

In an acoustically untreated room (where the boundary surfaces are finished in hard, sound-reflecting surfaces) multiple sound reflections persist audibly for prolonged periods of time, even, in some instances, as long as 10 to 15 sec. In such a room, also, the sound levels will be high.

The fundamental purpose of acoustical materials is to reduce either the average sound-pressure level or the reverberation time in a room, or both. In fact, as we show in Appendix A to this chapter, the reverberation time halves and the sound-pressure level decreases by 3 db for each doubling of the total *sound absorption* in a room. Acoustical materials are generally applied to wall and ceiling surfaces and occasionally are suspended freely in the room volume. In spaces where speech and music listening are important, a certain prolongation of sound is desirable. The application of acoustical materials in these instances is referred to as "reverberation control." In spaces where noise may be a problem, any prolongation of sound is generally undesirable. The application of acoustical materials in these instances is referred to as "noise control." Thus, the designer must consider the use to which the space will be put and the amount of sound absorption required to meet satisfactorily the varying degree of room performance requirements.

15.2 Methods of Sound Absorption

Acoustical materials exhibit a wide range of sound-absorbing efficiency over the audible spectrum, depending upon their surface treatment, composition, method of mounting, and thickness.

Nearly all the more common acoustical materials are of a porous composition with a surface finish designed to permit entrance of the sound wave into the interior. Their porous composition consists of a labyrinth of capillary-like tunnels which are formed by the interstices of fibrous mineral, cellulose, or aerated plaster products. When the material is exposed to the sound wave, the air molecules contained in that part of the sound wave near the surface of the material partially migrate into the material in a rapid to-and-fro motion.

Here a part of the acoustic energy is converted into heat through the frictional drag or viscous resistance encountered by the air molecules against the walls of the capillary tunnels. In this manner, the absorption of an incident acoustic wave can be as high as 95 per cent. The fraction of incident energy absorbed (α, *absorption coefficient*) in the process will vary with frequency for a given material of a specified thickness and type of mounting.

Almost without exception, porous materials provide a greater percentage of sound absorption at the mid and high frequencies than at the lower frequencies. The lower curve of Fig. 15.1 illustrates a typi-

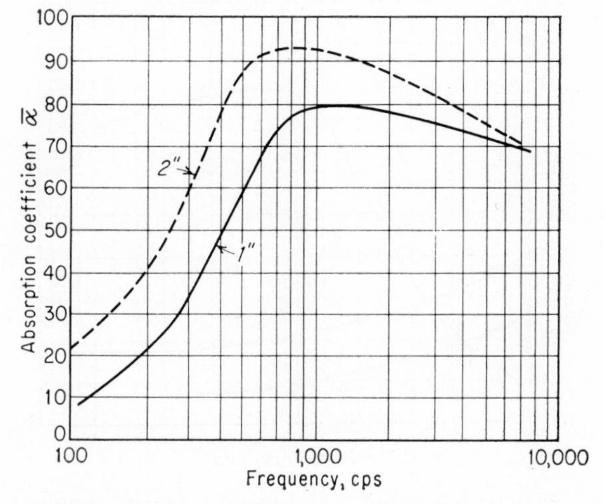

FIG. 15.1 Curves showing the increase in sound absorption resulting from a doubling in thickness of a typical homogeneous porous sound-absorbing material.

cal variation of absorption coefficient vs. frequency for a 1-in.-thick "unfaced" low-density porous blanket with no air-space backing. The upper curve of Fig. 15.1 shows that the same material having a 2-in. thickness exhibits a significant increase in sound-absorbing efficiency, particularly at the lower and mid frequencies. While this is in fact a typical characteristic of porous acoustical materials, it should be pointed out that the increased sound-absorbing efficiency resulting from the increased thickness may not always be as dramatic as that shown in Fig. 15.1.

Another material for sound absorption often used in architectural acoustics is the flexible impervious panel. When a panel is set into motion by an impinging sound wave, part of the energy contained in the wave is removed through internal viscous damping due to panel flexure. This is particularly true for frequencies in the region of the resonances of the panel. Panel materials such as plywood and composition hardboards of the sizes normally used for interior finishes provide useful sound absorption, but we must note that they exhibit their maximum absorption in the lower-frequency regions. On the other hand, ½- to 1-in.-thick porous acoustical materials are somewhat less effective at these lower frequencies. Figure 15.2 shows the sound absorption as a function of frequency for two thicknesses of plywood panels in comparison to that of a typical 1-in.-thick porous material. Maximum absorption is obtained with panels that are suitably damped and do not produce a ringing tone when struck. An

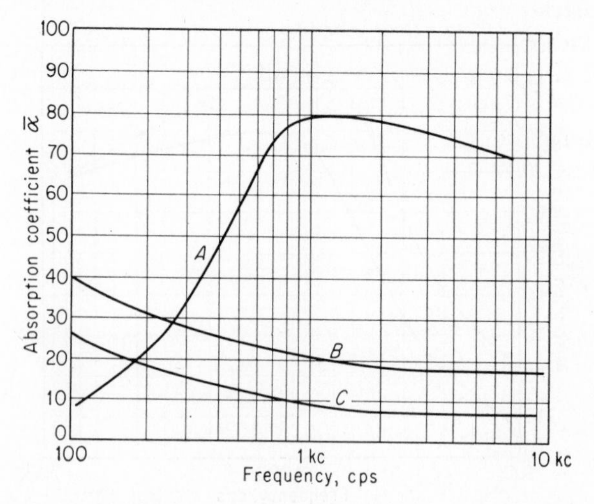

FIG. 15.2 Absorption coefficients for: (*A*) 1-in.-thick porous acoustical material; (*B*) two sheets of ⅛-in.-thick plywood, separated by tufts of randomly glued felt causing ⅛-in. spacing between sheets; (*C*) one sheet of ⅜-in.-thick randomly braced plywood. The measurements of *B* and *C* were made in small broadcasting studios.

example of an unsuitable panel material would be a thin, undamped sheet of steel which is actually capable of reemitting sound for a protracted period after the air-borne source of acoustic excitation has been removed.

Still another method of achieving sound absorption is the acoustic resonator of the Helmholtz type which consists of a vessel of any shape containing a volume of air. This air volume is in direct communication with the air within the room through an interconnecting tunnel which may be long or short and of any cross-sectional shape. An example of a resonator of this type might be a gallon jug. When a sound wave impinges on the aperture of the tunnel (neck of the jug), the air in the tunnel will be set into vibration which periodically expands and compresses the air within the vessel. The resulting amplified motion of the air particles in the tunnel due to phase cancellation between the air plug in the tunnel and the air volume in the vessel causes energy absorption due to frictional drag in and around the tunnel. This type of absorber can be designed to produce maximum absorption over a very narrow frequency range or even a wide frequency range as shown in Fig. 15.3.

The above brief descriptions serve to illustrate that a number of materials and even combinations of methods are available for absorbing sound over a broad or narrow frequency range and selectively at low, middle, and high frequencies.

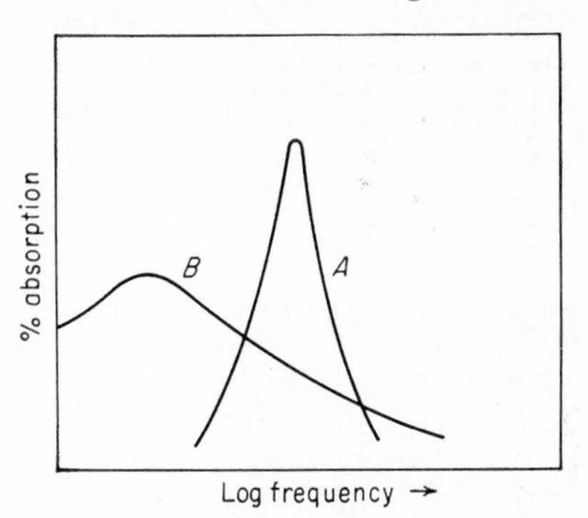

FIG. 15.3 Graphs illustrating absorption by acoustic resonators: (*A*) narrow-band type; (*B*) wideband type.

15.3 Prefabricated Acoustical Materials

In recent years there has been an ever-increasing demand for architectural acoustical materials. Competition among manufacturers has resulted in a significant increase in the number of products. The abundance of products which are now available in a wide variety of basic materials, thicknesses, tile sizes, board sizes, and surface finishes makes the selection of an appropriate material somewhat difficult. Figure 15.4 shows a sampling of a number of surface finishes.

Fortunately, nearly all the prefabricated acoustical materials now available have been typed in accordance with their general appearance and composition by the Acoustical Materials Association (AMA[1]). This association is composed of a number of nationwide organizations engaged in the manufacture of architectural acoustical products. Particular emphasis has been directed by the association toward maintaining or improving the absorption characteristics of various materials while simultaneously offering a greater variety of decorative features. Table 15.1 offers a list of the materials as classified by the AMA.

As mentioned earlier, the thickness and method of mounting are of equal importance in determining the sound-absorption coefficient of a material at various frequencies. The prefabricated products have been standardized essentially into ½-in., ¾-in., and 1-in. thicknesses by the AMA. The methods of application or types of mountings used in conducting tests to determine sound-absorption coeffi-

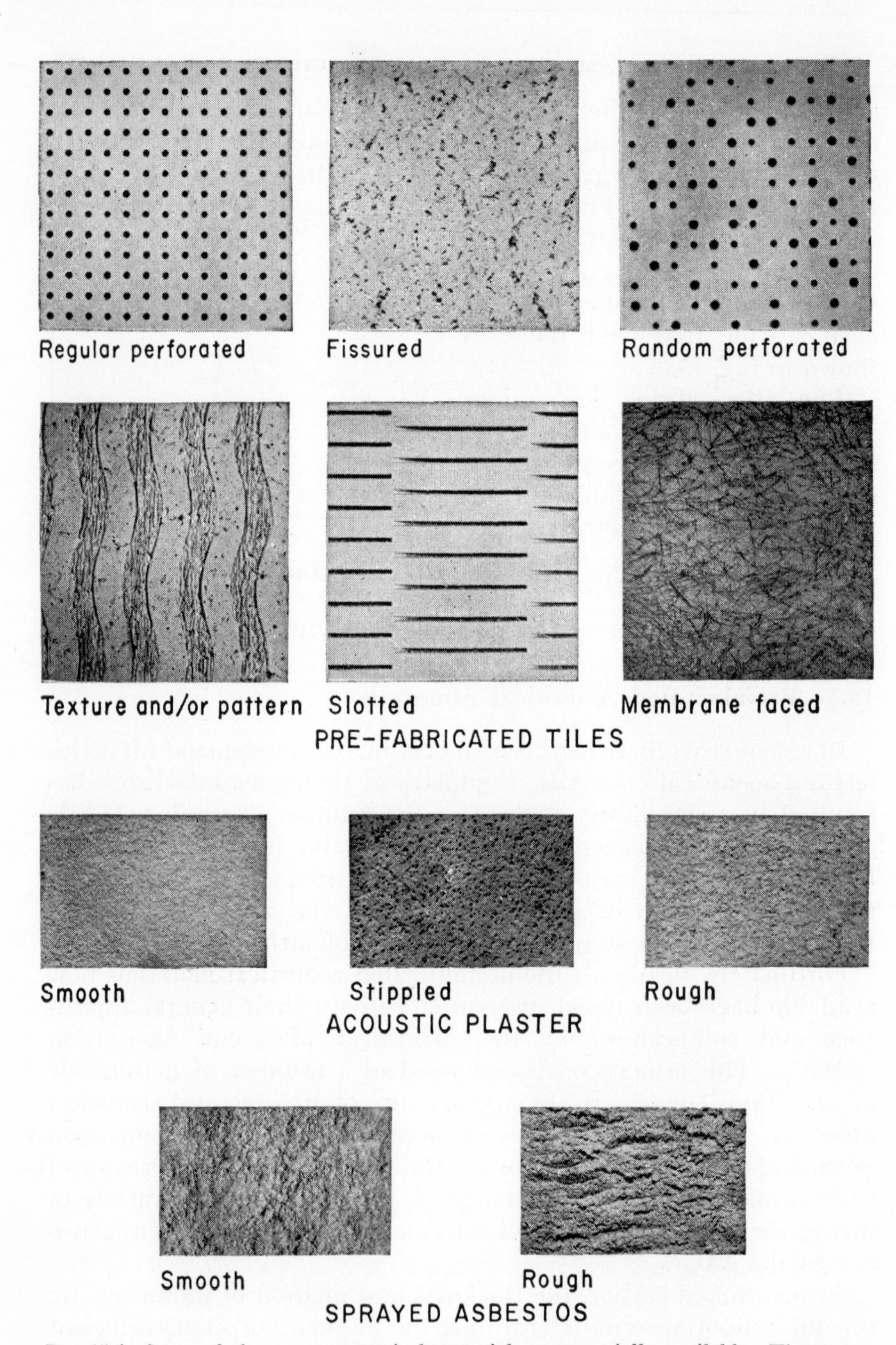

Fig. 15.4 Some of the many acoustical materials commercially available. The upper six are prefabricated tiles; the middle three are acoustic plasters; and the lower two are sprayed asbestos. (*Photographs courtesy of National Gypsum Company, Dant and Russell, Inc., Celotex Corporation, U.S. Gypsum Company, Owens-Corning Fiberglas Corporation, Tiger Products Division of Basic Corporation.*)

Table 15.1

AMA Classification of Materials

Type	*Material*
I	Regularly perforated cellulose fiber tile
II	Randomly perforated cellulose fiber tile
III	Slotted cellulose fiber tile
IV	Textured or fissured or simulated-fissured cellulose tile or board
V	Perforated mineral fiber tile
VI	Fissured mineral fiber tile
VII	Textured or smooth mineral fiber tile
VIII	Membrane-faced mineral fiber tile or board
IX	Perforated metal panels with mineral fiber pads
X	Perforated asbestos board panels with mineral fiber pads
XI	Sound-absorbent duct lining
XII	Special acoustical panels and systems

cients of acoustical materials have also been standardized as illustrated in Fig. 15.5.

15.4 Measurement of Sound Absorption

When selecting an acoustical material, it is important that the designer have an understanding of the test methods used in determining the absorption coefficient since the reported coefficient is likely to vary for a given material under different test conditions. The two techniques most frequently used in determining the absorption coefficients of materials are the reverberation-chamber and the impedance-tube methods.

In the reverberation-chamber method [2,3] a large rectangular sample of acoustical material mounted in one of the manners shown in Fig. 15.5 is placed in a highly reverberant room. Measurements are made at frequencies of 125, 250, 500, 1,000, 2,000, and 4,000 cps. The effect of the sample on the time rate of sound decay of the room at each frequency is observed and compared with that of the time rate of sound decay of the room without the sample at each frequency. From these comparative tests, the absorption coefficients are calculated for the sample and are reported as characterizing the material at the specified mounting and frequency. It is important to note that in this test procedure, the sound wave is incident upon the material from all angles (random incidence) which most nearly approximates the conditions to be found in a fairly reverberant architectural enclosure.

In the impedance-tube method a small sample, generally 3 to 4 in.

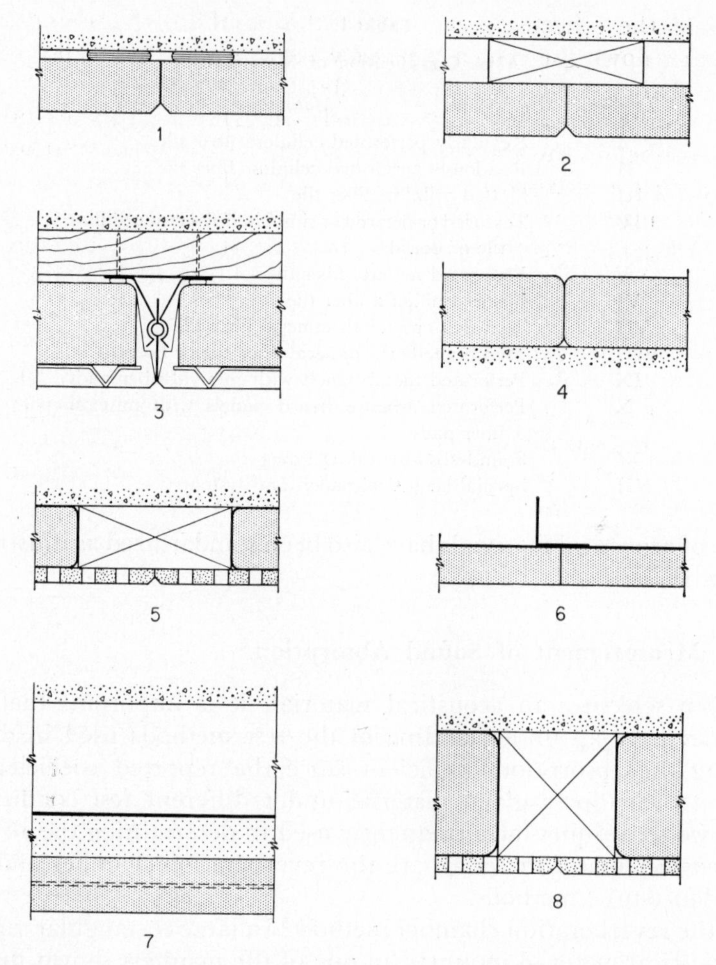

FIG. 15.5 Types of mounting used in laboratory tests of acoustical materials. These mountings are typical of actual installation methods used in the field. The sound-absorption coefficients of most materials vary with the type of mounting: (1) cemented to plaster board with 1/8-in. air space—considered equivalent to cementing to plaster or concrete ceiling; (2) nailed to 1- by 3-in. wood furring 12-in. on-centers unless otherwise indicated; (3) attached to metal supports applied to 1- by 3-in. wood furring; (4) laid directly on the laboratory floor; (5) furred 1 in., furring 24-in. on-centers, 1-in. mineral wool between furring; (6) attached to 24-gauge sheet iron, supported by metal angles; (7) mechanically mounted on special metal supports; (8) furred 2 in., furring 24-in. on-centers, 2-in. mineral wool between furring.

in diameter, is inserted at one end of a long tube (8 to 10 ft) with a sound source at the opposite end. A pure-tone acoustic wave is generated, and the acoustic field is explored, by means of a probe, to determine the difference in sound pressure between the incident wave and the reflected wave. With this method, the sound wave propagates down the tube as a plane wave at normal incidence to the surface of the sample. Tests are made at many frequencies.

The results of tests by both methods (reverberation-room and impedance-tube) on the absorption coefficients of a number of materials indicate that a consistent comparison is not always obtainable. One may generalize, however, and say that at low frequencies, the absorption coefficient of a given porous material at random incidence is approximately twice that measured at normal incidence, while at the higher frequencies they are approximately equal.

The sound-absorption coefficients for all acoustical materials reported by the AMA are based on the reverberation-room method and are, in fact, measured in the same reverberation chamber. In this manner, those discrepancies are minimized that might arise from differences among test laboratories. Thus, a satisfactory comparison may be made of the products of the member organizations of the AMA. However, not all American manufacturers are members of the AMA.

In many instances, materials such as sprayed asbestos, acoustical plasters, or fabricated products in tile form are available through local suppliers. These materials are made also with a variety of surface finishes not listed by the AMA, such as pebble grain, stipple, whorl, etc. The sound-absorption coefficients for these materials are determined in other laboratories such as the NBS, universities, or privately owned organizations that are properly instrumented to conduct such tests. The coefficients so obtained are not always strictly comparable with those published by the AMA.

15.5 Noise-reduction Coefficient

The noise-reduction coefficient (NRC)[1] is the (arithmetical) average of the sound-absorption coefficients of a material at 250, 500, 1,000, and 2,000 cps. It is intended for use as a single-number index of the sound-absorbing efficiency of a material. Caution should be exercised in the selection of a product on the basis of its NRC, since this index is not always indicative of the material's capabilities. For example, the NRC does not include the absorption coefficient of the material at 125 cps, where the great majority of acoustical materials are less efficient. This is of particular importance since low-frequency absorption is not only difficult to obtain, but is almost always required in greater quantity than is normally available. As mentioned earlier, porous materials are at maximum efficiency in the mid and high frequencies. Thus incremental differences of 0.10 to 0.15 in the

NRC among several materials may be neglected since they tend to average out across the measured frequency range. This would not be true in the case of two products having the same NRC but a similar incremental difference (0.10 to 0.15) in absorption coefficient at 125 cps. Such a difference at low frequencies can often represent as much as 100 to 300 per cent greater sound-absorbing efficiency by one product compared with another.

15.6 Selecting an Acoustical Material

The fact that there are nearly 400 commercially available prefabricated architectural acoustical products manufactured by member organizations of the AMA in addition to products of numerous nonmember organizations often makes the selection of a specific material difficult. The problem in material selection may be minimized in some cases where precise reverberation or noise control is not required. Typical spaces of this type may include ground-level lobby areas of office buildings and large concourse areas of various types of transportation terminal structures where adequate measures for minimizing the intrusion of exterior noises have been taken in the design of the building envelope. In these spaces, the mounting method of the acoustical material is usually integrated with the structural design, the desired visual impact of the space dictates the surface finish, the building code spells out the fire-resistance classification, and the light reflection characteristics are governed by the lighting system and the corresponding visual task requirements. Once these limitations are imposed, it is apparent that the number of applicable products from which one can choose is considerably narrowed down. As a result (and only in these cases) the sound-absorption coefficient is of perhaps secondary importance.

In those cases where the noise-reduction or reverberation time is specified, the above selective process of a material is essentially altered to the extent where an equal or even greater emphasis is placed on its acoustical properties relative to its other properties.

If the problem, for example, is one involving noise reduction in a factory, an analysis of the spectrum characteristics of the noise produced by the machinery would dictate the selection of a material which has maximum sound absorption in the frequency regions where the noise would be most troublesome. Similarly, if the problem is one of reverberation control, the design criterion for optimum reverberation time at various frequencies dictates the selection of a material that will have the appropriate sound-absorption coefficient at the various frequencies.

Some difficulties in both cases may be encountered, however, when the building project is to be contracted for on an open competitive bidding system. In such cases, there can often be discrepancies between the design objectives and the final results. This is particularly true if the design has been based on a manufacturer's product having a specified sound-absorption coefficient at the design frequencies and that manufacturer either refuses to bid or may be rejected for some reason. The "or equal" clause introduced in the specification is seldom adequate protection, since some manufacturers' products (according to published data) cannot be equated acoustically. The NRC is also an ineffective rating for reasons already described.

In order to minimize these possible discrepancies, the accepted engineering procedure in such cases is to design around the average absorption coefficient of a group of materials having the same thickness, surface finish, and method of mounting with a specification of the minimum acceptable absorption coefficient at each of the design frequencies. Thus, a specification of this type might read as follows:

. . . to be ¾-in.-thick, type I, regularly perforated, cellulose fiber tile on a no. 7 mounting in accordance with AMA standards and having a coefficient of absorption that is equal to or greater than that specified at the following frequencies:

Frequency, cps	125	250	500	1,000	2,000	4,000
Coefficient	0.40	0.40	0.55	0.80	0.75	0.70

said coefficients to be certified by an acceptable laboratory using the procedures of test outlined in ASTM Test Code $C384\text{-}56T^2$ or its demonstrated equivalent.

For convenience, Table 15.2 lists the average sound-absorption coefficient for type I through type VIII acoustical materials as manufactured by all members of the AMA. These averaged data are grouped according to type number, thickness, and mounting. Reference should be made to published data for the performance of specific products.

15.7 Site Fabricated—Composite Materials

As previously shown, there are a number of types of prefabricated acoustical materials with various surface finishes that exhibit a wide range of sound-absorbing efficiency over the audible spectrum. In spite of the latitude of choice, there are a number of cases where none of the prefabricated materials (for one or more reasons) will meet either acoustical or visual design requirements. In these in-

Table 15.2

Average Absorption Coefficients for Type I Through Type VIII Acoustical Materials*

Type I—Regularly Perforated Cellulose-fiber Tile

Thickness, in.	Mounting	Frequency, cps					
		125	250	500	1,000	2,000	4,000
½	1	0.08	0.19	0.61	0.76	0.80	0.64
	2	0.15	0.64	0.60	0.75	0.87	0.80
	7	0.40	0.50	0.50	0.66	0.79	0.79
¾	1	0.12	0.31	0.74	0.88	0.83	0.65
	2	0.21	0.63	0.64	0.83	0.84	0.69
	7	0.44	0.44	0.61	0.84	0.86	0.69
1	1	0.15	0.38	0.87	0.95	0.80	0.66
	2	0.27	0.64	0.78	0.92	0.80	0.64
	7	0.39	0.51	0.72	0.87	0.83	0.68

Type II—Random-perforated Cellulose-fiber Tile

Thickness, in.	Mounting	125	250	500	1,000	2,000	4,000
½	1	0.14	0.25	0.61	0.68	0.72	0.68
	2	0.15	0.61	0.53	0.61	0.73	0.73
	7	0.43	0.45	0.47	0.57	0.72	0.75
¾	1	0.18	0.22	0.75	0.75	0.78	0.62
	2	0.24	0.66	0.63	0.77	0.79	0.63
	7	0.45	0.46	0.59	0.74	0.82	0.69
1	1	0.28	0.46	0.83	0.81	0.75	0.61
	2	0.33	0.72	0.66	0.75	0.76	0.62
	7	0.48	0.53	0.65	0.75	0.77	0.63

Type III—Slotted Cellulose-fiber Tile

Thickness, in.	Mounting	125	250	500	1,000	2,000	4,000
¾	1	0.18	0.29	0.72	0.82	0.84	0.71
	2	0.20	0.62	0.50	0.67	0.82	0.73

Type IV—Fissured or Textured Cellulose Tile or Board

Thickness, in.	Mounting	125	250	500	1,000	2,000	4,000
½	1	0.11	0.24	0.61	0.62	0.57	0.52
	2	0.20	0.67	0.48	0.53	0.53	0.40
9/16	1	0.11	0.22	0.65	0.69	0.74	0.70
	2	0.13	0.63	0.56	0.62	0.74	0.74
¾	1	0.22	0.34	0.78	0.72	0.80	0.77
	2	0.22	0.69	0.60	0.69	0.80	0.79
	7	0.46	0.49	0.54	0.68	0.80	0.81
1¼	1	0.03	0.12	0.52	0.62	0.41	0.46
	7	0.11	0.47	0.50	0.27	0.44	0.50

Table 15.2

Average Absorption Coefficients for Type I Through Type VIII Acoustical Materials* (*Continued*)

Thickness, in.	Mounting	Frequency, cps					
		125	250	500	1,000	2,000	4,000
Type V—Perforated Mineral-fiber Tile							
½	1	0.08	0.16	0.54	0.88	0.80	0.58
	2	0.05	0.53	0.60	0.74	0.76	0.52
⅝	1	0.07	0.20	0.64	0.93	0.81	0.61
	2	0.26	0.57	0.70	0.83	0.79	0.49
	7	0.50	0.65	0.74	0.80	0.84	0.66
¾	1	0.07	0.20	0.75	0.99	0.75	0.63
	7	0.67	0.83	0.79	0.86	0.82	0.58
⅞	1	0.11	0.35	0.90	0.93	0.72	0.53
	7	0.43	0.65	0.92	0.99	0.91	0.73
1	1	0.11	0.38	0.79	0.98	0.84	0.67
	7	0.41	0.61	0.67	0.91	0.89	0.74
Type VI—Fissured Mineral-fiber Tile							
⅝	7	0.34	0.56	0.72	0.83	0.73	0.67
¾	1	0.10	0.26	0.79	0.93	0.86	0.80
	7	0.64	0.70	0.72	0.83	0.90	0.85
Type VII—Textured or Smooth Mineral-fiber Tile							
½	1	0.14	0.22	0.66	0.74	0.72	0.56
⅝	1	0.14	0.25	0.73	0.81	0.76	0.67
	7	0.34	0.59	0.66	0.65	0.71	0.71
¾	1	0.12	0.29	0.75	0.87	0.84	0.78
	2	0.18	0.45	0.81	0.97	0.93	0.82
	7	0.68	0.73	0.75	0.85	0.87	0.85
1	7	0.70	0.77	0.83	0.91	0.94	0.91
Type VIII—Membrane-faced Mineral-fiber Tile or Board							
¾	1	0.12	0.33	0.72	0.88	0.80	0.56
	7	0.53	0.46	0.45	0.45	0.65	0.61

* See Fig. 15.5 for mounting details.

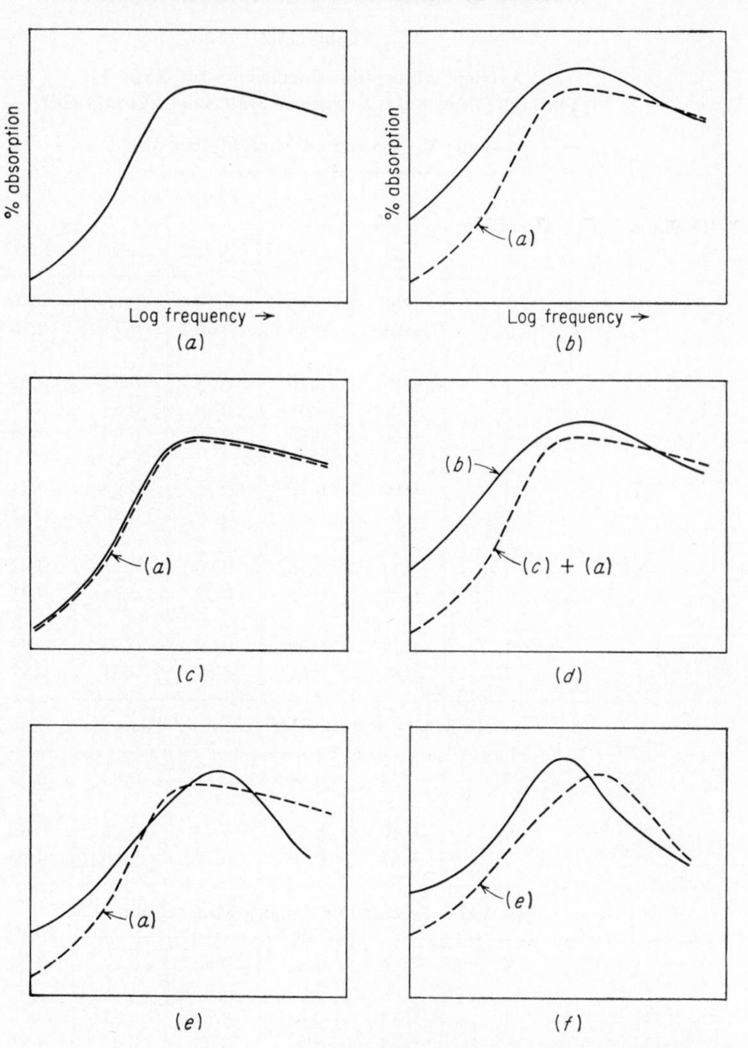

FIG. 15.6 Graphs showing the effects on space behind a porous acoustical material The metal facing is assumed to have 5 to air space; (*b*) unfaced—with air space; (*c*) open-weave fabric facing—with air space; (*f*) perforated metal facing with air space; air space; (*h*) four to six coats of sealer board" or gypsum lath—no air space; (*j*) with air space; (*k*) lightweight plastic facing —with air space. Some curves are duplicated appropriate letters.

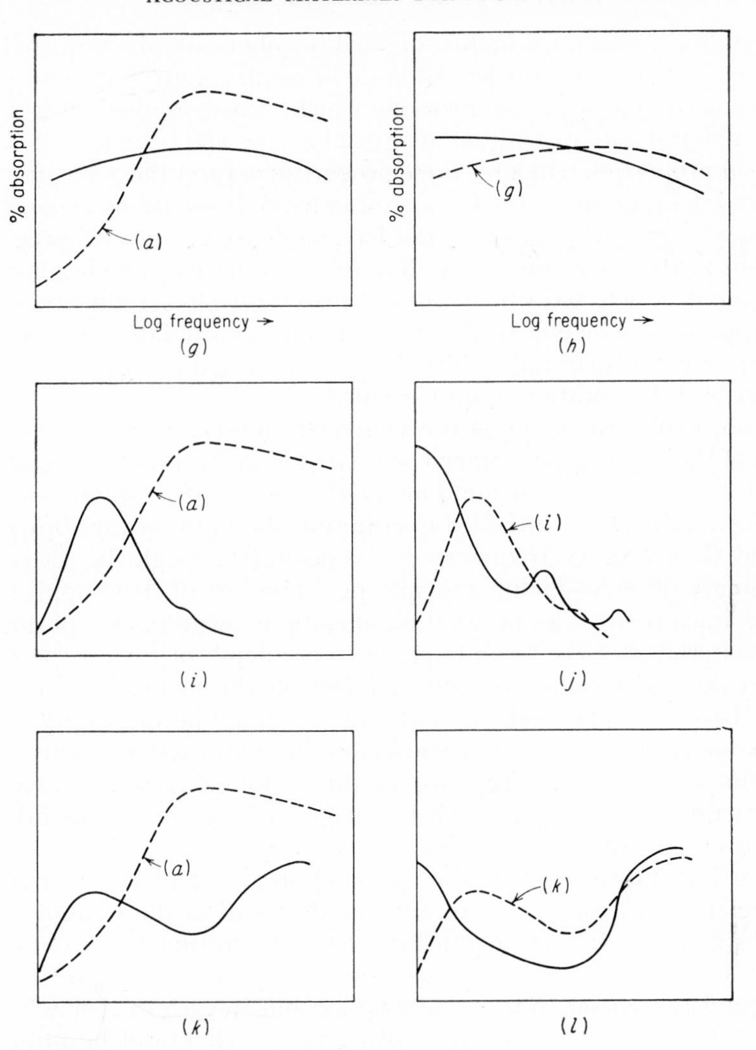

sound absorption with and without air
and of facings in front of the material.
10 per cent open area; (*a*) unfaced—no
open-weave fabric facing—no air space; (*d*)
(*e*) perforated metal facing—no air space;
(*g*) four to six coats of sealer paint—no
paint—with air space; (*i*) perforated "button
perforated "button board" or gypsum lath—
—no air space; (*l*) lightweight plastic facing
on other graphs as indicated by the

stances, other means must be investigated for obtaining the required sound absorption or appearance or both. The procedure is generally to select or design a facing material which, when applied over a porous material with or without air-space backing and having certain acoustical properties, will alter these properties to meet the acoustical and visual design objectives. On the other hand, it should be evident that where a certain porous material has the desired acoustical properties but at the same time an undesirable appearance, a facing material should be selected which will not materially change the acoustical properties. The end result in either case is a site fabrication of a combination of materials which have been designed to meet the visual as well as acoustical requirements.

It is not within the scope of this chapter to present a detailed discussion of the physical phenomena associated with the propagation of an acoustic wave by two or more materials when used in series combinations. Rather, it is intended to point out the range of variability of sound absorption vs. frequency that is possible through the use of two primary devices—facing and spacing. The first of these applies to the characteristics (e.g., perforation, membrane, woven, etc.) of the facing material, and the second applies to the depth of the air space between the porous material (directly behind the facing material) and the back-up surface such as a structural wall. The variability of sound absorption vs. frequency obtainable through these methods is indeed impressive and limited only by the combination of the number of facing materials which can be imagined, types of porous materials available, and possible air-space backings.

Figure 15.6 illustrates how the effects of spacing and facing may alter the acoustical properties of a material. In all of the examples, a porous material of a given thickness has been assumed, with and without an air-space backing. The curves are generalized as to frequency and percentage of sound absorption but in each case may be regarded as the parent of a family of curves which would be built around a generalized grouping of facing materials. Variations within the family of curves would result from major or minor differences in the material characteristics of proprietary products such as percentage open area, depth of perforation relative to perforation diameter and on-center spacing of perforation, surface weight of membrane, method of attachment, depth of air space, flow resistance and thickness of sound absorbing pad, etc.

15.8 Materials of Plastic Application

This category consists of porous acoustical materials which are applied in a wet semiplastic state, either by hand trowel or machine.

They consist of two basic products—plaster or mineral fiber—to which are added a binding agent and water at the time of application.

Acoustical Plasters. The acoustical plasters consist of an aggregate of Vermiculite or Perlite with either a "setting" type of binder, such as gypsum or lime, or a "nonsetting" type of binder, such as Bentonite. The acoustical plasters may be applied to a scratch or browncoat plaster base or directly to a concrete or masonry construction. In any case, they require a hard back-up surface. Application is generally made in two coats to a combined maximum practical thickness of 1 in. (not including the base-coat plaster). The final coat is hand or machine finished depending upon the surface texture desired. Surface porosity may be increased by wire brush or nail-roller stippling which aids in the retention of porosity after painting. Cleaning of acoustical plaster surfaces is accomplished by a brushing (such as camel hair) or a noncontact vacuuming preferably from a distance of approximately 1 in. from the surface. Refinishing the material requires the use of a nonbridging water-thinned resin emulsion or casein paint preferably by spray-gun application. Brush application of oil-base paints can seal the porous facing and essentially render the material "nonacoustic."

Since acoustical plasters require adhesion to a hard back-up surface, they seldom provide the very large low-frequency sound-absorption coefficients required for adequate control of reflections and focusing effects in spaces having curvilinear wall or ceiling surfaces or both. This problem may be further complicated if the material is machine applied.

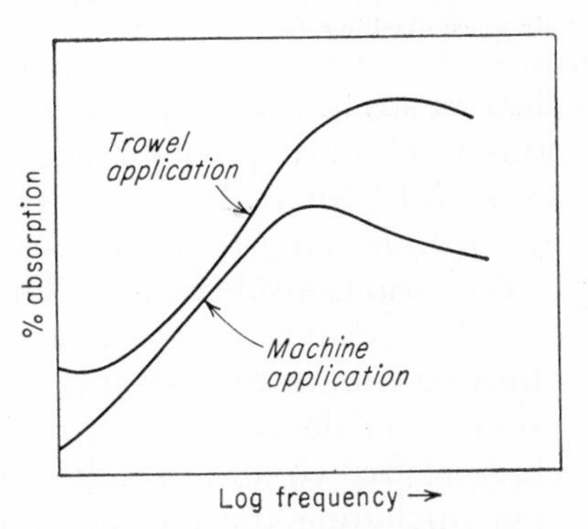

FIG. 15.7 Laboratory-determined sound absorption of acoustical plaster for two types of application.

Figure 15.7 shows laboratory-determined sound-absorption coefficients of an acoustical plaster as a function of frequency for a ½-in.-thick trowel application and for a ½-in.-thick machine application. The reduced efficiency of the machine-applied material is probably due to the compacting of the material under pressure of the applicator gun, reducing the surface and internal porosity.

Field measurements on the sound-absorption coefficient of a number of installations of acoustical plasters have often shown significant differences from laboratory data. In some cases, the field data are greater than the reported laboratory data, but more often it is less.

The differences may often be traced to variations from the specified thickness.

Mineral-fiber Products. The fibrous products consist of a mineral base, principally asbestos fibers, with a variety of binding agents and are referred to as "sprayed-on" materials. During the spray process, the binder, asbestos fibers, and water combine in air by means of a special spray gun to form a porous, lightweight, soft, acoustical, and fireproofing material which can be applied in thicknesses up to 3 in. The surface texture is a small-scale coarseness that may be varied to some extent depending upon the type of darby used in the hand-finishing process. If edge lighting of the material is proposed, a smooth finish should be specified. These materials may be applied directly to a wire-mesh or metal lath with an air-space backing. In addition, they may be applied to hard back-up surfaces similar to those required of acoustical plasters. The possibility of application of these materials to a metal lath with an air-space backing is a distinct advantage since it permits a very high sound absorption at the low- and mid-frequency range. Figure 15.8 shows the sound-absorption coefficient vs. frequency for a ¾-in.-thick sprayed asbestos with and without an air-space backing.

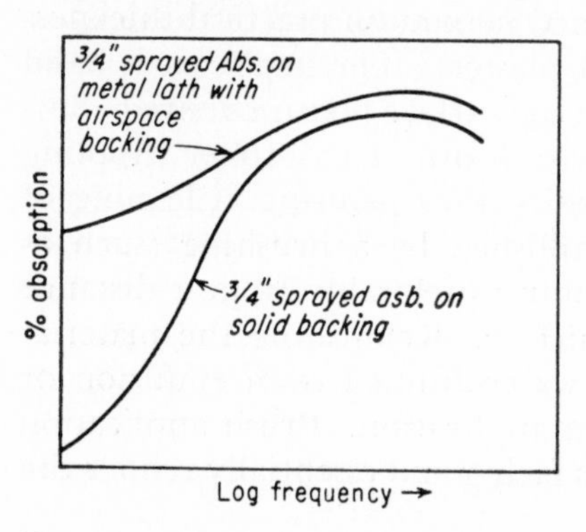

FIG. 15.8 Sound absorption of ¾-in.-thick sprayed asbestos acoustical material with and without 10-in.-deep air-space backing.

Practical Considerations. The use of prefabricated acoustical units usually demands a rigid modular pattern formed by the joints of abutting tiles or the carrying channels of the mechanical suspension system. On the other hand, the materials of plastic application are characterized by an unbroken surface continuity. This basic distinction (including the possibility of integral coloring combined with the advantage of the plastic materials to conform to structural surfaces of a compound curvature) makes these products particularly appealing for use in architectural interiors. However, as with all materials, there are limitations to both the acoustical plasters and sprayed asbestos which should be considered.

All porous acoustical materials are subject to damage when located within hand reach, unless they are covered with a protective facing. This is particularly true of sprayed asbestos products, which are quite fragile. Also, when supply or exhaust air registers are located in the plane of the treated surface, severe discoloration occurs around the register due to the accumulation of dirt particles deposited by the air stream. Cleaning of sprayed asbestos areas presents problems, particularly if the material is damaged in the process, because it is difficult to match the color of the remaining area. This difficulty may be offset to some extent by the machine application of a more dense surface treatment in the vicinity of the register which forms a protective crust over the normally soft material. Similar problems in maintaining the specified thickness are encountered with sprayed asbestos as with acoustical plaster.

15.9 Suspended Acoustical Units

This classification consists of a number of types of prefabricated space absorbers that are suspended within a room either from the ceiling or structural system or from a secondary suspension system such as stretched wires. They are manufactured in several geometrical configurations, e.g., flat baffles, cones, prisms, parallelepipeds, etc. The principal "nonacoustic" advantages of these products are their ease of installation, reusability, and maintainability. These advantages, combined with their relatively high acoustical efficiency, make these products particularly useful for acoustical correction in existing or new factories, loft spaces, arenas, gymnasiums, etc., where they may be spaced so as not to interfere with fire sprinklers, lighting systems, and air-handling units.

The conventional method of reporting sound-absorption coefficients of acoustical materials does not apply to these products, because the sound-absorbing efficiency varies per absorber unit, depending primarily upon the spacing between units. As the spacing between units is increased, their efficiency *per unit increases,* reaching a maximum stabilized value at a very wide spacing which is determined for the various products by experiment. The measurements are reported in sabins (unit of sound absorption equal to the absorption of 1 ft^2 of surface which is totally absorbent) per unit *for a given spacing in a specified geometric array.*

Although the efficiency per unit rises with increased spacing, it unfortunately does not do so linearly. Therefore, there will be less *total* room absorption with increased spacing (due to the limited number of units) than there would be with a smaller spacing (which permits a greater number of units). Thus, an optimum design for noise reduction would be one which, through proper spacing, would produce the required or maximum number of sabins with a minimum number of suspended units. A comparison may be made of the effectiveness of the suspended array with that of an over-all ceiling treatment of the same area by dividing the total number of sabins (the number of sabins for each unit times the number of units) by the area of ceiling. The result is comparable to the sound-absorption coefficient per unit area of an over-all acoustic ceiling treatment.

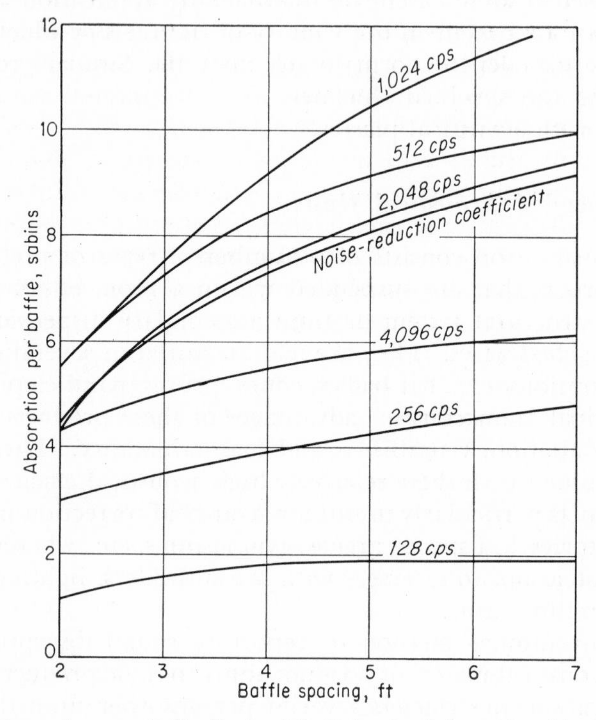

FIG. 15.9 Absorption per baffle (2- by 4-ft *Noise Stop*) in sabins (English units) at various frequencies for spacing of the baffles ranging between 2 and 7 ft. (*Courtesy of Owens-Corning Fiberglas Company.*)

Figure 15.9 shows the absorption per baffle (2- by 4-ft "Noise Stop") in sabins at various frequencies as a function of baffle spacing.

15.10 Paintability

One of the important considerations in the selection of an acoustical material is that its acoustical properties be maintained and essentially unaltered after painting.[4] As previously discussed, the mechanism of sound absorption takes place when the air particles of the sound wave are permitted to enter the material. If this means of access into the material is completely or partially blocked off, there can be a loss in sound-absorbing efficiency or a shift in the absorption-vs.-frequency curve. Either one of these effects will alter the room acoustics from the original acoustical design.

The results of repeated tests on the paintability of all types of acoustical materials have shown that no appreciable loss in the sound-absorbing properties will occur, provided the proper precautions are taken in accordance with manufacturer's recommendations. It is of importance to note, however, that there is a wide latitude in these precautions depending primarily upon the risk involved in the sealing of the surface of the material.

Perforated or slotted facings (AMA types I, II, III, V) in the prefabricated units usually present the least difficulty in retaining their acoustical properties after painting. Metal facings with very small perforations, however, can prove troublesome if the consistency of the paint causes bridging of the perforation. Perforated hardboard facings with a back-up sound-absorbing blanket can be a problem if the blanket is in direct contact with the facing since the paint has a tendency to "puddle" at the bottom of the perforation. This may be prevented by spacing the acoustical blanket at least ¼ in. from the facing material, thus removing the "bottom of the well."

Fissured tiles (AMA types IV and VI) may be painted without impairing the acoustical efficiency provided the area of application is restricted to the flat surfaces surrounding the fissures. Painting of the actual fissures can result in a sealing of the surface resulting in a loss of sound absorption. Since these products consist of a "high and low" surface the generally accepted technique for painting is a "damp" roller application to the high surface only. The simulated fissured products present some difficulties since the fissures are not actually depressed. In these cases spray-gun applications of well-thinned water-base paints are recommended.

The textured or smooth (AMA types IV and VII) prefabricated units, including acoustical plasters and sprayed asbestos, present the least satisfactory painting potential since these materials "breathe" through minute pores which essentially cover the entire surface of the material. Extreme care should be exercised in the painting of these surfaces and preferably with strict adherence to the manufacturer's recommendations. In no case should any attempt be made to bring the surface color up to the shade of the paint.

15.11 Light Reflection

Light-reflection values are reported in the AMA Bulletin for all acoustical materials manufactured by its member organizations. These values are established by the association using the "Baumgartner Sphere" method.[5] The values reported are the average of five tests on three different samples for a finish designated as white. All light reflection values are reported in the following ranges: (*a*) 0.75 or more, (*b*) 0.70 to 0.74 inclusive, (*c*) 0.65 to 0.69 inclusive, and (*d*) 0.60 to 0.64 inclusive.

Since the effects of aging and accumulation of dust and dirt can reduce light reflection by 0.10 to 0.20, there is no need for more definite values.

15.12 Flame Resistance

Because architectural acoustical materials are used as interior surface finishes, their flame-resistance characteristics are of extreme importance. The flame resistance of materials as reported by the AMA laboratory is based on a method essentially like that described in Federal Specification SS-A-118B.[6] This federal specification establishes specific criteria by which materials may be classified from *A* to *D*, depending on their test performance. No specific terms are given to describe these classes, but materials classified as *A* are usually considered incombustible, and those classified as *D* as combustible. Classes *B* and *C* represent materials of intermediate flame resistance.

It should be pointed out that while the AMA has adopted the designation for various degrees of flame resistance, as determined by a recognized standard test under a federal specification, there are certain localities in which these specifications are not necessarily recognized by local fire marshals. In a number of instances a material may have a class *A* rating which would imply that the material is flameproof, and yet the material may be rejected on a local basis. It is suggested that where acoustical materials are to be used in civic structures, schools, and places of public assembly, the designer obtain authorization through the local authorities for the use of specific materials. Another method of flame-resistance testing is given by the ASTM.[7]

15.13 Job Conditions

The following job conditions are recommended by the AMA[1] for the installation of acoustical materials to insure maximum performance.

Installations of acoustical materials should not be made when the building is excessively cold and damp or hot and dry. Temperature and humidity conditions closely approximating the interior condi-

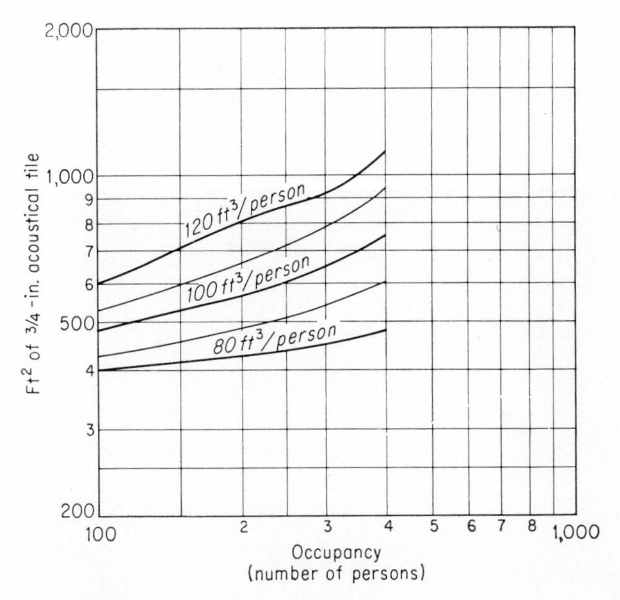

Fig. 15.10 Recommended areas of ¾-in.-thick acoustical tile, type I mounting, for reverberation control in speech rooms as a function of occupancy and volume per occupant.

tions which will exist when the building is occupied should be maintained before, during, and after installation.

All plastering, concrete, and terrazzo work (including grinding) should be complete and dry. All windows and doors should be in place and glazed. The heating system should be installed and operat-

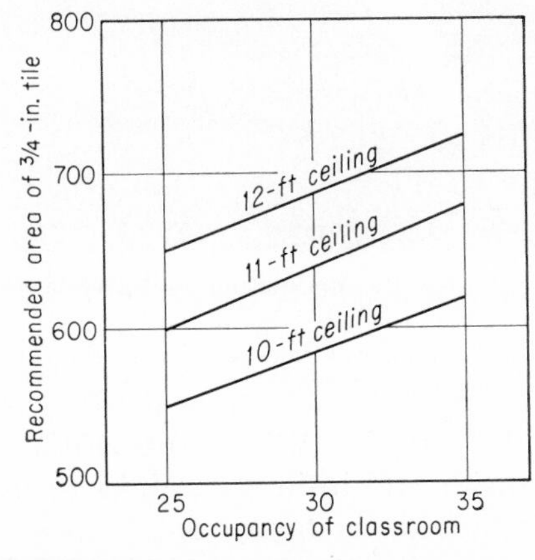

Fig. 15.11 Recommended areas of ¾-in.-thick acoustical tile, type I mounting, for typical grammar-school classrooms as a function of occupancy and ceiling height. A horizontal floor and rectangular construction are assumed.

ing where necessary to maintain proper conditions before, during, and after the acoustical work is in progress. Poured or precast concrete and gypsum or similar roof decks should be thoroughly dry and the space between such decks and suspended acoustical ceilings ade-

Fig. 15.12 Photograph showing the undulating wall and acoustic chambers in the sides of the MIT Chapel. This room is basically circular in shape. The undulations and the acoustical absorption render it free of focus in the center.

quately vented to the outside. Where substantial temperature differences between outside and inside of building occur at any season, acoustical materials should not be secured by cementing directly to the underside of a concrete, gypsum, or similar roof deck unless adequate thermal insulation is provided on the top side of such deck.

Where light from fixtures, cove lights, or high windows strikes the surface at a sharp angle, even slight unevenness of joints may result in unsatisfactory appearance. Under such conditions beveled materials should be used in preference to square-edge materials and installed with considerable care.

FIG. 15.13 The means for adding sound-absorbing material to a large arena through the use of hung flat baffles.

15.14 Amount of Acoustical Materials to Be Used

In the acoustical design of rooms, acoustical materials often need to be introduced to reduce noise, control echoes, and reduce reverberation. The design of large halls for speech or music is not within the scope of this book. The architect, however, is frequently faced with the acoustical treatment of conference and meeting rooms in hotels and convention halls and with the design of grammar-school classrooms (not of the university lecture-hall type). Guidance in the amount of material to be introduced into such rooms may be obtained from Figs. 15.10 and 15.11.

(*a*)

FIG. 15.14 (*a*) Sound-diffusing panels, auditorium, Aula Magna, Caracas, Venezuela. (*b*) Means for creating a highly absorbing surface at the rear of an auditorium with a long circularly curving wall with focal point on the stage. Auditorium, Aula Magna, Caracas, Venezuela.

15.15 Aesthetic Solutions to Specific Acoustical Problems

The solution to reverberation and noise-control problems in architectural acoustics need not be limited to the "obvious" use of prefabricated acoustical materials. Figures 15.12 through 15.17 show a number of examples of specific problems in auditorium, church, and arena design, taken from the files of Bolt Beranek and Newman Inc. The captions are self-explanatory.

15.16 Summary

This chapter discusses acoustical materials for architectural uses. Sound is absorbed in the interstices of porous materials, or by flexure of thin panels or by Helmholtz resonance. Porous materials are of two types: (1) prefabricated tiles, usually 0.5, 0.75, or 1.0 in. thick, and (2) acoustic plasters and sprayed-on mineral fibers.

Practical considerations include paintability, light reflection, flame resistance, and job conditions. These aspects are discussed here.

Recommended amounts of materials for meeting and conference rooms and for grammar-school classrooms are given in the figures.

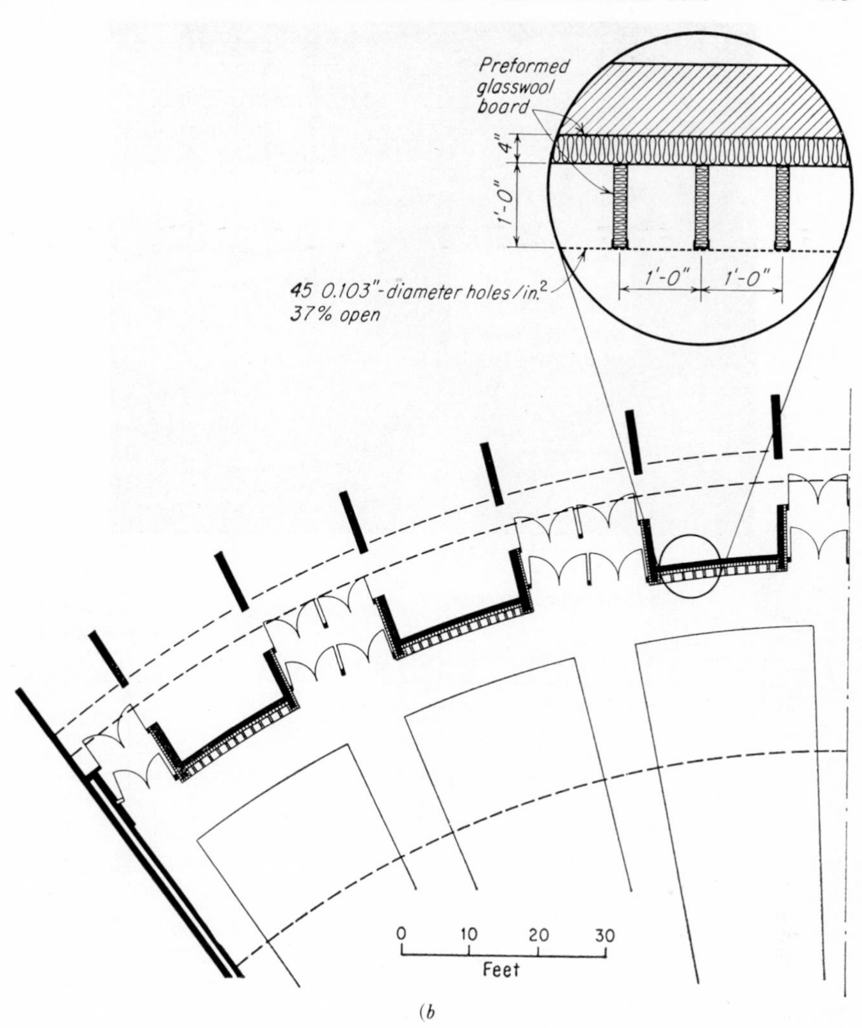

(b

Solutions of an interesting aesthetic nature are shown for several specific types of acoustical problems.

In Appendix A to this chapter, which follows, a review of the two fundamental formulas of room acoustics is made, and an example of their use is given.

Appendix A. Reverberation Time and Sound-pressure Level in a Room

The sound-pressure level, averaged by moving a microphone throughout a room, is calculated approximately from the formula

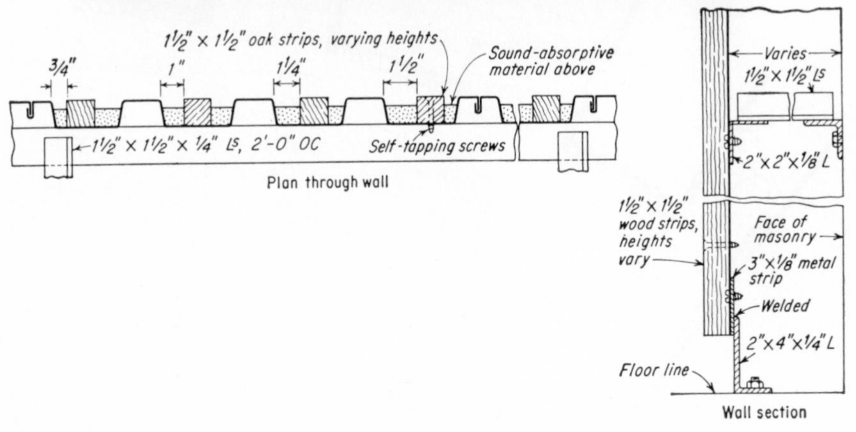

FIG. 15.15 An acoustical design for the sidewalls of an auditorium consisting of strips of standard perforated acoustical tile that are inserted between the ribs of metal deck. Due to the "edge" absorption effects the treated wall area absorbs sound very nearly as effectively as if the metal wraps were covered as well.

$$\mathrm{SPL_{av}} = \mathrm{PWL} - 10 \log_{10} S\bar{\alpha} + 6.5 \qquad \text{db} \qquad (15.1)$$

where PWL = sound-power level of source
$= 10 \log_{10} W + 130$ db
W = power radiated by source, watts
S = area of boundary surfaces of room, ft^2
$\bar{\alpha}$ = average absorption coefficient (see Eq. 11.1)
$S\bar{\alpha}$ = total absorption, sabins (units are ft^2)

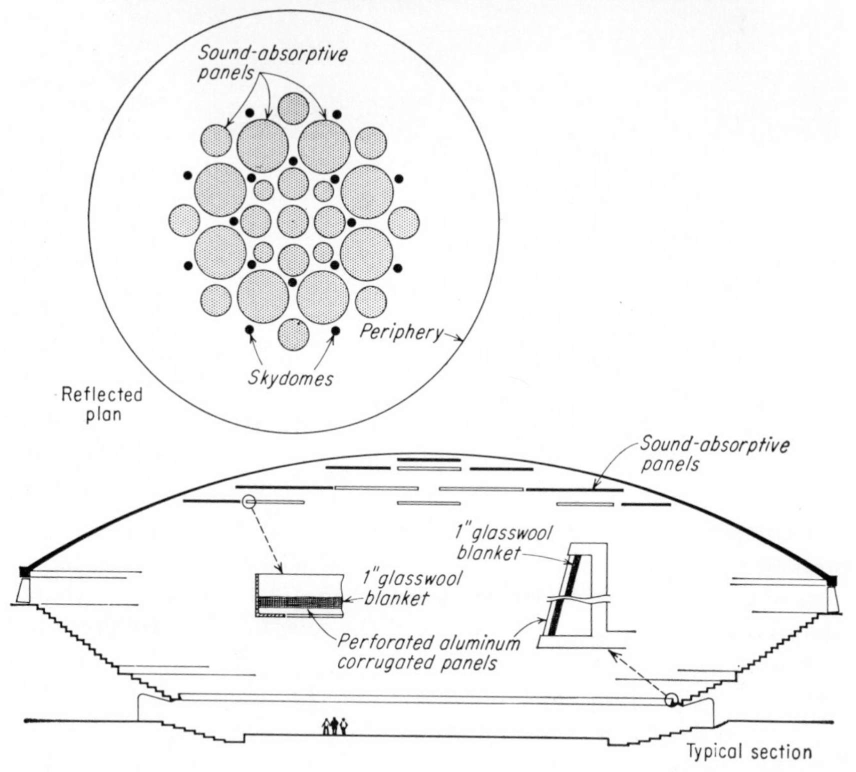

FIG. 15.16 Means for correcting the focusing effects of a spherical domed surface in an arena.

FIG. 15.17 Photograph showing means for covering absorbing and reflecting acoustical surfaces by means of an acoustically transparent grille work.

The reverberation time in an enclosure is given approximately (see Chap. 11 for more accurate formulas) by

$$T \doteq \frac{0.05V}{S\bar{\alpha}} \qquad \text{sec} \tag{15.2}$$

where T = length of time for sound to decay 60 db after source is turned off
V = volume of room, ft^3

Example 15.1. A source with a power output of 0.1 watt is located in a room with an average absorption coefficient of 0.15. If the dimensions of the room are 50 × 100 × 15 ft, what is the average sound-pressure level in the room and the reverberation time?

Solution

$$V = 75{,}000 \text{ ft}^3$$
$$S = 14{,}500 \text{ ft}^2$$
$$S\bar{\alpha} = 2{,}175 \text{ ft}^2$$
$$\text{PWL} = 120 \text{ db}$$
$$\text{SPL}_{av} = 120 - 10 \log 2{,}175 + 6.5 = 93 \text{ db}$$

$$T \doteq \frac{0.05 \times 75{,}000}{2{,}175}$$

$$\doteq 1.7 \text{ sec}$$

REFERENCES

1. "Sound Absorption Coefficients of Architectural Acoustical Materials," AIA No. 39b. Reissued annually by the Acoustical Materials Association, 335 E. 45th St., N.Y. 17, N.Y.

2. *ASTM Test Code C* 384–56*T*. Obtainable from American Society of Testing Materials, 1916 Race St., Philadelphia, Penna.

3. Beranek, L. L.: "Acoustic Measurements," John Wiley & Sons, Inc., New York, 1949.

4. Research Paper RP1298. *J. Research Natl. Bur. Standards,* vol. 24, May, 1940.

5. "Baumgartner Sphere," *Trans. Illum. Eng. Soc. N.Y.,* vol. 33, p. 379, 1938.

6. *Federal Specifications SS-A*-118*b*. Obtainable from the Superintendent of Documents, Washington, D.C., price 10 cents.

7. "Fire Hazard Classification of Building Materials," *ASTM Tentative Method E*-84-50-*T,* see Ref. 2 above.

SUGGESTIONS FOR FURTHER READING

Farrell, W. R.: "Acoustical Materials for Use in Monumental Spaces," *Noise Control,* vol. 4, p. 32, January, 1958.

Parkin, P. H., and H. R. Humphreys: "Acoustics, Noise and Buildings," Faber and Faber, Ltd., London, and Frederick A. Praeger, Inc., New York, 1958.

Sabine, H. J.: Acoustical Materials, chap. 18 of C. M. Harris (ed.), "Handbook of Noise Control," McGraw-Hill Book Company, Inc., New York, 1957.

Burris-Meyer, H., and L. S. Goodfriend: "Acoustics for the Architect," Reinhold Publishing Corporation, New York, 1957.

"Symposium on Acoustical Materials," *Special Tech. Publ.* 123, *ASTM,* 1952. Subjects include: H. J. Sabine: Measurement of Sound Absorption; W. Waterfall: Combustibility of Acoustical Materials; P. Chrzanowski and A. London: Maintenance of Acoustical Materials; L. Yerges: Erection of Acoustical Tile; W. Jack: Basic Physical Properties of Acoustical Materials.

Bolt, R. H., and R. B. Newman: Architectural Acoustics, *Architectural Record,* April, June, September, November, 1950.

Chapter 16

REACTIVE MUFFLERS

Peter A. Franken

16.1 General

A muffler is a special duct or pipe that impedes the transmission of sound while permitting the free flow of air. Successful muffler design requires that at least three criteria be satisfied simultaneously:

1. The *acoustical criterion* which specifies the minimum noise reduction required for the muffler as a function of frequency. The maximum permissible noise generated by the air flow through the muffler may also need to be specified.
2. The *aerodynamic criterion* which usually specifies the maximum permissible average pressure drop through the muffler at a given temperature and mass flow.
3. The *geometrical criterion* which specifies the maximum allowable volume and restrictions on shape.

Usually, the purchaser, when queried, asks for a muffler with an impossibly high noise attenuation, virtually no back pressure, and little or no volume. It is also important to him, of course, that the muffler be inexpensive, be very durable, and present no maintenance problems. Needless to say, in practice, these criteria for muffler design are modified somewhat. In this chapter we shall outline some of the analytical and empirical "tools" helpful in muffler design.

Ideally, we can divide mufflers into two categories, dissipative and reactive. For convenience we will define a *dissipative muffler* simply as a muffler whose acoustical performance is determined predominantly by the presence of flow-resistive material, and a *reactive muffler* as one whose acoustical performance is not greatly affected by the presence of flow-resistive material. In the interests of simplicity, we will relegate certain reactive elements (bends) to the following chapter on dissipative mufflers, since such elements frequently will contain some dissipative material in order to enhance their acoustical value. (Chapter 21 contains an alternate treatment of some dissipative and reactive devices, from the particular point of view of air-conditioning noise control.)

One step in the design of a muffler is the estimation of the acoustical behavior of the muffler. This may be expressed in terms of *insertion loss,* defined as the difference in the noise level at a particular point without and with the muffler. But in order to have meaning for the muffler manufacturer, these insertion-loss estimates must be accompanied by estimates of the change in mechanical performance due to the muffler. The details of mechanical operation involve complicated problems of superimposed steady and alternating flows which are not well understood. As a result, we cannot expect to predict performance changes in great detail, but we should at least investigate such changes qualitatively.

16.2 Acoustical Performance

The fundamental "equations of motion" governing linear acoustics are formally very similar to the equations of electrical circuit theory. This similarity leads to the convenient practice of studying acoustical systems with electrical analogs. Thus, we will often find it useful to take over the terminology and analytical results of electrical filter theory in our study of mufflers.

A block diagram (not an analogous electrical circuit) of our system is shown in Fig. 16.1. A source of sound pressure p, with internal

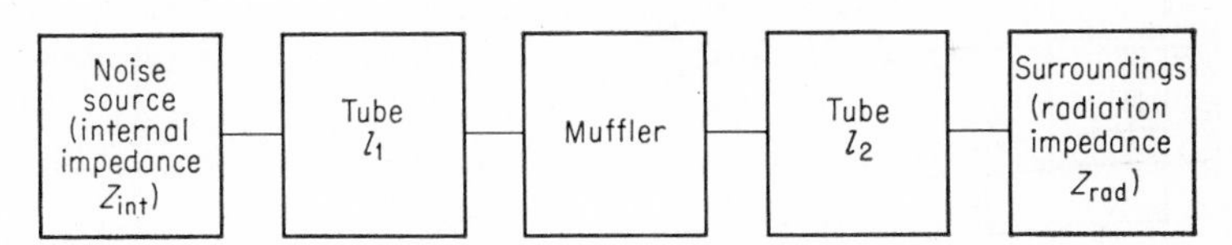

FIG. 16.1 Block diagram of muffler system, showing the noise source, the connecting tubes l_1 and l_2, the muffler, and the surroundings into which the sound is radiated.

impedance Z_{int}, is joined by a tube l_1 to the muffler, which in turn is joined by a tube l_2 to the surroundings. Radiation from l_2 to the surroundings is characterized by a radiation impedance Z_{rad}. It is clear that this system contains a complicated series of interactions between the component parts. In fact, the more general the system, the more complicated the interactions. It will therefore be valuable for us first to look at some idealized cases which are relatively simple to analyze. These simpler cases will often be sufficient to point out significant trends in the acoustical behavior of mufflers. After we become familiar with the more idealized cases, we will consider some of the complications of the more general system.

Transmission Loss of Mufflers. It would be extremely desirable if we could eliminate the source and radiation characteristics from our system in Fig. 16.1, and look only at some property of the muffler itself. Obviously, this simplification never occurs in practice, but we can readily perform the simplification in principle by defining a quantity called "transmission loss" (TL) as follows:

$$\text{Muffler TL} = \frac{\text{energy in incident wave at muffler inlet}}{\text{energy transmitted at muffler outlet into perfectly absorbing termination}} \tag{16.1}$$

Using a nonreflecting source and the incident energy only, we avoid any effects of source characteristics. We will discover later that TL, defined in this way, is a particular case of the quantity "insertion loss," which is of considerable interest in practical problems.

We will now consider the TL characteristics of three general muffler types.

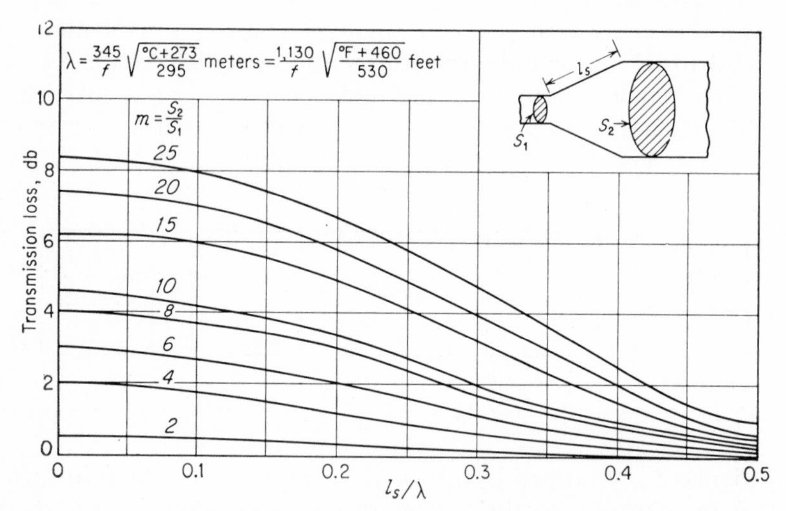

FIG. 16.2 TL of conical connectors. The units of l_s must be the same as those for λ. The wavelength λ is found either in meters or feet from the formulas shown.

Conical Connector. The first type is the conical connector, which joins a duct of cross-sectional area S_1, to another duct of cross-sectional area S_2.[1,2] The slant length of the connector is l_s. The TL for such a muffler is given in Fig. 16.2, as a function of l_s/λ. The wavelength of sound in the muffler is given by

$$\lambda = \frac{345}{f}\sqrt{\frac{\theta + 273}{295}} \qquad \text{m} \tag{16.2a}$$

$$\lambda = \frac{1{,}130}{f}\sqrt{\frac{T + 460}{530}} \qquad \text{ft} \tag{16.2b}$$

where f = frequency, cps
θ = muffler temperature, °C
T = muffler temperature, °F

The dimensionless parameter m in Fig. 16.2 is the area ratio

$$m = \frac{S_2}{S_1} \tag{16.3}$$

Expansion Chamber. The second type of muffler to be considered here is the expansion chamber shown in Fig. 16.3.[1,2] The behavior

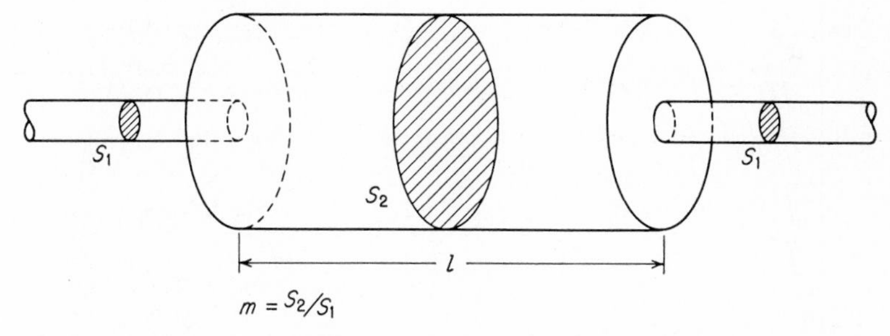

FIG. 16.3 Geometry of expansion chamber muffler.

of this unit can be described in terms of two dimensionless parameters m and kl, where

$$m = \frac{\text{muffler cross-sectional area}}{\text{duct cross-sectional area}} = \frac{S_2}{S_1} \tag{16.4}$$

$$k = \text{wave number} = \frac{2\pi}{\lambda} = \frac{2\pi f}{c} \qquad \text{m}^{-1} \text{ (or ft}^{-1}\text{)} \tag{16.5}$$

$$l = \text{muffler length} \qquad \text{m (or ft)} \tag{16.6}$$

So, in metric units,

$$kl = \frac{2\pi fl}{345}\sqrt{\frac{295}{\theta + 273}} \tag{16.7a}$$

or, in English units,

$$kl = \frac{2\pi fl}{1{,}130}\sqrt{\frac{530}{T + 460}} \tag{16.7b}$$

The TL for the expansion chamber muffler is given in Fig. 16.4. The equation describing the family of curves in Fig. 16.4 is

$$\mathrm{TL} = 10 \log \left[1 + \frac{1}{4}\left(m - \frac{1}{m} \right)^2 \sin^2 kl \right] \quad \text{db} \qquad (16.8)$$

From the form of Eq. (16.8) we see that the TL will be a periodic function in kl, repeating every π radians (180°). This result is valid

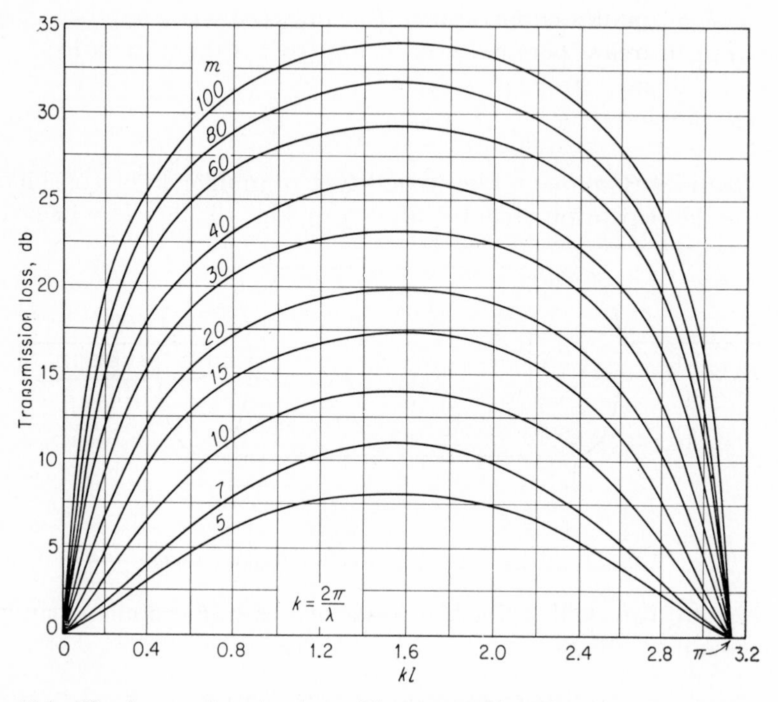

FIG. 16.4 TL of expansion chambers. The cross section of the muffler need not be round. The muffler width should not exceed about a quarter wavelength, for the graph to be valid. For values of kl between π and 2π, subtract π and use the graph. Similarly, for values between 2π and 3π, subtract 2π, etc.

for mufflers whose linear cross dimension (such as a diameter) is much less than one wavelength. This restriction is not serious, because wavelengths of interest are generally of the order of several feet.

Since m is proportional to the square of a linear cross dimension of the muffler, and kl is proportional to the length of the muffler, we can use Fig. 16.4 to "trade" muffler width against length in order to obtain the required TL, provided that the muffler width does not exceed about a quarter wavelength.

Example 16.1. A muffler is desired with a minimum TL of 10 db over the range of frequencies of 50 to 150 cps. The average temperature in the muffler is approximately 320°C (600°F).

(*a*) What is the minimum expansion ratio required if the muffler length is 0.305 m (1 ft)?

(*b*) With no restriction on muffler length, what is the minimum expansion ratio which will satisfy the TL requirements? What is the corresponding muffler length?

Solution. (*a*) From Eq. (16.7), the range of kl of interest corresponding to frequencies between 50 and 150 cps is from 0.2 to 0.6. Figure 16.4 shows that a value of m of approximately 36 will give 10 db TL at $kl = 0.2$ and 19 db at $kl = 0.6$. (*b*) Inspection of Fig. 16.4 shows that an expansion ratio of $m = 9$ provides 10 db or more of TL over the range of kl from 0.8 to 3×0.8. From Eq. (16.5) the value of k corresponding to 50 cps is 0.66 m^{-1} (0.20 ft^{-1}). The appropriate muffler length is then given by

$$l = \frac{kl}{k} = \frac{0.8}{0.66} = 1.2 \text{ m (4 ft)}$$

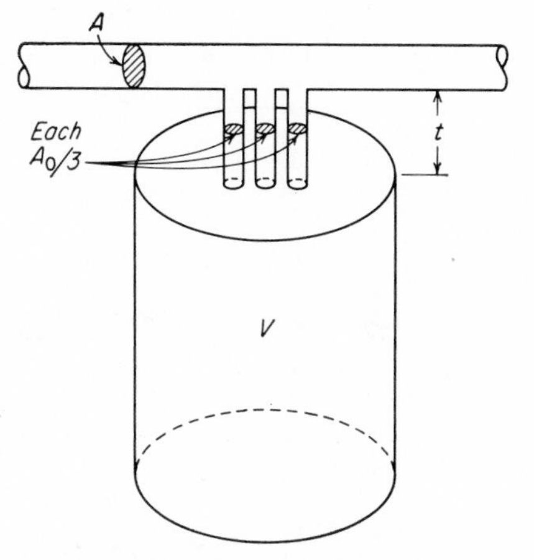

FIG. 16.5 Geometry of side-branch resonator. Volume V is joined to duct of cross-sectional area A by three tubes in parallel, each of length t and cross-sectional area $A_0/3$.

Side-branch Resonator. The third muffler type considered in this section is the side-branch resonator, in which flow does not pass through the resonator (see Fig. 16.5).* The equation for the TL of a side-branch resonator is

$$\text{TL} = 10 \log \left[1 + \frac{\alpha + \frac{1}{4}}{\alpha^2 + \beta^2(f/f_0 - f_0/f)^2} \right] \quad \text{db} \qquad (16.9)$$

where α = resonator resistance (dimensionless)
β = resonator reactance (dimensionless)
f_0 = resonance frequency, cps

Before we relate the dimensionless parameters α, β, and f/f_0 to the resonator dimensions, it will be interesing to look at the shape of the TL curves predicted by Eq. (16.9). At resonance ($f = f_0$) the TL is a function of α only, and can be written

* This treatment of resonators is based on material in Ref. 3.

$$\text{TL (at } f = f_0) = 20 \log \frac{\alpha + \frac{1}{2}}{\alpha} \qquad \text{db} \tag{16.10}$$

For α much less than $\frac{1}{4}$ and for frequencies much lower or higher than f_0, Eq. (16.9) becomes

$$\text{TL (for negligible } \alpha) \doteq 10 \log \left[1 + \frac{1}{4\beta^2 \, (f/f_0 - f_0/f)^2} \right] \text{ db} \tag{16.11}$$

Two families of TL curves are plotted in Figs. 16.6 and 16.7, as functions of f/f_0. The resistance parameter α has been set equal to the re-

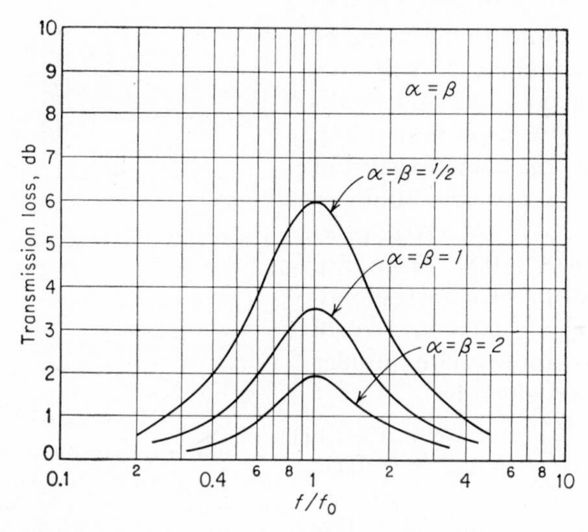

FIG. 16.6 TL of side-branch resonators. Resistance parameter α equal to reactance parameter β.

actance parameter β in the curves of Fig. 16.6. In the curves of Fig. 16.7 α has been set equal to $\frac{1}{2}\beta$.

We shall now relate the parameters to the dimensions. The following new symbols are involved in Fig. 16.5 and in the procedure described below:

A = duct cross-sectional area, m^2 (or ft^2)
V = resonator volume, m^3 (or ft^3)
n = number of aperture holes
A_0 = total aperture hole area, m^2 (or ft^2)
t' = equivalent neck length $\doteq t + 0.8\sqrt{A_0/n}$, m (or ft)
t = neck length, m (or ft)
c = speed of sound, m/sec (or ft/sec). [The speed c may be found from Eq. (16.2) by multiplying both sides by f, because $f\lambda = c$.]

q = "length" quantity convenient for use in charts, m (or ft)
ρ = density of air, kg/m^3 (or slugs/m^3)*

A procedure for choosing the appropriate resonator dimensions is itemized as follows:

1. Starting with the desired TL as a function of frequency, pick, from Figs. 16.6 and 16.7, the values of α, β, and f_0. For example, if the TL must exceed 4 db over a frequency range of 100 to 160 cps, then from Fig. 16.7 we see that $\alpha = \frac{1}{2}$, $\beta = 1$, and $f_0 = 125$ cps. If the curves of Figs. 16.6 and 16.7 do not contain the proper values of

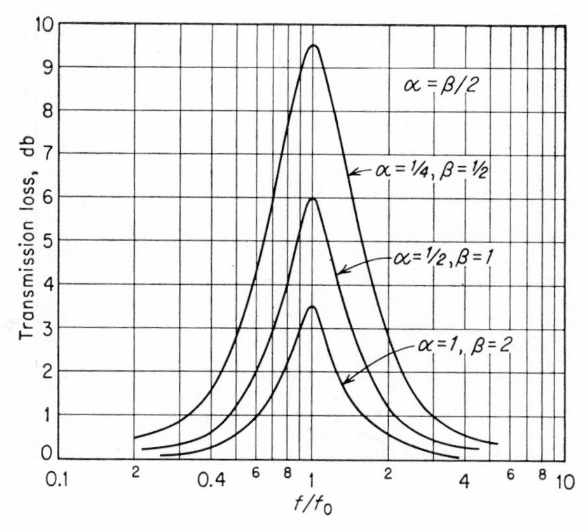

Fig. 16.7 TL of side-branch resonators. Resistance parameter α equal to $\frac{1}{2}$ reactance parameter β.

α and β, use Eqs. (16.10) and (16.11) to investigate other combinations of α and β. Smaller values of α lead to more TL at resonance, and smaller values of α/β lead to sharper TL curves. (In order for the theory to be valid, β should exceed approximately $\frac{1}{2}$, or if N resonators are in parallel, β should exceed approximately $\frac{1}{2}N$. These requirements are equivalent to the assumption that the resonator dimensions be small compared to a wavelength.)

2. Using the known duct area A and the resonance frequency f_0, determine the resonator volume from

$$V = \frac{Ac}{2\pi f_0 \beta} \qquad \text{m}^3 \text{ (or ft}^3\text{)} \tag{16.12}$$

* See Appendixes B and C for discussion of units.

3. Calculate the "length" quantity

$$q = \frac{2\pi f_0 A}{c\beta} \qquad \text{m (or ft)} \tag{16.13}$$

Choose an appropriate neck length t and number of holes n. Compute the dimensionless quantity q/nt. Then enter Fig. 16.8 and from

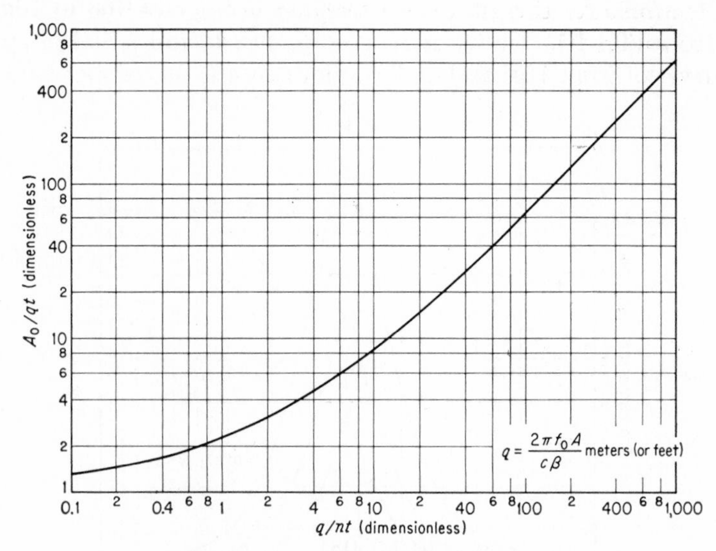

FIG. 16.8 Design chart for resonator aperture area A_0.

it obtain the value of the dimensionless quantity A_0/tq from the value of q/nt just computed. Determine A_0. If the resulting area of the apertures A_0 is unsatisfactory, vary the choices of n and t. Figure 16.8 indicates that larger values of n and t lead to smaller values of A_0. It is desirable to allow for some variation in the hole size of a prototype muffler to permit optimum tuning.

4. Choose a screen or other flow-resistive material for the aperture in line with the desired resistance α. The acoustic resistance R_s required in the aperture is

$$R_s = \frac{A_0}{A} \alpha \rho c \qquad \text{newton-sec/m}^3$$

where ρ = density of gas in resonator, kg/m³
c = speed of sound in resonator, m/sec

The resistance unit "newton-second per cubic meter" is also called the "mks rayl." (See Tables 12.6 and 12.7 for values of R_s for various cloths and screens.)

The equation for f_0 is implicitly contained in Fig. 16.8, and is

$$f_0 = \frac{c}{2\pi} \sqrt{\frac{A_0}{Vt'}} \quad \text{cps} \tag{16.14}$$

Example 16.2. Design a resonator with a TL of 6 db at a resonance frequency of 50 cps and a TL of 1 db at 100 cps. The duct diameter is 1.82 m (5.95 ft), there are four aperture holes, and the neck length is 0.0305 m (0.1 ft). The duct is at room temperature (22°C or 70°F).

Solution. From Fig. 16.7 or Eq. (16.9)

$$\alpha = 0.5$$

$$\beta = 1.0$$

The duct area is $A = \pi(0.91)^2 = 2.6$ m² (28 ft²). From Eq. (16.12)

$$V = \frac{2.6(345)}{2\pi(50)1} = 2.86 \text{ m}^3 \text{ (101 ft}^3\text{)}$$

From Eq. (16.13) calculate

$$q = \frac{2\pi(50)(2.6)}{345(1)} = 2.37 \text{ m (7.8 ft)}$$

Also calculate

$$\frac{q}{nt} = \frac{2.37}{4(0.0305)} = 19.4$$

From Fig. 16.8

$$\frac{A_0}{qt} = 15$$

$$A_0 = 15\,\text{qt} = 15(2.37)(0.0305) = 1.08 \text{ m}^2 \text{ (11.6 ft}^2\text{)}$$

$$\frac{A_0}{n} = 0.27 \text{ m}^2 \text{ (2.9 ft}^2\text{)}$$

The necessary aperture resistance is

$$R_s = \frac{A_0}{A}\alpha\rho c = \frac{1.08}{2.6}(0.5)(406) = 84 \text{ mks rayls}$$

After working through this procedure several times, it will be seen that the use of side-branch resonators is generally limited to frequencies below about 200 cps. Resonance frequencies higher than 200 cps require impractical dimensions. For example, the requirements that $\alpha = \beta = \frac{1}{2}$, $f_0 = 2{,}500$ cps, $A = 0.028$ m² (0.3 ft²), $n = 1$, $t = 0.0305$ m (0.1 ft), and normal room temperature, lead to $V = 0.0012$ m³ (0.04 ft³) and $A_0 = 4.3$ m² (47 ft²). Such an area violates the basic assumption that the lateral dimension be small compared to a wavelength. A lined duct or similar treatment is more suitable than a resonator at these higher frequencies (see Chap. 17).

Insertion Loss of Mufflers. We are now ready to consider some of the complications that are brought on by the nature of the noise source and the surroundings. The acoustical behavior of a muffler will be expressed in terms of "insertion loss" (IL), the difference in the noise levels measured at some external point with and without the muffler in the system. The TL defined earlier in the chapter is just the insertion loss for nonreflecting source and termination.

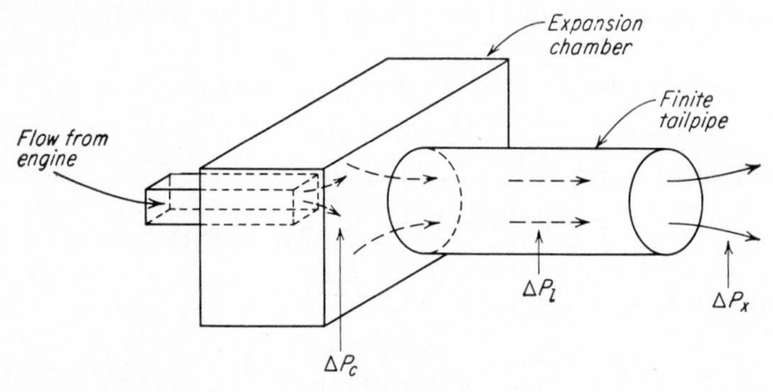

FIG. 16.9 Schematic view of a straight-through resonator, showing the components of steady-pressure drop, $\Delta P_{\text{total}} = \Delta P_c + \Delta P_l + \Delta P_x$.

A reactive muffler configuration of frequent interest is the straight-through resonator, in which the flow passes through the muffler (see Fig. 16.9). The straight-through resonator consists of an expansion chamber and a finite tailpipe. At some resonance frequency f_0 the effects of the cavity and the tailpipe effectively cancel each other, and the insertion loss then depends only on the dissipation in the muffler.

In order to make a quantitative study of the IL of the resonator, we must make some reasonable assumptions about the noise source and the surroundings. Let us consider that this muffler will be attached to a small 2-cycle gasoline engine. Such engines are commonly found on boats, garden tools, motorcycles, and some of the smaller foreign cars. We argue that a piston-driven engine of this type approximates a constant-volume-velocity (high-impedance) source, because the engine exhausts about the same volume of air, no matter what is hung on the tailpipe provided that the back pressure thus induced is small. Let us consider also that the tailpipe exit is unflanged and exhausts directly to the atmosphere.

We can set up an analogous electrical circuit for our cavity muffler and investigate its behavior. At high frequencies, there will be wave motion within the tailpipe and the cavity, so that the analog must be arranged to take these effects into account. Without setting up the analog, the results of such a study for a particular geometry of muffler are shown in Fig. 16.10, plotted as the insertion loss as a function of frequency.

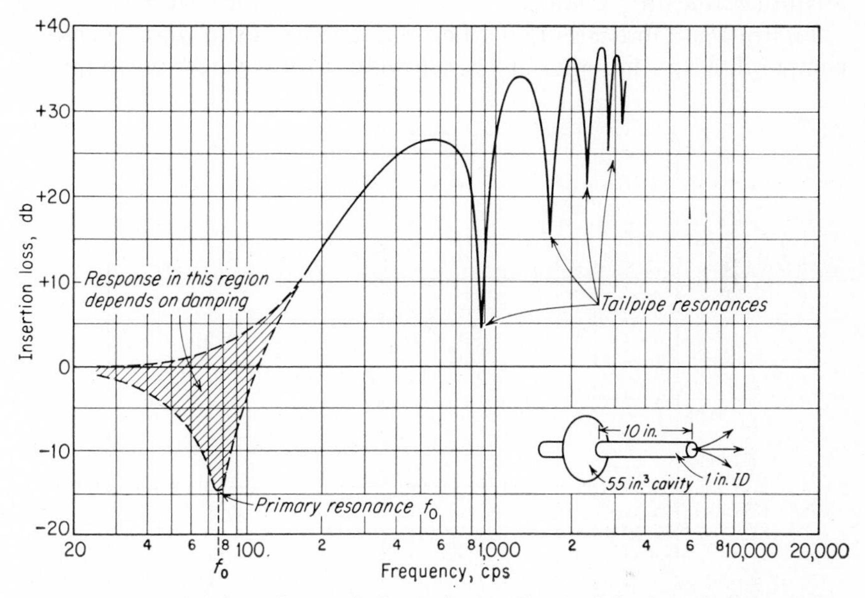

FIG. 16.10 Insertion loss of a particular cavity muffler (straight-through resonator) as determined from an analog study.

The resonance frequency f_0 is related to the muffler configuration by

$$f_0 = K\left(\frac{d^2\phi}{Vl\phi_0}\right)^{1/2} \qquad \text{cps} \tag{16.15}$$

where V = cavity volume
l = tailpipe length
d = tailpipe diameter
K = constant = 48 if units are meters
= 1.9×10^3 if units are inches
ϕ_0 = reference absolute temperature $\doteq$ 295°K (Kelvin) or 530°R (Rankine)
ϕ = absolute temperature in muffler, °K (or °R), i.e., $(273 + \phi)$°K or $(460 + T)$°R

At frequencies below f_0 there is practically no attenuation. Well above f_0 the attenuation increases rapidly. In the vicinity of f_0 the attenuation is determined by the acoustic dissipation of the muffler. This dissipation depends in a complicated way on the steady and alternating flows through the muffler. Thus, if a muffler is tested with no air flow, that is, with a loudspeaker as a noise source, then the muffler may *increase* the sound radiation at f_0 by improving the match between the loudspeaker and the air. But if the same muffler is tested on an engine, then the attenuation near f_0 will usually be significantly higher.

Figure 16.10 indicates that when the wavelength of sound becomes comparable to the muffler dimensions, "passbands" (dips in the

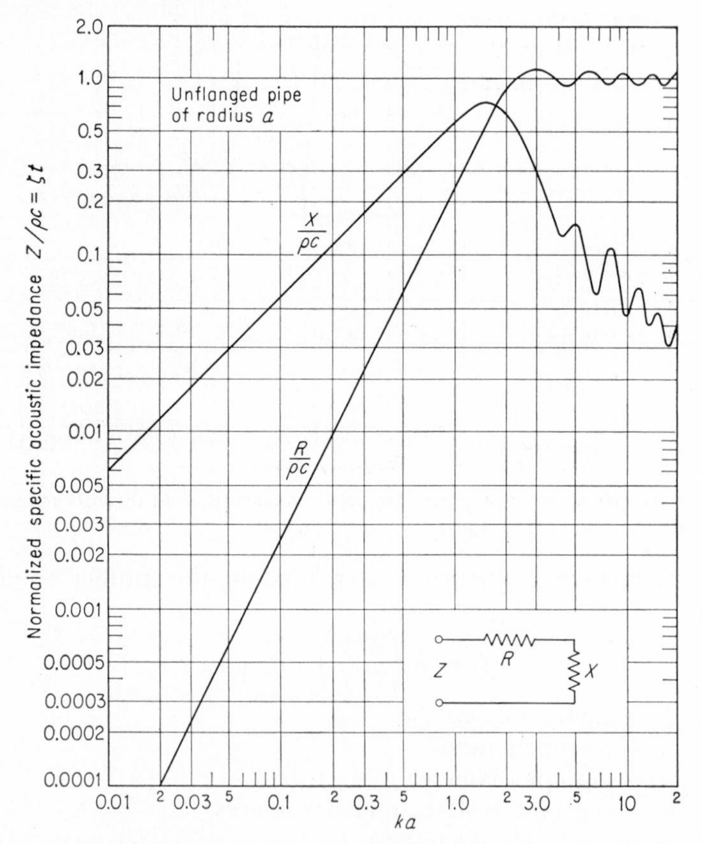

FIG. 16.11 Normalized specific acoustic impedance of a plane piston of radius a radiating outward from the end of a long unflanged tube. $R/\rho c$ is the real part of this normalized impedance and $X/\rho c$ is the imaginary part. The quantity ka is equal to the ratio of the circumference of the piston to the wavelength of the sound being radiated.

attenuation curve) appear. These passbands are located at frequencies satisfying the relation

$$f = n\frac{c}{2l} \qquad \text{cps} \tag{16.16}$$

where l and c are in consistent units.

We usually wish to locate the resonance frequency f_0 low enough so that there will be sufficient attenuation for low-frequency components of the noise. Equation (16.15) shows us that we need a large cavity volume and a tailpipe which is long or thin, or both.

For the cavity muffler, we have assumed that the termination impedance is that of an unflanged pipe. This termination is very common, and it is helpful to present this information in a general form for other applications.

Figure 16.11 plots the real and imaginary parts of the normalized specific acoustic impedance ζ_t of a plane piston of radius a mounted in the end of a long tube. Frequency is plotted in terms of ka, where

$$ka = 2\pi\frac{fa}{c} = \frac{\text{circumference}}{\lambda} \tag{16.17}$$

We have determined the insertion loss for one particular muffler of interest (cavity muffler), with an assumed source impedance (constant volume velocity) and termination impedance (unflanged pipe). We can now look at the problem more generally. The following notation will be employed:

$Z_i = \rho c \zeta_i$ = specific acoustic impedance of source
$Z_m = \rho c \zeta_m$ = specific acoustic impedance of muffler
$Z_t = \rho c \zeta_t$ = specific acoustic impedance of termination
$l = l_1 + l_2$ = total length of pipe between source and exhaust
$\eta = 1/\zeta$ = nondimensional admittance

Note that the letter ζ designates a dimensionless quantity equal to the ratio of the specific acoustic impedance to ρc. For a sound source with constant volume velocity, $\zeta_i = \infty$; for a sound source with constant pressure, $\zeta_i = 0$; for a nonreflecting sound source, $\zeta_i = 1$.

Figure 16.12 shows a schematic diagram of a side-branch muffler. The insertion loss for this configuration is*

$$\text{IL} = 20 \log \left| 1 + \frac{\zeta_i \eta_m (\cos kl_1 + j\eta_i \sin kl_1)(\cos kl_2 + j\eta_t \sin kl_2)}{\cos kl + j\zeta_i \sin kl + \eta_t(\zeta_i \cos kl + j \sin kl)} \right| \tag{16.18}$$

Equation (16.18) readily simplifies for special cases of interest. For example, for a nonreflecting source and termination we obtain the expression for transmission loss

* A portion of this material is based upon results in Refs. 2 and 3.

$$\text{IL}_{\zeta_i=\zeta_t=1} = \text{TL} = 20 \log \left| 1 + \frac{\eta_m}{2} \right| \tag{16.19}$$

For the case of a resonator muffler, where

$$\zeta_m = \frac{1}{\eta_m} = \alpha + j\beta \left(\frac{f}{f_0} - \frac{f_0}{f} \right) \tag{16.20}$$

Eq. (16.19) becomes our previous expression for a side-branch resonator,

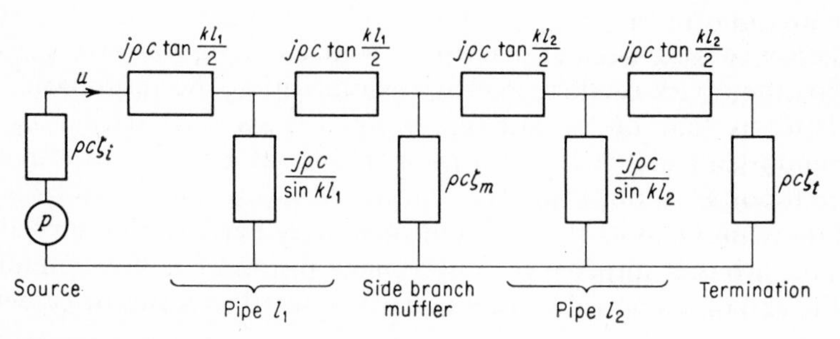

FIG. 16.12 Generalized schematic diagram of a side-branch muffler. $\zeta\rho c$ represents the specific acoustic impedance of the various elements.

Eq. (16.9). For a muffler consisting of a closed pipe of length L

$$\zeta_m = -j \cot kL \tag{16.21}$$

For an open pipe of length L (low frequencies only)

$$\zeta_m = jkL \tag{16.22}$$

Equation (16.18) and the special cases derivable from it can be used to determine the optimum placing of a side-branch resonator.

Example 16.3. A side-branch muffler is to be installed in a duct with a perfectly absorbing termination. The noise source has constant volume velocity. What should be the distance from the source to the resonator, in order to have a high IL over a relatively wide frequency range? Assume that the real part of the muffler admittance is never negative. (This assumption is not very restrictive.)

Solution. From Eq. (16.18)

$$\text{IL}_{\zeta_i=\infty,\zeta_t=1} = 20 \log |1 + \eta_m e^{-jkl_1} \cos kl_1|$$

This indicates that we have 0 db IL when $\cos kl_1 = 0$, or when

$$kl_1 = (2n+1)\frac{\pi}{2} \qquad \text{or} \qquad \frac{l_1}{\lambda} = \frac{2n+1}{4} \qquad \text{where } n = 0, 1, 2, \ldots$$

We see also that IL is greater than 0db for small values of kl_1. We conclude that l_1 should be as small as possible to avoid zero values of IL over a broad range of frequencies.

It is important to realize that many simplifying assumptions are implicit in this entire discussion on the acoustic performance of reactive mufflers. We may itemize the more important assumptions:

1. It has been assumed that flow through the system does not affect the sound propagation.

2. It has been assumed that temperature variations in the system do not affect the sound propagation. The effect of the *average* temperature on the velocity of sound and wavelength has been taken into account in Eq. (16.2).

3. It has been assumed that sound pressures are small compared with absolute pressures, so that nonlinear effects are negligible.

4. It has been assumed that muffler wall surfaces do not conduct or transmit sound.

5. It has been assumed that sound is propagated in plane waves, unattenuated by viscosity or heat conduction.

The first assumption, the neglect of flow effects, appears to be the most serious item in this list. Theoretical explanations of flow-sound interactions are not available at present. A descriptive account of some of the complicated effects of flow observed experimentally is given in Ref. 4.

Despite this impressive list of qualifications, the foregoing material serves as a useful guide in estimating muffler behavior.

Conventional mufflers for automotive exhausts are good examples of devices involving a combination of reactive elements such as those described in this chapter. Expansion chambers and side-branch resonators are tuned to eliminate predominant frequency components of the exhaust noise. By using a series of such tuned elements, one may achieve attenuation over a fairly broad frequency range. However, such combinations will also exhibit less attenuation than might be predicted in certain narrow bands of frequency where the individual elements interact with each other.

A preliminary estimate of the performance of an automotive exhaust muffler can be obtained as illustrated in the examples of this chapter, by using Fig. 16.4 for each expansion-chamber element and Eqs. (16.9), (16.12), and (16.14) for each side-branch resonator in the muffler. If improved accuracy is desired, the effects of interactions between the elements and the impedances of the source and the surroundings must be included. In an automotive muffler the source impedance is determined by the cylinder and manifold geometry, which may involve several parallel paths. Mathematical analysis in this case leads to complicated expressions such as Eq. (16.18). As we mentioned earlier in the chapter, a very convenient alternative technique is the use of analogous electrical circuits for the muffler system. Each element in the mechanical system (piece of tubing, resistance, noise generator, etc.) can be associated with simple electrical elements.* Analysis of the muffler system is then replaced by study of the electrical circuit, which is usually not difficult to perform experimentally. Thus, for example, the methods given in this chapter may first be used to choose the approximate dimensions for a series of automotive muffler resonators. Fine adjustments on the elements of an analogous electrical circuit will then permit improved design, including interaction effects among the resonators. Additional complications of flow or nonlinear effects must be studied in an actual mechanical model of the system.

16.3 Mechanical Performance

We mentioned earlier that it is important to be able to assess the changes in mechanical performance created by a muffler, but that our present understanding of the factors influencing this performance is very inadequate. We can, however, make some qualitative statements about back pressure, which influences mechanical performance.

Consider a muffler to consist of a chamber of some kind and a finite tailpipe. The steady pressure drop through such a muffler is made up of three principal components, as shown in Fig. 16.9.

1. A pressure drop (ΔP_c) caused by expansion and contraction of the flow in the chamber
2. A pressure drop (ΔP_t) caused by turbulent flow in the tailpipe
3. A pressure drop (ΔP_x) caused by expansion of the flow at the tailpipe end

ΔP_c is a function of the cavity dimensions and the relative size of the inlet and exhaust ducts of the cavity. The following design recommendations will tend to reduce ΔP_c:

1. Avoid square corners at the cavity exhaust.
2. Use slits perpendicular to the flow, rather than circular holes, as connections joining a side-branch muffler to the main duct. [The use of slits will bring the value of the equivalent neck length of a side-branch resonator closer to the value of the actual neck length (t' approaches t). This change requires care in tuning the resonator experimentally. Fig. 21.14 of Ref. 2 is a theoretical plot showing this effect.]
3. In straight-through muffler design, locate the inlet and exhaust ducts in line, and make the exhaust duct somewhat larger than the inlet in order to allow for expansion of the jet flow.

* The subject of acoustical analogs is discussed fully in Chap. 3 of Ref. 5.

ΔP_l, the pressure drop due to turbulent flow in the tailpipe, may be expressed as

$$\Delta P_l = 4 \frac{F_m l}{d} \times \frac{1}{2} \rho v^2 \tag{16.23}$$

where l = tailpipe length, in same units as diameter
d = tailpipe diameter
ρ = average gas density
F_m = Fanning friction factor for tailpipe
v = *instantaneous* linear velocity in tailpipe

For turbulent flow in small-diameter commercial steel pipe, or other pipe of comparable roughness, F_m is approximately 8×10^{-3}.*

It is important to note that the instantaneous value of ΔP_l depends on the total instantaneous particle velocity in the tailpipe. This means that the presence of an alternating particle velocity (due to a sound wave) will tend to increase ΔP_l. In the case of a noise source, such as a reciprocating engine, significant alternating velocities are generated at the fundamental firing frequency and its harmonics. Thus, it may be worthwhile to use a muffler with appreciable attenuation at the fundamental firing frequency and harmonics to reduce the alternating velocities and thereby reduce ΔP_l. Of course, care must be taken not to make l too long or d too small, since Eq. (16.23) indicates that these changes tend to increase ΔP_l.

Preliminary experiments indicate that ΔP_x, the pressure drop due to end expansion, is approximately

$$\Delta P_x \doteq 0.4 \times \tfrac{1}{2} \rho v^2 \tag{16.24}$$

Thus, reduction of any important alternating component in the flow will aid in reducing ΔP_x also. This behavior has been observed experimentally: a cavity muffler with a long tailpipe has *improved* the mechanical performance of a small reciprocating engine, compared with the unmuffled engine or with a muffler with a shorter tailpipe. The following example illustrates this unexpected change.

Example 16.4. A small piston engine has an average exhaust volume velocity V_0(m³/sec). Fourier analysis of the unmuffled exhaust time pattern shows that the volume velocity V_1(m³/sec) at the fundamental engine frequency f_1 is $2V_0$. Two cavity mufflers are to be tested on this engine. Muffler *A* does not attenuate sound of the engine fundamental; muffler *B* attenuates sound of that frequency by 6 db. The dimensions of the mufflers are given in Table 16.1.

* See, for example, pp. 124–129 of Ref. 6.

Table 16.1
Muffler Dimensions for Example 16.4

Dimension	Muffler A	Muffler B
Cavity volume, m^3	0.001	0.001
Tailpipe length, m	0.08	0.32
Tailpipe diameter, m	0.02	0.02

(*a*) What is the ratio of the resonance frequency of muffler A to that of muffler B? (*b*) What is the ratio of the steady pressure drops $\Delta P_l + \Delta P_x$ for the two mufflers?

Solution. (*a*) From Eq. (16.15), we see that f_0 is proportional to $[d^2/l]^{1/2}$. Therefore $(f_0)_A/(f_0)_B = 2$. (*b*) From Eqs. (16.23) and (16.24),

$$\Delta P_x + \Delta P_l \sim \frac{\rho}{2}\left[0.4 + 0.032\,\frac{l}{D}\right]\left\{(v_0 + v_1 \sin 2\pi f_1 t)^2\right\}$$

In order to compute the average pressure drop, we must take a time average of the quantity in the braces. Thus

$$[(v_0 + v_1 \sin 2\pi f_1 t)^2]_{\text{av}} = v_0^2 + \frac{v_1^2}{2}$$

since

$$[\sin 2\pi f_1 t]_{\text{av}} = 0$$

$$[\sin^2 2\pi f_1 t]_{\text{av}} = \tfrac{1}{2}$$

Also, since muffler B attenuates sound at frequency f_1 by 6 db, v_1 in muffler B is one-half the value of v_1 in muffler A.

By assumption $\qquad (v_1)_A = 2(v_0)_A$

So $\qquad (v_1)_B = \tfrac{1}{2}(v_1)_A = (v_0)_A$

Then
$$\frac{(\Delta P_x + \Delta P_l)_A}{(\Delta P_x + \Delta P_l)_B} = \frac{[0.4 + 0.032(4)]}{[0.4 + 0.032(16)]} \cdot \frac{v_0^2 + \dfrac{[2(v_0)]^2}{2}}{v_0^2 + \dfrac{(v_0)^2}{2}}$$

$$= \frac{0.528}{0.912}\,\frac{3}{1\tfrac{1}{2}} = 1.16$$

Muffler A with the shorter tailpipe has 16 per cent *higher* steady pressure drop than muffler B with the longer tailpipe.

16.4 Summary

This chapter reviews information on the acoustical performance of several common reactive muffler elements. The distinction is made between "transmission loss" and "insertion loss" as descriptions of acoustical behavior. Transmission-loss characteristics are given for conical connectors, expansion chambers, and side-branch resonators. Insertion-loss characteristics are given for cavity resonators and arbitrary side-branch mufflers. Mechanical performance is discussed in terms of the back pressure developed in common muffler elements.

REFERENCES

1. Davis, D. D., Jr., G. M. Stokes, D. Moore, and G. T. Stevens, Jr.: *Natl. Advisory Comm. Aeronaut. Rept.* 1192, 1954.

2. Davis, D. D., Jr.: Acoustical Filters and Mufflers, chap. 21 in C. M. Harris (ed.), "Handbook of Noise Control," McGraw-Hill Book Company, Inc., New York, 1957.

3. Ingard, U.: "Side Branch Resonators in Ducts," Bolt Beranek and Newman Inc. (unpublished).

4. Meyer, E., F. Mechel, and G. Kurtze: Experiments on the Influence of Flow on Sound Attenuation in Absorbing Ducts, *J. Acoust. Soc. Am.*, vol. 30, pp. 165–174, 1958.

5. Beranek, L. L.: "Acoustics," McGraw-Hill Book Company, Inc., New York, 1954.

6. Knudsen, J. G., and D. T. Katz: "Fluid Dynamics and Heat Transfer," University of Michigan Press, Ann Arbor, Mich., 1954.

Chapter 17

DISSIPATIVE MUFFLERS

Norman Doelling

17.1 Introduction

The distinction between dissipative and reactive mufflers is necessarily arbitrary. A purely reactive or purely dissipative muffler is a useful concept, but all real mufflers achieve noise reduction by both dissipation and reflection of energy.

Dissipative mufflers usually have relatively wideband noise-reduction characteristics. The sharp peaks (and valleys) in the NR curves which are always found in reactive mufflers are not present. Dissipative mufflers are most useful for noise-control problems associated with continuous noise spectra such as fan noise, jet-engine noise, and speech. Dissipative mufflers may also be useful if the center frequency of a narrow-band noise varies over a wide range as the operating conditions of the source fluctuate.

In this chapter, we consider the NR characteristics of lined ducts, bends, and lined plena which are the building blocks of all dissipative mufflers. Special emphasis is placed on the importance of "finite length" effects and on the differences between the noise reduction for plane axial sound waves, and for random-in-space waves at the input to the dissipative muffler. The analysis of Davis[1] is utilized to show how to calculate finite length effects. Empirical data have been obtained which allow reliable estimation of noise reduction for the random-in-space case from well-known plane-wave theories. With this information in hand, the reader will be able to estimate the noise reduction for almost any type of dissipative muffler.

17.2 Terminology and Definitions

Measures of Acoustical Performance. Insertion loss, transmission loss, SPL differences, noise reduction, end differences, and attenuation are but a few of the terms which are used in the literature of acoustics to describe the effectiveness of a noise-control component in an air passage.

Unfortunately, standard nomenclature* has not yet been adopted in the field of noise control for air passages. As a result authors frequently use different words to describe the same measure of acoustical effectiveness and, conversely, the same word may be used to describe different measures of acoustical effectiveness. We, therefore, temporarily digress to discuss some of the generally accepted meanings of these terms as they have been used in the previous chapter and as they will be used in this chapter.

Insertion loss is generally defined as the difference between two sound-pressure levels (or power levels or intensity levels) which are measured at the same point in space before and after a muffler is inserted between the measurement point and the noise source.

Transmission loss (TL) is defined as the ratio of sound power incident on the muffler to the sound power transmitted by the muffler. The ratio is expressed in decibels. Transmission loss is a useful analytic concept, but its measurement is hampered by the lack of an acoustic wattmeter. As noted in Chap. 8, sound power can be derived from measurements of sound pressure only under certain limited circumstances. Furthermore, one must devise ways of separating the sound power *incident* at the input to the muffler from the sound power *reflected* back toward the source by the muffler. The separation can be accomplished only by indirect techniques.

SPL difference, Noise Reduction, and *end differences* all have the same meaning and refer to the *difference* between the sound-pressure levels measured at the *source side* (input) of a muffler and the sound-pressure levels measured at the *receiving side* (output) of the muffler. "End differences" is perhaps the most descriptive term, but the author prefers "Noise Reduction" (NR) because it is parallel to the familiar terminology used to describe the acoustical effectiveness of walls. Furthermore, the differences between the Noise Reduction and transmission loss for a muffler are the same as the differences between NR and TL for a wall. The term *Noise Reduction*† will be used throughout this chapter to describe this measure of acoustical effectiveness for mufflers.

* An American Standards Committee, S-1-X-42, is presently exploring the feasibility of writing such standards.

† We capitalize Noise Reduction to avoid confusion with the nonspecific term "noise reduction" which may refer to a decrease in noise level accomplished by any means, such as, increasing the distance between source and receiver, lowering the noise output of the source, etc.

Attenuation is the decrease of *sound power* in decibels between two points in an acoustical system. The concept of attenuation is used by most writers in describing wave propagation in lined ducts, and we shall use it for this purpose in later sections of this chapter. In many cases, the attenuation can be measured by determining the decrease in sound-pressure level (in decibels) per unit length of duct—measured away from the ends of the duct—and multiplying by the total length of the duct.

Attenuation is a useful concept for describing the loss of power of a sound wave traveling in a muffler. However, the transmission loss, the insertion loss, and the Noise Reduction of the mufflers are not uniquely determined by the attenuation. Put in another way, attenuation is a useful concept for describing what is going on inside of a muffler, but it does not by itself tell the entire story of how effective a muffler is in a given system.

It is important to note also that insertion loss, transmission loss, and Noise Reduction are not uniquely related to the physical properties of a muffler. As illustrated by specific examples in the previous chapter, the acoustical effectiveness of all mufflers depends on the source and/or termination impedance. Thus, each of these parameters (TL, NR, or IL) must be considered as measures of the interaction between the muffler and its acoustical environment.

In noise-control engineering, insertion loss is perhaps the most useful measure of acoustical effectiveness. One generally measures or calculates the sound-pressure level at some point for a system which has no acoustical treatment. The difference between this sound-pressure level and the criterion sound-pressure level is just the insertion loss required for a muffler. The NR (sound-pressure-level difference from input to output) is, however, easier to measure than insertion loss or TL, so that most measured data are Noise Reductions.

The relations among insertion loss, Noise Reduction, and transmission loss can be found—provided the input impedances of the muffler, the output impedance of the source, and the input impedance of the termination are known. In the general case of arbitrary impedances one cannot say which of the functions is largest and which is smallest. However, for most dissipative mufflers the input and output impedances are not too different from ρc (the characteristic impedance of air). Furthermore, dissipative mufflers are usually used with low-impedance sources.* If in addition the termination impedance is nearly ρc, the following relations will hold:

* As pointed out in Chap. 16, reciprocating engines are a notable exception.

$$\text{Insertion loss} \approx \text{TL} < \text{NR}$$

In the rest of the chapter we shall assume insertion loss is equal to transmission loss. NR can usually be assumed to be about 3 db greater than TL. These relationships are realistic under the impedance conditions described above.

Description of Noise Fields. It is frequently assumed that a sound wave impinging on the input to a muffler travels parallel to the longitudinal axis of the muffler and that the wavefront is plane.* In almost all noise-control problems, the noise input is not such a *plane axial* wave. In fact, one must carefully design experiments in the laboratory to assure that only plane axial waves are obtained. When the wavelength of sound is less than the height or width of the duct, the sound wave generally has radial components as well as axial components.

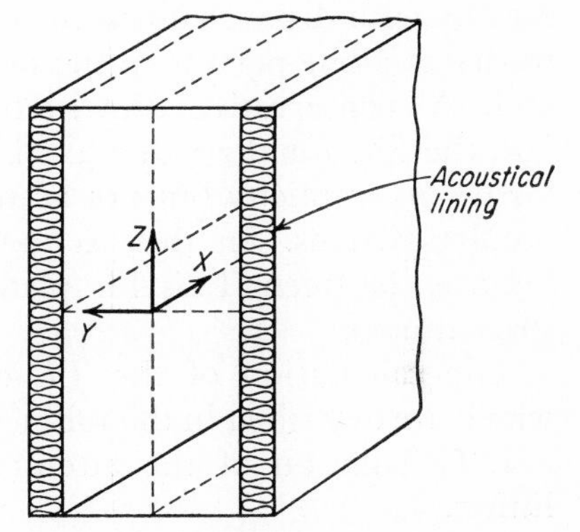

FIG. 17.1 An isometric sketch of a lined duct, showing the nomenclature used in describing the direction of a sound wave in the duct.

The axial component lies in the x direction shown in Fig. 17.1. The y component in Fig. 17.1 is perpendicular to the lined sides of the duct and shall be called the *normal component* of the sound wave. The z component in Fig. 17.1 is parallel to the lined sidewalls and shall be called the *grazing component* of the sound wave. If the noise input to a muffler has equal components in all directions, the input is said to be *randomly incident.*

17.3 Lined Ducts and Parallel Baffles

Introduction. A lined duct is an air passage with one or more interior surfaces covered with a porous acoustical material. Parallel baffles are merely a series of side-by-side ducts that generally have a rectangular cross section and may be considered as a special type of lined duct.

The TL of lined ducts arises primarily from the *attenuation* of sound by the absorbing lining. When the total cross-sectional area

* A truly plane wave cannot exist in a duct. Because of losses at the boundary, the wavefront is somewhat inclined toward the sidewalls.

of the lining becomes as great as the cross-sectional area of the free air passage or when there is a significant change in area at the ends of the ducts, reflected waves are produced which cause increased losses over and above the total attenuation. The combined loss is called the "transmission loss." For plane waves traveling along the longitudinal axis of a duct, the TL can be calculated from well-known theories. But, in practice, the noise input usually consists of many waves that impinge on the input to the muffler at all angles of incidence (random-in-space incidence). It is found that for such random-in-space inputs the oblique waves are attenuated much more rapidly with distance down the duct than the plane axial component so that it alone persists in the muffler a distance from the beginning end. A "conservative" (low) estimate of the TL would be to calculate the loss for the plane axial wave alone. This approach has the virtue of simplicity but not accuracy, and results in "overdesigned" muffler systems. In this section, therefore, we emphasize the differences between TL's for plane axial inputs and for random-in-space inputs.

The calculation of the TL of lined ducts involves three steps which are described in the following three sections:

1. Calculation of the attenuation of plane axial waves by the lining.

2. Calculation of the plane-axial-wave TL. This calculation takes into account the wave-reflection effects that occur at the ends of the ducts. This increase in attenuation would be unimportant for very long ducts, but for "finite" length ducts it is often very significant.

3. Calculation of the corrections to the plane-axial-wave TL for the case when the incident sound field is random-in-space.

Attenuation of Lined Ducts for Plane, Axial Inputs. The attenuation of a lined duct is dependent primarily on the duct length, the thickness of the lining, the flow resistance of the lining, the width or widths of the air passage, and the wavelength of sound. Morse's classic paper,[2] which appeared in 1939, shows how to calculate the loss of sound energy in a lined duct when the required parameters are known. Unfortunately, the calculations required are somewhat tedious. Cremer[3,4]* has developed charts which have greatly simplified the calculation procedures.

Unfortunately, neither of these works is convenient to use when solving the inverse problem of designing a duct to meet given NR requirements. Ingard [5] has obtained a series of approximate solu-

* Reference 4 contains a condensed English version of the original German work given in Ref. 3.

tions to the wave equation which are valid over fairly wide frequency ranges. These approximate solutions, which have been reduced to chart form, are more useful for designing a duct because they show explicitly the relations between the various parameters and attenuation. One can see immediately from his charts, for example, the effect on attenuation of varying the flow resistance of the lining while holding all other variables constant.

We present here some ways of rapidly estimating the attenuation of a lined duct. The data presented will be useful for the engineer concerned with the day-to-day routine noise-control problems. However, familiarity with the works of Morse, Cremer, and Ingard is essential to anyone doing research and development of "optimum" dissipative mufflers.

The acoustical effectiveness of many types of lined ducts has been experimentally determined.[6–14] Some of the data were obtained by Noise-Reduction measurements, some by insertion-loss measurements, and others are measurements of attenuation. The measurements of attenuation are most useful for design purposes. Several types of linings are included in the data in the references noted: flow-resistive linings, such as porous acoustical materials mounted on the duct wall with no facing or a facing with a large open area; flow-resistive linings mounted on the duct wall with "resonant" facings; flow-resistive linings separated from the duct wall by an air space. The first type is the most commonly used.

The attenuation characteristics of ducts that have flow-resistive linings mounted directly on their walls with no facing or a facing with relatively large (25 per cent to 35 per cent) open area are given in Fig. 17.2.

The attenuation parameter (Al_y) is the attenuation per length of duct equal to the duct width l_y. The attenuation for a duct of length l_e is therefore $Al_y(l_e/l_y)$. The frequency parameter l_y/λ is the ratio of the width between the linings to the wavelength of sound in air at the frequency of interest. To facilitate the use of this graph, a frequency scale for a duct 1 ft wide at an ambient temperature of about 70°F has been added. To use this chart for a duct 2 ft wide, for example, the frequency scale shown for a 1-ft duct must be halved. The change in the frequency scale caused by changes in temperatures may be neglected for the temperature range encountered in heating, ventilating, and air-conditioning problems. If very high temperatures are encountered, the value of λ should be calculated for the appropriate temperature and frequency.

The flow resistance R_1t of the duct linings for which these data are applicable is given on the graph in a dimensionless form. The attenuation, however, is not very sensitive to changes in the flow-resistance parameter, and this graph may be applied over an appreciable range of values of the flow-resistance parameter. The data are not seriously affected if the flow-resistance parameter is varied from one-half to twice the nominal value given. The flow resistance of many lining materials is given in Chap. 12.

If the duct is lined on four sides, the total attenuation may be obtained by adding, arithmetically, the attenuation of the "sides"

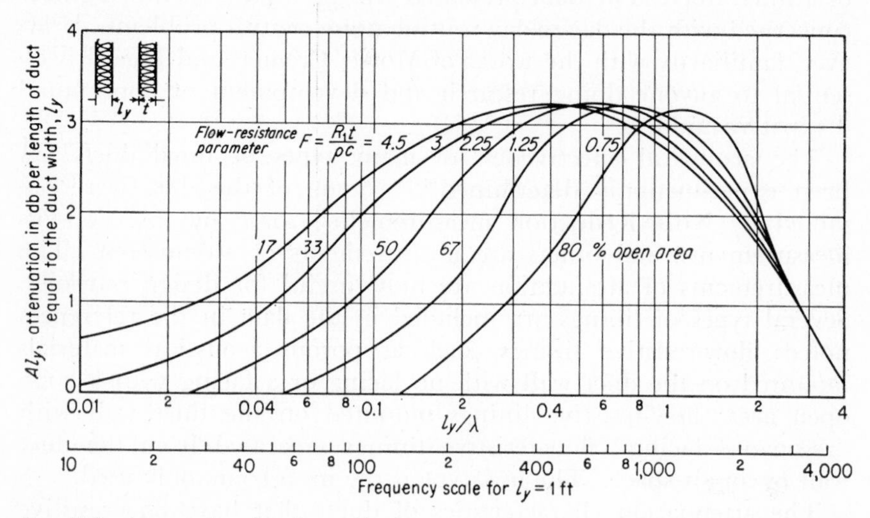

FIG. 17.2 Attenuation for ducts lined on two sides with a flow-resistive material. The flow-resistance parameter F is nondimensional so that R_1 (the flow resistance per unit length), t (the thickness of the lining), and ρc (the characteristic impedance of air) must be in consistent units. A is the attenuation in decibels per unit length of duct. To find the attenuation for a duct length l_e multiply the ordinate by l_e/l_y. To achieve the attenuations shown, for each percentage open area indicated, the flow resistance of the lining should be chosen to yield the proper flow-resistance parameter.

to the attenuation of the "top and bottom" of the duct. A square duct lined on four sides will have about twice the attenuation of a duct lined on only two opposite sides.

If the duct is lined on only one side, the attenuation is approximately equal to the attenuation calculated for an equivalent duct which is lined on two sides and which is twice as wide as the actual duct.

At very low frequencies ($l/\lambda < 0.1$) Sabine's approximation for attenuation[12] may be used even though the flow resistance is other than that shown in Fig. 17.2. Sabine found empirically that the attenuation of a lined duct could be expressed as:

$$A' = 12.6\,\bar{\alpha}^{1.4}\,\frac{P}{S} \qquad \text{db/ft} \tag{17.1}$$

in which A' = attenuation, db/ft
$\bar{\alpha}$ = statistical sound-absorption coefficient
P = acoustically lined perimeter of duct, in.
S = cross-sectional open area of duct, in.^{2}

This formula becomes increasingly inaccurate as l/λ increases, and its use must be restricted to small values of l/λ (low frequencies). Another restriction is that the impedance of the lining be much greater than ρc. For most acoustical materials, the restriction that l/λ be less than 0.1 will assure a sufficiently large impedance.

Calculation of the Transmission Loss of a Lined Expansion Chamber. The TL of a finite length of a lined duct can be found by considering the lined section of the duct as a lined expansion chamber. In this way one can account for the reflective effects of the change in cross section at the beginning and end of the lining. The results of an analysis based on this consideration are interesting. They show that the losses associated with the changes in cross section and the losses associated with the lining interact in an interesting and, to the noise-control engineer, favorable way. Only in the limit of very large attenuation by the lining can the TL be calculated by considering separately the attenuation of the lining and the reflections at the input and output. In most cases the TL is found to be greater than the sum of the attenuation and the reflection effects considered separately.

The effects of adding dissipation to an expansion chamber can be derived from Davis's[1] analysis by changing the phase-propagation exponent from kl_e to $(kl_e + j\sigma l_e)$. This procedure leads to an equation that looks formidable. However, we shall simplify the results by presenting the equation in the form of charts separately applicable to various frequency regions. One finds, by making the substitution of $(kl_e + j\sigma l_e)$ for kl_e and carrying out some algebra, that the ratio of the incident sound power to the transmitted sound power is:

$$\text{TL} = 10\log_{10}\left\{\left[\cosh\sigma l_e + \frac{1}{2}\left(m + \frac{1}{m}\right)\sinh\sigma l_e\right]^2\cos^2 kl_e + \left[\sinh\sigma l_e + \frac{1}{2}\left(m + \frac{1}{m}\right)\cosh\sigma l_e\right]^2\sin^2 kl_e\right\} \qquad \text{db} \tag{17.2}$$

in which TL = $10\log_{10} 1/\tau$
$1/\tau$ = ratio of incident sound power to transmitted sound power

σ = attenuation per unit length for lined duct, nepers (neglects end losses)

m = ratio of area of expanded or lined sections to area of initial (and final) sections of duct. The presence of the sound-absorbing lining is ignored in determining m. Implicitly, therefore, we have assumed that the thickness of the lining is small compared with the wavelength of sound λ_m in the material comprising the lining.

$k = \omega/c = 2\pi f/c$

l_e = muffler length

It is interesting to note that substitution of $1/m$ for m in no way affects Eq. (17.2). Thus a "contraction" chamber has the same TL as an expansion chamber.

For comparison with the TL, the total attenuation, as defined in Sec. 17.2, expressed in decibels is

$$\text{Total attenuation} = 10 \log_{10} e^{2\sigma l_e} = 8.68\sigma l_e \tag{17.3}$$

We shall now investigate the TL for small and large values of attenuation for the lined duct.

Small Attenuation ($\sigma l_e \ll 1$). With small attenuation, sinh σl_e nearly equals σl_e, and cosh σl_e nearly equals 1. The TL for a lined expansion chamber for $kl_e = 0, \pi, 2\pi, \ldots, n\pi$ is

$$\text{TL} = 10 \log_{10} \left[1 + \tfrac{1}{2}\left(m + \frac{1}{m}\right)\sigma l_e\right]^2 \quad \text{db} \quad (\sigma l_e \ll 1) \tag{17.4}$$

The TL of an unlined expansion chamber ($\sigma l_e = 0$) is zero at these values of kl_e. Thus the transmission loss arises solely because of the sound attenuation σl_e in the expansion chamber. The TL for a lined expansion chamber for $kl_e = \pi/2, 3\pi/2, \ldots, (2n + 1)\pi/2$ is

$$\text{TL} = 10 \log_{10} \left[\sigma l_e + \tfrac{1}{2}\left(m + \frac{1}{m}\right)\right]^2 \quad \text{db} \quad (\sigma l_e \ll 1) \tag{17.5}$$

Since σl_e is small, the TL of the duct results almost entirely from the reflections at the ends.

Large Attenuation ($\sigma l_e \gg 1$). When the attenuation of the lined duct becomes large, the hyperbolic functions both approach $\tfrac{1}{2}e^{\sigma l_e}$, and the TL becomes:

$$\text{TL} = 10 \log_{10} (e^{2\sigma l}) + 1 \log_{10} \frac{1}{4m}(1 + m)^2 + 10 \log_{10} \frac{1}{4} m \left(1 + \frac{1}{m}\right)^2 \quad \text{db} \quad \sigma l_e \gg 1 \tag{17.6}$$

This equation differs from the TL for an unlined expansion chamber [Eq. (16.8)] in two significant ways. First, it is not explicitly dependent on frequency ($k = 2\pi f/c$), although it is implicitly dependent on frequency because σ is a function of frequency. Second, the equation can be conveniently factored into terms which depend only on σl_e or m but not on both. This indicates that there are

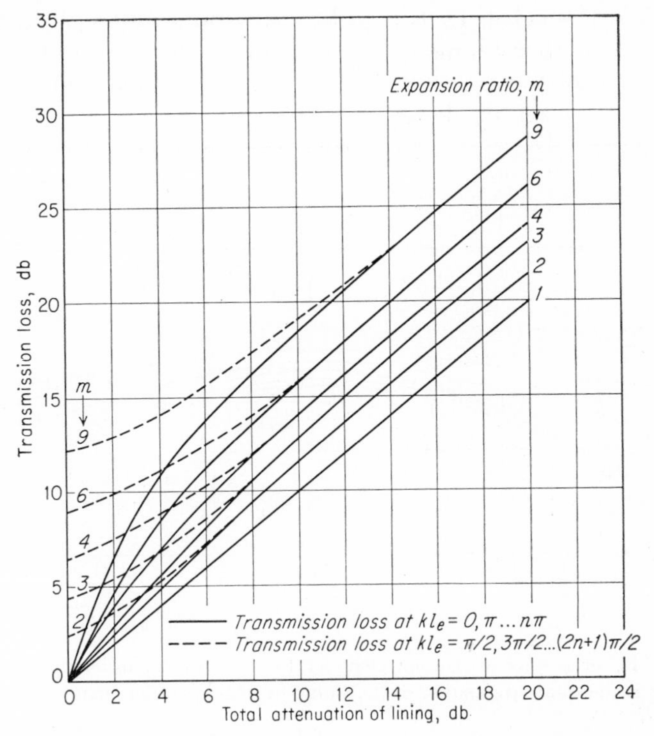

FIG. 17.3 The TL of a lined expansion chamber as a function of the area expansion ratio m and the total attenuation ($8.68\sigma l_e$ db) of the lining. The solid curves show TL for $kl_e = 0, \pi, \ldots, n\pi$, and the dotted curves show TL for $kl_e = \pi/2, 3\pi/2, \ldots, (2n+1)\pi/2$.

three "loss" mechanisms which operate independently and which can be calculated independently. The first term in this expression is just the attenuation provided by the lining ($10 \log_{10} e^{2\sigma l_e} = 8.68\sigma l_e$). The second term is the reflection loss at the "input" end of the expansion chamber, and the third term is the reflection loss at the "output" end of the expansion chamber.*

Intermediate Attenuation ($0.1 < \sigma l_e < 2$) with $kl_e = 0, \pi/2, \pi, \ldots, n\pi/2$. When finding the TL of lined ducts, it is usually not necessary

* See Ref. 15, p. 140.

to evaluate Eq. (17.2) at all values of the frequency parameter kl_e. It is usually sufficient to find the TL at the minima ($kl_e = 0$) and the maxima ($kl_e = \pi/2$). The TL at $kl_e = 0, \pi, \ldots, n\pi$ and at $kl_e = \pi/2, 3\pi/2, \ldots, (2n+1)\pi/2$ is given as a function of the total attenuation of the lining and the expansion ratio m in Fig. 17.3.

For large attenuations, the TL at $kl_e = 0$ approaches the TL at $kl_e = \pi/2$. Then the total TL is equal to the attenuation plus 20 $\log_{10} [\frac{1}{4}(1+m)(1+1/m)]$. [This relation follows directly from Eq. (17.6)]. Figure 17.3 may be used to estimate the TL for mufflers with an arbitrary expansion ratio and known attenuation character-

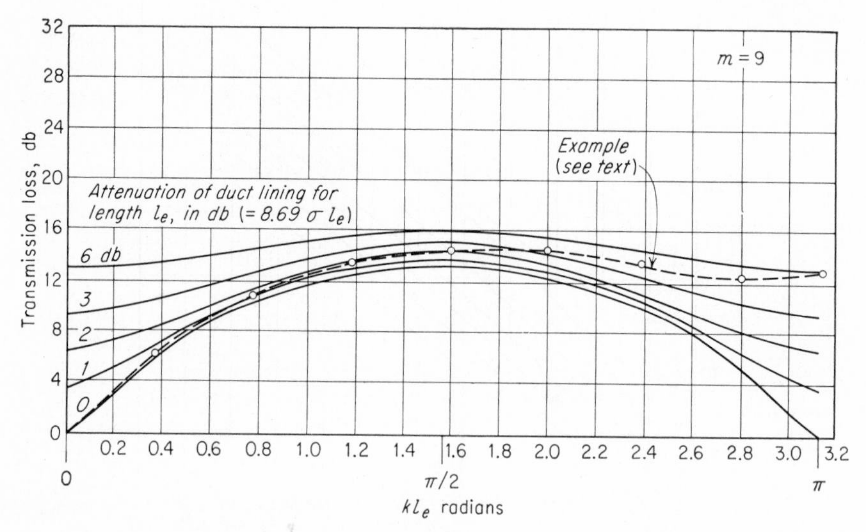

FIG. 17.4 TL of a lined expansion chamber for an area expansion ratio of 9. The parameter is the total attenuation of the lining of the expansion chamber (8.69 σl_e db).

istics. The TL for other frequencies (values of kl_e) will lie between the value at $kl_e = \pi/2$ and the value at $kl_e = 0$. An example illustrating the use of Fig. 17.3 is given in the summary paragraph below.

Comparison of Lined and Unlined Expansion Chambers. Although Fig. 17.3 gives all of the information needed to deal with the end effects for lined ducts, it is of interest to compare the TL characteristics of lined and unlined expansion chambers.

Figures 17.4 and 17.5 show TL's of lined expansion chambers (σl_e greater than 0) and of unlined expansion chambers ($\sigma l_e = 0$) for $m = 9$ and $m = 25$ respectively. The total attenuation for a sound wave propagating through the duct of length l_e is given as a parameter in both figures. The attenuation for any real lining will vary with frequency, and the TL for any real lined muffler must be found by entering the graph at the appropriate value of kl_e and reading off the TL at the appropriate value of total attenuation. All the important effects of adding dissipation are illustrated; a large* increase of TL with total attenuation at $kl_e = 0$; a smaller increase at $kl_e = \pi/2$; and for large σl_e a decrease in dependence on kl_e.

We see that the addition of dissipation to an expansion chamber (1) increases the frequency range in which a certain TL may be achieved, (2) lowers the expansion ratio required to achieve a given TL in a specified frequency range, and (3) decreases the length of muffler needed to obtain a given TL in some frequency range for a fixed expansion ratio.

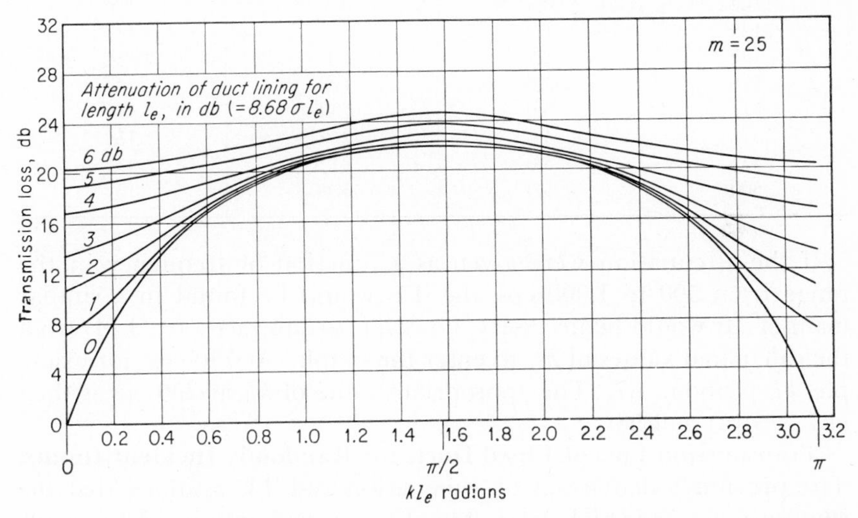

FIG. 17.5 TL of a lined expansion chamber for an area expansion ratio of 25. The parameter is the total attenuation of the lining of the expansion chamber (8.69 σl_e db).

Example 17.1. Assume an expansion chamber with an expansion ratio of 9 and a length l_e of about 1.1 ft. The frequency and the total attenuation of the lining are given respectively in columns 1 and 2 of Table 17.1. This information may be determined experimentally or analytically. It must be known before the TL can be found.

Solution. The TL is found by calculating the value of kl_e for each frequency and then entering Fig. 17.4 at the appropriate attenuation. For example, at 500 cps kl_e is equal to 3.14, and the attenuation of

* Note from Fig. 17.4 (for $m = 9$) that when the *attenuation* (due to adding a dissipative lining to the expansion chamber) is only 2 db, the *transmission loss* at $kl_e = 0$ is 6 db. The TL of an *unlined* expansion chamber is zero for $kl_e = 0$. Thus, the increase in TL (6 db) is three times as great as the increase in attenuation (2 db).

the lining is 6 db. The TL, found from Fig. 17.4, is about 13 db. The TL at other frequencies is plotted in Fig. 17.4 and tabulated in column 4 of Table 17.1.

Table 17.1

Steps for Calculating the Transmission Loss of the Lined Expansion Chamber Given in Example 17.1

Frequency, cps	Attenuation of Lining, db ($8.69\ \sigma l_e$)	kl_e, radians	Transmission Loss, db (From Fig. 17.4)
(1)	(2)	(3)	(4)
62	0.5	0.39	6
125	1.0	0.79	11
187	1.5	1.17	13
250	2.0	1.57	14
312	3.0	1.96	14
375	4.0	2.36	13
437	5.0	2.75	13
500	6.0	3.14	13

If the attenuation were given as a function of frequency in the range from 500 to 1,000 cps, the TL would be found in a similar manner. It would be necessary, however, to subtract π (= 3.14) from the calculated values of kl_e to enter the graph. At 750 cps, for example, kl_e is about 4.7. The appropriate value of kl_e at 750 cps is then (4.7 − 3.14) or 1.56.

Transmission Loss of Lined Ducts for Randomly Incident Inputs. The previous calculations of attenuation and TL assumed that the input was a plane axial wave. The TL for randomly incident inputs is now to be considered.

First let us consider cases where the frequency is high enough that the wavelength of sound is much less than the smallest duct width. Assume a wave that enters the duct at an angle ϕ and travels in the x,y plane of Fig. 17.1. The wave reflected from a lined side is reduced to $[1 - \alpha(\phi)]$ times the incident wave [$\alpha(\phi)$ is the absorption coefficient of the duct lining at an angle of incidence ϕ]. The total number of reflections n is approximately

$$n = \frac{l_e}{l} \tan \phi \tag{17.7}$$

in which l = duct width

l_e = total length of duct

ϕ = angle of incidence of sound wave at lined side, measured between duct axis and direction of wave travel

The noise reduction of the duct is $[1 - \alpha(\phi)]^{(le/l) \tan \phi}$. As ϕ approaches 90°, tan ϕ and the noise reduction become very large. As ϕ approaches 0°, the wave travels down the axis of the duct, and the noise reduction approaches zero because the frequency has been assumed to be so high (and the wavelength so short) that the wave travels like a light beam. In that case, the wave is attenuated very little by the sound-absorbing boundaries.

Consider next a sound wave with a component parallel to the lined sides in the x,z plane of Fig. 17.1. A sound wave moving at an angle $\pm\theta$ to the "top" or "bottom" of the duct goes $(1/\cos\theta)$ times as far in going through the duct as a wave traveling along the axis ($\theta = 0$). If the noise reduction is N db for $\theta = 0$, then noise reduction will be $N/(\cos\theta)$ for a wave at an angle θ. Again the noise reduction becomes very large as the angle θ approaches 90°.

Sound waves in a duct, of course, will generally have both parallel (z) and normal (y) components. The noise reduction characteristics for these cases, however, will be similar in certain respects to the cases described above. In particular, the sound waves with radial components are attenuated more rapidly than the purely axial sound waves. As a result, only the axial component will persist at some distance into the muffler. The noise reduction per unit length is then equal to the value given by the axial-plane-wave theory.

With this qualitative description of the random incidence attenuation phenomena, we can now estimate how attenuation per unit length varies with length of the muffler. Assume, for example, that the attenuation for a duct is 6 db/ft for sound waves incident at 90° to 70° from the longitudinal axis, 3 db/ft for waves between 50° and 70°, 2 db/ft for waves between 30° and 50°, and 1 db/ft for waves from 0° to 30°. Furthermore, let the relative intensity level in each of these angular ranges be 100, 95, 90, and 80 db, respectively, at the input plane. The total sound intensity at any point in this structure is obtained by a summation of the sound energy from 0° to 90°. Figure 17.6 shows the sound energy in each angular range as a function of distance into the structure. The total sound intensity at the input plane is about 102 db, and the major contribution comes from the energy in the 70° to 90° region. At a distance of 10 ft into the structure, the level is about 73 db, with approximately equal contributions from the 30° to 50° and 0° to 30° angular regions. At 20 ft only the 0° to 30° component remains of importance.

The attenuation per unit distance (the slope of the curve) varies significantly as a function of length into the treatment. In the first 1 ft of distance from the entrance, the attenuation is about 6 db/ft.

From 10 ft to 11 ft, the attenuation is about 1½ db/ft. At 20 ft, the attenuation is 1 db/ft.

Detailed field measurements[13] bear out the qualitative arguments described above. It has been found that the attenuation per unit length is large near the input and approaches the constant axial, plane wave value about 4 duct widths into the duct. The noise reduction for ducts longer than four duct widths can be approximated by an attenuation of the form of $(a + bl_e)$, where a is an end correc-

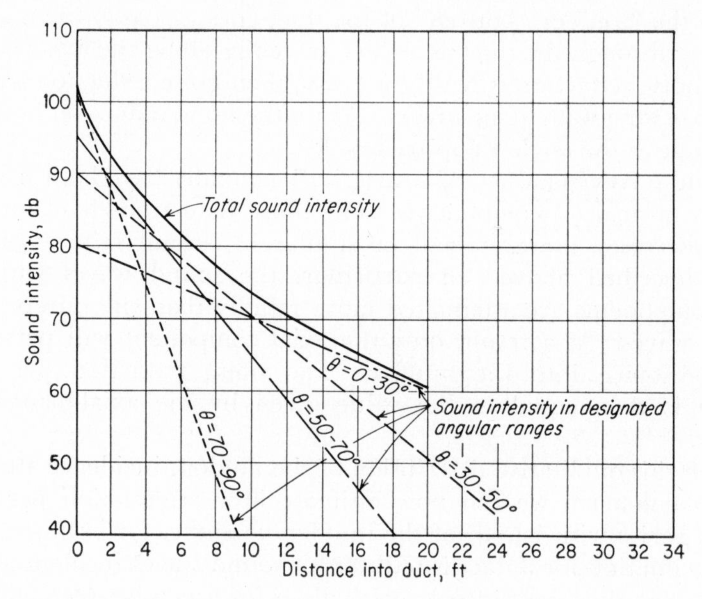

Fig. 17.6 An example of the decrease of sound intensity as a function of distance into a lined duct for a case in which the attenuation increases with increasing angle of incidence. Note that the change in total sound intensity per unit distance is not constant.

tion which measures the loss of the noise energy from the higher order modes, b is the attenuation per foot, and l_e is the length of the muffler in feet. In Fig. 17.6, for example, a is 22 db, and b is 1 db/ft.

The frequency at which the random incidence end correction a becomes important depends both on the width between the lined sides of the duct and the height of the duct (even though the top and bottom may not be lined). The magnitude of these end effects has been found for several widely differing geometries of ducts and baffles.[13] The magnitudes do *not* (1) correlate well with the ratio of l/λ and (2) do *not* have a frequency dependence similar to that of the plane-axial-wave attenuation. The magnitude of a does correlate rather well with $\sqrt{S}/\lambda$, in which S is the area of the duct. Furthermore, since a is essentially a measure of the ratio of the energy in higher order modes to the energy in the plane axial component, a is almost independent of the attenuation of the duct.

Figure 17.7 shows the values of a as derived from experimental data. There is an argument* which indicates that a ought to con-

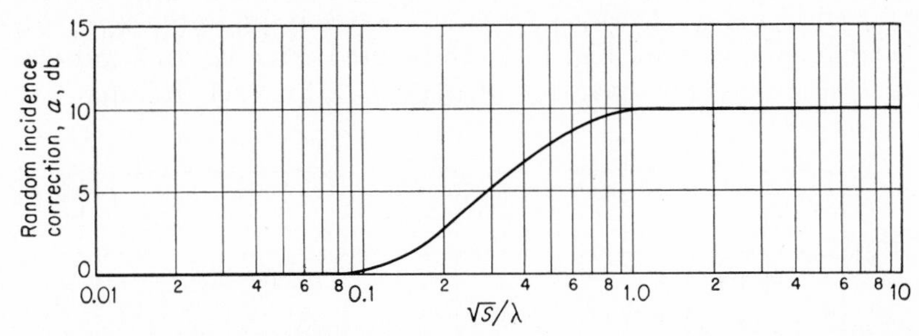

Fig. 17.7 Correction a to be added to plane-axial-wave TL of a duct to obtain the TL for a random incidence input. S = open area of duct; λ = wavelength of sound.

tinually increase with frequency; however, experimental data indicate an upper limit of about 10 db.

Example and Summary. We now have the necessary information to derive the TL of a lined duct with a large expansion ratio (or as it is usually expressed in relation to lined ducts, a small percentage open area). All of the information in the preceding paragraphs is most easily summarized by use of an example.

Example 17.2. An expansion chamber, 5 ft in length, is inserted in a large ventilating duct. The chamber is lined with acoustical material on two opposite sides. The width of the lined section is 5 ft, and the height is 3 ft, which is comprised of one side lined with 1 ft of acoustical material, a 1-ft air passage, and the other side lined with 1 ft of acoustical material. The ducts preceding and following the expansion chamber are 5 ft wide and 1 ft high. The open cross-sectional area of the expansion chamber is 33 per cent of the total cross section, so that the expansion ratio is 3.

Solution. The total TL is found in three steps. First, the attenuation of the duct is obtained from Fig. 17.2. Next, the plane-wave TL is calculated by considering the duct as a lined expansion chamber.

* The argument, briefly, is as follows: For random excitation of a duct, the sound energy is equally divided among all of the modes. The number of modes in a given frequency interval increases with increasing frequency above the second mode. Therefore the ratio of energy in all modes to the energy in the first mode (axial plane wave) must also increase with frequency.

(This step will be essential only for m greater than 2. The TL is approximately equal to the attenuation for m less than 2.) Finally, the random incidence end correction is added to the plane-wave TL to find the total TL.

The attenuation of the lining as a function of frequency can be found very simply if the total flow resistance $R_1 t$ is about 3 ρc as required in Fig. 17.2. Since the width between the lining is one foot, the frequency scale on Fig. 17.5 can be used directly. In a more general case one would have to calculate l_y/λ for each frequency. At

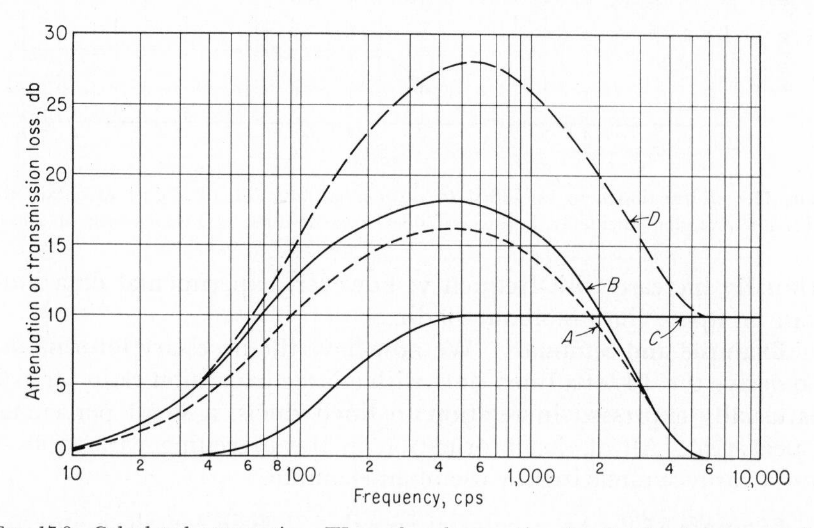

FIG. 17.8 Calculated attenuation, TL, and end correction for the lined duct used in Example 17.2. (*A*) Axial-plane-wave attenuation of duct; (*B*) TL of duct for plane axial waves; (*C*) random-incidence end correction; (*D*) total TL for random-incidence, curve *B* plus curve *C*.

$l_y/\lambda = 1$, about 1,100 cps, the attenuation per duct width Al_y is 2.8 db. The total length of the muffler is five duct widths, so the total attenuation is 5×2.8 or 14.0 db at 1,100 cps. The attenuation of the muffler at other frequencies for axial plane waves has been found from Fig. 17.2 and has been plotted as a function of frequency in Fig. 17.8 as curve *A*.

The next step is to calculate the TL from the attenuation. A complete set of curves similar to those in Figs. 17.4 and 17.5 has not been calculated for $m = 3$. However, the TL characteristics can be estimated from Fig. 17.3, which gives TL at $kl_e = 0$, and $kl_e = \pi/2$ as a function of attenuation and expansion ratio m.

First, the frequency for which $kl_e = \pi/2$ is calculated

$$kl_e = \frac{2\pi f l_e}{c} = \frac{\pi}{2}$$

$$f = \frac{c}{4l_e} = \frac{1{,}100}{4 \times 5} = 55 \text{ cps}$$

From Fig. 17.8 one finds that the attenuation of the lining is 6 db at 55 cps. Entering Fig. 17.3 at an attenuation of 6 db and reading up to the dotted curve ($kl_e = \pi/2$) for $m = 3$, one finds the TL is almost 9 db. For small attenuations and large values of m, it would also be necessary to calculate the frequencies for which $kl_e = \pi$, 2π, . . . , $n\pi$ and $kl_e = 3\pi/2$, . . . , $(n + 1)\pi/2$ in order to find the TL as a function of frequency. In this case ($m = 3$), however, one can see from Fig. 17.3, that the TL at $kl_e = \pi/2$ is almost equal to the TL at $kl_e = \pi$ for all attenuations greater than 6 db. Furthermore, one finds from Fig. 17.3 that the TL for $m = 3$ is just about 2.5 db greater than the total attenuation for all total attenuations greater than 6 db.

At frequencies above which the wavelength is less than half the duct width (about 2,000 cps for this example), the expansion chamber no longer behaves as an effective reactive muffler. The effect of the change in cross-sectional area, which is the source of the difference between TL and attenuation, will tend toward zero above 2,000 cps, and the TL will be about equal to the attenuation rather than 2.5 db greater.

At low frequencies the attenuation and TL both approach 0 db. The TL for the muffler has been estimated from the considerations in the previous paragraphs and is given in Fig. 17.8 as curve B.

The random incidence end correction of Fig. 17.7 is now added to the axial-plane-wave TL to find the TL for random incidence. The area of the duct S is 5 ft^2, and $S^{1/2}$ is 2.24 ft. Now we find the frequency for which $S^{1/2}/\lambda = 1$.

$$\frac{S^{1/2}}{\lambda} = S^{1/2}\frac{f}{c} = 1$$

$$= \frac{c}{S^{1/2}} = \frac{1{,}100}{2.24} = 490 \text{ cps}$$

For all frequencies above 490 cps, the random incidence end correction is 10 db (see Fig. 17.7). The end correction below 490 cps has been obtained from Fig. 17.7 and is plotted in Fig. 17.8 as curve C. The total attenuation, curve D, is found by adding curve C to curve B.

17.4 Lined and Unlined Bends

Discussion. There are many data available pertaining to the transmission of sound around bends. The data are difficult to compare because many different measurement techniques are used and there is no consistent nomenclature for the noise reduction of bends. The state of the art is also hampered by a lack of an exact theoretical treatment of the problem. Miles[16] has obtained a solution to the wave equation which is useful for frequencies below which the wavelength is about one half of the duct width. Experimental data[17,18,19] are in good agreement with the results obtained by Miles.

In the high-frequency range, where the wavelength is less than a duct width, the experimental data are notable for their lack of consistency.[7,13,16–19] Part of the inconsistency arises from confusion of nomenclature, and part arises because the effectiveness of a bend depends very strongly on the mode structure (randomness) of the incident sound field.

To illustrate the important differences between the various definitions of acoustical effectiveness of bends, consider the lined bend illustrated in Fig. 17.9*a*. Noise is propagated down the duct in the direction of the arrow. At plane *B* just before the bend only the axial incident wave exists because any oblique components which may have been introduced at plane *A* have been suppressed by the lining. However, the diffracted and reflected waves which are transmitted around the bend are primarily waves with oblique components. Between plane *C* and plane *D*, the oblique components are suppressed and only axial waves remain at plane *D*.

The usual definitions of TL (or Noise Reduction) require that a difference between the noise levels at plane *B* and plane *C* be used to determine the effect of the bend. Such definitions are not reasonable. They do not reflect the very important fact that the noise reduction between planes *C* and *D* is increased because the bend scatters energy into higher-order modes.

Obviously, the most useful way to evaluate a bend in a lined duct is to measure (at some point, such as plane *D*, well beyond the bend) the insertion loss. In the example given, one would measure the loss of a straight lined duct about $8\frac{1}{2}l_z$ long and subtract the loss from the loss measured between planes *A* and *D* in Fig. 17.9*a*. Brittain[20] obtained an equivalent result by measuring sound-pressure level along the center line of the duct and plotting the SPL's as a function of distance. The plot is a straight line in front of and also beyond the bend, but there is a sharp drop going around the bend. If the straight SPL vs. distance lines before and after the bend regions are extrapolated, they are found to be parallel. The "height" between them is the *insertion loss of the bend.*

The *noise reduction of bends* for random incidence waves may be very small under certain circumstances. The behavior of bends for random incidence waves can be illustrated by the optical analogy where the hard sidewalls are considered as mirrors. As indicated in

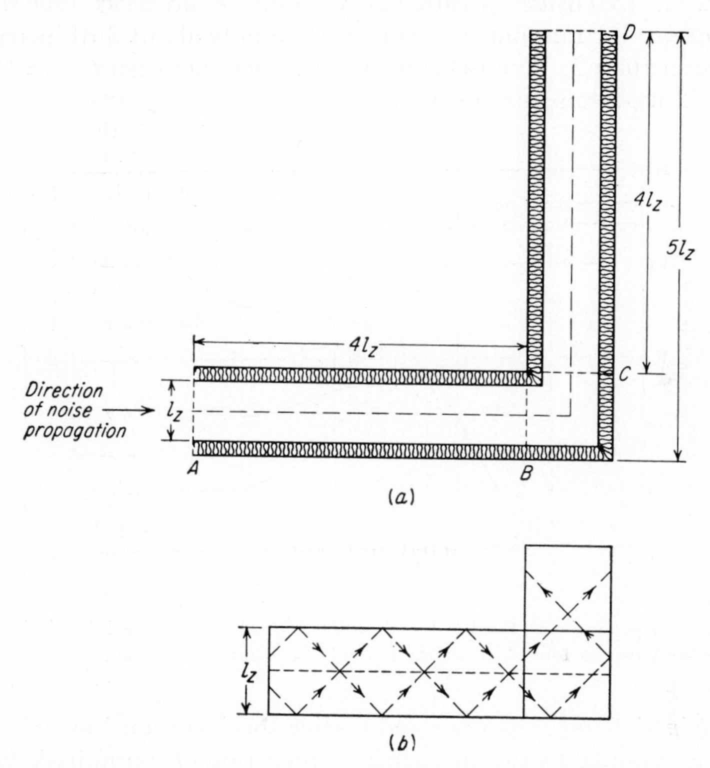

FIG. 17.9 Sketches of bends used in the discussion of their noise-reduction characteristics: (*a*) a typical lined bend with a plane-axial-wave input; (*b*) a typical unlined bend with a higher-order mode wave in the input duct.

Fig. 17.9*b* the higher-order modes in the *z* direction will easily travel around the bend. Note also that the very oblique waves in front of the bend tend to become the axial modes beyond the bend. Therefore, to assure a large insertion loss for the bend at high frequencies, the top and bottom of the duct should be lined for a short distance (one duct width) before the bend as well as beyond the bend.

The modes with large components in plane *B* (perpendicular to the paper) will tend to be reflected back toward the noise source at high frequencies even if the bend is not lined.

Calculation of the Insertion Loss of Bends. 1. *Unlined Bends.* The insertion loss for unlined bends for the plane axial wave is given by the upper dashed line in Fig. 17.10. These data are based on Lippert's analysis up to l_z/λ equal to about $\frac{1}{2}$ and on empirical data[18] at higher frequencies. For random incidence inputs (see lower dashed line) the insertion loss is the same as for axial waves up to $l_z/\lambda = \frac{1}{2}$. Extensive measurements[13] indicate, however, that the insertion loss for random incidence will be only about 3 db when l_z/λ is greater than 2. The value of insertion loss between $l_z/\lambda = \frac{1}{2}$ and $l_z/\lambda = 2$ has been interpolated.

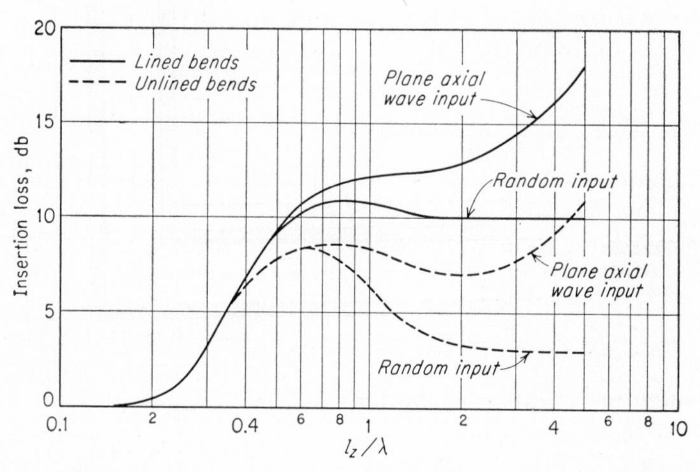

FIG. 17.10 Insertion loss for lined and unlined bends. In a lined bend the lining must extend two to four duct widths beyond the bend for the data shown to apply.

2. *Lined Bends.* As discussed earlier the insertion loss of a lined bend is obtained by two mechanisms, reflection of sound back toward the source and scattering of sound energy into higher-order modes which are rapidly attenuated by the lining beyond the bend. Higher-order modes will be attenuated by even an unlined duct for frequencies below which $l_z = \lambda$. Therefore, the insertion loss of a lined bend will be the same as the insertion loss of an unlined bend for frequencies below which $l_z/\lambda = \frac{1}{2}$. At frequencies well above $l_z/\lambda = \frac{1}{2}$ the insertion loss of a lined bend might be expected to be comparable to the reverberant-field end correction derived for ducts (about 10 db). Field measurements bear this out. As indicated previously, the end correction is merely a measure of the ratio of sound energy in the higher modes to the sound energy in the plane or axial modes. Thus the end correction is essentially independent of the attenuation char-

acteristics of the lining provided the lining extends 2 to $4l_z$ beyond the bend. The insertion losses of lined bends for plane axial inputs and random incidence inputs are given by the solid lines in Fig. 17.10.

The total TL, or insertion loss, of a lined bend will be given by the insertion loss of the bend given in Fig. 17.10 plus the *plane axial wave* TL of the length of duct lining used beyond the bend (at least $2l_z$ and preferably $4l_z$).

Very little data exist concerning the insertion loss of bends greater or less than 90°. Some data are given by Beranek[15] and Brittain[20] for bends less than 90°. As a rough engineering approximation, one may assume that insertion loss is proportional to the angle. For example, one might estimate the insertion loss of a 30° bend to be one-third that of a 90° bend. In jet engine test cells[13] one finds that the random incidence Noise Reduction for 180° bends is about 1½ times the Noise Reduction for a 90° bend. Comparable relations probably hold for insertion loss. To the author's knowledge, data for the insertion loss for 180° bends having a plane-wave normal incidence input are not available.

17.5 Plenum Chambers

A plenum chamber is similar in many ways to a lined expansion chamber. The most significant difference is that the inlet and outlet of a plenum chamber are not located directly opposite one another. They are generally offset to minimize the direct transmission of sound. At high frequencies, at least, almost all of the sound energy must reflect off the lined sides several times to get from the inlet to the outlet. The geometry and nomenclature for a single plenum are given in Fig. 17.11.

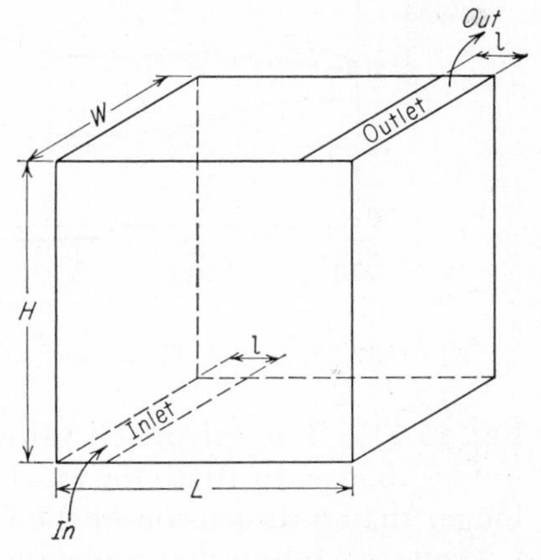

FIG. 17.11 A single chamber plenum.

Wells[21] has experimentally determined the TL of several acoustical plenum chambers and has determined an approximate method of calculating TL in the general case. The TL of a single plenum chamber can be found approximately from the expression:

$$\mathrm{TL} = -10 \log_{10} \left[S \left(\frac{\cos \phi}{2\pi d^2} + \frac{1}{R} \right) \right] \quad \mathrm{db} \qquad (17.8)$$

in which TL = transmission loss, db
$S = lW$ = area of inlet or exit opening, ft²
$d^2 = (L - l)^2 + H^2$
d = slant distance from entrance to exit, ft
$\cos \phi = H/d$
$R = a/(1 - \bar{\alpha})$ ft²
a = total absorption of lining, sabins—ft² (total lined area in chamber times absorption coefficient)
$\bar{\alpha}$ = statistical absorption coefficient of plenum lining

This formula has been found by Wells to be in reasonably good agreement at high frequencies and not too large values of l/L (see

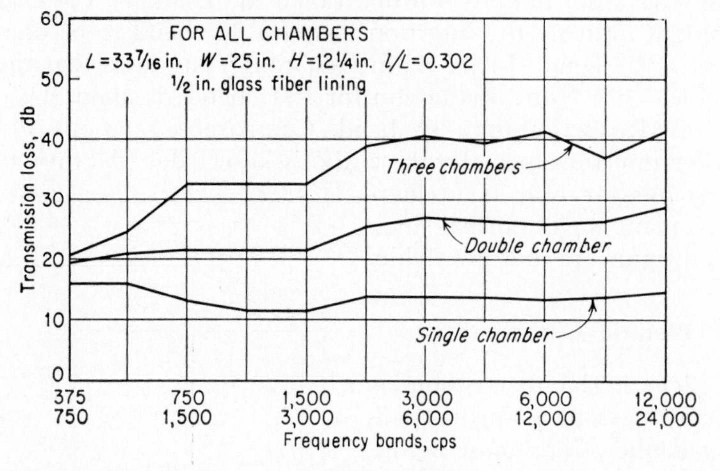

FIG. 17.12 Measured TL for one to three plenum chambers. (*After Wells.*)

Fig. 17.11). The calculated values are lower than the measured values by about 5 to 10 db for low frequencies, where the wavelength is longer than a dimension of the chamber.

Wells has found that a plenum with n chambers does not have an attenuation n times as large as a single chamber. However, this assumption may be used provided that a low-frequency end correction is added only once. If, for example, the calculated TL for a single chamber is N db and the low-frequency end correction is taken to be 10 db, then the attenuation of three chambers would be $(3N + 10)$ db.

Figure 17.12 shows the measured TL of plenums with one, two, and three chambers of the type illustrated in Fig. 17.11. Figure 17.14 indicates the TL of a special plenum (see Fig. 17.13) designed by Wells. Comparison of the measured data shows that this design is much more effective in the high frequencies than the ordinary single-chamber plenum. At the highest frequency the TL of the special plenum (40 to 50 db) is about three times the TL of the ordinary plenum (15 db). Unfortunately pressure-drop data are not available for these plenums so that one cannot estimate the penalty paid for the additional attenuation obtained.

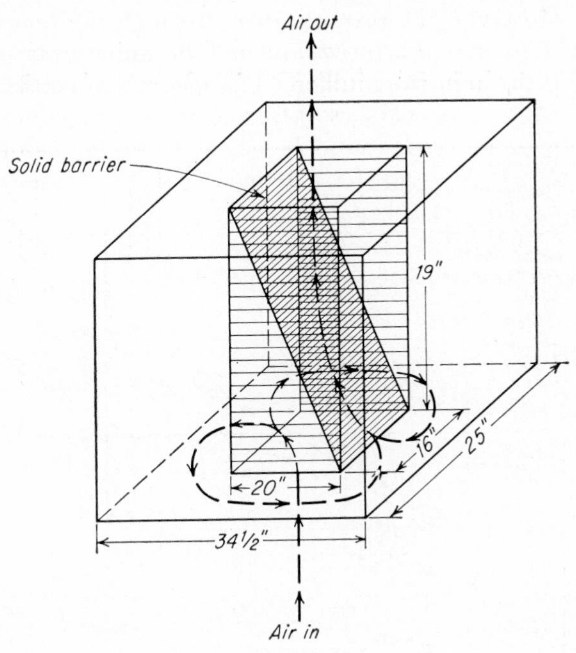

FIG. 17.13 Sketch of a high TL plenum. Note that there is no direct path from entrance to exit and no path with less than three or four reflections. (*After Wells.*)

17.6 Prefabricated Mufflers for Ventilating Systems

There are many types of prefabricated muffling devices for noise control in air passages. Standard items are available for controlling noises in ducts from 6 by 6 in. up to 50 by 50 ft. Mufflers for the very large air passages encountered in wind tunnels and large jet engine test cells are of interest to a limited audience and are not discussed here. Measured data for many commercially available test-cell systems are reported in Appendix *C* of Ref. 13.

Many different types of mufflers are available for applications in ventilating systems.* Most of the prefabricated mufflers for ventilating systems are available in a wide range of lengths, NR charac-

* Among the companies manufacturing such items are Industrial Acoustics Company, New York, N.Y., and Koppers Company, Inc., Baltimore, Md. The products of these companies and other companies do not perform alike.

teristics, cross-sectional areas, and airflow characteristics. The primary advantage of prefabricated units is that a noise-reduction requirement can be satisfied in a much shorter length than is possible by use of lined ducts and bends. The pressure drop through such systems is usually somewhat higher than through an equal length of lined duct. However, the pressure drop through the *length of lined duct required for equal attenuation* may be much greater than the pressure drop through the muffler. The use of a prefabricated muf-

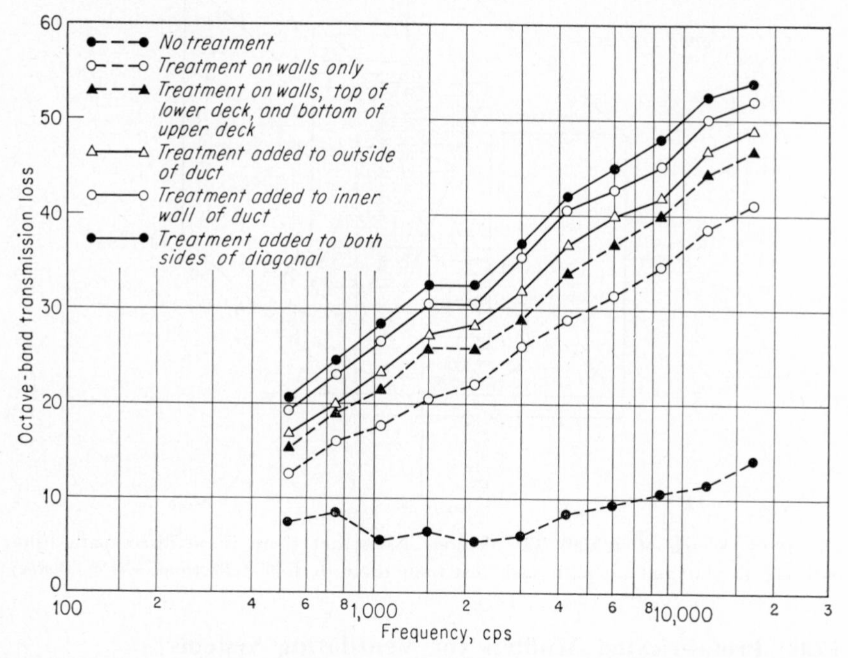

FIG. 17.14 Measured TL for plenum in Fig. 17.13 with no absorptive lining and for various partial treatments. (*After Wells.*)

fler may be mandatory if the available space is severely limited or if there are large noise-reduction requirements.

Several types of prefabricated muffling devices have been measured at Bolt Beranek and Newman Inc. All the data have been obtained by an end-difference method and hence the values given in Tables 17.2, 17.3, and 17.4 below are Noise Reduction values. Details of the measurement procedure are given elsewhere.[22] The approximate cross sections of the units tested are included as the product lines are sometimes redesigned without a change in the product name.

These mufflers are constructed on a module basis. The Manufacturer *A* units are 12 by 24 in. The Manufacturer *B* units have a

12- by 12-in. module size. The Manufacturer *C* units are designed on 6- by 14-in. module size. These mufflers may be used in larger duct systems by using parallel combinations of the basic module sizes. Some larger module units are prefabricated by the manufacturers.

Table 17.2

The Octave-band Noise Reduction Characteristics of Some Commercial Prefabricated Package Attenuators—Manufacturer *A* (1959 Product)

Unit Designation	Length, in.	Cross Section	Frequency Bands 20–75	75–150	150–300	300–600	600–1,200	1,200–2,400	2,400–4,800
A	24		2	6	6	9	19	23	16
B	34	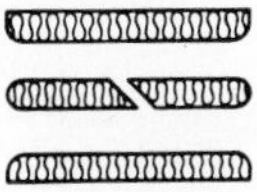	4	5	5	8	16	22	12
C	34	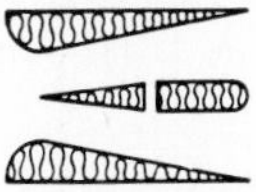	6	6	6	10	20	21	14
D	34	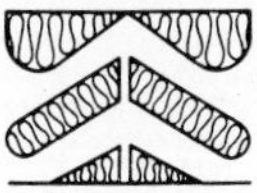	4	5	9	13	25	35	32
E	34	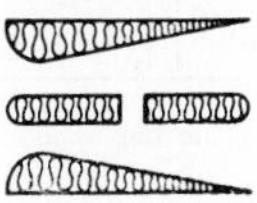	7	7	9	12	24	29	22

When the mufflers are used in parallel, the noise-reduction characteristics are approximately the same as those for a single unit. Stacking units in parallel or increasing module size decreases the pressure drop for a given volume flow since the pressure drop depends essentially upon the linear velocity through the units.

The pressure-drop vs. volume-flow characteristics are shown for representative Manufacturer *C* units as sold in 1959 in Fig. 17.15.* The 3-ft-, 5-ft-, and 8-ft-long half-open models differ only in the length of the straight section in the center. The pressure drop is almost entirely attributable to the change in direction of the airflow

* These data, based on tests performed by U.S. Testing Laboratories, Inc., are taken from Koppers' 1959 sales literature.

at the ends. Thus the pressure drops for these three lengths are all about equal. The pressure drops for the full-open models are somewhat higher than for the half-open models because a contraction and expansion of the airflow take place as well as a change of direction of the airflow (see Table 17.4).

Table 17.3

The Octave-band Noise Reduction Characteristics of Some Commercial Prefabricated Package Attenuators—Manufacturer *B* (1959 Product)

Unit Designation	Length, in.	Cross Section	Frequency Bands						
			20–75	75–150	150–300	300–600	600–1,200	1,200–2,400	2,400–4,800
A	28		6	11	12	12	25	42	40
B	46	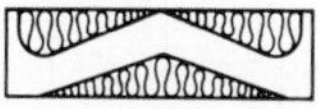	7	13	13	18	30	47	36
C	60	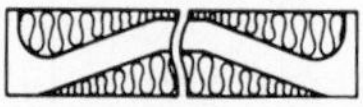	10	13	18	30	56	60	56
D	84	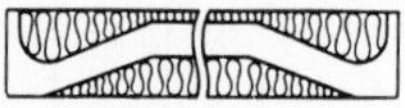	14	14	28	43	66	67	68
*E**	120		16	24	39	49	63	50	46

* The glass fiber in this muffler was enclosed in a polyethylene bag which lowered the high-frequency noise reduction. It is not known if all 120-in. units are so designed.

17.7 The Determination of Octave-band Transmission Loss

Most noise problems are resolved to a set of *octave-band* TL or insertion-loss requirements which must be achieved to reduce the *octave-band* noise levels at some point to a certain *octave-band* criterion level. A remaining problem, then, is to relate the continuous noise reduction functions to octave-band values of TL.

The general nature of the solutions to the problem is intuitively obvious. Suppose, for example, that the TL function increases through the 300- to 600-cps octave band from 10 db at 300 cps to 20 db at 600 cps. The TL at the geometric mean frequency, 425 cps, is

15 db. If the spectrum of the noise input has a very steep negative slope, almost all of the energy will be concentrated near 300 cps, and one might expect the TL to be very near the value at 300 cps or about 10 db. Conversely, if the spectrum of the noise input has a very steep positive slope, then the octave-band TL is expected to be

Table 17.4

The Octave-band Noise Reduction Characteristics of Some Commercial Prefabricated Package Attenuators—Manufacturer *C* (1959 Product)

Unit Designation	Length, ft	Cross Section	Frequency Bands						
			20–75	75–150	150–300	300–600	600–1,200	1,200–2,400	2,400–4,800
2S--*	2	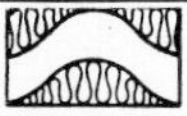	9	9	11	15	25	37	28
3S--	3	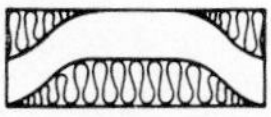	11	11	16	19	30	44	38
4S--F	4	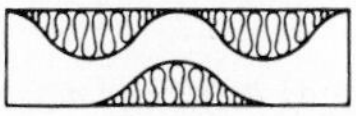	7	9	11	19	29	40	28
5S--	5	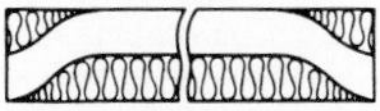	11	12	20	26	45	55	49
6S--F	6	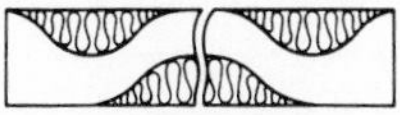	10	10	17	25	45	55	44
8S--F	8	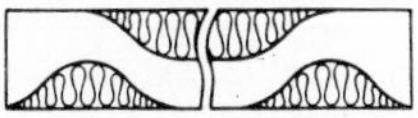	11	14	25	33	56	62	52
8S--	8	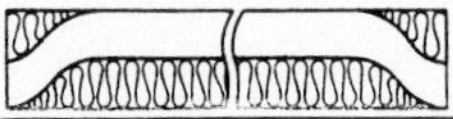	13	15	25	34	59	63	62

* The last two digits indicate the height (perpendicular to the section shown) of the units. The noise reduction is essentially independent of the height.

near 20 db. If the spectrum of the noise is flat (the noise energy is equally distributed among all frequencies), then the octave-band TL might be expected to be near the value at the geometric mean frequency.

The octave-band TL for an arbitrary input spectrum can be written as:

$$\mathrm{TL}_{\text{octave}} = -10 \log_{10} \frac{\text{output power}}{\text{input power}}$$

$$= -10 \log_{10} \left[\int_{f_1}^{2f_1} I(f)\tau(f)\, df \Big/ \int_{f_1}^{2f_1} I(f)\, df \right] \quad \text{db} \qquad (17.9)$$

in which TL_{octave} = octave-band value of TL
$I(f)$ = intensity of input as function of frequency
$\tau(f)$ = transmission coefficient as function of frequency

Note that

$$\text{TL spectrum} = -10 \log \tau(f) \qquad \text{db}$$

and

$$\text{Input spectrum} = 10 \log I(f) \qquad \text{db}$$

The function of Eq. (17.9) has been evaluated for input spectra and

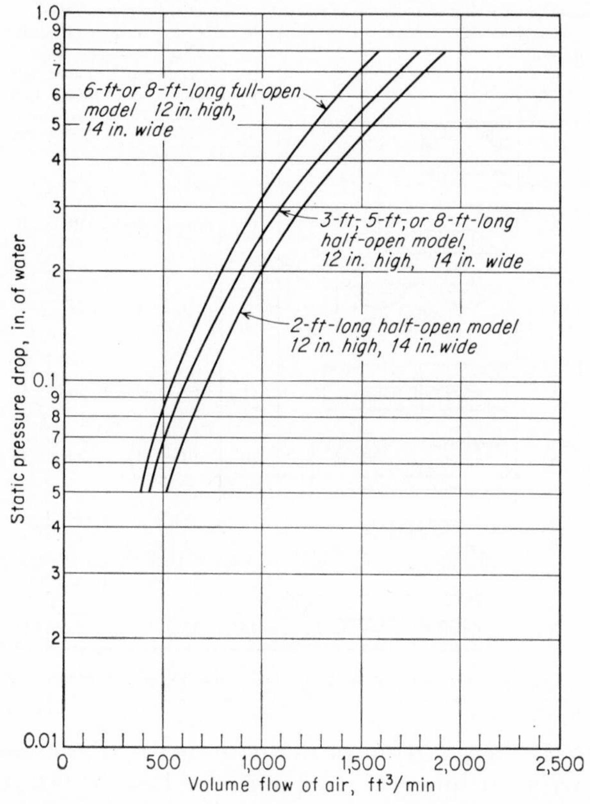

FIG. 17.15 Static pressure loss as a function of volume flow for three manufacturer *C* units (1959 models).

TL spectra which vary linearly on a db vs. log f scale. The results are summarized in Fig. 17.16.

Example 17.3. Use of Fig. 17.16 is best illustrated by a specific example. The following conditions are assumed:

(*a*) The TL curve has a slope of +20 db/octave.

(*b*) The TL at the geometric mean frequency of the octave band is 18 db.

(*c*) The octave-band spectrum of the input noise has a slope of -12 db/octave.

The octave-band TL is to be found for these conditions.

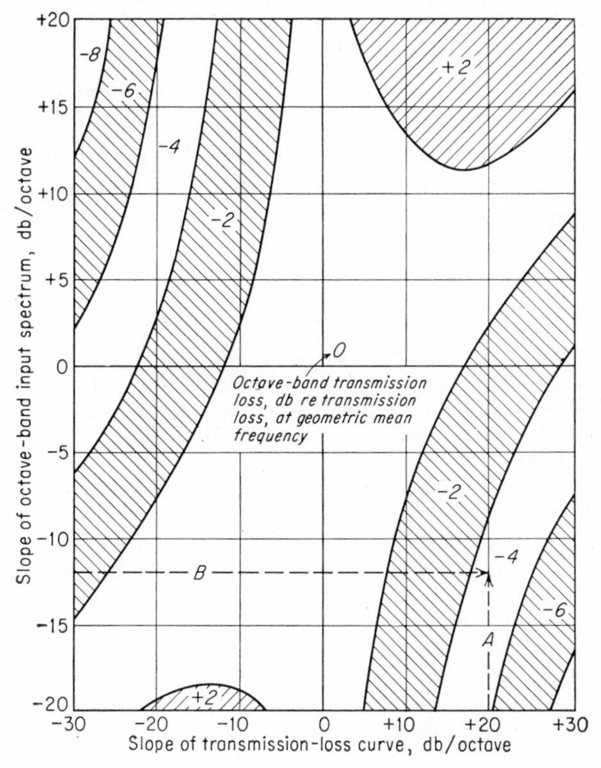

FIG. 17.16 A graph for finding the octave-band TL (for wideband noise) from the spectrum of the TL (measured with pure tones) and the spectrum of the noise input.

Solution. We enter Fig. 17.16 at a TL slope of $+20$ db/octave (line *A*) and spectrum slope of -12 db/octave (line *B*). At the intersection of these two lines, one finds the octave-band TL to be 4 db less than the TL at the geometric mean frequency, or 18 db $-$ 4 db $=$ 14 db.

17.8 Summary

The noise-reduction characteristics of many dissipative mufflers have been presented. The importance of "finite length" effects has been shown by extending the theoretical analysis of purely reactive expansion chambers to include dissipation. Data obtained in the field have been used to estimate the magnitude of finite-length effects associated with the space distribution (mode structure) of the noise input to the muffler. Examples have been given to show how to derive the noise reduction characteristics of a dissipative muffler of finite length by considerations of (1) the attenuation of an infinite length for the dissipative section, (2) reactive effects of the change in cross section at the beginning and end of the dissipative section, and (3) the spatial distribution of sound energy at the input.

Data are given on the noise reduction of lined and unlined bends, lined ducts, expansion chambers, and typical commercial mufflers for ventilating systems.

REFERENCES

1. Davis, D. D., Jr., G. M. Stokes, D. Moore, and G. L. Stevens, Jr.: *Natl. Advisory Comm. Aeronaut. Ann. Rept.* 1192, 1954, or D. D. Davis, Jr.: Acoustical Filters and Mufflers, chap. 21 in C. M. Harris (ed.), "Handbook of Noise Control," McGraw-Hill Book Company, Inc., New York, 1957
2. Morse, P. M.: Transmission of Sound inside Pipes, *J. Acoust. Soc. Am.,* vol. 11, p. 205, 1939.
3. Cremer, L.: Theory of Damping of Airborne Sound in a Rectangular Duct with Absorbing Walls and the Resulting Maximum Attenuation Constant, *Acustica,* vol. 3, p. 249, 1953.
4. Lukasik, S. J., A. W. Nolle, and the Staff of Bolt Beranek and Newman Inc.: Physical Acoustics, vol. 1, suppl. 1 of "Handbook of Acoustic Noise Control," pp. 252–261, *WADC Tech. Rept.* 52–204, April, 1955.
5. See Ref. 4, pp. 217–240.
6. See Ref. 4, pp. 240–249.
7. Beranek, L. L., S. Labate, and U. Ingard: Noise Control for NACA Supersonic Wind Tunnel, *J. Acoust. Soc. Am.,* vol. 27, p. 85, 1955.
8. Brittain, C. P., C. R. Maguire, R. A. Scott, and A. J. King: Attenuation of Sound in Lined Air Ducts, *Engineering,* p. 97, Jan. 30, 1948.
9. King, A. J.: Attenuation of Sound in Lined Air Ducts, *J. Acoust. Soc. Am.,* vol. 30, no. 6, p. 505, 1958.
10. Doelling, N.: Noise Control for Aircraft Engine Test Cells and Ground Run-up Suppressors, vol. 2, Design of Noise Control in Aircraft Engine Test Cells and Ground Run-up Suppressors, *WADC Tech. Rept.* 58–202(2).
11. Dyer, I.: Noise Attenuation of Dissipative Mufflers, *Noise Control,* vol. 2, no. 3, p. 50, May, 1956.
12. Sabine, H. J.: The Absorption of Noise in Ventilating Ducts, *J. Acoust. Soc. Am.,* vol. 12, p. 53, 1940.
13. Doelling, N.: Noise Control for Aircraft Engine Test Cells and Ground Run-up Suppressors, vol. 3, An Engineering Analysis of Measurement Procedures and of Design Data, *WADC Tech. Rept.* 58–202(3), August, 1958.
14. Meyer, E., F. Mechel, and G. Kurtze: Experiments on the Influence of Flow on Sound Attenuation in Absorbing Ducts, *J. Acoust. Soc. Am.,* vol. 30, no. 3, pp. 165–174, March, 1958.

15. Beranek, L. L.: "Acoustics," McGraw-Hill Book Company, Inc., New York, 1954.

16. Miles, J. W.: The Diffraction of Sound due to Right-angled Joints in Rectangular Tubes, *J. Acoust. Soc. Am.,* vol. 19, pp. 572–584, 1947.

17. Lippert, W. K. R.: The Measurement of Sound Reflection and Transmission at Right-angled Bends in Rectangular Tubes, *Acustica,* vol. 4, p. 313, 1954.

18. Sound Control, chap. 40 of "Heating, Ventilating, and Air Conditioning Guide," American Society of Heating and Air-conditioning Engineers, New York, 1957.

19. Watters, B. G., S. Labate, and L. L. Beranek: Acoustical Behavior of Some Engine Test Cell Structures, *J. Acoust. Soc. Am.,* vol. 27, p. 449, 1955.

20. Brittain, C. P., C. R. Maquire, R. A. Scott, and A. J. King: Attenuation of Sound in Lined Air Ducts, *Engineering,* p. 145, Feb. 13, 1948.

21. Wells, R. J.: Acoustical Plenum Chambers, *Noise Control,* vol. 4, no. 4, p. 9, July, 1958.

22. Doelling, N.: "On the Noise Attenuation Characteristics of Package Mufflers for Ventilating Systems," paper presented at the meeting of the Acoustical Society of America, Cleveland, October, 1959.

Chapter 18

ISOLATION OF VIBRATIONS

*D. Muster and R. Plunkett**

18.1 Introduction

Vibration isolation is a means of decreasing transmission of vibratory motions or forces from one structure to another. It involves interposing a relatively flexible element between the two structures. The driven structure will have a vibration amplitude controlled by its inertia resisting the vibratory forces; the other structure will have little force transmitted to it by the flexible support. Adding damping to a structure for the purpose of reducing its response at resonance has a different purpose; damping actually tends to decrease vibration isolation.

There are two types of vibration-isolation applications: (1) those in which we seek to prevent transmission of unbalanced forces from a machine to its foundation, (2) those in which it is desirable to reduce the motion transmitted from a substructure to a device mounted on it. Reciprocating engines and other types of rotating equipment (such as electric motors, fans, turbines, etc.) are mounted on vibration isolators to accomplish the former. As an example of the latter, electronic instruments used in aircraft are mounted with resilient mountings so that the motion they experience will be less severe than that of the airframe.

The elastic device which accomplishes this effect is called a "vibration isolator." It may use a metal spring, some rubberlike material, cork, or felt as its resilient element; the more resilient it is, the more effective it is. "Vibration dampers," which are normally characterized by high damping, serve to reduce the vibration response of a system only at or near a resonant frequency and must always be placed in parallel with another element of the system. This chapter

* Both at General Engineering Laboratory, General Electric Company, Schenectady, N.Y.

is concerned only with a discussion of the uses and properties of vibration isolators.

18.2 Vibration Isolation

Transmissibility: Force and Displacement. The effectiveness of an isolator is measured by its transmissibility ϵ, which is defined separately for the two cases mentioned earlier. *Force transmissibility* ϵ_f is defined as the ratio of the force transmitted through the isolator to the exciting force applied to a mass on it. *Displacement* (or motion) *transmissibility* ϵ_d is defined as the ratio of the displacement transmitted through the isolator to the exciting displacement applied to it. Obviously, in both cases the transmissibility depends not only upon the characteristics of the isolator but also upon the properties

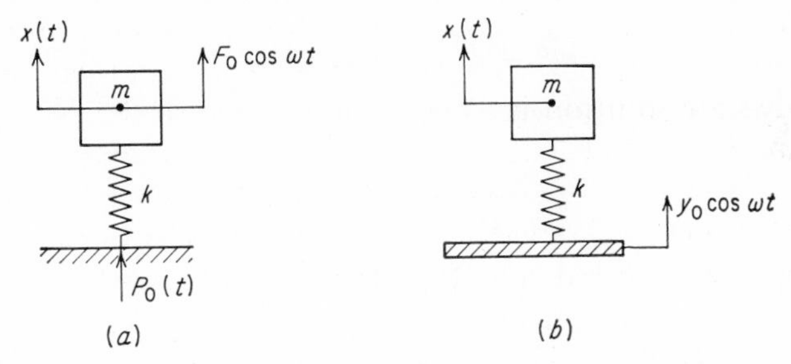

FIG. 18.1 Undamped, single-degree-of-freedom system: (*a*) force driven; (*b*) displacement driven.

of the support used in conjunction with it. If, for the present, we limit our discussion to the single-degree-of-freedom systems shown in Fig. 18.1, then the transmissibilities ϵ_f and ϵ_d are functions of the isolator properties only.

Undamped, Single-degree-of-freedom System. Consider the case of Fig. 18.1*a*, a simple undamped mass-spring system, under the action of a periodic applied force $F_0 \cos \omega t$. In addition to any other motion, such as a static deflection due to its own weight, the equation of motion for the mass m due to the force F is:

$$m\ddot{x} + kx = F_0 \cos \omega t \tag{18.1}$$

where k is the stiffness of the isolator. The steady-state solution of Eq. (18.1) is

$$x = \frac{F_0/k}{1 - (\omega/\omega_n)^2} \cos \omega t \tag{18.2}$$

where $\omega_n^2 = k/m$. The force P_0 transmitted to the foundation is

kx, and, by definition, force transmissibility is the ratio of transmitted force to applied force, or

$$\epsilon_f = \frac{kx}{F_0 \cos \omega t} \tag{18.3}$$

which by Eq. (18.2) becomes

$$\epsilon_f = \frac{1}{1 - (\omega/\omega_n)^2} \tag{18.4}$$

For the case in Fig. 18.1*b*, displacement transmissibility is defined as the ratio of the displacement of the mass m to the applied displacement $y_0 \cos \omega t$, or

$$\epsilon_d = \frac{x}{y_0 \cos \omega t} \tag{18.5}$$

The equation of motion for the mass m due to the displacement y is

$$m\ddot{x} = k(y_0 \cos \omega t - x) \tag{18.6}$$

The steady-state solution is

$$x = \frac{y_0}{1 - (\omega/\omega_n)^2} \cos \omega t \tag{18.7}$$

which when substituted into Eq. (18.5) gives

$$\epsilon_d = \frac{1}{1 - (\omega/\omega_n)^2} \tag{18.8}$$

Thus, we see that the expressions for force and displacement transmissibility are identical.

An interesting aspect of Eqs. (18.4) and (18.8) is that force (or displacement) transmissibility becomes negative for $\omega/\omega_n > 1$. This sign change is associated with a change in phase between the direction of the applied force and the motion of the mass, that is, for $\omega/\omega_n < 1$, m moves downward as $F_0 \cos \omega t$ is directed *downward;* for $\omega/\omega_n > 1$, m moves downward as $F_0 \cos \omega t$ is directed *upward.* This implies that the force transmitted to the foundation is either in phase or 180° out of phase with the applied force according as ω/ω_n is less or greater than unity. A plot of transmissibility as a function of frequency ratio for which this phase relation is ignored is shown in Fig. 18.2.

The principle underlying all vibration isolation is contained in Fig. 18.2. For very low frequency ratios ($\omega/\omega_n \approx 0$), the base in Fig. 18.1 senses almost the same force as though the mass m were fastened directly to it. As the frequency ω of the exciting force increases and approaches the natural frequency ω_n, the force sensed by the base increases, until at $\omega = \omega_n$ (since we have assumed there is no damping in the system) the force becomes infinitely large.

As the forcing frequency ω becomes greater than the natural frequency ω_n, the force applied to the base through the spring decreases rapidly, until at $\omega = \sqrt{2}\omega_n$, the applied force and the force sensed by the base are the same. For exciting frequencies greater than $\sqrt{2}\omega_n$, the transmitted force is less than the applied force and, in general, the higher the frequency the more effective will be the action of the isolator. There are limits to this generalization which are discussed in a later section, but if the assumptions of the linear, lumped theory are valid for the conditions of a given problem, the greater the frequency ratio, the more effective the isolator.

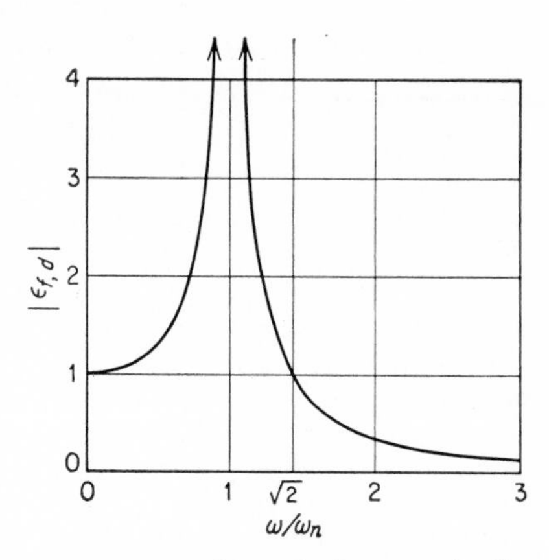

FIG. 18.2 Undamped force and displacement transmissibility as a function of frequency ratio.

From the curve in Fig. 18.2, it is clear that a resilient mounting can make the situation worse if its natural frequency is incorrectly selected. Only for $\omega > \sqrt{2}\omega_n$ is an isolator effective.

The natural frequency ω_n of an undamped, single-degree-of-freedom system is given by

$$\omega_n = \sqrt{\frac{k}{m}} \qquad \text{radians/sec} \tag{18.9}$$

or

$$f_n = \frac{1}{2\pi}\,\omega_n = \frac{1}{2\pi}\sqrt{\frac{k}{m}} = \frac{1}{2\pi}\sqrt{\frac{k'g}{W}} \qquad \text{cps} \tag{18.10}$$

where k = spring constant, lb_f/ft (or lb/ft or newtons/m) (see Appendix B for a discussion of units)
m = mass, slugs (or lb/32.2 or kg)
g = acceleration of gravity (386 in./sec^2)
k' = spring constant, lb/in.
W = weight of the mass m, lb

The static deflection δ_{st} of the isolator is given by $\delta_{st} = W/k'$, which when used in Eq. (18.10) shows that

$$f_n = 3.13\sqrt{\frac{1}{\delta_{st}}} \tag{18.11}$$

where δ_{st} is the deflection, in inches, of the mass under its own weight. Thus, the natural frequency of a simple isolator is a function only of its static deflection.

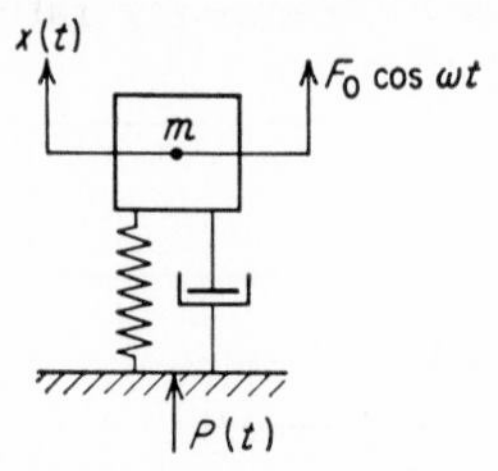

Fig. 18.3 Damped, single-degree-of-freedom system—force driven.

Damped, Single-degree-of-freedom System. In Fig. 18.3, a damped, single-degree-of-freedom system is shown under the action of an applied force $F_0 \cos \omega t$. We recall that the governing equation of motion is

$$m\ddot{x} + C\dot{x} + kx = F_0 \cos \omega t \tag{18.12}$$

where C is the coefficient of viscous damping.

The solution of Eq. (18.12) is

$$x = e^{-(C/2m)t} (A \cos \omega_d t + B \sin \omega_d t) + X \tag{18.13}$$

where the "damped natural frequency" ω_d is $\omega_d^2 = \omega_n^2 - (C/2m)^2$, and X is a particular solution of Eq. (18.12). It is clear that the first group of terms in Eq. (18.13) will become negligibly small for large values of time and that X (the steady-state solution) will then govern the motion of the mass m. It can be shown that

$$X = \frac{F_0/k}{\left[\left(1 - \frac{\omega^2}{\omega_n^2}\right)^2 + \left(2\frac{C}{C_c}\frac{\omega}{\omega_n}\right)^2\right]^{1/2}} \cos (\omega t - \phi) \tag{18.14}$$

where $C_c = 2m\omega_n$ is the "critical damping," and the phase angle is defined by

$$\tan \phi = \frac{2\dfrac{C}{C_c}\dfrac{\omega}{\omega_n}}{1 - \dfrac{\omega^2}{\omega_n^2}}$$

The force $P_0 \cos (\omega t - \phi)$ transmitted to the foundation is $C\dot{x} + kx$, and the magnitude of the force transmissibility ϵ_f of the system in Fig. 18.3 is

$$\epsilon_f = \left|\frac{P_0}{F_0}\right|$$

or

$$\epsilon_f = \left\{\frac{1 + \left(2\dfrac{C}{C_c}\dfrac{\omega}{\omega_n}\right)^2}{\left[1 - \left(\dfrac{\omega}{\omega_n}\right)^2\right]^2 + \left(2\dfrac{C}{C_c}\dfrac{\omega}{\omega_n}\right)^2}\right\}^{1/2} \tag{18.15}$$

A plot of this equation presents a rather accurate picture of the behavior of resilient mountings for a single-degree-of-freedom system. Force transmissibility (Eq. 18.15) as a function of frequency ratio for several damping ratios is shown in Fig. 18.4.

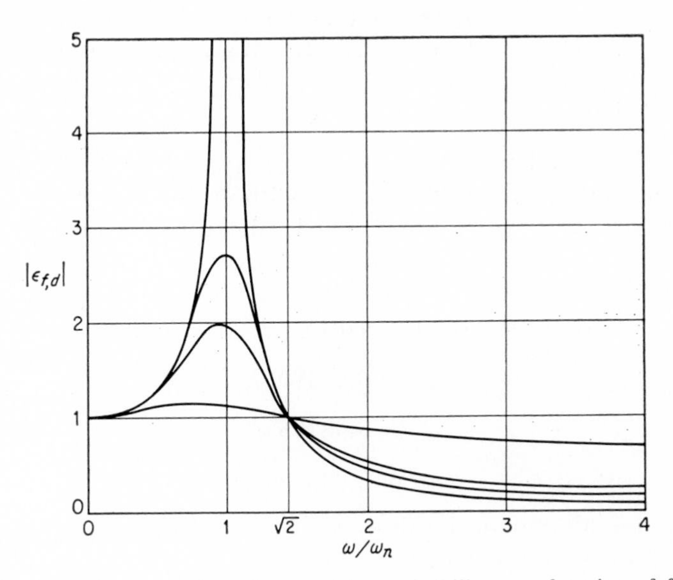

FIG. 18.4 Damped force and displacement transmissibility as a function of frequency ratio.

The shift of the resonance peak with damping is evident. Since we are plotting transmissibility vs. frequency, it is logical to ask at what frequency value is ϵ_f a maximum and how is this value related to other resonant frequencies?

From Eq. (18.15), it can be shown that the *frequency of maximum transmissibility* ω_ϵ is

$$\omega_\epsilon = \frac{\omega_n}{2\zeta}\left[(8\zeta^2 + 1)^{1/2} - 1\right]^{1/2} \tag{18.16}$$

where $\zeta = \dfrac{C}{C_c}$ and $\omega_n = \sqrt{k/m}$. Equation (18.14) implies that the *frequency of maximum forced amplitude* ω_a is

$$\omega_a = \omega_n\sqrt{1 - 2\zeta^2} \tag{18.17}$$

Thus, the four important frequencies are

$$\omega_n = \sqrt{\frac{k}{m}}$$

$$\omega_d = \omega_n \sqrt{1 - \zeta^2}$$

$$\omega_a = \omega_n \sqrt{1 - 2\zeta^2}$$

$$\omega_\epsilon = \frac{\omega_n}{2\zeta} \sqrt{\sqrt{8\zeta^2 + 1} - 1}$$

where ω_d is defined as the *damped natural frequency*.

Commercial resilient mountings without external dashpots or friction devices have relatively little damping—normally it corresponds to $\zeta < 0.25$ and in most cases $\zeta < 0.10$. For $\zeta = 0.10$, that is, the damping present is 10 per cent of critical damping, the frequencies listed above take the values

$$\omega_d = 0.99499\ \omega_n$$

$$\omega_a = 0.98995\ \omega_n$$

$$\omega_\epsilon = 0.99038\ \omega_n$$

Thus, for all practical values of damping, the four frequencies can be assumed to be equal.

The curves of Fig. 18.4 show that for $\omega/\omega_n < \sqrt{2}$, the force transmitted through the mount to the base is actually magnified for any degree of damping. When $\omega/\omega_n = \sqrt{2}$, the transmitted force is equal to the applied force. In the region of isolation ($\omega/\omega_n > \sqrt{2}$), we can see from the curves of Fig. 18.4 that viscous damping affects isolation adversely.

In the case of rotating machinery, a secondary problem makes it mandatory that there be some damping present. As a rotating device is started or stopped, it will, in most cases, pass through a resonant range—the range of frequencies where the system response is magnified by a resilient mounting. If the machine can be accelerated rapidly, it will pass through this region quickly and the amplitude of the transmitted force does not have time to build up to the steady-state levels indicated by Fig. 18.4. If, on the contrary, the machine accelerates slowly through the resonant range, then the transmitted force may become very large. In this case, a large damping factor (say, $\zeta = 0.5$) may be required to prevent excessive vibration as the machine passes through resonance. An external damper can be installed to accomplish this; but even in the presence of such large damping values, the amount of isolation it is possible to obtain at higher frequencies will, in most cases, be adequate.

The previous analysis considered only viscous damping, that is, systems in which the damping force is directly proportional to velocity. It has been shown that other forms of damping can be reduced to an "equivalent viscous damping," using energy dissipation per cycle as the criterion of equivalence. Jacobsen[1] demonstrated that such a treatment is practical. However, in such cases, the frequency effects of damping are quite different than are shown in Fig. 18.4.

For hysteresis or structural damping, the damping term depends on displacement instead of velocity. In this case there is little or no frequency effect. The equation for ϵ_f is:

$$\epsilon_f = \left[\frac{1 + \delta^2}{\left(1 - \frac{\omega^2}{\omega_n^2}\right)^2 + \delta^2} \right]^{1/2} \tag{18.18}$$

where $\delta = 1/Q$ and Q is the amplification factor at resonance. Generally, δ is less than 0.2. A few calculations will show that Eq. (18.18) is closely approximated by Eq. (18.4) if ω is very much larger than ω_n. In actuality, wave effects (Fig. 18.17) occur at higher frequencies so that Eq. (18.18) is a good approximation if a suitable value for δ is used. At $\omega > 10\ \omega_n$, the interaction of these various factors makes the best value for damping very much smaller than would be found at resonance ($\omega = \omega_n$).

Isolation at Acoustic Frequencies. The theory that has been shown so far is that which is customarily presented in books and papers dealing with vibration isolation.[2,3] With modifications to take care of the three-dimensional nature of the machine and the fact that several mounts are used, it gives excellent results for relatively low frequencies. This infrasonic frequency range is where we are concerned about physical damage or fatigue failure. Unfortunately, results calculated by Eq. (18.15) predict attenuations for the audio-frequency range which are apt to be very much higher than can be achieved in practice.

Since there is nothing wrong with the mathematical development, we must look to the assumptions for the reasons for this discrepancy. Figure 18.3 shows a rigid mass mounted on an isolator which in turn rests on an absolutely rigid foundation. If the foundation were actually rigid, the isolation problem would be of no interest since the foundation could not be moved. In actuality, almost any foundation and any reasonable machine will have many resonances in the acoustic frequency range.

Let us consider first the foundation. Since for small motions it is a linear system, if a force $F = F_0 \cos \omega t$ is applied to it, the vibratory velocity at the point of application is given by

$$v = UF_0 \cos (\omega t - \phi)$$

where U and ϕ will be functions of frequency ω. This may also be written in the form

$$v = UF_0 \cos \omega t \cos \phi + UF_0 \sin \omega t \sin \phi$$

If we use the Euler equation

$$e^{j\theta} = \cos \theta + j \sin \theta$$

we may redefine the force F as the real part of

$$F = F_0 e^{j\omega t}$$

and the velocity at the point of application of the force as the real part of

$$\begin{aligned} v &= UF_0 e^{j(\omega t - \phi)} \\ &= (Ue^{-j\phi})(F_0 e^{j\omega t}) \\ &= M_f F \end{aligned} \tag{18.19}$$

In Eq. (18.19) we have introduced the quantity M_f, called *mobility*, which is the complex number whose modulus is the magnitude of the velocity resulting from the application of a unit force and whose argument is the phase angle between them.

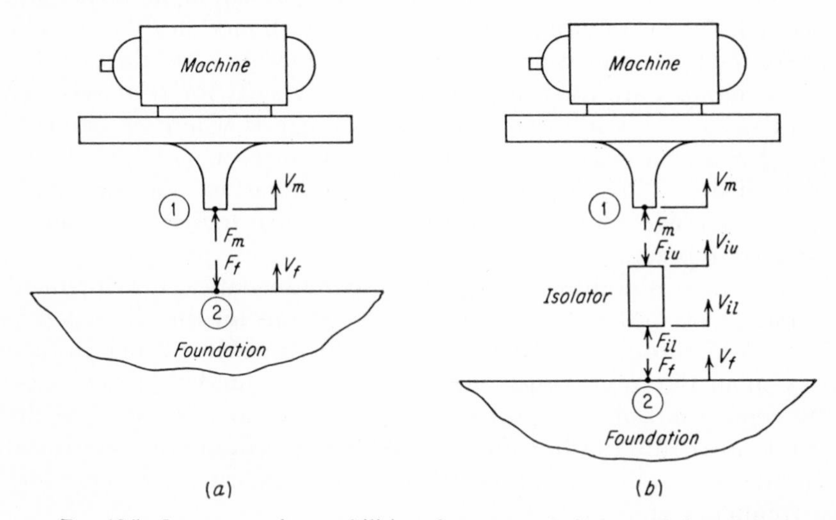

FIG. 18.5 Interconnection mobilities of a motor, isolator, and foundation.

The internal mobility of the machine M_m can be defined in a similar manner. By superposition, the net motion of that point on the machine base where the machine is fastened to the foundation is the sum of the motion it would have due to internal forces only (that is, the point of attachment is free) and the motion due to the resisting force (Fig. 18.5*a*), or

$$v_m = v_0 + M_m F_m \tag{18.20}$$

It should be noted that v_0 is not necessarily a rigid-body motion of the machine and may be modified appreciably by internal resonances and antiresonances. It is not always easy to calculate but can be measured with reasonable accuracy if the machine is supported on mountings (isolators), which are very soft in comparison to M_m (that is, $M_m \ll M_i$).

Let us consider that point 1 on the machine is attached directly to point 2 on the foundation (Fig. 18.5*a*); then for motion of the combined system (which we denote by a tilde)

$$\tilde{v}_m = \tilde{v}_f$$

$$\tilde{F}_m = -\tilde{F}_f$$

From Eqs. (18.19) and (18.20)

$$\tilde{F}_f = \frac{v_0}{M_m + M_f} \tag{18.21}$$

If a massless isolator is interposed between the machine and its foundation (Fig. 18.5*b*), then for motion of the combined system

$$\begin{aligned} \tilde{v}_m &= \tilde{v}_{iu} \\ \tilde{v}_f &= \tilde{v}_{il} \\ \tilde{F}_m &= -\tilde{F}_{iu} = +\tilde{F}_{il} = -\tilde{F}_f \end{aligned} \tag{18.22}$$

We define the force in and relative velocity of the isolator by, respectively,

$$\tilde{F}_i \equiv \tilde{F}_{iu} = -\tilde{F}_{il}$$

$$\tilde{v}_l \equiv \tilde{v}_{iu} - \tilde{v}_{il}$$

which, with Eqs. (18.19), (18.20), and (18.22), becomes

$$\tilde{v}_i = M_i \tilde{F}_i = v_0 + M_m \tilde{F}_m - M_f \tilde{F}_f$$

Since $\qquad F_i = \tilde{F}_f = -\tilde{F}_m$

we obtain

$$\tilde{F}_f = \frac{v_0}{M_m + M_i + M_f} \tag{18.23}$$

If we now redefine ϵ_f as the ratio of the force on the foundation with the isolator to that without

$$\epsilon_f = \frac{\tilde{F}_f}{F_f} = \frac{M_m + M_f}{M_m + M_i + M_f} \tag{18.24}$$

It will be found that this definition is the same as the previous one if the previous assumptions are used. In addition, it may be shown that ϵ_d will have the same value as ϵ_f.

From Eq. (18.24) we find that an isolator does no good unless it is more flexible (has greater mobility) than the sum of the mobilities

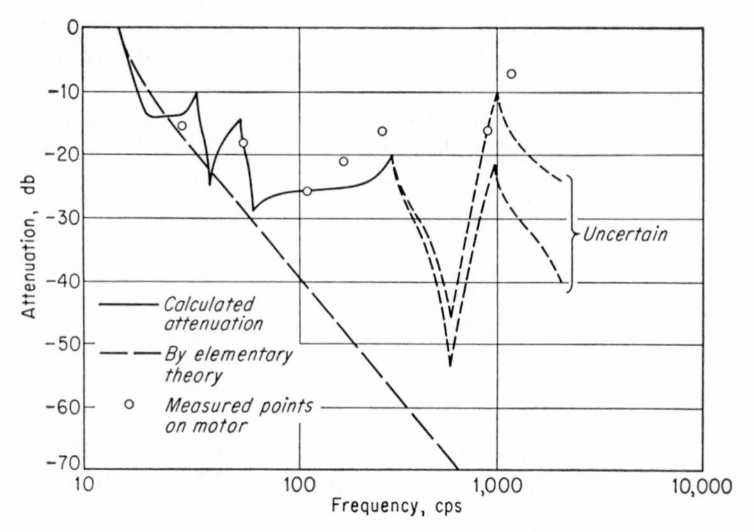

Fig. 18.6 Predicted and measured attenuation of an isolator.[4]

of the machine and foundation. The mobilities of simple structures may be calculated, and those of any structures may be measured. At present, we have very few measured values, but those that are available indicate that M_m and M_f are very much larger than might be anticipated. It is rare to get more than 20-db attenuation at acoustic frequencies ($\epsilon = 0.1$) with mounts of reasonable stiffness, and it is not uncommon to get no attenuation at all. For this reason, very soft mounts ($f_n = 5$ to 6 cps) are being offered, and special constructions, such as pneumatic mounts, are beginning to be marketed. In any case, we may see from Eq. (18.24) that if a mount is effective at all, a softer one (M_i larger) will be that much more effective.

In Fig. 18.6 a predicted curve and some measured values of ϵ_f are shown for an experimental setup.[4] In Fig. 18.7 a measured value for M_m of a typical base plate is plotted.[5]

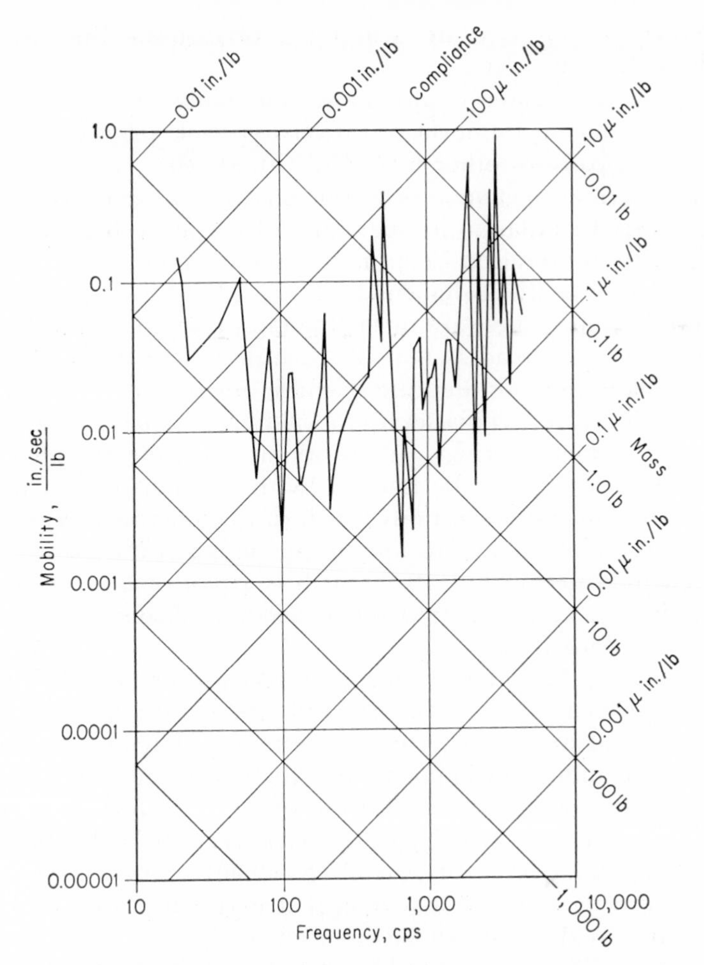

Fig. 18.7 The jagged line shows the measured mobility (in./sec-lb_f) as a function of frequency of a ribbed motor-generator base plate ⅜ in. thick weighing 60 lb (or lb_m). Note that the lines running from upper left to lower right on the graph give the mobility as a function of frequency for the masses indicated. For example, 100 lb_m (or a weight of 100 lb) at 100 cps has a mobility of 0.0062 in./sec-lb_f. The lines running from upper right to lower left give the mobility as a function of frequency for the compliances (reciprocals of the spring constants) indicated. For example, a spring with a compliance of 10^{-5} in./lb_f has, at 1,000 cps, a mobility of 0.062 in./sec-lb_f. This graph shows us that below 100 cps this motor base behaves on the average like a mass of 40 lb_m (or a weight of 40 lb). Above 1,000 cps it behaves on the average like a spring with a compliance of 3×10^{-6} in./lb_f. (See Appendix B for a discussion of English units.) (*After Plunkett.*[5])

18.3 Vibration Isolators and Isolator Materials

Four resilient materials are most commonly used in vibration isolators: Metals in the form of various types of springs are used as well as mesh pads (or other shapes) of rubber, cork, and felt. Other materials, such as steel-mesh pads, pneumatic sacks, and even a gelatinous material similar to hectograph pad gel, have been used, but the majority of vibration isolators use one or more of the first four materials as the resilient element.

Metal Springs. These are by far the most common, since the field of their use is as broad as that of machine design itself. They are used to isolate the most delicate scientific instruments from foundation vibrations, and yet masses up to 450 tons have been satisfactorily isolated with metal springs. In theory, at least, the complete spectrum of frequencies can be isolated by them. This is due, in part, to the large range of deflections which can be obtained by changing the dimensions and materials used in the design of the springs. This has been indicated in Fig. 18.8, which is a plot of Eq. (18.11) with the nominal maximum permissible deflections of some elastic media superimposed upon it.

Metal springs have the advantages of ready interchangeability and several beneficial chemical characteristics (they resist corrosion by oil and water and are not affected by extremes of temperature). An advantage of industrial importance is that they can be produced in large quantities with only small variation in their individual characteristics. Inherently, metal springs have very little damping; the damping is about 0.001 of critical ($\zeta \approx 0.001$). However, external damping can be added to a system, if it is required by a particular application. A dashpot can be inserted in parallel with the metal spring (Fig. 18.3). Recoil mechanisms are sometimes built employing this principle—a metal spring for elasticity, a separate oil dashpot for a large amount of viscous damping.

Some resilient mountings are fabricated with a metal spring inside a rubber sack with a calibrated orifice which regulates the flow of air in and out of the sack and, thus, furnishes viscous damping. Coulomb (friction) damping can also be supplied by external means. In leaf springs damping is caused by the rubbing action between individual leaves; in certain resilient mountings with steel mesh inserts it is caused by the rubbing of the wires on each other. Although it is incidental rather than intentional in many cases, this form of damping occurs also in nonrigid mechanical joints where one part can slip relative to another.

For purposes of analysis, it is convenient to reduce coulomb and other nonviscous forms of damping to an equivalent viscous damping which uses energy dissipated per cycle as the basis of equivalence.[1]

Metal springs have the practical disadvantage of transmitting high frequencies very readily. For example, although the low natural frequency of an internal combustion engine (say, 15 cps) can be isolated easily, the higher frequencies present are transmitted through the metal of the spring to the foundation. These higher frequencies

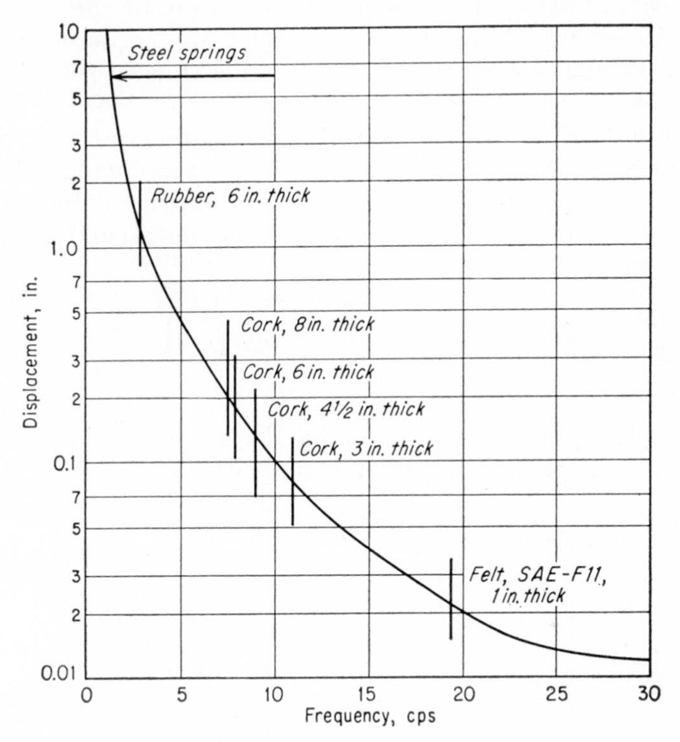

FIG. 18.8 Relation between static deflection and natural frequency. Typical natural frequencies of some practical isolators are shown.

may range from two hundred to several thousand cycles per second and are due to detonation in the cylinders, local resonances at the mounts and other sources. Transmission of these frequencies is minimized by insuring that there be no direct contact between the spring and the supporting structure. In commercial mountings, this is accomplished by inserting rubber or felt pads between the ends of the spring and the surfaces to which it is fastened.

1. *Helical Springs.* For most applications, the design of helical compression or tension springs (e.g., coil springs) has been reduced to nomographs or other handbook methods. Treatises have been

written on the subject,[6] and there is extensive commercial and technical literature available. Some design considerations of importance in certain applications, such as buckling and lateral stiffness of helical springs, have been subjects of recent investigations, and these are summarized here.

Timoshenko[7] has reviewed the general subject of the buckling of helical springs. More recently, Haringx[8] has reported on several phases of an analytical and experimental study of the mechanical properties of helical springs and rubber rods in compression. In the first of these papers, he considers the buckling of helical springs with various end conditions.

If the unloaded length of a steel spring is denoted by l_0 and its diameter by D_0, for the case of a helical spring of circular wire cross section with ends hinged or constrained to remain parallel, the relative compression ($\xi = \Delta l/l_0$) at which buckling occurs is given by Haringx as

$$\xi = 0.8125\left\{1 - \left[1 - 6.87\left(\frac{D_0}{l_0}\right)^2\right]^{1/2}\right\} \tag{18.25}$$

where l_0/D_0 is the slenderness ratio.

This relationship is derived under the following simplifying assumption: the spring is idealized as an elastic prismatic rod made of

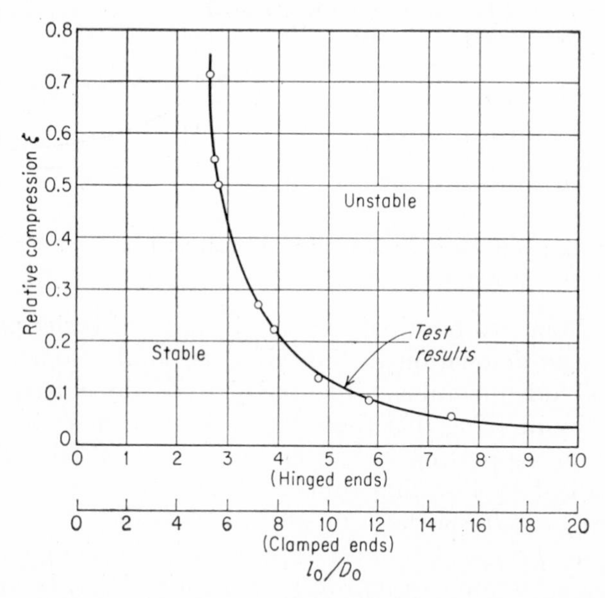

FIG. 18.9 Stable and unstable values of relative compression for coil springs.[8]

a hypothetical material with stiffness in bending, compression, and shear which correspond to those of the original spring. The experimental evidence appears to justify this assumption.

A plot of the relation given in Eq. (18.25) is shown in Fig. 18.9. The curve delineates a region wherein all combinations of spring slenderness and relative compressions lead to instability. It shows that there is a certain critical slenderness ratio ($l_0/D_0 = 2.62$ if the spring ends are hinged or constrained to remain parallel; $l_0/D_0 = 5.24$ if the spring ends are clamped) below which the spring does not buckle, above which it does. Haringx has measured the relative compressions at which buckling occurs for the springs of several slenderness ratios. The results of these experiments are shown superimposed on the curve in Fig. 18.9.

Burdick et al.[9] and Haringx[8] have studied lateral stiffness of axially loaded helical compression springs for several end conditions. The results of both studies can be compared qualitatively, although the quantitative results of Haringx's more carefully controlled tests are more useful to us here. The details of Haringx's analysis will not be reproduced here. He assumes that the spring behaves as an elastic prismatic rod and that the lateral load L and applied end moment M_l (shown in Fig. 18.10) can be expressed as the linear combinations

$$
\begin{aligned}
L &= k_1\delta - k_2\psi_l \\
M_l &= -k_2\delta + k_3\psi_l
\end{aligned}
\qquad (18.26)
$$

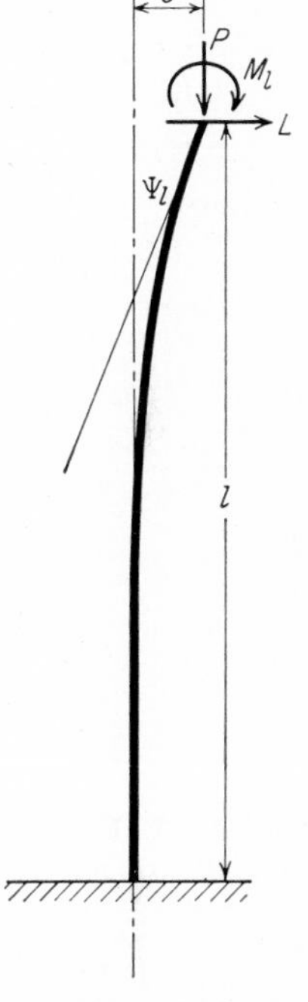

Fig. 18.10 End moment, load, and deflections for coil spring [Eq. (18.26)].

where δ and ψ_l are the lateral displacement and slope of the free end of the rod, respectively, and k_1, k_2, and k_3 are stiffness coefficients of appropriate dimensionality.

As an example, let us consider the case of steel, helical compression springs with circular wire cross section and ends constrained to remain parallel (that is, $\psi_l = 0$). For this case, Haringx demonstrates that the ratio of lateral stiffness k_1 to axial stiffness k_a as a function of relative compression ξ is given by the family of curves shown in Fig. 18.11. The intercepts of these curves and the ξ axis correspond to the values of relative compression for which the spring end is free to deflect without a lateral force being supplied by an external agency. The spring is then in an unstable condition and buckles. In a sense, then, buckling of helical springs under compression is a special case of the problem of lateral stiffness.

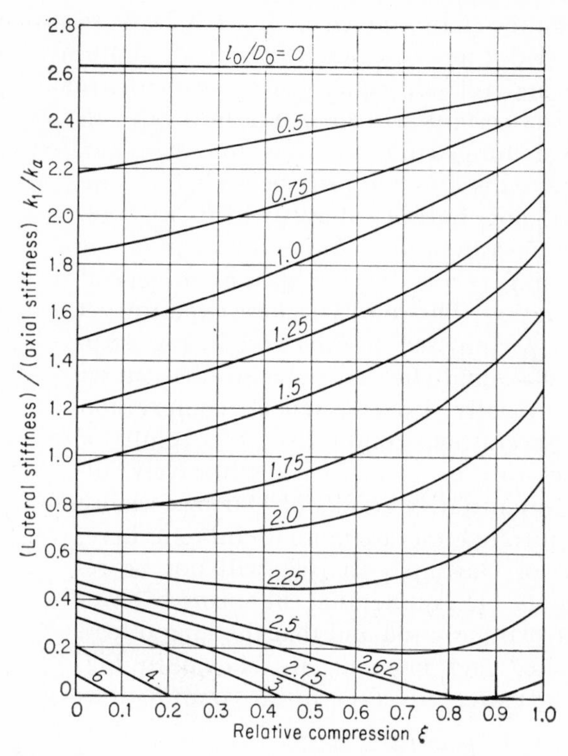

FIG. 18.11 Lateral stiffness as a function of relative compression for various diameters of coil springs.

It is interesting to note that whereas the axial stiffness of a helical spring remains constant for all loads, such is not the case with lateral stiffness (see Fig. 18.11). Further, Haringx's theory predicts a limit of $k_1/k_a = 2.62$, as the slenderness ratio approaches zero.

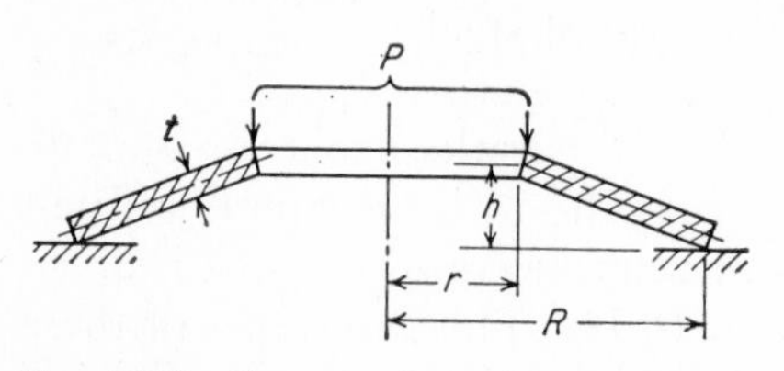

FIG. 18.12 Cross section of Belleville or disk spring.

2. *Belleville Springs.* Coned disk springs, perhaps more widely known as "Belleville springs," consist essentially of circular disks dished to a conical shape as indicated in Fig. 18.12. A manual dealing with their design and use is available,[10] so that here we will only sketch their general characteristics.

Load-deflection characteristics of Belleville springs can be obtained in considerable variety by changing the ratio h/t (Fig. 18.12). In

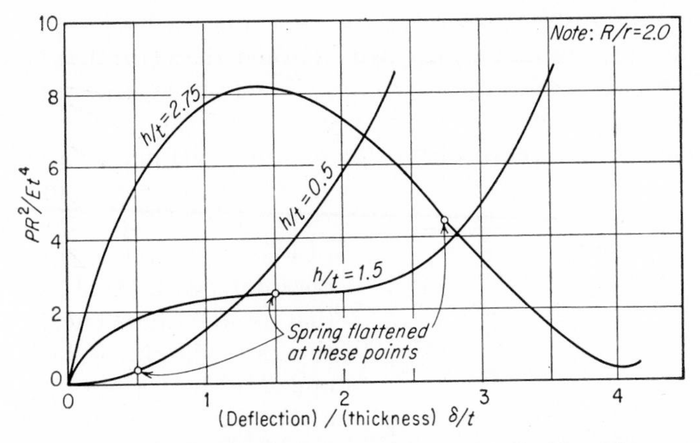

FIG. 18.13 Load-deflection characteristics of Belleville springs.

Fig. 18.13, the effect of changes in this ratio upon the load-deflection diagram is shown for three cases. For $h/t = 1.5$, we note that the spring has practically zero stiffness for a considerable range of deflection, that is, the load is almost constant over this range.

Belleville springs may be used singly or may be stacked in parallel (Fig. 18.14*a*) or in series (Fig. 18.14*b*). By stacking in series, the deflection for a given load is increased directly with the number of disks; in parallel, the load for a given deflection increases in proportion to the number of disks.

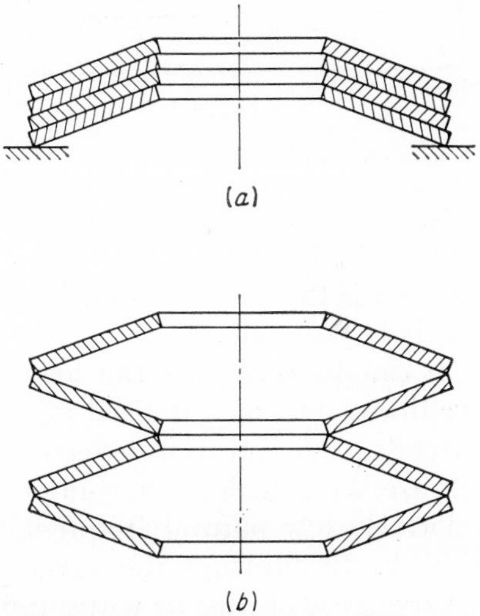

(*a*)

(*b*)

FIG. 18.14 Stacking of Belleville springs for increasing stiffness (parallel) or deflection (series): (*a*) stacked in parallel; (*b*) stacked in series.

Advantages of Belleville springs include:

1. High energy storage for a relatively small space requirement in the direction of load application
2. Controlled nonlinear stiffness characteristics (which can be obtained by varying the ratio h/t)
3. Coulomb damping (obtained from parallel stacking)

4. Variable stiffness (obtained by series or parallel stacking)

On the other hand,

1. The springs fatigue easily.
2. Guides must be provided to prevent sidewise buckling.
3. Clearance must be provided between the central tube or rod.
4. There is a possible "snapping" action as the springs pass through a flat shape.

In practice, the ratios of outside to inside diameter vary between 1.5 to 3.5. The optimum value of this ratio will depend on the type of loading as well as other design considerations. Analyses indicate that for maximum energy storage, the ratio should be between 1.8 and 2.0. In springs of this type at stresses which are commonly used for static loading conditions, energy storage capacities may be as high as 500 to 700 in.-lb_f/lb_m (in.-lb/lb) of active material.

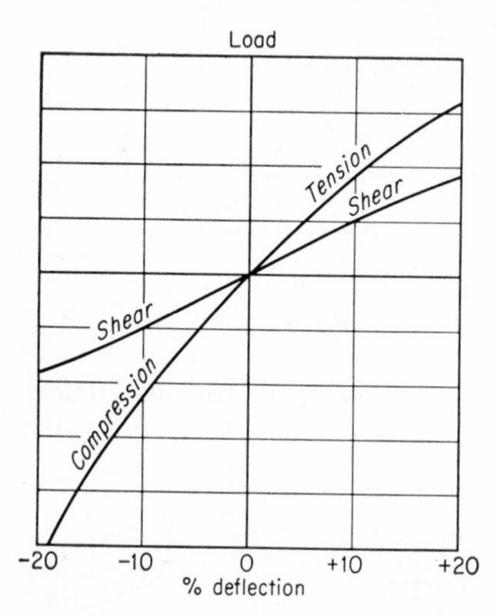

FIG. 18.15 Load-deflection curves for rubber in shear, tension, and compression.

Rubber Mountings. Rubber is used very effectively to isolate small machinery and mechanical devices—engines, motors, instruments, electronic gear, etc. In Fig. 18.15 a typical set of static load-deflection curves for rubber is shown. The exact values will vary with the composition, but the qualitative behavior will be essentially the same.

It can be seen that the load-deflection curves for all three cases are linear over only limited ranges, although that for rubber in shear extends to the greatest deflections. The soft characteristics of the latter are used widely in commercial resilient mountings to obtain relatively low natural frequencies at reasonable stress levels in the rubber. In addition, the advantages of minimal effect of shape and a very great ability to withstand repeated variations in stress* outweigh the disadvantages of small energy storage capacity and the more complex molding and bonding problems associated with fabricating an actual mounting.

* Useful life of actual mountings in service at normal loads is 1.5×10^9 cycles or greater.

Rubber in compression is used widely for applications which require high energy-storage capacity and an ability to support heavy loads without failure. The stiffness of a compressive rubber pad depends on the end restraint against lateral bulging. The *shape factor* of a compressive pad is defined as the ratio of the area of one loaded surface to the total area of free surface. Roughly, the shape factor varies linearly with the greatest dimension of the loaded surface and inversely with the undeflected height of the pad. For a given load, both the compressive and shear moduli increase with increasing shape factor.[11,12,13] The maximum service life of rubber in compression is less than that for rubber in shear.

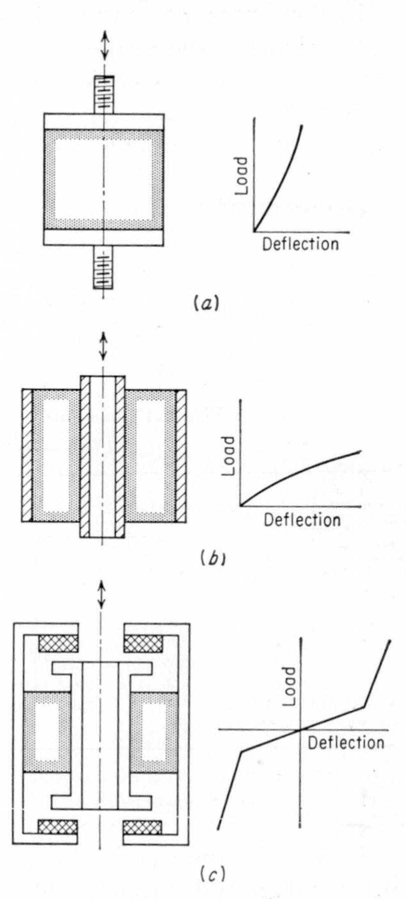

Fig. 18.16 Typical load-deflection curves for three common types of rubber mounts: (*a*) compression; (*b*) shear (section); (*c*) shear compression (section).

Rubber can seldom be used in tension to support the main load because of its tendency to tear or separate at bonded edges under tensile loading. There are no commercial resilient mountings which use rubber in this manner, although it may be used rarely in custom installations.

In current design practice, it is common to find that mountings use rubber in shear as the primary elastic element and rubber in compression as a secondary element which furnishes a snubbing action if the mounting is subjected to an overload. Typical designs are shown in Fig. 18.16 with their appropriate load-deflection curves.

Most resilient mountings are designed to operate primarily in a preferred direction, usually an axis of symmetry. The stiffness characteristics in other directions are usually not the same. For example, in a shear mounting, the stiffness in the transverse directions is greater because displacements in these directions tend to compress the resilient element (Fig. 18.16*b*). Mountings have been designed which are equally stiff in an axisymmetric direction and in the transverse directions as well.

In the design of compression mountings, account must be taken of the shape of the resilient element and, since the volume compressibility of solid rubber is extremely low,* it cannot be used successfully as a compression spring, unless space is provided for lateral expansion. Ribbed construction will provide the space into which the rubber can bulge.

Detailed information on the properties of rubber and rubberlike materials is available in commercial and technical literature. In Table 18.1, some representative engineering properties are given for various hardness rubber stocks.†

Table 18.1

Various Physical Constants of Some Representative Rubber Stocks*
(Temperature of rubber 70°F)

Durometer Hardness, Shore *A*	Moduli (psi)† Static			% Volume Compression at 1,000 psi Hydrostatic Pressure	Damping (Log Decrement)	Specific Gravity	Specific Heat	Velocity of Sound in Rubber Rods
	Compression	Tension	Shear					
30	185	175	50	0.34	0.041	1.01	0.47	115
40	270	230	70	0.31	0.055	1.06	0.43	165
50	375	305	95	0.29	0.14	1.11	0.40	210
60	550	450	140	0.27	0.23	1.18	0.38	345
70	750	610	195	0.24	0.35	1.25	0.35	750
80	1,200	1,025		0.22	0.47	1.31	0.33	

* "Some Physical Properties of Rubber," U.S. Rubber Company, Akron, 1941.
† psi ≡ pounds (lb_f) per square inch.

Under dynamic conditions, the apparent stiffness of rubber is greater than under comparable static conditions. The work of many observers has shown that this phenomenon is a function of (1) the ambient temperature, (2) the frequency (forced or free) at which the observations are made, (3) the static strain in the specimen being tested, (4) the amplitude of vibration, and (5) the kind of rubberlike material of which the specimen is made.

It should be noted that when interpreting results in the literature

* Almost all rubbers have less volume compressibility than water, e.g., at a hydrostatic pressure of 30,000 psi, the percentage decrease in original volume of water is 7.3 per cent, of a 30-durometer (soft) rubber 6.5 per cent, and of an 80-durometer (hard) rubber 4.7 per cent.

† A Pearl pencil eraser has hardness roughly equivalent to 30-durometer rubber; a good Vulcanite pipe stem has that roughly equivalent to 80-durometer rubber.

obtained from dynamic tests it is important to distinguish between the behavior of a material in thin strips or rods or in bulk. For example, the velocity of a bulk compressional wave in a rubberlike material is greater than 3,000 ft/sec, while that of a wave in a thin strip is from 100 to 1,000 ft/sec. Resilient mountings are almost always constructed in such a manner as to use rubberlike materials in bulk, so that in the discussion that follows, the dynamic properties in bulk only will be mentioned.

The dynamic modulus of the visco-elastic mathematical models which are used to characterize the properties of rubberlike materials is less dependent upon frequency as temperature is increased. This has been verified experimentally, although there is considerable variation in the degree of temperature dependence of dynamic modulus with the type of rubberlike material being tested. Usually, for a given frequency value, synthetic rubbers display a greater variation of dynamic modulus with temperature than does natural rubber. In general, dynamic modulus increases with increasing frequency.

Both static and dynamic moduli increase with increasing strain in the specimen. In addition, for any value of strain, the dynamic modulus is always greater than the corresponding static modulus.

With increasing amplitude of vibration, it has been shown that the dynamic modulus of most rubberlike materials decreases. Despite the many observations of this phenomenon, there is no agreement among workers in the field as to the reason for the dependence of modulus upon amplitude. Gehman[14] found that with decreasing modulus, there is an associated temperature increase, which might be attributed to higher hysteresis losses. However, the temperature rise is too small to account for the entire effect. Other tests indicate that if an initially stressed specimen is caused to vibrate, increasing amplitude of vibration causes a decrease in apparent dynamic modulus measured from the hysteresis loop. That is, at the original static strain, the slope of the principal axis of the hysteresis loop is less than that of the static stress-strain curve at that point. Whatever the mechanism, the over-all effect is to reduce modulus with increasing amplitude of vibration.

There are other second-order dynamic effects on the properties of rubber which will not be mentioned here. The reader is referred to the extensive technical literature available.

Wave Effects in Resilient Mountings. It has been shown experimentally that the efficiency of resilient mountings as vibration isolators is reduced appreciably at certain frequencies associated with a resonance in the elastic element of the mounting. In helical wire springs of the most commonly used sizes, the frequency at which the resonance occurs is high (several thousand cps or higher), but it has a more marked effect on the isolator efficiency. In most rubber resilient mountings, the so-called "wave effects" first occur at much lower frequencies (about 200 to 350 cps), with only moderate increases in transmissibility. As an example, in Fig. 18.17, a plot of decibel change (transmissibility*) as a function of frequency is shown for a shear-type rubber resilient mounting under a nominal static load of 58 lb.

The relative peaks in decibel change which occur at 880, 1,940, 2,710, . . . , cps can be related to a wave resonance phenomenon in

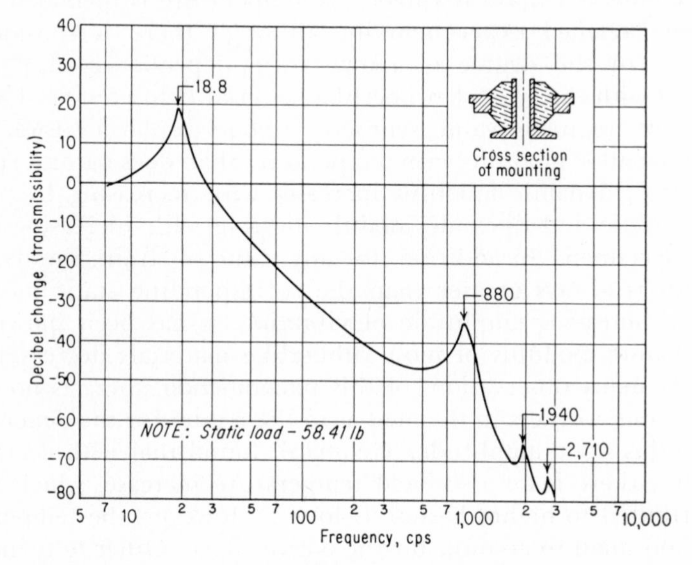

Fig. 18.17 Measured transmissibility curve for a standard rubber mount.

the rubber element of the mounting. The frequency at which this resonance occurs is a function of the dimensions of the rubber element and the hardness of the rubber. In general, a rubber resilient mounting which displays poor sound attenuation properties is characterized by (1) a soft rubber resilient element, (2) relatively large dimensions, and (3) a light static load.

Cork. One of the oldest materials used for vibration isolation is cork. It is used in compression or in a combination of compression and shear. Unlike a rubberlike material in compression, cork becomes less stiff at high loadings, displaying the same type of stress-

* For a single-degree-of-freedom system, it can be shown that transmissibility ϵ and decibel change n are related by the relationship $n = 20 \log_{10} \epsilon$. For $f > 30$ cps, the modulus of M_i is approximately $|M_i| \approx \omega m \epsilon$.

strain curve as copper. The dynamic properties of cork are also very much dependent upon frequency.

Generally, in order to obtain sufficiently large deflections, the machine to be isolated is mounted on large concrete blocks which are separated from the surrounding foundation by several layers of cork slabs 1 to 6 in. thick. The recommended pressure to which the cork should be subjected for optimum performance is between 7 and 20 psi.

Oil, water, and moderate temperatures have little effect upon the operating characteristics of cork; but cork does tend to compress with age under an applied load. At room temperature its effective life may be measured in decades; at 200°F it is reduced to a fraction of a year. Figure 18.18 shows why cork must have sufficient load to give a reasonable resiliency. It is not a very effective isolator in the low-frequency range, since great thicknesses are needed to achieve the correspondingly large deflections which are required. Unless the slabs are properly spaced, this can lead to an unstable condition.

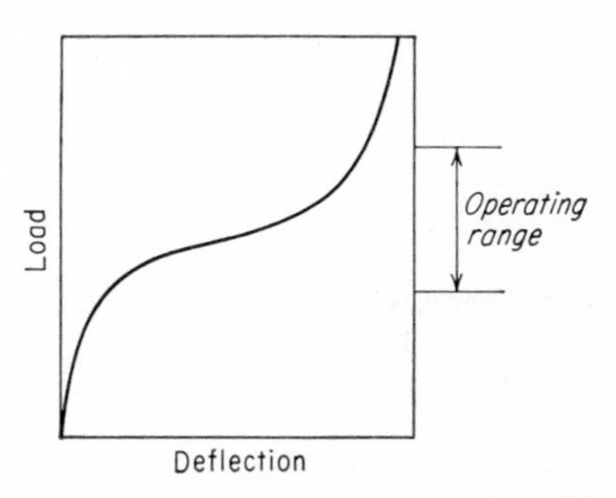

FIG. 18.18 Load-deflection curve for cork.[15]

Felt. When felt is used as a vibration isolation material, the greatest isolation efficiency is obtained by using the smallest possible area of the softest felt, in maximum thickness, under a static load that the felt will resist without excessive compression or loss of structural stability. It has a high damping factor and thus is particularly useful in reducing amplitude of vibration at resonance. The amplification factor at resonance is almost independent of amplitude and load and is about 4 for soft felt. For general purposes, felt mountings of ½- to 1-in. thicknesses are recommended, with an area of 5 per cent of the total area of the base if the machine has a flat bed. In installations where vibration is not excessive, no bonding is necessary between the felt and the machine.

Resonance curves of loaded felt pads in compressional vibration show the nonlinear character of the damping and stiffness properties of felt. The compliance of a felt pad increases somewhat more slowly than the thickness of the pad and there is a decrease in stiffness with increasing amplitude of vibration. The increase of stiffness with pressure is so large that the ratio of elastic modulus to pressure (stiffness to mass) does not vary greatly between pressures of 3 to

100 psi. In this range, the natural frequency of a mass mounted on a felt pad is determined by the thickness of the pad rather than its area and static load (Fig. 18.19). In most cases, the effectiveness of felt in reducing vibration transmission is limited to frequencies above 40 cps. Felt is particularly useful in reducing vibration transmission in the audio-frequency range since it offers a large impedance mismatch to most engineering materials.

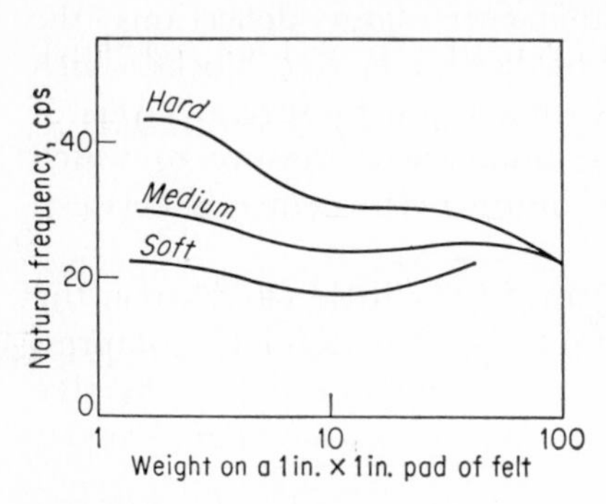

FIG. 18.19 Natural frequency as a function of weight (lb_m or lb) loading on different grades of felt. (*After Tyzzer and Hardy.*[16])

REFERENCES

1. Jacobsen, L. S.: Steady Forced Vibration as Influenced by Damping, *Trans. ASME,* vol. 52, p. APM169, 1930.

2. Crede, C. E.: "Vibration and Shock Isolation," John Wiley & Sons, Inc., New York, 1951.

3. Myklestad, N. O.: "Vibration Analysis," McGraw-Hill Book Company, Inc., New York, 1944.

4. Plunkett, R.: Interaction between a Vibrating Machine and Its Foundation, *Noise Control,* vol. 4, no. 1, pp. 18–22, January, 1958.

5. Plunkett, R.: Experimental Measurement of Mechanical Impedance or Mobility, *J. Appl. Mechanics,* vol. 21, pp. 250–256, 1954.

6. Wahl, A. M.: "Mechanical Springs," Penton Publishing Company, Cleveland, 1944.

7. Timoshenko, S.: "Theory of Elastic Stability," McGraw-Hill Book Company, Inc., New York, 1936.

8. Haringx, J. A.: On Highly Compressible Helical Springs and Rubber Rods and Their Application for Vibration-free Mountings, *Philips Research Repts.,* vol. 3, pp. 401–449, 1948; vol. 4, pp. 49–80, 206–220, 261–290, 375–400, 407–448, 1949.

9. Burdick, W. E., F. S. Chaplin, and W. L. Sheppard: Deflection of Helical Springs under Transverse Loadings, *Trans. ASME,* vol. 61, p. 623, 1939.

10. "Manual on Design and Manufacture of Coned Disk Springs or Belleville Springs," Society of Automotive Engineers, January, 1950.

11. "Some Physical Properties of Rubber," U.S. Rubber Company, Akron, 1941.

12. "Rubber in Engineering," Chemical Publishing Company, New York, 1946.

13. "Handbook of Molded and Extruded Rubber," Goodyear Tire and Rubber Company, Inc., Akron, Ohio, 1949.

14. Gehman, S. D.: Rubber in Vibration, *J. Appl. Phys.,* vol. 13, p. 402, 1942.

15. Dart, S. L., and E. Guth: Elastic Properties of Cork, *J. Appl. Phys.,* vol. 17, no. 5, pp. 314–318, 1946; vol. 18, no. 5, pp. 470–478, 1947.

16. Tyzzer, F. G., and H. C. Hardy: Properties of Felt in the Reduction of Noise and Vibration, *J. Acoust. Soc. Am.,* vol. 19, pp. 872–878, 1947.

SUGGESTIONS FOR FURTHER READING

General Works

Alfrey, T.: "Mechanical Behavior of High Polymers," Interscience Publishers, Inc., New York, 1948.

Houwink, R.: "Elasticity, Plasticity, and the Structure of Matter," Cambridge University Press, London, 1953.

Periodicals

Crede, C. E., and J. P. Walsh: The Design of Vibration-isolating Bases for Machinery, *J. Appl. Mechanics,* vol. 14, p. 7, 1947.

Harris, C. O.: Some Dynamic Properties of Rubber, *J. Appl. Mechanics,* vol. 9, pp. 129–135, 1942.

Haushalter, F. L.: The Mechanical Characteristics of Rubber, *India Rubber World,* vol. 99, pp. 39–42, 1939.

Hull, E. H.: Influence of Damping in the Elastic Mounting of Vibrating Machinery, *Trans. ASME,* vol. 53, p. 155, 1941.

Hull, E. H.: The Use of Rubber in Vibration Isolation, *J. Appl. Mechanics,* vol. 4, p. 109, 1937.

Lewis, R. C., and K. Unholtz: Simplified Method of Design of Vibration Isolating Supports, *Refrig. Eng.,* vol. 53, p. 291, April, 1947.

Madden, B. C.: Effectiveness of Shear-stressed Rubber Compounds in Isolating Machinery Vibrations, *Trans. ASME,* vol. 65, p. 617, 1943.

Smith, J. F. Downie: Rubber for the Absorption of Vibration, *Iowa Eng. Expt. Station Eng. Rept.* 2, Ames, Iowa, July, 1950.

Soderberg, C. R.: Vibration Absorbers, *Elec. J.,* vol. 21, p. 160, 1924.

Part Three

CRITERIA FOR NOISE AND VIBRATION CONTROL

Chapter 19

DAMAGE-RISK CRITERIA FOR HEARING*

K. D. Kryter

19.1 Need for a Criterion

It is well known that a person exposed to a sound of sufficient intensity may suffer a degree of permanent hearing loss. By "degree" is meant that the hearing acuity of the exposed person will be somewhat less than it was prior to his exposure to the noise; by "permanent" we mean that his loss will not disappear as the result of rest away from noise.

There is a definite need to specify the physical acoustical characteristics of noises that are *safe* with regard to their effect upon hearing acuity and noises that are potentially *unsafe*. In this regard a number of educated guesses have been made of the maximum level, for a given sound frequency, that can be tolerated for a given period of time without risk of damage to the ear. These maximum tolerable levels have come to be called "damage-risk criteria" for hearing.

Damage risk criteria are useful in several ways. They serve as guides for: (1) hearing-conservation programs in industry and in the military services, (2) noise-control procedures and techniques as applied to machinery and work environments, and (3) equitable rulings with respect to workmen's compensation cases.

19.2 Factors Involved in the Derivation of Damage-risk Criteria

There are a number of things that make difficult the derivation of damage-risk criteria from existing industrial and laboratory data on hearing loss. Among them are factors which may contribute to

* This chapter was prepared in part under Research Grant B-1727 from the National Institutes of Health, Public Health Service.

an overestimation of the damage effects of noise. For example, there is the question of whether too much weight has been given to the losses suffered by "tender-eared" persons who suffer hearing loss at sound levels that are not, by the usual standards, very intense.

A second factor that may lead to overestimation of the effects of industrial noise is that the level to which a given worker is said to be exposed is possibly underestimated for various environments. For instance, in a recent study performed by Rosenwinkel and Stewart[1] the average hearing loss suffered by workers in a large machine shop was related to the average level of the room noise. In spite of the fact that there were a large number of machines involved in this study, there is a considerable probability that the noise level at the operator's position of some of the machines was greater than the general level of room noise.[2] Inasmuch as the hearing losses of all the workers were averaged regardless of the particular noise levels in which they worked, it follows that the losses reported may be due largely to those workers exposed to levels considerably above the level of the general room noise.

On the other hand, there are factors that may mask the harmful effect of industrial noise. One possibility is, for instance, that only persons who have the "toughest" ears (ears that are more resistant than the average ear to injury from noise) remain in jobs that involve exposure to excessive levels of noise; accordingly the hearing losses these workers suffer are less than what would be experienced by persons with "average" ears.

19.3 The Measurement of Hearing

To evaluate the effects of an industrial noise on hearing, we need to compare a worker's preemployment audiogram* with audiograms taken after his employment or exposure to noise. However, it has not been until recent years that these audiometric measurements have been obtained on workers *prior* to exposure to industrial noise, so that for the most part our population of deafened industrial workers is without preexposure audiograms. Therefore, to determine the effect of industrial noise on hearing it has been the practice

* Auditory acuity is commonly determined by having the subject listen to a pure tone at several frequencies (usually 250, 500, 1,000, 2,000, and 4,000 cps). The tone is generated electronically and presented via an earphone to the listener seated in a quiet room; the intensity of the tone is decreased until the listener signals that the tone is inaudible. The minimum intensity level (called threshold) at which the listener is able to reliably hear each tone is recorded in decibels *re* the intensity required to reach threshold for that frequency by young adults with normal hearing. The plot of these values for the different frequencies tested is called an "audiogram."

to compare the hearing of industrial workers to that of young adults with presumably normal hearing.

Whether the hearing loss for a person exposed to noise is determined by comparing his postexposure audiogram to his preexposure audiogram or to a standard, "normal" audiogram, it is necessary to

Table 19.1

Presbycusis for Men and Women*

(In decibels below the mean hearing loss of the 20- to 30-year group)

Age Group, years	Survey †	440–512 cps	880–1,024 cps	1,760–2,048 cps	3,520–4,096 cps	No. of Ears	No. of People
20–30	B	0	0	0	0	322	161
	WF	0	0	0	0	7,495	7,495
	SDF	0	0	0	0	1,576	788
	WSF	0.2	2.1	3.15	10.6	1,258	629
	NHS	1.5	1.4	1.6	5.9	2,420	1,210
						13,071	10,283
		0.50‡	0.46	0.60	2.11		
30–40	B	0.5	0	1.3	4.0	348	174
	WF	2.0	1.95	2.6	5.3	7,175	7,175
	SDF	1.95	2.2	2.45	3.5	1,480	740
	WSF	2.0	4.55	6.7	20.15	1,320	660
	NHS	3.65	4.05	5.5	13.85	3,212	1,606
						13,535	10,355
		2.35‡	2.68	3.64	8.55		
40–50	B	2.5	1.8	4.0	8.2	346	173
	WF	4.85	5.15	6.85	12.75	8,897	8,897
	SDF	3.3	3.35	5.6	6.95	1,048	524
	WSF	4.7	7.7	10.65	28.0	1,326	663
	NHS	7.55	8.65	11.65	26.55	3,076	1,538
						14,693	11,795
		5.24‡	5.91	8.04	16.49		
50–60	B	12.5	3.5	9.5	15.8	270	135
	WF	8.55	8.75	11.55	19.7	4,473	4,473
	SDF	5.55	6.1	11.15	16.15	664	332
	WSF	6.85	10.8	16.85	38.0	1,266	633
	NHS	12.4	14.8	21.7	38.45	1,930	965
						8,603	6,538
		9.06‡	10.04	14.51	26.20		

* In the B, WF, and SDF surveys the mean hearing loss of the 20 to 30 age group found in each study was used as the measure of normal acuity ("audiometer zero"). The audiometer zero established by the "normal" group (average of medians for young men and women) in the NHS survey was taken in the WSF and the NHS studies as the reference or "normal" acuity. The mean acuity was less than the median or "normal" acuity in the NHS survey.

† How the studies were made. B—Bunch:[3–5] group mean; monaural listening, both ears tested—scores of right and left ears averaged. WF—World's Fair:[7] group mean; monaural listening—one ear tested. SDF—San Diego Fair:[6] group mean; binaural listening—both ears tested together. WSF—Wisconsin State Fair:[9] group median; monaural listening, both ears tested—scores of right and left ears averaged. NHS—National Health Survey:[8] group median; monaural listening, both ears tested—scores of right and left ears averaged.

‡ Average over all groups, weighted by number of ears.

make allowance for the loss that normally occurs in hearing as a person grows older. This decline in acuity is called *presbycusis* and is attributed to the normal processes of aging. The amount of hearing loss that remains for a given individual after correcting for

Table 19.2

Presbycusis in Men*

(In decibels below the mean hearing loss of the 20- to 30-year group)

Age Group, years	Survey †	440–512 cps	880–1,024 cps	1,760–2,048 cps	3,520–4,096 cps	No. of Ears	No. of People
20–30	B	1.0	−1.0	0	1.5	180	90
	WF	0	−0.2	−0.1	2.0	3,287	3,287
	SDF	1.3	−0.7	0.4	3.8	830	415
	WSF	1.0	3.4	4.5	13.9	638	319
	NHS	0.9	1.3	2.0	8.3	944	472
						5,879	4,583
		0.48 ‡	0.34	0.81	4.54		
30–40	B	0	−1.0	1.0	8.0	182	91
	WF	1.4	1.3	2.3	8.2	3,197	3,197
	SDF	0.3	1.3	2.2	7.7	766	383
	WSF	2.4	5.4	7.8	28.7	694	347
	NHS	2.6	3.5	5.8	19.2	1,238	619
						6,077	4,637
		1.58 ‡	2.15	3.59	12.71		
40–50	B	1.5	0.5	4.0	12.5	210	105
	WF	3.7	4.5	7.0	17.7	4,528	4,528
	SDF	1.3	2.4	6.0	13.2	542	271
	WSF	5.4	9.4	12.4	40.8	602	301
	NHS	5.7	7.8	12.4	37.2	1,312	656
						7,194	5,861
		3.96 ‡	5.24	8.27	22.70		
50–60	B	2.0	4.0	9.5	18.5	166	83
	WF	6.8	7.7	12.1	25.6	1,935	1,935
	SDF	4.2	4.9	11.9	25.0	378	189
	WSF	7.3	12.2	20.9	52.8	638	319
	NHS	10.1	13.7	24.6	49.5	908	454
						4,025	2,980
		7.18 ‡	9.35	16.19	34.95		

* In the B, WF, and SDF surveys the mean hearing loss of the 20 to 30 age group found in each study was used as the measure of normal acuity ("audiometer zero"). The audiometer zero established by the "normal" group (average of young men and women) in the NHS survey was taken in the WSF and the NHS studies as the reference or "normal" acuity.

† How the studies were made. B—Bunch:[3–5] group mean; monaural listening, both ears tested—scores of right and left ears averaged. WF—World's Fair:[7] group mean; monaural listening—one ear tested. SDF—San Diego Fair:[6] group mean; binaural listening—both ears tested together. WSF—Wisconsin State Fair:[9] group median; monaural listening, both ears tested—scores of right and left ears averaged. NHS—National Health Survey:[8] group median; monaural listening, both ears tested—scores of right and left ears averaged.

‡ Average over all groups, weighted by number of ears.

presbycusis is presumably due to exposure to noise or to some otological disease or abnormality.

Presbycusis has been measured in a number of studies of the hearing acuity of large groups of people. The results of these investi-

Table 19.3

Presbycusis in Women*

(In decibels below the mean hearing loss of the 20- to 30-year group)

Age Group, years	Survey †	440–512 cps	880–1,024 cps	1,760–2,048 cps	3,520–4,096 cps	No. of Ears	No. of People
20–30	B	1.0	1.0	0	−1.5	142	71
	WF	0	0.2	0.1	−2.0	4,208	4,208
	SDF	1.3	0.7	−0.4	−3.8	746	373
	WSF	−0.6	0.8	1.8	7.3	620	310
	NHS	2.1	1.5	1.2	3.5	1,476	738
						7,192	5,700
		0.53 ‡	0.59	0.42	−0.25		
30–40	B	1.0	1.0	1.5	0	166	83
	WF	2.6	2.6	2.9	2.4	3,978	3,978
	SDF	3.6	3.1	2.7	−0.7	714	357
	WSF	1.6	3.7	5.6	11.6	626	313
	NHS	4.7	4.6	5.2	8.5	1,974	987
						7,458	5,718
		3.13 ‡	3.23	3.69	4.44		
40–50	B	3.5	3.0	4.0	4.0	136	68
	WF	6.0	5.8	6.7	7.8	4,369	4,369
	SDF	5.3	4.3	5.2	0.7	506	253
	WSF	4.0	6.0	8.9	15.2	724	362
	NHS	9.4	9.5	10.6	15.9	1,764	882
						7,449	5,934
		6.51 ‡	6.54	7.68	9.87		
50–60	B	3.0	3.0	9.5	13.0	104	52
	WF	10.3	9.8	11.0	13.8	2,538	2,538
	SDF	6.9	7.3	10.4	7.3	286	143
	WSF	6.4	9.4	12.8	23.2	628	314
	NHS	14.7	15.9	18.8	27.4	1,022	511
						4,578	3,558
		10.37 ‡	10.80	12.92	17.73		

* In the B, WF, and SDF surveys the mean hearing loss of the 20 to 30 age group found in each study was used as the measure of normal acuity ("audiometer zero"). The audiometer zero established by the "normal" group (average of young men and women) in the NHS survey was taken in the WSF and the NHS studies as the reference or "normal" acuity.

† How the studies were made. B—Bunch:[3–5] group mean; monaural listening, both ears tested—scores of right and left ears averaged. WF—World's Fair:[7] group mean; monaural listening—one ear tested. SDF—San Diego Fair:[6] group mean; binaural listening—both ears tested together. WSF—Wisconsin State Fair:[9] group median; monaural listening, both ears tested—scores of right and left ears averaged. NHS—National Health Survey:[8] group median; monaural listening, both ears tested—scores of right and left ears averaged.

‡ Average over all groups, weighted by number of ears.

gations are given in Tables 19.1 to 19.3.* It will be noted in Table 19.1 that the data obtained by Bunch[3,4,5] (B) and those obtained at the San Diego Fair[6] (SDF) and World's Fair[7] (WF) are in fairly close agreement; also these data show considerably less presbycusis than the data from the National Health Survey[8] (NHS) and Wisconsin State Fair[9] (WSF). The results of the latter two studies are very similar to each other. Why there is the systematic difference between these two groups of data is not known; the use of "means" (Bunch, San Diego Fair, and World's Fair) rather than medians (National Health Survey and Wisconsin State Fair) would not account for the discrepancies. For example, the use of medians

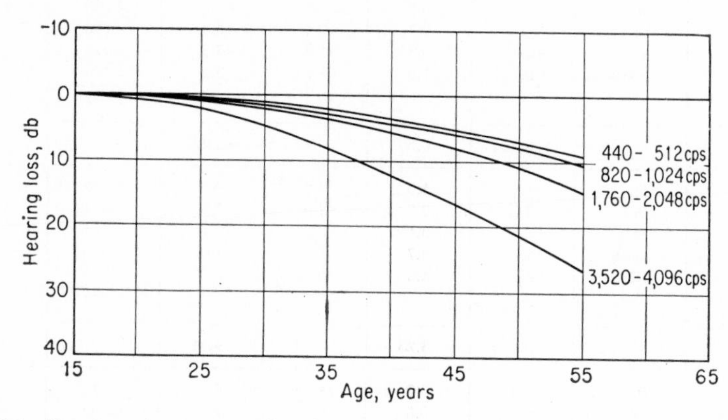

FIG. 19.1 Showing the average loss in hearing acuity for the combined results of men and women as a function of age. The approximate frequency of the test tones is indicated on the individual curves. The data are taken from the weighted averages presented in Table 19.1.

for the Bunch, San Diego Fair, and World's Fair studies presumably would indicate even less apparent hearing loss than the means which were used.

The medians of each group should be an appropriate measure of hearing losses inasmuch as extreme hearing losses due to otological disease, intense noise, etc., would not affect its value. Since medians were used with the National Health Survey and Wisconsin State Fair data it would appear that the hearing losses for persons with "normal" hearing are properly shown for those groups.

Until more measurements are made of presbycusis or some satis-

* Since the preparation of Tables 19.1 to 19.3, additional data on presbycusis have been published: J. F. Corso, Age and Sex Differences in Pure Tone Thresholds, *J. Acoust. Soc. Am.*, vol. 31, pp. 498–507, 1959. Corso's results are in close agreement with the averages presented in Tables 19.1 to 19.3.

factory explanation of the differences among these studies is found, it is perhaps proper to average the results of the investigations and let this average represent "normal" hearing of the general public according to age; this is done in Fig 19.1 for the combined results for men and women, in Fig. 19.2 for men, and in Fig. 19.3 for

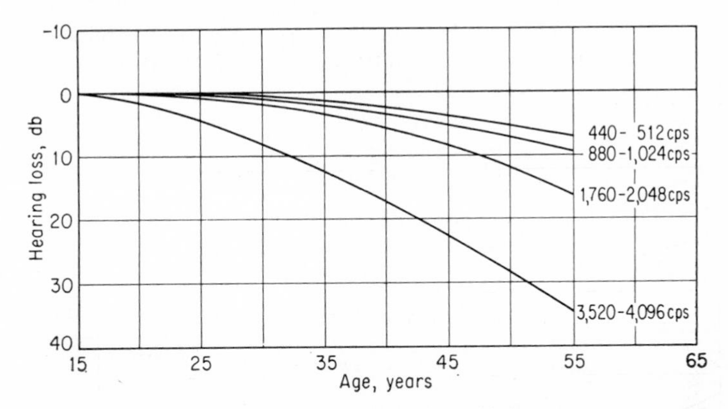

Fig. 19.2 Showing the average loss in hearing acuity for men as a function of age. The approximate frequency of the test tones is indicated on individual curves. The data are taken from the weighted averages presented in Table 19.2.

women. The importance of the curves in Figs. 19.1, 19.2, and 19.3 is, of course, that the appropriate hearing losses for a given age and frequency are to be subtracted from the hearing losses experienced by, say, an industrial worker in order to determine how much of his hearing loss is presumably induced by exposure to noise.

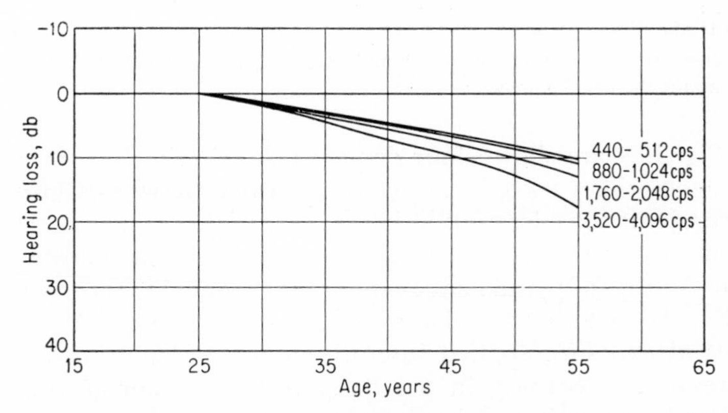

Fig. 19.3 Showing the average loss in hearing acuity for women as a function of age. The approximate frequency of the test tones is indicated on individual curves. The data are taken from the weighted averages presented in Table 19.3.

19.4 Present Damage-risk Criteria

Sterner[10] in 1952 summarized a large number of standards or criteria for noise tolerance. These criteria, and there are many of them, differ widely. They represent estimates based on various industrial and laboratory studies on the deafening effects of noise. No one of these studies is sufficient to provide us with a single criterion that all investigators will agree upon (within 10 db or so) as being valid.

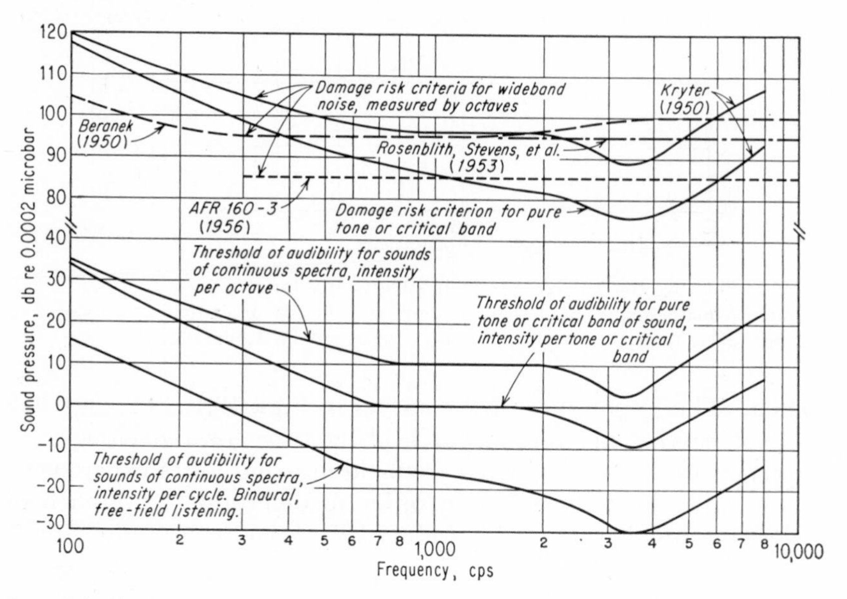

FIG. 19.4 Showing various thresholds of audibility curves and damage-risk criteria. The damage-risk criteria of Beranek[14] and Rosenblith and Stevens[15] are the same except at the higher frequencies.

In Fig. 19.4 are shown four damage-risk criteria that are, among others, sometimes used in noise-control work. Kryter's[11] guess was based on joint consideration of laboratory studies of temporary deafness due to pure tones and bands of noise as well as industrial and military studies of permanent deafness in personnel exposed to noise. In brief, this criterion was that a sound should not exceed ("for long periods of exposure applied intermittently over months or years") its threshold of hearing in the quiet by more than 85 db. This criterion is not quite as simple as it may first appear because the threshold of hearing varies not only as a function of the frequency of sound but is also dependent upon the bandwidth of a complex sound and the density of the frequency components within the band.

The concept of "critical bandwidth of the ear" was used in determining what the threshold level of a complex noise would be. In essence, the "critical band" notion is that (1) the ear "integrates" the acoustic power over a relatively narrow band of frequencies for a "unit" of loudness or masking or (presumed) damage effect, and (2) energy more than one critical bandwidth away from a given frequency or narrow band will not contribute significantly to the response of the auditory system to that given frequency or narrow band. The width of these bands as a function of frequency is shown in Fig. 19.5. This concept of a critical bandwidth was developed

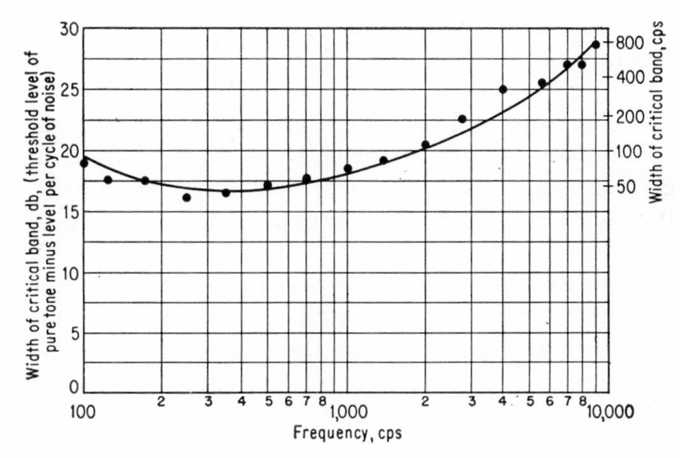

FIG. 19.5 Ratio between the monaural masked threshold of a pure tone and the level per cycle of the masking noise measured at the frequency of the pure tone. This ratio is the so-called "critical band" width. (*From Hawkins and Stevens.*[13])

primarily on the basis of data pertaining to the masking of pure tones by noise* and, while useful in the present context of the deafening effects of noise, has not been adequately tested in that regard.[12,13]

Beranek[14] and Rosenblith and Stevens[15] modified the "85 db *re* threshold criterion" for continuous spectrum and for pure tones or very narrow-band noises as shown in Fig. 19.4. These latter curves have been widely used for noise-control purposes.

Finally, there is a curve on Fig. 19.4 to show limits for lifetime exposure to broad-band noise as set forth in Air Force Regulation 160–3.[16] This regulation states that "the risk of damage to hearing

* A "critical bandwidth" is here defined as the width in cycles per second a narrow band can attain, keeping total acoustic power over the band constant, before its effectiveness in masking a pure tone at the center frequency of the band starts to diminish.

is slight when the pressure level of a (octave) band does not exceed 85." According to Air Force Regulation 160–3, if a band contains "pure-tone" components the criterion is to be lowered by 10 db. The difference in 10 db between the criteria is based on the fact that there are roughly 10 critical bands for masking in most octave bands and, according to the critical band concept, one pure tone or critical band would be one-tenth as damaging as ten such bands each containing the same amount of energy. This assumption is probably conservative; that is, the ear is perhaps somewhat less sensitive to narrow bands of noise than the "critical band for masking concept" would indicate. The supporting argument, of course, is that energies present simultaneously in a number of adjacent "critical" bands are to some degree mutually inhibitory or interfering and must add less to deafness or "damage" than bands taken one critical band at a time.

Zwicker, Flottrop, and Stevens[17] recently found that the critical band for loudness was approximately 2½ times as wide as that for masking. This suggests that the critical band for damage risk is probably wider than ⅒ octave; perhaps it is, on the average, about ⅓ octave wide.

As we shall see in the next few paragraphs, our fundamental measured effect of noise on permanent deafness is the result of what might be called widebands—or at the least octave-wide bands—of industrial noise. Hence our pure-tone or narrow-band criterion is derived from the observed effect of wideband noise. Perhaps the difference between damage-risk criteria for wideband vs. pure tone or narrow bands of noise should be more like 5 db rather than 10 db. We believe that additional research data will be required before this matter can be decided with any degree of confidence.

The study that must be given the greatest weight in deciding where on a scale of acoustic energy the damage-risk criterion should be placed is that conducted by Sub-committee Z24–X–2 of the American Standards Association under the chairmanship of W. A. Rosenblith.[18] This study provides suggestive evidence that a continuous spectrum noise that is 80 db *re* 0.0002 dyne/cm^2 or less in any octave band above 300 to 600 cps will cause negligible damage to persons exposed for 25 years (8 hr/workday). Noise of greater intensity, according to that study, may cause some hearing losses. Although the Z24–X–2 group did not specify a damage-risk criterion in their report, their findings and conclusions represent the best single basis for such a criterion that is available to date.

The rationale, as suggested in the Z24–X–2 report, behind a possible damage-risk criterion can be illustrated by Fig. 19.6. We see in Fig. 19.6 that a damage-risk criterion rests upon the concept of a statistical distribution of hearing acuities and the acceptance as insignificant of some arbitrary increase in the degree of hearing losses in some fraction of the total population. Figure 19.6 was

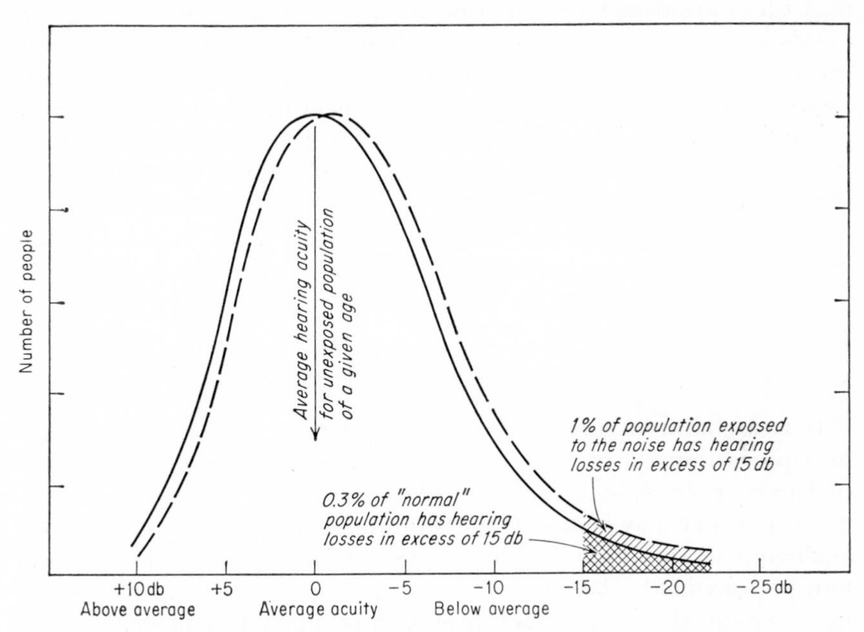

Fig. 19.6 An illustration of the statistical model underlying damage-risk criteria. It is presumed that a noise exposure at a specified damage-risk criterion would result in a slight and, practically speaking, an insignificant shift in the hearing acuity of the total population of persons so exposed. Solid curve shows distribution of "normal" hearing for a 2,000-cps tone by people who are 55 years old and who have had no damage to hearing from disease and have not been exposed to excessive amounts of noise. The dashed curve shows the distribution of hearing that would be expected from the same population of persons shown by the solid curve if they had been exposed 8 hr per working day for 25 years to either an octave band of noise centered at 1,200 cps and having an SPL of 80 db or a narrow band of noise centered at 1,200 cps and an SPL of 70 db. A loss of 15 db or more is usually required before the loss is considered abnormal or significant. Thus, about 0.3 per cent of the normal population has a hearing loss in excess of 15 db and about 1 per cent of the population exposed to noise can be expected to have a hearing loss in excess of 15 db. The values indicated should not be interpreted as the result of precise measurements but are based on interpretations and estimates of data from various sources.

constructed on the assumption (1) that the noise exposures described in the legend will increase the percentage of people having a hearing loss of greater than 15 db, from 0.3 per cent to 1 per cent, and (2) that this increase should be considered unimportant for practical purposes.

The findings of the Z24–X–2 report, which were published subsequent to all but Air Force Regulation 160–3, indicate that some of the criteria in Fig. 19.4 may possibly be too high, particularly if applied to situations involving persons exposed to continuous spectra noises for periods as long as 25 years.

However, a damage-risk criterion based on noise exposure lasting for 25 years at 8 hr/workday, while useful as a limiting case, may require refinement and modification to meet other situations. Two major questions to be asked in this regard are:

1. Should the damage-risk criterion be the same for work periods that are to be shorter than 25 years?

2. What intensity levels of noise can the ear safely tolerate for periods of time shorter than 8 hr?

19.5 The Susceptibility of the Ear to Damage from Noise as a Function of Years of Exposure to Noise

It is a common conception that the damage effects of noise on hearing add over a number of years of exposure to a significant impairment, even though for any 1-year period of exposure the decrement in acuity may be hardly measurable. This opinion is somewhat analogous to that held for the accumulation of radiation or of certain poisons in the body: a series of doses of the agent, in amounts which singly are not harmful, results in permanent damage.

There is, however, another notion of how noise damages the ear that is somewhat in opposition to that outlined in the preceding paragraph. This other concept is perhaps readily understood if we think of a "model" of the inner ear in which the structure is such that an exceedingly intense sound "breaks" or injures irreparably the weakest receptors in the cochlea. In our "model" this damage does not make the ear more susceptible to succeeding auditory insult; to injure the ear further the noise intensity must now exceed the level of resistance of the remaining structures to damage.

To make this "model" of the ear behave realistically like the human ear, we must give it another property or characteristic. This property is that the structures of the ear can become weaker with changes in the physiological state of the organism or as the result of aging. This theory would say that the *increase* found in the deafness of workers as a function of years of exposure to a steady-state noise could be due to increased susceptibility of the ears of the workers to damage as the result of aging; this is in contrast to the notion that the *increase* in deafness as a function of years of exposure to noise is the result of an accumulative, additive effect of noise.

It would follow from considerations of our "model" that the deafness measured in 50-year-old men working in a given intense industrial noise should be about the same for all those men regardless of the number of years they have been exposed to the same noise. Because there is such a strong correlation between age and years of service, the industrial data as now published do not readily permit the separation of the effects of years of exposure from the effects of aging and thereby do not permit a test of our "model." The decline of acuity with age, presbycusis, in this theory could result from an increased susceptibility, as a person grows older, to injury from the noise of everyday living.

Of course, it is likely that the ear is made more susceptible to damage as the result of *both* noise exposure and aging, but it is essential to consider the possible contribution to the over-all hearing loss of each of these factors. For example, if a person's auditory "life" is indeed not shortened by exposure to sounds which do not, when he is young, reduce his auditory acuity, it might be desirable for the military services and industry to have damage-risk criteria based on age as well as years of exposure.

To make use of this line of thinking we must decide upon how much sound intensity the ear can tolerate at different age levels. There are, to our knowledge, no data available that bear directly on this question, but perhaps a reasonable approximation can be achieved by considering the general trend of the presbycusis curves of Fig. 19.1, particularly for the important frequencies around 2,000 cps. For the sake of the argument, let us assign a value of 0 to the normal acuity of persons 25 years old. From Fig. 19.1, we see that by increasing the age of our reference 25-year-old group by 15 years (to 40 years) the acuity for tones in the 2,000-cps region has decreased by 6 db. Doubling this age step (adding 30 instead of 15 years to our reference age of 25 years) takes us to the age of 55 and a decrease relative to our 25-year-old group of roughly 12 (actually 14) db.

Now let us make (with due hesitation) two assumptions: (1) that the "presbycusis" shown in Figs. 19.1, 19.2, and 19.3 is due to exposure to everyday noise; the older persons show greater hearing loss because their ears become progressively susceptible to auditory insult as they grow older; and (2) that a loss in hearing acuity of 1 db is indicative, for frequencies around 2,000 cps, of a change in tolerance that can be compensated for by a 1-db decrease in the sound-pressure level of noise to which a person might be exposed.

The above rationalizations lead to Fig. 19.7, which shows the decrease in tolerance of the ear to sound-pressure level as a function of age.

19.6 Proposed Damage-risk Criteria for Long-duration (8-hr) Exposures

Using the function given in Fig. 19.7 and a damage-risk criterion derived primarily from the Z24–X–2 report and our assumption regarding the relation between the shape of the damage risk and auditory threshold curves as a function of frequency, a family of damage-risk criterion curves for octave bands of noise are drawn on the upper portion of Fig. 19.8 for each decade age group from 20 to 60 years.

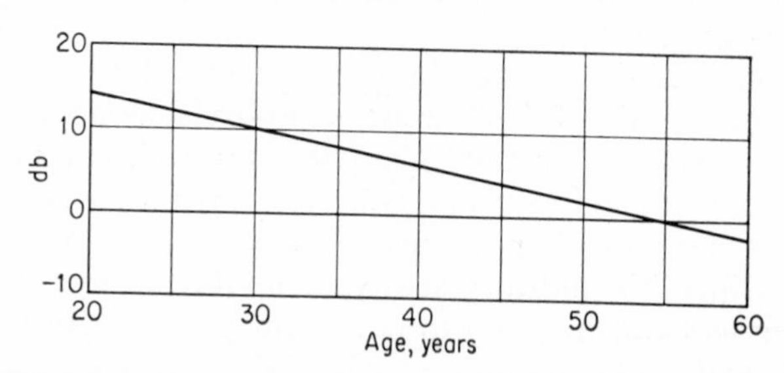

Fig. 19.7 Decibel corrections to be added to proposed damage-risk criteria for 50- to 60-year-old group to obtain criteria for persons of different ages, based on trends of presbycusis curves, particularly for frequencies between 1,760 and 2,048 cps. The slope of this curve shows in terms of sound-pressure level the change in susceptibility of the ear to damage as a function of age.

In addition, comparable damage-risk criteria are shown in Fig. 19.8 for pure tones or narrow "critical" bands of noise. These latter criteria are based upon the aforementioned critical-band concept, modified for the sake of simplicity, to the extent that 10 critical bands are allowed for each octave regardless of frequency.

The criteria given in Fig. 19.8 are based upon the report of Subcommittee Z24–X–2 in several ways. For one thing, that report suggests that the octave band of noise from 300 to 600 cps should not exceed a sound-pressure level of 80 db for 25 years of workday exposures. This relationship is shown by the part of the curve for persons 50 to 60 years old in the upper portion of Fig. 19.8.

The Z24–X–2 study also included some data on the effects of exposure to a predominantly low-frequency noise which had a sound-pressure level of about 100 db in the band below 150 cps. The data showed a negligible amount of hearing loss after 12 years of exposure, the longest exposure time for which there were data available. Considering this finding, along with the upward trend of the curve for the acuity of hearing at low frequencies (see Fig. 19.4), the damage-

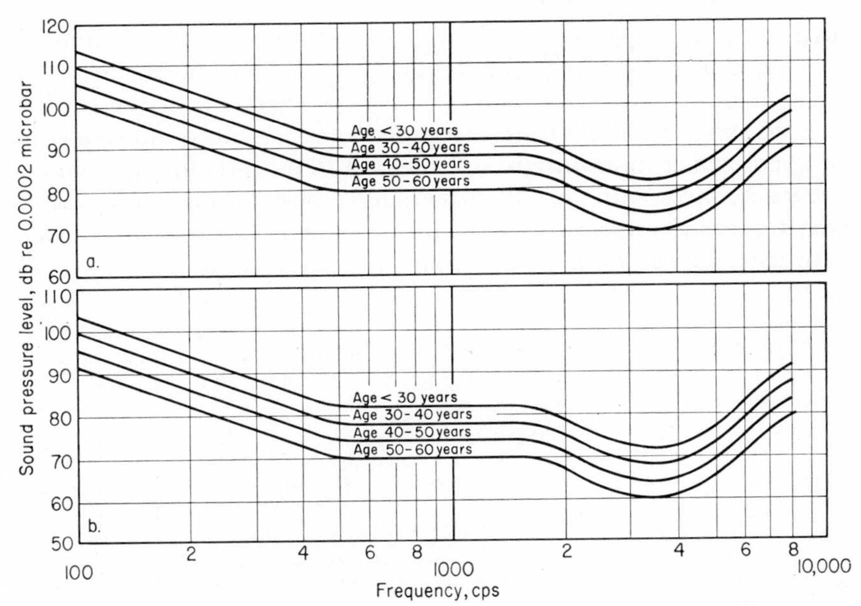

FIG. 19.8 Proposed damage-risk criteria for (*a*) wideband noise measured by octave, 8 hr continuous exposure; and (*b*) pure tones or critical bands of noise. The parameter is age.

risk criterion for 50- to 60-year-old people was modified for frequencies below 300 cps as shown in Fig. 19.8.

19.7 Tolerable Levels for Brief Exposures

It is logical to believe that listening to a noise for a few minutes will be less harmful than listening to the same noise for as long as 8 hr, and in 1955 Eldred, Gannon, and von Gierke[19] proposed that a "constant-energy" principle should be followed in estimating the damage effects of sounds of brief duration. Their concept is that the amount of energy absorbed by the ear (sound power multiplied by length of exposure) uniquely determines the amount of damage to hearing; the lower curve in Fig. 19.9 illustrates their notion.

Davis et al.[20] conducted some research with pure tones and wideband noise that bears on this matter. They found in general that for all the sounds they tested, the same degree of temporary deafness was experienced if the sound-pressure level was decreased by about 6 db while the duration of exposure was doubled. Their results are shown by the solid lines in Fig. 19.9.

Spieth and Trittipoe[21] and Ward, Glorig, and Sklar[22] recently conducted studies on the effect of different intensities and durations of noise exposure upon temporary hearing loss. Their findings for relatively brief exposures are in substantial agreement with the earlier work of Davis et al. Thus, research data would seem to indicate that for exposures of short durations, from 5 to 10 minutes to a few hours, the ear behaves with regard to temporary hearing loss more like a pressure-sensitive device than an intensity or power-sensitive device as proposed by Eldred, Gannon, and von Gierke.

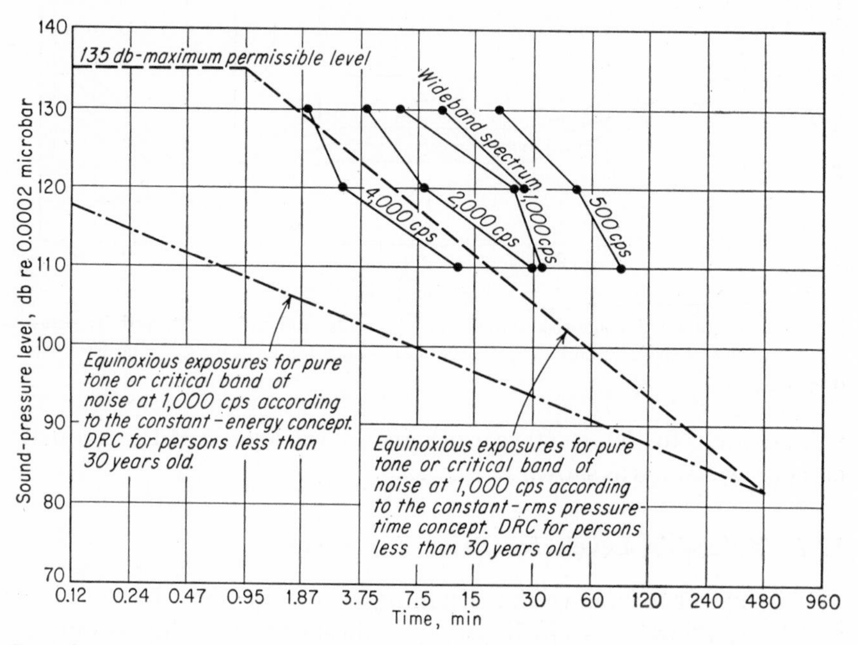

FIG. 19.9 Showing equinoxious exposure curves based on (1) the "constant energy concept"; (2) the constant rms pressure time constant; and (3) data obtained by Davis et al.[20] Thirty decibels of temporary hearing loss (average over 2 octaves) about 10 min after exposure to sounds is indicated with full recovery within 24 hr.

However, Spieth and Trittipoe[21] as well as Ward, Glorig, and Sklar[22] found that the relation between sound intensity and duration, for equal temporary hearing loss, is curvilinear, and that a simple trading between time and sound pressure tends to overestimate the deafening effect of exposures of less than 10 min or so per day and to underestimate the deafening effect of exposures exceeding one hour or so per day. Figure 19.10 shows for the octave band of noise from 1200–2400 cps the relation between duration of exposure and sound-pressure level with a temporary hearing loss of 20 db measured 2 minutes after exposure.

Figure 19.10 can be used in conjunction with Fig. 19.8 to determine the risk to hearing experienced by persons working under varying noise conditions. For an example, let us take a young engine mechanic who is required for 10 min per day to be in the presence of a noise that peaks in the 1200–2400 cps band and has a sound pressure level of 105 db. This is equivalent (causes the same fatigue) to being exposed to a level of 89 db for 480 min. Such an exposure

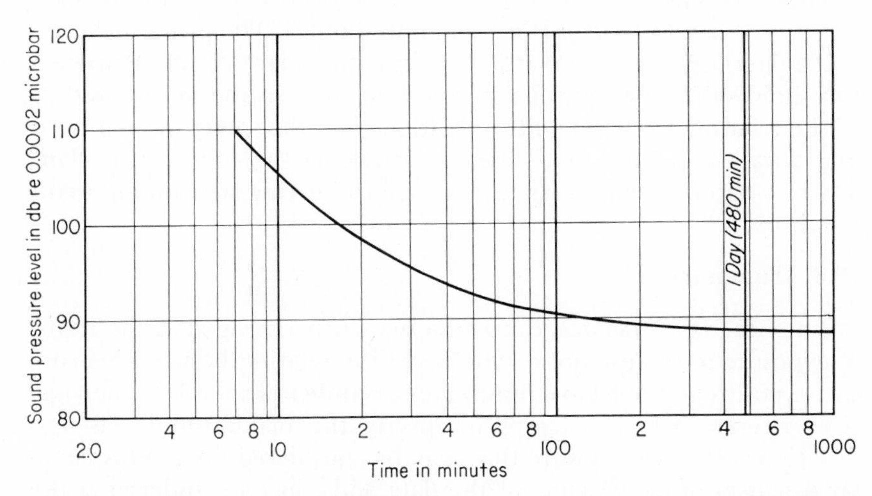

FIG. 19.10 Showing for octave band 1200–2400 cps the relation between duration of exposure in minutes and sound-pressure level. The parameter of the contour is 20 db Temporary Threshold Shift, measured 2 minutes following exposure. After Ward, Glorig, and Sklar.[22]

meets the damage risk criterion suggested in Fig. 19.8 for the 30–40 year age group. An additional 10 min exposure per day to a level of but 95 db would be equivalent to an exposure of 30 min according to Fig. 19.10; in this way we can convert the durations and levels of his exposures to other noises throughout the day into equivalent durations. If the equivalent exposure durations when added together exceed 480 min (as would be the case in our example), we conclude that the person is in danger of suffering damage to his hearing. If the equivalent exposure duration equals, or is less than 480 min his hearing is safe.

It is to be noted that the curve of Fig. 19.10 is for the 1200–2400 cps band. To a first approximation the shape of the contour in Fig. 19.10 is representative of similar functions found for wide-band noise and other octave bands by Ward, Glorig, and Sklar and can, therefore, be used to compute the equivalent exposure time for bands of noise other than the 1200–2400 cps band. To do so, however, one must first adjust upward or downward, depending on the band, the vertical ordinate of Fig. 19.10. The amount of adjustment is determined by comparing the sound-pressure level at 1800 cps for a given age group in Fig. 19.8 with the sound-pressure level at the center frequency of the band in question. For example, the sound-pressure levels in Fig. 19.10 should be increased by 12 db when one is finding the equivalent exposure durations of an octave band of noise centered at 200 cps, and the levels should be decreased by 10 db when the octave band of noise is centered at about 3,200 cps.

The recent work of Ward, Glorig, and Sklar[22–25] on temporary threshold shift represents a significant advance in the understanding of the relations between auditory fatigue and the spectrum of a noise and duration of exposure. The application of their findings to damage risk criteria for permanent damage to hearing will undoubtedly continue.

19.8 Summary

The hearing of persons can be permanently damaged as the result of exposure to intense noise. Studies of the relation between hearing and exposure to noise have resulted in a number of so-called "damage-risk criteria," which attempt to specify the maximum levels and durations of noise exposure that may be considered safe. This chapter discusses in detail some of the data and concepts underlying the more widely accepted of these damage-risk criteria.

REFERENCES

1. Rosenwinkel, N. E., and K. C. Stewart: The Relationship of Hearing Loss to Steady State Noise Exposure, *Am. Ind. Hyg. Assoc. Quart.*, vol. 18, pp. 227–230, 1957.

2. Karplus, H. B., and G. L. Bonvallet: A Noise Survey of Manufacturing Industries, *Am. Ind. Hyg. Assoc. Quart.*, vol. 14, 1953.

3. Bunch, C. C.: Age Variations in Auditory Acuity, *AMA Arch. Otolaryngol.*, vol. IX, pp. 635–636, 1929.

4. Bunch, C. C.: Further Observations on Age Variations in Auditory Acuity, *AMA Arch. Otolaryngol.*, vol. XIII, pp. 170, 1931.

5. Bunch, C. C., and T. S. Raiford: Race and Sex Variations in Auditory Acuity, *AMA Arch. Otolaryngol.*, vol. XIII, p. 423, 1931.

6. Webster, J. C., H. W. Hines, and M. Lichtenstein: San Diego Fair Hearing Survey, *J. Acoust. Soc. Amer.*, vol. 22, pp. 473–483, 1950.

7. Steinberg, J. C., H. C. Montgomery, and M. B. Gardner: Results of the World's Fair Hearing Tests, *J. Acoust. Soc. Amer.*, vol. 12, pp. 291–301, 1940.

8. Beasley, W. C.: Normal Hearing for Speech at Each Decade of Life, *Natl. Health Survey Hearing Study Ser. Bull.* 3, U.S. Public Health Serv., 1938.

9. Glorig, A., D. Wheeler, R. Quiggle, W. Grings, and A. Summerfield: 1954 Wisconsin State Fair Hearing Survey, *Trans. Am. Acad. Ophthalmol. Otolaryngol.*, Los Angeles, Calif., 1957.

10. Sterner, J. H.: Standards of Noise Tolerance. Noise—Causes, Effects, Measurement, Costs, Control, *School Public Health & Inst. Ind. Health, Univ. Mich.,* Ann Arbor, Mich., 1952.

11. Kryter, K. D.: The Effects of Noise on Man, *J. Speech & Hearing Disorders,* monograph suppl. 1, pp. 1–95, 1950.

12. Fletcher, H.: Loudness, Masking, and Their Relation to the Hearing Process and the Problem of Noise Measurement, *J. Acoust. Soc. Am.,* vol. 9, pp. 275–293, 1938.

13. Hawkins, J. E., Jr., and S. S. Stevens: The Masking of Pure Tones and of Speech by White Noise, *J. Acoust. Soc. Am.,* vol. 22, pp. 6–13, 1950.

14. Beranek, L. L.: Noise Control in Office and Factory Spaces, *Trans. Chem. Engring. Confs., Ind. Hyg. Foundation Am.,* Nov. 19, 1950.

15. Rosenblith, W. A., and K. N. Stevens: Noise and Man, vol. II of "Handbook of Acoustic Noise Control," *WADC Tech. Rept.* 52–204, Wright-Patterson Air Force Base, 1952.

16. Hazardous Noise Exposure, *Air Force Regulation* 160–3, Dept. Air Force, Oct. 29, 1956.

17. Zwicker, E., G. Flottrop, and S. S. Stevens: "The Critical Bandwidth in Loudness Summation," *J. Acoust. Soc. Am.,* vol. 29, pp. 548–557, 1957.

18. Rosenblith, W. A. (chairman): The Relations of Hearing Loss to Noise Exposure, *A Report by Sub-committee Z24–X–2,* American Standards Association, Inc., 70 E. 45th St., N.Y. 17, N.Y., 1954.

19. Eldred, K. M., W. J. Gannon, and H. von Gierke: Criteria for Short Time Exposure of Personnel to High Intensity Jet Aircraft Noise, *WADC Tech. Note* 55–355, Wright-Patterson Air Force Base, 1955.

20. Davis, H., C. T. Morgan, J. E. Hawkins, Jr., R. Galambos, and L. W. Smith: Temporary Deafness Following Exposure to Loud Tones and Noise, *Acta Oto-Laryngol.,* suppl. 88, pp. 1–57, 1950.

21. Spieth, W., and W. J. Trittipoe: Intensity and Duration of Noise Exposure and Temporary Threshold Shifts, *J. Acoust. Soc. Am.,* vol. 30, pp. 710–713, 1958.

22. Ward, W. D., A. Glorig, and D. L. Sklar: Dependence of Temporary Threshold Shift at 4 kc on Intensity and Time, *J. Acoust. Soc. Am.,* vol. 30, pp. 944–954, 1958.

23. Ward, W. D., A. Glorig, and D. L. Sklar: Temporary Threshold Shift from Octave-band Noise: Applications to Damage-Risk Criteria, *J. Acous. Soc. Am.,* vol. 31, pp. 522–528, 1959.

24. Ward, W. D., A. Glorig, and D. L. Sklar: Relation between Recovery from Temporary Threshold Shift and Duration of Exposure, *J Acous. Soc. Am.,* vol. 31, pp. 600–602, 1959.

25. Ward, W. D., A. Glorig, and D. L. Sklar: Temporary Threshold Shift Produced by Intermittent Exposure to Noise, *J. Acous. Soc. Am.,* vol. 31, pp. 791–794, 1959.

Chapter 20

CRITERIA FOR NOISE AND VIBRATION IN BUILDINGS AND VEHICLES

Leo L. Beranek

20.1 Introduction

An essential aspect of noise-control design is the establishment of acceptable noise levels for the space being quieted. The purpose of quieting may be to protect hearing. That subject was discussed in the previous chapter. The purpose may be to protect a machine or electronic component from noise levels that will cause malfunction. That subject is beyond the scope of this book. Or, the purpose may be to provide an environment in offices favorable to those occupying them, in concert halls favorable to the listening of music, in vehicles that will permit riding comfort, and so forth.

Earlier references on this subject are given at the end of this chapter.[1–5] The principal data for this chapter are based on five recent papers[6–10] and on some unpublished data. Discussed here are criteria for noise control in offices, broadcast studios, concert halls, legitimate theaters, musicrooms, schoolrooms, television studios, apartments, and hotels, assembly halls, homes, motion picture theaters, hospitals, churches, courtrooms, libraries, restaurants, coliseums for sports, aircraft, trains, and automobiles. Some information is given on criteria for vibration control.

Speech-interference Levels (Units = db). SIL, as it is commonly known, is the arithmetic average of the readings, in decibels, in the three octave frequency bands, 600 to 1,200, 1,200 to 2,400, and 2,400 to 4,800 cps. The resulting number, in decibels, is a handy guide to the interfering effect of the noise on speech.

Loudness Level (Units = Phons). The LL of a particular sound or noise, in phons, is defined as the sound-pressure level of a 1,000-cps tone that sounds equal in loudness to the sound or noise being rated. In this chapter, however, the loudness level is a quantity computed from octave-band levels according to a procedure given in Appendix A.

Perceived-noise Level (Units = PNdb). The PNL of a particular sound or noise, in PNdb, is defined as the sound-pressure level of a band of noise from 910 to 1,090 cps that sounds as "noisy" as the sound or noise under comparison. The principal difference between loudness level and perceived-noise level is that "equal-annoyance contours" in place of "equal-loudness contours" were used in its derivation. As proof of its validity for judging the "noisiness" of aircraft fly-over noise, a series of experiments was conducted in which people judged the relative acceptability of the fly-over noise made by various jet- and piston-engined aircraft. The perceived-noise level predicted the acceptability of the jet- and piston-aircraft noises more accurately than did loudness level, speech-interference level, or sound-pressure level. Perceived-noise levels in PNdb are calculated from octave-band levels according to a procedure given in Appendix B.

20.2 Criteria for Noise in Buildings

Criteria for Office Spaces. The results of a series of noise ratings made by office workers are plotted against speech-interference level and loudness level in Figs. 20.1 and 20.2—Fig. 20.1 for executive offices and small conference rooms, Fig. 20.2 for stenographic and large engineering drafting rooms (commonly found in aircraft companies) where many people are conversing simultaneously. The subjective ratings did not correlate as well with the over-all sound-pressure levels as did either the speech-interference level or the loudness level. The correlation of the subjective ratings was generally better with loudness level (LL) than with the speech-interference level (SIL). Because Figs. 20.1 and 20.2 are summary charts, they need a bit of explaining.

The left-hand ordinate of Fig. 20.1 shows the rating scale used on the questionnaires given to the office workers. On the average the personnel in executive offices thought that the noise should not exceed a rating just below "moderately noisy" if their work was not to suffer. More than half of those questioned said that if the room were more than moderately noisy, their ability to telephone and to converse would be affected. Nearly all of them said that they used the telephone "often" to "very often" and that they conversed "often" with coworkers as far away from them as 8 to 10 ft. It follows, then, that the SIL in executive offices should not exceed about 40 db if conversations are to be carried on in a more-or-less normal tone of voice over distances up to 8 to 10 ft.

In some offices the subjective ratings correlated better with the computed LL than with measured SIL. Examination of the octave-band spectra measured in these offices revealed that the low-frequency noise levels were higher relative to the levels in the speech range than

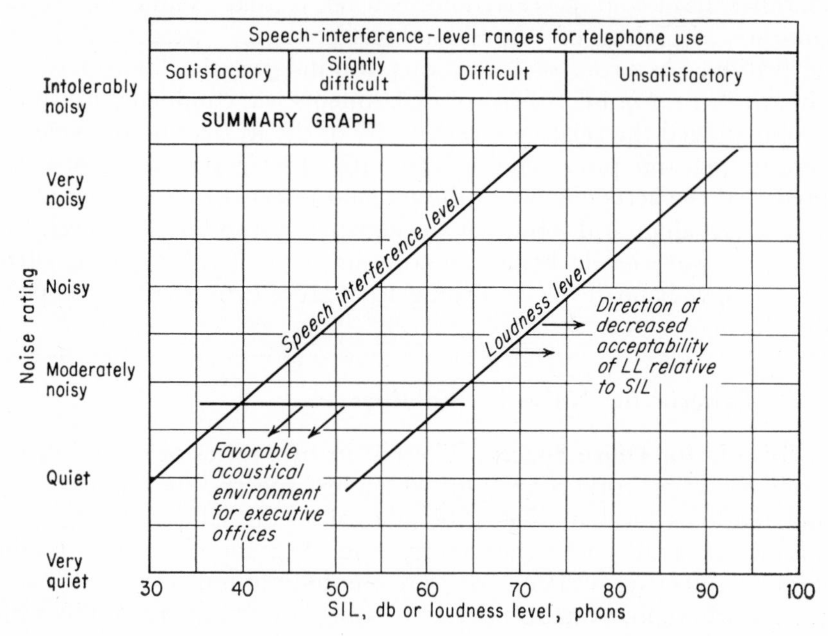

FIG. 20.1 The relation of the subjective noise ratings of executive-office personnel to SIL and LL. For a favorable acoustical environment for executive-office personnel who must converse "often" to "very often" at 8 to 10 ft, the SIL should be below 40 db. For the noise to have a balanced spectrum, the LL should not exceed the SIL by more than about 22 units. Office-personnel's ratings of ease of telephone use (at the top of the graph) are related to the SIL (bottom scale on the graph).

they were in offices in which the subjective ratings correlated well with both the SIL and the LL.

Where the correlation was good with both LL and SIL, the LL in phons was not more than 22 units higher than the SIL in decibels. Whenever the difference between LL and SIL became substantially greater than 22 units, say 30 units, the subjective rating was higher than would be predicted by the SIL alone. In addition, when the LL minus the SIL difference exceeded 30 units, there were usually complaints, even when the SIL met the requirement of being lower than 40 db.

For the reasons just given the acceptable LL as plotted in Fig. 20.1 is about 22 units to the right of the SIL. If the difference between the two curves exceeds about 22 units, the noise level is likely to be found less acceptable than would be predicted by the SIL alone. These studies, therefore, show that neither the SIL nor the LL alone is sufficient to characterize the criterion spectrum; both are needed.

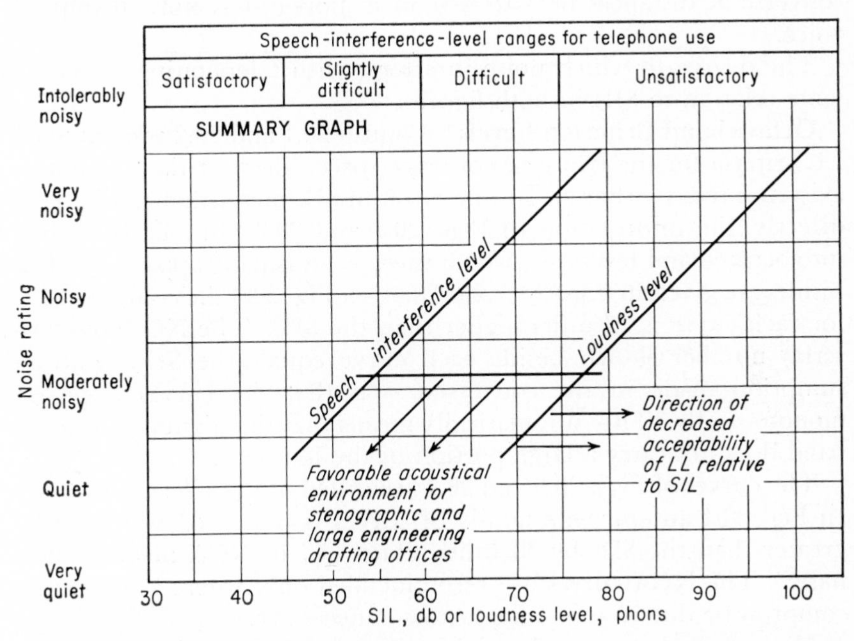

FIG. 20.2 The relation of the subjective noise ratings to the SIL and LL for stenographic and large engineering drafting rooms. For a favorable acoustical environment for those personnel who must converse "often" to "very often" at 3 to 4 ft, the SIL should be below 55 db. For the noise to have a balanced spectrum, the LL should not exceed the SIL by more than about 22 units. Office personnel's ratings of ease of telephone use (at the top of the graph) are related to the SIL (bottom scale on the graph).

Kryter[9] surmises that noise (including narrow bands placed anywhere in the audible frequency spectrum) ideally should not have an LL exceeding 62 phons, inasmuch as the "threshold of annoyance" is reached under certain listening conditions with narrow bands of noise having an LL of about 62 phons. However, he says, people apparently adapt themselves in a given work situation to higher noise levels, with a subsequent upward shift in their "thresholds of annoyance." This certainly is true for those personnel with less stringent speech communication requirements, as we see in Fig. 20.2.

The curves in Fig. 20.2 are like those of Fig. 20.1, except that the ratings of the noise were made in stenographic pools or large drafting rooms. In these areas, the average noise usually has a speech-interference level of 50 to 55 db. As we can see, the personnel say that if the SIL were to exceed about 55 db (or the LL exceed about 77 phons) their work would suffer. An SIL of 55 permits them to converse at distances of 3 to 4 ft in a more-or-less normal tone of voice.

The degree to which noise interferes with telephoning bears the same relation to SIL in both figures.

Octave-band Criterion Levels.[8] Figures 20.1 and 20.2 present SIL–LL criteria for the two types of office space. Because there is not in existence today either an LL meter or an SIL meter that can be read directly, the information in Figs. 20.1 and 20.2 must be converted into octave-band levels, for which there is measuring apparatus. The results are given in Figs. 20.3 and 20.4. In Fig. 20.3 the computed LL for each curve is 22 units higher than the SIL. The NC (noise criteria) number shown beside each curve equals the SIL. The assumption made in the conversion was that the curves slope off monotonically. This automatically means that the highest frequency band does not carry a large portion of the loudness.

The curves of Figs. 20.3 and 20.4 are called *noise-criterion curves.** In Fig. 20.4 an *alternate* family of curves is presented whose LL is greater than the SIL by 30 units. The "A" in NCA means "alternate." The NCA curves are recommended only where a maximum compromise due to economic factors is necessary.

In a particular room the eight octave-band levels should not exceed the values indicated by the NC or NCA curve selected in the specification for that office.

In choosing between the NC and NCA families of curves, the quality of the space being quieted and the character of the low-frequency noise should be considered. Both groups of office personnel prefer a noise spectrum that is shaped like the NC curves, i.e., an LL − SIL difference equal to 22 units; and experience indicates that negligible improvement in personnel comfort is achieved by still further reducing the LL − SIL difference. A majority of the personnel objects to a noise whose octave-band levels at the low frequencies exceed, for a given SIL, the values given by the NCA curves, even when the SIL is low enough for voice communication to be satisfactory. In other words, substantial objections occur whenever the LL

* In earlier publications[3,4,6] these were called *speech-communication* (SC) criteria. The new curves embody both speech and annoyance considerations and are given the more general name of *noise criteria* (NC).

— SIL difference exceeds 30 units. Objections are also likely to occur if the low-frequency noise fluctuates or if there are beats between low-frequency components. When such fluctuations or beats at low

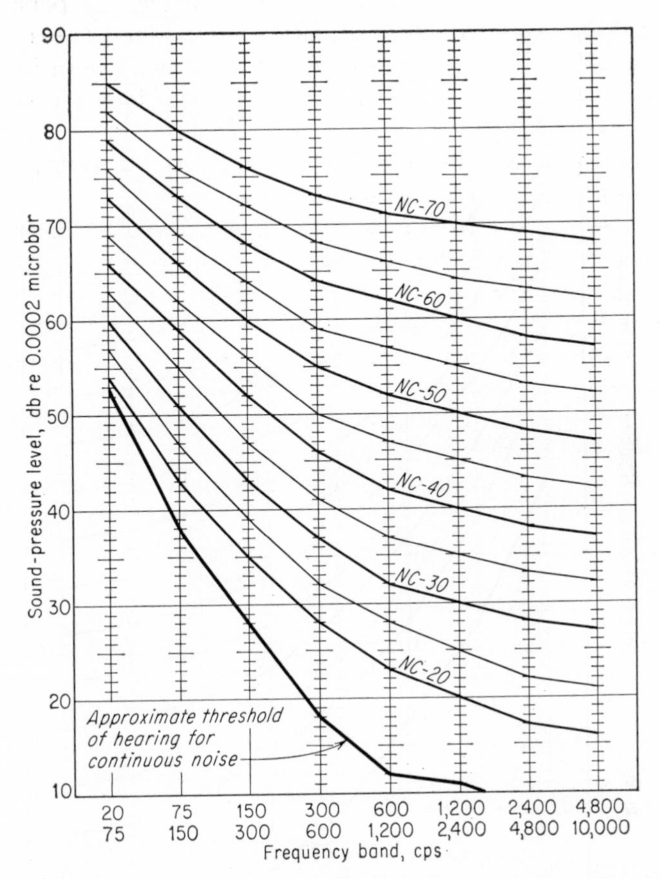

FIG. 20.3 Criterion curves for use with Tables 20.1 and 20.2 in determining the permissible (or desirable) SPL's in eight octave bands. Each NC curve has an LL in phons that is 22 units greater than the SIL in decibels (expressed by the NC number of the curve). These curves are recommended for specifications wherever a favorable relation between the low-frequency and the high-frequency portion of the spectrum is desired (see text).

frequencies exist, the LL — SIL difference cannot be as large as 30 units without risk of complaints.

The two families of curves in Figs. 20.3 and 20.4 represent, therefore, the lower and upper limits of the LL — SIL difference that is generally acceptable to office personnel. In writing specifications either an NC or NCA number (to the nearest 5 units) or the corresponding sound-pressure levels (to the nearest decibel) in the eight octave bands can be given. It is possible, of course, to interpolate between the NC and NCA curves.

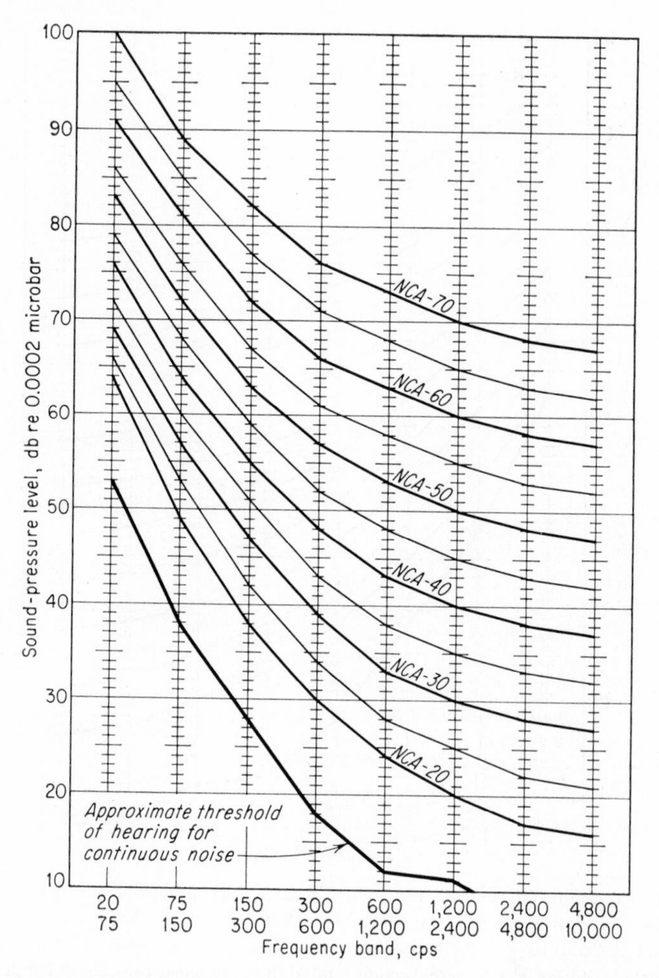

FIG. 20.4 Same as for Fig. 20.3 except that each NCA curve has an LL in phons that is 30 units greater than the SIL in decibels (expressed by the NCA number of the curve). These curves may be used in specifications in place of those of Fig. 20.3 wherever economy dictates a maximum compromise and where, in addition, the noise is steady and free of beats between low-frequency pure-tone components (see text).

Recommended Office Criteria.[8] Based on the results of the studies described above and on observations made recently in other industrial buildings, it is recommended that the NC curve for a particular office space be selected with the aid of Table 20.1. As was stated in detail above, an NCA curve should be substituted for an NC curve only in instances calling for extreme economy.

Table 20.1

Recommended Noise Criteria for Offices

NC Curve of Fig. 20.3, NC units	Communication Environment	Typical Applications
20–30	Very quiet office—telephone use satisfactory—suitable for large conferences	Executive offices and conference rooms for 50 people
30–35	"Quiet" office; satisfactory for conferences at a 15-ft table; normal voice 10 to 30 ft; telephone use satisfactory	Private or semiprivate offices, reception rooms and small conference rooms for 20 people
35–40	Satisfactory for conferences at a 6- to 8-ft table; telephone use satisfactory; normal voice 6 to 12 ft	Medium-sized offices and industrial business offices
40–50	Satisfactory for conferences at a 4- to 5-ft table; telephone use occasionally slightly difficult; normal voice 3 to 6 ft; raised voice 6 to 12 ft	Large engineering and drafting rooms, etc.
50–55	Unsatisfactory for conferences of more than two or three people; telephone use slightly difficult; normal voice 1 to 2 ft; raised voice 3 to 6 ft	Secretarial areas (typing) accounting areas (business machines), blueprint rooms, etc.
Above 55	"Very noisy"; office environment unsatisfactory; telephone use difficult	Not recommended for any type of office

Note: Noise measurements made for the purpose of judging the satisfactoriness of the noise in an office by comparison with these criteria should be performed with the office in normal operation but with no one talking at the particular desk or conference table where speech communication is desired (i.e., where the measurement is being made).

Application to Other Interior Spaces.[8] Although a detailed study has not been made of the reaction of people to noise in other spaces, fairly dependable conclusions can be drawn from general experience gained in studying noise-control problems generated by ventilating systems and exterior sources. One must also recognize that local situations and local attitudes affect the criticalness of people. What seems adequate in one location may evoke complaints in another. For example, in a convention hall with a near-perfect sound system, a designer may get away with an air-conditioning system that is 5 to

15 db noisier in the speech bands than that required in a hall with a mediocre sound system. In other words, when the articulation index (AI) of speech produced by an auditorium sound system drops as low as 60 per cent as a consequence of the sound system's restricted frequency response, the noise in the speech bands must be 20 db below the average speech levels at the listener's position if he is to understand what is being said. On the other hand, if the sound system is so very good that its articulation index is nearly 100 per cent, the noise need be only 5 to 10 db below the average speech levels for easy listening.[10]

The greatest variability in requirements for noise control exists in homes. Here the habits and temperaments of people are paramount. Except for the need to converse over the telephone—which requires only that the SIL be lower than 60—the activities in a home vary. In many homes, the noise of children, radios, television, and outside traffic creates so much din that noise at a loudness level of 80 phons would pass unnoticed. In other homes, where there are no children or where a writer or a scholar engages in creative thought, an LL may need to be as low as 40 phons to pass unnoticed.

Table 20.2 presents the NC or NCA curves (see Figs. 20.3 and 20.4) which are believed to provide acceptable noise levels for a variety of interior spaces. The noise levels referred to are to be measured with the particular room in question unoccupied but with other activities inside and outside the building proceeding normally. For comparison with the NC curves, the sound-level meter *A*-scale readings in dba, computed from the appropriate NC curve, are given. *It is not recommended that A-scale readings be used in specifications because the same A-scale reading may be obtained for a wide variety of shapes of spectra. Furthermore, the eight octave bands are necessary in the engineering design of noise-control measures, and no single number can substitute.*

If an *A*-scale number were to be used in a specification in place of an NC or NCA curve, it probably should be chosen as 5 dba less than the number given in the right-hand column of Table 20.2. Such a choice would partially ensure against the possibility of spectra shapes that are different from the NC and NCA curves.

Summary. Criteria for noise in buildings are presented in Tables 20.1 and 20.2 and in Figs. 20.3 and 20.4. The NC curves are recommended for use in all spaces. The NCA curves are to be used as alternates for the NC curves only in instances where extreme economy is needed and where the noise is steady and free of beats between low-frequency pure-tone components. In the specification of an NC or NCA curve, it is intended that the sound-pressure level in any

Table 20.2

Recommended Noise Criteria for Rooms

Type of Space	Recommended NC Curve of Fig. 20.3 or 20.4, NC units	Computed Equivalent SLM Readings* Weighting Scale *A*, dba
Broadcast studios	15–20	25–30
Concert halls	15–20	25–30
Legitimate theaters (500 seats, no amplification)	20–25	30–35
Musicrooms	25	35
Schoolrooms (no amplification)	25	35
Television studios	25	35
Apartments and hotels	25–30	35–40
Assembly halls (amplification)	25–35	35–40
Homes (sleeping areas)	25–35 †	35–45†
Motion-picture theaters	30	40
Hospitals	30	40
Churches (no amplification)	25	35
Courtrooms (no amplification)	25	30–35
Libraries	30	40–45
Restaurants	45	55
Coliseums for sports only (amplification)	50	60

* If there were relatively less noise in the low-frequency bands than indicated by the recommended noise criterion curve, the dba numbers would be lower by about 5 dba. All numbers in this column should be dropped by 5 dba if they are to be used to estimate the compliance of a normally encountered noise level with an NC criterion; these numbers should not be used for specification purposes.

† Room air conditioners manufactured prior to 1957 commonly produce levels of 40 to 55 dba in sleeping areas.

Note: Noise levels are to be measured in unoccupied rooms. Each noise criterion curve is a code for specifying permissible sound-pressure levels in eight octave bands. It is intended that in no one frequency band should the specified level be exceeded. The computed equivalent dba numbers in the right-hand column are presented for information only and are not recommended for use in specifications. Ventilating systems should be operating, and outside noise sources, traffic conditions, etc., should be normal when measurements are made.

one of the eight frequency bands not exceed that shown by the curve. It is recommended that NC and NCA designations replace older *A*-scale numbers in specifications.

The architect or consultant will have to use his own judgment in selecting a curve for a particular specification because of the wide range of attitudes toward noise and because of local customs and expectations in different locations. In some cases lack of funds for quieting may require that a calculated risk of complaint be taken. In others, previous experience may have conditioned the people exposed to the noise so that they are less, or even more, critical of their noise environment than Table 20.1 or 20.2 would predict. Just as in other fields involving human reactions, numbers on a page do not alone replace careful analysis of each noise problem and the taking into account of local differences.

Recent progress in criteria for noise in buildings reported since this chapter was sent to the publisher is given in Appendix C.

20.3 Criteria for Noise in Vehicles

The criteria that have been developed for noise in vehicles are largely empirical and are based in part on economic and practical limits. Nevertheless, there are clearly discernible parallels with the requirements for quiet in offices. These requirements are:

1. Conversation among members of a group sharing adjacent seats or the same compartment should be possible without straining the voice.

2. The spectrum balance should be such that the difference between LL and SIL not exceed about 30 units.

3. The noise should have no strong tonal components within a few cycles of each other in frequency which might produce a disturbing "beat" note.

4. There should be no prominent single component that is observable as a persistent "singing" tone.

5. There should be no rattles or clatter arising from loose objects or mechanisms.

Aircraft. *Propeller Aircraft.* In the period prior to 1959 when all aircraft were propeller driven, some clearly defined standards of acceptability evolved from observation of passenger complaints. These results as tabulated by Kerwin and Beranek in unpublished reports are shown in Fig. 20.5.

Such airplanes as the Convair Metropolitan model 440, and the Vickers Viscount have noise levels in the vicinity of the lower edge of the shaded region of Fig. 20.5. An unacceptable airplane from the noise standpoint was the Convair 340, whose levels fell above the shaded region of Fig. 20.5. Even though the lower edge of the shaded region of Fig. 20.5 has been judged "acceptable" by airline passengers and even though conversation is possible with passengers in adjacent seats, still passengers complain of fatigue after long journeys. This fatigue is partly a matter of vibration of the floor and the seat and partly a matter of the excessive low-frequency noise and consequent large difference (37 units) between LL and SIL. Practical considerations of weight and design precluded further reduction of the low-frequency noise by significant amounts.

Jet Aircraft. With the advent of the jet aircraft, it is believed that the NCA curves of Fig. 20.4 may be applied as design criteria. The suggested design criterion is NCA-60 (see Fig. 20.5). For example,

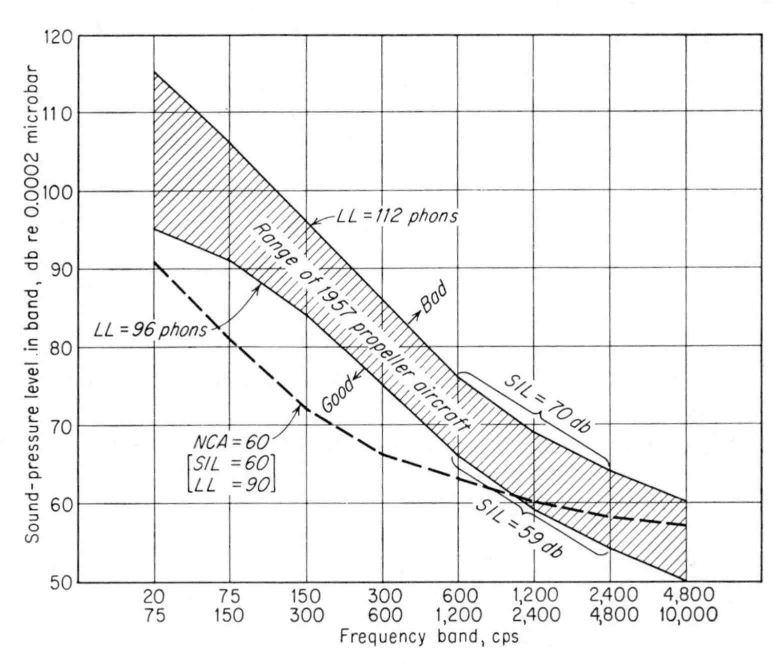

FIG. 20.5 Acceptability contours for propeller-driven aircraft as derived empirically by Beranek and Kerwin (unpublished data). Airplanes with levels near or below the lower edge of the shaded region are rated "noisy" but "acceptable" by passengers. Those with levels near or above the upper edge are rated "intolerably noisy" and "unacceptable" by passengers. *Accepted practice:* noise levels in 1957 propeller aircraft—below shaded region. *Suggested practice:* noise levels in jet aircraft below NCA-60 (see dashed curve).

reference to Table 20.1 shows that for conversation between two or three people using a raised voice at 3 to 6 ft, the noise level should lie between NC-50 and NC-55 criterion curves. Certainly, noise levels in excess of NCA-60 will meet with passenger complaints on trips of any significant length.

Trains. Similar curves to those given in Fig. 20.5 have been developed in unpublished reports by Dyer for use in guiding the design of quieting for railway passenger coaches. The results are shown in Fig. 20.6. Again, the shape and levels of these curves are dictated

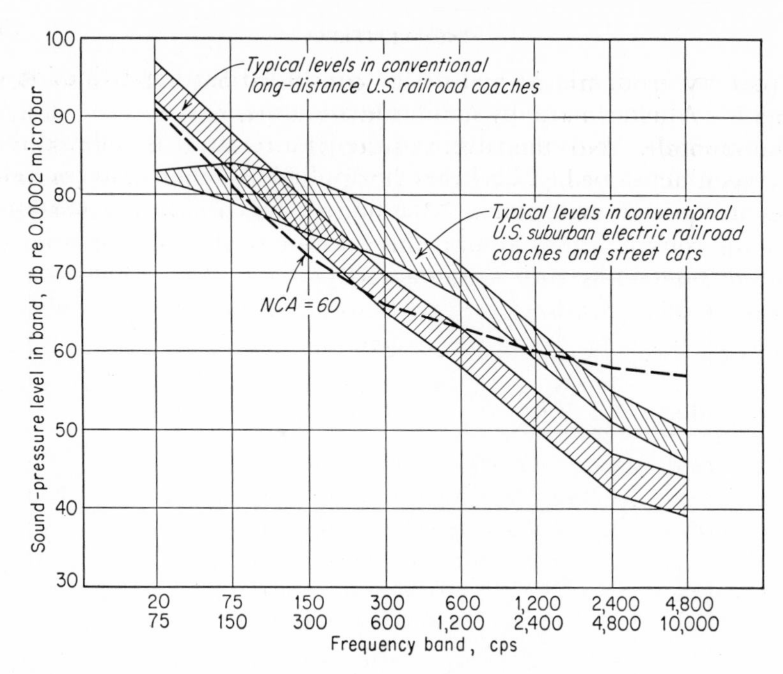

Fig. 20.6 Acceptability contours for modern railway Pullman cars as derived empirically by Dyer (unpublished data). An NCA-60 criterion is suggested as maximum permissible levels in design.

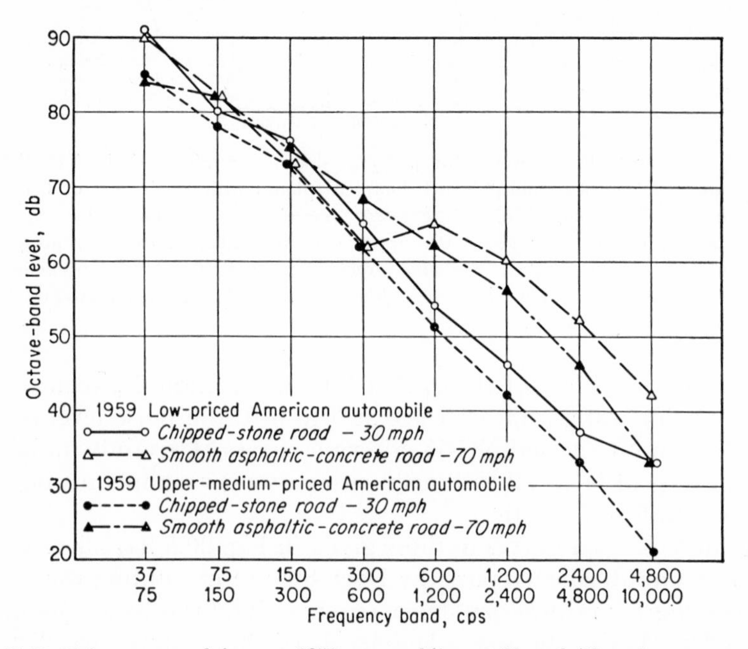

Fig. 20.7 Noise measured in two 1959 automobiles at 30 and 70 mph on two types of road. Measurements were made with all windows closed.

in part by economic and practical considerations. (See also Bonvallet.[11]) A criterion of NCA-60 or lower is recommended in design.

Automobiles. No standards of acceptability for automobiles have ever been developed. Passengers readily observe that some automobiles are quieter than others. The noise arises primarily from aerodynamic considerations around the exterior of the car and from vibration induced by road irregularities.

Some typical measured interior noise levels in 1959 cars are given in Fig. 20.7. These levels may be used as a guide in design.

20.4 Criteria for Vibration in Buildings and Vehicles

No recent definitive work has been done on the acceptability of buildings or vehicles from the standpoint of vibration. The problem is very difficult and would need the same sort of practical study that has gone into the formulation of criterion curves for office noise given in the first part of this chapter. For example, a building vibration that would be imperceptible in the daytime in a home might be quite bothersome at night when everything else is quiet.

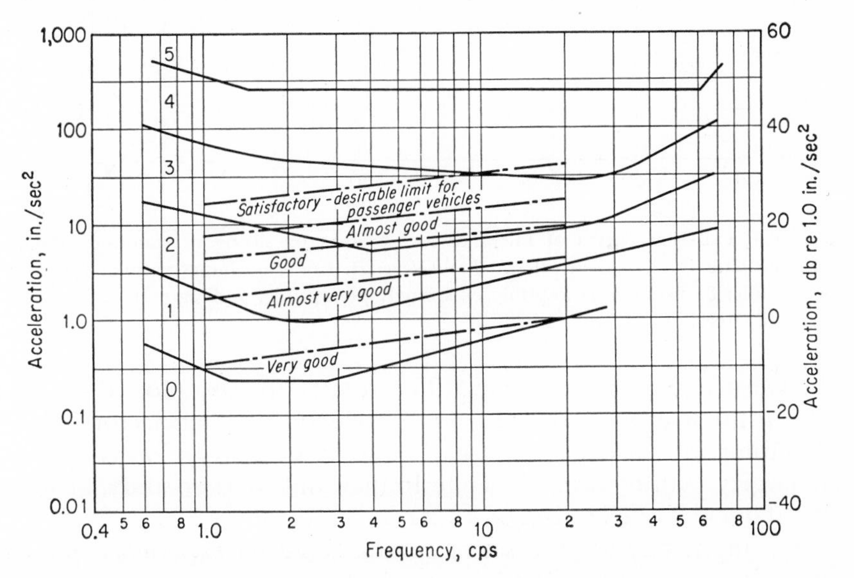

FIG. 20.8 Vibration sensitivity of seated and standing persons. The ordinate is rms acceleration in the units indicated. The regions are to be interpreted as follows: 0, imperceptible; 1, barely perceptible; 2, distinctly noticeable but not uncomfortable; 3, slightly disagreeable; 4, disagreeable; and 5, exceedingly disagreeable. (*From Reiber and Meister.*) The five dash-dot-dash lines labeled from "very good" to "satisfactory-desirable limit for passenger vehicles" were obtained by Helberg and Sperling for passenger comfort in railway vehicles. (*Graph after Koffman.*[12])

Koffman[12] has summarized the literature of up to 1949 and given as a summary graph the material shown in Fig. 20.8. This graph has two independent sets of data on it. First, the results of laboratory tests conducted by Reiher and Meister were used to develop the six regions (0 through 5) shown on the graph. Other tests by Helberg and Sperling were carried out to ascertain the effect of vibrations experienced in railway coaches. The final results have been used as a basis for the evaluation of the riding qualities of rolling stock of

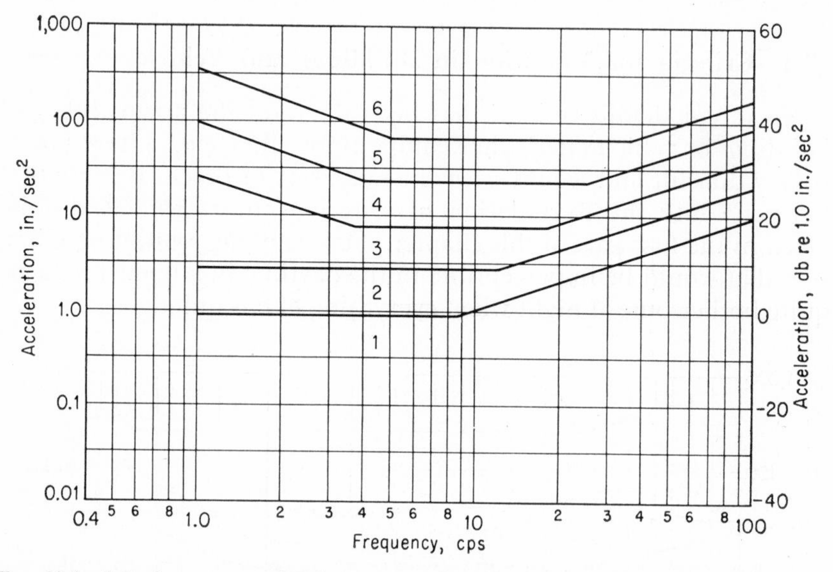

FIG. 20.9 Criterion curves of tolerance to vibration by sitting or standing persons. The ordinate is rms acceleration. (*From Dyer.*[13]) Curve 1, imperceptible; 2, barely perceptible; 3, distinctly perceptible; 4, strongly perceptible; 5, disagreeable; 6, exceedingly disagreeable.

the German Federal Railways. The results are also plotted in Fig. 20.8 as the five lines labeled from "very good" to "satisfactory desirable limit for passenger vehicles." The latter designation is seen to lie largely within region 3 of the Reiher and Meister study, at least for frequencies below 10 cps.

An empirically derived set of curves, based on experience gained from six engineering situations in which human responses to vibration were involved, has been advanced by Dyer[13] (see Fig. 20.9). In Dyer's interpretation, the rms acceleration in the resonances should be held below the tolerance limits. He believes that the stimulus basis for the response at low levels is somewhat different from the stimulus basis for the response at high levels, and this difference is reflected in the fact that the high-level boundaries bend upwards at low frequencies, whereas the low-level boundaries do not.

20.5 Summary

This chapter starts with definitions of speech-interference level (SIL), loudness level (LL), and perceived-noise level (PNL). Next it presents criteria for noise control in offices, broadcast studios, concert halls, legitimate theaters, musicrooms, schoolrooms, television

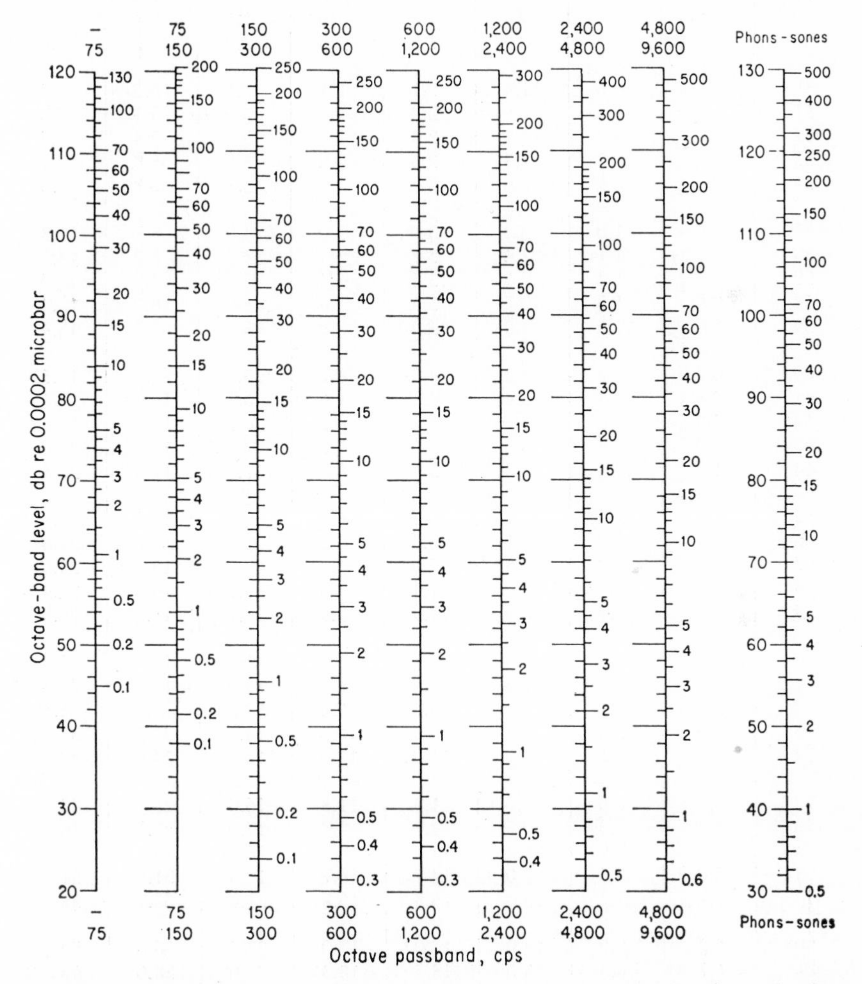

Fig. 20.10 Nomograms relating sound-pressure level in octave bands to loudness in sones. The calculated total loudness in sones can be converted to loudness level in phons by means of the nomogram at the right. (See Appendix A for computational instructions.) (*From S. S. Stevens.*[14])

Table

Conversion of Octave-band Sound-pressure Levels (db)

SPL, db re 0.0002 μbar	Frequency Band, cps							
	20–75	75–150	150–300	300–600	600–1,200	1,200–2,400	2,400–4,800	4,800–10,000
20							0.6	1.0
30				0.5	0.5	0.7	1.2	2.1
40			0.6	1.0	1.0	1.4	2.4	4.3
45			0.9	1.5	1.5	2.0	3.3	6.0
50		0.6	1.4	2.1	2.1	2.9	4.7	8.5
51		0.6	1.6	2.2	2.2	3.0	5.1	9.0
52		0.7	1.7	2.4	2.4	3.2	5.4	9.7
53		0.8	1.8	2.6	2.6	3.5	5.8	10.5
54		0.9	2.0	2.8	2.8	3.7	6.2	11.0
55		1.0	2.2	3.0	3.0	4.0	6.7	11.8
56		1.1	2.4	3.2	3.2	4.3	7.2	12.5
57	0.5	1.2	2.6	3.4	3.4	4.6	7.7	13.4
58	0.6	1.3	2.7	3.6	3.6	5.0	8.2	14.4
59	0.6	1.5	3.0	3.9	3.9	5.3	8.7	15.4
60	0.7	1.6	3.2	4.2	4.2	5.7	9.3	16.5
61	0.8	1.8	3.4	4.5	4.5	6.1	10.0	17.6
62	0.9	2.0	3.7	4.8	4.8	6.5	10.7	18.8
63	1.0	2.2	4.0	5.2	5.2	7.0	11.5	20.0
64	1.1	2.4	4.3	5.6	5.6	7.5	12.3	21.3
65	1.2	2.6	4.7	6.0	6.0	8.0	13.1	22.7
66	1.4	2.9	5.1	6.5	6.5	8.6	14.0	24.0
67	1.6	3.2	5.5	7.0	7.0	9.2	15.0	25.6
68	1.8	3.5	5.9	7.5	7.5	9.9	16.0	27.4
69	2.0	3.9	6.3	7.9	7.9	10.7	17.2	29.3
70	2.2	4.3	6.8	8.4	8.4	11.5	18.5	31.3
71	2.5	4.7	7.3	9.0	9.0	12.2	19.8	33.4
72	2.7	5.1	7.8	9.6	9.6	13.0	21.0	35.5
73	3.0	5.6	8.4	10.3	10.3	14.0	22.5	38.0
74	3.4	6.1	9.0	11.0	11.0	15.0	24.0	40
75	3.8	6.6	9.7	11.8	11.8	16.0	25.7	43
76	4.2	7.3	10.5	12.7	12.7	17.0	27.5	46
77	4.6	7.9	11.2	13.7	13.7	18.3	29.5	49
78	5.0	8.5	12.0	14.7	14.7	19.5	31.5	51
79	5.5	9.2	13.0	15.8	15.8	21.0	33.7	55
80	6.2	9.9	14.0	17.0	17.0	22.5	36.0	59
81	6.7	10.8	15.0	18.3	18.3	24.0	38.0	63
82	7.3	11.7	16.0	19.5	19.5	25.5	41	67
83	8.0	12.7	17.1	21.0	21.0	27.5	44	71
84	8.7	13.8	18.3	22.5	22.5	29.5	47	76

B20.1

to Octave-band Perceived "Noisiness" (noys)

SPL, db *re* 0.0002 μbar	Frequency Band, cps							
	20–75	75–150	150–300	300–600	600–1,200	1,200–2,400	2,400–4,800	4,800–10,000
85	9.4	14.8	19.5	24.5	24.5	31.0	50	81
86	10.2	15.9	21.0	26.0	26.0	34.0	53	86
87	11.1	17.0	22.5	28.0	28.0	36.0	57	92
88	12.1	18.3	24.0	30.0	30.0	38.0	61	97
89	13.1	19.5	26.0	32.0	32.0	41	65	103
90	14.2	21.0	28.0	34.0	34.0	44	70	110
91	15.3	22.0	30.0	37.0	37.0	47	75	117
92	16.5	24.0	32.0	40	40	50	80	124
93	17.0	26.0	34.0	43	43	53	85	132
94	19.5	28.0	36.0	46	46	57	91	141
95	21.0	30.0	39.0	49	49	61	97	150
96	22.5	32.0	42	53	53	66	104	160
97	24.0	34.0	45	57	57	71	111	170
98	26.5	37.0	48	61	61	75	119	180
99	28.0	40	51	65	65	80	127	191
100	30.5	42	55	70	70	86	135	203
101	32.5	45	59	75	75	92	144	215
102	35.0	49	63	80	80	98	159	228
103	38.0	53	67	86	86	105	165	242
104	41	56	71	92	92	113	175	256
105	43	60	76	98	98	120	187	273
106	46	64	82	106	106	128	200	290
107	50	68	88	113	113	136	213	310
108	53	73	94	120	120	145	226	328
109	57	78	100	129	129	155	240	348
110	61	84	106	139	139	165	260	370
111	66	90	114	150	150	177	275	395
112	71	96	122	166	166	190	293	420
113	76	103	130	170	170	203	315	445
114	82	111	139	181	181	215	335	470
115	87	119	148	192	192	230	360	500
116	93	125	158	205	205	245	380	
117	100	134	169	220	220	260	405	
118	106	145	180	237	237	280	435	
119	113	155	192	255	255	300	465	
120	120	165	204	275	275	320	500	
125	170	225	280	385	385	440		
130	235	315	390	510	510			
135	325	425						
140	450							

studios, apartments and hotels, assembly halls, homes, motion-picture theaters, hospitals, churches, courtrooms, libraries, restaurants, sports coliseums, aircraft, trains, and automobiles.

The chapter ends with three appendixes, one giving the procedure for calculating loudness level, another for calculating perceived-noise level, and the third giving recently developed information on speech privacy between adjacent rooms.

Appendix A. Computation of Loudness Level [14]

The loudness level is a single number computed from octave-band levels according to the following procedure. The unit is the phon.

To obtain the loudness level in phons, convert the sound-pressure levels in the eight octave bands to sones by means of the appropriate columns in Fig. 20.10 and sum the loudness in sones:

$$\Sigma S = S_1 + S_2 + S_3 + \cdots + S_8 \tag{1}$$

The loudness in *sones* of total noise is given by

$$S_t = S_m + 0.3(\Sigma S - S_m) \tag{2}$$

where S_m is the loudness of the loudest band, and ΣS is the sum of the loudness of all eight bands. The loudness in sones is converted to loudness level in phons by means of the nomogram at the right-hand side of Fig. 20.10.

Appendix B. Computation of Perceived-noise Level [15,16]

The perceived-noise level is a single number computed from octave-band sound-pressure levels according to the following procedure. The unit is the PNdb.

The first step is to convert the sound-pressure level in each of the eight octave bands to "noisiness" in noys by means of the appropriate column in Table B20.1. These values of octave-band "noisiness" are summed:

$$\Sigma N = N_1 + N_2 + N_3 + \cdots + N_8 \tag{1}$$

Then the "noisiness" in noys of the total noise is given by

$$N_T = N_m + 0.3(\Sigma N - N_m) \tag{2}$$

where N_m is the largest value of "noisiness" for any one of the eight bands and ΣN is the sum given by Eq. (1).

This total noisiness N_T (in noys) is finally converted to total perceived-noise level in PNdb by means of the formula

$$\text{Number of PNdb for total noise} = 40 + 33.3 \log_{10} N_T \qquad (3)$$

As an example, see Table B20.2 for octave-band levels for a certain jet aircraft flying overhead at 500 ft.

Table B20.2

Octave Band	*SPL, db re 0.0002 μbar*
20–75	97
75–150	96
150–300	101
300–600	101
600–1,200	102
1,200–2,400	103
2,400–4,800	102
4,800–10,000	95

From Table B20.1 we find that

$$\begin{aligned} \Sigma N &= 24 + 32 + 59 + 75 + 80 + 105 + 159 + 150 \\ &= 684 \text{ noys} \\ N_m &= 159 \text{ noys} \\ N_T &= 159 + 0.3 \times 525 = 316 \text{ noys} \end{aligned}$$

$$\text{Perceived-noise level} = 123 \text{ PNdb}$$

Appendix C. Speech-privacy Considerations*

Noises seem to fall into two categories: Those which, because of their character, are "acceptable," and those which are "unacceptable." Some examples of the former are: the rustle of wind in trees, the patter of rain on the roof, the rumble of distant traffic, the blur of a large number of voices such as a concert audience at intermission time. One important example of an "unacceptable noise" is speech intruding into what is ostensibly a private office, home, or dormitory room. Other examples are: the tinkle of breaking glass, the roar of your neighbor's power lawnmower, the whine of the neighbor's air-conditioning compressor, and the sound of your neighbor's television or Hi-Fi.

Often, diffuser noise and similar mechanical-equipment sounds are classified by the listener as "acceptable." In such cases the diffuser noise is usually blended with distant traffic noise or office activity noise. Other times, diffuser noise is definitely unacceptable. In those cases, the diffuser is usually the major source of noise and can be readily located by the listener.

* This appendix is a contribution by B. G. Watters.

When the noises in a space are of an "acceptable" character, the listener is usually not aware of their existence. Annoyance from such noises is usually small unless their level is great enough to interfere with conversation or in some other way force themselves upon the listener's attention. When an unacceptable noise intrudes into a space having a certain level of acceptable noise, the resulting annoyance is often determined not by the *level* of the unacceptable noise but rather by the magnitude of the difference between the unacceptable and the acceptable noise levels.

To understand the problem of speech privacy, let us look at a recent study.[17] Test subjects (listeners) were placed in a private office where there was an "acceptable" noise. They were then subjected to intruding speech sounds from an adjoining office. The level of the speech was variable and was controlled by the experimenter. The subjects were asked to respond when the level of the speech just reached the point where they considered it objectionable. A typical result is shown in Fig. 20.11. For this group of about 100 subjects, there was a well-defined variation of about 10 db between the speech levels tolerated by the most critical 10 per cent of the subjects and by the least critical 10 per cent.

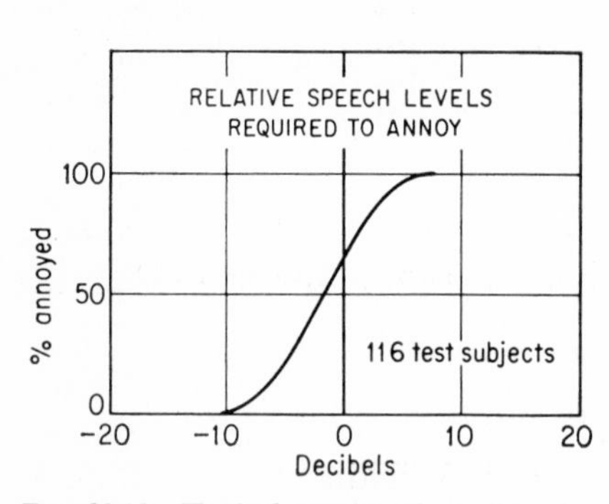

Fig. 20.11 Typical results of studies giving the percentage of persons judging speech intrusion "objectionable" in the presence of an "acceptable" noise as a function of the difference in levels between the "objectionable" and "acceptable" noises.

When the privacy conditions required by the most critical half of the subjects were further analyzed it was found that the intruding speech was called "objectionable" when it became *intelligible.* This is not to say that the subjects were necessarily trying to understand the speech. Each subject had a different although quite definite idea of what speech privacy meant for him. But for the most critical half of the subjects, this definition could be correlated with the intelligibility of the intruding speech.

Some of the consequences of this finding are as follows:

1. It is the numerical *difference* in decibels between (*a*) the intruding speech and (*b*) the "acceptable" noise levels, not the level of the intruding speech, that is important. Thus, in a "quiet" office the intruding speech levels need be only say, 20 db (in the 600- to

1,200-, 1,200- to 2,400-, and 2,400- to 4,800-cps bands) to be rated as objectionable while in a "moderately noisy" office the intruding speech levels (in the same bands) must increase to about 30 db before they become objectionable.

2. For the most critical half of the subjects, the noise becomes objectionable before the intruding speech levels contribute appreciably to the individual band levels in the listening space. For example, 20-db levels will often cause annoyance when the levels of "acceptable" background noise are about the same.

3. The maximum allowable difference between speech levels and "acceptable background noise levels" for a particular octave band is not a fixed quantity. Since speech intelligibility is an "integrated" effect, i.e., is contributed to by several octave bands, the permissible speech level in any one band depends on the levels in the other bands.

4. The relative importance of speech levels in the various frequency bands varies from virtually no importance for the 20- to 75- and 75- to 150-cps bands, small importance for the 150- to 300-cps band, maximum importance for the 1,200- to 2,400-cps band and slightly less importance for the 2,400- to 4,800-cps band.

These ideas form the basis of a design procedure for choosing sound-isolating structures such as partitions, doors, ceilings, and the like. Such a procedure should take into account the following aspects of the spaces or rooms to be isolated:

1. The noise reduction between the spaces
2. The speech effort (conversational speech, raised voice, loud voice, etc.) which is expected
3. The privacy requirements of the occupants (sufficient for everyday work, confidential work, etc.)
4. The level of background noise in the spaces
5. The desired probability that the occupants will be satisfied with the privacy provided (recognizing that some people are more critical listeners than others; some are louder talkers than others)

The result of the procedure of calculation is an *articulation index* (*AI*) that lies between 0 and 1.0. If the AI is less than 0.1, speech intelligibility is generally low. If it is above 0.6, speech intelligibility is generally high. For adequate speech privacy in executive offices we generally assume that the AI, calculated by the procedure below, should be less than about 0.03.

While this design procedure for determining speech privacy cannot be presented in its entirety, the principal ideas are illustrated in Example C20.1 and Table C20.1 below.

Example C20.1. *Example of Speech-privacy Design.* Item 1 of Ta-

Table C20.1

Computing the Articulation Index

Elements Used in Computation	Frequency, cps				
	150–300	300–600	600–1,200	1,200–2,400	2,400–4,800
1. Relative importance of band (multiplying factor to be used below)	0.7*	2.6*	3.8*	5.9*	5.1*
2. Speech levels exceeded 1% of time in room having $S\bar{\alpha}$ = 400 sabins of absorption (average talker)	65	70	68	63	51
Case 1:					
3. NC-30 Background Noise	43	37	32	30	28
4. Approximate NR of the Wall (2 in. solid plaster)	31	31	35	43	50
5. Item 3 plus item 4	74	68	67	73	78
6. Item 2 minus item 5		2	1		
7. Item 6 times item 1		5.2	3.8		

Summation of the values in item 7 = 9.0. Divide by 600 to get the articulation index; AI = 0.015. Privacy is probably satisfactory.

Case 2:					
8. NC-25 background noise	39	32	27	25	23
9. Item 8 plus item 4	70	63	62	68	73
10. Item 2 minus item 9		7	6		
11. Item 10 times item 1		18.2	22.8		

Summation of the values of item 11 = 41. Divide by 600 to get articulation index; AI = 0.068. Privacy is marginal.

Case 3:					
12. Approximate octave-band hearing threshold for continuous noise	28	18	12	11	8
13. Item 12 plus item 4	59	49	47	54	58
14. Item 2 minus item 13	6	21	21	9	
15. Item 14 times item 1	4.2	54.6	79.8	53.1	

Summation of the values in item 15 = 192. Divide by 600 to get articulation index; AI = 0.32. Privacy is inadequate.

* The speech frequency range has been divided by French and Steinberg into 20 bands of equal importance to intelligibility. The values here are the number of these bands per octave. See L. L. Beranek, The Design of Speech Communication Systems, *Proc. IRE*, vol. 35, pp. 880–890, 1947.

ble C20.1 gives the relative weighting of the various octave bands. Thus, the 150- to 300-cps band is fairly unimportant; the 1,200- to 2,400-cps band is most important.

Item 2 gives the peak speech levels to be expected in a moderately large room with typical sound-absorbing treatment for an average talker speaking in a normal tone of voice.

In case 1 of Table C20.1 an NC-30 background noise is assumed (item 3). The octave-band levels of item 3 are added to item 4, the NR of the wall. Item 2 minus item 5 yields the difference between the speech strength and the background noise. In the 300- to 600-cps band, for example, the peak speech levels are 2 db greater than the background noise; in the 600- to 1,200-cps band they are 1 db greater. In all other bands, the peak speech levels lie below the background noise and are thus assumed to be of no importance.

When the speech-noise level differences of item 6 are (1) multiplied by the weighting factors of item 1, (2) added together, and (3) divided by 600, the AI is obtained. We have found that in a private executive office, for example, nearly every occupant will be satisfied if the AI is less than about 0.02 to 0.04. A different value of AI applies when the privacy requirement is different.

In case 2, an NC-25 background noise is assumed. The result is an increase in AI to 0.068. In our experience, a significant percentage of executives will feel like complaining at this degree of privacy.

In case 3, the background noise is removed altogether, the AI increases to 0.32, and nearly every executive would be expected to complain about the privacy.

If the background noise is of an acceptable character, the conditions of case 1 are definitely to be preferred over cases 2 or 3.

Several important office buildings are being constructed in the United States with built-in masking noise set at NC-25 to NC-30. With this noise, it is expected that adequate speech privacy between offices will be achieved using modern lightweight office partition construction.

REFERENCES

1. Parkinson, J. S.: Control of Interior Noise, *Noise Control,* vol. 1, no. 1, pp. 54–62, January, 1955.

2. Kryter, K. D.: The Effects of Noise on Man, *J. Speech and Hearing Disorders,* monograph suppl. 1 (1950).

3. Beranek, L. L.: Noise Control in Office and Factory Spaces, *Trans. Ind. Hyg. Foundation, Bull.* 18, pp. 26–33, 1950.

4. Beranek, L. L.: "Acoustics," pp. 417–429, McGraw-Hill Book Company, Inc., New York, 1954.

5. Knudsen, V. O.: Acoustics in Comfort and Safety, *J. Acoust. Soc. Am.*, vol. 21, pp. 296–301, July, 1949.

6. Beranek, L. L.: Criteria for Office Quieting Based on Questionnaire Rating Studies, *J. Acoust. Soc. Am.*, vol. 28, pp. 833–852, 1956.

7. Miller, L. N., and I. Dyer: Noise Control for Offices, *Noise Control*, vol. 3, no. 2, pp. 70–75, March, 1957.

8. Beranek, L. L.: Revised Criteria for Noise in Buildings, *Noise Control*, vol. 3, pp. 19–27, January, 1957.

9. Kryter, K. D.: Noise Control Criteria for Buildings, *Noise Control*, vol. 3, pp. 14–20, November, 1957.

10. Beranek, L. L.: "Acoustics," pp. 408–417, McGraw-Hill Book Company, Inc., New York, 1954; articulation index is explained.

11. Bonvallet, G. L.: Levels and Spectra of Transportation Vehicle Noise, *J. Acoust. Soc. Am.*, vol. 22, pp. 201–205, March, 1950.

12. Koffman, J. L.: Vibration and Noise, *Automobile Engr.*, pp. 73–77, February, 1957.

13. Dyer, I.: Unpublished data prepared for a client of Bolt Beranek and Newman Inc., Cambridge, Mass.

14. Stevens, S. S.: Calculation of the Loudness of Complex Noise, *J. Acoust. Soc. Am.*, vol. 28, pp. 807–832, September, 1956.

15. Beranek, L. L., Karl D. Kryter, and Laymon N. Miller: Reaction of People to Exterior Aircraft Noise, *Noise Control*, pp. 23–31*ff.*, September, 1959.

16. Kryter, K. D.: Scaling Human Reactions to the Sound from Aircraft, *J. Acoust. Soc. Am.*, vol. 31, pp. 1415–1429, November, 1959.

17. Speech-privacy project sponsored by the Owens-Corning Fiberglas Corporation and performed by Bolt Beranek and Newman Inc. The results will be published in the Journal of the Acoustical Society of America.

Part Four
PRACTICAL NOISE CONTROL

Chapter 21

NOISE CONTROL IN VENTILATION SYSTEMS

Clayton H. Allen

21.1 General Approach

The use of a ventilation and air-conditioning system in a building often eliminates noise from traffic and other outside activities. But a noise problem may still remain because attention is then drawn to the noise of the ventilating equipment itself and to sounds from neighboring offices. The problem is made more difficult when there are lightweight walls between rooms or when high-velocity air-handling systems are installed. The solution of the over-all problem requires the systematic application of advanced acoustical engineering principles.[1]

Noise control is best begun in the early stages of building design. Then while there is still freedom in the location of quiet areas and noisy mechanical components, many problems can be eliminated directly. The modifications necessary for satisfactory noise control may add relatively little cost to the building when they are incorporated in initial construction. Similar or substitute modifications are likely to be expensive when installed later in the form of corrective measures. The most satisfactory solution of a noise-reduction problem in ventilating or air-conditioning systems requires the close cooperation of the architect, ventilation engineer, and acoustical engineer.

The aim in ventilation quieting is not to eliminate all noise but to produce a balanced noise environment. Generally, noise should be reduced only to the extent needed to allow normal conversation at a comfortable level for the anticipated activity. The quieting of a ventilation system much below the level of the noise created by normal activities is wasteful. Furthermore, excessive quieting is often detrimental since it reduces acoustical isolation of one area from another; some background noise masks conversation and other distracting noises in adjacent areas by overriding these weaker sounds and rendering them inaudible or unintelligible. Frequently the only acoustical privacy found in large industrial office spaces, drafting rooms, purchasing departments, etc., results from the masking effects of the ambient noise. Even where walls are provided, the degree of privacy and freedom from outside distractions may be governed largely by the steady background noise provided by an air-conditioning system. Therefore the activity to be conducted, the construction of the building, and the amount of privacy required determine the amount of background noise which is desirable in an occupied space. We shall call the desired amount of background noise the *noise criterion*. It is used in determining both the maximum and the minimum amount of noise control desired in the ventilating and air-conditioning system.

Methods are given in this chapter for predicting ventilation-noise levels in occupied spaces. Even where a building already exists, it is frequently economical to predict rather than to measure. The cost involved in adding a small margin of safety in the design of control measures to offset predictive uncertainties is generally less than the cost of making detailed acoustic measurements. Thus it is possible to make reliable engineering estimates of the amount and kind of noise reduction which may be required in a specified space, whether the space exists or is still to be constructed.

Methods are given in Chap. 23 for the control of air-conditioning-system noise and vibration transmitted through machinery-room walls and floors to adjacent office spaces.

Once the noise criterion has been chosen and the ventilation noise has been predicted (or measured) the ventilation quieting problem involves most of the noise-control principles discussed in previous chapters. The familiar outline

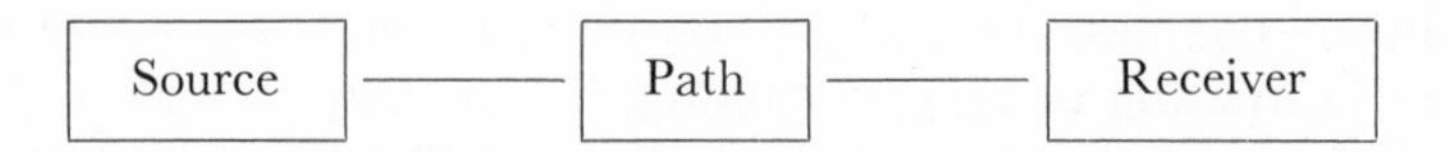

applies, but it is complicated by the fact that there are generally many sources, many paths, and many receivers, all interconnected, as illustrated in Fig. 21.1. In addition, the noise criterion may differ widely from one listener's location to another (e.g., greater quiet is necessary in an auditorium than in a mechanical equipment space). This multiplicity of problems means that frequently there is no single, straightforward procedure. The most direct approach is to determine successively:

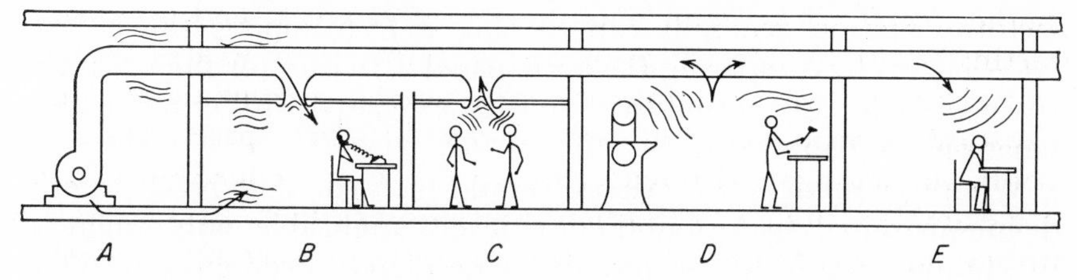

FIG. 21.1 Interconnecting sources, paths, and receivers. The fan in room *A* produces vibrations which enter room *B* through the floor and it produces noise which may enter room *B* through the air diffuser or by vibration of the duct walls. The noise may travel to all other rooms through the duct. The men talking in room *C* produce noise in room *B*. The noise from the shop *D* may travel through the ducts to rooms *B*, *C*, and *E*.

1. The noise criterion for each occupied space
2. The sound-power level of the noise produced by each source
3. The attenuation of the noise by walls, ducts, etc., between each source and the space in question
4. The noise levels at typical listener's positions in that space
5. The required additional attenuation (item 4 minus item 1)
6. The kind and placement of noise-reducing devices
7. Any special mountings of the devices necessary to control flanking noise
8. Any vibrating elements that may cause transmission of noise through building structure.

This order of approach simplifies the analysis because it places first those determinations that have the most influence on subsequent determinations. Furthermore, this order generally allows the acoustical engineer to work most efficiently with the architect and ventilation engineer during successive design stages.

21.2 Noise Criteria

Ventilation-noise control starts with the determination of a criterion for background-noise level in each occupied space. Acceptable levels are so dependent upon the activities to be carried on that it is essential to obtain as complete a description of these activities as possible. This includes knowing when the space is to be occupied and what other activities may be occurring simultaneously in other parts of the building. Only in this way can the most critical requirements be determined and a suitable background criterion selected.

Criteria for noise in occupied spaces have been discussed in detail in Chap. 20. Office noise criteria are illustrated by the NC and NCA curves of Figs. 20.3 and 20.4 of the previous chapter. A satisfactory criterion generally can be chosen from among the NC or NCA curves by comparing the anticipated activities for the occupied space with the descriptions of activities associated with criterion designations in Table 20.1.

Wherever possible NC criteria should be used. The NCA criteria define marginal background-noise levels applicable only where a maximum amount of compromise is necessary. Economic considerations occasionally may dictate the choice of a criterion intermediate between the NC and NCA criteria. Seldom is a noise level above the NCA criterion acceptable, and seldom is it economical or even desirable to design for a noise level below that NC criterion which provides the minimum required speech-interference level (SIL).

Noise levels in factories and mechanical spaces are frequently governed by hearing conservation criteria as discussed in Chap. 19. Usually the noise levels in these spaces are not controlled by air-conditioning or ventilating systems.

A noise criterion intended for the protection of delicate equipment or processes must be determined on an individual basis.

21.3 Sources of Noise

The acoustic output of a noise source may be described by the sound-power level in each of a group of contiguous frequency bands covering the audible frequency range and by the total sound-power level. For engineering purposes frequency bands an octave wide (see chap. 4) are generally found to be sufficient except where pure tones or unusual spectra are involved. Where a source is not confined, it may be necessary to specify its directivity also.

The noise sources of primary concern in the quieting of ventilating and air-conditioning systems are grouped below according to the principal way in which the noise from these sources reaches the occupied space:

1. Noise transmitted through or along ducts:
    a. Mechanical equipment noise, fans, motors, etc.
    b. Self-noise from air motion and turbulence within the distribution system
    c. Cross-talk from one room to another (i.e., noise that enters an air outlet in one room, travels through the duct, and then emerges in an adjacent room through another air outlet)
    d. Noise and vibration transmitted through duct walls
2. Noise transmitted through building structure resulting from:
    a. Machine vibration

   b. Transmission through walls, ceilings, etc., from mechanical equipment or other rooms
3. Noise transmitted from sources external to the building

All of these must be considered in order to avoid serious omissions.

Mechanical-equipment Noise. Fans are usually the principal sources of mechanical-equipment noise in ventilating systems. The intensity and character of the noise vary not only with size, speed, and load but also with type and manufacturer of the fan. The noise output of a fan can be estimated [2,3,4] within close limits when the static pressure is known, and either the driving-motor horsepower or the delivered air volume per minute is given.

The over-all sound-power level radiated into a duct system from *either* end of a fan can be determined from one of the following equations. The PWL radiated in *both* directions is 3 db higher.

$$\begin{aligned} \text{Over-all PWL} &= 100 + 10 \log hp + 10 \log p \\ \text{Over-all PWL} &= 65 + 10 \log q + 20 \log p \\ \text{Over-all PWL} &= 135 + 20 \log hp - 10 \log q \end{aligned} \tag{21.1}$$

where over-all PWL = sound-power level of total noise spectrum (20 to 10,000 cps), db *re* 10^{-13} watt
hp = rated motor horsepower
p = static pressure, in. H_2O
q = fan discharge, cfm (ft^3/min)

Any one of these equations should give the over-all PWL within ±4 db for conventional air-conditioning fans. Special or radical designs may fall outside these limits.

The spectral shape of the noise from fans varies, but in general, there are two types of spectra that are commonly encountered as shown in Fig. 21.2. Axial-flow fans have a spectrum that is nearly flat. Centrifugal fans, either with forward- or backward-curved blades, have a spectrum that slopes off approximately 5 db/octave with increasing frequency. Noise from centrifugal fans with forward-curved blades has a spectrum very nearly like that shown. The noise from centrifugal fans with backward-curved blades generally has more energy at high frequencies and less energy at low frequencies. The noise spectrum for large (several horsepower) backward-curved blades is sometimes predicted by shifting the spectrum to the right until the value shown for the first octave band lies in the octave band containing the blade passage frequency. Below the blade-passage frequency the spectrum usually falls at a rate of approximately 5 db/octave.

From a noise standpoint, forward-curved blades are best suited to unitary equipment or small short-run systems such as in household installations where there is little opportunity to attenuate high-frequency noise. Backward-curved blades are best suited to large installations where long duct runs provide natural reduction of high-frequency noise and where low-frequency noise may be troublesome because of the large physical size and driving power of the fan.

Even with the variations commonly encountered among fans of different manufacturer, quality, design, and workmanship, the values

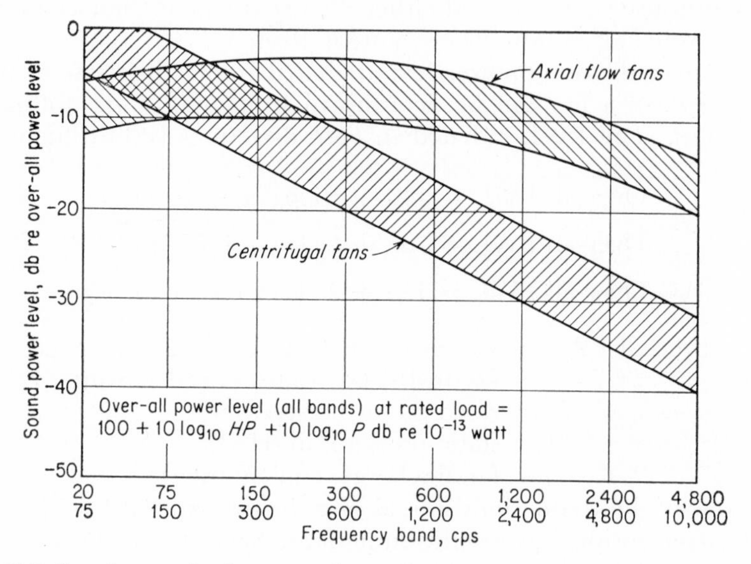

FIG. 21.2 Sound-power level spectra of centrifugal and axial-flow fans.[2,4] *Hp* is the rated motor horsepower; *P* is the static pressure, in. H_2O.

of the noise output estimated as indicated above are sufficiently close for general engineering purposes. Closer values might be obtained from fan manufacturers for particular units. When it is necessary to determine noise-reduction requirements for a ventilation system before the actual fan or manufacturer has been selected, then Eqs. (21.1) and Fig. 21.2 furnish the best available information.

Other mechanical-equipment noises are generally less important than fan noise. However, where the special noise sources do exist such as compressors, cooling towers, sprays, etc., they must be evaluated and their contribution added appropriately to the fan noise in calculating the total noise-power level.

Cooling towers, although generally located outdoors, generate noise which may be troublesome because it may enter the building through windows or may create a community or neighborhood noise problem. Results of measurements[5] on many cooling towers ranging from small fractional horsepower sizes to over 50-hp units are summarized in Eq. (21.2), which differs only slightly from Eqs. (21.1).

$$\text{Over-all PWL} = 105 + 10 \log hp \qquad \text{db } re\ 10^{-13} \text{ watt} \qquad (21.2)$$

The output-noise spectra are similar to those for centrifugal fans except that they fall with a slope of approximately 3 db instead of 5 db/octave (see Fig. 21.3).

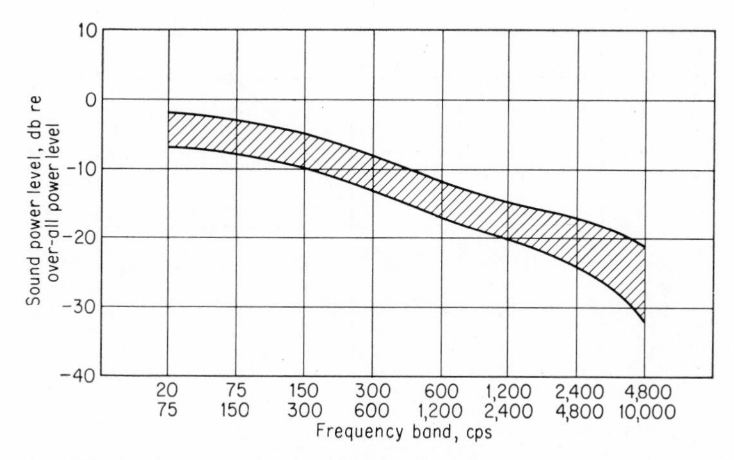

FIG. 21.3 Cooling-tower noise: This generalized spectrum, based on measurements, should apply to the noise generated by most induced-draft cooling towers since they are usually designed to be operationally similar, whatever the capacity. Equation (21.2) can be used to estimate over-all level from fan power.

Self-noise. Self-noise is generated by air traveling through the various sections of the distribution system. Grilles, pressure regulating valves, dampers, turning vanes, bends, branches, and other discontinuities or obstructions in the smooth flow of air create noise. With low airflow velocities and good duct design, these obstructions cause little noise and may be neglected. However, in high-velocity systems, even the flow of air in unobstructed ductwork may be so noisy that such ducts should be routed away from quiet areas.

Little engineering information is available on self-noise. Some manufacturers are beginning to evaluate their products, but this information is not yet well cataloged. The only work which now is available in engineering form is that on grilles and diffusers.

1. *Grille Noise.* Like other system self-noise, grille noise is caused by turbulence in the airflow around solid obstructions and deflecting vanes. Simple perforated register grilles are the quietest, those with

crossed deflecting vanes are the noisiest. Grille noise has a broad spectrum which is most important in the speech interference region of 600 to 4,800 cps.*

At the lower frequencies, fan noise usually predominates over grille noise. Therefore we are generally interested in grille-noise levels in the speech-interference range, and grille noise can be represented conveniently in terms of the SIL power level.* The curves in Fig. 21.4 are derived from manufacturers' data and from experi-

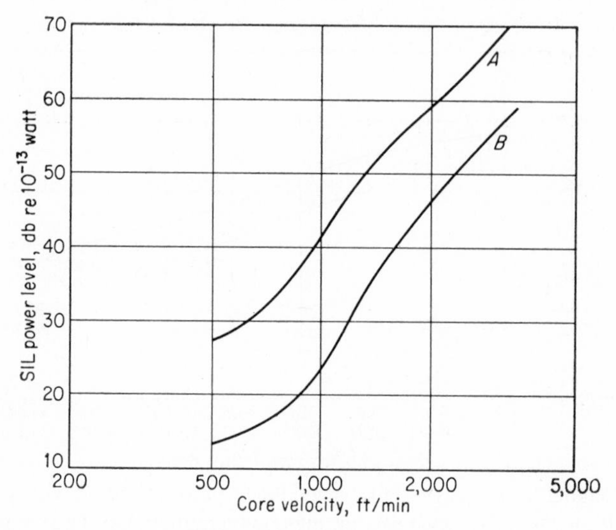

FIG. 21.4 Noise from simple grilles with core area = 1 ft². The *SIL power level* is defined as the simple average of the power levels (in decibels) in the 600 to 1,200, 1,200 to 2,400, and 2,400 to 4,800 cps bands. Curve (*A*) vertical or horizontal bar deflection type, all angles; (*B*) plain perforated grille. These graphs assume approximately a 20 per cent blockage by the grille members. For grilles with core area *A*, add, algebraically, $10 \log_{10} A$ to the ordinate.

mental measurements at MIT. They give the SIL power level radiated by a grille having a core area of 1 ft² (core area is the area of the duct opening into which the grille fits). The power level radiated by a grille having a core area of A ft² will be obtained by adding $10 \log A$ to the value given by Fig. 21.4.

2. *Diffuser Noise.* Diffuser noise is similar to grille noise. It is

* A designation of the intensity of the noise in this frequency region is the *speech interference level* (SIL). As discussed previously in Sec. 20.1, SIL is defined as the average of the sound-pressure levels (in decibels) in the three octave frequency bands, 600 to 1,200, 1,200 to 2,400, and 2,400 to 4,800 cps. By extension, we can also speak of the sound-power level (*re* 10^{-13} watt) in the SIL bands. For brevity, we shall call this the *SIL power level.*

higher in level for the same open area, perhaps because of the hornlike action of the diffuser flare. Recent measurements[6] on a limited number of diffusers from three manufacturers over a range of sizes from 8- to 18-in. neck diameter indicate that there is a broad peak in the spectrum which shifts toward the higher frequencies as the air velocity is increased. However, when the velocity is high enough to generate a significant amount of noise, it is the noise in the speech-interference region which is of concern.

The over-all sound-power level for the diffusers tested, in the region from 75 to 10,000 cps, is found to be described by the following empirical equation:

$$\text{Over-all PWL} = 13 \log A + 60 \log V \qquad \text{db} \tag{21.3}$$

where Over-all PWL = over-all sound-power level, db *re* 10^{-13} watt
A = minimum open area at damper, ft^2
V = air velocity through minimum area, ft/sec

The sound-power level in octave bands relative to the over-all PWL determined from Eq. (21.3) is given for a limited range of diffuser neck sizes in Fig. 21.5.

3. *Other Sources of Self-Noise.* Dampers and pressure regulators are known to be serious sources of noise, especially in high-velocity

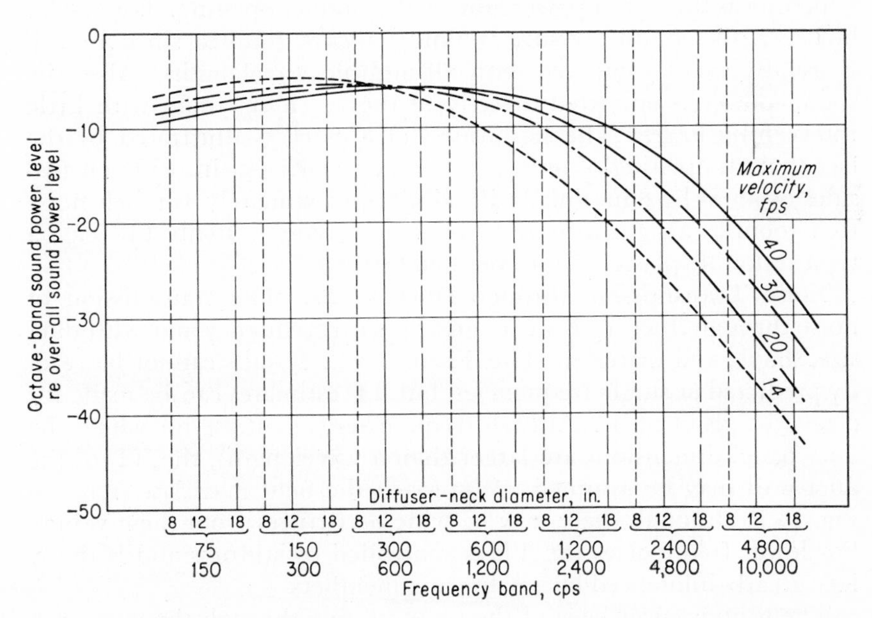

FIG. 21.5 PWL spectra of ceiling diffusers.[6] (The numbers 8, 12, and 18 refer to the diffuser neck diameter in inches.)

systems. Mixing boxes and other terminal devices are also noisy. Manufacturers should be consulted regarding the acoustic treatment recommended for the intended application.

Cross-talk. Cross-talk as applied to ventilation systems refers to direct transmission of sound through ducts from one duct opening to another or through unsealed cracks. A grille or opening not only allows air and noise to enter a room, but it allows sound from the room to enter the duct system. Thus the opening is a noise source when viewed from some other part of the duct system. The sound-power level of the noise *entering* the grille is found by determining the sound-pressure level (SPL) at the grille opening and adding 10 log A, where A is the core area of the grille in square feet. The SPL at the grille opening may be found by measurement or by estimating the sound-pressure level at the grille resulting from the known sources in the room. Note that when the grille dimensions are small compared to a wavelength there will be a pressure doubling (6-db increase in level) at the grille face due to sound reflection at the wall around the grille. When the grille dimensions are large compared to a wavelength, little increase in pressure occurs.

A crack from one room to another presents a similar cross-talk problem. It acts as a small section of wall with nearly zero TL. The sound power admitted to the receiving room can be estimated in the same way as the sound power admitted to a duct opening. For cracks, however, the sound pressure is almost always doubled since a crack is usually small compared with all audible wavelengths. Also, the sound power is admitted directly to the receiving room with little intervening losses. The seriousness of a crack is illustrated by the fact that an opening of only 1 in.^{2} (e.g., a crack $\frac{1}{16}$ in. wide on two sides of an 8- by 8-in. duct) will admit approximately as much noise to a room as an entire wall 12 by 12 ft having a 40-db TL (e.g., a typical double-plastered wall on metal studs).

Noise Transmission through Duct Walls. The transmission of noise through duct walls varies greatly with frequency and with duct size, shape, and material. The TL of duct sidewalls cannot be readily predicted at single frequencies, but TL estimates can be made for octave bands of noise. At high frequencies (i.e., those for which the duct panel dimensions are larger than a wavelength), the TL of the duct wall may be assumed to approach the field mass law (refer to Fig. 13.7) although there may be wide departures from these values. For lower frequencies the TL is controlled by stiffness and is therefore greatly influenced by any bracing members.

The sound-power level of the noise passing through the walls *out-*

ward to the surrounding space may be estimated from the following approximate equation:

$$PWL_T = PWL_D - TL + 10 \log \frac{S_w}{A} \tag{21.4}$$

where PWL_T = sound-power level of noise transmitted through duct walls in any octave frequency band
PWL_D = sound-power level transmitted along the inside of duct at beginning of section of duct considered (in same octave band)
S_w = total area of duct wall for section considered, ft^2
A = cross-sectional area of duct, ft^2

In cases where $10 \log S_w/A$ approaches or is larger than the TL of the duct wall this approximation no longer is valid since the PWL_T obviously cannot exceed the PWL_D which was originally inside the duct. For such a condition a reasonable approximation is to assume that PWL_T is 3 db less than PWL_D, i.e., half the sound is transmitted through the duct walls.

Similar considerations apply for transmission from a room *into* a duct, but here the sound-pressure level inside the duct (SPL_i) is calculated from the sound-pressure level existing outside the duct wall (SPL_o). Then the PWL of the resulting sound in the duct is calculated from the SPL_i. The equations and procedures are as follows:

$$SPL_i = SPL_o - TL + 10 \log \left(\frac{1}{4} + \frac{S_w}{R} \right) \tag{21.5}$$

$$R = \frac{\alpha_e (S_w + 2A)}{1 + \alpha_e} \tag{21.6}$$

$$\alpha_e = \frac{2A + \alpha S_w}{2A + S_w} \tag{21.7}$$

where TL = transmission loss of duct sidewall, db (see Fig. 13.7)
A = duct cross-sectional area, ft^2
S_w = duct wall area exposed to noise, ft^2
α = absorption coefficient of duct walls
α_e = effective absorption coefficient
R = quantity analogous to "room constant," ft^2 (see Chap. 11)

The absorption coefficient α for the duct wall is made up of an absorption due to any lining or wrapping of the duct plus an absorption due to any transmission of sound back into the room. The part of α due to lining can be found from the acoustical characteristics of the lining material. The part of α due to transmission is simply anti-$\log_{10}$ (TL/10). It is usually negligible for lined ducts except at low frequencies, but it is not negligible for unlined ducts when S_w is large compared with $2A$. Wrapping the outside of a duct with thermal insulation increases the absorption at low frequencies. For large ducts, wrapping may produce an α of 0.1 in the lowest two octave bands but produces negligible change at higher frequencies.

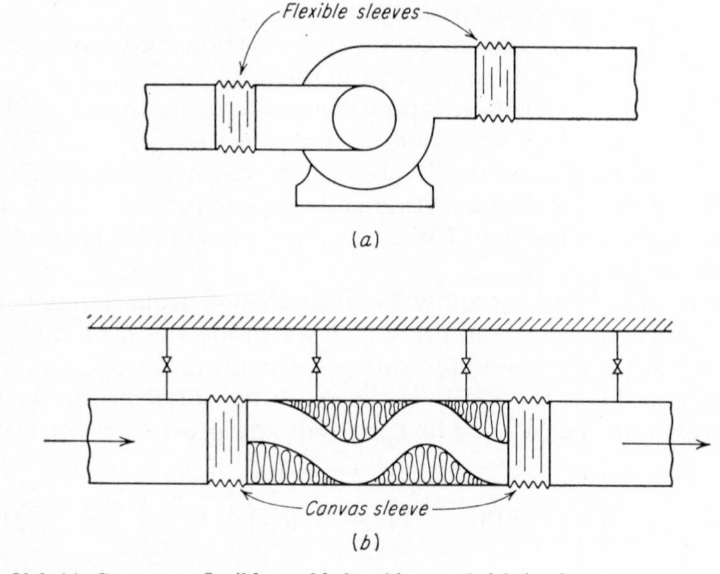

FIG. 21.6 (a) Canvas or flexible molded rubber and fabric sleeves serve as vibration breaks between fan and connecting ductwork. (b) Canvas or molded rubber connectors between ductwork and high-attenuation devices such as Aircoustat package sound attenuators prevent short-circuiting of noise through duct walls. Vibration-isolating hangers should be used where objectionable amounts of noise may short-circuit through supports and building structure.

The PWL entering a duct and traveling in each direction along the duct from the exposed area is computed from the SPL_i of Eq. (21.5) by adding 10 log A, where A is the cross-sectional area of the duct. This section of duct is then treated as a noise source having the calculated PWL. The noise-power level in each octave band of this source can be added to other noises at this point on a power-level basis when both noises are of a random nature. However, if the transmitted noise is speech or other intelligible noise, it must be handled separately as a cross-talk problem on the basis of privacy and speech intelligibility.

When ducts travel parallel to each other in close proximity, as in a duct chase, noise may pass from one duct to the other through the duct walls. This type of noise transfer may be particularly serious where large ducts that serve very quiet areas, such as an auditorium or broadcast studio, pass next to ducts serving lobbies or other noisy areas.

Noise from Duct-wall Vibration; Flanking. Noise may be "telegraphed" along the duct walls as vibration. The duct walls may resonate at some of the vibrational frequencies and act as new sources of noise when the vibrational energy converts back into sound at remote points. This type of noise source is much too complicated to predict quantitatively, but it can be eliminated by suitable vibration breaks. Typical examples of such vibration breaks are given in Fig. 21.6.

Where a large amount of air-borne noise reduction is provided in a short duct run, say by using an efficient duct lining, the vibrations traveling in the duct wall may cause serious flanking. Flanking of this sort sometimes can be controlled by adding vibration-damping material such as a heavy layer of mastic or an adhered covering of cork or glass-fiber board to the outside surface of the attenuating section of duct. Further reduction of the flanking problem may require spreading the noise-control treatment over a greater length of duct or dividing the treatment into two or more sections, each one separated by a length of duct equal to several duct diameters, with vibration breaks or damping material between each section of treatment.

21.4 Performance of Noise-reducing Components

The ducts of a ventilation system carry air-borne sound in much the same way as a speaking tube. There are several ways in which sound is reduced between the source and the receiver:

1. Absorption of sound due to transmission through duct walls into spaces outside the ducts
2. Absorption of sound in duct-wall linings
3. Division of sound among several branches
4. Reflection of sound back toward the source
5. Spreading of sound into the open space (room) at a duct end
6. Absorption of sound in the room where a duct ends.

There is always some natural noise reduction associated with the basic ventilation system, and it generally involves a combination of most of these processes. Necessary additional noise reduction involves the use of linings, plenums, or package units designed to obtain the required results with a minimum of cost and modification to the ventilation system.

In some instances the added noise-reducing "device" may be simply a rearrangement of the materials comprising the original system as, for example, applying thermal insulation inside rather than outside of duct walls. The difference between natural and added noise reduction is then primarily a matter of objective. Some elements of noise control in a simple duct system are illustrated in Fig. 21.7.

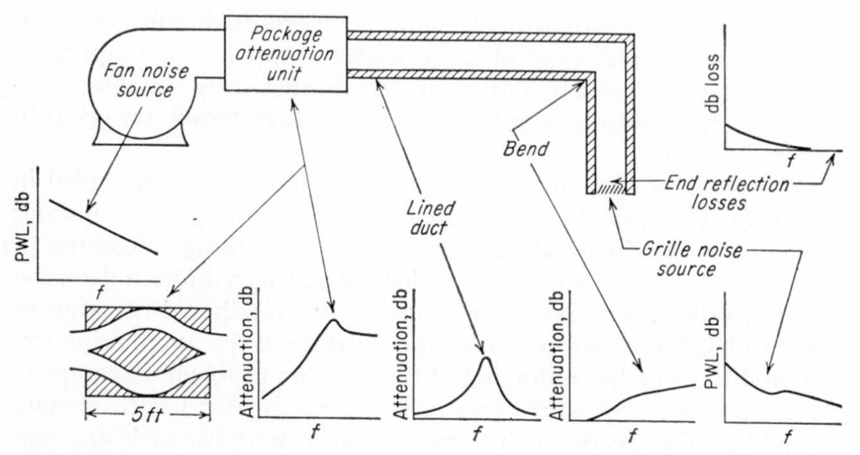

FIG. 21.7 Noise sources and attenuation in a simple duct system. The sources are the fan and the grille. Attenuation is provided by a package attenuation unit, a lined duct, a bend, and by reflection of low-frequency waves backwards at the end of the duct.

Unlined Sheet-metal Ducts. Rectangular sheet-metal ducts have large, flat, relatively flexible walls. Transmission through and absorption by such duct walls are greatly influenced by duct size and wall thickness and by the presence of any stiffening members. The attenuation of broad-band noise can be estimated very approximately from the figures given in Table 21.1.

The high-frequency attenuation in smooth round ducts may be estimated conservatively as equal to that given for rectangular ducts of the same cross-sectional area. Recent measurements* on small round, rigid, spiral, and rolled seam metal ducts 4 in. to 12 in. diameter show attenuations of approximately 0.03 db/ft for frequencies below 1,000 cps, rising irregularly at higher frequencies to approximately 0.1 db/ft.

Thermal insulation such as cork or glass-fiber board, applied either

* Carrier Corporation, Syracuse, New York, unpublished data.

inside or outside of rectangular ducts, damps vibrations in the duct walls. Limited experimental results[7] indicate that if the insulation is cemented to the duct so that it loads the walls, it may result in a doubling of the values given in Table 21.1 in the lowest three octave bands but has negligible effect at higher frequencies.

Table 21.1

Approximate Attenuation in Bare Rectangular Sheet-metal Ducts[7]

Duct	Size, in.	Frequency, cps			
		20–75	75–150	150–300	Above 300
		Attenuation, db/ft			
Small	6 × 6	0.2	0.2	0.15	0.1
Medium	24 × 24	0.2	0.2	0.1	0.05
Large	72 × 72	0.1	0.1	0.05	0.01

Resonances in the duct walls may provide high attenuation for limited-frequency regions. The use of duct wall resonances for obtaining attenuation requires careful attention to duct construction and mounting details, and in general is used only to meet special requirements which necessitate individual design calculations.

Lined Ducts. Acoustical material such as glass or mineral-fiber board, applied to the inside of either rectangular or round ducts, increases the sound attenuation along the duct, particularly in the frequency region for which the duct cross-sectional dimensions are approximately ½ wavelength. Methods for estimating the attenuation and noise reduction in lined ducts are given in Sec. 17.3. Data are also to be found in manufacturers' catalogs.

Lined and Unlined Bends. Noise is reflected at a square-cornered bend when the cross dimension of the duct is sufficiently large. The noise reduction provided by lined and unlined bends is given in Sec. 17.4. Turning vanes, if used, should be made small in width (less than ⅛ of the wavelength of the sound for which attenuation is required). The noise reduction provided by a bend may be limited by self-noise if the air velocities are sufficiently high.

Round bends provide considerably less attenuation than square bends. Conservative estimates which can be used for the attenuation of round unlined bends in either circular or rectangular ducts of various sizes are given in Table 21.2.

Reflection Loss Due to Change in Area. When a duct changes area abruptly, some sound is reflected back toward the source. The

Table 21.2

Approximate Attenuation Provided by Round Bends or Square Bends with Turning Vanes

Diameter, in.	Frequency, cps							
	20–75	75–150	150–300	300–600	600–1,200	1,200–2,400	2,400–4,800	4,800–10,000
	Attenuation, db							
5–10	0	0	0	0	1	2	3	3
11–20	0	0	0	1	2	3	3	3
21–40	0	0	1	2	3	3	3	3
41–80	0	1	2	3	3	3	3	3

reflection loss in decibels due to a change in area is given in Table 21.3. (The data in this table represent the low-frequency limits shown in Fig. 16.2.)

Table 21.3

Power Loss Due to Reflection at an Abrupt Change in Duct Cross-sectional Area

Ratios of Cross-sectional Area		Reflection Loss, db
S_1/S_2	S_2/S_1	
1	1	0.0
0.5	2	0.5
0.4	2.5	0.9
0.33	3	1.3
0.25	4	1.9
0.20	5	2.6

Branches. When a duct divides into two or more branches, the cross-sectional area of the branches is nearly equal to but generally somewhat greater than that of the single main duct. Noise traveling along the main duct divides approximately in proportion to the areas of the branches and continues with little loss due to reflection. The reduction in power level from a main duct to a branch duct, therefore, can be considered as an attenuation equal to:

$$\text{Attenuation at a branch} = 10 \log \frac{T}{B} \tag{21.8}$$

where B = cross-sectional area of branch duct in question
T = total cross-sectional area of all branches leaving junction

The use of the area of the main duct in place of the total area of all branches leaving the junction will generally result in only a small error at any one junction, but in a complicated duct system the cumulative error may amount to several decibels.

When noise arrives at a junction from a side branch, as may be the case with cross-talk from one room to another, there will be a division of power level in the same way, but there will also be a significant reflection of sound back along the branch because of the large change in area at the junction. The power-level reductions due to these two effects are additive. The power division loss can be found from Eq. (21.8); the numerator now contains the sum of the areas of all ducts except the branch from which the noise enters the junction. The reflection loss can be found from Table 21.3 by using the ratio of the area of the source branch to the combined area of all other ducts leaving the junction.

End Reflection Loss. Where a duct opens abruptly into a large space, a room, or large cavity with absorbing walls, end-reflection losses occur which vary with the frequency of the sound and with the size and shape of the duct opening and its position with respect to adjacent walls. Particularly at low frequencies, a large part of the noise does not radiate from the duct. The reflection loss for a square duct with width L is plotted in Fig. 21.8. If the duct opening is not

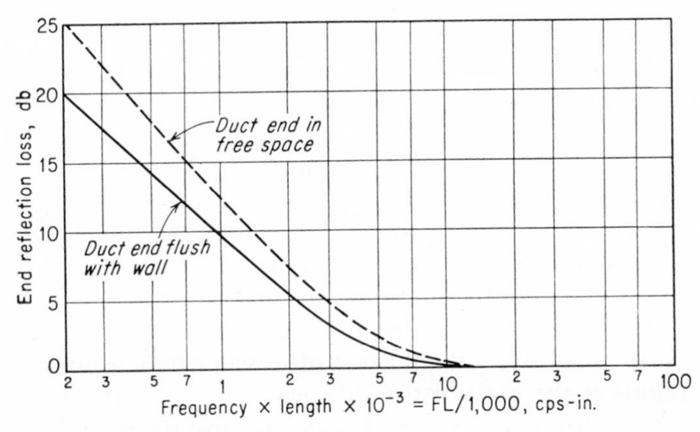

FIG. 21.8 End reflection loss at the open end of square ducts. The upper curve corresponds to position A of Fig. 21.10, and the lower curve corresponds to position B.

square, the effective size may be taken as approximately $\sqrt{L_1L_2}$, or for a round opening use approximately 0.9 of the diameter in place of L.

Plenum Chambers. Noise reduction in a duct system can also be provided by use of a plenum chamber (large expanded section of duct) which is lined with absorbing material. This is especially economical where a large number of small ducts must be fed by one main supply fan. The design of plenums and the estimation of their noise reduction are given in Sec. 17.5.

A plenum is particularly suited for reducing low-frequency noise. Other means are usually more economical for control of high-frequency noise. However, when a plenum is used, it will reduce high-frequency noise effectively if the inlet and outlet openings are not opposite each other.

Rectangular-cell- or Splitter-type Absorbers. When the available duct run is short, added attenuation can be gained at the expense of increased pressure drop by dividing the duct into an egg-crate type of sound-absorbing rectangular cells or by dividing the duct with

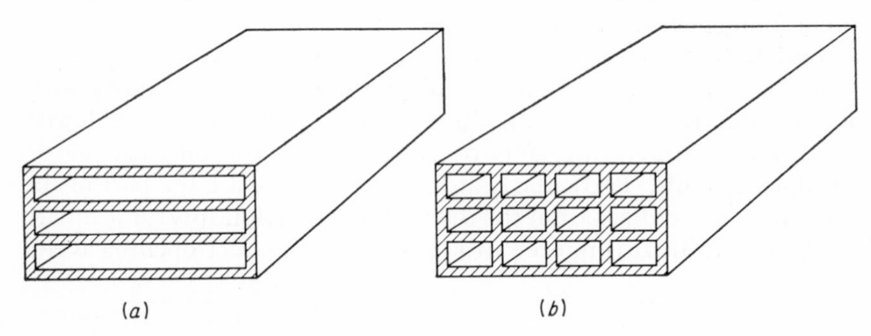

FIG. 21.9 Types of absorber for noise control in ducts: (*a*) splitter type; (*b*) rectangular cell type.

sound-absorbing plate splitters. To provide adequate sound absorption, all surfaces of the duct and splitters are lined as shown in Fig. 21.9. These arrangements provide a much larger lined surface and yield an attenuation per foot equivalent to that for a small lined duct equal in size to one individual section (see Sec. 17.3).

Where the number of splitters is large and the open area decreases to a fraction of its original size, an additional attenuation occurs due to reflections at the ends of the splitters. Although the area change aids in sound attenuation, it increases pressure drop, and therefore, where pressure drop must be minimized, the total cross-sectional area of the splitter type attenuator must be made greater than that of the duct. The reflection loss is calculated from the ratio of the open area of the splitter to the area of the ducts connected to the splitter. When the attenuation in the splitter section is more than 10 db, the reflection of sound from one end will be independent of the reflection at the other, and the total reflection loss due to area discontinuities at both ends will be double that at either end. The attenuation as a function of area ratio is given in Table 21.4.

Table 21.4

Attenuation Due to Area Change in an Attenuator
(Splitter, cell-type absorber, or packaged unit)

Open area of splitter / *Area of duct*	*Attenuation*, db, *total from both ends*
0.50	1
0.35	2
0.30	3
0.25	4
0.20	5

Package Attenuator Units. More efficient use of space and less pressure drop for a given amount of attenuation can be obtained from package units now available commercially. These units give a known amount of attenuation over a wide frequency range and thereby greatly reduce design costs and increase reliability. They also minimize assembly and inspection costs.

Various units are available giving different amounts of attenuation and pressure drop. The cost of package units is generally found to be less than duct lining where noise reduction is required over a wide range of frequencies and where lining is not needed for thermal insulation. These units are available in a range of sizes to fit ducts 6 by 12 in. and larger. Typical values of noise reduction and the range of static pressures associated with several types of recent (1959) units are presented in Sec. 17.6. More complete charts are available from manufacturers, including, also, recommended maximum air velocities when an upper limit for self-noise level is specified. In applying manufacturers' data, it is necessary to determine whether the self-noise values quoted are total power level, power level per unit area of duct cross section, or sound-pressure levels in a typical room used for test purposes.

21.5 Calculation of Room-noise Levels

The final step in determining room noise levels is the conversion from sound-power level (PWL) entering the room to the sound-pressure level (SPL) in the room. Sound-pressure levels in rooms were discussed in detail in Chap. 11. The present discussion reviews information particularly applicable to ventilation systems.

The preceding sections have presented the means for estimating the sound-power level produced by various sources normally encountered in ventilation systems. Also, procedures have been given for estimating the reduction in the sound-power level provided by various types of duct elements between the source of noise and a room. By subtracting in any one frequency band all of the values of noise reduction from the sound-power level of the source, the net sound-power level entering the room due to any source is obtained. This net sound-power level radiated into the room is used in the following calculations to determine the noise levels at a listener's position in a room.

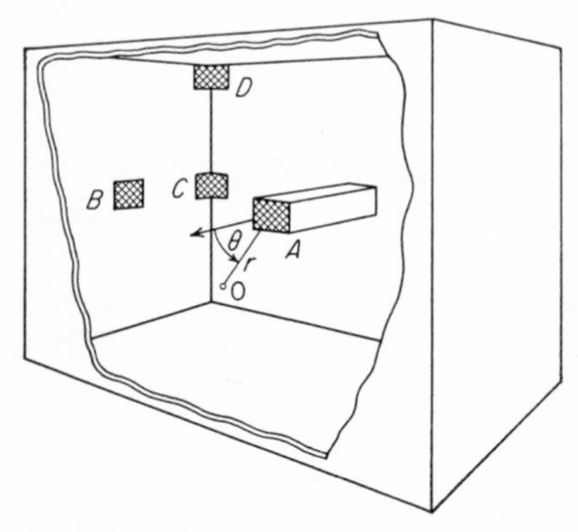

Fig. 21.10 Sketch showing duct opening positions A through D in a room and the meaning of θ and r. O is the position of a listener.

The sound-pressure level at any point in a room due to a source, say a grille, is made up of two parts: (1) direct sound from the source to the listener, and (2) reverberant sound arriving at the listener's position after one or more reflections from the surfaces of the room.

The direct sound is determined by the directivity factor of the source Q and the distance from the listener r (see Chap. 11). The reverberant sound is independent of the directivity factor and the distance and is determined only by the room constant R, which in turn is determined by the total absorption of the various surfaces of the room and its contents. The sound-pressure level at a listener's position at a distance r and at an angle θ from the axis perpendicular to a grille face is given by Eq. (21.9).

$$\text{SPL} \doteq \text{PWL} + 10 \log \left(\frac{Q}{4\pi r^2} + \frac{4}{R} \right) \quad \text{db} \tag{21.9}$$

where SPL = sound-pressure level at angle θ and distance r from duct opening, db *re* 0.0002 μbar (see Fig. 21.10)

PWL = power level radiated from duct opening, db *re* 10^{-13} watt

Q = directivity factor of duct opening, which varies with frequency, duct size, and angle θ (see Chap. 11)

r = distance from duct opening to listener, ft

R = room constant, ft^2—varies with frequency (see Chap. 11)

The directivity factor depends upon the position of the duct opening with respect to walls, edges, and corners of the room. Four duct positions are shown in Fig. 21.10. Approximate values of Q for these positions are given in Fig. 21.11 for $\theta = 0$ and $\theta = 45°$.

The room constant at various frequencies can be computed approximately from known values for acoustical treatment, room furnishings, and occupants. Alternatively, an estimate of room constant R can be obtained from Fig. 11.7.

Noise Level for a Single Source. The value of the sound-pressure level (SPL) relative to the sound-power level (PWL) of a source as

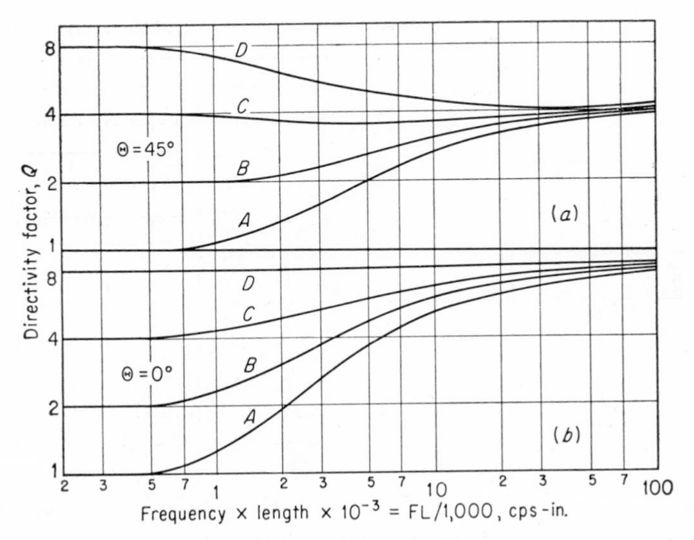

FIG. 21.11 Directivity factor Q for $\theta = 0$ and $\theta = 45°$ at the four positions of the duct opening shown in Fig. 21.10. $L = \sqrt{A}$, where A is the core area of the grille. Curve A, grille at center of room; B, grille at center of walls; C, grille at center of edge; D, grille at room corner.

calculated from Eq. (21.9) has been plotted in Fig. 21.12. To use the graph one must first know the directivity factor (Q) for the duct opening in the specified direction and the room constant (R) for the room at the frequency of interest. The graph is entered from the bottom at the distance r. Follow diagonally upward and to the left until the proper value of Q is reached, then move vertically to the value of R for the room. Finally read the *relative* sound-pressure level from the ordinate at the left. To obtain the actual sound-pressure level, add the sound-power level to the ordinate reading.

An example is shown by the dotted line in Fig. 21.12 for $r = 7$ ft, $Q = 2$, and $R = 2{,}000$ ft^2. The relative sound-pressure level is -22 db. Assume a sound-power level of 120 db *re* 10^{-13} watt. Then

120 db is added algebraically to the relative sound-pressure level to give the actual sound-pressure level at the listener's position. The result is 120 — 22 = 98 db *re* 0.0002 μbar.

Noise Levels for Multiple Sources. Where several duct openings serve one room, the noise-level computation must be modified. The reverberant noise level is determined by using the total sound-power

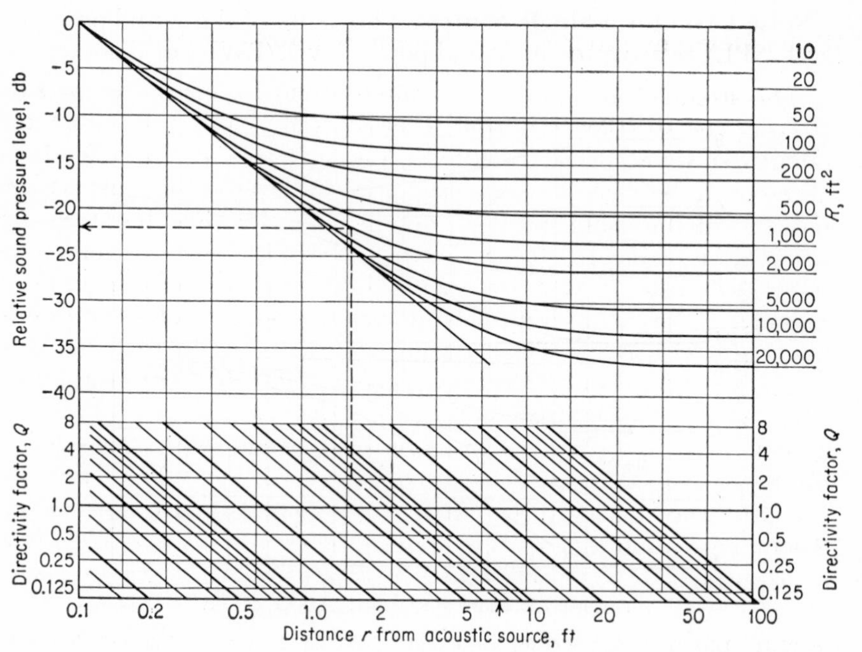

FIG. 21.12 Method for determining the sound-pressure level, given the sound-power level (PWL), directivity factor (Q), distance (r), and the room constant (R). The absolute SPL (in decibels *re* 0.0002 μbar) equals the sum of the ordinate and the PWL (in decibels *re* 10^{-13} watt) of the noise source at normal room temperature and pressure conditions. Relative sound pressure in a room $= 10 \log_{10} [(Q/4\pi r^2) + (4/R)]$ db. $R = S\bar{\alpha}/(1 - \bar{\alpha})$ = room constant, ft². S = total area of boundaries of room, ft². $\bar{\alpha}$ = average energy absorption coefficient of the surface of the room.

level from *all duct openings.* To get the total sound-power level, perform the operation

$$\text{Total PWL} = 10 \log_{10}\left(\text{antilog}_{10}\frac{\text{PWL}_1}{10} + \text{antilog}_{10}\frac{\text{PWL}_2}{10} + \cdots\right) \tag{21.10}$$

The noise level at a specified position near any particular duct outlet can be determined from Eq. (21.9) by replacing the room constant

R by an "effective" room constant R_e which applies for the duct in question. R_e is given by

$$R_e = R \frac{\text{antilog } (\text{PWL}_{\text{duct}}/10)}{\text{antilog } (\text{PWL}_{\text{total}}/10)}$$

An example of this procedure is worked out in Sec. 21.8.

21.6 Determination of Attenuation Required

The attenuation required in a duct system is determined by subtracting the specified criterion level in each octave band from the calculated sound-pressure level at the listener's position 0. It is necessary to be sure that the sound-pressure level at the most critical position in the room has been determined. In some instances the reverberant sound levels may be the controlling levels, in other instances the listener positions near grilles may be the most critical. Usually the levels in several positions in a room or auditorium must be compared with the criterion in order to ascertain that the most severe condition has been examined. Note that the applicable criterion may be different in different parts of the same room. This may be true particularly in an auditorium where higher noise levels can be permitted near the stage than at the rear of the hall.

Grille and Duct Noise; Additive Effects on Criterion Level. Room noise levels due to grille noise alone are usually calculated first to determine that the grille size is sufficiently large. Any necessary changes in grille sizes should be made at the outset because they may affect the duct sizing and necessitate a rebalancing of the distribution system.

If the noise level due to grilles is more than 10 db below the criterion level, it may be neglected in subsequent calculations. However, it is frequently advisable to adjust grille noise to a value only slightly below the criterion level at the most critical listener's position. Such an adjustment has not only the advantages of minimizing the duct size and maximizing the throw of the air but provides a control over the noise level in the speech-interference region which is necessary for maintaining adequate privacy. Normal attenuating means used in duct systems to reduce fan noise usually provide excessive attenuation at high frequencies resulting in a deficiency of noise in the speech interference bands. Noise in this frequency region is needed to provide privacy by masking. Noise generated by properly sized grilles may be used to fill in this frequency region.

If grille noise is adjusted to meet the criterion level in the SIL region, then the fan noise should be reduced at least 10 db below the criterion level in this region so that the sum of the two will not exceed the criterion level. Usually a compromise is made by adjusting the grille noise to a value 3 to 5 db below the criterion level. In this way the sum of the grille noise and the fan noise will not depart from the criterion by more than 3 to 5 db whether the fan noise by itself meets the criterion exactly or is completely attenuated.

Cross-talk. In order that intruding spoken words will be unintelligible, cross-talk noise levels must be held well below the criterion level and at least 5 db below the lowest anticipated background level if privacy is required. The calculation of attenuation needed to control cross-talk noise levels is handled in the same way as those for random noise once an acceptable level has been specified for the room or area in question.

Note that privacy will be greatly augmented by general activity noise which provides masking during working hours. After normal working hours this masking noise drops many decibels, and privacy will be greatly reduced.

21.7 Selection of Noise-control Devices

The choice of any one or more of the wide selection of noise-control devices depends upon the amount and frequency range of noise reduction required.

Duct lining is the most common noise-control device and is often used wherever the lining can serve the double purpose of noise reduction and thermal insulation. The transfer of thermal insulation from outside to inside the duct may in some cases provide all needed noise reduction at little additional cost if the transfer is made in the building design stage.

Package units, splitters, or egg-crate units may be used where insufficient noise reduction results from the lining required for thermal insulation or where a large amount of reduction is required in a short duct run. Larger freedom in the choice of the frequency region of high noise reduction exists with these units than with simple lined ducts. Since costs are roughly proportional to noise reduction and volume, any excess noise reduction provided in frequency regions where it is not needed is generally wasteful. Furthermore, excess noise reduction in some frequency regions may lower the background levels excessively and decrease privacy.

Where fibrous attenuating materials are used, suitable facings should be chosen to give adequate protection against erosion at the anticipated stream velocities. Where large amounts of noise reduction are required and where the resulting noise levels are low, it is always necessary to consider for the attenuator the self-noise of the attenuator due to air turbulence within its passages or at its termination. For low noise levels, such as are required for television studios or comparable spaces, the air velocities in ducts and in open areas of attenuators should not exceed 1,000 ft/min, and duct outlet velocities should be even lower.

21.8 Example

A typical ventilation-system noise-control problem is worked out here to illustrate the procedures described in this chapter.

Noise sources should be considered in proper sequence. As mentioned earlier, grille noise should first be determined to permit necessary changes in sizes and discharge velocities to be made. Fan noise should be calculated as soon as fan sizes and approximate layouts are determined, because minor changes in layout will not seriously affect these calculations. Early consideration of cross-talk and cross-transmission of noises frequently may influence the location of some noise-control devices. Self-noise may dictate the placement of package attenuators near outlets when very quiet areas are served.

The choice of a suitable combination of attenuation devices involves all of the above determinations. The selection and installation should be checked for possible flanking, and appropriate modifications should be made to correct deficiencies.

The problem chosen is illustrated schematically in Fig. 21.13. It involves the analysis and recommendation of the means for control of noise necessary for an air supply system serving a conference room with dimensions 30 by 33 by 10 ft. The ventilation schedule calls for a change of air every 5 min, i.e., 2,000 ft^3/min through the supply ducts. The air is distributed to the room through four louvered deflecting vane-type grilles, each 6 by 12 in. The air is supplied through a branch from a central supply system that employs a centrifugal fan operating at 1.5 in. static pressure and supplying a total of 10,000 cfm to the building. The motor driving the fan has a rating of 5 hp.

1. *Criterion.* The criterion which is considered satisfactory for background noise in a conference room of this size is NC–30 (see Table 20.1 and Fig. 20.3).

2. *Grille Noise.* Four grilles, each 6 by 12 in., supply a total of 2,000 cfm, thus the nominal velocity through the core area of each grille is 1,000 ft/min. From Fig. 21.4 we find for a deflecting-vane grille having a core area of 1 ft^2 and an outlet velocity of 1,000 ft/min that the SIL sound-power level is approximately 42 db *re* 10^{-13} watt.

For each grille being considered here (area = 0.5 ft²), the SIL power level is then found by adding algebraically the quantity 10 log 0.5 = −3 db, yielding 39 db *re* 10^{-13} watt. The SIL sound-power level for all four grilles equals

$$42 \text{ db} + 10 \log (2 \text{ ft}^2) = 45 \text{ db } re\ 10^{-13} \text{ watt}$$

3. *Room Levels.* For a room with a volume of 10,000 ft³ an average value of room constant is about 500 ft² (see Fig. 11.7). If the acoustical treatment and furnishings of the room are known in detail,

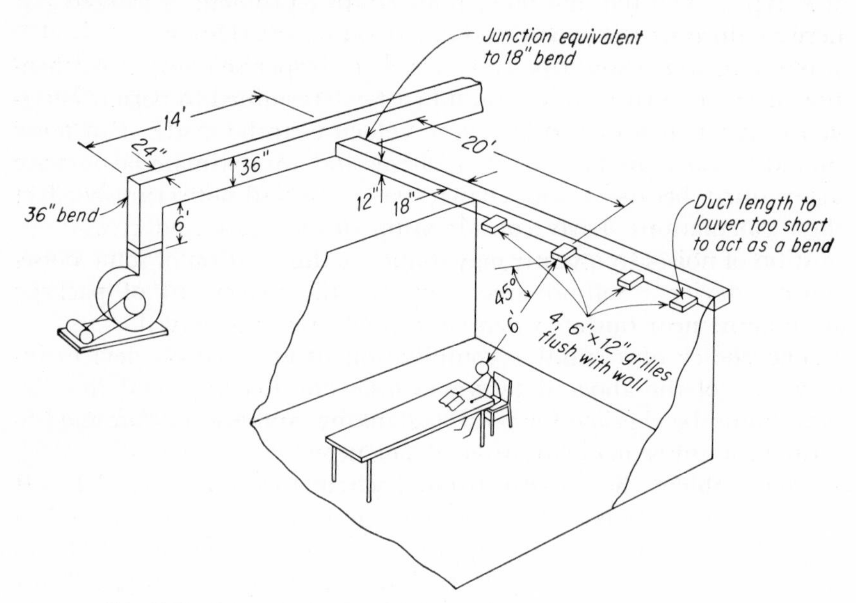

FIG. 21.13 Schematic duct arrangement for the example of Sec. 21.8.

more accurate values of room constant for each octave band may be determined.

It is necessary to look at both the reverberant field and near field of each outlet. First, let us consider the reverberant field. We see from Fig. 21.12 that for a room constant of 500 ft² the reverberant field is −20 db relative to the total sound-power level. Thus the SIL in the reverberant field due to all four grilles is 45 − 20 = 25 db. The reverberant field is below the NC–30 criterion.

For the near-field levels, because all grilles are equivalent, we need look only at the position of the listener nearest to any one of the grilles. To obtain the level in the near field of one grille in a room served by several, we must first find the "effective" room constant for the one grille. The effective room constant will be the actual room constant multiplied by the ratio of the power radiating from the single duct or grille in question to the total acoustic noise power. In this case since the four outlets are equivalent, the acoustic power from one outlet is one-fourth of the total, and the effective room constant for one grille is $\frac{1}{4} \times 500\ \text{ft}^2 = 125\ \text{ft}^2$.

Now if the grilles are located in the wall near the ceiling, the nearest person seated at a table is approximately 6 ft from one grille at a 45° angle. From Fig. 21.11, for a 6- by 12-in. grille in the frequency region from 600 to 4,800 cps, Q is approximately 4. Entering Fig. 21.12 from the bottom at the point labeled 6 ft and proceeding diagonally upward to the horizontal line labeled $Q = 4$, then proceeding vertically to a point for $R = 125$, which lies just below the curve labeled $R = 100$, we find from the corresponding ordinate at the left that the relative sound-pressure level is −13. Thus the SIL at the 6-ft position is $39 - 13 = 26$ db *re* 0.0002 μbar. This is again below the NC–30 criterion chosen for this room, and we may therefore consider the grille noise satisfactory.

If the grille noise were found to be too high, larger grilles giving lower exit velocities would have been required. If the grille noise were more than 5 db below the criterion, it might be desirable to decrease the grille size to raise the background noise closer to the criterion level for the purpose of increasing privacy.

4. *Fan Noise.* Fan-noise data generally are not available for a particular fan so that one form of the approximate Eqs. (21.1) must be used together with spectral information from Fig. 21.2. We have

$$\begin{aligned} \text{Over-all PWL} &= 100 + 10 \log hp + 10 \log p \\ &= 100 + 10 \log 5 + 10 \log 1.5 \\ &\doteq 109 \text{ db } re\ 10^{-13} \text{ watt} \end{aligned}$$

or

$$\begin{aligned} \text{Over-all PWL} &= 65 + 10 \log q + 20 \log p \\ &= 65 + 10 \log 10{,}000 + 20 \log 1.5 \\ &\doteq 108 \text{ db } re\ 10^{-13} \text{ watt} \end{aligned}$$

or

$$\begin{aligned} \text{Over-all PWL} &= 135 + 20 \log hp - 10 \log q \\ &= 135 + 20 \log 5 - 10 \log 10{,}000 \\ &\doteq 109 \text{ db } re\ 10^{-13} \text{ watt} \end{aligned}$$

Any one of these equations may be used depending upon the information available. They usually agree much closer than the estimated maximum error of ± 4 db. The spectrum for a centrifugal fan with 108-db over-all PWL as obtained from Fig. 21.2 is given as the first line of Table 21.5.

5. *Natural Attenuation of the System (See Fig. 21.13).* We will assume attenuation to result from:

a. A sound-power division due to branching which may be determined as the sum of attenuations calculated for each branch in accordance with the procedure outlined in Sec. 21.4 or may be estimated with somewhat less accuracy in a single step as 10 log of the ratio of the cubic feet per minute delivered to one outlet to the total fan cubic feet per minute, i.e., $10 \log \frac{2{,}000}{10{,}000} \times \frac{1}{4} = -13$ db

b. A 3-ft square-cornered bend (Fig. 17.10)
c. 20 ft of 24- by 36-in. unlined main duct (Table 21.1)
d. 20 ft of 12- by 18-in. unlined branch duct (Table 21.1)
e. An 18-in. unlined bend (Fig. 17.10)
f. End reflection loss, $L = \sqrt{6 \times 12} \doteq 8.5$ in. (Fig. 21.8)
g. Conversion from PWL to SPL in the room (Fig. 21.12)

Numerical values for these system losses are listed in Table 21.5, lines 2 through 8.

Note: In obtaining the difference in level between PWL and the SPL at the listener's position, we need consider here only the 6-ft distance because the near-field condition was found to be the most serious in the calculation of the grille noise for this room. Where several ducts entering a room may have different amounts of acoustic treatment, both the reverberant field and the near field of each grille must be examined. In this example all grilles are nearly equivalent and are considered to have the same noise output.

A particular duct system may involve several more items that may provide noise reduction. The extension of the procedure is obvious. The total losses in level are the sum of lines 2 through 8, as shown in line 9. The net SPL is the power level (line 1) minus the total losses (line 9) as shown in line 10. The required noise attenuation is then found by subtracting the criterion (in this instance NC–30) as listed in line 11 from the calculated noise levels in line 10; the result is shown in line 12.

From the various alternatives available, the noise-reduction measures which most economically meet the requirements should be chosen. Noise reductions offered by typical noise-control corrective measures are indicated in lines 13 through 17 of Table 21.5.

Note that where a corrective device replaces a section of the duct system, the increase in noise reduction is the NR of the new device minus the NR of the original section of the duct system. For example, if a 36-in. square-cornered lined bend is inserted in place of a 36-in. unlined bend, we have added attenuation values which are the differences between the lined bend and unlined bend as shown in line 3. A loss due to the absorption characteristic of length of lined duct is obtained in addition to the attenuation of the lined bend itself. Thus, if the duct was lined for a total of 10 ft past the bend,

Table 21.5

Typical Room-level Calculation Table

Octave Bands of Frequency	Frequency, cps							
	20–75	75–150	150–300	300–600	600–1,200	1,200–2,400	2,400–4,800	4,800–10,000
1. Fan PWL, db *re* 10^{-13} watt	107	102	97	92	87	82	77	72
	Losses, db							
2. Power division to one grille	−13	−13	−13	−13	−13	−13	−13	−13
3. Bend, 3-ft square-cornered	−1	−4	−8	−6	−3	−3	−3	−3
4. 20 ft of 2- by 3-ft duct	−4	−4	−2	−1	−1	−1	−1	−1
5. 20 ft of 12- by 18-in. duct	−4	−4	−3	−2	−2	−2	−2	−2
6. Bend, 18 in., unlined	0	−1	−4	−8	−6	−3	−3	−3
7. End Loss, L = 8.5 in.	−14	−10	−5	−2	0	0	0	0
8. PWL to SPL, r = 6 ft	−13	−13	−13	−13	−13	−13	−13	−13
9. Total losses PWL to SPL (sum of lines 2–8)	−49	−49	−48	−45	−38	−35	−35	−35
10. Net SPL (line 1 + line 9)	58	53	49	47	49	47	42	37
11. NC-30	60	51	43	37	32	30	28	27
12. Required NR (line 10 − line 11)		2	6	10	17	17	14	10
Typical noise-control measure, Alternate A: Line 6 ft of 2- by 3-ft duct between fan and bend on four sides (thickness of lining selected to give 80% open area).								
13. Random incidence attenuation of lined duct (36″ × 24″ × 6′)	0	7	14	23	22	13	11	10
14. Increased attenuation for 3-ft bend (difference between plane-axial-wave input and random input	0	0	1	3	5	10	10	10
15. Increased attenuation due to duct-wall damping by rigid acoustical lining	1	1	0	0	0	0	0	0
16. Total, lines 13 through 15	1	8	15	26	27	23	21	20
Typical noise-control measure, Alternate B: Add 2 ft Aircoustat unit in 2- by 3-ft duct.								
17. NR of 2-ft Aircoustat (approximate octave-band values derived from Table 17.4)	9	9	11	15	25	40	28	20

the lined duct absorption would be that given in line 14. Because of duct-wall damping provided by the lining, there would be additional absorption in the first three bands given in line 15.

The total attenuation for these items, line 16, is seen to be adequate to meet the requirements for NC–30.

As an alternative, a package unit such as a 2-ft Aircoustat might have been used to provide the attenuation shown in line 17. If the conference room were the only problem area in this building, a small package unit in the branch line might be the most practical solution. If several other problem areas exist, a larger sized acoustic treatment probably should be placed in a main duct to save duplication of parts and labor. It may be noted that these treatments provide significantly more than the necessary attenuation in the higher octave bands. In view of precautions given earlier about the possibility of excess attenuation causing a reduction in privacy, this excess attenuation might seem inexpedient; however, excessive high-frequency attenuation usually results when sufficient low-frequency attenuation is introduced. Therefore grilles should be sized to maintain the high-frequency noise near or a little below the criterion level.

21.9 Summary

Proper noise control in occupied spaces in buildings is not just noise reduction; it implies the adjusting and shaping of background noise to meet the needs for speech communication, privacy, and comfort (in that order). Procedures can be simplified and costs minimized by recognizing and treating noise problems in the early stages of a building design. The final success of noise-control treatment depends upon meticulous attention to detail.

This chapter presents methods by which noise levels in building spaces may be predicted in advance of construction. Means for reducing or modifying the predicted or measured background noise as required to meet applicable noise criteria are described and example computations are presented to illustrate the procedures introduced. Precautionary measures necessary to avoid common omissions and design errors are emphasized.

REFERENCES

1. Sound Control, chap. 40 of "Heating, Ventilating, and Air-conditioning Guide," American Society of Heating and Air-conditioning Engineers, New York, 1957 (and later editions).

2. Beranek, L. L., G. W. Kamperman, and C. H. Allen: Noise of Centrifugal Fans, *J. Acoust. Soc. Am.,* vol. 27, pp. 217–219, March, 1955.

3. Allen, C. H.: Noise from Air Conditioning Fans, *Noise Control,* vol. 3, pp. 28–34, January, 1957.

4. Beranek, L. L.: "Acoustics," McGraw-Hill Book Company, Inc., New York, 1954.

5. Dyer, I., and L. N. Miller: Cooling Tower Noise, *Noise Control,* vol. 5, no. 3, pp. 44–47, May, 1959.

6. Chaddock, J. B.: "Ceiling Diffuser Noise," to be published.

7. Chaddock, J. B., J. K. Nunnely, T. L. Moore, and H. H. Bell: Sound Attenuation in Straight Ventilation Ducting, *Refrig. Eng.,* vol. 67, pp. 37–42, 1959.

Chapter 22

CASE HISTORIES OF MACHINE AND SHOP QUIETING

Laymon N. Miller

22.1 Introduction

On the occasion of the twenty-fifth anniversary of the Acoustical Society of America Dean Vern Knudsen[1] of the University of California at Los Angeles observed that some of the noise levels encountered in our present-day living had increased approximately 1 db/year over the past 25 years. He wondered what this would lead to in another 75 years. Can we control this trend or will the trend control us?

We can certainly ask this question about noise in industry, for we have much credible evidence that high noise levels in work areas have already contributed to some hearing loss of some personnel. With no apparent letup in industrial growth, we must expect continued high noise levels in industry, and we must be prepared to work actively to bring about long-time hearing preservation of our entire society.

22.2 Review of Noise Criteria

Chapters 19 and 20 have been devoted to details of noise criteria for hearing preservation and speech communications. Without repeating the details, we present here in four charts the essential elements of those criteria as summarized for engineering noise-control purposes. Most factory noise falls within the range of noise levels given by these charts. Figure 22.1 gives the broad-band noise levels which suggest the need for hearing-preservation programs. Figure 22.2 presents noise criteria for hearing preservation in terms of the average length of exposure time each day to various average noise levels. Figure 22.3 shows a range of noise levels which will permit

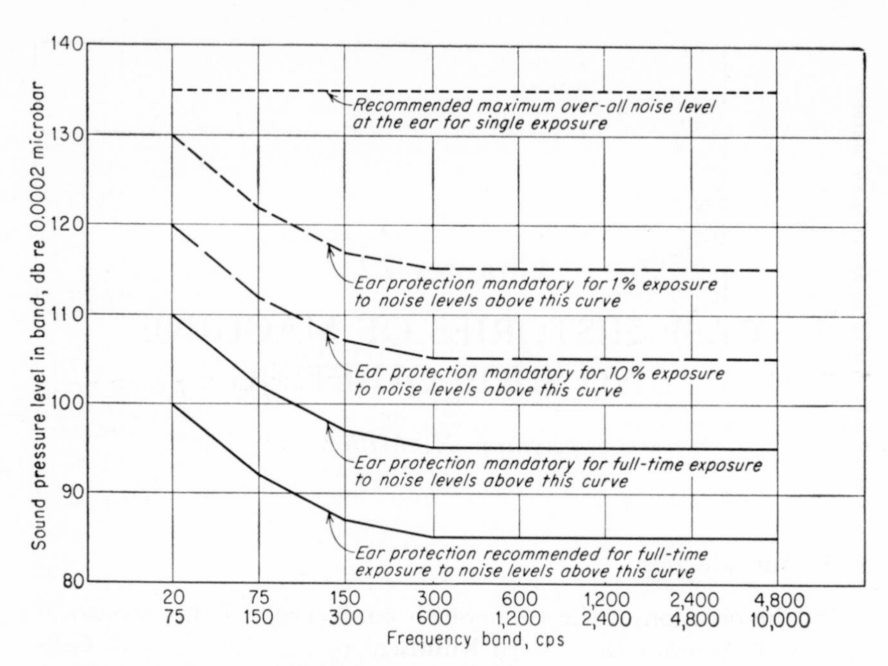

FIG. 22.1 Broad-band noise-level exposures for which conservation-of-hearing measures are recommended or mandatory. The parameter of the upper three curves is percentage of each working day. Exposure over a period of years is assumed.

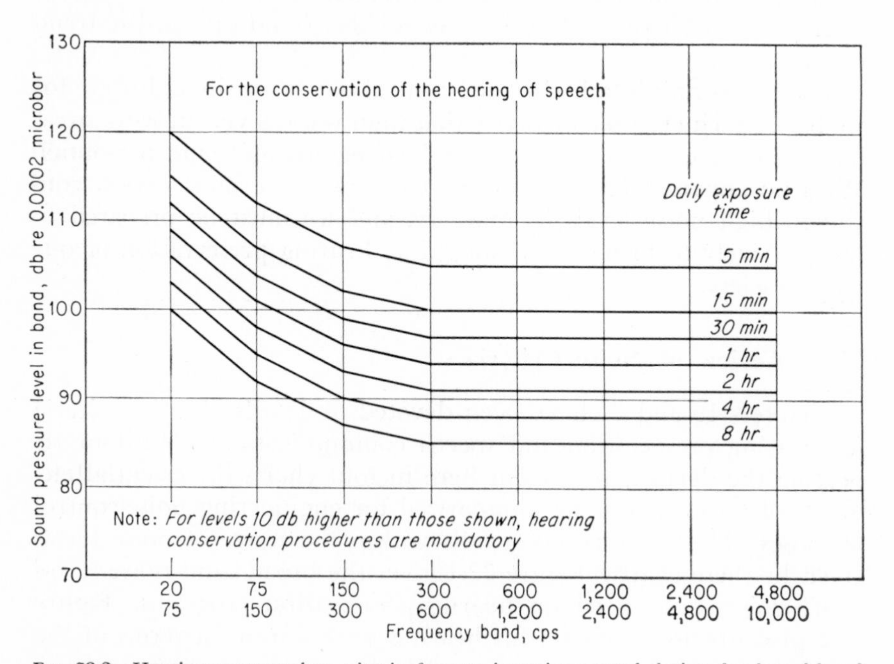

FIG. 22.2 Hearing-conservation criteria for use in noise-control design for broad-band noise levels of various average daily exposure times. These curves are built on the lowest curve of Fig. 22.1, assuming that a halving of the daily exposure permits 3-db-higher noise levels.

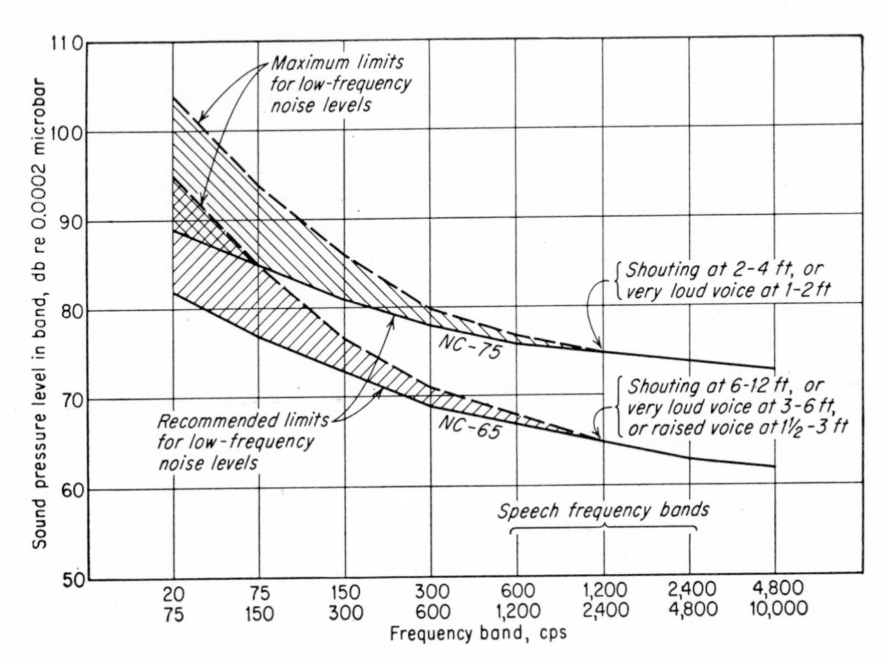

FIG. 22.3 Noise-level criteria and associated voice levels for acceptable *discontinuous* speech communication using *limited* vocabulary.

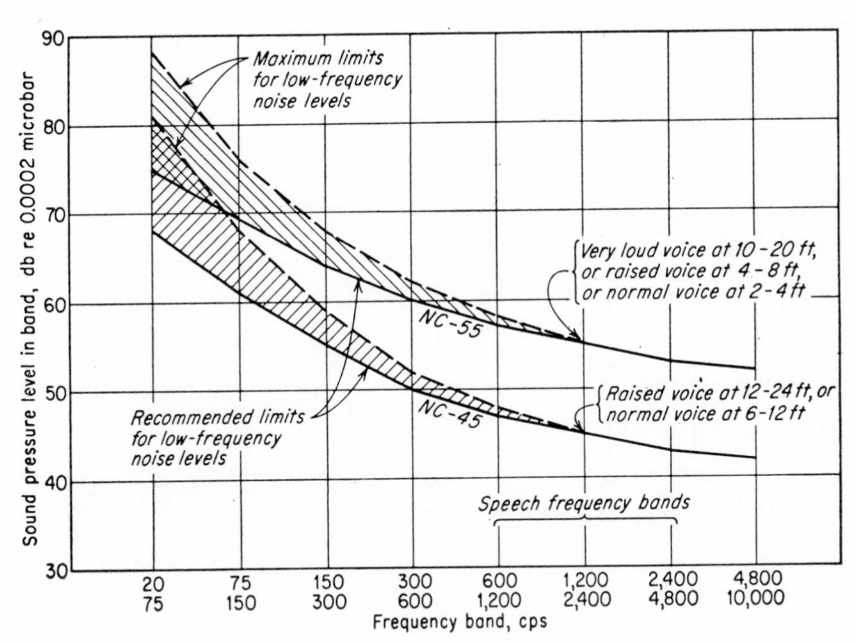

FIG. 22.4 Noise-level criteria and associated voice levels for reliable *continuous* speech communication using *unlimited* vocabulary.

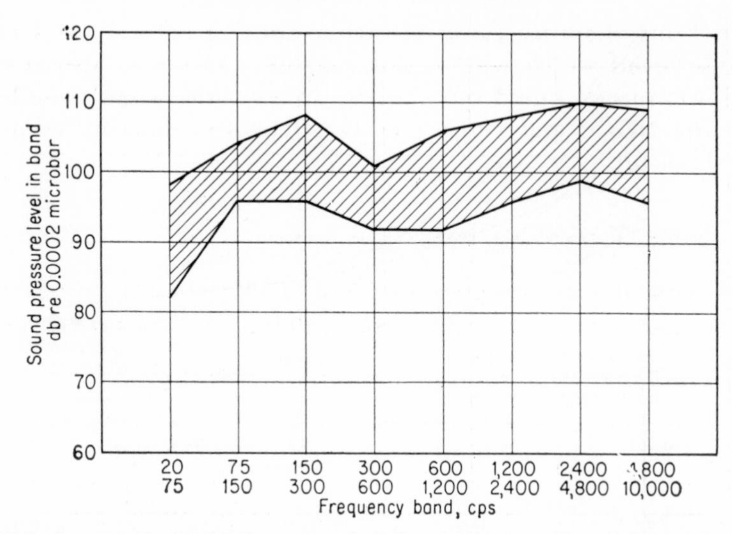

FIG. 22.5 Range of noise levels measured at the operator position for several riveting operations in an aircraft plant.

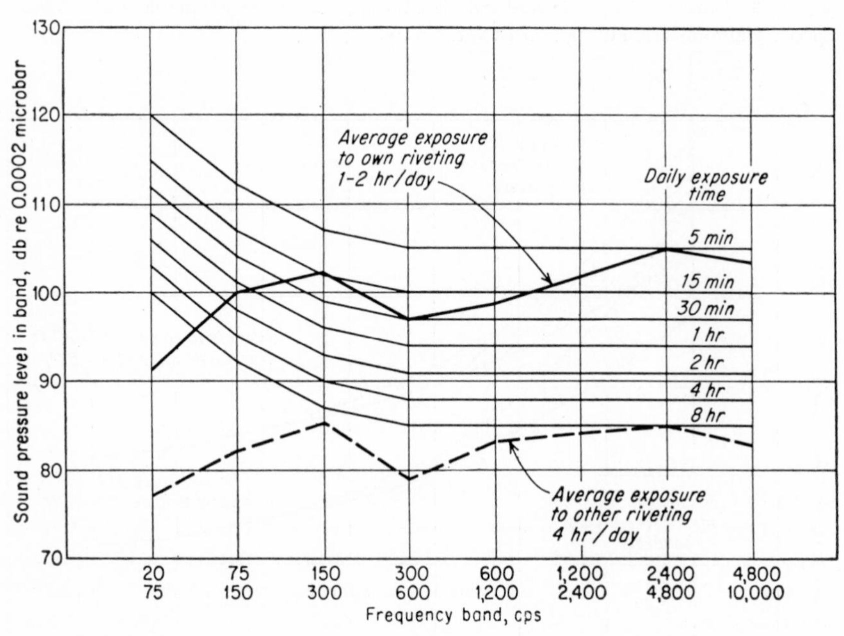

FIG. 22.6 Comparison of noise-exposure curves of riveting noise with criterion curves for recommended hearing-conservation programs.

shouted bits of speech communication, and Fig. 22.4 shows a range of noise levels which will permit reliable continuous speech communication using raised voice levels. In addition, when noise levels reach the NC–55 and NC–65 levels, telephone communication becomes difficult.

22.3 Some Typical Factory Noise Levels

The next few graphs give some typical noise levels measured in several work areas in various types of industries. Note some of the

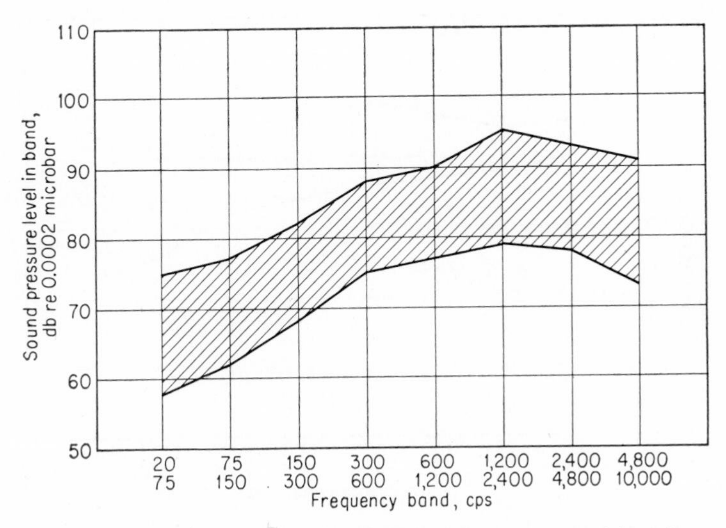

FIG. 22.7 Range of noise levels measured at several operator positions in a typical textile-weaving plant.

characteristic differences in the types of noise. The reader should refer to Chaps. 4 to 6 for information on noise measurement.

The impulsive, intermittent noise levels of Fig. 22.5 were measured in the riveting area of an aircraft plant. The riveters involved in these operations were exposed to their own riveting noise for approximately 10 to 20 per cent of the working day and to somewhat lower noise levels produced by their neighboring workers for approximately 50 per cent of the working day. Comparison of these noise levels with the criterion curve in Fig. 22.2 for 1 to 2 hr's daily exposure (10 to 20 per cent of the workday) reveals that a hearing-conservation program should be mandatory at that plant. Such a comparison is shown in Fig. 22.6.

The noise levels of Fig. 22.7 were measured at several operator positions in the weaving room of a textile plant. These noise levels are continuous throughout the entire workday. A hearing-preservation program would be recommended for this noise environment.

The noise levels at several locations in a rock-crushing and grinding plant are given in Fig. 22.8. Much of the processing in this particular plant is automatic and the employees are exposed to these noise levels only 5 to 10 per cent of the time. In addition, the cautious management of the plant uses the findings of periodic audiometric testing of employees as a guide in job assignments to various work areas. With the relatively low daily exposure time and the con-

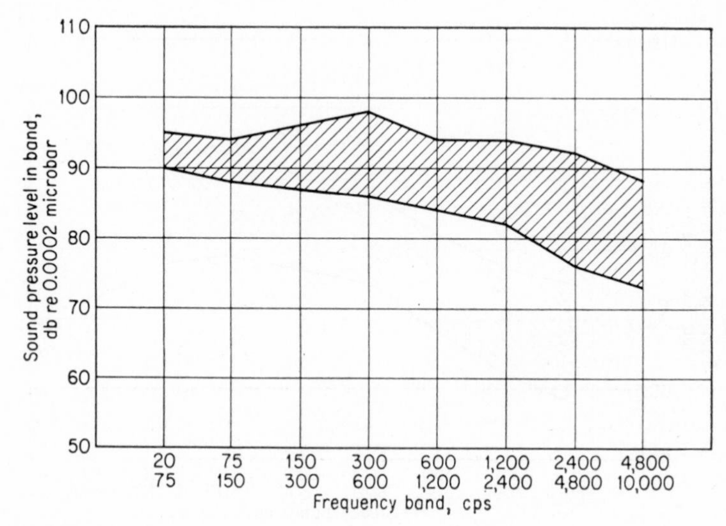

FIG. 22.8 Range of noise levels measured at several locations in a plant for crushing and grinding rocks.

scientious use of audiometric testing, this plant would provide little risk of noise-induced hearing damage to operating personnel.

Figure 22.9 shows the range of noise levels measured at the operator position of a wood-chipping machine when logs are being fed into the machine. When logs are not being fed to the machine, the high-speed cutting blades produce a siren-like single-frequency signal near 1,000 cps having a noise level of approximately 95 db. Thus, for the full work period, the operator is exposed to potentially hazardous noise levels.

The noise levels of Fig. 22.10 were measured at the operator position of a 60-ton multistage automatic punch press. This particular machine operates at a rate of three punchings per second for 60 to 80 per cent of the average workday. Several types of noise-level readings are given in the graph. The lower solid curve shows the av-

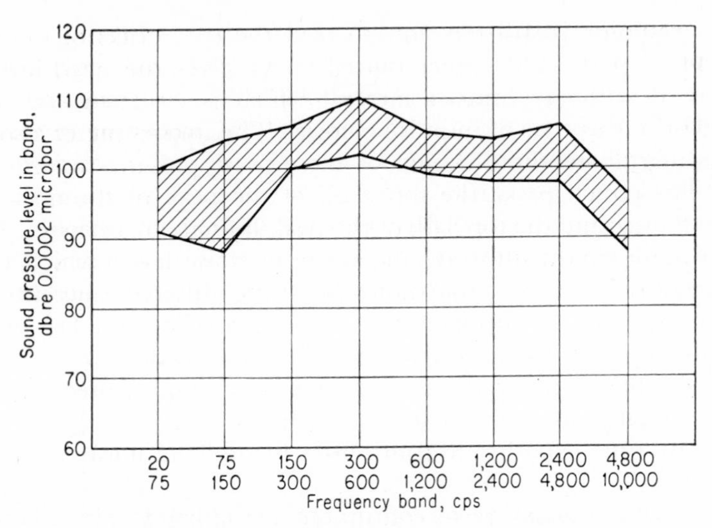

Fig. 22.9 Range of noise levels measured at the operator position of a wood-chipping machine in a wood-pulp plant.

erage values as read with the "slow" meter movement of an octave-band analyzer (OBA). The "slow" meter averages the noise over a 1- to 2-sec time interval and thus averages the noise level over several punching operations. The middle dashed curve represents the noise levels read with the "fast" meter movement of the OBA. The time constant of this movement is approximately ½ sec, but the

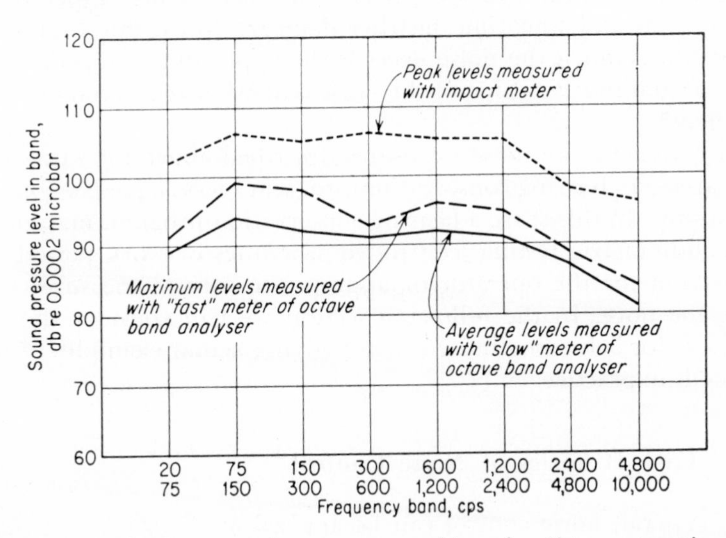

Fig. 22.10 Noise levels measured at operator position of a 60-ton automatic punch press in an automotive-parts plant.

meter response is still too slow to read the true peaks of the noise of each impact. The upper dotted curve gives the peak levels as measured with an "impact meter," which is designed specifically to measure this type of impulsive noise. The impact meter also tells something about the duration of these noise pulses. For example, with this punch press, the top 3 db of the peak of the noise lasts about 2 msec and the top 8 db of the peak lasts about 10 msec. These three noise curves illustrate the range of noise levels which might be measured for the same noise by using different measurement techniques. It is cautioned that hearing-preservation criteria for impulsive type noise have not yet been established.

In summary, the various noise exposures represented by these various plant noise environments include:

1. Full-time exposure to continuous broad-band noise (the textile plant)
2. Part-time exposure to continuous broad-band noise (the rock-crushing and -grinding plant)
3. Part-time exposure to intermittent noise (riveting in the aircraft plant)
4. Exposure to narrow-band noise (the whine of the wood-chipping machine without logs)
5. Exposure to impact noise (the noise of the punch press)

By comparing the noise levels measured in any shop or factory space with the noise criteria given in Figs. 22.1 to 22.4, it is possible to estimate the degree of severity of the noise as it affects speech communication and long-time hearing damage. Since this comparison reveals how much the noise exceeds the applicable criterion, it provides a quantitative value of the noise-reduction requirements of the problem.

In reviewing the noise exposures described above for various factory areas, a "hearing-conservation program" was suggested for some situations. In this sense a hearing-conservation program may encompass audiometric testing, controlled schedules of work periods, ear protection for the operator or appropriate control measures to reduce the noise. In the following section we will consider the basic methods for achieving noise control, giving actual examples of these methods in practice.

22.4 Case Histories of Noise Control

In general, noise control can be applied at the source, at the receiver, or in the transmission paths between the source and the receiver. Noise-control measures include:

1. Noise reduction design in the noise source
2. Use of vibration isolation methods
3. Use of enclosures or barriers
4. Use of mufflers or sound absorption devices in open passages
5. Suitable location of noise sources in the surroundings

We will discuss some typical examples of some of these noise-control measures.

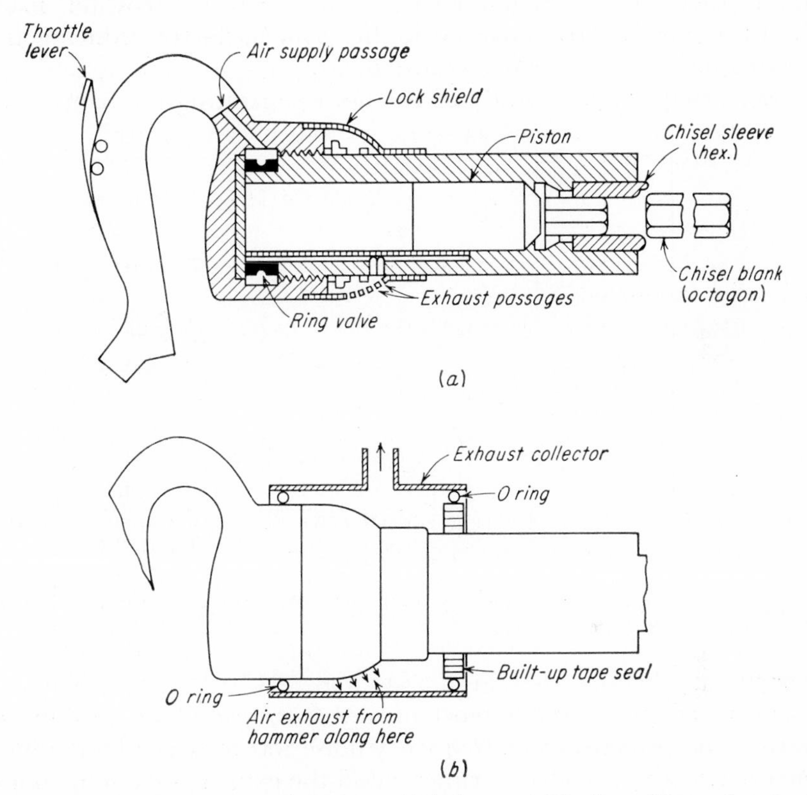

FIG. 22.11 Section of pneumatic hammer (*a*) before and (*b*) after adding exhaust collector ring to reduce air exhaust noise. (*After Plunkett.*[2])

Noise Reduction in Design. It makes good sense to try to reduce noise at its source before it has a chance to spread out to the surrounding areas. In many modern mechanical devices, reduced noise output gives greater public acceptance and sales appeal. This incentive for quieter machinery is beginning to be evident in factory equipment.

One power tool company specializing in air-driven screw drivers and nut setters has made design changes in its products to muffle the whining exhaust noise of the air motors.

A riveting machine company has designed a stationary spinning riveter which does away with the impact noise of a conventional hammer riveter.

One large plant has converted its gasoline engine fork lifts to propane operation with a significant reduction in noise inside the plant. Rubber tires are replacing metal rim wheels on dollies and platform trucks used for inside plant transportation. Another plant has replaced wood-block flooring with concrete strips for main shop routes taken by carts and dollies as a means of reducing the rattling noise of in-plant parts movements.

One group[2] made a study of the noise-generating mechanisms of

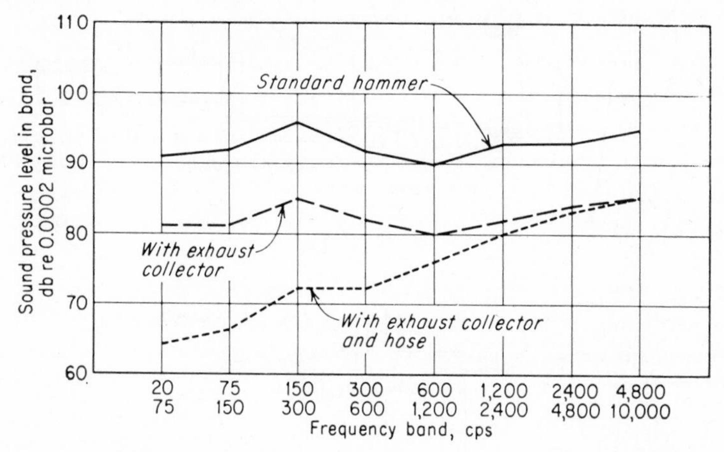

FIG. 22.12 Noise reduction achieved by addition of exhaust collector ring and hose to experimental pneumatic hammer.

a pneumatic hammer (see Figs. 22.11 and 22.12). With the hammer operating against a sample block of rubber instead of a typical metal casting, the principal high-frequency noise was found to be the hissing exhaust air. A collector ring around the exhaust gave some noise reduction, and a long rubber tube connected to the outlet of the collector ring further reduced the noise at the operator position. Although this was only an experimental study, it illustrates a systematic approach toward the noise reduction of a particular mechanical device.

Studies in reduction of noise radiated by chipping on certain large castings have been reported [3] by the industrial hygiene engineer of Bethlehem Steel Company. Pneumatic chipping on large propeller castings produces loud ringing sounds. Several thousand pounds of metal must sometimes be removed from a single casting; this may require several hundred hours of chipping. It was ultimately found that vibration dampers clamped around the edges of the propeller blades could reduce the noise down to that of the chipping tool itself. The vibration dampers were modified C clamps in which 6-in.-square steel plates surfaced with an industrial felt material were clamped tightly on either side of the propeller blade's edge. The Bethlehem engineer gives this simple clue to the possible effectiveness of vibration damping:

> If striking the structure which is ready to be chipped produces a ringing sound afterward, then the noise produced by chipping on that casting can be reduced by vibration damping. The longer a casting rings after being struck, the more noise it will produce during chipping and the greater the noise reduction which can be obtained through damping.

Another group[4] reports that "flame gouging" is more efficient than chipping in removing excess metal in some of their operations, and it produces 20 db less noise. They also find that high-speed burring is more effective than chipping in another operation, and this results in approximately 15 db less noise.

Another company has introduced an acoustically treated stock tube[5] for use with automatic screw machines. For screw machines, bar or rod stock rotates at high speed inside the stock tube, and the rattling and whipping of the stock against the stock tube create considerable high-frequency noise. The conventional stock tube is an undamped steel tube which radiates this noise directly into the shop area. A shop full of these automatic screw machines can produce noise levels throughout the shop ranging from 95 to 105 db in the high-frequency octave bands.

Figure 22.13 shows that the acoustically treated stock tube contains a helically wound metal strip as an internal lining. This strip is separated from the outside tube with a pad of textile webbing. The webbing serves to damp the vibrations of the helical lining and to mechanically isolate the lining from the outside tube.

This treated stock tube is said to reduce stock-tube noise by 10 to 30 db in the high-frequency bands. Actually, total noise reductions in factory installations will depend on how much other high-frequency noise is generated by the screw machines.

One company in the foundry industry reduced tumbling barrel noise 10 to 30 db in the high-frequency bands by placing a rubber lining inside the tumbling barrel walls.[6] The cost was about $200 per liner installation, but noise reduction was not the only gain. Maintenance cost on the barrels has been reduced, and the effectiveness of the tumbling process has not been impaired.

All the above examples illustrate noise control designed at the source. Such designs frequently result in new ways of doing things, as compared with conventional methods. For example, Fig. 22.14 shows a large shear which is operated hydraulically instead of by the more conventional mechanical gear drive. The manufacturer claims that this design essentially eliminates the impact noise normally associated with the operation of large shears, and in addition the new design reduces damage and wear on the cutting blade. This is an example of noise-control design and improved mechanical performance working together toward a better product.

It is, of course, highly desirable to design and engineer a new

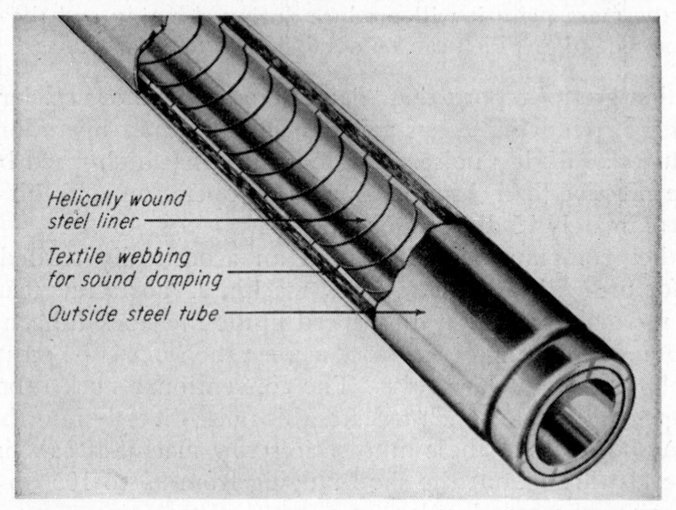

Fig. 22.13 Cutaway view of special noise-reducing stock tube for screw machines.

mechanical device to be relatively quiet in operation. Our factories are filled, however, with machines and devices which are already designed, manufactured, sold, and in use; and many of them are noisy. Thus, let us see how noise control can be applied to existing noisy factory machinery.

Vibration Isolation. Let us consider first the use of vibration-isolation methods to achieve noise reduction. An important aspect of diagnosing a noise or vibration problem is the separation of *air-borne* and *structure-borne* noise or vibration. Transmission of noise by these two types of paths can be evaluated with the use of noise- and vibration-measurement instruments and techniques. As the term implies, *air-borne* noise is noise radiated directly into the air by the noise source. *Structure-borne* noise is the transmission through supporting structures of "forced vibrations" caused by the vibrating machinery. Just as a vibrating tuning fork does not sound very loud until it is touched to a table or some other "sounding board," so a piece of vibrating machinery may not actually be very noisy until it is clamped to a floor or other supporting structure. Then the vibration of the machine sets the floor into motion, and the floor becomes the sounding board.

Fig. 22.14 Hydraulically operated shear has less impact noise than conventional mechanical drive shear. (*Courtesy of Pacific Industrial Manufacturing Company.*)

To solve structure-borne noise problems, it is customary to try to separate the vibrating source from its potential sounding board. A properly designed resilient mount, which takes into account the loading of the machine and the principal resonating frequencies and modes of vibration of the machine, can appreciably reduce the vibration transmitted from a vibrating source to its structural support. An improperly designed or improperly used mount, however, may give little improvement and under some circumstances can actually increase the amount of vibration transmitted to the mounting surface. Standard vibration-isolation mounts are available for many types of machines, and specially engineered mounts can be designed for unusual vibration problems. Refer to Chap. 18 for detailed design material on vibration isolation.

Recently a special mount was designed for use under the printing

FIG. 22.15 Vibration isolation of printing presses required to reduce noise in office on floor below.

press shown in Fig. 22.15.[7] When a room full of these printing presses were in operation, office personnel located on the floor beneath the printing shop were disturbed by the noise. It was determined by measurements that the noise was structure-borne, and appropriate vibration-isolation mounts were designed and tested at the printing shop. The noise levels in the office below were reduced the desired 10 to 15 db. Yet, the noise levels in the printing shop were not reduced a bit. This is an example of air-borne noise controlling the shop and structure-borne noise controlling the office. The vibration mounts, of course, were not expected to reduce the air-borne noise output of the press.

It is also important in the vibration isolation of a machine that there be no rigid mechanical ties between the machine and the building structure which would reduce the effectiveness of the mounting system. This means that flexible connections must be inserted in all pipes and conduits leading from the vibrating machine, and sometimes even the connecting pipes must be given special attention.

Figure 22.16 illustrates a noise problem in which a vibration-isolated reciprocating compressor was generating vibrations in the floor

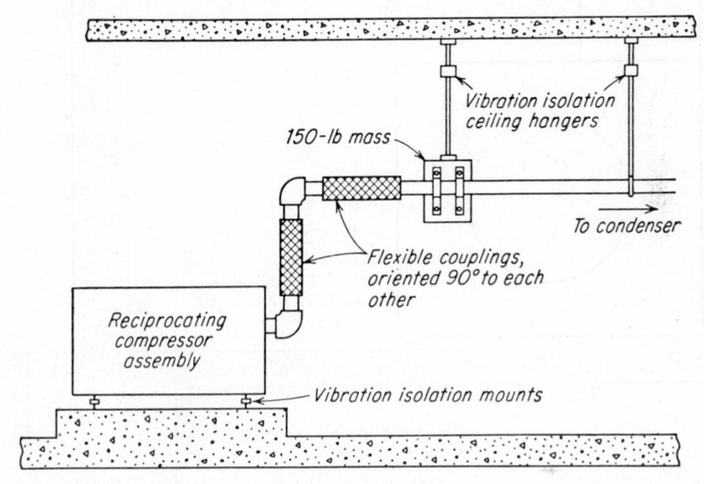

FIG. 22.16 Vibration-isolation mounting of the piping connected to a reciprocating compressor. Install in all pipes connecting to compressor.

overhead because the compressor piping was originally mounted rigidly to the overhead floor slab. To reduce the floor vibrations, the following steps were taken:

1. In the pipeline, two flexible couplings were inserted, one oriented 90° to the other.

2. A heavy mass of 150 lb was attached to the pipe beyond the second flexible coupling in order to increase the inertia of the pipe.

3. The pipe and the mass were supported from the floor slab by means of vibration-isolation ceiling hangers.

In many structure-borne vibration problems it is desirable to isolate one part of a building from the remainder of the building. Figure 22.17 shows a vibration break between the basement floor slab of a building and the columns and walls of the building. This structure break was required to reduce the transmission of textile machinery noise and vibration into the building proper.

Adequate structure breaks are even more important when heavy impact-type machinery is involved. In one problem the operation of a high-speed automatic punch press resulted in surface-grinding imperfections at a machine nearly 100 ft away. It was found that the punch press was mounted on a massive concrete inertia block imbedded in the ground. The concrete floor slab for the entire plant was one large poured section with no expansion joints and with no

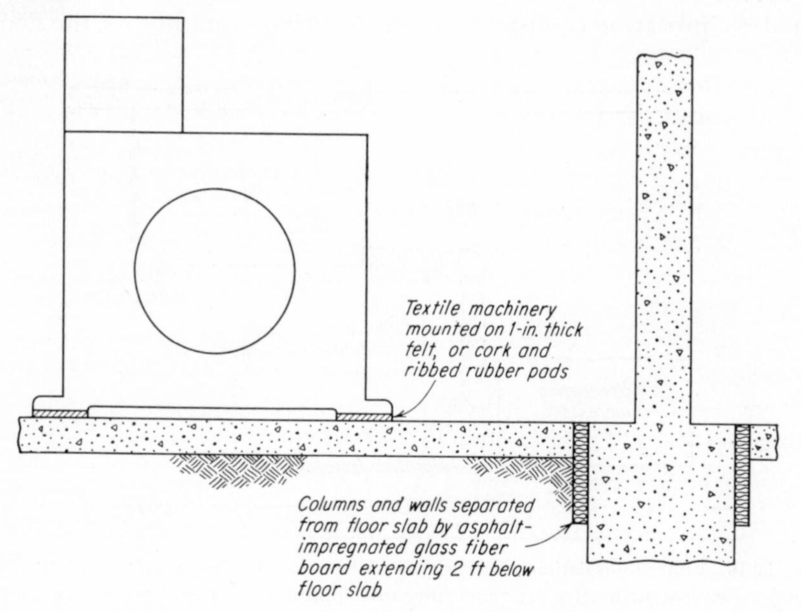

FIG. 22.17 Vibration break in building structure reduces transmission of noise and vibration to other parts of the building.

structure breaks between the inertia block supporting the punch press and the pad supporting the surface grinder. A saw-cut in the floor slab around the punch press and another one around the surface grinder was all that was required to eliminate the vibration-induced imperfections at the grinding machine.

Thus, with these examples we note the need for proper and complete vibration isolation of machinery when we have structure-borne noise problems. We offer some suggestions on vibration isolation of a few typical pieces of machinery. These are of a very general nature and must not be interpreted as solutions for every type of vibration problem.

High-speed rotational equipment, such as motor-generator sets and centrifugal-type compressors, have to be dynamically balanced by the manufacturer in order to perform well at high-rotation speeds. Thus, in general, these types of equipment usually require fairly simple vibration-isolation techniques—depending, of course, on the nature of the noise or vibration problem. For rotation speeds of about 600 rpm or higher a vibration-isolation mounting could consist of a group of rubber-in-shear mounts or steel spring mounts. Conventionally, rubber-in-shear mounts are designed to have approximately ¼-in. static deflection for the rated load. Steel springs may be selected to have any desired static deflection, but ½ to 1 in. would not be unusual for this type of installation. Steel springs transmit some high-frequency vibration, so it is desirable to place ribbed-rubber pads in series with the steel springs in order to provide high-frequency isolation.

For large air-moving fans and reciprocating pumps or compressors, whose shaft speeds fall in the range of about 600 to 1,800 rpm, it is sometimes necessary to install the machine on a heavy inertia block and then vibration-mount the assembly somewhat as indicated above. Depending on the nature of the problem, the weight of the inertia block might be one to four times that of the supported load.

Shop equipment containing numerous gears, cams, shuttles, and ratchets, such as automatic screw machines and textile weaving machines, generate considerable high-frequency noise and vibration. To isolate this type of high-frequency vibration, when there is no significant amount of low-frequency vibration, it is frequently sufficient to place the equipment on pads of cork or felt or ribbed rubber. In using any of these pads it is important to follow the manufacturers' recommended surface-loading rates. When these recommended loading rates are not followed, excessively large areas of pad materials are frequently used, resulting in isolation mountings having higher values of stiffness than desired. Most of these pad materials perform well when loaded at about 30 to 50 lb/in.^{2}

In heavy-machine shops, large punch presses and drop hammers typify the most serious kinds of vibration problem. Usually these large impact machines are mounted rigidly to very massive concrete inertia blocks having weights many times greater than the weight of the supported machines. These inertia blocks are usually isolated from the earth and from the building structure with large wood timbers and with thick pads of cork. It is cautioned that cork loses much of its resilience when immersed in water, so it is desirable to provide suitable drainage around these pads.

When designing vibration-isolation mountings for machines in critical installations, it is desirable to try to locate the resilient mounts in a plane which contains the center of gravity of the mounted assembly. It is also desirable to locate the mounts laterally as far away as possible from the center of the machine.

It must be kept in mind that vibration isolation is normally required as a means of reducing structure-borne noise transmitted by structural paths away from the machine. Thus, this type of isolation usually results in little or no reduction in air-borne noise radiated by the machine itself. To reduce air-borne noise we use other types of noise control.

Barriers and Enclosures. When it is not possible or practical to reduce the amount of air-borne noise *generated within* a machine, it is frequently desirable to place the machine in an enclosure or behind a barrier in order to reduce the amount of noise radiated. We use the term "barrier" to mean a partial-height wall or a partial enclosure which does not completely contain the noise but may change the amount of noise transmitted in certain directions. Figure 22.18 shows a number of examples of different types of barriers which have been devised [8] for certain factory machines and processes. Notice that these barriers offer noise shielding in certain directions and relatively free flow of stock and operating personnel in other directions. Barriers such as these usually afford adequate accessibility to the machines under most normal conditions.

The acoustic characteristics of barriers or partial enclosures vary considerably. In general, if a single large-area barrier wall with no openings is placed between the noise source and the observer, one might expect the order of 2- to 5-db noise reduction in the low-frequency bands and possibly 10- to 15-db noise reduction in the high-frequency bands. These are only rough approximations because dimensions, distances, room configurations, and the use of acoustic absorption materials on the barrier wall and in the room greatly influence the actual values of noise reduction.

A two-sided or three-sided barrier, with or without a top, will give still larger amounts of noise reduction opposite the closed walls of the barrier. However, since these barriers usually change the over-all radiation pattern of the noise source, they may produce higher noise levels in front of the opening of the barrier than if there were no barrier at all. This effect can be reduced if the interior walls of the barrier are covered with acoustic absorption material. For well-designed partial enclosures which provide easy accessibility to the machine inside, one should be able to achieve the order of 5- to 10-db noise reduction in the low-frequency bands and approximately 20-db noise reduction in the high-frequency bands in the direction of the closed walls of the barrier. Partial enclosures, such as the open-front acoustically lined telephone booth (Fig. 22.19), may also be used to achieve a relatively quiet environment in a noisy area.

Much larger amounts of noise reduction can be achieved with com-

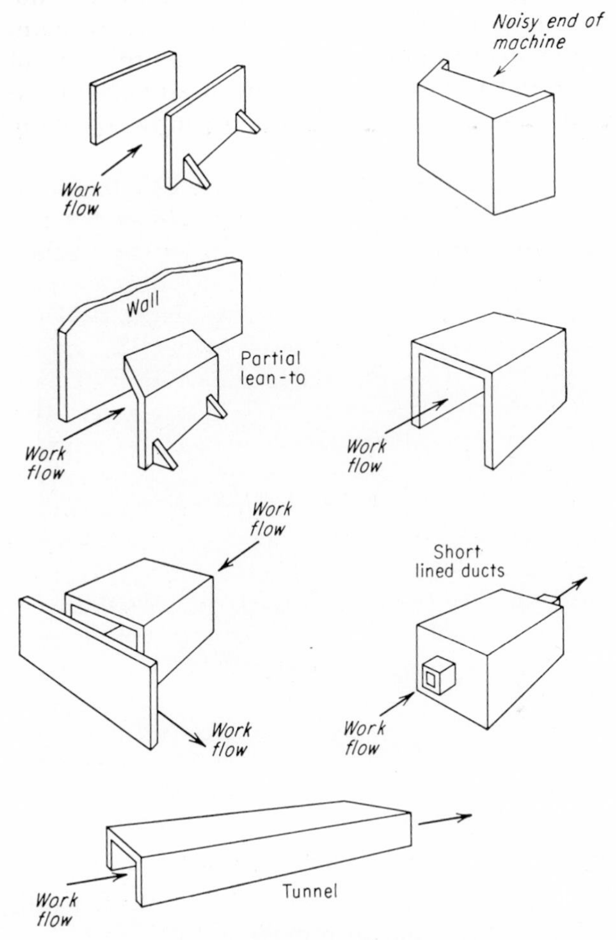

FIG. 22.18 Typical barriers for partial noise control in work areas. (*After Callaway.*[8])

plete enclosures. Complete enclosures for machines are of two general types. One type is a close-fitting box around the machine such that the operator performs his normal work outside the box and thus is not subjected to the high noise levels of the machine. Figure 22.20 shows a multislide punch press in its original installation in a factory shop at IBM. The noise from the punch press was sufficiently objectionable to require an enclosure. Figure 22.21 shows an enclosure for the machine which was designed and installed at IBM.[9] The operator works in a quieter working area, and yet he can maintain general control of the machine through the various observation windows and access panels.

Complete form-fitting enclosures such as the one shown in Fig. 22.21 are usually made of sheet steel. This one is made of approximately 18-gauge steel and provides noise reduction of approximately 10 to 15 db in the low-frequency bands and 25 db in the high-frequency bands. Special attention was given to this enclosure to pro-

FIG. 22.19 Acoustically absorbent partial enclosure for telephone booth provides localized quiet area in noisy surroundings. (*Courtesy of Erdle Perforating Company, Inc.*)

vide good seals of the windows and panels, to avoid rattling of sheet-metal sections, to provide forced ventilation to the machine through acoustically lined ducts, and to provide acoustic seals for the entry of the strip stock which was fed to the machine.

A significant improvement in the noise reduction provided by metallic covers can frequently be achieved by the addition of a vibration damping material to the sheet metal. Just as automobile under-coating is sprayed on the bottom of an automobile to reduce "road rumble," so an application of this mastic material will reduce the vibration and noise transmission of a sheet of metal. A handy rule of thumb is to spray the mastic to a thickness approximately twice that of the metal under treatment.

FIG. 22.20 Multislide punch press before addition of acoustic enclosure. (*After Engstrom.*[9])

FIG. 22.21 Multislide punch press inside acoustic enclosure.

Metal damping tape is a new product which has been developed to give a clean, fireproof method of applying vibration damping material. This thin aluminum tape is coated with an adhesive film which sticks firmly to the metal to be damped and which at the same time provides the damping action. A few layers of this tape can give reductions of possibly 3 to 6 db in the noise transmission of a sheet metal panel (see also Chap. 13).

If the size of the machine and the working area do not permit close-fitting enclosures such as the one described above, it may be necessary to house the machine (and sometimes also the operator) in a room of its own in order to achieve lower noise levels for other occupants of the area. For these types of complete enclosures we are usually interested in gaining large amounts of noise reduction, so we are interested in the "acoustic transmission loss" of the wall of the enclosure. The TL of various types of walls has been discussed in Chaps. 13 and 14.

When a noise source is enclosed, the noise levels inside the enclosure are generally higher than they would be at the same distance from the source if there were no enclosure (see Chap. 10). Thus, if a machine operator works inside the enclosure with his machine, consideration should be given to this increase in noise levels. This leads us to the use of acoustic absorption materials as a noise-control treatment.

Acoustic Absorption Devices. In Chaps. 10 and 11 the use of absorption materials in rooms has been discussed. In this section we will only summarize briefly the pertinent uses of this material in shops and near machine operators.

Acoustic absorption material may be placed on the ceiling and sidewalls of a shop to reduce the reverberant noise levels in the room. But if the machine operator must remain very close to his machine, say within 2 or 3 ft of it, he will receive the direct sound from the machine, and the use of acoustic tile on the ceiling will give him no significant noise reduction. However, if occupants of the room are located 10 ft or more from the machine, they hear not only the direct sound but also many reflected paths of sound which bounce around the room. For these occupants, the use of acoustic absorption material on the ceiling and sidewalls of the room will yield some noise reduction. The amount of reduction depends on many factors (distance from noise source, size and shape of the room, type and amount of absorption material, etc.), but as an order of magnitude one might achieve approximately 3- to 5-db reduction of low-frequency noise and 5- to 8-db reduction of high-frequency noise for an application of standard 1-in.-thick perforated acoustic tile on the entire ceiling and possibly on some of the sidewall area of a room. Acoustic absorption material has an important place in noise control and in architectural acoustics, but it is cautioned that perforated tile on the ceiling is not the solution to all noise problems.

It was mentioned earlier that a machine operator located only 2 or 3 ft from a noisy machine would benefit very little from absorption material located quite a distance away. Several new types of sound absorption units have been devised during recent years in an

FIG. 22.22 Functional sound absorbers can be located close to noise sources. (*Courtesy of Elof Hansson, Inc., and Sonosorber Corporation.*)

attempt to give more noise reduction near at hand. Figure 22.22 shows a shop equipped with one type of "functional sound absorber." These units can be clustered near a noisy machine or can be distributed at will in any desired pattern to obtain lower noise levels within the shop. Compared on the basis of equal total exposed surface areas, the functional sound absorbers have slightly higher noise-reduction coefficients than conventional acoustic tiles cemented directly to ceilings and walls.

Acoustic mufflers are another type of noise-control treatment which may use acoustic absorption material. In Chaps. 16 and 17 mufflers were described and evaluated in detail. Here we wish to remind the reader that mufflers may serve in other ways than merely reducing the noise of internal combustion engines and the like. A muffler or some type of sound-attenuating device may be required for any air opening or passageway between a noisy area and a quiet area. The opening may be a passageway for ventilation, for air, gas, or steam inlets or outlets, for movement of stock into a machine or finished parts out of a machine, or even for personnel movement between areas.

The January, 1956, issue of *Noise Control* quotes one interesting case history of a muffler problem taken from *Industry and Power* of August, 1955. Several large reciprocating compressors were connected together to a common header in the compressed-air supply system for an automobile plant. When two particular compressors were in operation at the same time, large amplitude vibrations were set up.

> Soon after operation began, the vibration caused welds in the piping to crystallize, resulting in pipe breakage.
>
> To prevent further breakage, the engineers rigidly anchored the header by welding it to the building columns. This transferred the vibration to the building walls.
>
> Fluorescent lights fell to the floor and bricks worked loose from the walls. Pipe anchors snapped, and welds in structural steel components broke. Then large cracks began to appear in the walls in which the header was anchored. The building itself became endangered.

The solution to this problem was the installation of snubbers (mufflers) between the output lines of the compressors and the header. "The snubbers effectively reduced the pressure pulsations from the compressors and eliminated the oscillations within the header which were the actual source of the excessive vibrations."

Noise Control by Location. Finally, we mention briefly noise control by suitable location of noise sources in the surroundings. Noise control by location can be considered both in the design of new layouts and in the use of existing layouts. A simple rule to try to follow is simply this: keep approximately equally noisy machines, processes, and work areas together, and separate particularly noisy and particularly quiet areas by buffer zones having intermediate noise levels.

Such an arrangement provides a minimum of annoyance or interference due to noise. Reasonable attention to building layout from an acoustical point of view will not eliminate all noise problems, but it will certainly minimize them and provide more favorable working conditions for large proportions of the personnel.

22.5 Summary

We conclude this discussion by pointing out that there is no fixed or stereotyped solution to noise problems. Each problem is solved on the basis of the amount of noise reduction required and the analysis of noise and vibration levels along the various paths from the source to the receiver. We point out that for many factory noise problems there are systematic engineering approaches and noise-

FIG. 22.23 Tumbling barrels are coated with vibration damping material and shielded by partial height barriers. (*Courtesy of Allis-Chalmers Manufacturing Company.*)

control devices which are currently available and are in reasonably widespread use.

To illustrate the varied nature of noise problems and to summarize partially the noise-control measures available, we note some of the work done in noise reduction at the La Crosse, Wisc., plant of Allis-Chalmers.[10] The results are not phenomenal, and the cost was not excessive; this merely illustrates an unpretentious but conscientious effort by the industrial hygiene group of that plant.

In the foundry cleaning room before noise treatment was started, an array of tumbling barrels, grinders, and grit blasters, and frequent loading and unloading of metal castings into carts and trays resulted in noise levels ranging from 80 to 102 db in all the frequency bands. The various steps taken to achieve noise reduction up to 20 db in some of the operator positions are shown in Figs. 22.23 to 22.25 and are described briefly as follows. The tumbling barrels were coated with a ½-in. layer of a rubber base sound deadener and were then shielded by partial height barriers. The grit blaster hoppers were lined with a three-layered construction of ¾ in. soft wood, ¾ in. oak and 10-guage perforated sheet steel to reduce the impact noise of parts dropped into the hopper. All metal trays and carts were sprayed

Fig. 22.24 Hopper and tray are lined and coated with vibration damping material.

on all outside surfaces with ⅛-in. to ¼-in. thicknesses of the sound-deadener material, and a 2-in. layer of sand in the bottom of each tray was used to reduce the impact noise of castings falling into the trays. Vibration-isolation pads were placed under the tumblers to reduce noise and vibration transmission to the floor slab, and a vibration-isolated tool rest was designed for use with the grinding machines. Glass-fiber acoustic baffles were hung vertically from the ceiling to achieve significant reduction of the reverberant noise about the room. Each of these steps provided only a few decibels of noise reduction, but the combined effect has been a significant improvement in the working environment. Many other plants have done as well with the ingenious application of basic noise-control principles.

We conclude this chapter with three reminders. First, if you have to shout to be heard 1 to 2 ft away in your shop or plant, you probably have sufficiently high noise levels to produce hearing damage for some workers. Second, even though noise-induced hearing damage

FIG. 22.25 Tool rests are vibration-isolated from grinders and ceiling has sound-absorbent glass-fiber baffles.

is a long-time process, it is not wise to wait a long time before doing something about it. Third, almost every noise problem has a practical and economical solution.

REFERENCES

1. Knudsen, Vern D.: Noise, the Bane of Hearing, *Noise Control,* vol. 1, no. 3, p. 11, May, 1955.

2. Plunkett, R.: Noise Reduction of Pneumatic Hammers, *Noise Control,* vol. 1, no. 1, p. 78, January, 1955.

3. Botsford, James H.: Noise Reduction at Chipping Operations on Large Castings, *Facts on Factory Noise,* no. 2, p. 15, Associated Industries of New York State, Inc., 1956.

4. Preussman, William: An Approach to the Noise Problem in a Large Machine Shop, *Facts on Factory Noise,* no. 2, p. 21, Associated Industries of New York State, Inc., 1956.

5. Schweitzer, B. J.: A Silent Stock Tube for Automatic Screw Machines, *Noise Control,* vol. 2, no. 2, p. 14, March, 1956.

6. Lake, James W.: Tumbling Barrel Noise Control, *Facts on Factory Noise,* no. 3, p. 44, Associated Industries of New York State, Inc., 1958.

7. Miller, L. N., and I. Dyer: Printing Machine Isolation, *Noise Control,* vol. 4, no. 4, p. 21, July, 1958.

8. Callaway, D. B.: Design of Noise Control Structures, *Noise Control,* vol. 1, no. 5, p. 49, September, 1955.

9. Engstrom, J. R.: Noise Reduction by Covers, *Noise Control,* vol 1, no. 2, p. 19, March, 1955.

10. Less Racket at Little Cost, *Factory Management and Maintenance,* vol. 114, no. 6, p. 108, June, 1956.

SUGGESTIONS FOR FURTHER READING

Machine and Shop Quieting

Facts on Factory Noise are the published proceedings of annual meetings of the Associated Industries of New York State, Inc. The Proceedings are also published in the *Am. Ind. Hyg. Assoc. Quart.,* or are available as reprints from Associated Industries of New York State, Inc., 30 Lodge St., Albany, N.Y. These booklets describe numerous industrial noise-control treatments devised by various hygiene, safety, and plant engineers.

Industrial Noise Manual, published by the American Industrial Hygiene Association, 1958, a valuable compilation of information on noise, noise measurements, and industrial noise control.

Noise Control, published by the Acoustical Society of America, is an important bi-monthly source of new and practical industrial noise-control treatments.

Harris, C. M. (ed.): "Handbook of Noise Control," McGraw-Hill Book Company, Inc., New York, 1957. This is a recent publication which will be found helpful in the analysis and solution of many noise problems.

Karplus, H. B., and G. L. Bonvallet: A Noise Survey of Manufacturing Industries, *Am. Ind. Hyg. Assoc. Quart.,* vol. 14, p. 235, December, 1953. Some of this material is reviewed briefly by Bonvallet in Noise Sources in Modern Industry, *Noise Control,* vol. 1, no. 3, p. 30, May, 1955.

Engstrom, John R.: Riveting Hammer and Paper Shredder, *Noise Control,* vol. 2, no. 2, p. 18, March, 1956.

Gould, H. A., and L. F. Yerges: Functional Sound Absorber Installations in Two Industrial Plants, *Noise Control,* vol. 2, no. 2, p. 26, March, 1956.

"Handbook of Acoustic Noise Control," vols. I and II, and suppl. I, *WADC Tech. Rept.* 52–204, Wright-Patterson Air Force Base, 1955.

Miller, L. N.: series of articles on noise and noise control, *Safety Maintenance and Production,* vol. III, June, July, and August, 1956.

Chapter 23
CASE HISTORIES OF NOISE CONTROL IN OFFICE BUILDINGS AND HOMES

Laymon N. Miller

23.1 Introduction

With the increased use of lightweight construction in buildings, the provision of more wide open living areas in homes, the increased use of mechanical devices in offices and homes, and an increased noise-consciousness on the part of office workers and home owners, the disturbing effects of noise have been given more attention during recent years and the application of noise-control measures has advanced from an art to a reasonably predictable engineering know-how. Although some unusual noise problems may have difficult or economically unfeasible solutions, most current noise problems have quite practical solutions involving standard materials and reliable engineering practices.

In this chapter we wish to review a number of noise problems in offices and homes which have been encountered during the last few years and to indicate some typical noise-control measures applicable to these problems.

23.2 Noise-level Criteria for Offices

In order to determine the amount of noise reduction desired, we recall the noise criteria for offices as discussed in Chap. 20. The NC curves of Fig. 20.3 and the office areas listed in Table 20.1 describe the types of noise environments generally considered desirable for various office or functional areas. The NCA curves given in Fig. 20.4 are alternate noise-level criteria used in certain situations where an excess of low-frequency noise is present and it is considered uneconomical or impractical to attempt to achieve the NC criteria.

A special comment on criteria is given in regard to mechanical equipment noise as heard in private and executive offices and conference rooms. We usually recommend approximately NC-25 to NC-30 background-noise-level criteria for mechanical equipment noise as heard in a nearby critical office area, even though somewhat higher (say NC-30 to NC-35) criteria for air supply noise or city traffic noise may be specified for that office area. The masking noise provided by an air diffuser or under-the-window air supply unit, for example, is usually a steady, unobtrusive, almost unidentifiable noise. Mechanical equipment noise, however, may contain such disturbing noises as the low-frequency hum of transformers, or the low-frequency rumble of ventilation fans, or the "grinding" sounds of pumps, or the high-pitched whine of refrigeration compressors or elevator motor-generator sets. We believe these noises, which are identifiable and objectionable, should be somewhat suppressed below the general masking noise levels of the office areas. Hence, we strive for NC-25 to NC-30 criteria for mechanical equipment noise, even in the presence of NC-30 to NC-35 criteria for masking noise backgrounds.

23.3 Types of Office-noise Sources

Noise problems in offices can usually be attributed to four general groupings of noise sources:

1. Industrial noise, such as that associated with factory machinery and manufacturing processes, which arises from areas located near offices
2. Vehicular noise, such as automobile, truck, subway, and train traffic noise, and aircraft noise
3. Office machinery and equipment, such as telephones, typewriters, business machines, and electronic data computing equipment
4. Air-conditioning systems, which include cooling towers, compressors, pumps, refrigeration machines, boilers, fans, ducts, air mixers, air valves, and diffusers

It would be difficult to give exhaustive discussions of all the problems and solutions encountered in these four categories; but we shall try to give enough case histories to indicate in a broad manner the types of solutions available to these problems. In many cases the solutions appear ridiculously simple to the trained acoustician, but the original problems are usually sufficiently complex that a moderate amount of detective work is required to determine the true sources and paths of noise.

Turning various pieces of equipment "on" and "off" is the easiest way to identify the most annoying noise sources (with or without the aid of noise-measurement and analysis equipment); and the noise paths must be correctly identified and separated into air-borne paths and structure-borne paths in order to know what kind of treatment must be applied. Acoustic tile on the ceiling will not have much effect on subway-induced vibration transmitted into a building through underground footings, columns, and walls. Similarly, vibration mounts under a circular saw will not have much effect on the control of the air-borne noise radiated by the whirring blade.

In the sections to follow, we will discuss various noise aspects of the four different types of noise sources enumerated above.

23.4 Industrial Noise in Offices

Much of our work is concerned with the design of new buildings to assure adequate acoustic isolation between noisy and quiet areas. We are frequently involved, however, with noise problems in existing buildings, many of which were not built to serve some of the present-day purposes for which they are being used. Acoustically this brings on problems which occasionally are simple to solve; but more often, the structures just do not afford practical solutions. Let us look at a few examples.

One office area was plagued by a repetitive but irregular thumping from a machine shop on the floor below. The noise had all the characteristic sounds of a heavy punch press, and the office occupants in the wood-floor building began to believe that nothing could be done to control such a tremendous vibration source. Ultimately, an investigation led to the discovery that a machine shop worker on the floor below was doing a manual hammering operation on a pedestal that was bolted directly to a column leading to the office floor above. The hammering pedestal was removed from the column and was replaced by a massive concrete inertia block which was set on felt pads on the wood floor. The hammering noise as heard upstairs was reduced to insignificance.

In another wood-floor building, an office on the floor above a shoe factory was being remodeled for rental. The building owner was distressed at the factory noise (and odors) which seemed to penetrate the office. A brief inspection of the site indicated that this was entirely an air-borne noise problem: large, open clearance holes around all the radiator pipes connected between the two floors. Filling up all the holes by packing the spaces around the pipes resulted in sufficient reduction of noise. A previously contemplated suspended impervious ceiling in the factory area was not necessary.

Most noise problems in old wood-floor buildings are not so easily handled. Wood flooring is generally so lightweight and so resilient that heavy vibrating machinery cannot be well isolated with vibration-isolation mounts. Thus, the noise of impact-type machines is radiated by the flooring into areas on the floor below. The most satisfactory solution for riveting hammers, punch presses, and printing presses on lightweight floors is not to locate them over areas that must have low noise levels. Heavy platforms or bases upon which these machines could be isolation-mounted are usually so heavy or so complex that the building structure cannot safely carry the increased loading.

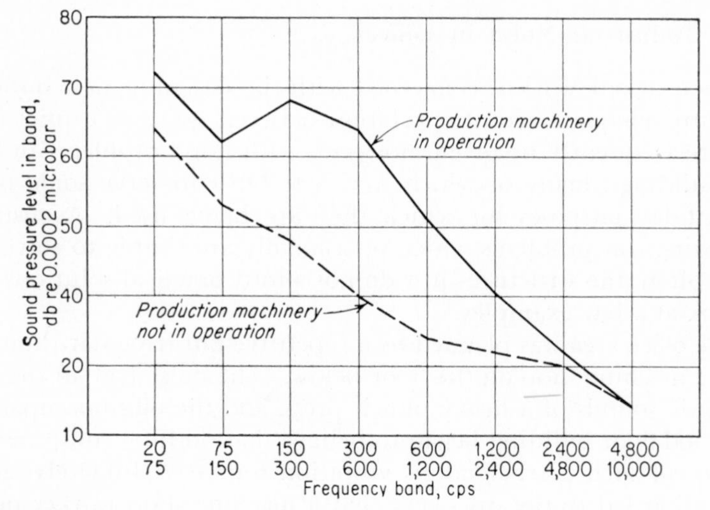

FIG. 23.1 Measured noise levels for an office area located immediately above a paper production plant.

In newer factory buildings with heavy dense concrete floors, there are fewer problems of floor-to-floor noise transmission. The problems that do arise, however, are usually quite severe. A printing shop located above an engineering office was described in Chap. 22. That problem was aggravated by the need for a vibration-isolation system that would not amplify the ½- to 2-cps driving frequency of the plattens, nor amplify the vibration of the already marginally loaded floor slab at its natural resonance in the region of 15 cps, yet give suitable vibration isolation for the audio frequencies and not interfere with register requirements for multicolor printing on the presses.

Another noise problem was found in an office floor above a paper production plant.[1] An 8-in.-thick concrete slab separated the two floors, yet occupants of the office area complained of annoyance and difficulty in communication. The noise levels in the office with the paper production equipment "on" and "off" are shown in Fig. 23.1. The noise levels in the production area were also measured; and knowing the approximate sound TL of the 8-in. concrete slab, it was possible to calculate the approximate noise levels in the office floor *if* the noise were primarily air-borne noise (that is, radiated by air paths from the production machinery, transmitted through the floor slab, and finally radiated by the floor slab into the office space). The resulting calculated noise levels were as much as 10 to 25 db lower

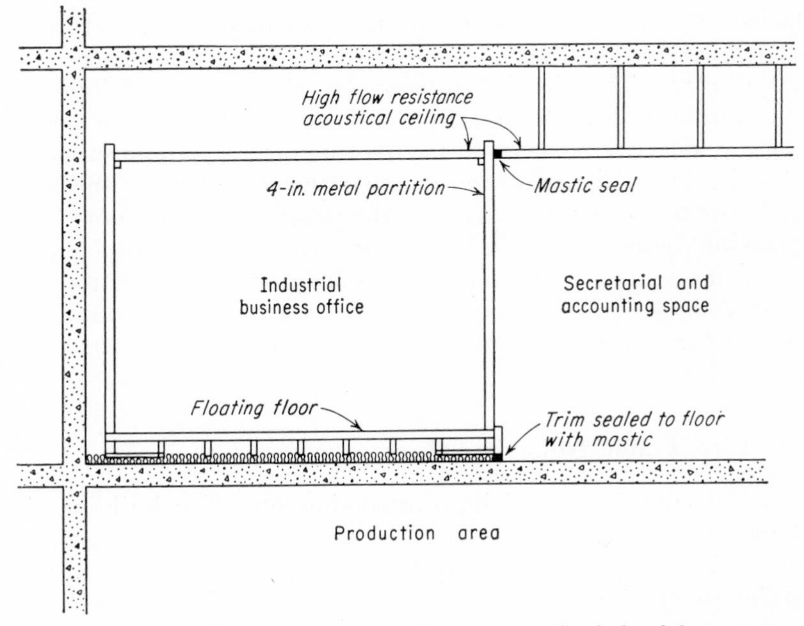

FIG. 23.2 Schematic arrangement of office enclosure vibration-isolated from surrounding structure.

than the noise levels actually measured in the office. Thus, it was concluded that the noise was not air-borne but was primarily structure-borne from the machinery to the office area.

The most direct solution would have been to provide vibration isolation for the machines following conventional techniques. In this case, however, the production line could not tolerate the misalignments that might result from the use of individual vibration mounts under the various machines, and extensive modifications and down-time of the machines for bases that would meet the alignment requirements were considered out of the question by the manufacturer.

The proposed solution is represented schematically in Fig. 23.2.

Each office is provided with a floating floor constructed of (1) wood sleepers resting on a resilient glass-fiber blanket, (2) rough flooring consisting of two layers of lapped and taped plywood, and (3) the finish flooring. The lightweight partitions rest entirely on this floor and do not contact the building proper. An acoustical ceiling of high flow resistance (having relatively high TL compared to lightweight, porous, perforated materials) is suspended from the partitions, so that each office becomes a closed box structurally independent of the remainder of the building. Note that this is an example of the use of a relatively lightweight structure to solve a noise problem. Electrical and ventilation connections to the offices are made through flexible conduits.

As illustrated by the above examples, the control of industrial noise transmission into office areas requires (1) the identification of the noise source (e.g., the man hammering on the column-supported pedestal), (2) the identification or separation of air-borne and structure-borne transmission paths (e.g., the paper-production noise), and (3) solving the respective air-borne and structure-borne problems by suitable measures (e.g., filling up the holes around the radiator pipes in the ceiling of the shoe factory to reduce air-borne noise, and proper vibration isolation of the printing press to reduce structure-borne noise).

23.5 City Traffic Noise in Offices

The advent of lightweight construction for office buildings has resulted in an interesting situation in regard to the intrusion of city traffic noise into buildings. On one hand, the use of all-glass walls and the use of glass and metal facing as exterior skins of office buildings have resulted in lower TL than formerly achieved with heavy exterior wall construction. On the other hand, the wide-scale use of complete air-conditioning systems in these buildings eliminates the open window as a path for city noise to enter an office. In the exchange, the new lightweight air-conditioned buildings are somewhat noisier than the older buildings with windows closed (as in winter) and somewhat quieter than the older buildings with windows open (as in summer). In any case, city traffic noise is to be reckoned with in new building designs.

We know of one manufacturer of air-conditioning system components who recently was placed in the position of having to defend the quietness of his equipment installed in a new modern glass-walled building. Noise measurements made one day in the office in question gave the levels shown by the upper curve of Fig. 23.3 when the air-conditioning equipment was in operation. In pursuit of the elusive noise problem, noise measurements were continued into the evening and repeat readings were taken in the same office. The resulting levels are shown by the lower curve of Fig. 23.3. There was no change in the operation of the equipment; only the city traffic noise decreased.

A range of typical New York City traffic noise is shown in Fig. 23.4. The cross-hatched area represents the range of maximum and minimum noise levels (as read on the "slow" meter of an octave-band

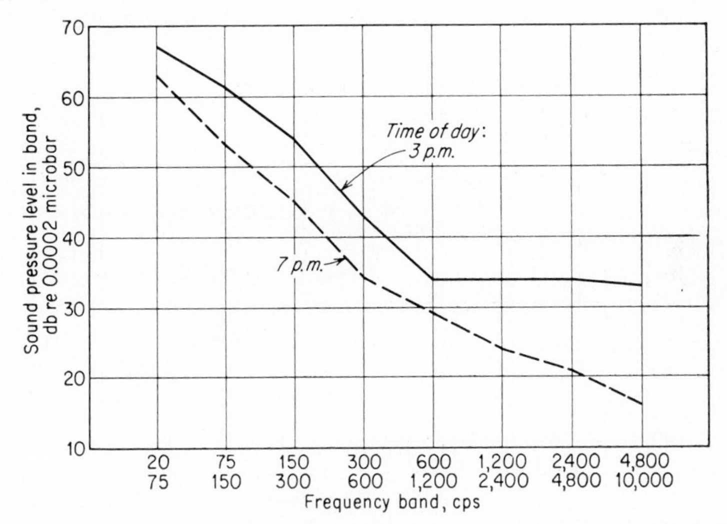

Fig. 23.3 Comparison of noise levels measured in glass-walled office at 3 P.M. and at 7 P.M. Afternoon levels attributed to traffic noise; evening levels represent air-conditioning noise and may include some traffic noise.

analyzer) measured between about 8 A.M. and 6 P.M. at three different locations. The measurements were made outside open windows on the 16th and 17th floors of the Biltmore Hotel (overlooking Vandebilt Avenue beside the Grand Central Station), the Commodore Hotel (overlooking 42nd Street near Lexington Avenue), and the Waldorf-Astoria Hotel (overlooking 50th Street between Park Avenue and Lexington Avenue). The low-frequency noise levels of Fig. 23.4 are due to trucks and buses, and the high-frequency noise levels are due to auto horns, police whistles, and squealing brakes. It is interesting to note that these noise levels do not drop off with the "inverse-square law" (6 db per double distance) for increased height above the street. This is because sound does not have a chance to spread out spherically. Instead, the close spacing of relatively tall buildings channels the sound in confined paths. Thus, fairly high traffic-noise levels may be expected to extend upward many floors around clusters of tall buildings surrounded by busy streets.

The measured "noise reduction" provided by two particular large-area windows in two modern glass-wall buildings is shown in Fig. 23.5. "Noise reduction" in this usage is the difference between the noise levels just outside the window and those just inside the window. The lower curve of Fig. 23.5 shows the noise reduction of a factory-sealed thermal-insulation window made up of two sheets of

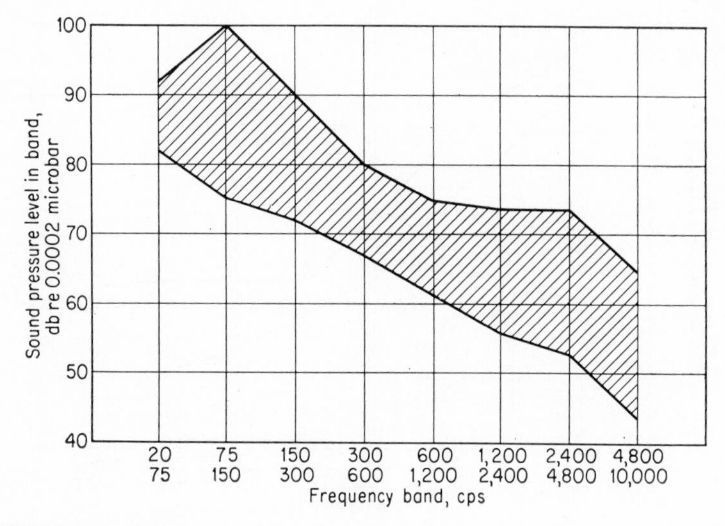

FIG. 23.4 Range of daytime traffic-noise levels measured outside the 16th and 17th floor windows of three New York City hotels.

¼-in.-thick glass separated by a ½-in.-thick air space. The upper curve shows the noise reduction measured for a double-glass window made up of two ⅛-in.-thick sheets of glass, set in rubber gaskets and separated with an air space of approximately 1½ in. The data given in Fig. 23.5 were obtained during crude field tests and are not intended to serve as authoritative performance measurements. Rather, the data are plotted here to illustrate two generalizations: (1) in double-window construction, the larger the air separation the higher the noise reduction, and (2) resilient mounting of the individual sheets of glass is essential to the achievement of high noise-reduction values. These two generalizations apply to lightweight and heavyweight double-wall construction as well as they do to double-window construction.

If we were to subtract the noise-reduction values of the windows of Fig. 23.5 from the range of noise levels of Fig. 23.4, we would find that the resulting inside noise levels would fall somewhat along the NC-20 to NC-35 curves. For single thickness of ⅛-in. to ¼-in. glass, the inside noise levels would be even higher. Thus, we see that in busy downtown offices inside the lower floors of glass-walled buildings, outside city traffic noise may raise the background levels approximately to the NC-25 to NC-35 criterion curves during periods of sustained noise, or even higher for short bursts of maximum noise.

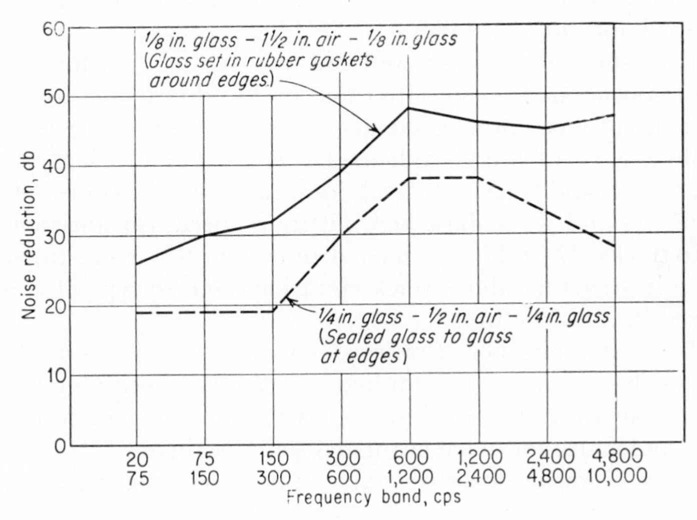

FIG. 23.5 Measured "noise reduction" provided by two types of double-glass windows. The curves are the differences in the noise levels measured just outside and just inside the windows.

This serves as a caution against setting very low noise-level criteria (NC-20 to NC-25) in offices subjected to high traffic noise from the outside.

It is apparent from this information that the window construction is essentially the controlling factor in setting background city traffic noise levels in offices. In cases where city noise is likely to be a serious disturbance, it is suggested that a noise-level survey be carried out at the site in order to determine the actual range of noise-level exposure. The noise levels thus determined may be used to influence the design of the windows and possibly the wall area that can be devoted to windows if the architects permit this latitude in the building design.

23.6 Structure-borne Railroad, Subway, and Highway Noise in Buildings

Another type of vehicular-noise problem which involves some buildings is the noise and vibration produced by nearby trains, subways, and automobile and truck expressways. In this discussion we will assume that the control of air-borne noise from these vehicles can be taken care of with correct selection of wall thicknesses and wall designs. We will present here some of the points of interest in control of structure-borne transmission. We discuss first the problem of railroad and subway noise.

Structure-borne Railroad and Subway Noise. Earth-borne and structure-borne noise and vibration in buildings due to railroad trains and subways moving on nearby tracks seem to be related to the following variables: (1) axle load of engines and cars, (2) smoothness of wheels and track, (3) track joints and switch points, (4) resilience of ties and the ballast bed, (5) train speed, (6) horizontal distance to tracks, (7) soil or earth conditions near the track, (8) vertical distance in structure above track elevation, and (9) type of construction used in the structure.

In the limited number of railroad and subway noise and vibration problems that we have encountered we have not been able to evaluate all of the above factors, but we have witnessed the effects of several of the factors and have evaluated some of them.

Noise or vibration measurements or both have been made at the following sites:

1. In a courtroom of the Philadelphia City Hall for the passage of subways a few floors beneath the courtroom
2. In Carnegie Hall in New York City for the passage of nearby subways (the closest subway track is about 40 ft from some seats in the hall)
3. In a hotel on Broadway in New York City near the site of the new Lincoln Center for the Performing Arts for passages of nearby subways
4. In the ground floor of the Waldorf-Astoria Hotel in New York City for trains immediately under the hotel on some of the tracks serving Grand Central Station
5. In the lower floors of the Queen Elizabeth Hotel in Montreal for train activity at the railroad station under the hotel
6. In a three-floor concrete structure in front of the Queen Elizabeth Hotel also above the railroad tracks.

Of the above sites, the Waldorf-Astoria Hotel and the Queen Elizabeth Hotel contain "lead-asbestos" vibration-isolation pads between the footings and the column bases of the buildings (these pads are described later). For the other structures in which measurements were made, there was no vibration-isolation material in the foundations of the buildings.

All the data are not completely consistent when intercompared, but we believe a few generalizations can be drawn.

1. Noise levels, measured in the vibration-isolated buildings in the one or two floors immediately above the railroad tracks, for trains at slow speed (approaching or departing from station platforms), fall in the range of 70 to 80 db in the 20- to 75-cps frequency band, in the range of 60 to 80 db in the 75- to 150-cps frequency band, and then drop off with increasing frequency somewhat following the shape of an NCA curve.

2. For horizontal distances away from the tracks, noise levels seem to decrease at the rate of about 5 db/50 to 70 ft. For vertical heights above track elevation, noise levels seem to decrease at the rate of about 5 db per 15 ft for building elevations higher than about 40 ft above the tracks.

3. Noise levels increase as the train speed increases, but the rate of change is not yet known. Possibly a rate of 3 to 5 db per doubling of speed could be assumed until more definite data are obtained. Noise levels also increase with axle load, but the rate is not fully established. In some measurements, engines were found to produce levels about 3 to 6 db higher than train cars. Axle loads of engines are about 2 to 4 times those of train cars; so this suggests a rate of about 3 db increase in noise for doubling of axle loads.

4. Higher noise levels are produced when wheels pass over the open gap of switch frogs than when they pass over conventional expansion joints between track sections. On this basis, we believe that long single-track sections without any track joints will produce noise levels possibly 3 to 6 db below those of conventional track sections with joints. We understand that ¼-mile track lengths have been delivered to some installations, and that several of the tracks in the western part of the country extend for several miles without track joints. Also portable rail welders are now available for welding adjoining rails on the site, thereby making it possible to eliminate at moderate expense the clicking of wheels passing over the track joints.

5. It appears that the "lead-asbestos" pads result in reductions of about 5 to 10 db in the transmission of low-frequency noise and vibration into the columns of buildings. These pads are about 1 in. thick and consist of a ⅛-in.-thick sealed jacket of lead sheet enclosing two ⅜-in.-thick pads of packed asbestos fibers which are separated by a sheet of 18- to 20-gauge steel. These pads are loaded at about 500

to 800 lb/in.^{2} in normal installations under the columns of buildings.

6. From the various subway measurements, it appears that the noise spectrum follows somewhat the shape of an NC curve; and for distances of about 40 to 80 ft from the subway tracks the radiated noise levels in unisolated buildings fall in the vicinity of the NC-35 to NC-50 criterion noise levels in the low- and mid-frequency bands.

7. The earth (bedrock) and normal building structures provide sufficiently high attenuation of high-frequency sound and vibration that there is no significant amount of transmitted and radiated noise above approximately 1,000 cps.

With the above information it is possible to make very rough estimates of the low-frequency noise levels to be expected in buildings located near railroad and subway tracks. We suggest conservative use of the material, however, because we do not have a large storehouse of data in support of these generalizations. The author will be pleased to learn of the experiences of others who have worked in this field.

As suggested by the material summarized above, a few reasonably positive noise-control measures are available. (1) When conditions justify, lead-asbestos pads may be used to provide vibration isolation in the bases of building columns. (2) Welded track joints will give noticeable reduction of the noise peaks produced by the impact of wheels crossing over open track joints. (3) Switches should be removed from critical locations if possible; if they cannot be removed, movable-point frogs or spring-loaded frogs may be designed for use near the critical locations. (4) Since speed is an important parameter, noise can be reduced by having low train speeds in the critical area. This, of course, requires the cooperation of the railroad or subway operators. (5) The tracks may be vibration-isolated from their ties or from the roadbed. We have devised a number of possible mountings, but none of these has been used to date; and as yet we have no acoustic information on the use of any other track isolation schemes.

Structure-borne Highway Noise. With the increasing use of high-speed expressways leading into and through the centers of our cities and with the rapidly decreasing amount of space available in the centers of our cities, it becomes more and more the practice to place high-speed, high-traffic-density, truck, bus, and automobile expressways over, under, or near occupied buildings.

Our limited data indicate that structure-borne noise levels in buildings, located beside or over medium-speed (25 to 40 mph) highways, fall in the range of 60 to 80 db in the 20- to 75-cps band and drop off with frequency at the rate of about 10 db/octave. Noise levels can range higher if there are expansion joints, drainage grilles, bumps, or rough sections in the road.

Some general suggestions for reduction of earth-borne and structure-borne noise and vibration due to automobile and truck traffic are as follows:

1. Provide smooth, fine-grained road finish.
2. Have no expansion joints in the roadway near critical buildings.
3. Have no drainage grilles running across the roadway.
4. Provide suitable vibration-isolation joints between the roadbed and the structural members of adjoining buildings.

23.7 Aircraft Noise in Buildings

Present and future aircraft noise is a factor of concern for some buildings, particularly those located near airports or along prescribed flight paths in the takeoff and landing patterns in the vicinity of airports. For in-town office buildings several miles by flight path from the nearest airport, aircraft fly-over noise is just another intermittent noise which does not cause any significant amount of interference with most normal office activities.

However, for office buildings located near (1) airports or military air bases, (2) prescribed flight paths close to the airport, or (3) in-town heliports, the noise levels are usually quite high and the frequency of flights is greater because of the concentration of activity within a relatively small area.

Some of the noise levels which might be encountered near an airport are given in Fig. 23.6. A distance of about 2,000 ft to the aircraft is assumed for all cases. Corrections to other distances to fit specific situations may be made by the reader from information given in Chap. 9. Figure 23.6 gives approximate outdoor noise levels for (1) a large passenger-carrying helicopter powered by piston engine, (2) a large four-engine propeller-driven commercial airliner (of the DC-6, DC-7, Constellation, and Super Constellation class) at climb power, and (3) three four-engine jet airliners (1959–1960 vintage) equipped with noise suppressors at reduced power following takeoff.[2] For these outdoor noise levels, the noise levels inside buildings having glass walls would fall approximately along the speech-frequency portions (600 to 4,800 cps) of the NC curves given in Table 23.1. Two types of glass walls are indicated: Column 2 of Table 23.1 represents single-thickness ¼-in.-thick glass; column 3 represents double glass construction of ⅛ in. glass, 1½ in. air, ⅛ in. glass as described by the top curve of Fig. 23.5.

For a large amount of jet aircraft activity, it would appear from this comparison that double windows would be desired for private and executive offices and conference rooms in buildings which are located closer than 2,000 ft from busy aircraft traffic lanes. As used here "double windows" means windows set in rubber gaskets so that they may be considered resiliently supported from each other.

Office areas in airport buildings should receive similar consideration.[3] In many cases the noise levels are even higher in the airport terminal buildings because aircraft engines at "idle" and "taxi" conditions may be only 50 to 200 ft from the outside walls of offices.[4] To

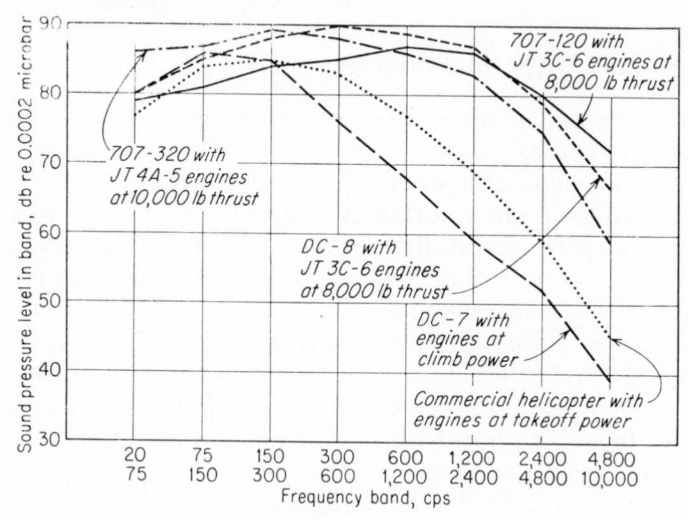

Fig. 23.6 Approximate noise levels due to various 1959–1960 commercial aircraft following takeoff, measured at a distance of 2,000 ft perpendicular to the flight path. The levels are the maximum recorded in each band during a fly-over. The thrusts for the jets are those generally used following a power cutback for noise-reduction purposes.

illustrate the need for acoustic design of the exterior walls of airport terminal buildings, we recall the case of the airline ticket agent who had to crawl under the ticket counter to talk on the telephone as a Viscount taxied in to the ramp.

The discussion given above is primarily based on present-day aircraft. With intercontinental jet airliners coming into use and with supersonic jet transports reputedly in design, we would be negligent in acoustic designs if we did not anticipate some increase in aircraft noise levels for the future. New advances in jet-engine designs may reduce the noise levels of future jets, but it is more likely that even higher thrusts will offset any reduction of noise by engine design. It does not appear unreasonable to assume that future jet engines in the thrust range of 20,000 to 30,000 lb will be about 5 to 10 db noisier than present-day engines in the thrust range of 10,000 to 15,000 lb. Changes in engine design and possible elimination or modification of present-day noise suppressors will probably also change the spectrum content of the noise from that presented in Fig. 23.6.

Table 23.1

Effects of Glass on Airplane Noise Levels inside Buildings

Type of Plane	Noise Criterion Curves, NC Units	
	¼-in.-Thick Glass	Double Glass
(1)	(2)	(3)
DC-7	30	15
DC-8	55	40
Helicopter	40	25
707-120	55	40
707-320	50	35

As with other problems involving external noise, the roof, wall, window, and door designs are the variables with which adequate noise reduction can be achieved for offices and buildings exposed to nearby aircraft noise.

23.8 Office-equipment Noise

The noise levels measured for a few specific types of office equipment are given in the next few plots. Figure 23.7 shows the noise levels at a distance of about 8 to 10 ft from a ringing telephone and an electric typewriter. These readings were taken with the "slow" meter of the octave-band analyzer. These levels typify the noise in a secretary's office which might adjoin a private office.

Figure 23.8 presents noise levels measured about 8 to 10 ft from an assortment of different types of office duplicating equipment. Included is a hectograph machine (liquid-process duplicator) and a table-type Bruning duplicating machine and a large Ozalid machine. Figure 23.9 gives noise levels near a multilith printing machine. These types of equipment are frequently found in or near stenographic areas or in special duplicating rooms.

New office buildings are now beginning to make provision for extensive installations of electronic data-computing centers. These centers are somewhat noisy, but they can be handled with no difficulty when they are anticipated in the building designs.

Figure 23.10 shows the noise levels measured by W. W. Stalker in the keypunch room of an industrial office.[5] The room has an acous tically treated ceiling and contains six keypunch machines, four verifiers, and other assorted office equipment. These noise levels may

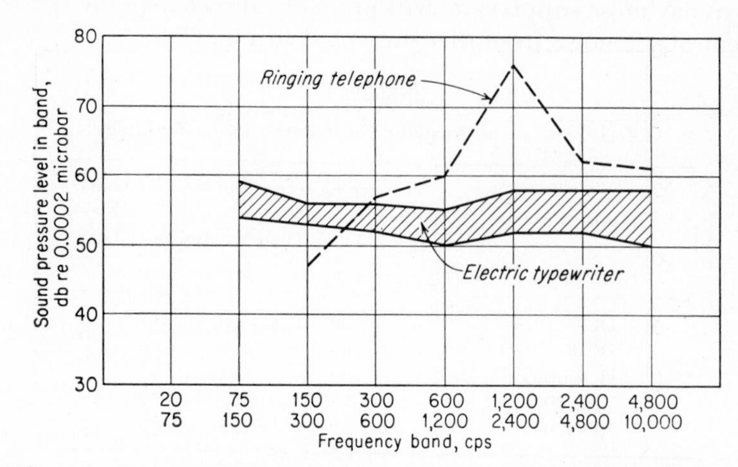

FIG. 23.7 Noise levels at distance of 8 to 10 ft from a ringing telephone and an electric typewriter.

persist almost continuously throughout the workday. Areas such as this may be found in many office buildings, with supervisory offices located about the floor at strategic positions.

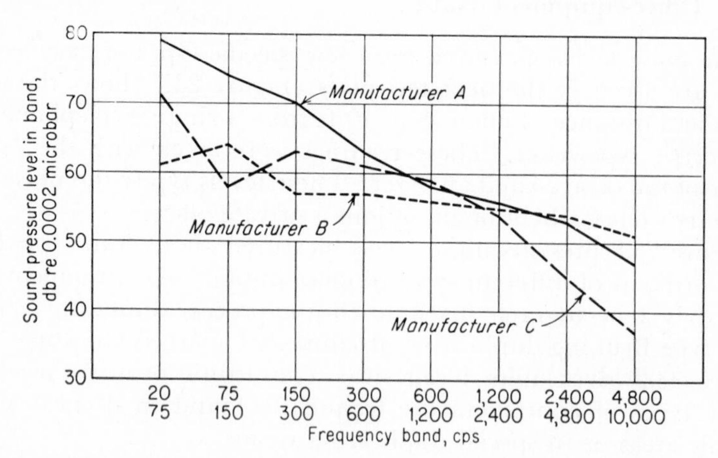

FIG. 23.8 Noise levels at distance of 8 to 10 ft from three different types of duplicating machines (1959 models).

Figure 23.11 gives the noise levels measured near a high-speed printer, one of the noisiest components of a data-computing center.

We go now from this brief survey of noise levels found in work areas containing office equipment to the treatments which can be used to help reduce the noise. There are a few good working prin-

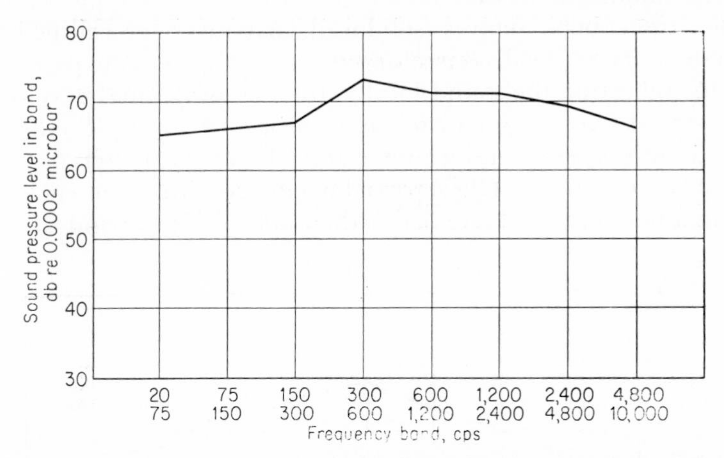

FIG. 23.9 Noise levels measured 6 to 10 ft from a multilith printing machine (1959 model).

ciples which can be applied to keep this noise from being bothersome to nearby office occupants.

First, if several people work in the noisy environment but do not

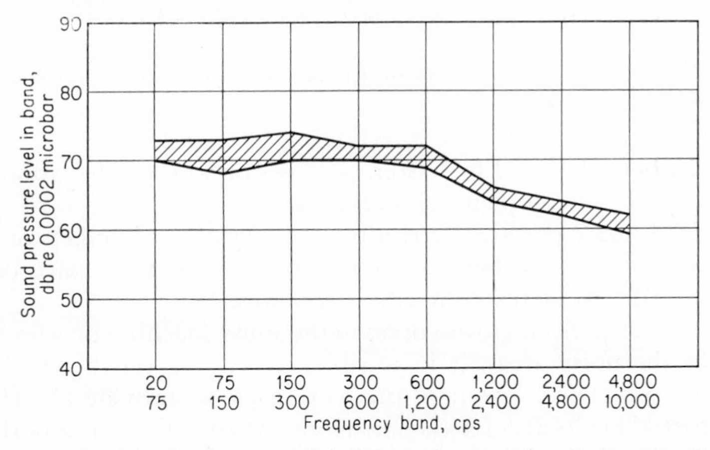

FIG. 23.10 Noise levels measured in keypunch room of an industrial office.

themselves make the noise, every effort should be made to separate them from the noise makers either by greater distance or by full or partial partitions. The annoyance of these machines is mostly due to their high-frequency noise content, and high-frequency noise is the easiest to control. If the work area must be a large open area, acoustic treatment of the ceiling is a necessity, and, as much as possible, tall obstacles should be scattered about the area. For example, banks of tall file cases, book shelves, solid-wall coat racks, or ceiling-height room dividers can be dispersed about the area to provide partial barriers for reflecting the sound locally, thereby providing more opportunity for the sound waves to be absorbed by multiple reflections from the absorbent ceiling and other absorptive materials in the nearby surroundings. Glass partial-height partitions or enclosures, although not very good, are better than nothing at all; and these can

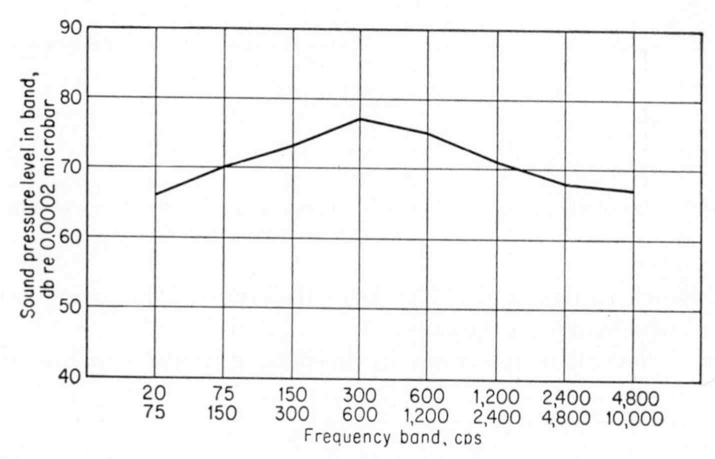

FIG. 23.11 Noise levels measured 6 to 10 ft from a high-speed printing machine of an electronic data system.

be used where the need for visibility exists or where ventilation and lighting prohibit complete solid enclosures.

If ceiling space permits, vertically hung absorbent blankets or unit functional sound absorbers can be located near the noisy parts of the room in order to increase the effective absorption.

Second, for personnel who operate the noisy machine or who must work in the noisy areas, it is possible, depending on the machines and the geometrical layout of the work area, to provide absorbent-lined partial enclosures for each machine or partial-height absorbent-lined separators between rows of machines and operators. As mentioned above, ceiling absorption can be enhanced with the use of vertically hung acoustic blankets or unit functional sound absorbers. Such devices usually will not reduce the noise of the operator's own machine but they may produce significant reduction of the noise from the neighboring machines.

Next, for enclosed offices bordering on noisy work areas, the office equipment noise can usually be brought down to a reasonably acceptable level. The common wall between the office and the noisy area, the door into the office, and the ceiling detail are the controlling factors. The wall should be as heavy as practical, but equally important, the wall should not have any air cracks or holes through it. Refer to Chaps. 13 and 14 for detailed discussions of the TL of walls. In addition, acoustical studies of the TL of walls should be followed in the *Journal of the Acoustical Society of America* and in *Noise Control*,[6] for many improvements are being made in lightweight wall designs.

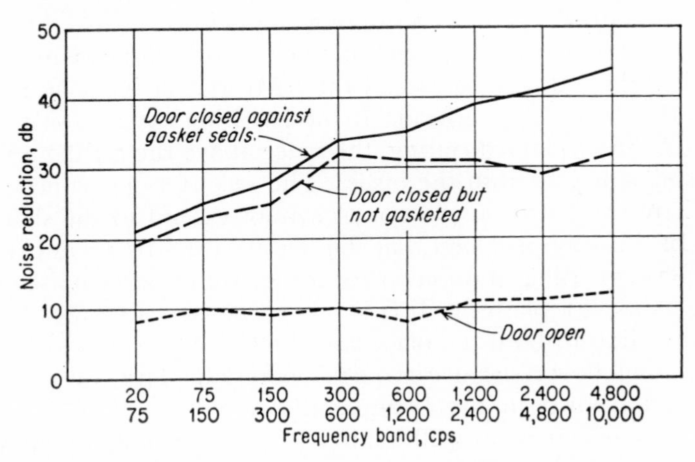

Fig. 23.12 Comparison of "noise reduction" for an open doorway, an ungasketed closed door, and a gasketed closed door.

The door is the most obvious weakness in the TL of a wall. It may not be obvious, however, just how much difference there is between a loose-fitting door and a snug-fitting gasketed door in keeping out unwanted noise. Figure 23.12 shows the noise reduction measured for a particular door-wall combination. Recall that *noise reduction* is the difference between noise levels measured on opposite sides of the door under study. In this case the microphone positions were about 1 ft from the door in the noisy area and about 6 ft from the door in the quiet office. These simple measurements suggest the following: (1) If there *is* a door, keep it closed, and (2) for maximum effectiveness of the door, install gaskets around the door frame.

This raises the question of ventilation through louvered doors. In some buildings, offices are designed to have louvered doors as the return path for ventilation air. For private offices this should certainly be avoided; it is a sure way to ruin a door which must otherwise serve as a sound barrier. If the door must provide a return air path and yet retain its function as a sound barrier, the door can be undercut about 1 in. and the resulting air gap treated with an "acoustic trap." The trap essentially consists of a channel insert in the bottom of the door. The channel is made as wide as the door will permit and about 2 in. deep and is filled with glass fiber. Design details for such a trap were presented by Paul Veneklasen at the May, 1959 meeting of the Acoustical Society of America[7] and will probably be published shortly. This acoustic trap is most useful in the high-frequency region of interest to us for office-equipment noise.

The ceiling detail is also an important consideration in maintaining noise control in offices adjoining noisy work areas. There are two items to note. First, where an office is to be located near noisy areas, it is good design to extend the partitions of that office all the way to the floor slab overhead. If the partition extends only to the height of the suspended ceiling, the space above the ceiling serves as a transmission path into the office. The second point concerns the case where the partitions do not or cannot extend to the slab overhead. In this case, it is best that the office ceiling be an impervious, dense plaster ceiling in order to reduce transmission of noise via the ceiling path. If a plaster ceiling is out of the question, then the suspended ceiling in both the office and the noisy area should be high-density, high-flow-resistance acoustic material. The use of such a ceiling was shown in the example illustrated by Fig. 23.2. Even here the use of return air grilles into a common air plenum above the ceiling can provide a flanking path for noise to travel from the work area to the office. An acoustically lined elbow on the ceiling side of the air grilles will decrease the feed-through of high-frequency noise.

A few precautions are in order for the installation of electronic data-computing centers. These precautions are intended to protect the floor below the computing center against excessive structure-borne noise. First, a lightweight metal deck floor with a suspended acoustic ceiling below is probably not a satisfactory floor and ceiling construction for the data-center floor. A heavy dense concrete floor (75 lb/ft^2 surface weight) with an acoustic ceiling below or a lightweight metal deck floor with a dense plaster ceiling below has been found to be adequate in two installations that we have measured. In some new installations not yet completed, we have suggested that with these minimum floor and ceiling constructions it may be necessary later to install vibration-isolation mounts under some of the noisier pieces of equipment, particularly those that contain large blower fans and those that operate with large impact forces, such as high-speed printers.

We know of several new bank buildings in which certain floors are being designed specifically to handle the data centers, and we know of one in which cooling air is being ducted into the bottom of several of the machines from a special ventilation system in the building in order to eliminate the extra heat and noise of the various blower fans normally contained in this equipment.

23.9 Air-conditioning-system Noise and Vibration Transmitted through Machinery-room Walls and Floors to Adjacent Office Spaces*

A trend in modern office-building design is to locate most of the air-conditioning equipment up at the top of the building. This

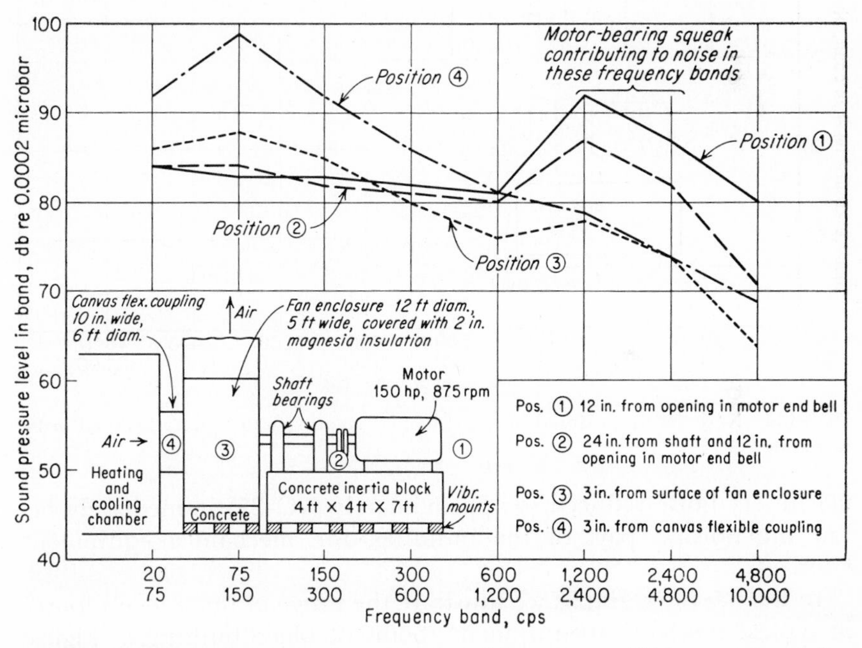

Fig. 23.13 Noise levels measured at several positions near a 150-hp centrifugal fan assembly.

places the heavy refrigeration compressors, pumps, and heat exchangers up near the cooling towers, which perform best at the

* The control of duct-borne fan noise and airflow noise is covered in Chap. 21.

highest building elevation. This permits large water pipes to and from the cooling towers to be of minimum length. It also places the fresh-air-supply fans up where the air is fresh and away from the polluted air at street levels. We have even seen a few new buildings recently in which the boilers are located at the top of the building.

Of course, it is traditional that the topmost occupied floors of buildings are premium floors for executive offices. Thus, we are faced with the situation of having to provide quiet office areas (with

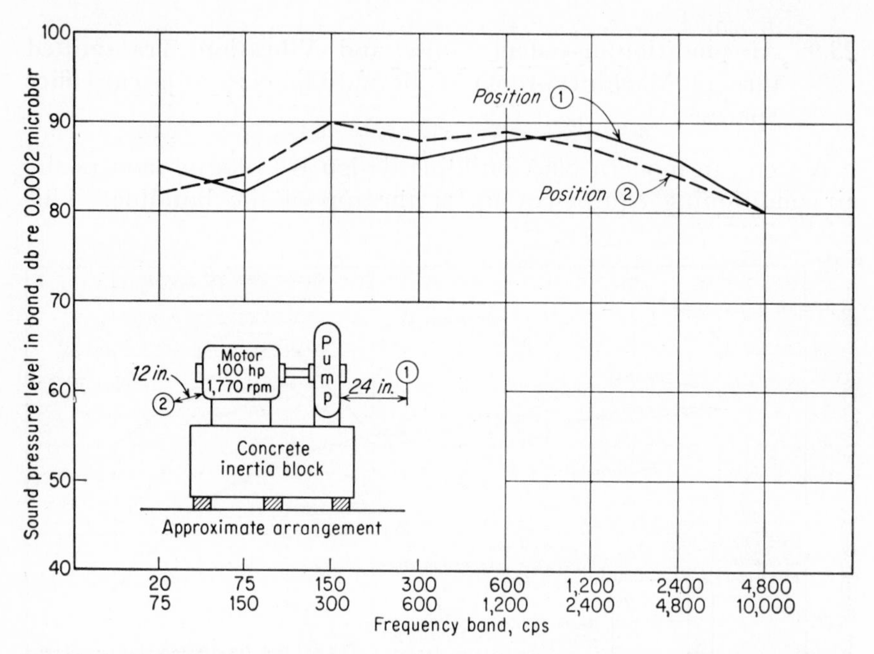

Fig. 23.14 Noise levels measured near a 100-hp circulating water pump believed to be cavitating.

machinery noise reduced to approximately NC–25) immediately below the noisiest part of the building, the mechanical-equipment floor.

In the next few graphs we illustrate the range of noise levels found in typical mechanical-equipment rooms of office buildings. Figure 23.13 shows the noise levels measured at several positions near a 150-hp centrifugal fan operating at 875 rpm at a static pressure of about 8 in. water. Figure 23.14 gives noise levels measured near a water-circulating pump driven at 1,770 rpm by a 100-hp motor. At the time of the measurement, the pump sounded as though it were cavitating. Figure 23.15 gives the noise levels measured at one end of a machine-room floor at a distance of about 30 ft from the fan

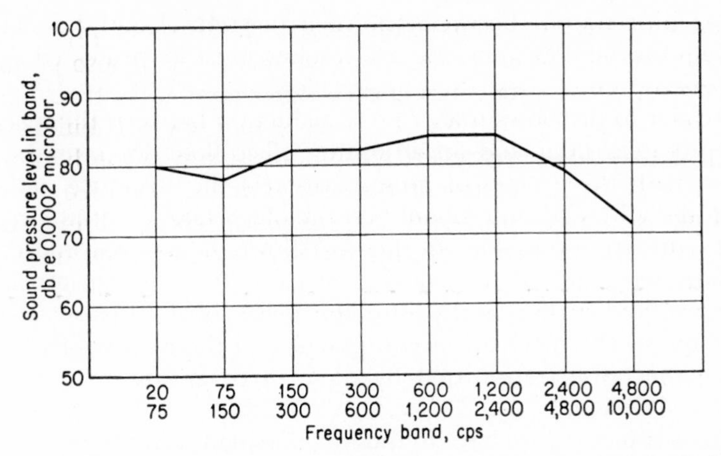

FIG. 23.15 Noise levels measured in a mechanical equipment room about 30 ft from the fan assembly of Fig. 23.13 and 10 to 15 ft from two pumps of Fig. 23.14.

system of Fig. 23.13 and about 10 to 15 ft from two pumps similar to that of Fig. 23.14.

The noise levels inside the inlet casing and the outlet duct of a 100-hp fan are shown in Fig. 23.16. In this installation, the intake side of the fan is enclosed with a casing which rests on the floor of

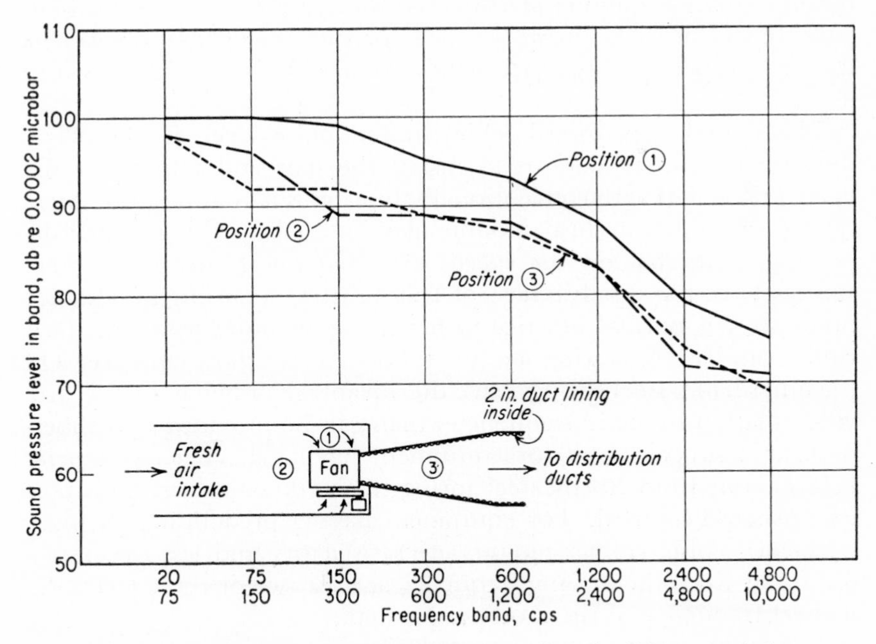

FIG. 23.16 Noise levels measured inside the casing of a 100-hp supply fan.

the machine room. Thus, the noise at positions 1 and 2 is actually incident on the floor; and the office below must be protected against the penetration of these high noise levels.

In order to provide suitable noise reduction between the mechanical-equipment floor and an adjoining office floor we must be concerned with three types of noise paths: (1) the structure paths between the vibrating equipment and the office areas, (2) the air-borne noise paths from one floor to the next through the floor and ceiling combination, and (3) the duct-borne paths for the transmission of fan noise and airflow noise into the office areas. Proper acoustic treatment of the mechanical equipment and the nearby offices must reduce the noise transmitted by these three groups of paths in a "balanced design."

The best method we have found for assuring low noise levels under a machine-room floor consists essentially of four steps. The first step is to install a "floated floor" over the entire machine-room area. This floated floor is a concrete slab floor, with suitable thickness and reinforcing to satisfy the loading requirements, supported off the normal structure floor with isolation mounts or pads. An effective isolation material is a thick resilient pad of glass fiber. A sketch of such a floated floor is shown in Fig. 23.17.

A glass-fiber pad is not the only way to achieve a floated floor. Vibration-isolation mounts of rubber-in-shear, stacks of ribbed-rubber pads, blocks of cork (of proper area) or steel-spring mounts can be used at proper spacings to isolate a floor slab above the structure floor.

The second step toward achieving low noise levels in the offices under a machinery floor is to install the fans and pumps on the floated floor slab with conventional vibration-isolation mounts. Steel springs or rubber-in-shear mounts may be used for this purpose depending somewhat on the speeds and frequencies involved. Steel springs have the disadvantage of transmitting some high-frequency noise, so it is normal practice to insert one or more pads of ribbed rubber under steel spring mounts to make them most effective. On the other hand, steel springs have the advantage of being able to provide isolation to lower frequencies than can be provided by rubber-in-shear mounts. Thus, for equipment having driving frequencies below about 10 to 20 cps, steel springs with ribbed-rubber pads normally should be used. For equipment having predominantly high-frequency noise, rubber mounts are satisfactory and less expensive. (*Caution:* Keep the driving frequency at least two or three times the natural frequency of the mounting system.)

The vibration isolation of large reciprocating compressors (50 hp

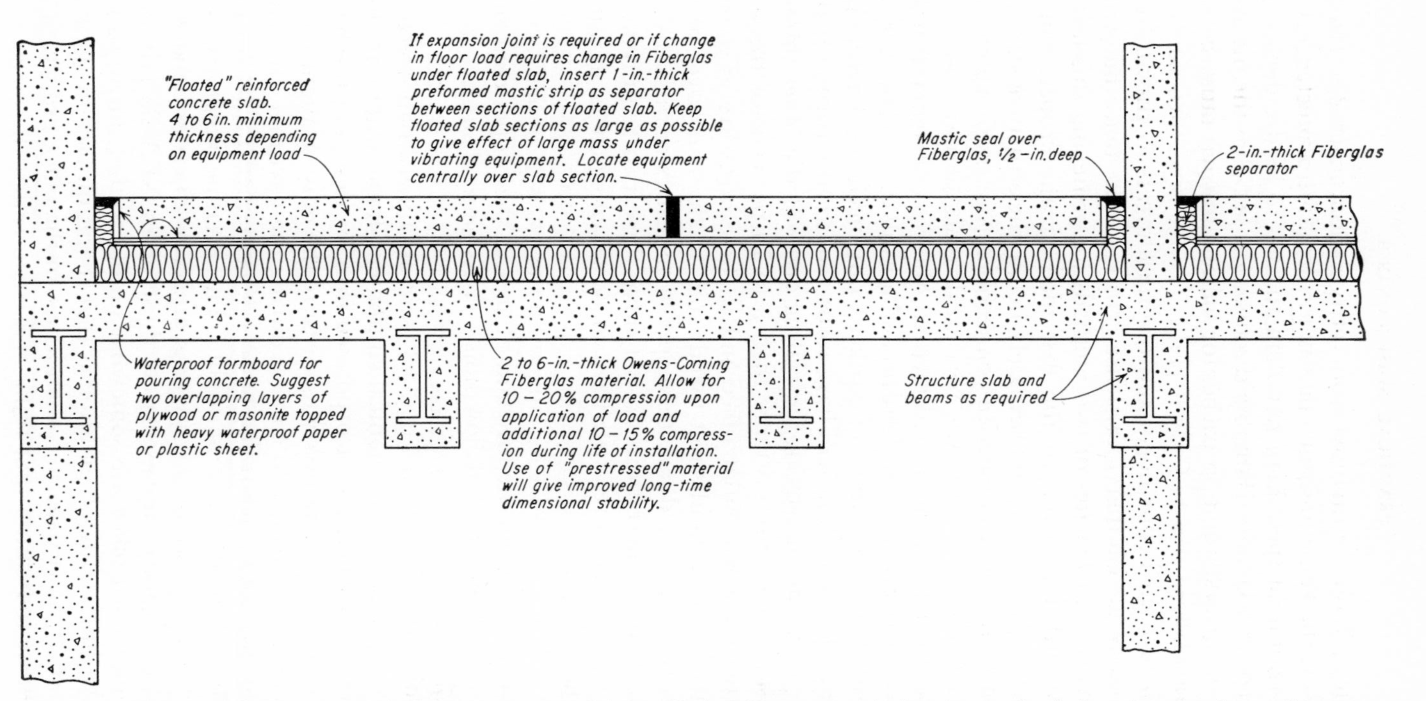
"Floated" reinforced
concrete slab.
4 to 6 in. minimum
thickness depending
on equipment load
If expansion joint is required or if change
in floor load requires change in Fiberglas
under floated slab, insert 1-in.-thick
preformed mastic strip as separator
between sections of floated slab. Keep
floated slab sections as large as possible
to give effect of large mass under
vibrating equipment. Locate equipment
centrally over slab section.
Mastic seal over
Fiberglas, 1/2-in. deep
2-in.-thick Fiberglas
separator
Waterproof formboard for
pouring concrete. Suggest
two overlapping layers of
plywood or masonite topped
with heavy waterproof paper
or plastic sheet.
2 to 6-in.-thick Owens-Corning
Fiberglas material. Allow for
10 – 20% compression upon
application of load and
additional 10 – 15% compression during life of installation.
Use of "prestressed" material
will give improved long-time
dimensional stability.
Structure slab and
beams as required

FIG. 23.17 Schematic arrangement of "floated" concrete slab installation for support of vibrating mechanical equipment above office areas.

and larger), large centrifugal-type refrigeration machines (weight 50,000 lb or more), and large cooling towers (25 hp and larger) becomes a somewhat specialized problem and has to be decided on the basis of specific layouts. For this discussion it is believed unwise to try to devise "standard" approaches for such mountings because there just are none.

The third step in this noise-control procedure is to provide a resiliently suspended impervious dense ceiling in the office below the machinery floor. Sketches of two methods for achieving this ceiling are shown in Fig. 23.18. In the first method, the *acoustic-barrier* plaster ceiling is installed immediately below the structure slab, and the finish ceiling and all service connections (lighting, piping, and air ducts) are below the acoustic barrier. If space permits, this is the simpler approach. In the second method, the *acoustic-barrier* plaster ceiling is penetrated by light fixtures and air ducts. Since we require the acoustic-barrier ceiling to help keep out mechanical-equipment noise, holes in this ceiling will have to be acoustically treated if they make up a significant part of the ceiling area. If the holes through the acoustic barrier are no more than 5 to 10 per cent of the total ceiling area for a particular office and if these holes are effectively closed by a piece of solid sheet metal (such as the rear cover of the light fixture or the air duct), then special measures are usually not required. If the penetrations do not meet these two requirements, however, special measures will have to be taken, such as shown in Fig. 23.18*b*.

The fourth step toward achieving low noise levels in an office, either close to or remote from the mechanical area, is the proper control of duct-borne fan noise and airflow noise. This general subject is covered in Chap. 21.

The noise levels given in Fig. 23.3 were obtained in an executive office directly under a mechanical-equipment floor, where all four steps described above were incorporated into the design of the building. Recall that it was necessary to measure the noise at night in that office because city noise held control during the daytime. In fact, it is possible that some city traffic noise was contained in the low-frequency measurements even at night. The evening noise levels in that office were approximately those of an NCA–25 criterion and the afternoon levels were approximately those of an NCA–35 criterion. The ceiling of that office was somewhat similar to that shown in Fig. 23.18*a*. An interesting observation is that it was quieter in the duct space between the two ceilings than it was in the office proper. This was a definite indication that traffic noise controlled the office.

By comparison, it is of interest to observe the noise levels that were

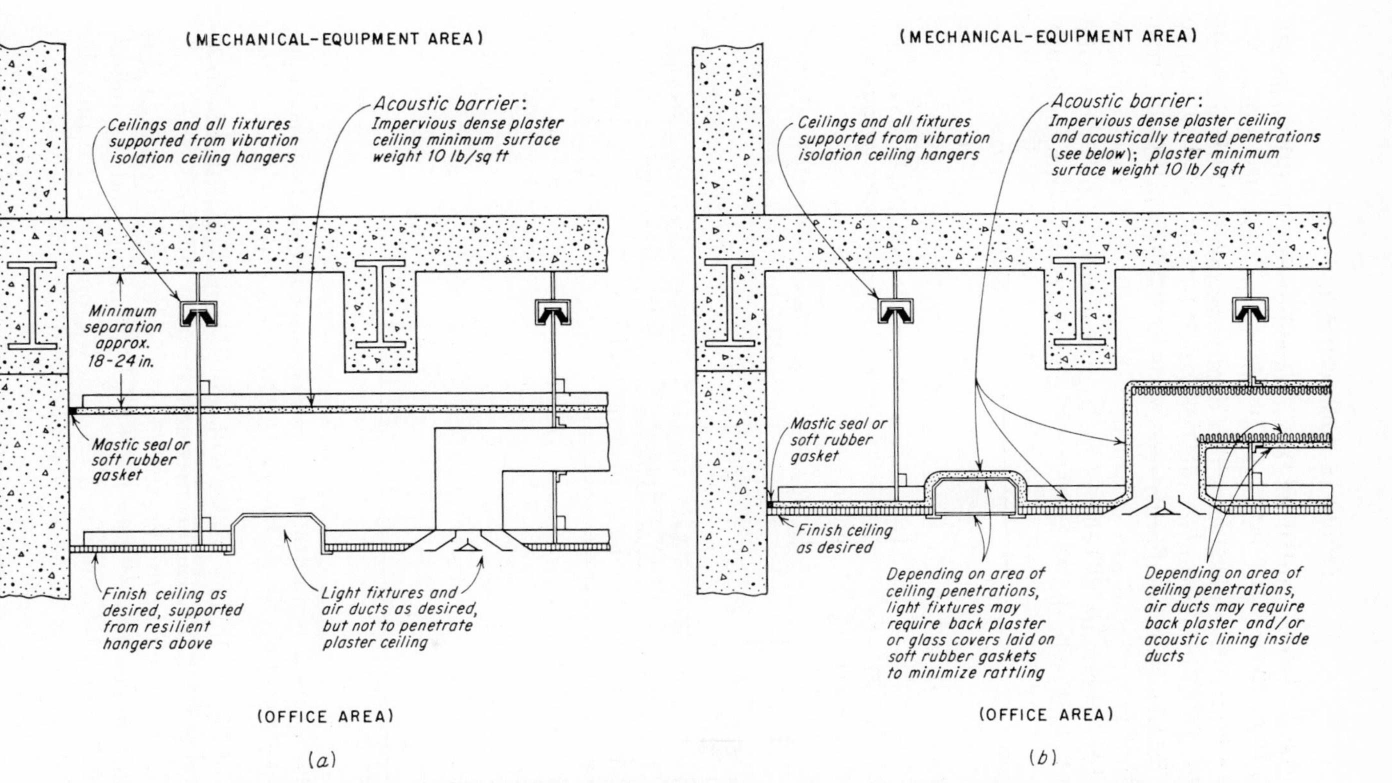

Fig. 23.18 Schematic arrangements of alternative ceiling treatments for office area beneath mechanical equipment area.

found to exist in a similar office without the special treatments described above. Figure 23.19 shows a conventional fan mounting above an executive office in which data were taken. The occupant complained of low-frequency audible "rumble" and some very obvious vibration associated with operation of the fan overhead. The measured noise levels are shown in Fig. 23.20. The low-frequency noise levels here are at least 14 db higher in the 20- to 75-cps band and 9 db higher in the 75- to 150-cps band than the evening measure-

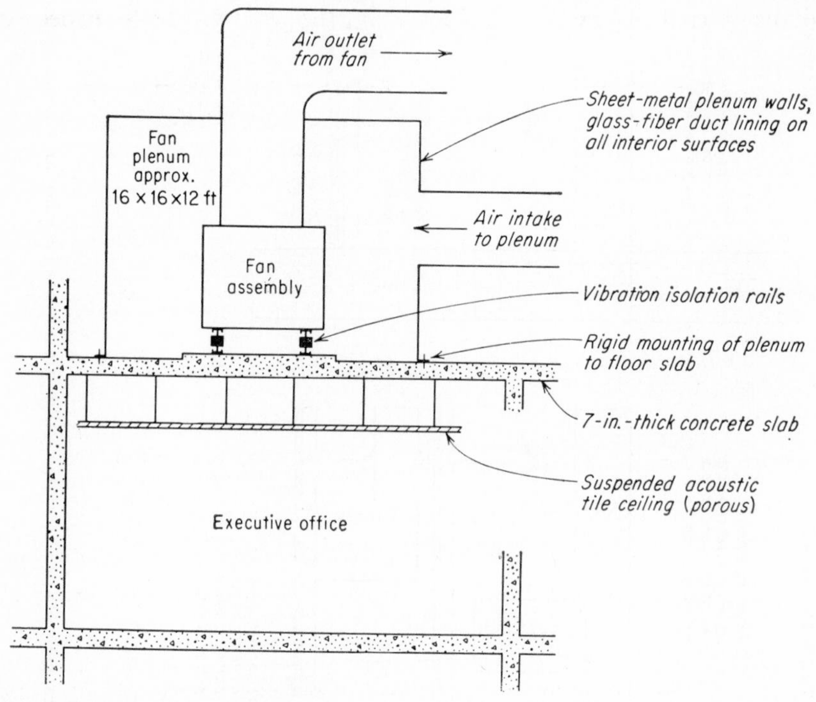

FIG. 23.19 Structural arrangement of executive office and overhead mechanical equipment area for which the data of Fig. 23.20 apply.

ments of Fig. 23.3. The reader will probably recognize that 9- to 14-db noise reduction in the low-frequency region does not come easily. The vibration felt in the Fig. 23.19 office would be eliminated with the addition of a floated floor for the fan mounting, and the audible "rumble" would be eliminated through the improved TL provided by the combination of the floated floor and the resiliently hung impervious dense ceiling.

There are undoubtedly some compromises from the four-step procedure which will still yield satisfactory installations. For example, in some mechanical-equipment areas it has been considered not feasible by the architects to provide a full-area floated floor. In such cases we have suggested smaller sections of floated slabs for individual fans or pumps or very heavy concrete inertia blocks that are vibration isolated from the floor. These fall into special designs and are not simply described and categorized.

The floated floor slab should not be dismissed lightly. Its major advantages are (1) it usually improves vibration isolation, (2) it definitely benefits the air-borne transmission problem, and (3) it provides a very real "safety factor" against inadvertent errors in isolation details carried out by the contractors on the job. We have inspected

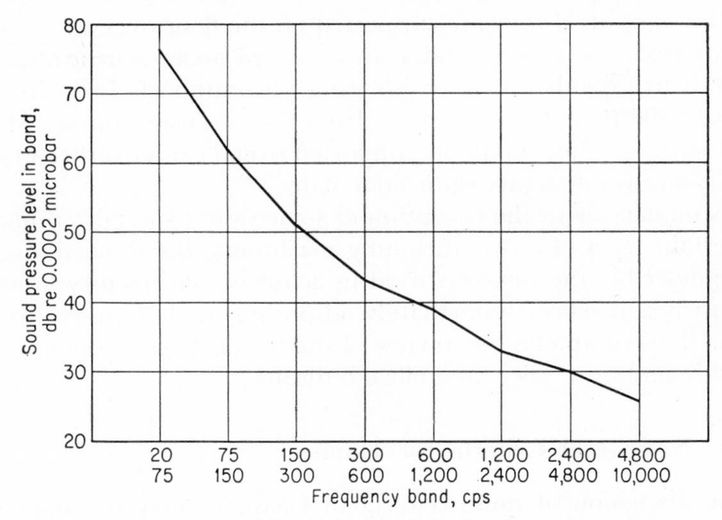

FIG. 23.20 Measured noise levels in office of Fig. 23.19 with overhead fan in operation.

machinery-room installations that have had some of the following mistakes in isolation details: unisolated floor-mounted pipe stands and duct supports, rigid electrical connections between isolated motors and the floor slab, rigid pipe drains connected to isolated pump bases, and several other examples in which rigid connections provided "short circuits" of the isolation mount. Many of these oversights on the job would not have shown up as problems if they had taken place on a floated slab rather than on the structure slab. It is easy to say that those oversights should not be permitted in construction. Unfortunately, we find, however, that there are usually several such mistakes on each machinery floor; and sometimes through complacency many of them are never found or even seriously sought. Therefore we believe that the use of a floated slab has considerable merit as a protection against many of these occurrences.

One final aspect of vibration isolation is deserving of attention. That is the need for eliminating all rigid ties between an isolated machine and the building proper. Electrical connections should be made with long "floppy" lengths of flexible conduit. Pipe connections to pumps should contain flexible rubber connections if the installation is known to be in a critical location; or at least the pipe should be laid out so that flexible connections can be added later if necessary. A rule-of-thumb estimate for a flexible rubber pipe connection is for the length to be about six times the outside diameter of the pipe.

Vibrating pipes and ducts should be supported with vibration-isolation mounts if they are supported on the floor over an occupied area or from the ceiling under an occupied area. A rule-of-thumb figure is to provide vibration-isolating pipe supports for a distance of about 100 pipe diameters away from the vibration source. Where pipes are large, they should be supported from beams overhead rather than from thin or lightweight floor slabs.

In summary, with the exception of some rather specialized mounts for certain types of air-conditioning machinery, the discussion given above describes the steps involved in achieving satisfactory noise environments in offices immediately adjoining mechanical-equipment floors. This completes our review of the major types of noise sources found in and near offices and office buildings.

23.10 Noise-level Criteria for Homes

The discussion of noise criteria in Chap. 20 may be adapted to home activities as desired for speech communication in various areas of the home. To apply these criteria to the design of homes is a more or less academic approach because when the automatic dishwasher or the vacuum cleaner is running one generally just "puts up with the noise" and shouts. These devices, although noisy, are usually of benefit in the home, and in any event, their noise is of rather short duration. As another example, you are glad to have the lawn mowed, so you tolerate the noise of the power lawn mower.

But when it's time to sleep, the noise criterion comes into its own. In most residential areas, an NC–20 criterion for indoor noise levels is satisfactory for sleeping. Highly sensitive persons might think that they need an even lower criterion, but this should be discouraged; for with a very low ambient, all intruding noises appear magnified in proportion. For busy urban areas containing significant amounts of nighttime traffic or industrial noise, an NC–25 or NC–30 may be acceptable for steady-state broad-band noise. Noise levels in bedrooms of city hotels often reach NC–35 to NC–40 with windows open, but these levels are not advocated as conducive to sleep for many people. And yet, large numbers of window-installed room air conditioners produce NC–40 to NC–45 noise levels in rooms, and we presume that many people sleep with these air conditioners running.

Thus there seems to be a wide range of tolerance in the noise levels that people can accept during sleeping. Some intermittent noises can and do exceed the above range of noise criteria considerably, depending upon the nature of the noise, the nature of the residential area, and the temperament of specific individuals in the area.

To produce NC–20 to NC–30 noise levels inside the house, outdoor

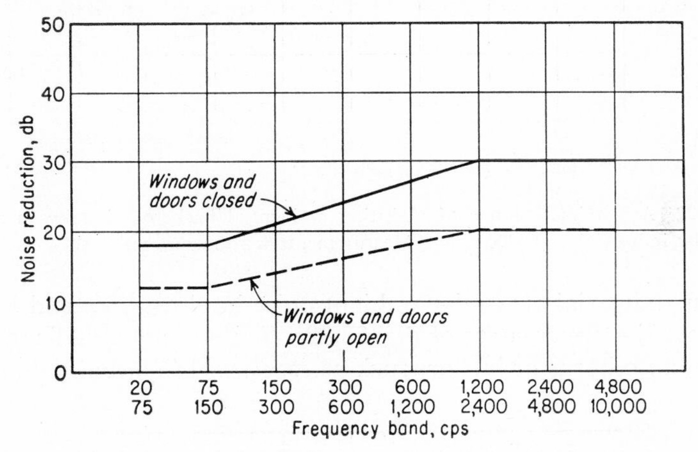

Fig. 23.21 Approximate noise reduction of outside noise provided by a typical one-family frame house.

noise levels may be somewhat higher because there is some reduction of noise in going from outside to inside, even for open windows in the house. Figure 23.21 shows the approximate reduction of outside noise provided by a typical one-family frame house These curves are the averages of several readings taken for several typical residences in which the window area was about 10 to 30 per cent of the outside wall area. For the curve labeled "windows and doors partly open," the windows ranged 10 to 50 per cent open. For different home-construction materials and for different areas of open and closed windows and for various room layouts, the noise reduction values will vary from those of Fig. 23.21, but the shape of the NR curves will generally follow the shape of these curves.

Based on noise measurements made in a number of residential areas, we plot in Fig. 23.22 a range of outdoor nighttime background levels found for typical residential areas. These are ambient levels recorded at times when nearby automobiles or intermittent noises are absent. In rural residential areas, the noise levels may range about 5 db below the lower curve; and for residences on or near busy streets

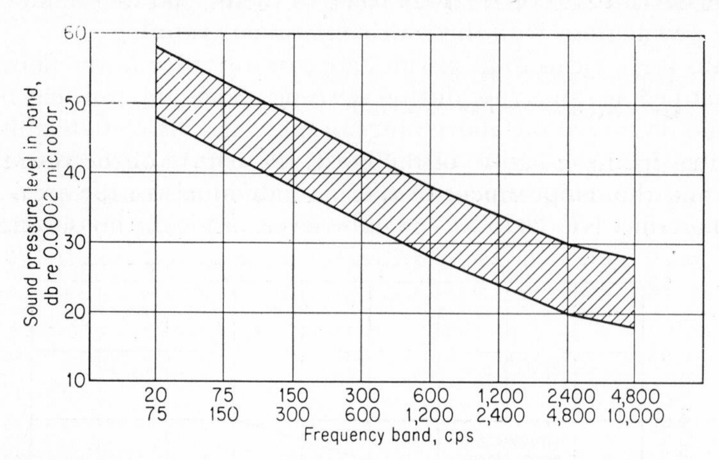

FIG. 23.22 Approximate range of nighttime residential background noise in absence of nearby industrial or traffic noise. Measurements were made outdoors.

or near industrial areas, the ambient noise levels may extend 5 to 10 db above the upper curve of Fig. 23.22. The noise of specific sources, such as auto horns or bus or truck passages, will be even higher.

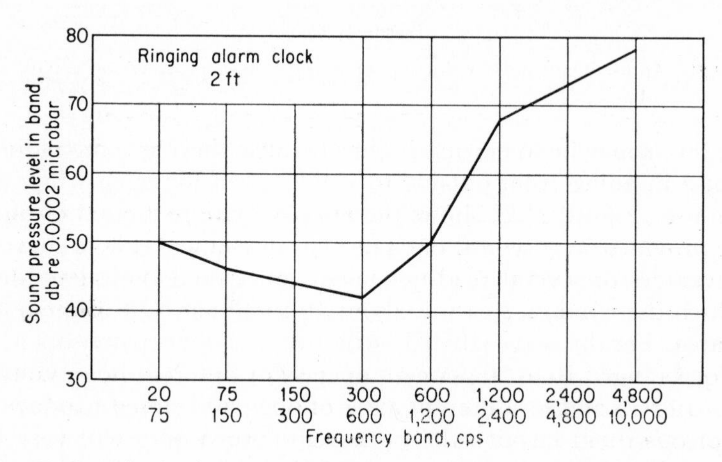

FIG. 23.23 Noise levels measured 2 ft from a ringing Baby Ben alarm clock.

If NC–20 to NC–30 criteria might be taken as applicable for sleeping, what noise levels will cause waking? This is a question with a complex answer, which is now under study by Karl Kryter and others.

Until we learn some of the acoustic and psychological factors which cause waking under noise stimulus, we may have to continue to rely on the alarm clock to induce wakefulness. Noise levels at a distance of 2 ft from a Baby Ben alarm clock are given in Fig. 23.23.

23.11 Home Noise Problems Caused by Outside Noise Sources

Noise in the home comes either from noise sources outside the home or from those inside the home. Outside noise sources are principally vehicular and industrial or commercial; but in some unfortunate cases noise sources may involve inconsiderate neighbors. Regardless of the source, the two basic solutions are (1) to try to reduce the noise at the source, and (2) if the noise persists, to try to protect the house against penetration of excessive amounts of the noise. For the purpose of this discussion we will assume that the first step has been carried out as far as possible but that some noise persists and the resident wishes to provide the desired noise control for his home.

If the home is already built, many control variables are already fixed and would be expensive or impractical to change. Thus, we caution that there are not many solutions available and the available ones may not produce very much noise reduction. If the house is not built, or if the site has not even been purchased, there is much more latitude in noise-control measures. It may seem rather elementary to warn about home sites being located too near airports or highways or public transportation routes, but it is surprising to learn how many people buy houses without knowing the flight patterns used at nearby airports or without visiting the site late at night to actually listen to the noise of nearby traffic. We even know of some acousticians who have done this!

In Fig. 23.6 some aircraft noise levels were presented. Figure 23.24 gives the noise levels of a few passing railroad freight trains at a distance of about 100 ft. These trains were pulled by diesel engines operating at slow speeds of 10 to 40 mph through a residential neighborhood.

Figure 23.25 gives the noise levels which would be produced by a new truck whose noise output satisfied the 125-sone specification adopted in 1954 by the Automobile Manufacturers Association. Although many new trucks may meet this specification on delivery, most trucks in current use exceed the specification.[8]

The curves in Fig. 23.26 show the noise levels associated with a typical automobile horn. The curves give the noise levels for various headings of the automobile relative to the recording microphone.

The noise levels produced by neighbors or neighboring industrial or commercial enterprises can be so varied that no attempt will be made to categorize them here. For specific noise problems, the noise levels should be measured in order to determine the type of treatments to be undertaken.

The following noise treatments will be discussed: (1) relocation and treatment of bedrooms, (2) improvement of windows, (3) use of masking noise, and (4) construction of fences and barriers.

Relocation and Treatment of Bedrooms. Since sleeping requires the lowest noise levels, we will limit our interest to bedrooms in this

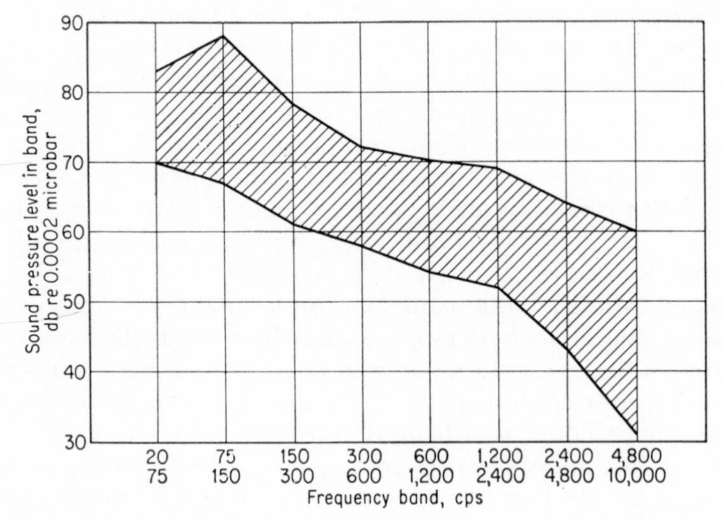

FIG. 23.24 Noise levels measured for several freight train passages (10 to 40 mph) at a distance of about 100 ft.

section. It is not a very ingenious suggestion, but where noise comes predominantly from one direction it is possible that some of the rooms of a house could be traded around to provide the lowest noise levels for the most sensitive sleepers. In one residence, the noise of passing rapid-transit trains was no disturbance at all to the two teen-agers in the house. So they were given the bedrooms facing the tracks. The parents were more sensitive sleepers and benefited from a move to the bedrooms on the opposite side of the house away from the tracks. The shielding provided by the house was good for 10 to 20 db of noise reduction. If you don't have teen-agers in your house, you're out of luck! Sometimes, teen-agers can provide enough masking noise that you could seldom hear the trains anyway.

For sounds coming from any direction, such as with airplane flyovers, there is little preference for bedroom location. In this case, a small amount of local noise reduction can be achieved with the use of acoustic absorption material on the ceiling and full-area thick

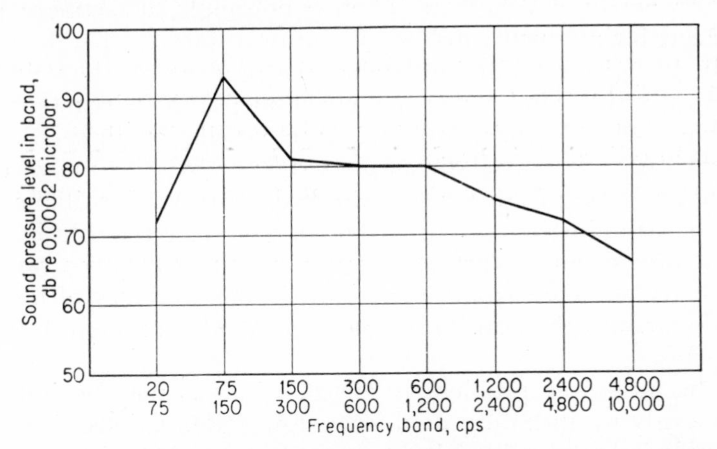

Fig. 23.25 Noise levels measured 50 ft from a typical new truck which will meet 1954 specification of 125 sones adopted by Automobile Manufacturers Association.

carpets on the floor. The benefits to be gained are not large, so we caution against undue optimism in applying this treatment.

For high-frequency noise, some reduction of noise levels can be

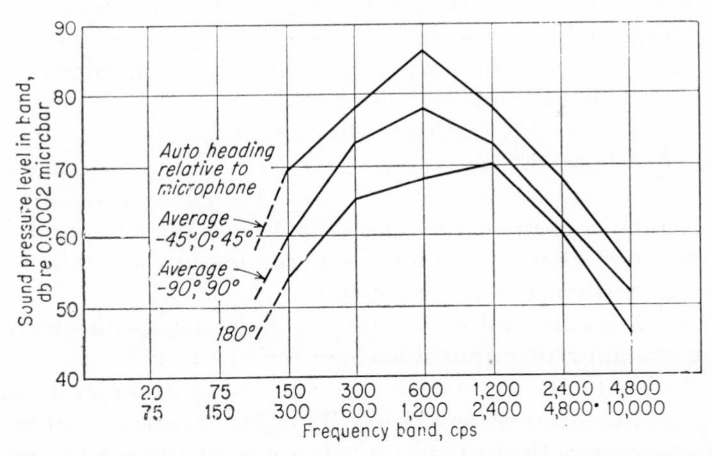

Fig. 23.26 Noise levels of automobile horn measured at distance of 80 ft (1956 lower-priced American model).

achieved with the use of portable screens (thin plywood or masonite, not cloth or porous fabric) placed near the window in such a way as to interrupt the direct sound path to the sleeper. Such screens, coupled with floor and ceiling absorption, can produce 5 to 10 db noise reduction in the high-frequency bands. The larger the area of the screen, the more effective it will be.

Improvement of Windows. There is no simple, attractive acoustic treatment for an open window. We must assume in this discussion that if the noise is really intolerable, the resident will be willing to close the outside windows. The mere closing of the windows may be enough to solve the noise problem. If so, ventilation air should then be provided by a central air-supply system or by some means of inducing or forcing fresh air into the bedroom. With suitable basement or attic ducting and small fans it would be possible in many cases to provide an inexpensive air supply without having a complete central system. One should not overlook the possibility of bringing outside air into the house through open windows remote from the bedrooms.

If the air-supply problem is solved, windows may be improved significantly by installing double-window combinations. First, the existing window should either be completely sealed (airtight) or replaced with a sealed-in window. Second, an additional window should be added. This window should also be sealed or gasketed in place and the air space between the windows should be made as large as possible (4 to 6 in. or more would be very good). Provision should be allowed for opening or removing one of the windows occasionally for window cleaning. Also, a small tray of dehydrated desiccant should be placed in the space between the windows to take away the moisture which would otherwise condense on the windows and obscure clear vision during periods of cold weather.

The approximate NR achieved by closed windows relative to open windows is shown in Fig. 23.21. The use of a double window will give several more decibels NR. To be most effective for the exclusion of all types of noise, the glass should be as thick as possible, and the glass or window edges should be installed in soft rubber gaskets to provide resilience of the mounts.

Use of Masking Noise. If ambient noise levels due to outside sources contain rather annoying characteristics or unpleasant sounds, it is possible to reduce some of the annoyance by superimposing a steady unobtrusive masking noise. This has been found to be effective in many practical situations. It is important that the masking noise not introduce a new cause of annoyance. For summer-time use, a window air-conditioner serves the purpose of a masking noise for it normally has no really unpleasant sound and it gives the benefit of fresh, cool air when needed. Also, in forced-air-supply systems the noise of air entering the bedroom through louvered openings or a series of small orifices will produce some masking noise. Small axial-flow air circulating fans also produce noise which could be used for masking purposes.

As a last resort, electronically produced "white noise" can be fed to one or more small loudspeakers inconspicuously placed around the room to artificially raise the ambient levels a small amount.

Fences and Barriers. A lot of walls have been built and a few papers have been written to demonstrate that solid fences or barriers provide worthwhile amounts of noise reduction for certain conditions. In general, if a solid wall can be built to extend about 5 to 10 ft above and beyond the line of sight between a noise source and the receiver, and if the wall can be located within 10 to 20 ft of either the noise source or the receiver, that wall will provide a noise reduction of about 5 to 10 db in the low-frequency bands and about 15 to 25 db in the high-frequency bands. We assume in this case that we are trying to protect a house or a garden or a window from the noise of passing automobiles, trucks or trains, and that the dimensions given are considered practical from this point of view. As the height of the wall increases and the distance from the wall to either the source or the receiver decreases, the wall becomes even more effective. For lower walls and greater distances, however, the wall becomes less effective. When the distance from the wall to either the source or the receiver becomes very large (several hundred feet), the wall loses some value because of the refraction effects of sound waves bending over the wall due to wind and thermal conditions.

For estimating the effective noise reduction of a solid wall for a range of possible dimensions, the reader is invited to study one or more of the indicated references.[9]

The wall should be of solid construction and should have a surface weight whose TL is not less than the NR estimated for the wall. In effect, this implies that the wall should be not less than about 5 lb/ft^2.

One factor of concern in setting up such a wall is to ascertain that other reflecting surfaces nearby will not reflect noise over or around the barrier wall in such a way that the full effectiveness of the wall will not be realized.

Another factor involved in setting the height of the barrier wall is to be sure of the actual location of the noise source which the wall is intended to protect against. As an example, on an expressway, tire noise originates at road level; auto and truck engine noise originates a few feet above road level; but the end of the exhaust pipe on many diesel trucks extends 7 or 8 ft into the air above road level. In the case of trains, wheel clicking takes place at track level, some of the wheel clicking noise is probably radiated by the lower side walls of the train cars, and the diesel engine noise escapes from vents along the top of the engine. Thus, we see that it is important that the wall height take into account the full height of the disturbing noise.

Residents frequently ask if trees and shrubs will help reduce noise. Occasional trees and shrubs will have no effect, and thick hedges that may serve as complete visual barriers will have almost no measurable effect. It takes large thicknesses (perhaps 50 to 100 ft thick) of dense growth in order to obtain any significant amount of noise reduction (possibly 5 to 10 db in the high-frequency region), and even then a solid wall would be better and much less expensive. Carl Eyring[10] gives some data on sound transmission through tropical jungle areas which indicate that very little noise reduction can be achieved with small strips of wooded areas. (Eyring gives TL's of 10 to 20 db in the low frequencies and 70 db or more in the high frequencies for 1,000 ft thicknesses of growth of different degrees of denseness.)

To conclude this section, we regret to advise that there are not many simple ways to achieve large amounts of reduction of external noise after a house is built. The changes required to achieve considerable noise reduction are expensive in that they usually involve drastic changes to the house. As we have seen, some of the available noise-control treatments are relocation of bedrooms, closing-up and improvement of windows (possibly resulting in an air-supply problem), introduction of masking noise to reduce the apparent severity of the intruding noise, and construction of barrier walls or screens to protect the house or parts of the house from certain kinds of noise sources.

23.12 Home Noise Problems Caused by Inside Noise Sources

Inside noise sources in homes are chiefly of three types: (1) automatic home appliances, (2) heating and air-conditioning systems, and (3) entertainment devices. Each of these types of noise makers serves the home-dweller in very positive and worthwhile ways and many of the devices have come to be identified with their characteristic noise output. Thus, the busy housewife may not be very disturbed by the noise of an automatic dishwasher if it frees her from a kitchen chore so she can watch television with the family. If the dishwasher is too noisy, just turn up the TV set!

The next few figures present the noise levels measured for a variety of automatic home appliances. Figures 23.27 and 23.28 show what might appear to be a progressive quieting of dishwashers and clothes washers over a 10-year period of manufacturing. But, look at

Fig. 23.29; the garbage disposal units seem to be getting noisier! The noise levels of typical clothes driers and tank-type vacuum cleaners are shown in Figs. 23.30 and 23.31. The manufacturers of home appliances appreciate the desire for low noise levels, but noise control

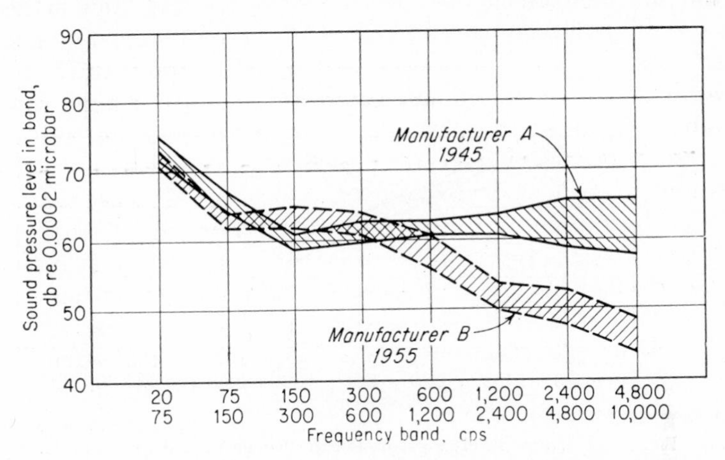

Fig. 23.27 Noise levels measured near two automatic dishwashers during washing cycle.

is only one of many design requirements, and the merchandising cost and the competitive pace ultimately decide how quiet the products will be.[11]

There is no doubt that many of these appliances can be made quieter by the manufacturer at some increase in cost, size, and weight.

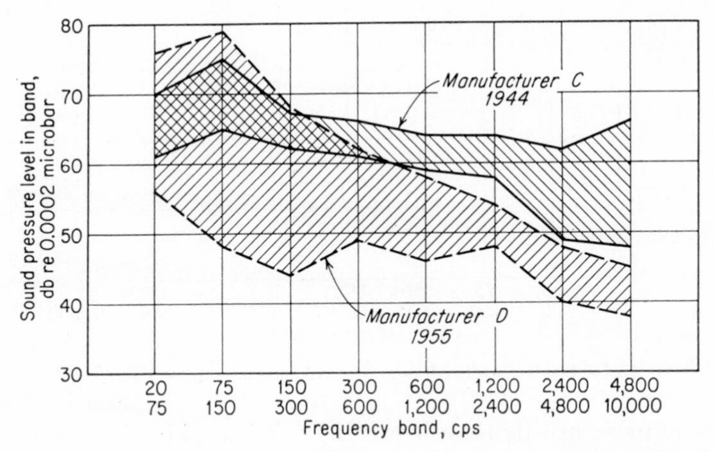

Fig. 23.28 Noise levels measured near two automatic clothes-washing machines during all active portions of washing cycle.

We are not in a position in this discussion to champion the cause of quiet products; that must come from large numbers of buyers. Also, since the completely assembled appliance is already filled to capacity

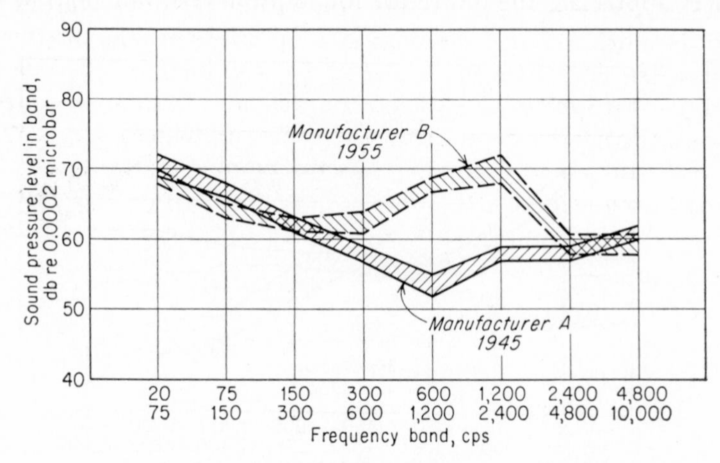

FIG. 23.29 Noise levels measured near two garbage-disposal units.

with the various components designed for the unit, there is usually very little space or opportunity for the do-it-yourself home engineer to make any significant noise-reduction improvements. Thus, the only remaining noise-control treatment is to insure that the unit op-

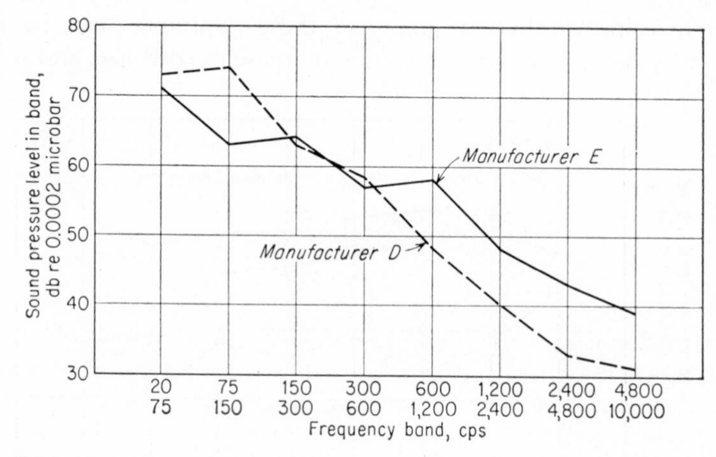

FIG. 23.30 Noise levels measured near two automatic clothes driers (1955 models).

erates no more noisily than it was intended to. Metal support straps, metal sidewalls, or metal cover lids sometimes rattle. The automatic washers and driers during active cycles sometimes vibrate enough to set other nearby appliances into vibration. The connecting pipes may vibrate and transmit noise through the house. These minor annoyances can usually be located and eliminated. The general treatments available, beyond the elimination of minor rattles, are to add as much acoustic absorption as possible in the work area around these appliances and to reduce as much as possible the "sounding-board effect," i.e., keep the machine from setting up vibration in the nearby floor, walls, and other appliances.

Heating and air-conditioning systems operate more nearly continuously, even at night; so there is more justification for quiet operation of this equipment. Hot-water-circulating pumps occasionally

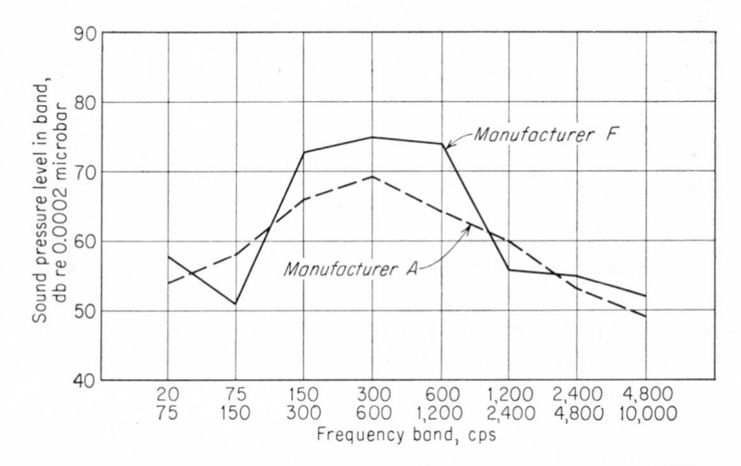

Fig. 23.31 Noise levels measured near two tank-type vacuum cleaners (1955 models).

transmit whines into the piping system which can be heard throughout the house. If the pump can't be replaced or made quieter, some noise might be reduced by removing the rigid pipe supports and resiliently hanging all exposed piping from the basement ceiling. Fan noise may be excessive in some forced-air heating systems. The addition of acoustically absorbent duct lining inside the available air ducts may be enough to reduce the noise to an acceptable level. Fan operation may also set up low-frequency vibration in the furnace proper, and this may be transmitted to the building by means of interconnecting pipes, conduits, ducts and tie-rods. These rigid connections can usually be broken and replaced with flexible connections.

Small-diameter water pipes in homes result in high water-flow velocities and this produces noise throughout the areas served by the pipes. If it is not worth while to install larger pipes, some noise improvement can probably be realized by detaching the pipes from their rigid supports along the basement ceiling and supporting these exposed lengths of pipe with resilient rubber-in-shear hangers or in fabric or metal straps isolated from the pipes with a 1- to 2-in.-thick

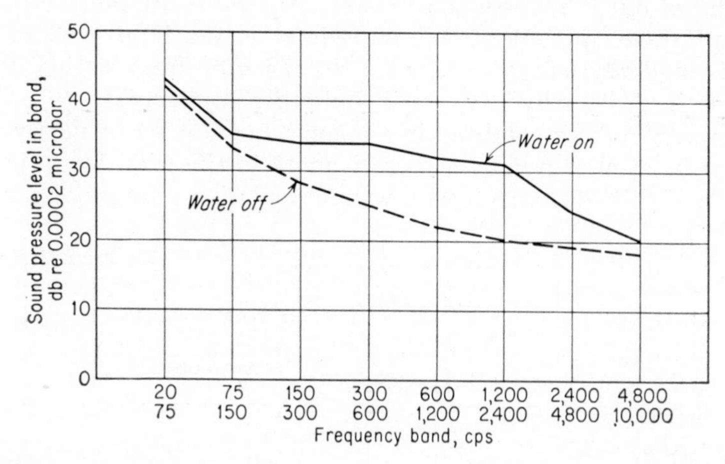

Fig. 23.32 Noise levels measured in dining room of residence with water flowing through water pipes in basement. Pipes are rigidly attached to floor joists.

wrapping of glass-fiber pad. Figure 23.32 shows the noise levels produced in a first floor room by water flow in unisolated pipes in the basement below.

Air-conditioning system noise is becoming of concern in many homes. Eugene Mikeska has discussed this and other home noise

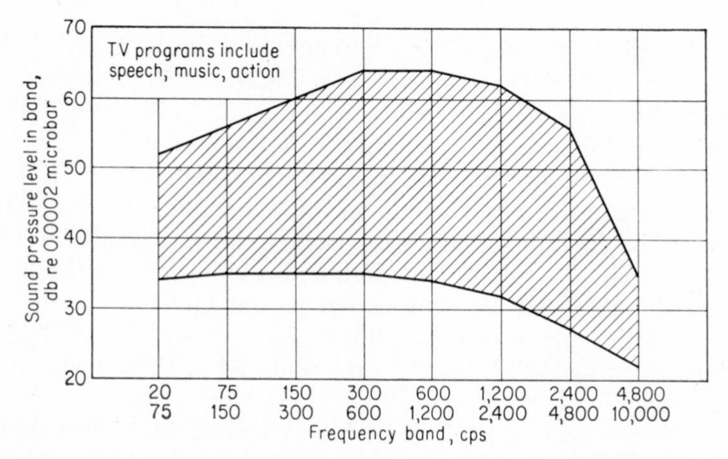

Fig. 23.33 Range of sound levels 12 ft from television set at typical operating level. (TV programs include speech, music, and action.)

problems in recent articles.[12] Adequate and complete vibration isolation of the compressor and fan, and proper attention to air-borne noise paths are essential to a quiet installation. The same principles discussed earlier for air-conditioning systems of office buildings are applicable for home equipment.

Figure 23.33 presents a typical range of sound levels measured for a television set. This is offered to illustrate that fairly high levels are produced even though the listeners may not be aware of this.

The above review of noise sources inside the home emphasizes the fact that we are being surrounded with devices (now being considered as necessities rather than luxuries) which produce intermittent noise levels ranging up to the NC–50 to NC–70 criterion levels. The length of time that these noise levels are "on" is a significant part of the day, so it is not unreasonable to begin to be concerned about how to escape from these noises into quiet environments for eating, sleeping, relaxing, reading, studying, and conversing.

From these examples of home noise sources, we see that for home construction, just as for office construction, we have the following worthwhile objectives:

1. Walls, floors, doors, and windows should have high TL.
2. Adequate vibration isolation should be provided in equipment mountings to reduce structure-borne transmission of noise.
3. Acoustic absorption materials are essential to noise control when used intelligently.

23.13 Summary

In this chapter we have noted first the need for setting noise criteria for specific office and home activities in order to determine the amount of noise reduction desired or required. We have then categorized the types of noise problems usually found in offices and homes. Noise in offices is usually attributable to nearby industrial activity, outside traffic (air-borne and sometimes structure-borne), railroad and subway movements, aircraft, office equipment, and ventilating and air-conditioning equipment. Noise in homes can be separated into two groups: noise from outside noise sources and that from inside noise sources.

Noise-control measures include suitable and thorough vibration isolation mounting of vibrational sources, use of barriers and partial enclosures for limited noise control, use of massive walls and floor structures for high-transmission-loss applications, proper attention to snug-fitting doors and windows, location of acoustically critical areas in the quieter parts of a building, use of masking noise for some special noise problems, and adequate use of acoustically absorbent material in rooms, passageways, and interconnecting ducts.

REFERENCES

1. Miller, L. N., and I. Dyer: Noise Control for Offices Located near Production Machines and Mechanical Equipment Spaces, *Noise Control,* vol. 3, no. 2, pp. 70–75, March, 1957.

2. Bolt Beranek and Newman Inc.: Studies of Noise Characteristics of the Boeing 707–120 Jet Airliner and of Large Conventional Propeller-driven Airliners, *Rept. for the Port of N.Y. Authority,* October, 1958; R. M. Gibbs and H. H. Howell: Noise Characteristics for the Boeing 707, *Noise Control,* vol. 5, no. 1, pp. 13–17, January, 1959.

3. Pietrasanta, A. C.: Aircraft Noise and Building Design, *Noise Control,* vol. 3, no. 2, pp. 11–18, March, 1957.

4. Miller, L. N., L. L. Beranek, and Karl D. Kryter: Airports and Jet Noise, *Noise Control,* vol. 5, no. 1, pp. 24–31, January, 1959. V. E. Callaway: Terminal Operation of 707, *Noise Control,* vol. 5, no. 3, pp. 42–43, May, 1959.

5. Stalker, W. W.: The Evaluation and Control of Noise in the Offices of an Industry, *Noise Control,* vol. 1, no. 4, pp. 34–36, July, 1955.

6. Mikeska, E. E.: Transmission Loss and Noise Reduction, *Noise Control,* vol. 4, no. 2, pp. 37–41, March, 1958. J. B. C. Purcell: Control of Airborne Sound by Barriers, *Noise Control,* vol. 3, no. 4, pp. 20–26, July, 1957. B. G. Watters: Transmission Loss of Some Masonry Walls, *J. Acoust. Soc. Am.,* vol. 31, pp. 898–911, July, 1959.

7. Veneklasen, P. S.: Sound Attenuating Gaps, *J. Acoust. Soc. Am.* (*Abstr.*), vol. 31, no. 6, p. 833, June, 1959.

8. Apps, D. C.: The AMA 125-sone New Vehicle Noise Specification, *Noise Control,* vol. 2, no. 3, pp. 13–17, May, 1956; Recent Developments in Traffic Noise Control, *Noise Control,* vol. 3, no. 5, pp. 34–36, September, 1957.

9. "Handbook of Acoustic Noise Control," vol. I, suppl. I, p. 295, *WADC Tech. Rept.* 52–204, 1955; Wesley Nyborg and David Mintzer: The Loss in Level Due to a Wall, *Noise Control,* vol. 3, no. 6, p. 55, November, 1957; M. Rettinger: Noise Level Reductions of Barriers, *Noise Control,* vol. 3, no. 5, pp. 50–52, September, 1957; R. O. Fehr: The Reduction of Industrial Machine Noise, *Proc. 2d Annual NNAS,* Chicago, October, 1951.

10. Eyring, Carl F.: Jungle Acoustics, *J. Acoust. Soc. Am.,* vol. 18, p. 257, 1946.

11. Burke, Richard S.: Merchandising Quiet Products, *Noise Control,* vol. 1, no. 2, pp. 24–26, March, 1955; G. Norman Sawyer: Noise Control of Appliances, *Noise Control,* vol. 1, no. 2, pp. 27–32, March, 1955; R. Plunkett: Noise Control of Domestic Appliances, *Metal Prods. Mfg.,* November, 1958.

12. Mikeska, E. E.: Air-conditioning Equipment Noise Levels in Homes, *Noise Control,* vol. 3, no. 3, pp. 11–14, May, 1957; Noise in the Modern Home, *Noise Control,* vol. 4, no. 3, pp. 38–41, May, 1958.

SUGGESTIONS FOR FURTHER READING

Parkin, P. H., and E. F. Stacy: Quieting of Apartments and Houses, *Noise Control,* vol. 1, no. 1, pp. 40–45, January, 1955.

Veneklasen, P. S.: City Noise—Los Angeles, *Noise Control,* vol. 2, no. 4, p. 14, July, 1956.

Salmon, Vincent: Surface Transportation Noise—A Review, *Noise Control,* vol. 2, no. 4, p. 21, July, 1956.

Newman, E. B.: Psycho-physical Effects of Noise, *Noise Control,* vol. 1, no. 4, p. 16, July, 1955.

Ashley, C. M.: Air-conditioning Noise Control, *Noise Control,* vol. 1, no. 2, p. 37, March, 1955.

Schuler, S.: As Quiet as a Christmas Mouse, *American Home,* December, 1958.

"Home Noise and What to Do about It," National Noise Abatement Council, 36 W. 46th St., N.Y. 36, N.Y.

The following articles appeared in *Noise Control,* vol. 3, no. 4, July, 1957: T. Mariner: Control of Noise by Sound Absorbent Materials; D. P. Loye: Noise Control in Hotels, Hospitals, and Multiple Dwellings; L. S. Goodfriend: Noise Control in Civic Buildings; W. W. Soroka: Noise Control in Office Buildings.

Chapter 24
JET NOISE

Peter A. Franken

24.1 Introduction

In recent years jet-noise sources have become important problems for noise control. In the case of turbojet engines, noise control may be necessary to avoid structural fatigue of the airframe or annoyance within the airplane or in the airport neighborhood. In the case of steam jets or high-pressure-ratio air jets, frequently used for industrial processes, noise control may be necessary to avoid hearing damage to workers.

The first section of this chapter will be devoted to a survey of engineering information presently available on the noise field associated with a jet. We will present procedures for estimating the sound-pressure levels at any point in this noise field. The second section of this chapter will discuss advances that have been made in jet noise control.

Before we start on a description of the jet-noise field, it will be useful to look at the geometry of a typical jet as shown in Fig. 24.1. The jet fluid emerges from the exit nozzle of diameter D at the far left. Distance downstream from the nozzle is measured in terms of the nondimensional quantity x/D, or number of diameters downstream from the nozzle. The portion of the jet stream right at the nozzle is the central core, in which flow is approximately sonic (Mach = 1.0) and very little turbulence is generated. This core comes to an end about $4\frac{1}{2}$ jet diameters downstream from the nozzle. At this point we are in the mixing region where the jet stream is being mixed with the ambient medium. This mixing region extends for many diameters downstream until the flow is fully turbulent.

The source of noise in a jet stream is not located right at the jet nozzle but is actually distributed over a considerable distance downstream from the jet nozzle. Qualitatively, it has been found that the very high frequencies of jet noise are produced close to the nozzle and the lower frequencies are produced further downstream from the jet nozzle. This distribution of noise sources along the jet stream leads to complications when we attempt to measure jet noise fairly close to the stream. Later we will discuss these near-field complications. (The term "near field" is described more fully in Chap. 8.)

The jet geometry shown in Fig. 24.1 is typical of present-day turbojet engines operating at or near *maximum rated power* (*100 per cent rpm*).* For the time being we will concentrate on this case. Later on we will discuss exceptions to this common case.

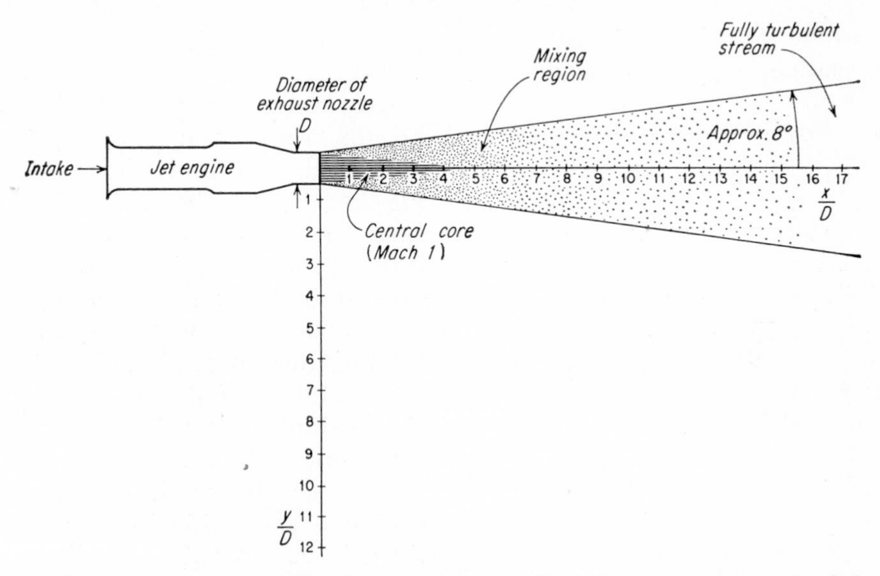

FIG. 24.1 Jet-engine geometry and typical jet-stream-flow pattern. Distance and radial distance from the jet center line are in terms of number of nozzle diameters (x/D and y/D, respectively).

24.2 Description of the Jet-noise Field

In previous chapters we have presented convenient methods of describing complex noise sources. We have seen that one important number needed to describe a noise source is the total sound power radiated by that source. This sound power may be given in terms of sound-power level (PWL), a logarithmic quantity with a reference of 10^{-13} watt. It is also important to describe the directional characteristics of the noise source and the frequency spectrum—a pure tone or a broadband, a high screech or a low rumble. Because a jet

* In United States military aircraft this is called "military power."

is a complex noise source, any description of the jet-noise field requires that we be able to give its sound-power level, its directivity, and its frequency-spectrum characteristics.

Turbojets at Maximum Rated Power (100 Per Cent rpm)—Far-field Noise. 1. *Stationary Conditions.* A general expression for the far-field sound-power level of turbojets may be written as:

$$\text{PWL} = 10 \log Av^8 + \text{constant} \qquad (24.1)$$

where A is the open area of the jet nozzle and v is the jet velocity relative to the ambient air.

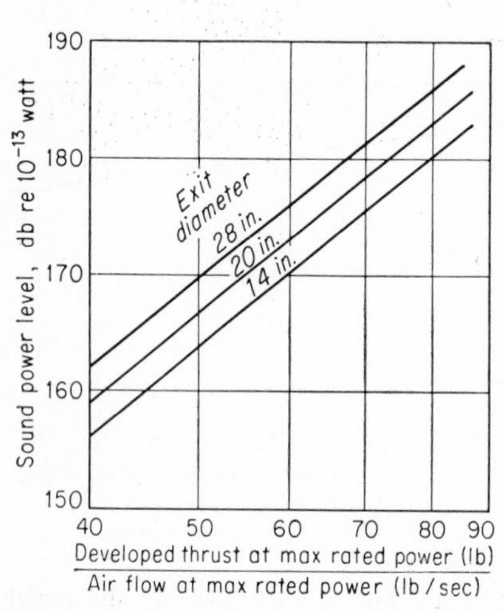

FIG. 24.2 Sound-power level in far field of turbojets at maximum rated power (100 per cent rpm). The abscissa is the ratio of thrust in lb to airflow in lb/sec. The exit diameter of the nozzle is the parameter which describes each of the lines. To convert from lb-thrust to newtons-thrust, multiply the former by 4.45. To convert from lb/sec to kg/sec, divide the former by 2.2. (See Appendix B for the English system of units.)

A convenient way of plotting Eq. (24.1) for turbojets at maximum rated power is shown in Fig. 24.2.[1] The abscissa on this figure represents the ratio of thrust produced by the engine divided by the airflow through the engine. The parameter describing the various lines is the exit diameter of the jet nozzle. For example, for a J-57 engine producing about 10,000 lb of thrust, the airflow through the engine is about 160 lb/sec and the exit diameter of the nozzle is about 22 in. Using these values in Fig. 24.2 we arrive at a power level for the J-57 operating at maximum rated power of 173 db, which is in good agreement with measurements.

A directivity curve for turbojets operating at maximum rated power is given in Fig. 24.3.[2] Here the abscissa gives the angle in degrees measured from the jet axis. The over-all sound-pressure level (SPL) is given relative to the space average SPL. This space average SPL is the SPL at any point on a hemisphere enclosing a nondirective source of the same PWL. We see that at about 40° from the jet axis there is a strong peak in power radiated by the turbojet. This directivity peak is typical of the noise field radiated by all jets.

A typical spectrum curve, in octave bands of frequency, for a turbojet at maximum rated power is shown in Fig. 24.4.[2] Since the turbojet noise is broadband in character, it is not generally necessary to analyze the noise in any greater detail than is given by octave bands. The jet-noise field can be divided into two angular regions, the first,

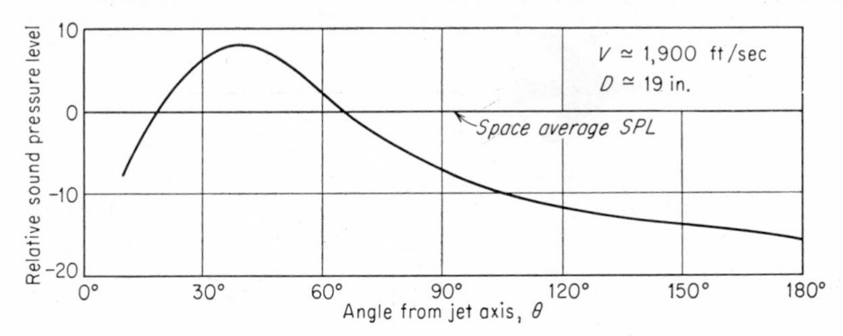

FIG. 24.3 Typical far-field turbojet directivity at maximum rated power. The abscissa is the angle θ measured from the jet axis. The curve gives the SPL relative to the space average SPL. The data are for a jet velocity of about 1,900 ft/sec and nozzle diameter of about 19 in.

20° to 50° from the jet axis, and the second, 60° to 180° from the jet axis. (Positions less than about 20° from the jet axis are generally too close to the jet stream to permit accurate noise measurements. Figure 24.3 indicates that the SPL at these positions makes a negli-

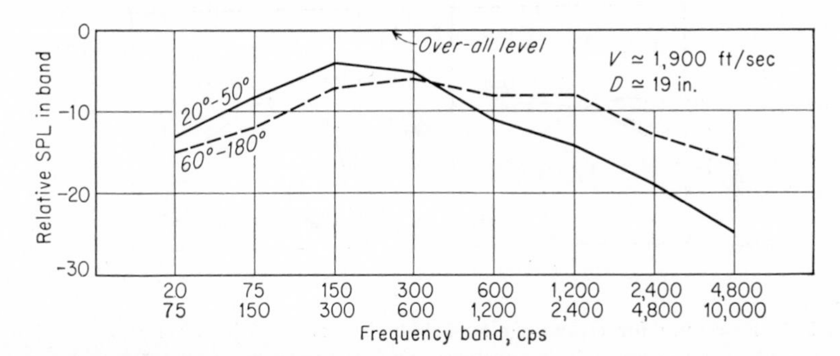

FIG. 24.4 Typical far-field turbojet spectra at maximum rated power. The curves give the SPL (approximately) in octave bands relative to the over-all level. The solid line is for 20° to 50° from the jet axis and the broken line is for 60° to 180° from the jet axis. The data are for a jet velocity of 1,900 ft/sec and nozzle diameter of about 19 in.

gible contribution to the PWL of the entire jet.) In these two regions the turbojet spectrum has the shapes shown in Fig. 24.4.

The information contained in Figs. 24.2, 24.3, and 24.4 may be combined into an estimation procedure for the far-field noise from stationary turbojets at maximum rated power, as shown schematically in Fig. 24.5. The sound-pressure level is desired at point P, which is located at distance r (measured in feet) from the exhaust nozzle of the turbojet engine, and at an angle θ from the jet axis as shown

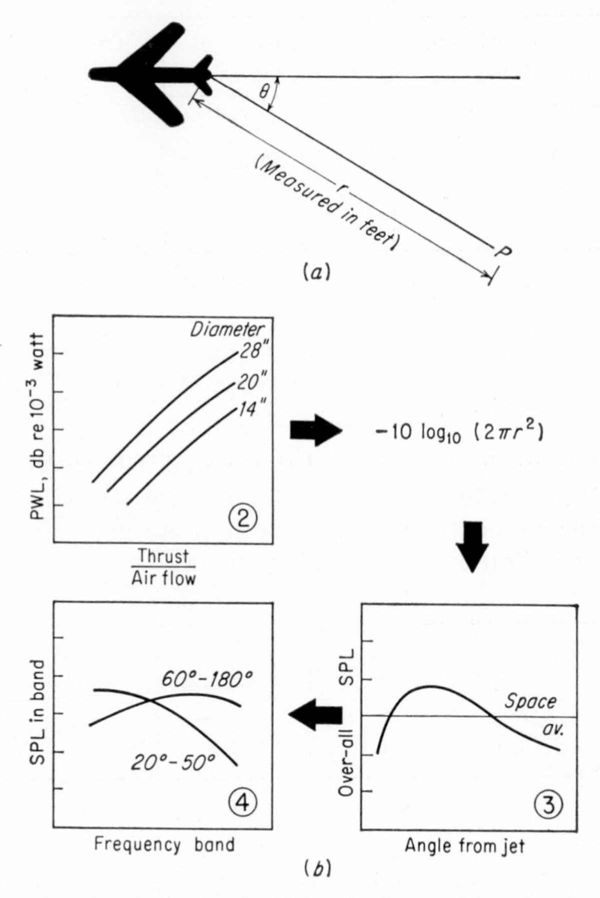

FIG. 24.5 Procedure for estimating far-field noise levels of jet aircraft at maximum rated power on the ground: (*a*) diagram showing observation point P at distance r from exhaust nozzle and angle θ from jet axis; (*b*) schematic representation of estimation procedure involving Figs. 24.2, 24.3, and 24.4.

in Fig. 24.5*a*. The power-level information from Fig. 24.2 is combined with the directivity information of Fig. 24.3 and the spectrum information of Fig. 24.4 as shown in Fig. 24.5*b*. The term 10 log $2\pi r^2$ accounts for hemispherical spreading of the jet noise. In other words, the jet on the ground can radiate only into a hemisphere. At a distance greater than about 400 ft, atmospheric effects and ground absorption become significant and must be taken into account.

Example 24.1. Find the sound-pressure level at a point 100 ft from the exhaust nozzle of the J-57 (PWL = 173 db) and at an angle of 90° with the jet axis.

Solution. The directivity correction from Fig. 24.3 is −7 db at 90°. The quantity $10 \log 2\pi(100)^2$ is 48 db. Therefore, the desired over-all sound-pressure level is

$$\begin{aligned} \text{SPL}_{\text{over-all}} &= \text{PWL} + \text{directivity} - 10 \log 2\pi r^2 \\ &= 173 - 7 - 48 = 118 \text{ db} \end{aligned} \tag{24.2}$$

and the sound-pressure level in octave bands is obtained by adding 118 db to the values given by the broken line in Fig. 24.4 which is applicable for the 90° position.

2. *Effects of Motion and Ambient-condition Changes.* We now ask how the above estimation procedure should be modified when the turbojet aircraft is in motion. We have stated earlier that the sound power radiated by the turbojets is proportional to the jet velocity raised to the eighth power. For a moving aircraft, the jet velocity relative to the ambient medium becomes the jet velocity minus the aircraft velocity, and thus the eighth power relation predicts a reduction in sound power with forward motion of the aircraft. The radiation pattern from a moving jet is less directional than that of a stationary jet.

Actually, the observed situation is considerably more complicated.[3] The low frequencies may be reduced by as much as 10 db over their static values. The high frequencies will be reduced considerably less or, in some cases, increased over their static values. These effects are explained qualitatively as the simultaneous decrease of jet engine exhaust noise, and the increase of noise from the rapid motion of the surrounding air into the wake of the aircraft. Also, at flight speeds higher than about Mach 0.5, the pressure fluctuations in the turbulent boundary layer set up motion in the aircraft skin, providing a new source of noise within the cabin.

It is important to consider the change in sound power which may result from a change in ambient conditions or from a change in the jet exhaust temperature. The power-level change is found to be[4]

$$\begin{aligned} \Delta\text{PWL} &= 10 \log \frac{m}{m_0} + 10 \log \frac{T_j}{T_{j0}} + 70 \log \frac{v}{v_0} - 35 \log \frac{T_a}{T_{a0}} \quad \text{db} \\ &= 10 \log \frac{p_a}{p_{a0}} + 40 \log \frac{T_j}{T_{j0}} - 35 \log \frac{T_a}{T_{a0}} \end{aligned} \tag{24.3}$$

where 0 subscripts refer to standard conditions

m = mass flow through engine
v = relative jet velocity
p_a = ambient pressure
T_j = static jet temperature
T_a = ambient temperature

Figure 24.6 presents Eq. (24.3) in graphical form. The variations in p_a and T_a relative to standard conditions are readily determined.

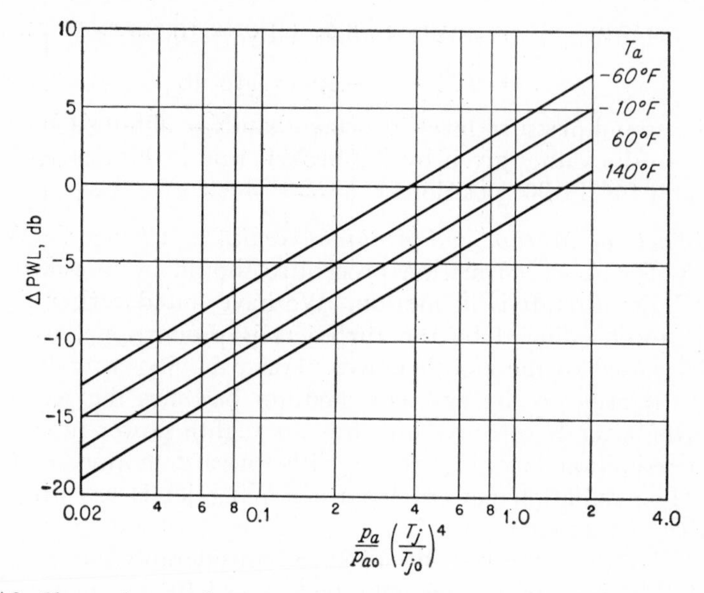

FIG. 24.6 Change in sound power with variations in ambient conditions and jet temperature. The subscripts a, j, and o refer to ambient conditions, jet conditions, and reference conditions, respectively. Each of the diagonal lines is described by an ambient temperature T_a. The reference ambient temperature is taken as 60°F. The ambient pressures and jet temperatures appearing in the abscissa scale are measured in absolute units such as pounds per square inch absolute, degrees Rankine, etc.

Variations in the tailpipe temperature will provide a satisfactory approximation to the variation in T_j.

The sound-pressure-level changes in the far field corresponding to the power-level changes are obtained from the relation

$$\Delta\text{SPL} = \Delta\text{PWL} - \Delta B \qquad \text{db} \tag{24.4}$$

ΔB is the number whose value depends on the change in the quantity $\rho_a c_a$, where ρ_a and c_a are the ambient density and sound velocity respectively. ΔB is plotted in Fig. 24.7. Figures 24.6 and 24.7 may then be used to calculate the changes in the sound-pressure level with changes in ambient conditions and jet temperature.

Turbojets at Maximum Rated Power (100 Per Cent rpm)—Near-field Noise. As pointed out earlier, the above estimation procedure assumes that the observation point is located in the far field, that is to say, sufficiently far from the jet exhaust so that the noise sources appear to originate from a point. The procedure for estimating the over-all sound-pressure level (SPL) and relative spectrum shape at

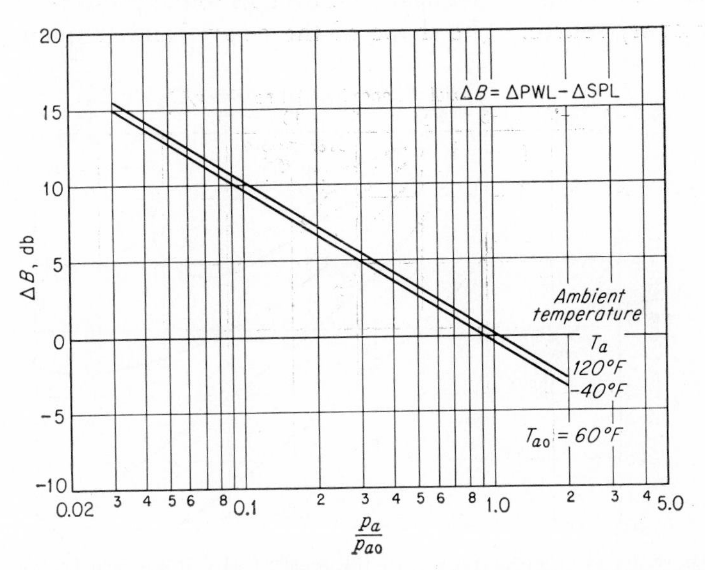

FIG. 24.7 Correction factor for converting power-level changes to sound-pressure-level changes. The subscripts a and o refer to ambient and reference conditions, respectively. The ambient pressures appearing in the abscissa scale are measured in absolute units such as pounds per square inch absolute.

points within 100 ft of a turbojet engine (near field) begins with the reference contour of Fig. 24.8. The geometry of the problem is shown by the sketch of Fig. 24.1 and the conditions are given in the caption to Fig. 24.8. The procedure is itemized as follows:

1. A correction factor to the over-all SPL of Fig. 24.8 is derived from the exit velocity V,

$$\Delta F = 80 \log_{10} \frac{V}{1{,}850} \quad \text{db} \tag{24.5}$$

where V is the relative expanded jet velocity in ft/sec and is computed by dividing the net thrust (in lb) by the mass flow (in slugs/sec).* For example, assume a net thrust of 10,000 lb and a mass flow of 10 slugs/sec. Then $V = 1{,}000$ ft/sec and $\Delta F = -21.4$ db.

2. The contour pattern shifts in angle away from the jet axis as the jet velocity relative to the surrounding medium increases. The angle of shift $\Delta\phi$ in the contour pattern of over-all SPL (Fig. 24.8) is given by Fig. 24.9.

3. When ΔF is applied to the reference contour values of Fig. 24.8 and the contours have been rotated about the exhaust nozzle by $\Delta\phi$, we have the predicted near-field over-all sound-pressure levels for a stationary source. The shape of the octave-band frequency spec-

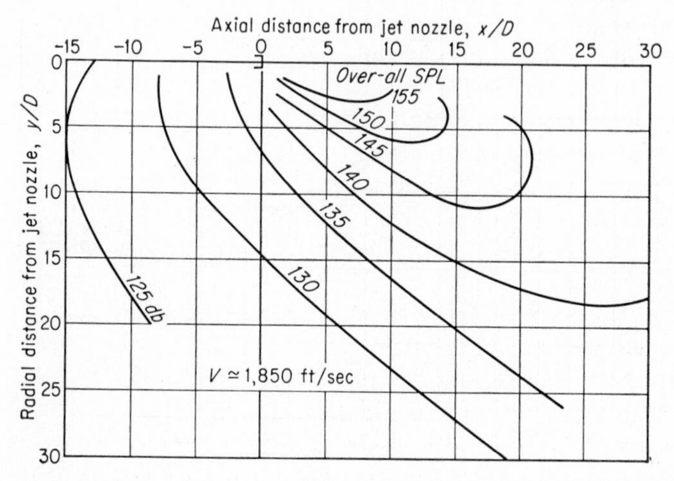

FIG. 24.8 Reference contours of over-all SPL *re* 0.0002 μbar in the near field of a turbojet ($r = \sqrt{x^2 + y^2}$ must be less than about $50D$). This graph assumes a contemporary (1959) turbojet engine operating at 100 per cent rpm with an exit velocity of 1,850 ft/sec and a nozzle diameter of 22 in. The measurements underlying these graphs were taken on turbojets producing about 10,000 lb thrust at maximum rated power.

trum at any point in space is found from Fig. 24.10. We note that for large values of x/D (far downstream of the jet) the frequency at which the octave-band spectrum is a maximum equals $0.2\ V/D$.

4. The near-field sound-pressure levels just computed are also functions of the forward speed and altitude of the aircraft. The dependence on altitude is the same as is given by Eqs. (24.3) and (24.4) or Figs. 24.6 and 24.7. The appropriate engine-operating conditions and ambient conditions are used with these equations or figures to obtain ΔSPL. As an example, ΔSPL has been computed from altitude operating data for a contemporary turbojet engine, and the results are shown in Fig. 24.11.

* See Appendix B for a discussion of English units.

This same procedure applies in determining the effect of forward speed, but there is also a compression of the near-field contours. The forward motion moves the receiver from its actual position R to an apparent position R' on a line parallel to the direction of motion. This change in position is shown in Fig. 24.12, for the receiver either upstream or downstream of the noise source S in the exhaust stream. The motion changes angle ϕ into angle ϕ'. A curve of ϕ' as a function of ϕ is given in Fig. 24.13 for several values of aircraft Mach number M. It may be seen that at supersonic speeds no jet noise propagates upstream of the noise source (positive values of ϕ). It should be noted that at very high speeds pressure fluctuations other than jet noise (such as boundary layer noise) may become significant.

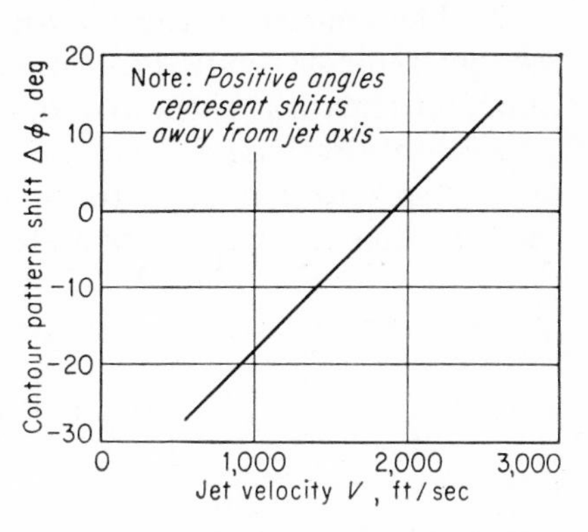

FIG. 24.9 Shift in angle of reference contour pattern of over-all sound-pressure level as a function of net jet velocity V. Positive angles represent a shift in the contour pattern of Fig. 24.8 away from the jet axis.

As an example, the angular transformation of Fig. 24.13 is applied to the near-field contours of Fig. 24.8, for the value of Mach number 0.8. Figure 24.14 shows the shape of typical over-all transformed contours. In obtaining Fig. 24.14 we have made the simplifying assumption that the sound sources are located near the jet exhaust nozzle. For greater refinement one may transform contours of noise meas-

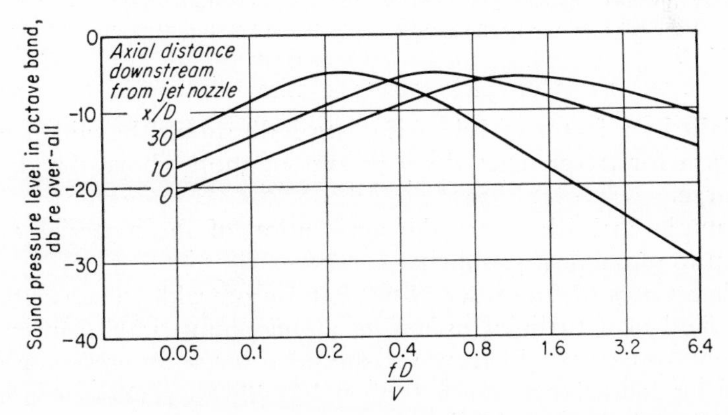

FIG. 24.10 Octave-band frequency spectrum of turbojet noise in the near field. The jet geometry is given in Fig. 24.1.

ured in bands of frequency, specifying the corresponding noise source location at some position downstream of the nozzle.

5. When an afterburner is used on a turbojet engine, another correction factor ΔG must be applied to the near-field contours of Fig. 24.8. This factor is given by Fig. 24.15 for full afterburner operation.

6. With afterburner operation, the peak in the octave-band frequency spectrum is one-half octave lower than that shown in Fig. 24.10 so that the frequency scale on the bottom must be shifted to

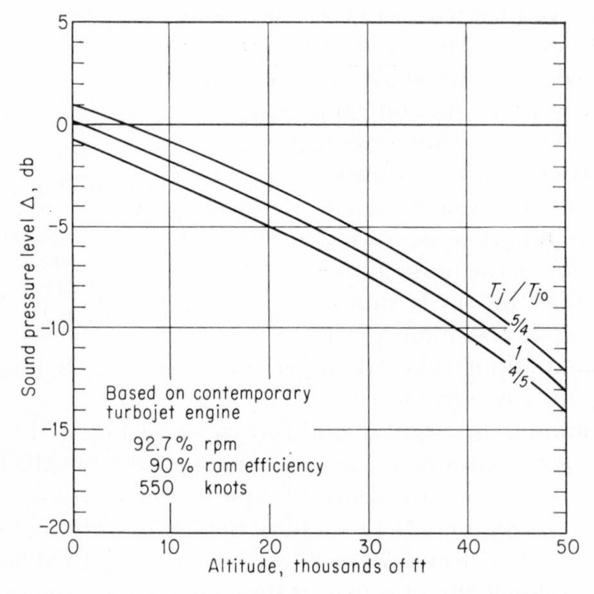

FIG. 24.11 Effect of altitude on near-field over-all sound-pressure level for a contemporary turbojet engine. The quantity T_j is the absolute temperature of the jet exhaust (averaged over the nozzle area) and T_{j0} is the absolute jet exhaust temperature at sea level.

the right by a factor of 1.4. Also 3 to 5 db should be added to the spectrum for frequencies three octaves or more above the peak in the curve.

Example 25.1 illustrates the application of the material of this section to a practical problem.

Other Cases of Interest. Now that we are able to describe the noise field of a turbojet operating at maximum rated power (100 per cent rpm), we will consider special cases of interest, for other operating conditions. We will then be able to generalize our discussion to other types of jets and will find certain interesting limiting cases.

1. *Turbojets Not at Maximum Rated Power.* Figure 24.16 shows the PWL corrections for turbojets operating below maximum rated power. The corresponding correction which should be applied for full afterburner operation was given in Fig. 24.15. As an engineering approximation we may say that between 90 and 100 per cent rpm the directivity and frequency characteristics of the jet noise are not appreciably changed. For afterburner operation, the peak of the far-field directivity curve moves out to about 50° and the frequency spectrum is shifted to lower frequencies by about half an octave compared to the spectrum for 100 per cent rpm conditions.

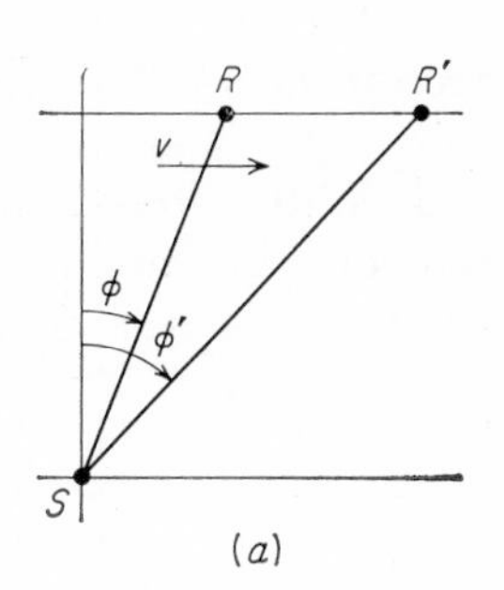

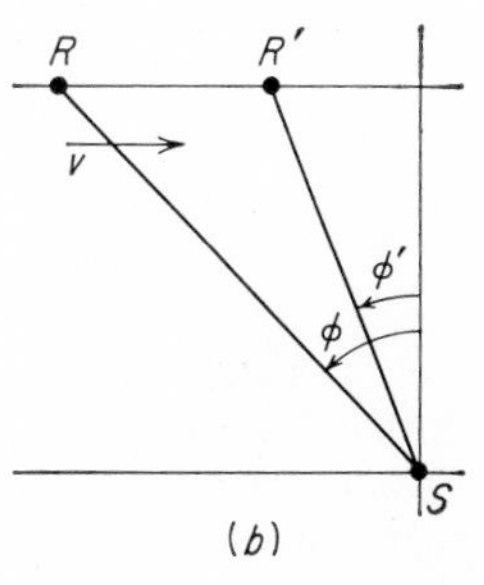

FIG. 24.12 The effect of forward motion on source and receiver geometry. The source is at position S and the entire system moves to the right at constant velocity V. The motion translates the receiver from position R to an apparent position R'. (*a*) Sketch for the case in which the source is downstream of the receiver; (*b*) sketch for the case in which the source is upstream of the receiver.

2. *Efficiency of Jet-noise Sources.* It is now convenient to introduce the concept of acoustic efficiency, denoted by the Greek letter η. Acoustic efficiency is defined as the ratio of the acoustic power to the mechanical power in the jet stream. Obviously, the acoustic efficiency can never exceed 1, that is, there can be no more sound power than mechanical power. However, there is evidence that the upper limit of the efficiency is the order of magnitude of $1/_{100}$. Using this limit on acoustic efficiency derived above, and the eighth power relation given in Eq. (24.1), we are now in a position to generalize our estimation procedure for the power level of jet noise. It is useful to describe the sound power radiated by a jet in terms of the acoustic efficiency, η, and this is done in Fig. 24.17. The abscissa of this figure is the expanded jet velocity V divided by the ambient sound velocity c_a, where the expanded jet velocity is defined as the ratio of thrust to mass flow. ρ and T are jet density and temperature, respectively, and ρ_a and T_a are ambient density and temperature, respectively. The information given by the diagonal lines is equivalent to that given in Fig. 24.2 for turbojets operating at military power. These diagonal lines in Fig. 24.17 reach an upper limit at the value of 10^{-2}, as we have just discussed. We actually expect a gradual transition between the diagonal lines and the upper limit

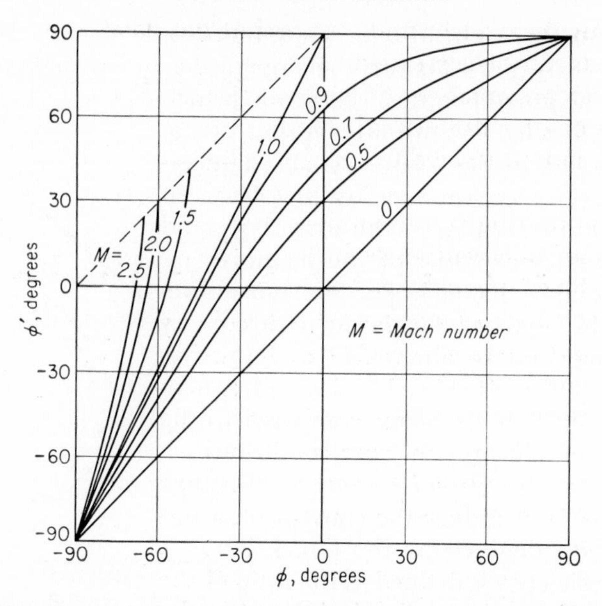

FIG. 24.13 The effect of forward motion on apparent receiver position. The geometry of the system is shown in Fig. 24.12. The aircraft moving at velocity V (Mach number times the speed of sound) changes angle of receiver position ϕ into angle ϕ', where the angles are measured from the normal to the direction of motion. These angular changes are to be applied to the near-field SPL contours.

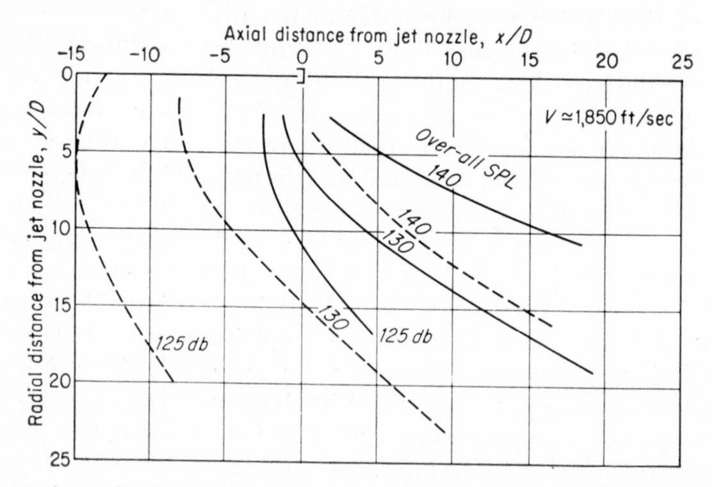

FIG. 24.14 The effect of forward motion on near-field SPL contours. The vehicle is moving at a Mach number of 0.8. The broken lines are the reference contours of Fig. 24.8, and the solid lines are the contours transformed according to Fig. 24.13.

as shown by the broken lines. Experimental data on many different types of jets can be correlated by this general presentation. Typical ranges of η for different types of jets are indicated by the cross-hatching.

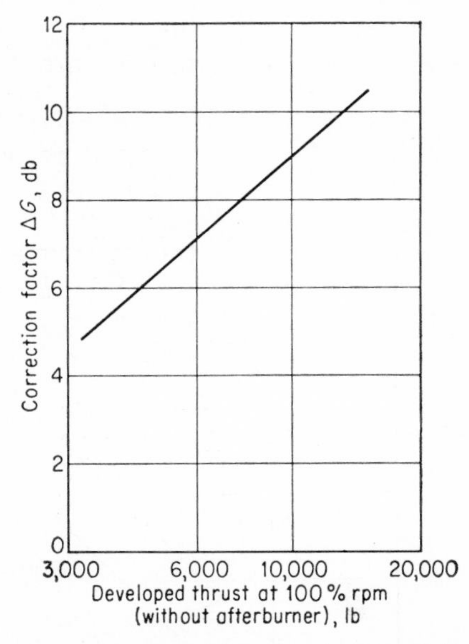

FIG. 24.15 Correction to the near-field contours of Fig. 24.8 for full afterburner operation.

Another limiting factor which occurs in our observations of the noise produced by large sound sources is the distortion which occurs in sound waves of finite amplitude. Some distance from the source is required for this distortion to build up to a significant level, so that the distortion may not be observed near the source but will be observed somewhat farther away. Qualitatively, finite amplitude effects are most important in the frequency range of the maximum SPL. These effects tend to transfer energy to higher frequencies, and to eventual absorption by molecular causes. Consequently, the observed maximum frequency would tend to be lower than the true maximum frequency because of finite amplitude effects. It is not possible at present to obtain engineering estimates of the importance of finite amplitude distortion in jet noise. This is a significant area for future study.

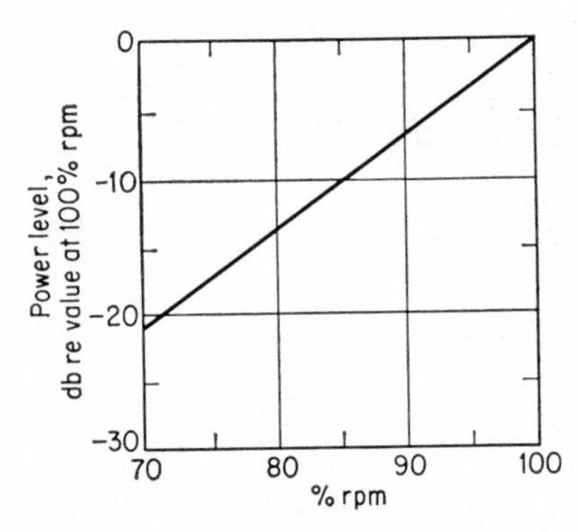

FIG. 24.16 Power-level corrections for a turbojet operating below 100 per cent rpm relative to PWL at 100 per cent rpm.

3. *Generalized Directivity and Spectrum.* We can also generalize our information on the directivity and spectrum characteristics of jets. It has been observed that jets whose geometry is similar to that shown in Fig. 24.1 have a far-field directivity pattern similar to that shown in Fig. 24.3. But the peak of this directivity curve may be shifted somewhat in angle, as has been shown in Fig. 24.9.[6] In general, most of the directivity curve of Fig. 24.3 will be shifted as prescribed by Fig.

24.9; however, parts of the directivity curve near 180° from the jet axis will be shifted slightly less than the rest of the curve.

In order to describe the jet spectrum in a general way, we introduce a nondimensional quantity that describes spectrum, similar to

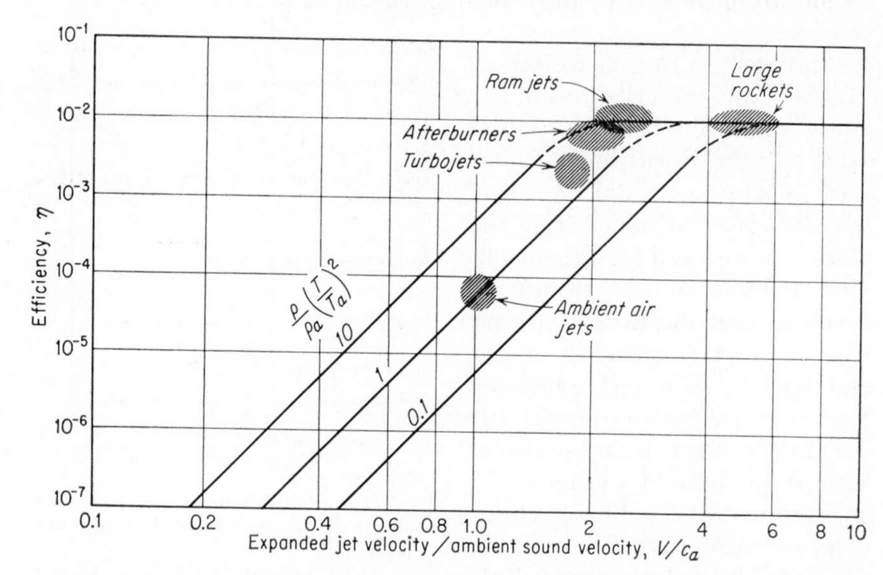

FIG. 24.17 Acoustic efficiency for jet noise. The abscissa is the ratio of the expanded jet velocity to the ambient sound velocity. The ratio $\rho/\rho_a(T/T_a)^2$ of jet and ambient densities and temperatures describes each of the diagonal lines. The upper limit of the efficiency η is 10^{-2}.

the nondimensional quantity η that describes acoustic power. This new quantity is called the *peak Strouhal number* S_0 and is defined as

$$S_0 = \frac{f_{\max} D}{V} \tag{24.6}$$

where $f_{\max}$ = peak of power-level spectrum
D = diameter of nozzle
V = jet exit velocity

From our earlier information on octave-band turbojet spectra we have seen that $f_{\max}$ occurred at about 200 cps, and this yields, for typical values of D and V, a peak Strouhal number of about 0.2. Experimentally, it is observed for all noise from jets similar to that shown in Fig. 24.1 that S_0 is approximately constant, and therefore, we can scale this result to any jet. For a jet breathing ambient air, the typical octave band SPL is given in Fig. 24.18. This spectrum is seen to be somewhat narrower than the spectra of Fig. 24.4. S_0 can be taken equal to 0.2 for an ambient air jet.

Figures 24.17, 24.9, 24.3, 24.4, and 24.18 and Eq. (24.6) now provide us with a general estimation procedure for obtaining the far-field power level, directivity, and spectrum of jets.

Example 24.2. Let us find the SPL spectrum at 45° from the jet axis and 20 ft from the nozzle of a 5-in.-diameter jet. The jet fluid is ambient air, and the expanded jet velocity 1,100 ft/sec. In English units the density of air at room temperature is 2.3×10^{-3} slugs/ft³, and there are 1.36 watts in 1 ft-lb/sec. We calculate the mechanical stream power, which is $(\frac{1}{2}\rho V^2)(\pi D^2 V/4)$ and which equals 2.8×10^5

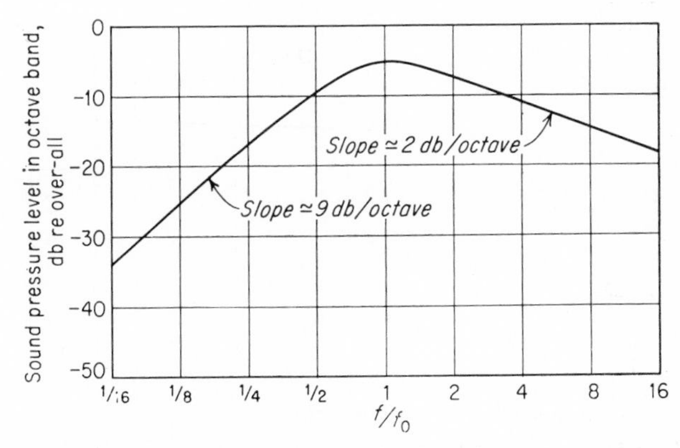

Fig. 24.18 SPL spectrum shape for an ambient air jet. The SPL in octave bands, relative to the over-all SPL, is given as a function of the ratio of the frequency to the peak frequency.

watts. The ratios $(\rho/\rho_a)(T/T_a)^2$ and V/c_a are both unity, and therefore Fig. 24.17 gives an acoustic efficiency η of 6×10^{-5}. The resultant over-all PWL is 142 db. For a jet velocity of 1,100 ft/sec, Fig. 24.9 indicates that the directivity maximum is shifted 16° closer to the jet axis than is shown by Fig. 24.3. When Fig. 24.3 is adjusted in this way, the directivity correction at 45° is found to be +2 db. Taking spherical spreading of the jet noise at 20 ft into account, we obtain an over-all SPL at 45° of $142 + 2 - 10 \log [4\pi(20)^2] = 107$ db. With the peak Strouhal number equal to 0.2, we obtain a spectrum peak of 530 cps. The spectrum shape is then given by Fig. 24.18.

4. *Choked Jet Noise.* There exists one important case of jet noise which has not yet been covered in our discussion. This case occurs when the pressure ratio (the static pressure upstream from the nozzle divided by the ambient pressure) exceeds 1.89. In this case, we have

a "choked jet." Flow through the nozzle is sonic and expands to a supersonic velocity downstream from the nozzle. A shock formation occurs outside of the nozzle, and a new source of noise appears, in addition to the usual jet noise described above. This is the situation which occurs in choked control valves.

The total sound power radiated by a choked jet is given in Fig. 24.19. The nondimensional ratio of sound power to the quantity ½ mc^2 is plotted in this figure, as a function of the nozzle pressure ratio defined in the previous paragraph. In this figure m is the mass flow of air through the nozzle, and c is the velocity of sound at the nozzle. Choked jet noise is radiated strongly at 90° to the jet axis,

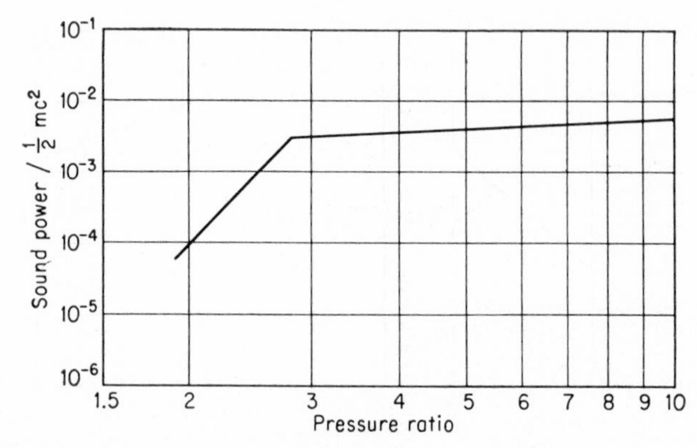

Fig. 24.19 Sound power of high-pressure-ratio jets breathing ambient air. The ordinate is the ratio of sound power to the quantity ½mc^2, where m is the mass flow of air and c is the sound velocity at the nozzle.

so that the sum of the choked jet noise and the usual jet noise yields an over-all sound field which is almost nondirectional up to about 120° from the jet axis.[7] In a particular frequency band, however, the directional characteristics of the choked or of the usual jet noise will be apparent.

The spectrum curve of choked jet noise is somewhat more peaked than the usual jet noise described above. The peak of this curve is found by the Strouhal number S_0 plotted in Fig. 24.20.[7] The abscissa on this curve is the excess pressure ratio, i.e., the pressure ratio minus 1.89. It is seen that for increasing pressure ratio the Strouhal number is a decreasing function. For values of the excess pressure ratio exceeding about 6, the shape of this curve is not determined at present. It is expected that for very high pressure ratios the line may level off at a constant value of Strouhal number, or perhaps may increase with increasing pressure ratio.

We may ask why we do not observe this choked noise in present-day turbojets. In such cases, the choked noise is masked by the usual jet exhaust turbulence noise, described earlier. However, at higher pressure ratios and altitude operating conditions, it is possible that

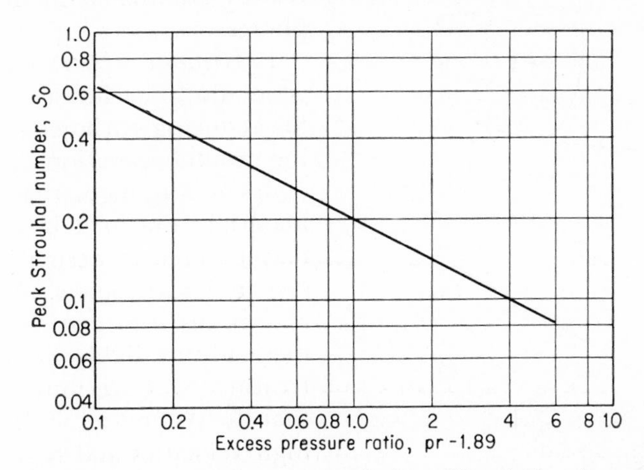

FIG. 24.20 Spectrum peak of high-pressure-ratio jets. The abscissa is excess pressure ratio (pressure ratio minus 1.89). The ordinate is the Strouhal number S_0, which equals $(f_{max}D/V)$, where f_{max} is the power-level spectrum peak, D is the nozzle diameter, and V is the jet velocity.

this choked noise may become a significant part of the sound power radiated by a turbojet.

24.3 Attempts at Noise Control

We now have available estimation procedures for describing the noise field of various jet sources. Therefore, we may consider the work being done on the noise control of such sources.

Since the production of the "choked-jet" noise is associated with regular shock formation in the stream, a device or technique which breaks up or disturbs the shocks should reduce the noise. This has been carried out successfully on choked air jets by means of a wire-gauze nozzle-extension screen, a convergent-divergent nozzle, auxiliary air injection, etc.

Once the choked noise has been eliminated, however, the usual jet noise remains, and the reduction of this latter noise poses a considerably more difficult problem. Equation (24.1) suggests that, since the jet PWL depends on the velocity to a high power, a promising method for reducing jet noise will be to reduce jet velocity. Since thrust is directly proportional to the square of the jet velocity, this method has obvious practical limitations. A shift in the jet spectrum, moving the acoustic energy to higher frequencies, may also aid in noise control, because the higher frequencies are more readily attenuated in the atmosphere.

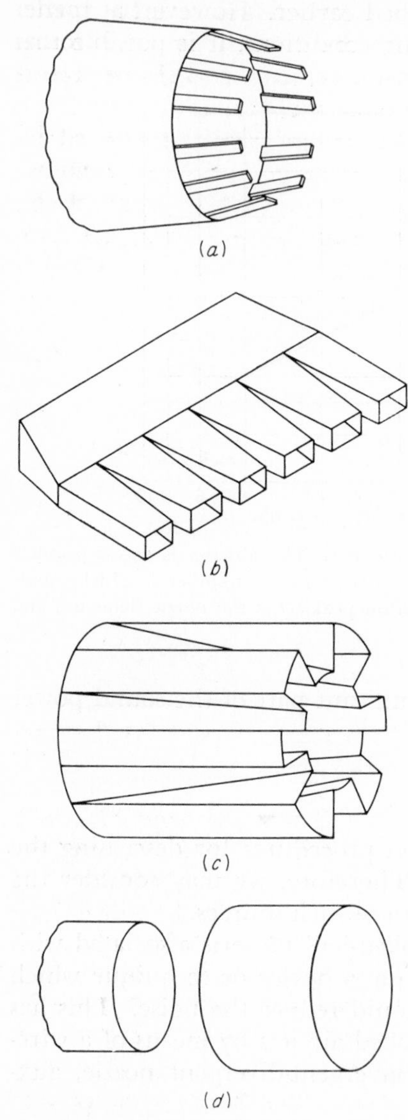

Fig. 24.21 Typical devices for in-flight noise reduction: (*a*) toothed nozzle, (*b*) rectangular segmented nozzle, (*c*) circular corrugated nozzle, (*d*) ejector.

1. *Ground Operation.* Two general approaches are available for reducing turbojet noise during ground operation. The principles of muffling with absorptive materials, outlined in Chap. 17, may be used to design a muffler system for turbojets. Reference 8 contains information on commercially available noise-control treatments. Care must be taken that the system can withstand the high velocities and temperatures encountered in a turbojet exhaust.[9] Wire screens placed transversely across the jet exhaust also provide a useful method of attenuating noise on the ground.[10] But, mufflers and screens bring with them high losses in the thrust produced by the jet engine, and it is necessary to look for other means of obtaining noise suppression in flight.

2. *In-flight Operation.* Work on in-flight noise reduction has been concentrated largely on modifications of the exhaust nozzle configuration. Typical devices which have been developed along this line are shown in Fig. 24.21. These include the toothed nozzle, the circular corrugated nozzle, the rectangular segmented nozzle, and the ejector.[11]

A simple analysis of jet-noise reduction due to combination with an induced secondary airflow may provide a partial explanation of the noise reduction brought about by such devices.[12] According to this theory, the modified nozzle shape may allow more secondary air to be induced than is the case for a conventional circular nozzle. When the secondary air combines with the primary air jet, it forms a new jet stream of larger area and lower velocity, and the net result of this new lower velocity jet may be noise reduction.

Figure 24.22 shows the power-level change ΔPWL predicted by this analysis, as a function of the ratio of the jet area after combination A_2 to the nozzle open area A_1. T_1 is the temperature of the primary jet, and T_0 is the ambient temperature. Thus, for contemporary turbojets at military power, T_1/T_0 is approximately 3.

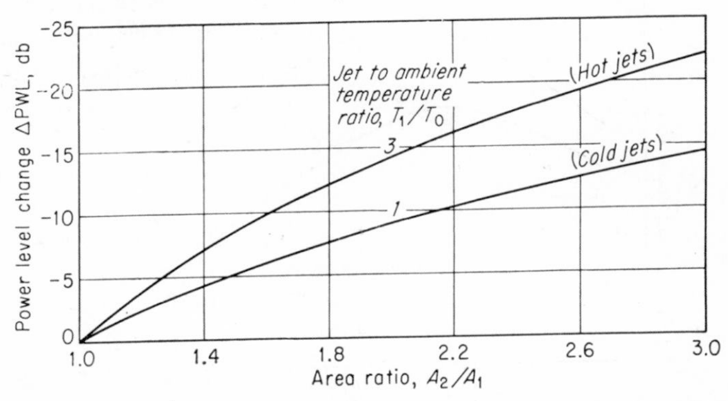

FIG. 24.22 Change in over-all power level due to induction of secondary air. A_2 is the jet cross-sectional area after complete combination of primary and secondary flows, and A_1 is the exhaust nozzle open area.

Actually, the primary and secondary flows do not combine right at the exhaust nozzle. The noise radiated before complete combination may make an important contribution to the total noise radiated by a modified jet and sets a limiting value on the noise reduction provided by inducing secondary air. The situation is shown schematically in Fig. 24.23, which gives the PWL spectrum as a function of frequency. The peak of the power-level spectrum has gone down in frequency from f_1 with the standard nozzle to f_2 with the modified nozzle, because the new jet has a lower velocity and a larger diameter than the original jet. There is also a secondary peak in the power-level spectrum, which is observed experimentally at a higher frequency, f_c. The noise in the high frequencies is produced near the exhaust nozzle, and these sound sources are unaffected by the mixing process which occurs further downstream. Thus, the ΔPWL values of Fig. 24.22 are upper limits of noise reduction.

It is known that the devices of Fig. 24.21 can induce secondary air into the jet stream. In the case of ejectors, much work has been performed to determine the optimum geometry for inducing secondary air.[13] Some similar work has been carried out for corrugated and segmented nozzles.[14] The results of these aerodynamic studies can now be used to investigate this proposed mechanism of noise reduction and thereby improve silencer design.

Noise reduction may also be accomplished by means of basic redesign of the turbojet engine. The principles involved in the bypass engine and low-temperature engine potentially promise noise reduction. However, such engines are considerably different in construc-

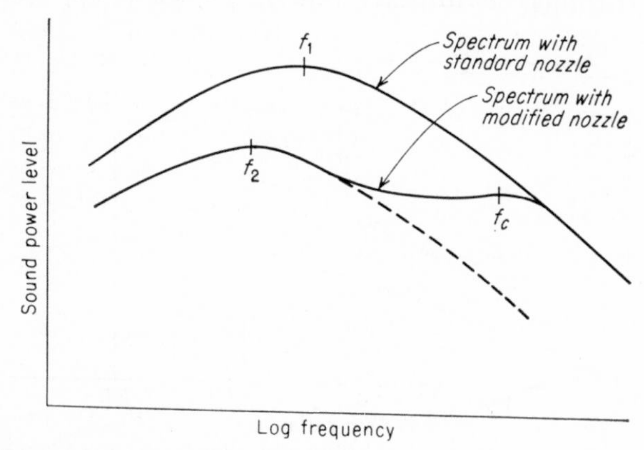

FIG. 24.23 Change in power-level spectrum due to induction of secondary air. The PWL spectrum with standard nozzle peaks at f_1. The PWL spectrum with a modified nozzle peaks at f_2, with a secondary peak at f_c.

tion from present-day military engines, and the new designs have not yet been fully exploited.

It is possible through proper placement to obtain reduction of jet-engine noise in specific areas. For example, placing the engine at a slight angle to the axis of the aircraft may reduce noise levels encountered in certain regions of the aircraft cabin.

At present, much of the developmental effort in noise reduction is being devoted to the type of nozzle modifications shown in Fig. 24.21. At best, the PWL reductions afforded by such devices appear to be limited to between 5 and 10 db.

24.4 Summary

The information now available on the noise field associated with a jet is reviewed in this chapter. The geometry and flow pattern common to many jets is described. Because the turbojet engine is a familiar source of jet noise, it is convenient to describe this case in some detail. Estimation procedures are outlined for obtaining the near- and far-field PWL's, directivity, and frequency spectra of the noise radiated by a stationary turbojet. The effects of motion and of ambient condition changes on the jet-noise field and the nature of the noise field near the jet stream are reviewed.

The noise from jets other than turbojets is then described by a generalized estimation procedure. Sound powers, directivity, and spectra are given in terms of nondimensional parameters. The noise radiated by a choked jet is also considered.

Attempts to control jet noise are described. The noise-control problem in flight is, in general, more difficult to solve than the noise-control problem for ground operations. A simple picture of jet-noise reduction is outlined. This picture provides a possible explanation of the reductions brought about by various exhaust nozzle configurations. About 5- to 10-db reduction in sound-power level appears likely with such devices.

REFERENCES

1. Pietrasanta, A. C.: Aircraft Noise and Building Design, *Noise Control,* vol. 3, no. 2, pp. 11–18, March, 1957.

2. Clark, W. E., A. C. Pietrasanta, W. J. Galloway, and the staff of Bolt Beranek and Newman Inc.: *WADC Tech. Note* 56–60, 1957. See also N. Doelling, D. M. A. Mercer, and the staff of Bolt Beranek and Newman Inc.: *WADC Tech. Rept.* 54–401, 1956.

3. von Gierke, H. E.: Aircraft Noise Sources, chap. 33 of C. M. Harris (ed.), "Handbook of Noise Control," McGraw-Hill Book Company, Inc., New York, 1957.

4. Doelling, N., I. Dyer, D. M. A. Mercer, and the staff of Bolt Beranek and Newman Inc.: *WADC Tech. Rept.* 55–477, 1956. (Certain equations in this report contain an error that has been corrected here.)

5. The data of Fig. 24.8 are based on measurements of W. L. Howes and H. R. Mull, *Natl. Advisory Comm. Aeronaut. Tech. Note* 3763, 1956, and unpublished data of USAF, Convair, Martin, and Bolt Beranek and Newman Inc. See also P. A. Franken, E. M. Kerwin, Jr., and the staff of Bolt Beranek and Newman Inc.: *WADC Tech. Rept.* 58–343, October, 1958. Obtainable from Wright-Patterson Air Force Base, Ohio; ASTIA document number AD-205-776.

6. Figure 24.17 contains data on model air jets (E. E. Callaghan and W. D. Coles: *Natl. Advisory Comm. Aeronaut Tech. Note* 3590, 1955), model helium jets (L. W. Lassiter and H. H. Hubbard: *Natl. Advisory Comm. Aeronaut. Tech. Note* 2757, 1952), and turbojets at military power and with afterburner (see Ref. 2 above).

7. Powell, A.: *Aeronaut. Quart.,* vol. IV, p. 103, 1953; J. M. Tyler and E. C. Perry: *SAE Reprint* 287, 1954; and E. E. Callaghan and W. D. Coles: *Natl. Advisory Comm. Aeronaut. Tech. Note* 3590, 1955.

8. Dyer, I.: Noise Attenuation of Dissipative Mufflers, *Noise Control,* vol. 2, no. 3, p. 50, May, 1956.

9. Labate, S.: Porous Materials for Noise Control, *Noise Control,* vol. 2, no. 1, p. 15, January, 1956.

10. Coles, W. D., and W. J. North: *Natl. Advisory Comm. Aeronaut. Tech. Note* 4033, 1957.

11. A typical sampling of the large literature on modified nozzles is: F. B. Greatex: *Inst. Aeronaut. Sci. Preprint* 559, 1955; E. J. Richards and D. J. Evans: *Aeronaut. Research Council Rept.* 18017, 1955; W. J. North and W. D. Coles: *Natl. Advisory Comm. Aeronaut. Tech. Note* 3573, 1955; M. M. Miller: *SAE Preprint* 818, 1956; H. W. Withington: Silencing the Jet Aircraft, *Noise Control,* vol. 2, no. 5, p. 46, 1956; W. D. Coles and E. E. Callaghan: *Natl. Advisory Comm. Aeronaut. Tech. Note* 3974, 1957.

12. Dyer, I., P. A. Franken, and P. J. Westervelt: Jet Noise Reduction by Induced Flow, *J. Acoust. Soc. Am.,* vol. 30, p. 761, 1958.

13. Kochendorfer, F. D., and M. D. Rousso: *Natl. Advisory Comm. Aeronaut. Research Mem.* E51E01, 1951; D. P. Hollister and W. K. Greathouse: *Natl. Advisory Comm. Aeronaut. Research Mem.* E52K17, 1953; S. C. Huntley and H. Yanowitz: *Natl. Advisory Comm. Aeronaut. Research Mem.* E53J13, 1954; R. D. Lemmerman and H. J. Lockwood: *Aeronaut. Eng. Rev.,* vol. 14, p. 37, 1955; and others.

14. Laurence, J. C., and J. M. Benninghoff: *Natl. Advisory Comm. Aeronaut. Tech. Note* 4029, 1957.

Chapter 25

NOISE CONTROL IN TRANSPORTATION

P. A. Franken and L. L. Beranek

25.1 Introduction

The noise control of a complex system, such as a train or airplane, involves many factors which must be integrated into a satisfactory and practical over-all solution. In order to arrive at detail design recommendations, the acoustical engineer must work toward a specific design goal and must have knowledge of the basic noise generators associated with the vehicle. The detailed acoustical recommendations should be taken into account in the vehicle structure, and these recommendations should subsequently be tested during the vehicle construction and evaluated in the completed vehicle. Ideally, then, a complete noise-control program of this type would fall into three general phases:

Phase 1: Preliminary Design. The acoustical engineer, jointly with his management, should: determine and agree upon a satisfactory working criterion for maximum permissible noise levels in the completed vehicle; calculate (or measure if the vehicle exists) the noise levels at appropriate points in the vehicle; establish the noise-reduction (NR) requirements for the various vehicle components; and prepare a series of preliminary component designs satisfying these NR requirements.

Phase 2: Detailed Design. The acoustical engineer should provide detailed design of the NR components determined in phase 1. The follow-up activity would include: the checking of all working drawings; the measurement of the acoustical behavior of models or mockups of the structure; the testing of prototype vehicles during construction; and the correction of any deficiencies in noise-control measures. This phase would of course require close liaison between the acoustical engineer and the manufacturing department to insure a balanced and economical engineering solution.

Phase 3: Design Evaluation. The acoustical engineer should perform noise surveys during actual performance tests of the "quieted" vehicles, in order to verify the appropriateness of the noise-control measures adopted. If necessary, modifications should be made at this point. At the completion of these tests, the entire program should be reviewed, and generalized engineering information should be organized for noise control of future vehicles of this type.

This book clearly cannot provide detailed answers to any particular noise-control problem as complex as those encountered in transportation vehicles. Rather, it is intended to aid the acoustical engineer in performing the first phase of his duties, the preliminary design. In general this chapter should:

1. Point out the predominant noise sources generally associated with interior noise levels in several contemporary vehicles. The methods of estimation given in this chapter are the best known to the writers, but many details are not fully verified. The user should not be surprised, therefore, if future investigations reveal better methods of prediction. In any specific problem, the acoustical engineer will probably be required to measure the noise levels in and around vehicles similar to the one he is considering, in order to obtain reliable estimates of the various noise source intensities.
2. Call attention to the need for carefully selecting the noise-control criteria appropriate to the quieting job at hand.
3. Suggest general noise-control measures that may be employed in these vehicles in order to satisfy the desired criteria. Here again, however, the acoustical engineer will be required to consider the design features of his particular configuration rather than to apply directly the general discussions included in this chapter.

25.2 Rail Vehicles

Important Sources of Noise in Rail Vehicles. It is possible to name vehicle noises in terms of their transmission paths between the sources and the observer within the vehicle. Thus we may speak of "air-borne sound" and "structure-borne sound." In most cases the noise is both air- and structure-borne. The terms *air-borne* and *structure-borne* are only relative, and describe the major part of the transmission path. For example, the noise of a train locomotive may be propagated in the air around a passenger coach, where it excites the coach walls which in turn reradiate noise to the interior. This example is considered air-borne sound because the major part of the path is in air. But, if locomotive vibrations are transmitted through the coupling structure to the coach wall which, in turn, reradiates the noise, this is a case of structure-borne sound, even though the noise source is also the locomotive.

The propagation of sound in complex structures is not nearly as well understood at present as is propagation in liquids and gases (such as air). As a result, there is little general engineering information available on structure-borne sound in vehicles, although structure-borne sound can be controlled by appropriate techniques based on specific design information and noise and vibration measurements. In the remaining discussion on rail-vehicle noise control, we will consider quantitatively only air-borne sound, although qualitatively important structural transmission paths will be mentioned. Several important noise sources exist in a rail vehicle:

1. *Interactions between Wheels and Rails.* The striking or sliding motion between a wheel, trolley, or third-rail pickup and a rail or wire will generate both air-borne and structure-borne sound.

2. *Aerodynamic Pressure Fluctuations on the Vehicle Surface.* Rapid motion of the vehicle through the air generates a turbulent boundary layer over the surface, and causes vortex shedding in regions of flow separation. Vortex shedding results in pressure fluctuations over the surface of the vehicle and these pressure fluctuations induce vibrations in the vehicle structure that are in turn radiated inside as noise.

3. *Power-plant Noise.* Intake and exhaust airflows, turbines, combustion process, gears, etc., act as sources of air-borne noise. In addition, each unit of mechanical equipment produces structure-borne vibration.

4. *Whistles and Other Audible Signals*

5. *Adjacent Vehicles*

6. *Air Conditioning*

The production and control of air-conditioning noise in rail vehicles are covered fully in Chap. 21, and will not require additional discussion here.

Unfortunately, there is little or no generalized information now available on the detailed characteristics of all of these sources, and the complexity of rail vehicles makes it unlikely that each source can be isolated and studied separately from every other source. However, empirical scaling procedures can be used with noise measurements on similar vehicles to obtain estimates of the noise under the other conditions. Two scaling parameters that are useful over some limited range of extrapolation are mechanical power and vehicle speed. The source information on air-conditioning fan noise, given

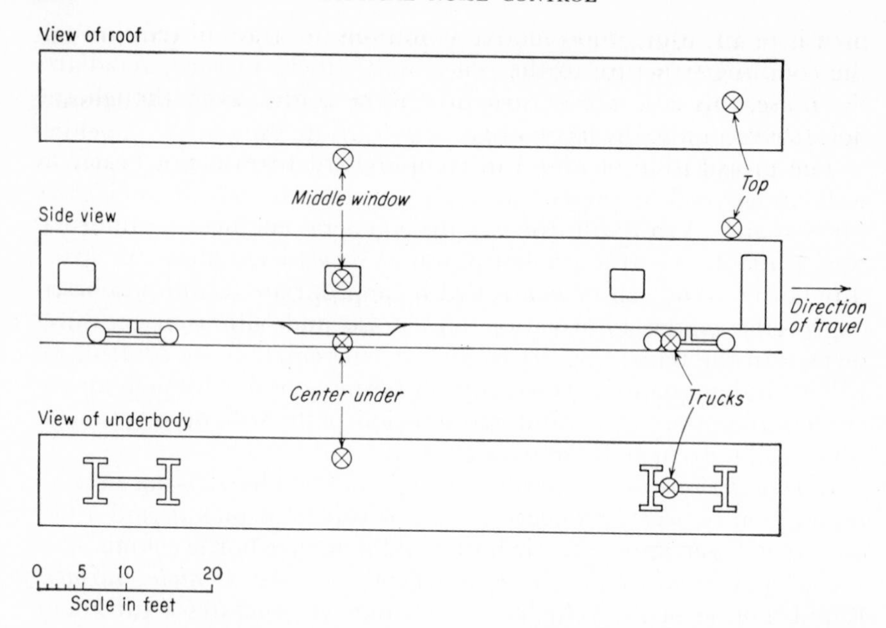

FIG. 25.1 Noise measurement positions outside a contemporary long-distance American railroad coach. The forward speed of the train ranged from 30 to 80 mph. An approximate scale of length is indicated.

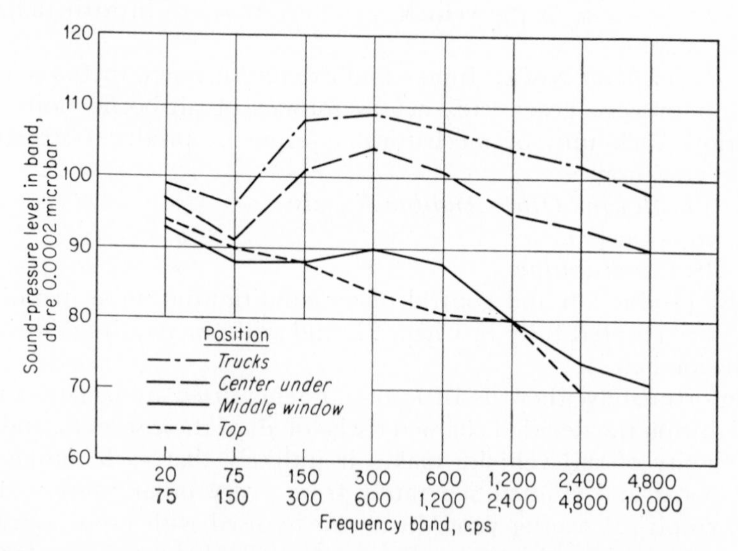

FIG. 25.2 Average octave-band noise levels measured at four positions outside an American railroad coach moving at a speed of 50 mph. These levels and the values of the speed exponent n may be used in Eq. (25.2) to obtain estimates of external noise levels at speeds ranging from 30 to 80 mph.[1]

in Chap. 21, exemplifies a scaling procedure based in part on mechanical power [see Eq. (21.1)].

In order to scale noise levels by vehicle speed, the sound-pressure levels (SPL) measured at any one point outside the vehicle at vehicle speeds v_1 and v_2 are assumed to be related in the following manner:

$$\mathrm{SPL}_{v_2} - \mathrm{SPL}_{v_1} = (10n) \log_{10} \frac{v_2}{v_1} \qquad \text{db} \tag{25.1}$$

where n is a function of measurement position and of frequency.

Noise measurements have been made outside a long-distance Amer-

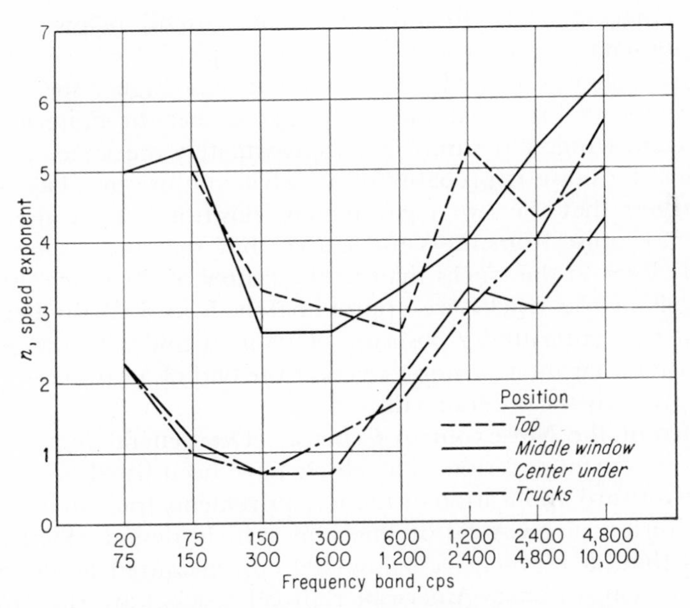

FIG. 25.3 Average variation with speed of noise levels at four positions outside an American railroad coach. These values of the speed exponent n and the average noise levels measured at a speed of 50 mph may be used in Eq. (25.2) to obtain estimates of external noise levels at speeds ranging from 30 to 80 mph.[1]

ican railroad passenger coach moving at speeds ranging from 30 to 80 mph.[1] Location of the measurement positions is shown in Fig. 25.1. Proper windscreen protection was provided to ensure that airflow disturbances around the microphones were not significant in determining the SPL. No noise from railroad signals or adjacent vehicles was included in the measurements, and the coach was sufficiently far from the locomotive to make engine noise negligible. The results of applying Eq. (25.1) to the measured data are shown in Figs. 25.2 and 25.3. Figure 25.2 gives the average exterior noise levels at four different measurement positions, for a speed of 50 mph. Figure 25.3 gives average values of the speed exponent n. In terms of these results, Eq. (25.1) may be rewritten to provide a procedure for estimating external noise levels at vehicle speed v mph:

$$\text{SPL}_v = \text{SPL}_{(50\text{ mph})} + (10n)\log_{10}\frac{v}{50} \quad \text{db} \qquad (25.2)$$

At speeds below about 40 mph, there is some experimental indication that the nature of some of the noise sources may change. Thus the procedure of Eq. (25.2) may lead to slight underestimates in the range of 30 to 40 mph and is not useful below 30 mph. Adequate data are not available to check Eq. (25.2) above 80 mph; therefore it should be used with caution at these higher speeds.

Inspection of Figs. 25.2 and 25.3 suggests that the noise levels at the top and window positions are approximately the same. At these positions the noise is probably of aerodynamic origin. The figures also suggest that the two measurement positions under the coach experience noise from the same sources but that these sources are located closer to the trucks than to the center of the coach, because the levels at the trucks are approximately 5 to 8 db higher than those at the center under position. This is in line with the general observation that interior noise levels at the end of a car exceed those at the center by the order of 5 db.

Choice of the Noise-control Criteria. The general requirements of a noise-control criterion in a vehicle have been listed in Sec. 20.3. Straightforward application of these requirements leads to the choice of a criterion of NCA–60 or lower in vehicle design. Figure 20.6 showed that, at frequencies below 300 cps, measured levels in conventional long-distance American railroad coaches are the order of 5 db above this criterion. We conclude that people accept more noise in present-day surface rail vehicles than they do in offices with the same SIL (speech interference level).

It is clear that the public attitude toward rail-vehicle noise may change from that reflected in the "accepted" levels in Fig. 20.6. For instance, a large reduction in scheduled service could conceivably lead to increased tolerance of noise. Whenever possible, the acoustical engineer should determine his own criterion by measuring noise levels known to be acceptable in a similar vehicle, rather than by attempting to interpret the "accepted" levels in Fig. 20.6 in terms of his particular vehicle configuration and intended use. In the absence of such measurements, the criterion of NCA–60 should be used.

Noise Reduction of Railroad Structures: Design Considerations. The noise-attenuating properties of a vehicle structure depend on many complicated factors:

1. They depend on the TL of the structure, which is related to the fraction of sound energy transmitted by the structure under ideal laboratory considerations.

2. The amount of acoustical absorption within the vehicle is important; this determines the reverberant character of the internal noise field.

3. The character of the external noise sources must be considered. For example, sound and turbulent pressure fluctuations differ in nature and will excite a vehicle structure in different ways. Therefore, the NR of the structure for sound will in general be different from that for turbulent fluctuations.

4. The structural connections to other elements of the vehicle also constitute a factor. Such connections will transmit structure-borne sound, which in turn reradiates sound to the internal space.

Because of these complications, it is clear that there may be no simple relation between the TL measured in an ideal laboratory experiment and the NR, i.e., the actual difference in levels between the external and internal spaces. Practically, we are interested in the NR; therefore, we must look at case histories to determine the NR achieved under field conditions.

Sidewalls. Transmission of sound through the sidewalls can be estimated from the general description given in Chap. 13. In modern, lightweight trains the TL's are similar to those for airplane fuselages which are discussed a few pages hence and in Chap. 14.

No details of sidewall construction are presented in this chapter as the sidewall transmission of most modern trains is determined by the windows, because of the large areas that they occupy. It is generally more difficult to obtain a high TL for the windows than for the sidewalls.

Windows. Figure 25.4 gives average NR's of double windows in present-day American railroad cars.[1] Data included in Fig. 25.4 were obtained from measurements just outside the windows and at window-seat positions half way between the ends of the cars. The window construction was approximately two panes of ¼-in. plate glass, separated by ¼ in. air space. Each pane is about 5 by 3 ft in size, and the panes are mounted with rubber gaskets around the edges in a rigid frame. The values obtained in long-distance coaches and in Pullman roomettes are given. The lack of data at the higher frequencies represents an instrumentation difficulty typical of railroad-noise measurements: the frequency spectrum of the noise within

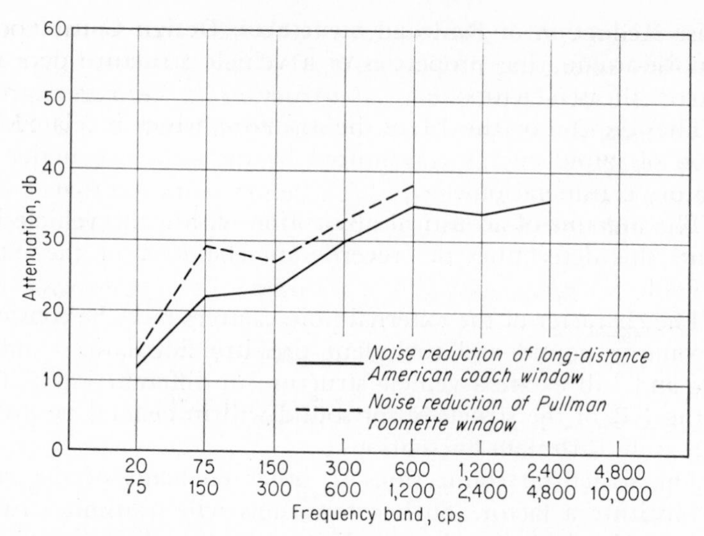

FIG. 25.4 Average noise reduction of double windows in contemporary American railroad cars. Data are shown for a long-distance coach and a Pullman roomette.[1]

the railroad cars has a very steep slope, and conventional instrumentation is noise-limited at the higher frequencies. Special high-pass filters must be used to obtain the levels at the higher frequencies.

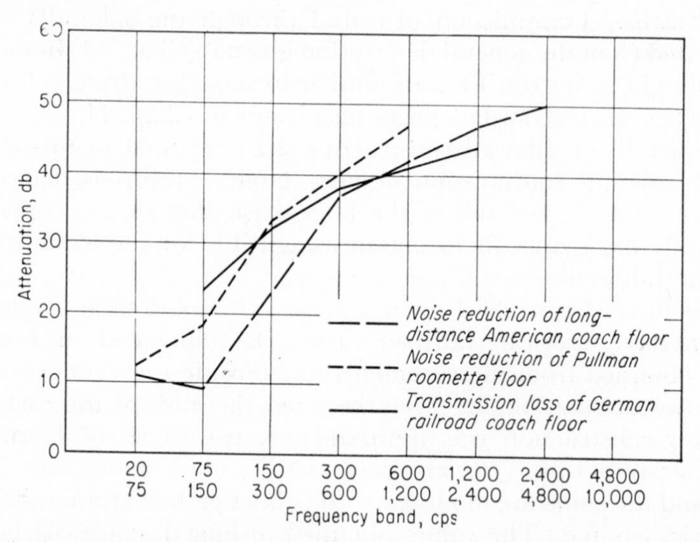

FIG. 25.5 Average noise reduction of floors in contemporary American railroad cars. Data are shown for a long-distance coach and a Pullman roomette.[1] The TL of a typical German railroad floor construction is also given.[2] See Fig. 25.6 for details of the floor constructions considered here.

Floors. Figure 25.5 gives average NR's measured at the floors of present-day American railroad cars.[1] Also shown in this figure is the measured TL of a typical floor construction in German railroads.[2] Cross sections of the floor constructions considered in Fig. 25.5 are given in Fig. 25.6.

Discussion and Design Details. We should point out that at speeds up to 80 mph the floor and window NR's given in Figs. 25.4 and

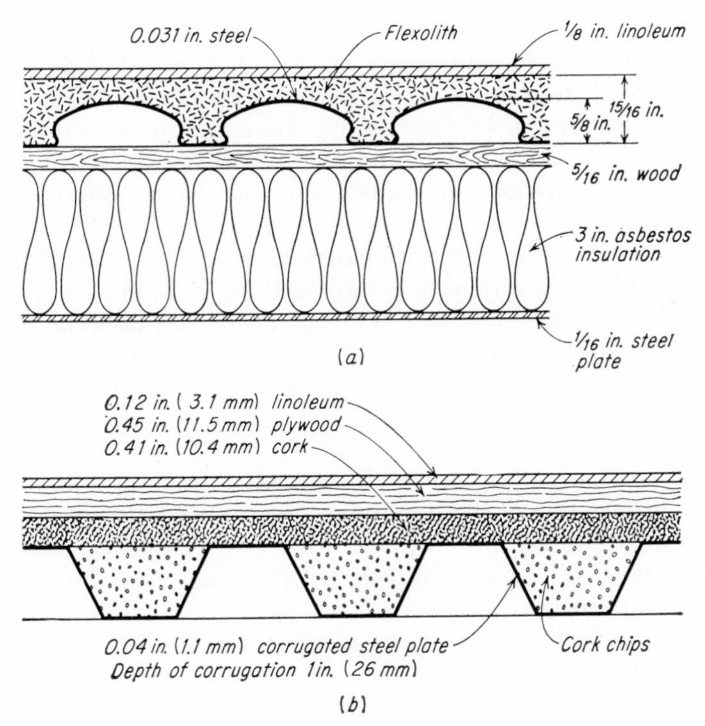

FIG. 25.6 Cross-sectional views of typical railroad-vehicle floors: (*a*) a floor used in American railroad cars; (*b*) floor studied in Ref. 2.

25.5 provide interior noise levels that in general fall within the range of typical levels in conventional long-distance American railroad coaches given in Fig. 20.6. We conclude that the noise environments provided by these structures are considered acceptable at present.

A number of design items may be considered to improve the NR of a rail-vehicle structure:

1. Penetrations or weak points in the structure can be found and eliminated. Such problems may be caused by ventilation holes, electrical conduits, drinking-cup receptacles, door gaskets, etc. When

it is not feasible to seal such penetrations, it may be necessary to provide a duct lined with acoustical material to reduce the noise transmitted through the penetration.

2. Providing a second impervious panel with a minimum of structural connections to the main structure creates a double wall. Chapter 13 has discussed theoretical aspects of this matter. Stüber[2] has measured improvements of the order of 15 db at all frequencies with the addition of a 1.1-mm-thick steel plate placed 40 cm from the structure of a railway coach.

3. Introduction of additional acoustical absorption in the interior spaces of the vehicle may reduce the reverberant field level in this space. In general because of the absorption normally provided by clothing, seat, carpet, etc., additional absorption will provide only a few decibels of reduction. Some discussion of these effects is given in Chap. 11.

4. "Ringing" of the structural panels can be reduced by additional damping. This damping can be provided by an applied viscoelastic layer or by covering a surface with an insulating material applied in strips along the panel surfaces.

25.3 Aircraft

The subject of noise control in aircraft cannot be covered adequately in one chapter. It might even be considered somewhat presumptuous to discuss the subject at all. It is introduced for the explicit purpose of illustrating one of the most complex types of noise-control problem that acousticians have to deal with today. The difficulties are highlighted by the recent remarks of one of the most experienced noise-control engineers in the aircraft industry. He says, "Calculations of the in-flight interior noise made from exterior sound-pressure-level measurements on the ground have been very disappointing. Even after taking all known items into account, errors of many decibels still may occur. We must not pretend that a calculation procedure will give us a good in-flight answer with the kind of basic data available to us today." An equally experienced noise-control engineer in another company says, "The trouble lies in properly evaluating all the various paths by which the noise can come into the cabin (sidewalls, windows, floors, partitions, hat racks, seats, ventilators, etc.). If, in flight, we measure sound directly outside and inside a point on the sidewalls or on a window using a special type of shielded microphone, the difference between the two sound-pressure-level measurements agrees well with the on-ground measurements of the sidewall or window transmission loss."

With these words of caution as to the reliability of what follows, we proceed to illustrate one of the most difficult types of noise-control problem.

Selection of Noise Criteria for Passenger Aircraft. As we have said before in this book, every noise-control project can be analyzed into three parts: (1) the source, (2) the transmission path, and (3) the receiver. Noise control is done for the benefit of the receiver—the passenger in the case of aircraft. Hence, the starting point in design is a knowledge of his requirements.

Actually the criterion chosen for the noise-control design in an aircraft is the result of a management decision that weighs passenger comfort and ability to hear speech vs. accepted standards in competitive aircraft vs. cost and weight considerations. Historically, however, the quietest aircraft have had great passenger appeal as witness the Convair Metropolitan (440), Douglas DC–7C, and the Britannia. On the other hand, passenger complaints were heard frequently about the noise levels in the Convair 340, the Douglas DC–7, and some British aircraft.

Stated simply, the passenger is generally satisfied with the noise and vibration conditions in an aircraft, if conversation is easy with adjacent passengers, if the loudspeaker announcements are easy to understand, and if vibration is not too obvious.

In Chap. 20, we discussed criteria for noise and vibration in aircraft. Essentially, we said that the noise should not exceed an NCA–60 curve measured at the ears of the passengers. Furthermore, the acceleration should not exceed about 4 in./sec^2 (0.1 m/sec^2) on the chair, floor, or armrests. We saw that aircraft today do not meet this criterion curve, partly because of cost and weight considerations and partly because competitive considerations have not yet emphasized quiet as being as important as speed.

Once the criteria for noise and vibration have been chosen, the source receives consideration next.

Exterior Noise Sources—Jet-driven Aircraft. The noises that require the greatest amount of control in high-speed jet-driven aircraft arise from the engines and from the turbulence of the air flowing over the surfaces of the aircraft. Both of these subjects are discussed in a lengthy report prepared by the staff of Bolt Beranek and Newman Inc. for the U.S. Air Force.[3] The reader is also referred to the excellent material by von Gierke.[4]

Those parts of Ref. 3 that deal with the near-field sound of turbojets were summarized in pp. 651 to 654. By that procedure one determines the SPL's that would exist over the surface of the fuselage if it were acoustically transparent. In other words we say that for each engine we can map out the free-field sound pressure over the surface of the fuselage. At each point on the fuselage, the levels from all engines in each octave band are summed on an intensity basis. If the engines are hung on pylons below the wing, shielding effects

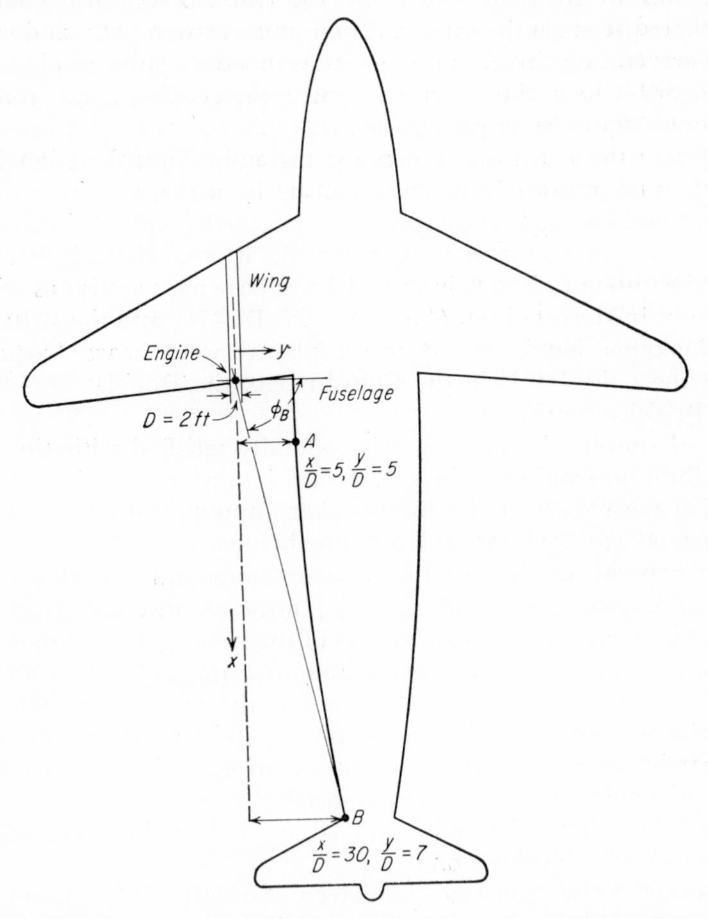

FIG. 25.7 Geometry of turbojet aircraft considered in working Examples 25.1 and 25.2.

must be taken into account. For example, for engine mountings similar to those used by 1960-vintage American jet-driven aircraft, the levels produced on the fuselage by a given engine in the shadow of the wing should be reduced by about 5 db at frequencies below 600 cps, and about 10 db at frequencies above 600 cps.

The question will be asked as to whether the sound pressure buildup normally associated with the reflection of sound from a surface should be considered on the outside of the fuselage. Whether it is or not is a matter of how the noise reduction of the fuselage sidewalls are handled in the computations. Let us elaborate on this question.

The TL of a wall is defined as the ratio of (*a*) the sound power transmitted through the wall to (*b*) the sound power that would be transmitted if the wall were removed from between the source and the observer. The receiving space is assumed to be anechoic.

The SPL's over the surface of the fuselage ($SPL_{incident}$) that we computed from the procedure of Chap. 24 for near-field noise are consistent with *b* in the preceding paragraph. Therefore, the transmitted sound-pressure levels (SPL_{inside}) are found from

$$SPL_{inside} = SPL_{incident} - TL \qquad \text{db} \tag{25.3}$$

Of course, the TL of a structure is a function of the angle at which the sound wave strikes the structure, and the SPL inside the fuselage is dependent on how reverberant the space is. These two factors will be discussed later in this chapter. If one wishes for some reason to estimate the SPL outside a fuselage surface, he must add about 3 db to the $SPL_{incident}$ levels to take into account the pressure buildup at the curved fuselage surface.

The use of the material of Chap. 24 for determining the near-field, free-field sound produced over the surface of a fuselage is illustrated by the example below.

Example 25.1. Figure 25.7 shows a portion of the wing and fuselage of a turbojet aircraft. The engine is buried in the wing. Dimensions and operating conditions are as follows:

Engine exhaust diameter	2 ft
Engine exhaust velocity (stationary aircraft)	2,000 ft/sec
Engine exhaust velocity (relative to moving aircraft)	2,000 ft/sec
Aircraft velocity	450 mph = 660 ft/sec
Aircraft altitude	35,000 ft

Static jet temperature for a moving aircraft equal to static jet temperature for a stationary aircraft

Find the SPL at the two points on the fuselage indicated in Fig. 25.7.

Solution. The near-field corrections for a moving aircraft, given in Sec. 24.2, can be applied in three steps:

Step 1. The values of near-field contours of Fig. 24.8 are adjusted by the quantity $\Delta F + \Delta PWL - \Delta B$. From Eq. (24.5).

$$\Delta F = 80 \log_{10} \frac{2{,}000}{1{,}850} = +3 \text{ db}$$

For an altitude of 35,000 ft

$$\frac{p_a}{p_{a0}} = 0.24 \qquad T_a = 394°\text{R} = -65°\text{F}$$

and by assumption

$$T_j = T_{j0}$$

Therefore, from Fig. 24.6,

$$\Delta\text{PWL} = -2 \text{ db}$$

and from Fig. 24.7

$$\Delta B = +6 \text{ db}$$

The total SPL adjustment is —5 db.

Step 2. The adjusted contours are rotated through the angle $\Delta\phi$, based on the exhaust velocity relative to the surrounding medium. This velocity is 2,000 — 660 = 1,340 ft/sec; therefore from Fig. 24.9, $\Delta\phi = -11°$. The adjusted and rotated contours are shown in Fig. 25.8. (See Fig. 24.8 for comparison.)

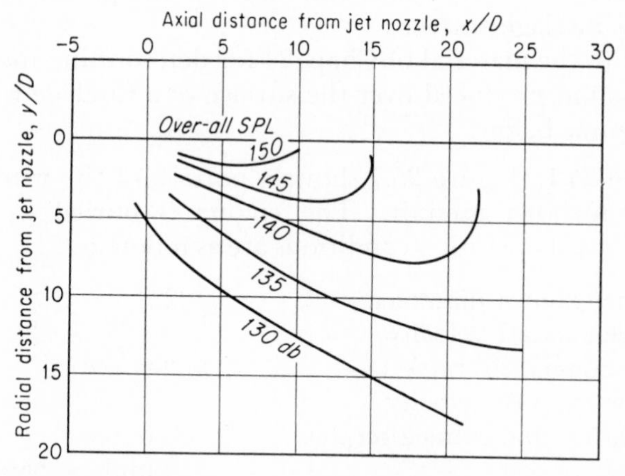

FIG. 25.8 Near-field contours used in working Example 25.1. The contours have been adjusted and rotated as described in the example.

Step 3. The forward motion of the aircraft distorts the near-field contours. This distortion is described in Fig. 24.13, and the results for the two points of interest are listed in Table 25.1. The aircraft Mach number at 35,000 ft is 0.67. Then obtain f_{max} from Fig. 24.10.

Note that the velocity occurring in the abscissa of Fig. 24.10 is the velocity relative to the surrounding medium, or 1,340 ft/sec, and that the "apparent axial position" is used in estimating the spectrum shape from Fig. 24.10.

Table 25.1

Results for Two Points Discussed in Example 25.1

Point	Actual Angle ϕ	Apparent Angle ϕ' (from Fig. 24.13)	Apparent Position		Over-all SPL, db (from Fig. 25.8)	f_{max}, cps (from Fig. 24.10)
			x'/D	y'/D		
A	−45°	−15°	1.3	5	132	670
B	−77°	−65°	15	7	140	270

Boundary-layer Noise.[3] The high-speed flow of air over an aircraft surface generates a turbulent boundary layer in a region near the surface. The fluctuating pressures exerted on the surface by this layer are convected along with the flow and set up vibration of the aircraft skin.

The procedure for estimating this noise is as follows:

1. Determine the over-all SPL of the pressure fluctuations at the surface of the fuselage using Fig. 25.9. The Mach number used in

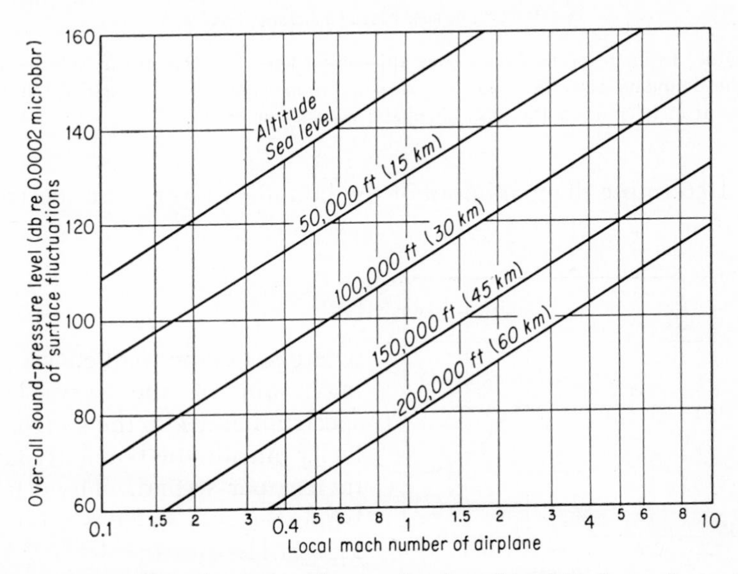

FIG. 25.9 Over-all boundary-layer noise as a function of local Mach number and altitude of the airplane.

the calculations is that existing at the part of the fuselage under consideration just outside the boundary layer. This is called the "local" Mach number. For example, some local numbers may be subsonic for a supersonic airplane and vice versa.

2. Estimate the boundary-layer thickness δ from Fig. 25.10. For a typical 1959 commercial jet transport, the value of δ might be 0.3 ft. See Example 25.2 for a value of ν at 35,000 ft.

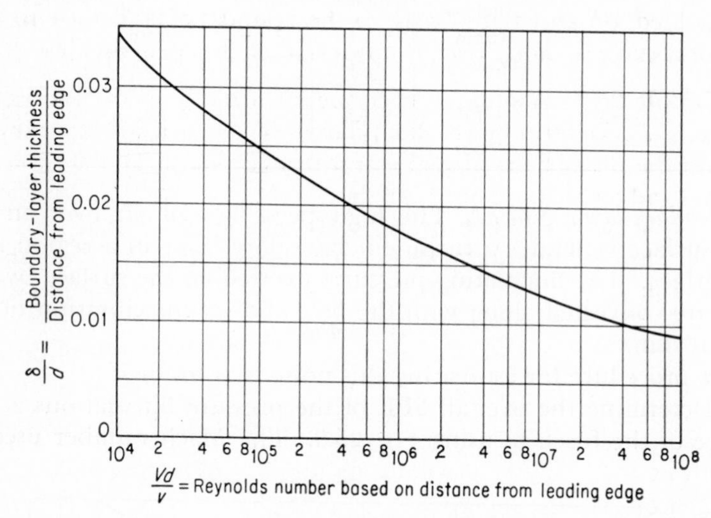

FIG. 25.10 Estimate of boundary-layer thickness where V = free-stream velocity outside the boundary layer in ft/sec; d = distance from leading edge of skin to observation point in ft; and ν is the local kinematic viscosity in ft²/sec.

3. Determine the maximum of the boundary-layer noise spectrum f_{max} from:

$$f_{max} = 0.4 \frac{V}{\delta} \quad \text{cps} \qquad (25.4)$$

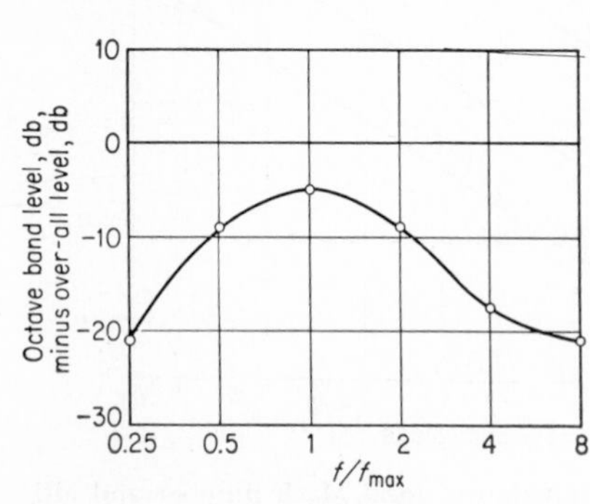

FIG. 25.11 Octave-band frequency spectrum relative to the over-all level. The abscissa equals the frequency divided by f_{max} as determined from Eq. (25.4).

where f_{max} is the frequency at the maximum of the octave-band spectrum and V is the stream velocity outside the boundary layer in feet per second. The octave-band-spectrum shape is given by Fig. 25.11.

The boundary-layer noise is affected by aircraft geometry, altitude, protuberances near the local position of measurement, and similar items. The procedure given above represents an average of the known information. The data underlying this procedure were taken at subsonic speeds only. Until organized data are taken at supersonic speeds the method is the only one available for calculations above Mach 1.0.

The boundary-layer noise is mapped out over the surface of the fuselage. The levels already mapped out for the turbojet noise are then added, on an energy basis, to the boundary-layer noise to yield the total exterior noise over the fuselage of the airplane.

Example 25.2. Assume a jet-propelled aircraft of the type shown in Fig. 25.7. Determine the boundary-layer noise outside a point 30 ft from the nose of the airplane at a true air speed of 450 mph (725 km/hr) and an altitude of 35,000 ft.

Solution. (*a*) At 35,000 ft, the speed of sound in standard atmosphere is about 670 mph (1080 km/hr). Hence, the Mach number of the aircraft is

$$\text{Mach No.} = \frac{450}{670} = 0.67$$

(*b*) From Fig. 25.9 we see that the over-all SPL of the surface fluctuations is 130 db.

(*c*) The boundary-layer thickness is found from Fig. 25.10. First we calculate,

$$\frac{Vd}{\nu} = \frac{660\ (30)}{4.11 \times 10^{-4}} = 4.8 \times 10^{7}$$

$$\frac{\delta}{d} = 0.01$$

$$\delta = 0.3 \text{ ft}$$

(*d*) Calculate f_{max} from Eq. (25.4).

$$f_{max} = 0.4\,\frac{660}{0.3} = 880 \text{ cps}$$

This is approximately the geometric mid-frequency of the 600- to 1,200-cps octave band.

(*e*) From Fig. 25.11 and items *b* and *d* of the solution above we obtain the octave-band levels listed in Table 25.2.

Engine Vibration. No numerical data are in the literature on vibration induced into the wing structure by a jet engine. In one airplane design where the jet engines were imbedded in the wing it was found that engine vibrations caused serious sound radiation from the floor of the passenger compartment. These levels were in excess of NCA–65 in the middle octave bands. To be certain of achieving the full value of the NR built into the sidewalls, the points of engine mounting should contain resilient material.

Table 25.2

Octave-band Levels Obtained in Example 25.2

Band frequency, cps	*SPL, db*
37.5–75	85
75–150	97
150–300	109
300–600	121
600–1,200	125
1,200–2,400	121
2,400–4,800	112

Exterior Noise Sources—Propeller Aircraft. In propeller aircraft, the exterior noise sources arise from the propeller, the exhaust, and the engine vibration. Detailed information on the propeller and exhaust noise is given in Ref. 3. Much of the material there and here is based on the work and reports of the National Advisory

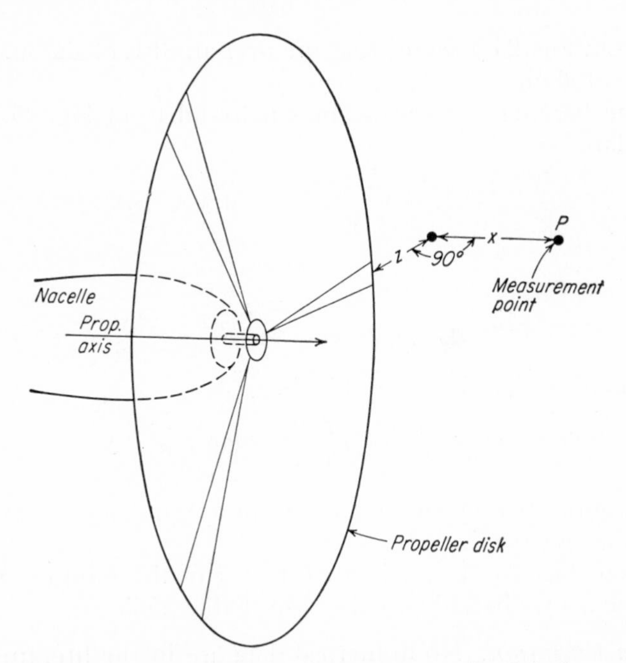

Fig. 25.12 Acoustically, the position of a measurement point P in free space outside the propeller cylinder is fully defined by two coordinates: one is a distance measured in the propeller plane and the other is a distance measured perpendicular to the propeller plane. The coordinate z, measured in the propeller plane, is a projection in the propeller plane of the line connecting the point P to the nearest point on the edge of the propeller disk. The coordinate x, measured perpendicular to the propeller plane, is the distance that the point P lies fore or aft of the extended propeller plane. The determination of x and z for two points on the surface of an airplane fuselage is shown in Fig. 25.19 accompanying Example 25.3.

Committee for Aeronautics (now the National Aeronautics and Space Administration). Other material is drawn from measurements made by Bolt Beranek and Newman Inc. and by Harvard University's wartime Electro-Acoustic Laboratory. Some of the concepts have not been fully verified. They represent the best judgment of the writers of Ref. 3.

Near-field Propeller Noise. The propeller noise in the near field can be estimated by the procedure given below. It consists of two parts, the rotational and the vortex noise. By definition, the near

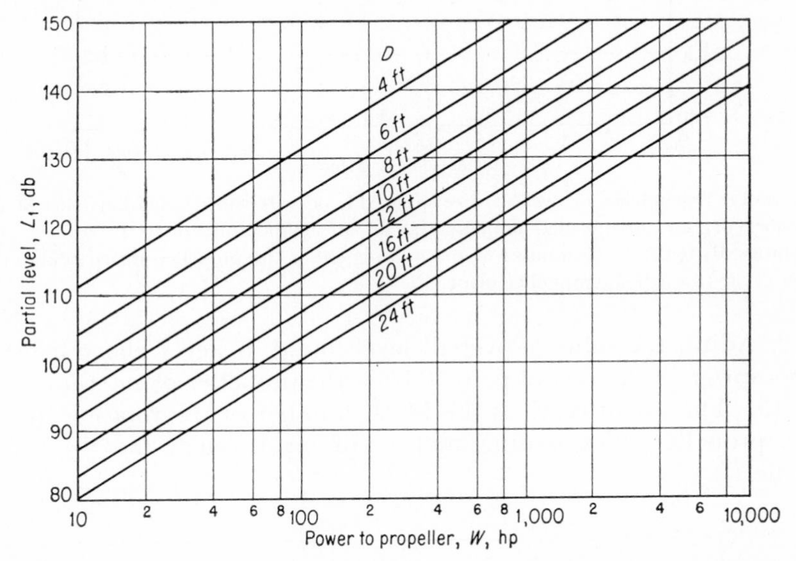

FIG. 25.13 Dependence of over-all, free-space sound-pressure level on horsepower W delivered to propeller and on diameter D, for three-bladed propellers at an air temperature of 68°F (20°C) (1 hp equals 746 watts).

field consists of those points of measurement that are located one propeller diameter or less from the propeller plane. The geometry of interest is given in Fig. 25.12. The near-field prediction given here applies only outside of the propeller cylinder. That is to say, the analysis is not valid for predicting noise levels directly in front of or directly behind the propeller disc.

1. Find the over-all SPL of the rotational noise L_{RN} (db) at the point P by adding algebraically the partial levels (in decibels) L_1 and L_2 given respectively by Figs. 25.13 and 25.14. For other temperatures subtract 10 log [(°F + 460)/528] from the over-all level just determined. For other numbers of blades m add 20 log $(3/m)$ to the over-all level just determined.

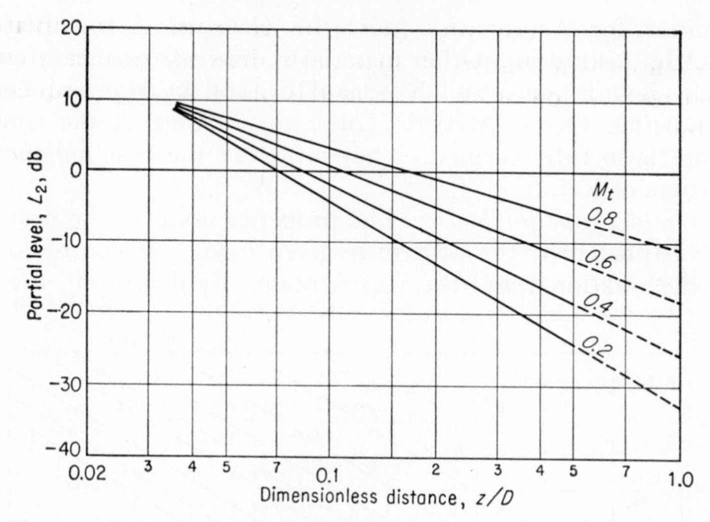

FIG. 25.14 Dependence of over-all, free-space SPL on z/D and on rotational tip Mach number M_t for three-bladed propellers at an air temperature of 68°F (20°C). The quantity M_t is the Mach number of the rotating tip of the propeller, disregarding motion of air through the propeller plane.

2. Adjust the value of over-all level found in step 1 above for the appropriate distance x of point P from the propeller plane using Fig. 25.15. The quantity M_t is the Mach number of the rotating tip of the propeller, disregarding motion of air through the propeller plane.

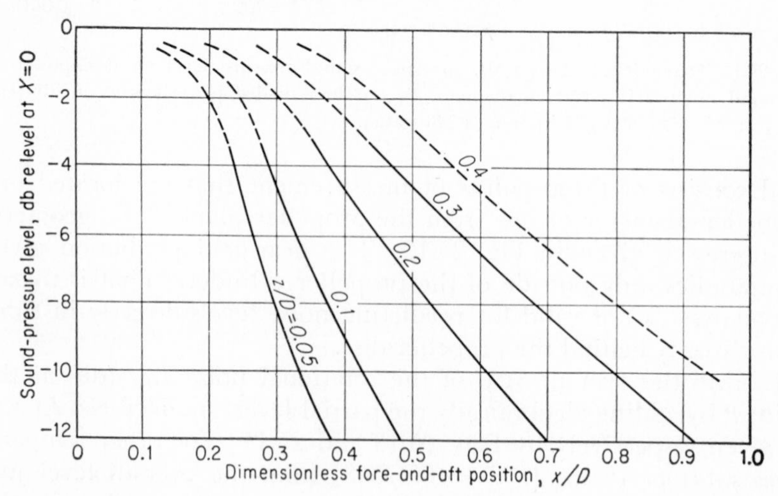

FIG. 25.15 Variation of over-all, free-space propeller-noise levels with axial position x/D fore and aft of propeller plane. $M_t = 0.6$.

3. The over-all level of 1 and 2 above includes the rotational noise alone, which consists of noise at discrete, harmonically related frequencies, the lowest of which is the blade passage frequency f_1 given by

$$f_1 = \frac{mRG}{60} \qquad \text{cps} \tag{25.5}$$

where m = number of blades
R = engine rpm
G = fractional gear ratio between engine and propeller, i.e., equals ratio of propeller rpm to engine rpm

4. The levels of the first eight harmonics (occurring at frequencies f_1 through f_8) of the rotational noise relative to the over-all level L_{RN} are found from Fig. 25.16. The rms SPL's in the octave bands containing more than one of the eight frequency components should be determined on an intensity basis as discussed in Chap. 3.

5. A second part of the propeller noise is the vortex (blade-wake) noise. This noise has a continuous spectrum with a peak at a frequency $f_{\max}$ determined by

$$f_{\max} = 0.13\,\frac{V_H}{L_{0.7}} \qquad \text{cps} \tag{25.6}$$

where V_H = helical propeller tip speed

$$= \sqrt{V_t^2 + V_F^2}, \qquad \text{ft/sec} \tag{25.7}$$

V_t = tangential tip speed, ft/sec
V_F = forward aircraft speed, ft/sec
$L_{0.7}$ = section of blade, ft, at 0.7 blade radius and given by insert on Fig. 25.18

6. The over-all vortex noise L_{VN} (db) is found from Fig. 25.17. For example, assume a four-blade propeller, a propeller loading of 10 hp/ft², and an over-all rotational noise level of 110 db; Fig. 25.17 shows that $L_{VN} - L_{RN} = -17$ db. Hence, $L_{VN} = 93$ db.

7. The octave-band levels of the vortex noise (relative to the over-all level, L_{VN}) are given by Fig. 25.18 where f is the geometric mean frequency of the octave band and $f_{\max}$ comes from step 5 above.

8. Add, on an intensity basis, in each octave band, the level of the rotational noise and the level of the vortex noise (see Chap. 3). The eight octave-band levels so obtained then give the spectrum of the noise to be expected at the measurement point P.

Example 25.3. A propeller-driven aircraft has the configuration shown in Fig. 25.19. Find the total SPL (rotational and vortex noise) in the usual eight octave frequency bands at two points on the sur-

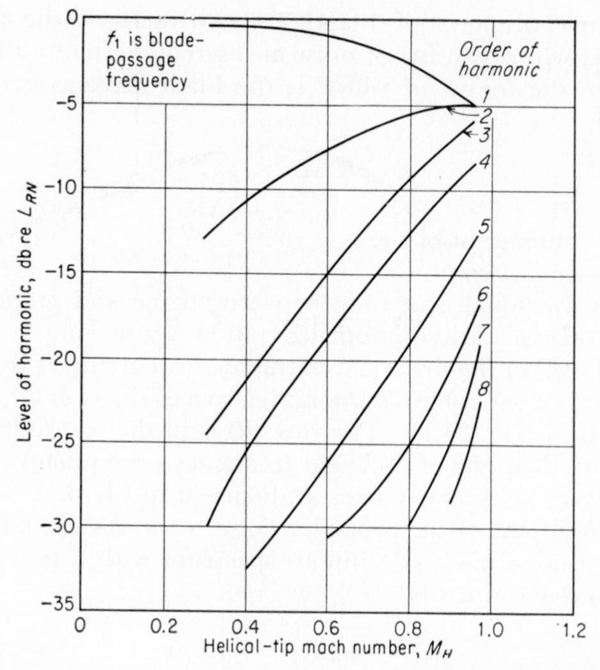

FIG. 25.16 Rotational noise level (in db relative to the over-all noise level L_{RN}) for each of eight harmonics as a function of the helical tip Mach number M_H.

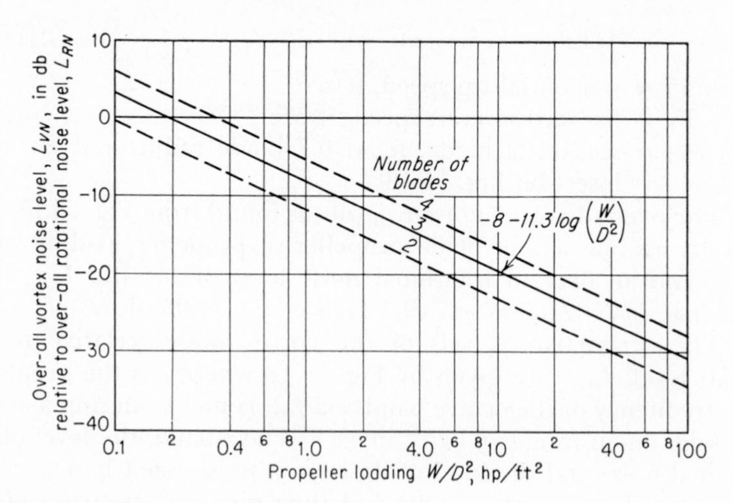

FIG. 25.17 Chart giving the over-all level of the vortex noise L_{VN} relative to the over-all level of the rotational noise L_{RN}.

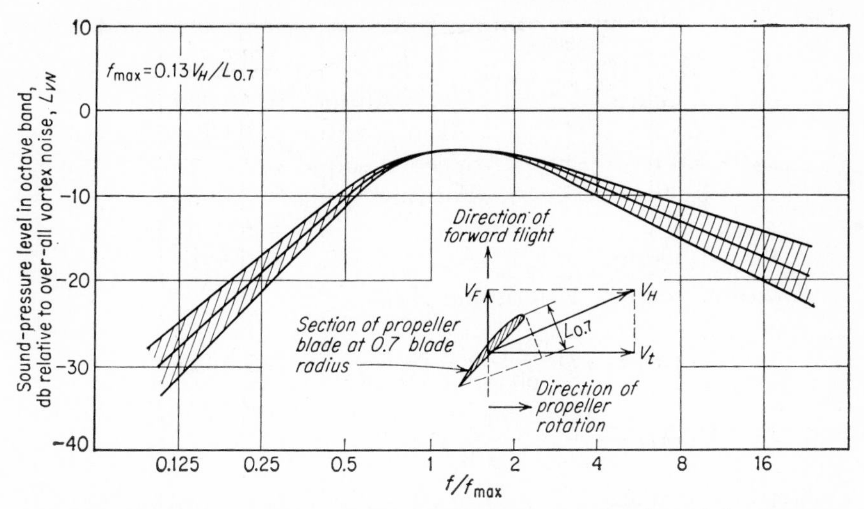

FIG. 25.18 Octave-band SPL's (*re* over-all level L_{VN}) as a function of the frequency ratio f/f_{max}. f is the geometric mean frequency of the octave band. The insert sketch gives the definition of $L_{0.7}$ in ft. V_H is the helical tip velocity in ft/sec.

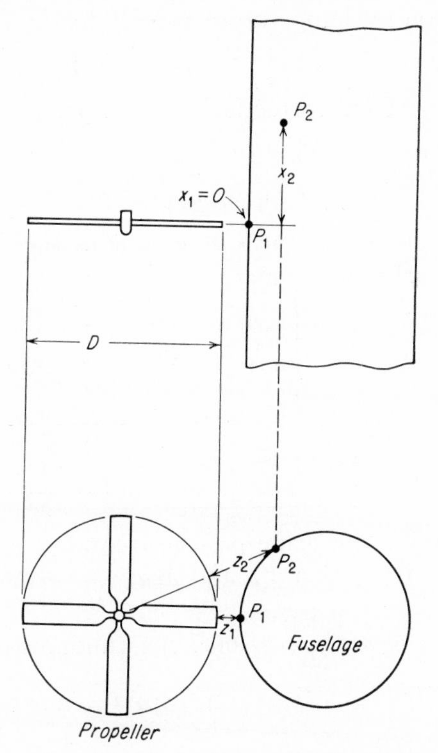

FIG. 25.19 Geometry and positional coordinates for measurement points P_1 and P_2 of Example 25.3.

face of the fuselage x_1 and x_2 as shown in Fig. 25.19 and below:

$D = 15$ ft
$m = 4$ blades
$z_1 = 1.5$ ft
$z_2 = 5.0$ ft
$x_1 = 0$
$x_2 = 7.5$ ft
$W = 1{,}500$ hp

Prop rpm = 1,000
Altitude = 10,000 ft
Temperature = 23°F
Speed of sound = c = 1,078 ft/sec
Forward speed = 350 mph
= 514 ft/sec
$L_{0.7} \doteq 0.5$ ft

Solution. Step 1. Determine:

$$V_t = \pi D \frac{\text{rpm}}{60} = \pi\, 15 \frac{1{,}000}{60} = 785 \text{ ft/sec}$$

$$M_t = \frac{V_t}{c} = \frac{785}{1{,}078} = 0.73$$

Step 2. Calculate:

$$\text{Correction for no. of blades} = +20 \log \tfrac{3}{4} = -2.5 \text{ db}$$

$$\text{Correction for temperature} = -10 \log \frac{23 + 460}{528} = +0.4 \text{ db}$$

$$\text{Total correction} \doteq -2 \text{ db}$$

Step 3. Tabulate. (See Table 25.3.)

Table 25.3

Results for Two Points Discussed in Example 25.3

Point	L_1,db (Fig. 25.13)	Correction for Number of Blades and Temperature, db (see step 2)	z/D	L_2, db (Fig. 25.14)	x/D	Correction for x/D (Fig. 25.15)	L_{RN}, db
P_1	132	−2	1.5/15 = 0.1	+3	0	0	132 + 3 − 2 = 133
P_2	132	−2	5/15 ≐ 0.33	−6	7.5/15 = 0.5	−4 (approximately)	132 − 6 − 4 − 2 = 120

Step 4. Determine the blade passage frequency f_1 and the helical tip speed and Mach number V_H and M_H:

$$f_1 = m \frac{\text{rpm}}{60} = 4 \times \frac{1{,}000}{60} = 66.7 \text{ cps (fundamental frequency)}$$

$$V_H = \sqrt{V_t^2 + V_F^2} = \sqrt{(785)^2 + (514)^2} = 938 \text{ ft/sec}$$

$$M_H = \frac{V_H}{c} = \frac{938}{1{,}078} = 0.87$$

We find from Fig. 25.16 the results in Table 25.4.

Table 25.4

Harmonic	f, cps	Harmonic Level, db *re* L_{RN}
1	66.7	−4
2	133	−5
3	200	−7
4	267	−10
5	333	−16
6	400	−21
7	467	−26
8	533	−30

The corresponding octave-band levels in decibels *re* the over-all octave-band level are given in Table 25.5.

Table 25.5

Octave band, cps	*Band level, db re* L_{RN}
20–75	−4
75–150	−5
150–300	−5
300–600	−14
600–1,200	−30 (extrapolated)

Step 5. Determine W/D^2, $L_{VN} - L_{RN}$, and $f_{\max}$:

$$\frac{W}{D^2} = \frac{1{,}500}{(15)^2} = 6.7 \text{ hp/ft}^2$$

From Fig. 25.17

$$L_{VN} - L_{RN} = -15 \text{ db}$$

$$f_{\max} \doteq 0.13 \frac{V_H}{L_{0.7}} = 0.13 \frac{938}{0.5} = 244 \text{ cps}$$

A frequency of $f_{\max} = 244$ cps is near the geometric mean frequency of the 150- to 300-cps band. Hence, in Fig. 25.18, $f/f_{\max} \doteq 1$ corresponds to the 150- to 300-cps band.

Step 6. Tabulate. (See Table 25.6.)

It is important in propeller-driven aircraft design to note that the noise levels produced at the outside of a fuselage drop off rapidly with distance from the point of closest passage of the propeller tip to the fuselage. This means that in design, the transmission of sound

Table 25.6

SPL Results of Example 25.3

Item	Frequency, cps								
	OA	20–75	75–150	150–300	300–600	600–1,200	1,200–2,400	2,400–4,800	4,800–10,000
	Sound-pressure Levels, db								
1. Rotational noise, band levels at $P_1(L_{RN} = 133$ db)	133	129	128	128	119	103			
2. Vortex noise, band levels at P_1 *re* (L_{RN}); see Fig. 25.18 and step 5	−15	−34	−25	−20	−20	−24	−28	−32	−36
3. Vortex noise, band levels at P_1	118	99	108	113	113	109	105	101	97
4. Total noise at P_1 (combine items 1 and 3)	133	129	128	128	120	110	105	101	97
5. Rotational noise, band levels at P_2 ($L_{RN} = 120$ db)	120	116	115	115	106	90			
6. Vortex noise, band levels at P_2	105	86	95	100	100	96	92	88	84
7. Total noise at P_2 (combine items 5 and 6)	120	116	115	115	107	97	92	88	84

along the fuselage structure from the point of closest passage to more distant points must be small, or the level of the noise nearest the propeller will set the levels throughout the length of the fuselage.

Exhaust Noise of an Internal-combustion Engine. Under certain conditions, the exhaust noise of internal-combustion engines may be a significant contributor in determining the noise levels in an aircraft. For estimating the noise levels, the following procedure is followed:

1. Obtain the over-all sound-power level of the engine exhaust noise from

$$\text{PWL} = 125 + 10 \log hp \qquad \text{db} \tag{25.8}$$

where hp is the total horsepower delivered by the engine shaft to the propeller, auxiliary pumps, generators, and other equipment. The experimental data used to obtain this formula showed a spread of ± 5 db.

2. Use Fig. 25.20 to obtain the octave-band noise spectrum of the exhaust noise. The frequency f_a is the average cylinder firing frequency of the engine.

3. The SPL at a point P a distance l from the engine exhaust can be found from the relation:

$$\text{SPL} = \text{PWL} - 10 \log (4\pi l^2) \qquad \text{db} \tag{25.9}$$

Spherical divergence of the sound has been assumed.

Vibration in Reciprocating Engines. Because of irregularities in the combustion, reciprocating engines are a source of serious vibration. Methods have been devised by the manufacturers of aircraft engines for suspending aircraft engines on rubber-in-shear mounts in such a way as to decouple the six modes of vibration (translational and rotational) so as to cause a significant reduction in the vibrational forces driving the fuselage. The detailed treatment of this subject is beyond the scope of this book.

Interior Sources. Interior sources of noise include: hydraulic

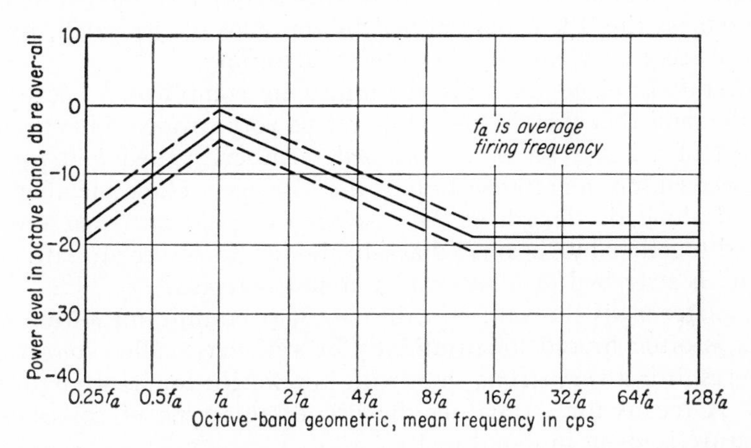

FIG. 25.20 Chart for determining the octave-band PWL's of the exhaust noise. The mid-frequency of the octave band to go with the spectrum shown is determined by f_a, the average cylinder firing frequency of the engine. It is assumed here that the exhaust port of each cylinder is connected to the outside air through cylindrical pipes equal in length, say, to about 1/10 wavelength at the frequency f_a, and joined together in one common exhaust opening. Obviously, if different distances obtain between the exhaust ports and the common exhaust opening, some noise components will be suppressed and others amplified according to the relative phases of the individual exhaust pulses at the time they combine in the exhaust system.

pumps for control of the aircraft in flight, air-conditioning compressors, ventilating fans and their associated hydraulic motors, ventilating outlets in the cabin, and cabin pressurization pumps and valves. The piping from hydraulic pumps to hydraulic motors may also be noisy.

It is not possible to discuss noise control of these items in detail. Furthermore, unless an operating mock-up is made before the plane is built and noise-control measures are planned, it may be virtually impossible to make corrections later. Pumps, fans, compressors, and so forth, should be bought with a noise specification consistent with the degree of quiet desired. The principles of noise reduction in air-handling systems have been discussed in Chap. 21.

Sound Transmission Loss through the Fuselage Walls.[1,3-5] In Chap. 13 sound transmission through solid walls and plates is discussed. We saw that at low frequencies a panel may not vibrate freely under excitation of a sound wave because of its stiffness. In this region we say that the TL is stiffness controlled. At somewhat higher frequencies, the panel resonates and the TL is small. At the lower middle frequencies the TL is controlled principally by the mass of the panel. At upper middle frequencies wave coincidence occurs and the TL either drops or levels off with increasing frequency. Finally at high frequencies the TL is determined by the mass of the panel, by the critical frequency, and by the internal damping.

Aircraft fuselage sidewalls are unusually complicated because of the demand that they be light but yet be very strong. They usually consist of a heavy set of ring-shaped members (called *belt frames*) that determine the cross section of the fuselage. These members are spaced along the length of the fuselage every 20 in. or so and are joined together by "beams" that run the length of the aircraft. The "skin" is stretched over this skeleton and is riveted or welded to it. The material of the skin is a thin sheet of aluminum alloy. This sheet is often braced longitudinally by stiffeners (called *longerons*). The result is a very strong shell with low weight for its strength and size. A means for estimating the TL of this type of complicated structure is given in detail in Ref. 3.

In the present chapter we shall simply look at the difference in noise levels measured outside and inside the fuselages of several typical aircraft in order to obtain an average value of transmission loss. The data are corrected to remove the effects of sound absorption of the sidewalls and of reverberation time. The average results for two typical structures, with details of construction given for guidance, are shown in Fig. 25.21.

The flattening off of the curves at frequencies above 1,000 cps is due to coincidence effects and is predicted from Fig. 13.9 of Chap. 13. The relatively high values of TL at low frequencies occur in the stiffness-controlled region (see Fig. 13.4).

The data of Fig. 25.21 must be used with care and only as a guide in design. The acoustical effects of the bracing provided by the longerons and belt frames are not fully understood. Actual data should be obtained on fuselage structures in flight in order to accurately plan internal noise control.

Vibration Transmission along the Fuselage. In connection with Example 25.3 we pointed out the need for reducing the transmission of sound along the length of the fuselage. In several propeller-driven aircraft it was found that the vibrational waves in fuselages of 1946 construction at frequencies below 600 cps decreased in amplitude at a rate of about 1 db for each 4 ft behind the plane of the propellers.[6] No new data have been reported in the literature since that time.

Regardless of how well the fuselage walls transmit a vibrational wave along their length, it is important to keep the wave from being induced in the fuselage to begin with. Some propeller-driven aircraft were built with extra heavy bracing in the vicinity of the propeller plane. This bracing reduced vibrational levels in the forward part of the fuselage by up to 10 db.

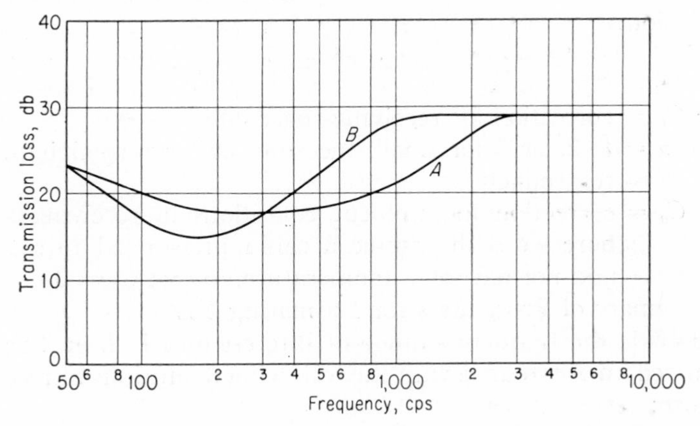

Fig. 25.21 Typical TL values for two representative fuselage constructions. It is not known how reliable these curves are for aircraft of various types of manufacture. The authors' experience and the available literature are confined to only a few examples of commercial airliners. Curve *A:* Fuselage diameter, 10 ft; skin 0.04 in. aluminum alloy; belt-frame spacing, 19 in.; longeron spacing, 6 in.; *Z* section, 0.75 in. web, 0.75 in. flanges. Curve *B:* Fuselage diameter, 11.5 ft; skin, 0.10 in. aluminum alloy; belt-frame spacing, 21 in.; only a few longerons.

Added Vibration-damping Materials. One method commonly used for reducing the transmission of sound along a fuselage other than by increasing its structural complexity and weight is to utilize vibration-damping materials. Such materials can reduce vibration (particularly above 300 cps) in a fuselage structure by 5 to 20 db if they are highly efficient for their weight. It is not within the scope of this text to treat vibration damping quantitatively. Some papers of recent issue are given in Refs. 7 through 11.

Cabin Configuration and Added Sound-absorbing Materials. In Chap. 14, the transmission of sound through structures containing porous materials was discussed. The properties of structures were stated there in terms of their TL. Although the TL of a structure is an important characteristic, it is by no means the whole story. In order to estimate the sound-pressure levels inside the fuselage, we must take account of the effects of the receiving space on the cabin noise levels.

The final result toward which we are working here is the development of a modified transmission loss which will include both the panel TL and the effects of the receiving space. Our modified value, denoted as TL′, can then be used directly to describe the level difference between the $SPL_{incident}$ and the sound-pressure levels in the receiving space.

Our procedure for determining TL′ in terms of TL is the following relation:

$$TL' = TL + C_n + C_a \qquad \text{db} \tag{25.10}$$

where C_n = correction for receiving-space effects
n = 1, 2, or 3 for small, medium, or large receiving space, respectively
C_a = correction for ambient conditions in receiving space;[3] here we shall assume a cabin pressurized to sea level and normal room temperature, so that $C_a = 0$

I. Summary of Procedures for Estimating TL′

A. Divide the frequency range of interest into 1, 2, or 3 regions, as required, in accordance with the following definitions of receiving-space size:

Small receiving space (low frequencies) $\frac{L}{\lambda} < \frac{1}{6}$

Medium-sized receiving space (mid-frequencies) $\frac{1}{6} < \frac{L}{\lambda} < 10$

Large receiving space (high frequencies) $10 < \frac{L}{\lambda}$

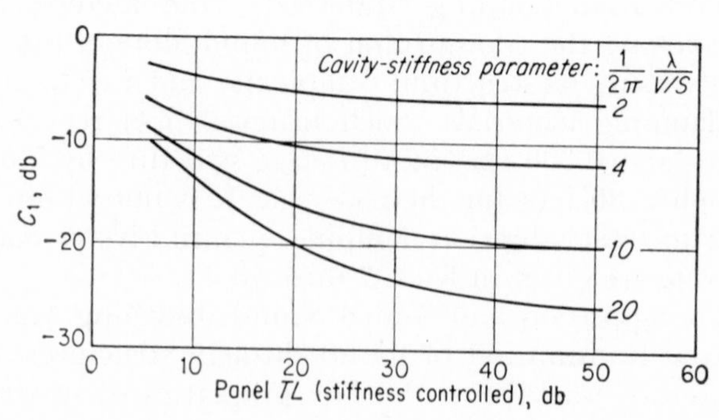

FIG. 25.22 C_1, correction to TL for stiffness effect of small receiving space.

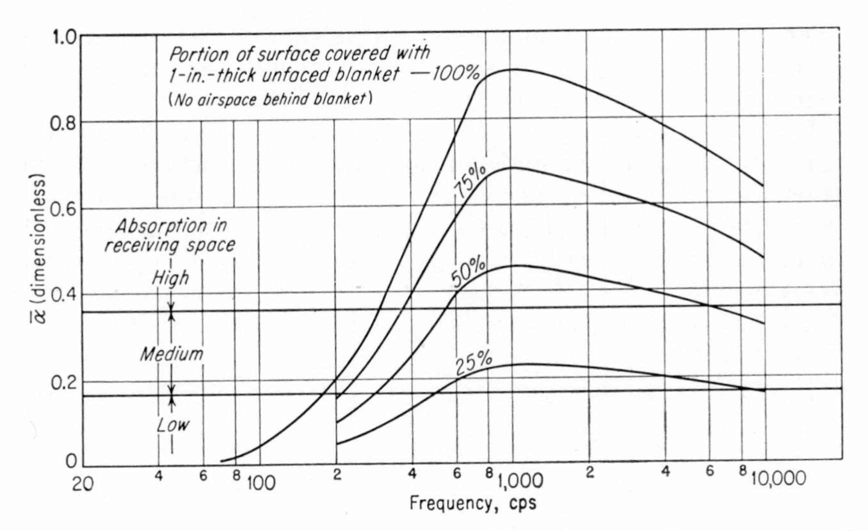

FIG. 25.23 Approximate average statistical coefficient $\bar{\alpha}$ for receiving spaces treated with 1-in.-thick aircraft-type acoustical blanket.

where λ is the sound wavelength within the receiving space and L is a typical dimension of the receiving space, usually the width of the space, or if the length is shorter, the length.

B. For a small receiving space, obtain the correction C_1 to TL from Fig. 25.22. The cavity stiffness parameter in this figure is $(1/2\pi)$

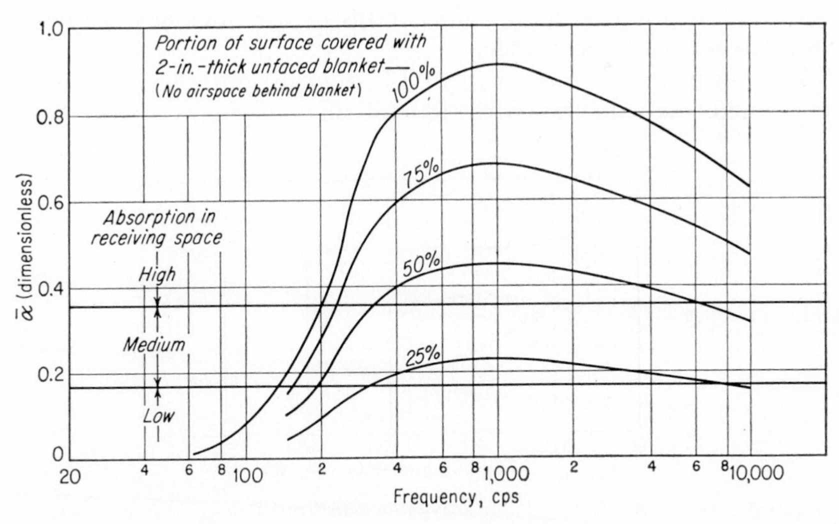

FIG. 25.24 Approximate average statistical coefficient $\bar{\alpha}$ for receiving spaces treated with 2-in.-thick aircraft-type acoustical blanket.

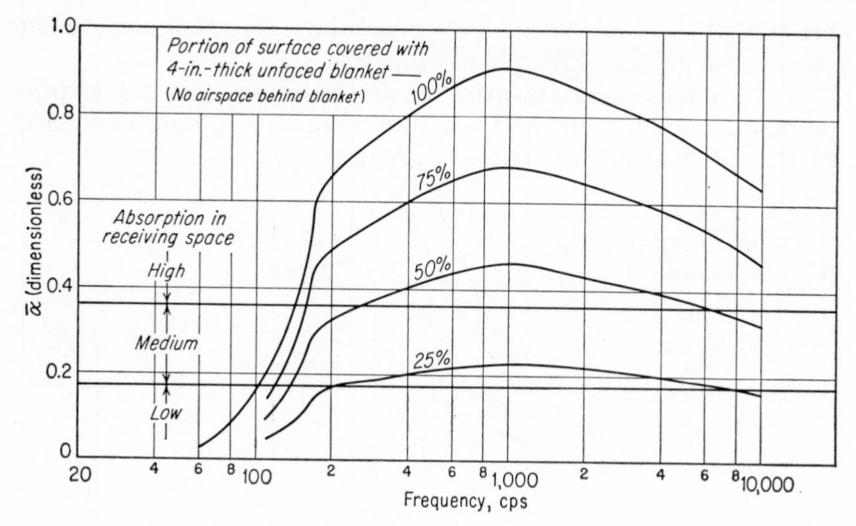

FIG. 25.25 Approximate average statistical coefficient $\bar{\alpha}$ for receiving spaces treated with 4-in.-thick aircraft-type acoustical blanket.

$\left[\frac{\lambda}{(V/S)}\right]$ where V and S are cavity volume and transmitting panel area, respectively.

C. For a medium-sized receiving space, estimate the amount of absorption in the space as "high," "medium," or "low" according to Figs. 25.23, 25.24, and 25.25. These figures describe the average statistical absorption coefficient $\bar{\alpha}$ in terms of the thickness of blanket and percent of wall coverage.

Determine the correction C_2 to TL from the appropriate curve in Fig. 25.26. As with the correction for small receiving spaces C_2 de-

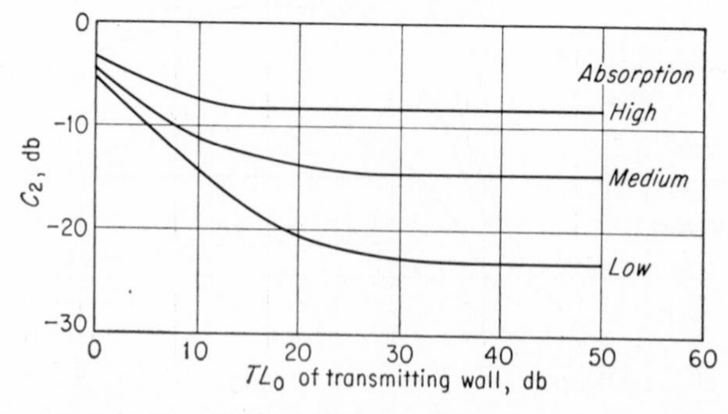

FIG. 25.26 C_2, correction to TL for maximum effect on standing waves in medium-sized receiving space.

pends on TL. Note that C_2 is the maximum correction. One should expect a range of effective corrections from 0 to C_2.

D. For a large receiving space determine the amount of absorption (high, medium, low) at the appropriate frequencies from Figs. 25.23, 25.24, and 25.25 as in *C* above. Calculate S_w/S_2

where S_w = area of the transmitting panel
S_2 = total surface area in the receiving space

Read the correction C_3 to TL from Fig. 25.27.

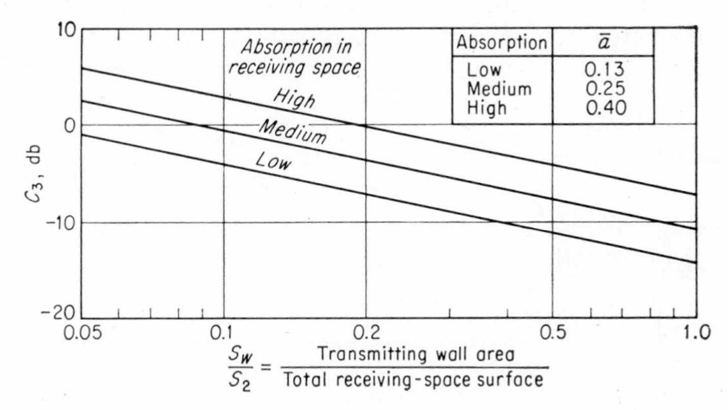

FIG. 25.27 C_3, correction to TL for reverberation in large receiving spaces.

E. Find the modified transmission loss (TL′) as follows:

1. Small receiving space:

$$\mathrm{TL}' = \mathrm{TL} + C_1 \tag{25.11}$$

2. Medium-sized receiving space:

$$\mathrm{TL}' = \mathrm{TL} + C_2 \tag{25.12}$$

3. Large receiving space:

$$\mathrm{TL}' = \mathrm{TL} + C_3 \tag{25.13}$$

II. Notes. The average absorption coefficient in a receiving space depends both on the type of surface treatment used and on the fraction of the total surface that is treated.

When an absorbing blanket is faced with an impervious septum (say, for example, a flexible plastic sheet weighing about 0.05 lb/ft²) the absorption changes appreciably. We may say qualitatively that the low-frequency absorption is enhanced as though the blanket absorption curve were simply shifted downward. (The downward shift is about one octave for a 1.5-in. blanket with a septum weighing 0.05 lb/ft^2.)

The effect of the septum is to reduce the high-frequency absorption sharply, so that the absorption coefficient reaches its peak value and falls off (on a logarithmic scale) just about as fast above the peak as it does below.

Figure 25.26 presents C_2, the maximum correction to TL giving TL′ *at space resonances,* for varying amounts of receiving-space absorption. At frequencies other than resonance frequencies we can expect corrections less than those shown in Fig. 25.26. In any case throughout this frequency range there will be a distribution of levels

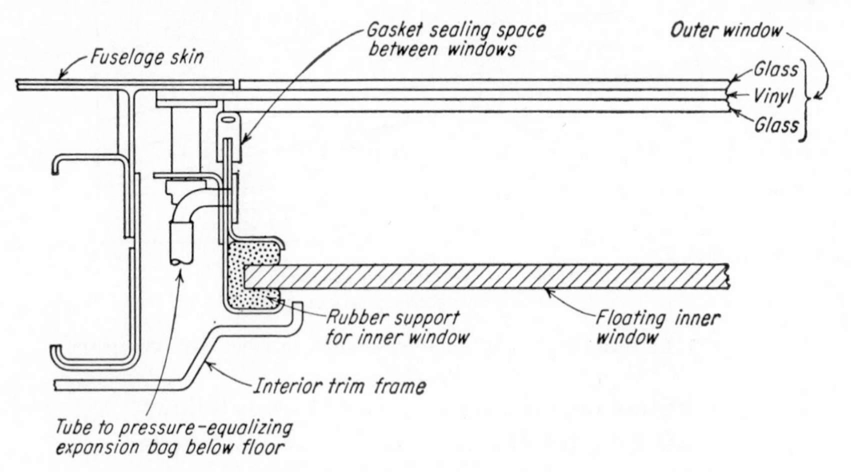

FIG. 25.28 Details of double windows. (*Courtesy of Convair.*)

ranging from approximately the levels for free-field receiving conditions at the standing-wave minima up to the maximum values calculable from the information in Fig. 25.26. Therefore, we apply a range of values in our estimate of TL′.

At resonance the maximum pressures are to be expected at the surfaces and corners in the receiving space as well as along the pressure antinode lines. In a standing-wave pattern the pressure maxima are considerably broader than the pressure minima. Therefore, the space-average sound-pressure level in a cavity at resonance will be closer to the maximum values than to the minimum or free-field values.

Design Details. Successful noise control means attention to details. Sound enters the passengers' cabin through walls, windows, floor, ventilation systems, bulkheads, seats, and baggage racks. External noise sources set the structure of the airplane into vibration. Any vibrating surface radiates sound which may be heard if it is high enough in frequency or may be felt if it is low enough in frequency. In noise control, one must either isolate a surface from a source that would otherwise set it into vibration or else build a barrier between that surface and the interior of the passenger cabin.

A noise-reduction job is no better than its weakest link. It does no good to provide a 40-db TL in the walls and only 20-db TL in the

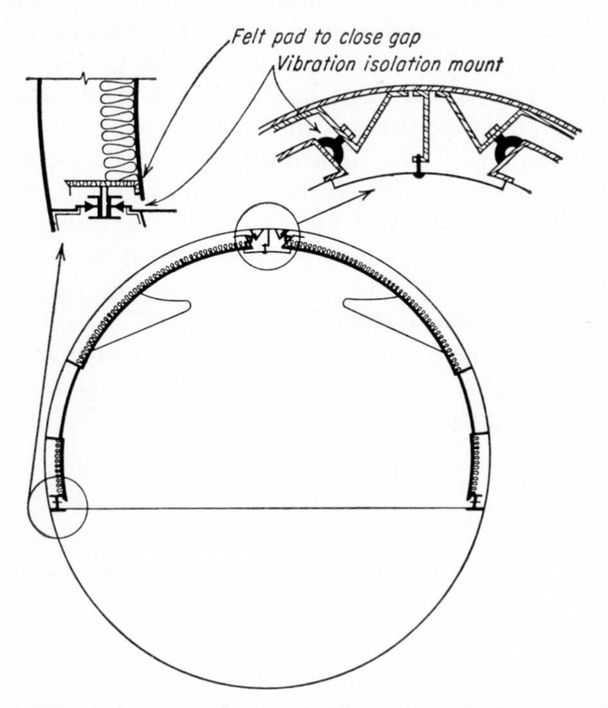

Fig. 25.29 Floating inner sidewalls. The weight of the walls is carried on vibration-isolation mounts. The windows are triple, and the outer two panes are designed according to Fig. 25.28. The third pane is carried on the floating structure. All joints (gaps) are sealed with felt pieces (not shown).

windows. The acoustically weaker windows will establish the noise levels inside the cabin. Let us take up some of the more important details one at a time. (We assume that the sidewall transmission has been properly provided for as discussed in the preceding paragraphs.)

Windows. A single pane of glass or transparent plastic may provide adequate TL in modern aircraft at frequencies below 500 cps if the surface density is sufficiently high. Above 500 cps, the TL requirements are great enough that from weight considerations alone, double or even triple panes are necessary. In order that the acoustical performance of a double-paned window exceed the performance of a single-paned window, two things must be achieved: (1) the spacing between the panes should be 3 in. or more, and (2) the inner pane should be floated in very soft rubber (see Fig. 25.28). The third pane of a triple-paned window, if needed, would be added to the second pane in the same manner as shown in Fig. 25.28.

Vibration Isolation of Baggage Racks, Window Frames, etc. Every surface that connects to the outside of the airplane can be vibrated by the motion of the exterior surfaces of the airplane. Such surfaces include coat racks, window frames, cabin-dividing bulkheads, floors, light fixtures, ventilation ducts, and the seats themselves. To prevent excitation of these surfaces the equivalent of an inner floating lining to the cabin may have to be built. A design detail for sidewalls and bulkheads, which can also carry the coat racks, light fixtures, and ventilation ducts, is shown in Fig. 25.29. In this design, two of the three panes of a triple window would be carried by the outer structure as seen in Fig. 25.28. The third window would be carried in the inner structure. A floating floor can be designed in a similar manner.

In a floating interior design, care must be taken to seal joints between adjacent panels, windows, and the floor. Special gaskets are necessary for this purpose.

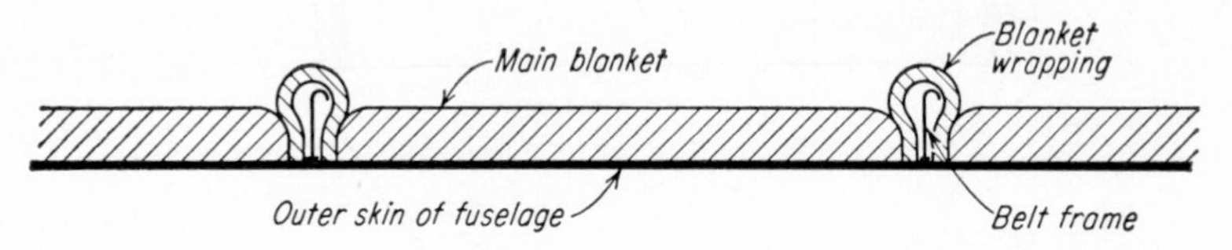

Fig. 25.30 Details of installation of sound-absorbing blanket in the fuselage.

When aircraft sound-absorbing blanket is used against the outer skin, as shown in Fig. 25.30, the belt frames must also be covered, and a tight joint must be formed between the wrapping and the main blanket. This prevents leakage at the joints and the wrapping reduces radiation of sound by the belt frames.

25.4 Summary

The two parts of this chapter deal with noise control in rail vehicles and in aircraft. The scaling of noise levels by vehicle speed is described. Design considerations in railroad-noise control are discussed with special emphasis on windows, floors, sidewalls, and details.

In the control of aircraft noise, the means for first determining the noise produced over the exterior surfaces of the airplane by the engines, propellers (if any), exhausts, and boundary-layer turbulence are treated. The transmission of vibrational waves along the fuselage is discussed. The required noise reduction between the outside of the fuselage and the passengers' ears inside is determined in each part of the airplane by subtracting the criterion from the exterior noise. The noise reduction provided by the sidewalls, windows, added damping materials, and added acoustical blankets is then presented. The influence of cabin configuration and added sound-absorbing materials on the noise levels inside the cabin is discussed next. The chapter concludes by discussing some design details.

REFERENCES

1. Data from files of Bolt Beranek and Newman Inc.

2. Stüber, C.: Luftschalldämmung von Fussboden-Konstruktionen für Eisenbahnwagen, *Larmbekampfung,* **1,** p. 1, March, 1958.

3. Franken, P. A., E. M. Kerwin, Jr., and the staff of Bolt Beranek and Newman Inc.: Methods of Flight Vehicle Noise Prediction, *WADC Tech. Rept.* 58–343, October, 1958. (Obtainable from the Aircraft Laboratory, WADD, Wright-Patterson Air Force Base, Ohio, *ASTIA Document AD*-205-776.)

4. von Gierke, H. E.: Aircraft Noise Sources, chap. 33 of C. M. Harris (ed.), "Handbook of Noise Control," McGraw-Hill Book Company, Inc., New York, 1957.

5. von Gierke, H. E.: Aircraft Noise Control, chap. 34 of C. M. Harris (ed.), "Handbook of Noise Control," McGraw-Hill Book Company, Inc., New York, 1957.

6. Rudmose, Wayne H., and L. L. Beranek: Noise Reduction in Aircraft, *J. Aeronaut. Sci.,* vol. 14, pp. 79–96, February, 1947.

7. Kurtze, G.: Bending Wave Propagation in Multi-layer Plates, *J. Acoust. Soc. Am.,* vol. 31, pp. 1183–1201, September, 1959.

8. Kerwin, E. M., Jr.: Damping of Flexural Waves by a Constrained Viscoelastic Layer, *J. Acoust. Soc. Am.,* vol. 31, pp. 952–962, 1959.

9. Dyer, I.: Response of Plates to a Decaying and Convecting Random Pressure Field, *J. Acoust. Soc. Am.,* vol. 31, pp. 922–928, 1959.

10. Oberst, H.: Werkstoffe mit extrem hoher innerer Dämpfung, *Acustica,* vol. 6, pp. 144–153, 1956 (*Akustische Beihefte,* no. 1). (This paper contains 61 entries in its bibliography.)

11. Liénard, P.: *Ann. Télécomm.,* vol. 12, pp. 359–366, 1957.

Appendix A
DECIBEL CONVERSION TABLES*

It is convenient in measurements and calculations in electroacoustics to express the ratio between any two amounts of electric or acoustic power in units on a logarithmic scale. The *decibel* (1/10th of a *bel*) on the briggsian or base 10 scale is in almost universal use for this purpose.

Since voltage and sound pressure are related to power by impedance, the *decibel* can be used to express voltage and sound-pressure ratios, if care is taken to account for the impedances associated with them.

Tables A.1 and A.2 have been prepared to facilitate making conversions in either direction between the number of *decibels* and the corresponding power, voltage, and sound-pressure ratios. All numbers have been verified by digital computing machine.

To Find Values Outside the Range of Conversion Tables. Values outside the range of either Table A.1 or A.2 can be readily found with the help of the following simple rules:

Table A.1: Decibels to Voltage, Sound-pressure, and Power Ratios

NUMBER OF DECIBELS POSITIVE (+). Subtract +20 db successively from the given number of decibels until the remainder falls within range of Table A.1. *To find the voltage (sound-pressure) ratio,* multiply the corresponding value from the right-hand voltage-ratio column by 10 for each time you subtracted 20 db. *To find the power ratio,* multiply the corresponding value from the right-hand power-ratio column by 100 for each time you subtracted 20 db.

Example. *Given:* 49.2 db

$$49.2 \text{ db} - 20 \text{ db} - 20 \text{ db} = 9.2 \text{ db}$$

Voltage (sound-pressure) ratio: 9.2 db →

$$2.884 \times 10 \times 10 = 288.4$$

* Courtesy of the General Radio Company, Cambridge, Mass.

Power ratio: 9.2 db →

$$8.318 \times 100 \times 100 = 83{,}180$$

NUMBER OF DECIBELS NEGATIVE (—). Add +20 decibels successively to the given number of decibels until the sum falls within the range of Table A.1. *For the voltage (sound-pressure) ratio,* divide the value from the left-hand voltage-ratio column by 10 for each time you added 20 db. *For the power ratio,* divide the value from the left-hand power-ratio column by 100 for each time you added 20 db.

Example. *Given:* —49.2 db

$$-49.2 \text{ db} + 20 \text{ db} + 20 \text{ db} = -9.2 \text{ db}$$

Voltage (sound-pressure) ratio: -9.2 db →

$$0.3467 \times \tfrac{1}{10} \times \tfrac{1}{10} = 0.003467$$

Power ratio: -9.2 db →

$$0.1202 \times \tfrac{1}{100} \times \tfrac{1}{100} = 0.00001202$$

Table A.2: Voltage and Sound-pressure Ratios to Decibels

FOR RATIOS SMALLER THAN THOSE IN TABLE. Multiply the given ratio by 10 successively until the product can be found in the table. From the number of decibels thus found, subtract +20 db for each time you multiplied by 10.

Example. *Given:* Voltage (sound-pressure) ratio = 0.0131

$$0.0131 \times 10 \times 10 = 1.31$$

From Table A.2, 1.31 →

$$2.345 \text{ db} - 20 \text{ db} - 20 \text{ db} = -37.655 \text{ db}$$

FOR RATIOS GREATER THAN THOSE IN TABLE. Divide the given ratio by 10 successively until the quotient can be found in the table. To the number of decibels thus found, add +20 db for each time you divided by 10.

Example. *Given:* Voltage (sound-pressure) ratio = 712

$$712 \times \tfrac{1}{10} \times \tfrac{1}{10} = 7.12$$

From Table A.2, 7.12 →

$$17.05 \text{ db} + 20 \text{ db} + 20 \text{ db} = 57.05 \text{ db}$$

Table A.1

Decibels to Voltage, Sound-pressure, and Sound-power Ratios*

[*To account for the sign of the decibel.* For positive (+) values of the decibel: Both voltage and power ratios are greater than unity. Use the two right-hand columns. For negative (−) values of the decibel: Both voltage and power ratios are less than unity. Use the two left-hand columns. Use the voltage columns for sound pressures.]

Example. *Given:* ±9.1 db. *Find:*

Decibels	Power Ratio	Voltage Ratio
+9.1	8.128	2.851
−9.1	0.1230	0.3508

Negative		Deci-bels	Positive		Negative		Deci-bels	Positive	
Voltage Ratio	Power Ratio		Voltage Ratio	Power Ratio	Voltage Ratio	Power Ratio		Voltage Ratio	Power Ratio
1.0000	1.0000	0	1.000	1.000	0.7079	0.5012	3.0	1.413	1.995
0.9886	0.9772	0.1	1.012	1.023	0.6998	0.4898	3.1	1.429	2.042
0.9772	0.9550	0.2	1.023	1.047	0.6918	0.4786	3.2	1.445	2.089
0.9661	0.9333	0.3	1.035	1.072	0.6839	0.4677	3.3	1.462	2.138
0.9550	0.9120	0.4	1.047	1.096	0.6761	0.4571	3.4	1.479	2.188
0.9441	0.8913	0.5	1.059	1.122	0.6683	0.4467	3.5	1.496	2.239
0.9333	0.8710	0.6	1.072	1.148	0.6607	0.4365	3.6	1.514	2.291
0.9226	0.8511	0.7	1.084	1.175	0.6531	0.4266	3.7	1.531	2.344
0.9120	0.8318	0.8	1.096	1.202	0.6457	0.4169	3.8	1.549	2.399
0.9016	0.8128	0.9	1.109	1.230	0.6383	0.4074	3.9	1.567	2.455
0.8913	0.7943	1.0	1.122	1.259	0.6310	0.3981	4.0	1.585	2.512
0.8810	0.7762	1.1	1.135	1.288	0.6237	0.3890	4.1	1.603	2.570
0.8710	0.7586	1.2	1.148	1.318	0.6166	0.3802	4.2	1.622	2.630
0.8610	0.7413	1.3	1.161	1.349	0.6095	0.3715	4.3	1.641	2.692
0.8511	0.7244	1.4	1.175	1.380	0.6026	0.3631	4.4	1.660	2.754
0.8414	0.7079	1.5	1.189	1.413	0.5957	0.3548	4.5	1.679	2.818
0.8318	0.6918	1.6	1.202	1.445	0.5888	0.3467	4.6	1.698	2.884
0.8222	0.6761	1.7	1.216	1.479	0.5821	0.3388	4.7	1.718	2.951
0.8128	0.6607	1.8	1.230	1.514	0.5754	0.3311	4.8	1.738	3.020
0.8035	0.6457	1.9	1.245	1.549	0.5689	0.3236	4.9	1.758	3.090
0.7943	0.6310	2.0	1.259	1.585	0.5623	0.3162	5.0	1.778	3.162
0.7852	0.6166	2.1	1.274	1.622	0.5559	0.3090	5.1	1.799	3.236
0.7762	0.6026	2.2	1.288	1.660	0.5495	0.3020	5.2	1.820	3.311
0.7674	0.5888	2.3	1.303	1.698	0.5433	0.2951	5.3	1.841	3.388
0.7586	0.5754	2.4	1.318	1.738	0.5370	0.2884	5.4	1.862	3.467
0.7499	0.5623	2.5	1.334	1.778	0.5309	0.2818	5.5	1.884	3.548
0.7413	0.5495	2.6	1.349	1.820	0.5248	0.2754	5.6	1.905	3.631
0.7328	0.5370	2.7	1.365	1.862	0.5188	0.2692	5.7	1.928	3.715
0.7244	0.5248	2.8	1.380	1.905	0.5129	0.2630	5.8	1.950	3.802
0.7161	0.5129	2.9	1.396	1.950	0.5070	0.2570	5.9	1.972	3.890

Table A.1 (*Continued*)

Negative		Decibels	Positive		Negative		Decibels	Positive	
Voltage Ratio	Power Ratio		Voltage Ratio	Power Ratio	Voltage Ratio	Power Ratio		Voltage Ratio	Power Ratio
0.5012	0.2512	6.0	1.995	3.981	0.3350	0.1122	9.5	2.985	8.913
0.4955	0.2455	6.1	2.018	4.074	0.3311	0.1096	9.6	3.020	9.120
0.4898	0.2399	6.2	2.042	4.169	0.3273	0.1072	9.7	3.055	9.333
0.4842	0.2344	6.3	2.065	4.266	0.3236	0.1047	9.8	3.090	9.550
0.4786	0.2291	6.4	2.089	4.365	0.3199	0.1023	9.9	3.126	9.772
0.4732	0.2239	6.5	2.113	4.467	0.3162	0.1000	10.0	3.162	10.000
0.4677	0.2188	6.6	2.138	4.571	0.3126	0.09772	10.1	3.199	10.23
0.4624	0.2138	6.7	2.163	4.677	0.3090	0.09550	10.2	3.236	10.47
0.4571	0.2089	6.8	2.188	4.786	0.3055	0.09333	10.3	3.273	10.72
0.4519	0.2042	6.9	2.213	4.898	0.3020	0.09120	10.4	3.311	10.96
0.4467	0.1995	7.0	2.239	5.012	0.2985	0.08913	10.5	3.350	11.22
0.4416	0.1950	7.1	2.265	5.129	0.2951	0.08710	10.6	3.388	11.48
0.4365	0.1905	7.2	2.291	5.248	0.2917	0.08511	10.7	3.428	11.75
0.4315	0.1862	7.3	2.317	5.370	0.2884	0.08318	10.8	3.467	12.02
0.4266	0.1820	7.4	2.344	5.495	0.2851	0.08128	10.9	3.508	12.30
0.4217	0.1778	7.5	2.371	5.623	0.2818	0.07943	11.0	3.548	12.59
0.4169	0.1738	7.6	2.399	5.754	0.2786	0.07762	11.1	3.589	12.88
0.4121	0.1698	7.7	2.427	5.888	0.2754	0.07586	11.2	3.631	13.18
0.4074	0.1660	7.8	2.455	6.026	0.2723	0.07413	11.3	3.673	13.49
0.4027	0.1622	7.9	2.483	6.166	0.2692	0.07244	11.4	3.715	13.80
0.3981	0.1585	8.0	2.512	6.310	0.2661	0.07079	11.5	3.758	14.13
0.3936	0.1549	8.1	2.541	6.457	0.2630	0.06918	11.6	3.802	14.45
0.3890	0.1514	8.2	2.570	6.607	0.2600	0.06761	11.7	3.846	14.79
0.3846	0.1479	8.3	2.600	6.761	0.2570	0.06607	11.8	3.890	15.14
0.3802	0.1445	8.4	2.630	6.918	0.2541	0.06457	11.9	3.936	15.49
0.3758	0.1413	8.5	2.661	7.079	0.2512	0.06310	12.0	3.981	15.85
0.3715	0.1380	8.6	2.692	7.244	0.2483	0.06166	12.1	4.027	16.22
0.3673	0.1349	8.7	2.723	7.413	0.2455	0.06026	12.2	4.074	16.60
0.3631	0.1318	8.8	2.754	7.586	0.2427	0.05888	12.3	4.121	16.98
0.3589	0.1288	8.9	2.786	7.762	0.2399	0.05754	12.4	4.169	17.38
0.3548	0.1259	9.0	2.818	7.943	0.2371	0.05623	12.5	4.217	17.78
0.3508	0.1230	9.1	2.851	8.128	0.2344	0.05495	12.6	4.266	18.20
0.3467	0.1202	9.2	2.884	8.318	0.2317	0.05370	12.7	4.315	18.62
0.3428	0.1175	9.3	2.917	8.511	0.2291	0.05248	12.8	4.365	19.05
0.3388	0.1148	9.4	2.951	8.710	0.2265	0.05129	12.9	4.416	19.50

Table A.1 (*Continued*)

Negative		Decibels	Positive		Negative		Decibels	Positive	
Voltage Ratio	Power Ratio		Voltage Ratio	Power Ratio	Voltage Ratio	Power Ratio		Voltage Ratio	Power Ratio
0.2239	0.05012	13.0	4.467	19.95	0.1496	0.02239	16.5	6.683	44.67
0.2213	0.04898	13.1	4.519	20.42	0.1479	0.02188	16.6	6.761	45.71
0.2188	0.04786	13.2	4.571	20.89	0.1462	0.02138	16.7	6.839	46.77
0.2163	0.04677	13.3	4.624	21.38	0.1445	0.02089	16.8	6.918	47.86
0.2138	0.04571	13.4	4.677	21.88	0.1429	0.02042	16.9	6.998	48.98
0.2113	0.04467	13.5	4.732	22.39	0.1413	0.01995	17.0	7.079	50.12
0.2089	0.04365	13.6	4.786	22.91	0.1396	0.01950	17.1	7.161	51.29
0.2065	0.04266	13.7	4.842	23.44	0.1380	0.01905	17.2	7.244	52.48
0.2042	0.04169	13.8	4.898	23.99	0.1365	0.01862	17.3	7.328	53.70
0.2018	0.04074	13.9	4.955	24.55	0.1349	0.01820	17.4	7.413	54.95
0.1995	0.03981	14.0	5.012	25.12	0.1334	0.01778	17.5	7.499	56.23
0.1972	0.03890	14.1	5.070	25.70	0.1318	0.01738	17.6	7.586	57.54
0.1950	0.03802	14.2	5.129	26.30	0.1303	0.01698	17.7	7.674	58.88
0.1928	0.03715	14.3	5.188	26.92	0.1288	0.01660	17.8	7.762	60.26
0.1905	0.03631	14.4	5.248	27.54	0.1274	0.01622	17.9	7.852	61.66
0.1884	0.03548	14.5	5.309	28.18	0.1259	0.01585	18.0	7.943	63.10
0.1862	0.03467	14.6	5.370	28.84	0.1245	0.01549	18.1	8.035	64.57
0.1841	0.03388	14.7	5.433	29.51	0.1230	0.01514	18.2	8.128	66.07
0.1820	0.03311	14.8	5.495	30.20	0.1216	0.01479	18.3	8.222	67.61
0.1799	0.03236	14.9	5.559	30.90	0.1202	0.01445	18.4	8.318	69.18
0.1778	0.03162	15.0	5.623	31.62	0.1189	0.01413	18.5	8.414	70.79
0.1758	0.03090	15.1	5.689	32.36	0.1175	0.01380	18.6	8.511	72.44
0.1738	0.03020	15.2	5.754	33.11	0.1161	0.01349	18.7	8.610	74.13
0.1718	0.02951	15.3	5.821	33.88	0.1148	0.01318	18.8	8.710	75.86
0.1698	0.02884	15.4	5.888	34.67	0.1135	0.01288	18.9	8.810	77.62
0.1679	0.02818	15.5	5.957	35.48	0.1122	0.01259	19.0	8.913	79.43
0.1660	0.02754	15.6	6.026	36.31	0.1109	0.01230	19.1	9.016	81.28
0.1641	0.02692	15.7	6.095	37.15	0.1096	0.01202	19.2	9.120	83.18
0.1622	0.02630	15.8	6.166	38.02	0.1084	0.01175	19.3	9.226	85.11
0.1603	0.02570	15.9	6.237	38.90	0.1072	0.01148	19.4	9.333	87.10
0.1585	0.02512	16.0	6.310	39.81	0.1059	0.01122	19.5	9.441	89.13
0.1567	0.02455	16.1	6.383	40.74	0.1047	0.01096	19.6	9.550	91.20
0.1549	0.02399	16.2	6.457	41.69	0.1035	0.01072	19.7	9.661	93.33
0.1531	0.02344	16.3	6.531	42.66	0.1023	0.01047	19.8	9.772	95.50
0.1514	0.02291	16.4	6.607	43.65	0.1012	0.01023	19.9	9.886	97.72
					0.1000	0.01000	20.0	10.000	100.00

Table A.1 (*Continued*)

Negative		Decibels	Positive	
Voltage Ratio	Power Ratio		Voltage Ratio	Power Ratio
3.162×10^{-1}	10^{-1}	10	3.162	10
10^{-1}	10^{-2}	20	10	10^{2}
3.162×10^{-2}	10^{-3}	30	3.162×10	10^{3}
10^{-2}	10^{-4}	40	10^{2}	10^{4}
3.162×10^{-3}	10^{-5}	50	3.162×10^{2}	10^{5}
10^{-3}	10^{-6}	60	10^{3}	10^{6}
3.162×10^{-4}	10^{-7}	70	3.162×10^{3}	10^{7}
10^{-4}	10^{-8}	80	10^{4}	10^{8}
3.162×10^{-5}	10^{-9}	90	3.162×10^{4}	10^{9}
10^{-5}	10^{-10}	100	10^{5}	10^{10}

* To find decibel values outside the range of this table, see pp. 705–706. Use the voltage-ratio columns for sound-pressure ratios.

Table A.2

Voltage and Sound-pressure Ratios to Decibels*

(*Voltage and sound pressure ratios:* Use the table directly. *Power ratios:* To find the number of decibels corresponding to a given power ratio, assume the given power ratio to be a voltage ratio and find the corresponding number of decibels from the table. The desired result is exactly one-half of the number of decibels thus found.)

Example. *Given:* a power ratio of 3.41. *Find:* 3.41 in the table.

$$3.41 \rightarrow 10.655 \text{ db} \times \tfrac{1}{2} = 5.328 \text{ db}$$

Voltage Ratio	0.00	0.01	0.02	0.03	0.04	0.05	0.06	0.07	0.08	0.09
1.0	0.000	0.086	0.172	0.257	0.341	0.424	0.506	0.588	0.668	0.749
1.1	0.828	0.906	0.984	1.062	1.138	1.214	1.289	1.364	1.438	1.511
1.2	1.584	1.656	1.727	1.798	1.868	1.938	2.007	2.076	2.144	2.212
1.3	2.279	2.345	2.411	2.477	2.542	2.607	2.671	2.734	2.798	2.860
1.4	2.923	2.984	3.046	3.107	3.167	3.227	3.287	3.346	3.405	3.464
1.5	3.522	3.580	3.637	3.694	3.750	3.807	3.862	3.918	3.973	4.028
1.6	4.082	4.137	4.190	4.244	4.297	4.350	4.402	4.454	4.506	4.558
1.7	4.609	4.660	4.711	4.761	4.811	4.861	4.910	4.959	5.008	5.057
1.8	5.105	5.154	5.201	5.249	5.296	5.343	5.390	5.437	5.483	5.529
1.9	5.575	5.621	5.666	5.711	5.756	5.801	5.845	5.889	5.933	5.977
2.0	6.021	6.064	6.107	6.150	6.193	6.235	6.277	6.319	6.361	6.403
2.1	6.444	6.486	6.527	6.568	6.608	6.649	6.689	6.729	6.769	6.809
2.2	6.848	6.888	6.927	6.966	7.005	7.044	7.082	7.121	7.159	7.197
2.3	7.235	7.272	7.310	7.347	7.384	7.421	7.458	7.495	7.532	7.568
2.4	7.604	7.640	7.676	7.712	7.748	7.783	7.819	7.854	7.889	7.924
2.5	7.959	7.993	8.028	8.062	8.097	8.131	8.165	8.199	8.232	8.266
2.6	8.299	8.333	8.366	8.399	8.432	8.465	8.498	8.530	8.563	8.595
2.7	8.627	8.659	8.691	8.723	8.755	8.787	8.818	8.850	8.881	8.912
2.8	8.943	8.974	9.005	9.036	9.066	9.097	9.127	9.158	9.188	9.218
2.9	9.248	9.278	9.308	9.337	9.367	9.396	9.426	9.455	9.484	9.513
3.0	9.542	9.571	9.600	9.629	9.657	9.686	9.714	9.743	9.771	9.799
3.1	9.827	9.855	9.883	9.911	9.939	9.966	9.994	10.021	10.049	10.076
3.2	10.103	10.130	10.157	10.184	10.211	10.238	10.264	10.291	10.317	10.344
3.3	10.370	10.397	10.423	10.449	10.475	10.501	10.527	10.553	10.578	10.604
3.4	10.630	10.655	10.681	10.706	10.731	10.756	10.782	10.807	10.832	10.857
3.5	10.881	10.906	10.931	10.955	10.980	11.005	11.029	11.053	11.078	11.102
3.6	11.126	11.150	11.174	11.198	11.222	11.246	11.270	11.293	11.317	11.341
3.7	11.364	11.387	11.411	11.434	11.457	11.481	11.504	11.527	11.550	11.573
3.8	11.596	11.618	11.641	11.664	11.687	11.709	11.732	11.754	11.777	11.799
3.9	11.821	11.844	11.866	11.888	11.910	11.932	11.954	11.976	11.998	12.019
4.0	12.041	12.063	12.085	12.106	12.128	12.149	12.171	12.192	12.213	12.234
4.1	12.256	12.277	12.298	12.319	12.340	12.361	12.382	12.403	12.424	12.444
4.2	12.465	12.486	12.506	12.527	12.547	12.568	12.588	12.609	12.629	12.649
4.3	12.669	12.690	12.710	12.730	12.750	12.770	12.790	12.810	12.829	12.849
4.4	12.869	12.889	12.908	12.928	12.948	12.967	12.987	13.006	13.026	13.045

Table A.2 (*Continued*)

Voltage Ratio	0.00	0.01	0.02	0.03	0.04	0.05	0.06	0.07	0.08	0.09
4.5	13.064	13.084	13.103	13.122	13.141	13.160	13.179	13.198	13.217	13.236
4.6	13.255	13.274	13.293	13.312	13.330	13.349	13.368	13.386	13.405	13.423
4.7	13.442	13.460	13.479	13.497	13.516	13.534	13.552	13.570	13.589	13.607
4.8	13.625	13.643	13.661	13.679	13.697	13.715	13.733	13.751	13.768	13.786
4.9	13.804	13.822	13.839	13.857	13.875	13.892	13.910	13.927	13.945	13.962
5.0	13.979	13.997	14.014	14.031	14.049	14.066	14.083	14.100	14.117	14.134
5.1	14.151	14.168	14.185	14.202	14.219	14.236	14.253	14.270	14.287	14.303
5.2	14.320	14.337	14.353	14.370	14.387	14.403	14.420	14.436	14.453	14.469
5.3	14.486	14.502	14.518	14.535	14.551	14.567	14.583	14.599	14.616	14.632
5.4	14.648	14.664	14.680	14.696	14.712	14.728	14.744	14.760	14.776	14.791
5.5	14.807	14.823	14.839	14.855	14.870	14.886	14.901	14.917	14.933	14.948
5.6	14.964	14.979	14.995	15.010	15.026	15.041	15.056	15.072	15.087	15.102
5.7	15.117	15.133	15.148	15.163	15.178	15.193	15.208	15.224	15.239	15.254
5.8	15.269	15.284	15.298	15.313	15.328	15.343	15.358	15.373	15.388	15.402
5.9	15.417	15.432	15.446	15.461	15.476	15.490	15.505	15.519	15.534	15.549
6.0	15.563	15.577	15.592	15.606	15.621	15.635	15.649	15.664	15.678	15.692
6.1	15.707	15.721	15.735	15.749	15.763	15.778	15.792	15.806	15.820	15.834
6.2	15.848	15.862	15.876	15.890	15.904	15.918	15.931	15.945	15.959	15.973
6.3	15.987	16.001	16.014	16.028	16.042	16.055	16.069	16.083	16.096	16.110
6.4	16.124	16.137	16.151	16.164	16.178	16.191	16.205	16.218	16.232	16.245
6.5	16.258	16.272	16.285	16.298	16.312	16.325	16.338	16.351	16.365	16.378
6.6	16.391	16.404	16.417	16.430	16.443	16.456	16.469	16.483	16.496	16.509
6.7	16.521	16.534	16.547	16.560	16.573	16.586	16.599	16.612	16.625	16.637
6.8	16.650	16.663	16.676	16.688	16.701	16.714	16.726	16.739	16.752	16.764
6.9	16.777	16.790	16.802	16.815	16.827	16.840	16.852	16.865	16.877	16.890
7.0	16.902	16.914	16.927	16.939	16.951	16.964	16.976	16.988	17.001	17.013
7.1	17.025	17.037	17.050	17.062	17.074	17.086	17.098	17.110	17.122	17.135
7.2	17.147	17.159	17.171	17.183	17.195	17.207	17.219	17.231	17.243	17.255
7.3	17.266	17.278	17.290	17.302	17.314	17.326	17.338	17.349	17.361	17.373
7.4	17.385	17.396	17.408	17.420	17.431	17.443	17.455	17.466	17.478	17.490
7.5	17.501	17.513	17.524	17.536	17.547	17.559	17.570	17.582	17.593	17.605
7.6	17.616	17.628	17.639	17.650	17.662	17.673	17.685	17.696	17.707	17.719
7.7	17.730	17.741	17.752	17.764	17.775	17.786	17.797	17.808	17.820	17.831
7.8	17.842	17.853	17.864	17.875	17.886	17.897	17.908	17.919	17.931	17.942
7.9	17.953	17.964	17.975	17.985	17.996	18.007	18.018	18.029	18.040	18.051
8.0	18.062	18.073	18.083	18.094	18.105	18.116	18.127	18.137	18.148	18.159
8.1	18.170	18.180	18.191	18.202	18.212	18.223	18.234	18.244	18.255	18.266
8.2	18.276	18.287	18.297	18.308	18.319	18.329	18.340	18.350	18.361	18.371
8.3	18.382	18.392	18.402	18.413	18.423	18.434	18.444	18.455	18.465	18.475
8.4	18.486	18.496	18.506	18.517	18.527	18.537	18.547	18.558	18.568	18.578

Table A.2 *(Continued)*

Voltage Ratio	0.00	0.01	0.02	0.03	0.04	0.05	0.06	0.07	0.08	0.09
8.5	18.588	18.599	18.609	18.619	18.629	18.639	18.649	18.660	18.670	18.680
8.6	18.690	18.700	18.710	18.720	18.730	18.740	18.750	18.760	18.770	18.780
8.7	18.790	18.800	18.810	18.820	18.830	18.840	18.850	18.860	18.870	18.880
8.8	18.890	18.900	18.909	18.919	18.929	18.939	18.949	18.958	18.968	18.978
8.9	18.988	18.998	19.007	19.017	19.027	19.036	19.046	19.056	19.066	19.075
9.0	19.085	19.094	19.104	19.114	19.123	19.133	19.143	19.152	19.162	19.171
9.1	19.181	19.190	19.200	19.209	19.219	19.228	19.238	19.247	19.257	19.266
9.2	19.276	19.285	19.295	19.304	19.313	19.323	19.332	19.342	19.351	19.360
9.3	19.370	19.379	19.388	19.398	19.407	19.416	19.426	19.435	19.444	19.453
9.4	19.463	19.472	19.481	19.490	19.499	19.509	19.518	19.527	19.536	19.545
9.5	19.554	19.564	19.573	19.582	19.591	19.600	19.609	19.618	19.627	19.636
9.6	19.645	19.654	19.664	19.673	19.682	19.691	19.700	19.709	19.718	19.726
9.7	19.735	19.744	19.753	19.762	19.771	19.780	19.789	19.798	19.807	19.816
9.8	19.825	19.833	19.842	19.851	19.860	19.869	19.878	19.886	19.895	19.904
9.9	19.913	19.921	19.930	19.939	19.948	19.956	19.965	19.974	19.983	19.991
	0	1	2	3	4	5	6	7	8	9
10	20.000	20.828	21.584	22.279	22.923	23.522	24.082	24.609	25.105	25.575
20	26.021	26.444	26.848	27.235	27.604	27.959	28.299	28.627	28.943	29.248
30	29.542	29.827	30.103	30.370	30.630	30.881	31.126	31.364	31.596	31.821
40	32.041	32.256	32.465	32.669	32.869	33.064	33.255	33.442	33.625	33.804
50	33.979	34.151	34.320	34.486	34.648	34.807	34.964	35.117	35.269	35.417
60	35.563	35.707	35.848	35.987	36.124	36.258	36.391	36.521	36.650	36.777
70	36.902	37.025	37.147	37.266	37.385	37.501	37.616	37.730	37.842	37.953
80	38.062	38.170	38.276	38.382	38.486	38.588	38.690	38.790	38.890	38.988
90	39.085	39.181	39.276	39.370	39.463	39.554	39.645	39.735	39.825	39.913
100	40.000									

* To find ratios outside the range of this table, see p. 706.

Appendix B
ENGLISH SYSTEM OF UNITS

The English system of units is inherently confusing. In everyday American life, the "pound" may have the units of force or it may have the units of mass (force/acceleration). Indeed, most physics courses in secondary schools teach the concept of pound force (lb_f) and pound mass (lb_m). On the other hand, mechanical engineers commonly use the "pound" (abbreviated "lb") either as a force or as a weight, both having the units of force. To further complicate matters, many technical writers seek to provide a system parallel to the metric system by using separate quantities for force and mass. Some adopt the "pound" as the unit of force and define a "slug" as the unit of mass. Others define a "poundal" as the unit of force and adopt the "pound" as the unit of mass. But neither the "slug" nor the "poundal" has found general acceptance in the technical literature.

As one well-traveled American acoustician said, "By experiment I have determined that 1 kg of butter bought in Zurich is exactly the same amount of butter as 2.2 lb bought in New York. Whenever I wish to solve a scientific problem in America without confusion, I immediately divide the number of pounds by 2.2 to obtain the equivalent number of kilograms. Then I work in the mks system, where force and mass are clearly distinguished."

Consistent Systems of Units Used in This Text

Three consistent systems of units are used in this text, the *mks*, the *cgs*, and the *fss* systems. To describe them, let us start with Newton's second law.

$$\text{Force} = \text{mass} \times \text{acceleration} \qquad \text{(B.1)}$$

In the *meter-kilogram-second* system

$$\text{No. of newtons} = \text{no. of kg} \times \text{no. of m/sec}^2 \qquad \text{(B.2)}$$

In the *centimeter-gram-second* system

$$\text{No. of dynes} = \text{no. of gm} \times \text{no. of cm/sec}^2 \qquad \text{(B.3)}$$

In the *foot-slug-second* system

$$\text{No. of pounds force (lb}_f) = \text{no. of slugs} \times \text{no. of ft/sec}^2 \qquad \text{(B.4)}$$

The relations among the magnitudes of the units are

$$1 \text{ kg} \equiv 0.0685 \text{ slug} \equiv 1{,}000 \text{ gm}$$

$$1 \text{ newton} \equiv 0.225 \text{ lb}_f \equiv 10^5 \text{ dynes}$$

Inconsistent Systems of English Units Used in This Text

Two inconsistent systems of English units are used in this text, the *fps* and *ips* systems.

In the *foot-pound-second* system

$$\text{No. of pounds force (lb}_f) = \frac{\text{no. of pounds mass (lb}_m)}{k_u} \times \text{no. of ft/sec}^2 \qquad \text{(B.5)}$$

where k_u is a *dimensionless* quantity equal to 32.2.

In the *inch-pound-second* system

$$\text{No. of pounds force (lb}_f) = \frac{\text{no. of pounds mass (lb}_m)}{k_u} \times \text{no. of in./sec}^2 \qquad \text{(B.6)}$$

where k_u is a *dimensionless* quantity equal to 386.

The relations among the magnitudes of the units are

$$1 \text{ kg} \equiv 2.2 \text{ lb}_m \equiv 1{,}000 \text{ gm}$$

$$1 \text{ newton} \equiv 0.225 \text{ lb}_f \equiv 10^5 \text{ dynes}$$

Other Systems of Units Used in America

In the *foot-pound-second* system (inconsistent system)

$$\text{No. of pounds force (lb)} = \frac{\text{no. of pounds weight (lb)}}{g} \times \text{no. of ft/sec}^2 \qquad \text{(B.7)}$$

where g is the acceleration due to gravity in units of ft/sec², that is, in this case 32.2 ft/sec².

In the *inch-pound-second* system (inconsistent system)

$$\text{No. of pounds force (lb)} = \frac{\text{no. of pounds weight (lb)}}{g} \times \text{no. of in./sec}^2 \qquad \text{(B.8)}$$

where g is the acceleration due to gravity in units of in./sec², that is, in this case 386 in./sec².

A working group of the American Standards Association has tentatively adopted the *in.-lb-sec* system of Eq. (B.8) as the standard in the field of shock and vibration.

The relations among the magnitudes of the units are

$$1 \text{ kg} \equiv 2.2 \text{ lb} \equiv 2.2 \text{ lb}_m \equiv 1{,}000 \text{ gm}$$

$$1 \text{ newton} \equiv 0.225 \text{ lb} \equiv 0.225 \text{ lb}_f \equiv 10^5 \text{ dynes}$$

One must observe that the *magnitude* of pounds mass (lb_m) and the *magnitude* of pounds weight (lb) are alike wherever $g = 32.2$ ft/sec². The difference is that pounds mass has dimensions of pounds force *divided* by acceleration, whereas pounds weight has the dimensions of pounds force. Obviously the two quantities lb (pounds) and lb_f (pounds force) have the same units and magnitudes.

In the *foot-poundal-second* system (consistent system)

$$\text{No. of poundals} = \text{no. of pounds mass } (\text{lb}_m) \times \text{no. of ft/sec}^2 \qquad \text{(B.9)}$$

In the *inch-poundal-second* system (inconsistent system)

$$\text{No. of poundals} = \text{no. of pounds mass } (\text{lb}_m) \times \frac{\text{no. of in./sec}^2}{k_u} \qquad \text{(B.10)}$$

where k_u is a *dimensionless* quantity equal to 12.

The relations among the magnitudes of the units are

$$1 \text{ kg} \equiv 2.2 \text{ lb}_m \equiv 1{,}000 \text{ gm}$$

$$1 \text{ newton} \equiv 7.23 \text{ poundals} \equiv 10^5 \text{ dynes}$$

$$1 \text{ pound force } (\text{lb}_f) \equiv 32.2 \text{ poundals} \equiv 1 \text{ lb}$$

Example B.1. One kilogram is accelerated 5 m/sec². Find the force necessary to do this in newtons, dynes, pounds force (lb_f), pounds (lb), and poundals. Assume $g = 32.2$ ft/sec².

Solution

$$1 \text{ kg} \equiv 2.2 \text{ lb}_m \equiv 2.2 \text{ lb} \equiv 0.0685 \text{ slug}$$

$$5 \text{ m/sec}^2 \equiv 16.4 \text{ ft/sec}^2$$

$$F \text{ (newtons)} = 1 \times 5 = 5 \text{ newtons}$$

$$F \text{ (dynes)} = 1{,}000 \times 500 = 5 \times 10^5 \text{ dynes}$$

$$F\ (\text{lb}_f) = 0.0685 \times 16.4 = 1.12 \text{ lb}_f$$

$$F\ (\mathrm{lb}_f) = \frac{2.2}{32.2} \times 16.4 = 1.12\ \mathrm{lb}_f$$

$$F\ (\mathrm{lb}) = \frac{2.2}{g} \times 16.4 = 1.12\ \mathrm{lb}$$

$$\boldsymbol{F}\ (\text{poundals}) = 2.2 \times 16.4 = 36.1\ \text{poundals}$$

Appendix C
CONVERSION FACTORS

The following values for the fundamental constants were used in the preparation of the factors:

$$
\begin{aligned}
1 \text{ m} &= 39.37 \text{ in.} = 3.281 \text{ ft} \\
1 \text{ lb}_m &= 0.4536 \text{ kg} \\
1 \text{ slug} &= 14.594 \text{ kg} \\
1 \text{ lb}_f &= 4.448 \text{ newtons} \\
1 \text{ poundal} &= 0.13825 \text{ newton} \\
\text{Acceleration due to gravity} &= 9.807 \text{ m/sec}^2 \\
&= 32.174 \text{ ft/sec}^2 \\
\text{Density of } H_2O \text{ at } 4°C &= 1.0 \text{ gm/cm}^3 \\
\text{Density of Hg at } 0°C &= 13.595 \text{ gm/cm}^3 \\
1 \text{ pound weight (lb)} &= 1 \text{ pound mass (lb}_m\text{) at point on earth} \\
&\quad \text{where } g = 32.177 \text{ ft/sec}^2
\end{aligned}
$$

Table C.1

Conversion Factors

To Convert	Into	Multiply by	Conversely, Multiply by
acoustic ohms	mks acoustic ohms	10^5	10^{-5}
acres	ft^2	4.356×10^4	2.296×10^{-5}
	m^2	4,047	2.471×10^{-4}
atm	mm Hg at 0°C	760	1.316×10^{-3}
	ft H_2O at 4°C	33.90	2.950×10^{-2}
	in. H_2O at 4°C	406.80	2.458×10^{-3}
	in. Hg at 0°C	29.92	3.342×10^{-2}
	newtons/m^2	1.013×10^5	9.872×10^{-6}
	kg/m^2	1.033×10^4	9.681×10^{-5}
	lb_f/in.^{2}	14.70	6.805×10^{-2}
°C	°F	(°C × 9/5) + 32	(°F − 32) × 5/9
cm	m	10^{-2}	10^2
cir mils	cm^2	5.067×10^{-6}	1.974×10^5
cm^3	m^3	10^{-6}	10^6
deg (angle)	radians	1.745×10^{-2}	57.30
dynes	lb_f	2.248×10^{-6}	4.448×10^5
	poundals	7.233×10^{-5}	1.383×10^4
	newtons	10^{-5}	10^5
dynes/cm^2	newtons/m^2	10^{-1}	10
ergs	ft-lb_f	7.376×10^{-8}	1.356×10^7
	joules	10^{-7}	10^7
ergs/sec	watts	10^{-7}	10^7
fathoms	ft	6	0.16667
ft	cm	30.48	3.281×10^{-2}
	m	0.3048	3.281
ft^2	m^2	9.290×10^{-2}	10.76
ft^3	liters	28.32	3.532×10^{-2}
	m^3	2.832×10^{-2}	35.31
ft H_2O at 4°C	in. Hg at 0°C	0.8826	1.133
	newtons/m^2	2,989	3.345×10^{-4}
	kg/m^2	304.8	3.281×10^{-3}
	lb_f/ft^2	62.43	1.602×10^{-2}
gal	m^3	3.785×10^{-3}	264.2
gal (liquid U.S.)	gal (liquid Brit. Imp.)	0.8327	1.201
gm	oz_m	3.527×10^{-2}	28.35
gm	lb_m	2.205×10^{-3}	453.6
hp (550 ft-lb_f/sec)	ft-lb_f/min	3.3×10^4	3.030×10^{-5}
	kw	0.746	1.341
in.	cm	2.540	0.3937
in.^{2}	cm^2	6.452	0.1550
in.^{3}	ft^3	5.787×10^{-4}	1,728
	cm^3	16.39	6.102×10^{-2}
	m^3	1.639×10^{-5}	6.102×10^4
in. Hg at 0°C	lb_f/in.^{2}	0.4912	2.036

Table C.1 (*Continued*)

To Convert	Into	Multiply by	Conversely, Multiply by
kg	lb_m	2.205	0.4536
	gm	10^3	10^{-3}
kg/m^2	lb_m/ft^2	0.2048	4.882
liters	m^3	0.001	1,000
	$in.^3$	61.03	1.639×10^{-2}
	gal (liquid U.S.)	0.2642	3.785
	pints (liquid U.S.)	2.113	0.4732
$\log_e n$, or $\ln n$	$\log_{10} n$	0.4343	2.303
mechanical ohms	mks mechanical ohms	10^{-3}	10^3
m	yd	1.094	0.9144
	ft	3.281	0.3048
	in.	39.37	0.0254
	cm	10^2	10^{-2}
m^2	ft^2	10.764	9.290×10^{-2}
m^3	ft^3	35.31	2.832×10^{-2}
	yd^3	1.308	0.7646
microbars	$newtons/m^2$	10^{-1}	10
miles (nautical)	ft	6,080.2	1.645×10^{-4}
	km	1.853	0.5396
miles (statute)	ft	5,280	1.894×10^{-4}
	km	1.609	0.6214
$miles^2$ (statute)	km^2	2.590	0.3861
mph	ft/min	88	1.136×10^{-2}
	km/min	2.682×10^{-2}	37.28
	km/hr	1.609	0.6214
nepers	db	8.686	0.1151
newtons	dynes	10^5	10^{-5}
	lb_f	0.225	4.448
$newtons/m^2$	$dynes/cm^2$	10	10^{-1}
lb_f	newtons	4.448	0.225
lb_m	kg	0.4536	2.205
lb_m H_2O (distilled)	ft^3	1.602×10^{-2}	62.43
	gal (liquid U.S.)	0.1198	8.346
lb_m/ft^3	kg/m^3	16.02	6.243×10^{-2}
$lb/in.^3$	lb/ft^3	1,728	5.787×10^{-4}
lb/ft^2	$lb/in.^2$	6.944×10^{-3}	144
lb_m/ft^2	gm/cm^2	0.4882	2.0482
$lb_m/in.^2$	kg/m^2	703.1	1.422×10^{-3}
poundals	dynes	1.383×10^4	7.233×10^{-5}
	lb_f	3.108×10^{-2}	32.174
rayls	mks rayls	10	10^{-1}
slugs	lb_m	32.174	3.108×10^{-2}
	kg	14.594	0.06852
tons, short (2,000 lb)	tonnes (1,000 kg)	0.9072	1.102
watts	ergs/sec	10^7	10^{-7}
	hp (550 ft-lb_f/sec)	1.341×10^{-3}	745.7

NAME INDEX

SUBJECT INDEX